Scientific Computation

C. Canuto M. Y. Hussaini
A. Quarteroni T. A. Zang

Spectral Methods

Evolution to Complex Geometries
and Applications to Fluid Dynamics

With 183 Figures and 11 Tables

 Springer

Claudio Canuto

Dipartimento di Matematica
Politecnico di Torino
Corso Duca degli Abruzzi, 24
10129 Torino, Italy
e-mail: claudio.canuto@polito.it

Alfio Quarteroni

SB-IACS-CMCS, EPFL
Station 8
1015 Lausanne, Switzerland
and
MOX, Politecnico di Milano
Piazza Leonardo da Vinci, 32
20133 Milano, Italy
e-mail: alfio.quarteroni@epfl.ch

M. Yousuff Hussaini

School of Computational Science
Florida State University
Tallahassee, FL 32306-4120, USA
e-mail: myh@scs.fsu.edu

Thomas A. Zang, Jr.

NASA Langley Research Center*
Mail Stop 449
Hampton, VA 23681-2199, USA
e-mail: Thomas.A.Zang@nasa.gov

* *This does not constitute an endorsement of this work by either the U.S. Government or the NASA Langley Research Center.*

Cover figures: Upper left: Structural dynamics analysis of the Roman Colosseum: spectral element calculation of unitary dilatation (Fig. 5.61). *Upper right*: Computational surface mesh in the vicinity of an engine for a sixth-order spectral discontinuous Galerkin computation of noise propagation and scattering off an aircraft surface (Fig. 5.58). *Lower left*: Velocity contours for a Taylor-Green vortex simulation using a Fourier spectral method (Fig. 3.4). *Lower right*: Spectral element approximation of an eigenfunction of the Laplace operator with Dirichlet boundary conditions in a square (Fig. 5.18).

Library of Congress Control Number: 2007924823

ISSN   1434-8322

ISBN   978-3-540-30727-3   Springer Berlin Heidelberg New York

Springer is a part of Springer Science+Business Media

springer.com

© Springer-Verlag Berlin Heidelberg 2007

Typesetting and production: LE-TeX Jelonek, Schmidt & Vöckler GbR, Leipzig, Germany
Cover design: WMXDesign GmbH, Heidelberg

SPIN: 11584735   55/3180/YL – 5 4 3 2 1 0   Printed on acid-free paper

Preface

Two decades ago when we wrote *Spectral Methods in Fluid Dynamics* (1988), the subject was still fairly novel. Motivated by the many favorable comments we have received and the continuing interest in that book (which will be referred to as CHQZ1), and yet desiring to present a more modern perspective, we embarked on the project which resulted in our recent book (Canuto et al. (2006), referred to as CHQZ2) and the present new book (referred to as CHQZ3). Our objectives with these two new books are to modernize our thorough discussion of classical spectral methods, accounting for advances in the theory and more extensive application experience in the fluid dynamics arena, while summarizing the current state of multidomain spectral methods from the perspective of classical spectral methods. While the two new books draw extensively from portions of our earlier text, CHQZ1, much of CHQZ2, and most of CHQZ3 is new. The added content has necessitated our publishing this new work as two separate books. The rationale for the division of the material between the books is that we furnished in the first new book a comprehensive discussion of the fundamental aspects of classical spectral methods in single domains. This second new book focuses on applications to fluid dynamics and on multidomain spectral methods.

The historical evolution of spectral methods from their initial (now classical) versions to the contemporary multidomain versions has been covered in some detail in the Preface of CHQZ2. In short, both the theory and the algorithms of classical spectral methods for smooth problems were reasonably mature already in the mid-1980s, and singular progress has been made over the past two decades in extending spectral methods to arbitrary geometries, enabling what some would consider the mathematical nirvana of a method of arbitrarily high-order capable of application to problems on an arbitrary geometry. In this respect, the trajectory of spectral methods over the past 20 years has been approaching that of hp finite-element methods. This process of migration from single-domain to multidomain spectral methods has required the injection of novel mathematical tools and stimulated original investigation directions. Mathematics has had a profound impact on the correct design and interpretation of the methods, and in some cases it has inspired the development of discontinuous spectral methods even for problems with continuous solutions. On the other hand, since in general a geometrically complex computational domain is split into polygonal or polyhedral subdomains (or elements), tensor-product domains are no longer a prerequisite for spectral methods, with the development of spectral bases on such elements as triangles and tetrahedra. Moreover, the splitting into many subdomains has motivated the use of moderate polynomial degrees in every subdomain—small from the perspective of classical spectral methods, but large from the perspectives of finite-volume and finite-element methods. In spite of this major change of perspective, the new multidomain spectral methods still enjoy some of the most distinguishing (and desirable) features of "classical" spectral methods—Gaussian integration formulas, low dispersion, and ease of preconditioning by low-order discretization matrices.

The main contents of the current text are as follows: Chapter 1 surveys the relevant equations of fluid dynamics, including the historical background and the underlying assumptions of the mathematical models. Chapter 2 presents spectral algorithms for solution of the boundary-layer equations and for analyses of linear and nonlinear stability of fluid flows. Chapter 3, on single-domain algorithms for incompressible flows, focuses its discussion of spectral methods for problems with at most one nonperiodic direction on those algorithms that remained in reasonably extensive use post-1990. Furthermore, it surveys the basic spatial discretization schemes for problems with two or more nonperiodic directions; these schemes are the foundation of many multidomain spectral methods for incompressible flows. Chapter 4, on classical spectral methods for hyperbolic systems and compressible flows, emphasizes algorithms for enforcing boundary conditions, methods for computing smooth compressible flows, and spectrally accurate solution of discontinuous compressible flows using the shock-fitting approach. Chapter 5 introduces the main strategies for constructing spectral approximations in complex domains, in particular, the spectral element method, the mortar element method, the spectral discontinuous Galerkin method, as well as the more traditional patching collocation method. Their theoretical properties are analyzed and their algebraic aspects are investigated. Some of the perspectives and results coincide with those of the hp version of finite-element methods (apart from the difference in notation). We arrive at them from the "spectral method" road, i.e., being coherent with the spectral technology within each element while reducing the size of the elements (the alternative road starts from the h version of FEM and conserves the typical global/local interplay while increasing the polynomial degree in each element). Chapter 6 illustrates solution strategies based on domain decomposition techniques for the spectral discretizations investigated in Chap. 5. Both Schur-based and Schwarz-based iterative solvers and preconditioners are considered, and their computational advantages (in particular, their property of scalability with respect to the number of subdomains) are illustrated. Chapter 7 caps our discussion of multidomain spectral methods with its survey of the modern approach to computing incompressible flows in general geometries using high-order, spectral discretizations.

Having CHQZ2 at hand is not necessary for reading CHQZ3, as all essential formulas are repeated here (in Chap. 8) in a short "primer for classical spectral methods", and we have also included the key elements of Appendices A–D from CHQZ2. (These appendices cover material on (A) some basic notations and theorems from functional analysis, (B) the Fast Fourier Transform, (C) iterative methods, and (D) temporal discretizations.) However, CHQZ2 does contain considerable material necessary for a better understanding of the subject, and we do make copious references to specific and more detailed material in CHQZ2. The complete Table of Contents from CHQZ2 is listed right after that of the present book, as a convenience to those readers wishing to locate specific background material. Also, the new Appendix D.3 contains some useful technical details for, but peripheral to, some of the algorithms discussed in the text.

In contrast to other recent expositions on the subject, such as the books by Karniadakis and Sherwin (2005) and Deville et al. (2002), this work certainly contains fewer implementation details of multidomain spectral algorithms. On the other hand, we accompany the algorithmic description with the theoretical framework necessary for the a priori prediction of their performance, which in turn supports informed choices for their use and provides some criteria for code verification. We trust that our copious numerical examples, and especially the large-scale applications, provide convincing empirical evidence that spectral methods have indeed evolved sufficiently that even in complex geometries they provide efficient, highly accurate numerical simulations of diverse physical phenomena. We hope that this effort, together with the other books and the ever-expanding literature on the subject, will foster the wider penetration of higher-order ideas and methods in computational engineering and science. The level of diffusion and appreciation that is by now granted to higher-order methods in certain communities, such as the structural mechanics one, is certainly a positive trend.

Whereas with our first text we made a valiant effort to provide comprehensive coverage of all available spectral methods (at least for fluid dynamics applications) and to provide a bibliography that encompassed all extant references to spectral methods, here we acknowledge the practical impossibility of such an ambition in the face of all the work that has since transpired in the field. We still aim to provide comprehensive coverage of general methodology. However, our discussion of particular algorithms is necessarily representative rather than complete. We focus on those single-domain algorithms that have seen the widest use in fluid dynamics applications and on a broad representation of multidomain algorithms. However, our knowledge in this area is certainly not exhaustive, and others would no doubt have made somewhat different choices. In our citations we enforce a strong preference for archival publications. We recognize that many developments appeared earlier (in some cases many years earlier) in pre-prints or conference publications, but we only cite non-archival sources when no archival reference is available.

The many numerical examples produced expressly for this book have been run on desktop computers in 64-bit arithmetic unless otherwise noted.

Nowadays, considerable software for spectral methods is freely available on the Web, ranging from libraries of basic spectral operations all the way to complete spectral codes for Navier–Stokes (and other) complex applications. Due to the highly dynamic nature of these postings, we have chosen not to list them in the text (except to acknowledge codes that we have used here for numerical examples), but to maintain a reasonably current list of such sources on the Web site (http://www.dimat.polito.it/chqz/) for this and the companion text. There is always the possibility that this site itself may need to be moved due to unforeseen circumstances; in this case, one should check the Springer site for the link to the detailed Web site of the book.

The authors are grateful to Dr. Wolf Beiglböck, Dr. Ramon Khanna and the Springer staff for their encouragement and facilitation of this project. The technical support of Paola Gervasio and Marco Discacciati in performing numerical tests, preparing figures and tables, typing, and editing much of the manuscript was indispensable. The authors are pleased to acknowledge the many discussions and helpful comments on the manuscript that have been provided by colleagues such as Erik Burman, Paola Gervasio, David Kopriva, Mujeeb Malik, Robert Moser, Luca Pavarino, David Pruett, Jamie Quirk, and Andrea Toselli. We appreciate the assistance provided by David Pruett, David Kopriva, and Benjamin Stamm for several of the numerical examples. Thanks are also due to Stefano Berrone, Dilek Dustegor, Giuseppe Ghibò, Nicola Parolini, and Svetlana Poroseva for providing additional technical support. We are also indebted to the many individuals who have graciously given us permission to reprint figures from their work. The authors are grateful to the Politecnico di Torino, Florida State University (esp. Lawrence Abele), the Ecole Polytechnique Fédérale de Lausanne, the Politecnico di Milano, and the NASA Langley Research Center for their facilitation of this endeavor. Finally, we are most appreciative of the support and understanding we have received from our wives (Manuelita, Khamar, Fulvia, and Ann) and children (Arianna, Susanna, Moin, Nadia, Marzia, and Silvia) during this project.

Torino, Italy   *Claudio Canuto*
Tallahassee, Florida   *M. Yousuff Hussaini*
Lausanne, Switzerland and Milan, Italy   *Alfio Quarteroni*
Hampton, Virginia   *Thomas A. Zang*

June, 2007

Contents

List of Figures

List of Tables

Contents of the Companion Book
Spectral Methods –
Fundamentals in Single Domains

References

Index

1. Fundamentals of Fluid Dynamics

1.1 Introduction

As pointed out in the Preface, the major thrust of this text is to extend our discussion of spectral methods from the simple model problems and single-domain methods described in the companion book, CHQZ2, to practical applications in fluid dynamics and to methods for complex domains. We refer the reader to Chap. 8 for a short introduction to spectral methods.

As preparation for the applications that are the focus of Chaps. 2–4, in this chapter we provide a survey of the fundamentals of fluid dynamics. We begin with some general background material on fluids on fluids, then concentrate on the basic equations of fluid dynamics (in Eulerian form). These equations are first stated in their full generality for compressible flow. Subsequent expositions include various special cases, especially those to which spectral methods have most commonly been applied. These special cases include Euler equations, incompressible Navier–Stokes equations, boundary-layer equations, and equations for both linear and nonlinear stability analysis. We also include some historical background on the various systems of equations that have been used over the years in the study of fluid dynamics.

The reader interested in detailed derivations and physical interpretations of these equations should consult standard references such as Courant and Friedrichs (1976), Howarth (1953), Liepmann and Roshko (2001), Landau and Lifshitz (1987), Serrin (1959a), Moore (1989), Batchelor (2000), Schlichting and Gersten (1999), and the various texts on more specialized topics cited below. (Most of these texts have been published in several editions. We have listed them here in the chronological order of their first editions but provide a reference to the most recent edition as of the publication of this text; we use this convention for ordering references throughout this book.)

1.2 Fluid Dynamics Background

Gases, such as air, and liquids, such as water, are both fluids. Before proceeding to the partial differential equations that describe the motion of such fluids, we provide the basis for the distinction between these two types of fluids and then summarize the relevant thermodynamic relationships.

1.2.1 Phases of Matter

The phases of matter may be broadly categorized into solids and fluids. Simply stated, solids resist deformation and retain a rigid shape; in particular, the stress is a function of the strain. Fluids do not resist deformation and take on the shape of its container owing to their inability to support shear stress in static equilibrium. More precisely, the stress is a function of the strain rate. The distinction between solids and fluids is not so simple and is based upon both the viscosity of the matter and the time scale of interest. The inorganic polymer-based toy known as Silly Putty™ in the U.S. (and Pongo™ in Italy), for instance, can be considered either as a solid or as a fluid, depending on the time period over which it is observed. Fluids include liquids, gases and plasmas. A liquid is a fluid with the property that it conforms to the shape of its container while retaining its constant volume independent of pressure. A gas is a compressible fluid that not only conforms to the shape of its container but also expands to occupy the full container. Plasma is ionized gas. The various phases of a substance are represented by a *phase diagram*, which marks the domains in pressure–temperature where each phase exists. Figure 1.1 illustrates the well-known phase diagram for water.

A rigorous mathematical description of the states of matter appears implausible. Goodstein (1975) states: "Precisely what do we mean by the term liquid? Asking what is a liquid is like asking what is life; we usually know it when we see it, but the existence of some doubtful cases makes it hard to define precisely." In his book, Anderson (1963) asks: "(2) How does one describe a solid from a really fundamental point of view in which the atomic nuclei as well as electrons are treated truly quantum mechanically? (3) How and why does a solid hold itself together? I have never yet seen a satisfactory fully quantum mechanical description of a solid." Phase diagrams would seem to require well-defined criteria, but there are always states at the boundaries of the so-called "phases" that cannot be precisely defined.

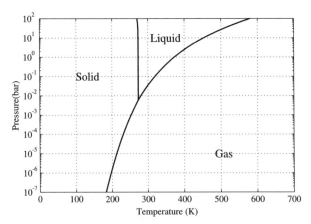

Fig. 1.1. Phase diagram for water

To be extremely precise, one should speak of "fluid-like" and "solid-like" behavior; their distinction depends upon the time and the length scales of interest. Glaciers and the Earth's mantle flow are thus fluid-like if the time scale is appropriately large, while they can be considered as solids if the time scale is appropriately small. However, it is not merely a matter of the time scale. Even if the time scale is on the order of months and the length scale on the order of miles, the Earth's mantle can be modeled as a solid, while if the length scale is the order of nanometers or less it may be appropriate to model it as a fluid. This suggests the need for appropriate nondimensional numbers. In rational mechanics, the nondimensional Deborah number (see Huilgol (1975)) is the ratio of a process time to an inherent relaxation time. The same lump of borosiloxane can be observed to flow as a fluid if the time scale is sufficiently large, to deform like an elastic solid, or to shatter like brittle glass.

Perhaps, the most accurate and lucid discussion of the distinction between solid-like and fluid-like behavior was given by Maxwell (1872): "What is required to alter the form of a soft solid is a sufficient force, and, this when applied produces its effect at once. In the case of viscous fluid it is time that is required, and if enough time is given, the very smallest force will produce a sensible effect, such as would require a very large force if suddenly applied. Thus a block of pitch may be so hard that you cannot make a dent in it by striking it with your knuckles; and yet it will in course of time, flatten itself by its own weight, and glide down hill like a stream of water." Maxwell is clearly referring to the importance of time scales and to a material that seems to have two time scales. However, materials can have more than two time scales.

1.2.2 Thermodynamic Relationships

General Gases. A complete description of a single-phase, homogeneous fluid is available if, as a function of space and time, the velocity $\mathbf{u} = (u, v, w)^T$, any two thermodynamic variables, and an equation of state are known. (The superscript T denotes the transpose of a vector or matrix.). The basic thermodynamic variables of most interest in fluid dynamics are the density ρ (or specific volume $V = 1/\rho$), the pressure p, the temperature T, the specific internal energy e, the specific entropy s, and the specific enthalpy h. (A *specific* quantity is one per unit mass.) Of these state variables, only two are independent, and the others can be expressed as functions of the two independent variables. In the classical Gibbs axiomatic formulation, the equation of state is

$$e = e(V, s) , \qquad (1.2.1)$$

and the variables *pressure* and *temperature* are defined by

$$p = -\frac{\partial e}{\partial V} \qquad \text{and} \qquad T = \frac{\partial e}{\partial s} , \qquad (1.2.2)$$

respectively, where p and T are always positive. By forming the total differential of the relation $e = e(V, s)$, we obtain the fundamental thermodynamic relation

$$T \, ds = de + p \, dV \; . \tag{1.2.3}$$

This may be considered a corollary of the second law of thermodynamics, and it defines the *specific entropy*, denoted by s.

The (specific) *total energy* is defined as

$$E = e + \frac{1}{2}(\mathbf{u} \cdot \mathbf{u}) \; , \tag{1.2.4}$$

the (specific) *enthalpy* is defined as

$$h = e + p/\rho \; , \tag{1.2.5}$$

and the (specific) *total enthalpy* is defined as

$$H = h + \frac{1}{2}(\mathbf{u} \cdot \mathbf{u}) \; . \tag{1.2.6}$$

The various relations among p, V, T, e, and s are known as *equations of state*. Relations giving p as a function of ρ (or V) and s (or T) occur with the greatest frequency in the theory of fluids:

$$p = f_1(\rho, T) \quad \text{or} \quad p = f_2(V, s) \quad \text{or} \quad p = f_3(\rho, s) \; . \tag{1.2.7}$$

Such relations are usually referred to as *caloric equations of state*. A unit mass of fluid in thermodynamic equilibrium has two important heat capacities

$$C_v = T \left. \frac{\partial s}{\partial T} \right|_V \quad \text{and} \quad C_p = T \left. \frac{\partial s}{\partial T} \right|_p \; . \tag{1.2.8}$$

They are called, respectively, the *specific heat at constant volume* and the *specific heat at constant pressure*. Experimental observations support the assumption that both specific heats are always positive. Using (1.2.3) and the corresponding total differential of the enthalpy, one obtains

$$C_v = \left. \frac{\partial e}{\partial T} \right|_V \quad \text{and} \quad C_p = \left. \frac{\partial h}{\partial T} \right|_p \; . \tag{1.2.9}$$

From (1.2.3) and the first part of (1.2.9), we have

$$C_p - C_v = \left(p + \left. \frac{\partial e}{\partial V} \right|_T \right) \left. \frac{\partial V}{\partial T} \right|_p \; . \tag{1.2.10}$$

Differentiating (1.2.3) with respect to T and V, respectively, we have

$$\left. \frac{\partial s}{\partial T} \right|_V = \frac{1}{T} \left. \frac{\partial e}{\partial T} \right|_V , \quad \left. \frac{\partial s}{\partial V} \right|_T = \frac{1}{T} \left(p + \left. \frac{\partial e}{\partial V} \right|_T \right) \; . \tag{1.2.11}$$

Then, equating the derivative of the former with respect to V with the derivative of the latter with respect to T, we have

$$p + \frac{\partial e}{\partial V}\bigg|_T = \frac{\partial p}{\partial T}\bigg|_V . \tag{1.2.12}$$

Hence, using (1.2.12) in (1.2.10), we have for the difference between the specific heats:

$$C_p - C_v = -\frac{T}{\rho^2} \frac{\partial p}{\partial T}\bigg|_\rho \frac{\partial \rho}{\partial T}\bigg|_p . \tag{1.2.13}$$

(See Howarth (1953) for more details.)

Another important positive quantity is

$$c^2 = \frac{\partial p}{\partial \rho}\bigg|_s , \tag{1.2.14}$$

where c is called the *sound speed*, which is the speed at which sound waves travel in the fluid (see Sect. 4.3).

Perfect Gases. The preceding equations are the general thermodynamic relationships. A number of special cases are of practical interest. A *thermally perfect gas* (also known as an *ideal gas*) is a fluid that obeys *Boyle's Law*, which is expressed by the equation of state

$$p = \rho \mathcal{R} T , \tag{1.2.15}$$

where the gas constant $\mathcal{R}$ is the ratio of the universal gas constant to the effective molecular weight of the particular gas.

In many applications the fluid is assumed to be calorically perfect in addition to being thermally perfect. For a thermally perfect gas one can show that both the specific internal energy and the specific enthalpy are functions of temperature only:

$$de = C_v(T)dT \qquad \text{and} \qquad dh = C_p(T)dT . \tag{1.2.16}$$

A thermally perfect gas is called *calorically perfect* if the specific heats are independent of temperature:

$$e = C_v T \qquad \text{and} \qquad h = C_p T , \tag{1.2.17}$$

assuming of course that both e and h vanish if T vanishes. Equations (1.2.4) and (1.2.6) become, respectively,

$$E = C_v T + \frac{1}{2}(\mathbf{u} \cdot \mathbf{u}) \tag{1.2.18}$$

and

$$H = C_p T + \frac{1}{2}(\mathbf{u} \cdot \mathbf{u}) . \tag{1.2.19}$$

It can be proven that an ideal gas that is calorically perfect has the *polytropic equation of state*

$$p = f(\rho, s) = K\rho^\gamma \,, \tag{1.2.20}$$

where K depends on the entropy only and γ (called the *adiabatic exponent*) is the ratio of the specific heats, C_p/C_v. Its value lies in the interval $1 < \gamma < 5/3$ for most fluids, and it is usually assumed to be 1.4 for air at moderate temperatures ($< 800\,°\mathrm{K}$).

For a calorically perfect gas (1.2.14) becomes

$$c^2 = \frac{\gamma p}{\rho} \,, \tag{1.2.21}$$

and (1.2.13) reduces to

$$C_p - C_v = \mathcal{R} \,. \tag{1.2.22}$$

The latter equation leads to the relations

$$C_p = \frac{\gamma}{\gamma - 1}\mathcal{R} \quad \text{and} \quad C_v = \frac{1}{\gamma - 1}\mathcal{R} \,, \tag{1.2.23}$$

and the calorically perfect gas equation of state

$$p = \frac{\rho e}{\gamma - 1} \,. \tag{1.2.24}$$

Furthermore, we have that

$$\gamma = \frac{C_p}{C_v} \,. \tag{1.2.25}$$

1.2.3 Historical Perspective

Before commencing our summary of the basic equations of fluid dynamics, we provide a brief historical perspective on the evolution of the mathematical description of fluid motion. Navier (1827) must be credited with the first attempt at deriving the equations for homogeneous incompressible viscous fluids on the basis of considerations involving the action of intermolecular forces. Poisson (1831) derived the equations for compressible fluids from a similar molecular model. The first mathematical description of the motion of an "ideal" (inviscid and incompressible) fluid was, however, due to Euler (1755), who applied Newton's second law of motion to a fluid moving under an internal force known as the pressure gradient. D'Alembert (1752) was the first to point out that this mathematical model leads to the eponymous paradox that a body at rest in a uniform stream of ideal fluid suffers no drag: "a singular paradox which I leave to geometricians to explain". Although Navier's (1827) intermolecular interaction model did include a viscous term, which was proportional to the Laplacian of the velocity field, he did not recognize the physical significance of viscosity and considered the viscosity coefficient to be a function of molecular spacing. The continuum derivation of the Navier–Stokes equations is due to Saint-Venant (1843) and Stokes (1845). Saint-Venant published a derivation of the equations on a more physical basis that applied not only to the so-called laminar flows considered by Navier but also to turbulent flows. However, it was Stokes who, under the sole assumption that the stresses are linear functions of the strain rates, derived the equations in the form that is currently in use. Interesting details of the history of fluid dynamics can be found in Truesdell (1953), Darrigol (2002), and in the texts by Rouse and Ince (1963) and Grattan-Guinness (1990).

As noted above, Euler (1755) was the first to provide a mathematical description of the motion of inviscid fluids. The Euler equations are often used to simulate vortical flows such as those resulting from shocks or breakaway separations whose basic mechanisms are not dominated by viscous effects. The matched asymptotic expansion theories (van Dyke (1975), Moore (1989)) have established the relevance of the Euler solution as the first term of the outer expansion to be matched with the first term of the inner expansion, which describes the relatively thin boundary layer abutting the solid streamlined boundary. Thus, the Euler solution coupled with corrections from the boundary-layer equations are sometimes used as an alternative to solving the relatively expensive full Navier–Stokes equations.

Prototypes of the concept of a boundary layer associated with the no-slip condition on a solid body had existed since the derivation of the equations of motion of a viscous fluid (Stokes (1845), Rankine (1864), Froude (1872), Lorenz (1881)). However, it was Prandtl (1904) who first derived what are now known as the boundary-layer equations based on the dimensional argument that in the thin layer adjacent to a body in a stream, the viscous stress in the streamwise direction is much larger than the normal stress and is of the same order of magnitude as the streamwise inertial force. The formal mathematical basis for deriving the boundary-layer theory from the Navier–Stokes equations was achieved by Lagerstrom and associates (see van Dyke (1969)), who developed the singular perturbation theory or method of matched asymptotic expansions. Simply stated, this consists of constructing the so-called outer and inner expansions by iterating the Navier–Stokes equations between an outer inviscid solution (that does not satisfy the no-slip condition on the body) and an inner solution (that satisfies the no-slip condition on the body) and matching them in their overlapping region of validity. Tani (1977) provides an excellent historical perspective on boundary-layer theory.

1.3 Compressible Fluid Dynamics Equations

This section begins with the most general form of the fluid dynamics equations considered in this text, namely, the compressible Navier–Stokes equations. Then various special or limiting forms of the compressible equations are presented. The following section focuses on the incompressible versions of these equations.

1.3.1 Compressible Navier–Stokes Equations

The constitutive equations are essential to the description of a viscous fluid. These are equations that relate the viscous stress tensor $\underline{\tau}$ at a given point $\mathbf{x} = (x, y, z)$ of the medium to the strain tensor $\underline{\mathbf{S}}$ at that point. The term *Stokesian fluid* is used for fluids that satisfy the conditions that the stress tensor (i) is a continuous function of the strain tensor (and independent of all other kinematic quantities), (ii) is spatially homogeneous, (iii) is spatially isotropic, and (iv) is equal to $-p\underline{\mathcal{I}}$ (where $\underline{\mathcal{I}}$ is the unit tensor and p is a scalar) if the strain vanishes. These imply (see, e. g., Serrin (1959a)), that the stress tensor is a linear combination of the identity matrix, the strain tensor, and the square of the strain tensor. *Classical fluids* are those for which the additional condition that the stress tensor is a linear function of the strain tensor is imposed. (The term Newtonian fluid is sometimes used for classical fluids, although this term is more properly applied to just the incompressible situation.) This yields the classical constitutive equation for the total stress tensor $\underline{\sigma}$:

$$\underline{\sigma} = -p\underline{\mathcal{I}} + \underline{\tau} \,, \tag{1.3.1}$$

with the viscous stress tensor

$$\underline{\tau} = 2\mu\underline{\mathbf{S}} + \lambda(\nabla \cdot \mathbf{u})\underline{\mathcal{I}} \,, \tag{1.3.2}$$

and the strain-rate tensor

$$\underline{\mathbf{S}} = \frac{1}{2}[\nabla\mathbf{u} + (\nabla\mathbf{u})^T] \,. \tag{1.3.3}$$

Here, μ and λ are called the *first* and *second coefficients of viscosity*, respectively. The first coefficient of viscosity μ is also called the *shear viscosity* and often simply the viscosity. The symbols $\nabla\cdot$ and ∇ denote the divergence and gradient operators, respectively, with respect to the spatial coordinate $\mathbf{x}$.

The *viscous dissipation function* Φ, which represents the work done by the viscous stresses on a particle, is defined as

$$\Phi = \underline{\tau} : \nabla\mathbf{u} \,, \tag{1.3.4}$$

with the colon denoting the contraction of the double-index tensors. The constraint that the dissipation function is always nonnegative requires that

$$\mu \geq 0 \quad \text{and} \quad \left(\lambda + \frac{2}{3}\mu\right) \geq 0 \,. \tag{1.3.5}$$

The expression $(\lambda + \frac{2}{3}\mu)$ is called the *bulk viscosity*. The limiting case corresponding to $\mu = \lambda = 0$ yields a simplified model, the Euler equations, discussed in Sect. 1.3.4.

When the effect of thermal conduction is significant, one uses the *Fourier law*

$$\mathbf{q} = -\kappa \nabla T \tag{1.3.6}$$

to express the *heat flux* $\mathbf{q}$ as a function of the temperature gradient through the *thermal conductivity* κ.

The relationship between κ and μ is customarily written in terms of the *Prandtl number*, defined as

$$\mathrm{Pr} = \frac{C_p \mu}{\kappa} . \tag{1.3.7}$$

This is the nondimensional ratio of the viscous diffusion to the thermal conductivity. The length scale for viscous diffusion is greater than or less than the length scale for thermal conductivity depending upon whether Pr is less than or greater than 1, respectively. In general, the Prandtl number is not constant, although the simplifying assumption of a constant Prandtl number is very common in applications.

The coefficients of viscosity and the thermal conductivity are dependent upon the thermodynamic variables, primarily the temperature. For gases, the shear viscosity is commonly taken from *Sutherland's formula*

$$\frac{\mu}{\mu_r} = \frac{T_r + S_1}{T + S_1} \left(\frac{T}{T_r} \right)^{3/2} , \tag{1.3.8}$$

where μ_r is the viscosity at the reference temperature T_r, S_1 is a constant, and the temperature T is given in degrees Kelvin. For air, $S_1 = 110.4\,^\circ\mathrm{K}$, and for the reference values $T_r = 273.1\,^\circ\mathrm{K}$ and $\mu_r = 1.716 \times 10^{-4}$ gm/cm-sec, (1.3.8) becomes

$$\mu = \frac{1.458\, T^{3/2}}{T + 110.4} \times 10^{-5}\ \mathrm{gm/cm\text{-}sec} , \tag{1.3.9}$$

and the bulk viscosity is taken to be zero, yielding

$$\lambda = -\frac{2}{3}\mu . \tag{1.3.10}$$

The algebraic form of Sutherland's formula is based on molecular forces considerations (Sutherland (1893)), and the constants are determined by fitting experimental data for a given gas (Anon. (1949, 1953)); for air, it is quite accurate for temperatures between $100\,^\circ\mathrm{K}$ and $1890\,^\circ\mathrm{K}$. Although empirical formulas for the thermal conductivity are also available, κ is often just taken to be proportional to μ, i.e., the Prandtl number Pr is taken to be strictly constant in (1.3.7). For example, Pr $= 0.7$ is usually used for air.

The *Navier–Stokes equations* are a differential form of the three conservation laws that govern the motion in time t of classical fluids. The conservation laws are first expressed in integral form as the appropriate conservation statement for any arbitrary volume that moves with the fluid, accounting for any relevant surface effects. The partial differential equations then follow under suitable continuity and differentiability conditions. The *equation of mass conservation*, known as the continuity equation, is

$$\frac{\partial \rho}{\partial t} + \nabla \cdot (\rho \mathbf{u}) = 0 . \tag{1.3.11}$$

The *equation for momentum conservation* is

$$\frac{\partial (\rho \mathbf{u})}{\partial t} + \nabla \cdot (\rho \mathbf{u} \mathbf{u}^T) + \nabla p = \nabla \cdot \underline{\boldsymbol{\tau}} , \tag{1.3.12}$$

and the *equation for total energy conservation* is

$$\frac{\partial (\rho E)}{\partial t} + \nabla \cdot (\rho E \mathbf{u}) + \nabla \cdot (p \mathbf{u}) = -\nabla \cdot \mathbf{q} + \nabla \cdot (\underline{\boldsymbol{\tau}} \mathbf{u}) . \tag{1.3.13}$$

(In (1.3.12) the divergence of a tensor appears. The symbol $\nabla \cdot \underline{\boldsymbol{\tau}}$ indicates the vector whose components are $(\nabla \cdot \underline{\boldsymbol{\tau}})_i = \sum_j \frac{\partial \tau_{ij}}{\partial x_j}$.) The momentum conservation statement that leads to (1.3.12) utilizes Newton's second law via the Cauchy principle that allows one to express the surface forces in terms of the total stress (volumetric forces have been neglected). The Navier–Stokes equations (1.3.11)–(1.3.13) must be supplemented with an equation of state, e. g., (1.2.15) for ideal gases or (1.2.24) for calorically perfect gases.

When volumetric forces $\mathbf{f}$ need to be taken into account, the momentum and total energy conservation laws become balance laws, whereas the mass conservation equation holds unchanged. In this more general case the Navier–Stokes equations read

$$\frac{\partial \rho}{\partial t} + \nabla \cdot (\rho \mathbf{u}) = 0 , \tag{1.3.14}$$

$$\frac{\partial (\rho \mathbf{u})}{\partial t} + \nabla \cdot (\rho \mathbf{u} \mathbf{u}^T) + \nabla p = \nabla \cdot \underline{\boldsymbol{\tau}} + \mathbf{f} , \tag{1.3.15}$$

$$\frac{\partial (\rho E)}{\partial t} + \nabla \cdot (\rho E \mathbf{u}) + \nabla \cdot (p \mathbf{u}) = -\nabla \cdot \mathbf{q} + \nabla \cdot (\underline{\boldsymbol{\tau}} \mathbf{u}) + \mathbf{f} \cdot \mathbf{u} . \tag{1.3.16}$$

Alternative forms of the energy equation in terms of specific (internal) energy e, entropy s, temperature T, and pressure p, are

$$\rho \left(\frac{\partial e}{\partial t} + \mathbf{u} \cdot \nabla e \right) + p \nabla \cdot \mathbf{u} = -\nabla \cdot \mathbf{q} + \Phi , \tag{1.3.17}$$

$$\rho T \left(\frac{\partial s}{\partial t} + \mathbf{u} \cdot \nabla s \right) = -\nabla \cdot \mathbf{q} + \Phi , \tag{1.3.18}$$

$$\rho \frac{\partial (C_v T)}{\partial t} + \rho \mathbf{u} \cdot \nabla (C_v T) + p \nabla \cdot \mathbf{u} = -\nabla \cdot \mathbf{q} + \Phi , \tag{1.3.19}$$

$$\frac{\partial p}{\partial t} + \mathbf{u} \cdot \nabla p + \gamma p \nabla \cdot \mathbf{u} = (\gamma - 1)[-\nabla \cdot \mathbf{q} + \Phi] . \tag{1.3.20}$$

Equations (1.3.11)–(1.3.18) apply even for non-ideal gases. The equation for the temperature (1.3.19), however, is specialized to thermally perfect gases, while the equation for the pressure (1.3.20) is restricted even further to calorically perfect gases.

The alternatives (1.3.17)–(1.3.20) hold even in the presence of volumetric forces, which appear explicitly in (1.3.16) but in none of these alternative equations. For the most part, in the remainder of this chapter our discussion presumes that volumetric forces are neglected, although elsewhere in this text they are quite often included.

Numerical computations of compressible Navier–Stokes problems are most commonly based on the conservation forms, (1.3.11)–(1.3.13). The alternative forms for the energy equation, (1.3.19)–(1.3.20), are used occasionally for stability analysis and for computations of compressible flows that have no discontinuities.

The previous equations hold at any point $\mathbf{x}$ of the region Ω occupied by the fluid at a given time. If the boundary $\partial\Omega$ of Ω consists of a solid, stationary wall, then the boundary conditions on $\partial\Omega$ are no-slip for velocity,

$$\mathbf{u} = \mathbf{0} \, ,$$

and some appropriate temperature condition, commonly either constant temperature,

$$T = T_{\text{wall}} \, ,$$

or adiabatic conditions,

$$\kappa \frac{\partial T}{\partial n} = \kappa \nabla T \cdot \mathbf{n} = 0 \, ,$$

where $\mathbf{n}$ is the outward (with respect to Ω) unit vector normal to the boundary.

The Navier–Stokes equations are an incompletely parabolic system (Belov and Yanenko (1971), Strikwerda (1976)); the absence of a diffusion term in the continuity equation prevents them from being fully parabolic. Consequently, there are only four boundary conditions on solid walls, but there needs to be a fifth boundary condition at open boundaries at those points for which $\mathbf{u} \cdot \mathbf{n} < 0$. The appropriate choice of mathematical boundary conditions on inflow and outflow boundaries is rather subtle. As discussed by Gustafsson and Sundstrom (1978) and Oliger and Sundstrom (1978), one desires not just that the problem be mathematically well-posed, but also that there be no nonphysical boundary layers near the inflow and outflow boundaries. For example, consider a flow in a two-dimensional channel with solid boundaries on the top and the bottom, an inflow boundary on the left and an outflow boundary on the right. There will be viscous boundary layers next to the two walls, but there should not be any boundary layers at the inflow and outflow boundaries other than those near the walls. This will be achieved if the imposed boundary conditions yield a well-posed problem in the inviscid limit. See the above references for further details.

The mathematical analysis of the compressible Navier–Stokes equations can be found in Serrin (1959b), Matsumura and Nishida (1979), Valli (1983), P.-L. Lions (1998), and Hoff (2002, 2006).

1.3.2 Nondimensionalization

The mathematical statement of a particular flow problem involves specification of geometric, flow and fluid parameters—some characteristic (or reference) length L_{ref} (such as the distance between two enclosing boundaries as in channel flow, or the maximum linear dimension of a solid body), characteristic length velocity U_{ref} (such as the free-stream velocity), characteristic temperature T_{ref}, pressure p_{ref} and density ρ_{ref} (based, for example, on free-stream conditions and related by $p_{\text{ref}} = \rho_{\text{ref}} R T_{\text{ref}}$), viscosity μ_{ref} and conductivity κ_{ref}.

Let us introduce nondimensional quantities identified with superscript $*$:

$$\mathbf{x}^* = \frac{\mathbf{x}}{L_{\text{ref}}}, \qquad t^* = \frac{U_{\text{ref}}\, t}{L_{\text{ref}}}, \qquad \mathbf{u}^* = \frac{\mathbf{u}}{U_{\text{ref}}}, \qquad p^* = \frac{p}{p_{\text{ref}}}, \qquad \rho^* = \frac{\rho}{\rho_{\text{ref}}},$$

$$E^* = \frac{E}{U_{\text{ref}}^2}, \qquad \mu^* = \frac{\mu}{\mu_{\text{ref}}}, \qquad \lambda^* = \frac{\lambda}{\mu_{\text{ref}}}, \qquad \kappa^* = \frac{\kappa}{\kappa_{\text{ref}}},$$

$$(1.3.21)$$

and let ∇^* denote differentiation with respect to $\mathbf{x}^*$.

The key dimensionless parameters that arise are the reference Mach number, the *Reynolds number* and the reference Prandtl number parameters:

$$\mathrm{M}_{\text{ref}} = \sqrt{\frac{\rho_{\text{ref}} U_{\text{ref}}^2}{\gamma p_{\text{ref}}}}, \qquad \mathrm{Re} = \frac{\rho_{\text{ref}} U_{\text{ref}} L_{\text{ref}}}{\mu_{\text{ref}}}, \qquad \mathrm{Pr}_{\text{ref}} = \frac{\mu_{\text{ref}} C_p}{\kappa_{\text{ref}}}. \quad (1.3.22)$$

(The Mach number and the Prandtl number also have the more local meanings given in (1.3.46) and (1.3.7), respectively. As we only consider flows with constant Prandtl number in this text, we henceforth drop the subscript $_{\text{ref}}$ on the Prandtl number in the nondimensional equations. We do, however, retain this subscript to ensure the distinction between the reference Mach number used for nondimensionalization and the local Mach number of the flow.)

In terms of these nondimensional coordinates, variables and parameters, the compressible Navier–Stokes equations (1.3.11)–(1.3.13) for a calorically perfect gas are

$$\frac{\partial \rho^*}{\partial t^*} + \nabla^* \cdot (\rho^* \mathbf{u}^*) = 0, \qquad (1.3.23)$$

$$\frac{\partial (\rho^* \mathbf{u}^*)}{\partial t^*} + \nabla^* \cdot (\rho^* \mathbf{u}^* \mathbf{u}^{*T}) + \frac{1}{\gamma \mathrm{M}_{\text{ref}}^2} \nabla^* p^* = \frac{1}{\mathrm{Re}} \nabla^* \cdot \boldsymbol{\tau}^*, \qquad (1.3.24)$$

$$\frac{\partial (\rho^* E^*)}{\partial t^*} + \nabla^* \cdot (\rho^* E^* \mathbf{u}^*) + (\gamma - 1) \nabla^* \cdot (p^* \mathbf{u}^*)$$

$$= \frac{\gamma}{\mathrm{Pr} \mathrm{Re}} \nabla^* \cdot (\kappa^* \nabla^* T^*) + \gamma(\gamma - 1) \frac{\mathrm{M}_{\text{ref}}^2}{\mathrm{Re}} \nabla^* \cdot (\boldsymbol{\tau}^* \mathbf{u}^*); \qquad (1.3.25)$$

the nondimensional (ideal gas) equation of state is

$$p^* = \rho^* T^* \,, \tag{1.3.26}$$

and the nondimensional viscous stress tensor and strain tensor are, respectively,

$$\underline{\tau}^* = 2\mu^* \underline{\mathbf{S}}^* + \lambda^* (\nabla^* \cdot \mathbf{u}^*) \underline{\mathcal{I}} \,, \tag{1.3.27}$$

$$\underline{\mathbf{S}}^* = \frac{1}{2} [\nabla^* \mathbf{u}^* + (\nabla^* \mathbf{u}^*)^T] \,. \tag{1.3.28}$$

Various limiting cases of these equations have important applications. The resulting simplified systems of equations are discussed later in this section and in the following section.

1.3.3 Navier–Stokes Equations with Turbulence Models

The Navier–Stokes equations were derived from first principles. They are appropriate for use in numerical computations for which all relevant spatial and temporal scales are resolved. This is usually feasible for laminar flow, but rarely feasible for turbulent flow. See CHQZ2, Sect. 1.3 for a detailed discussion of temporal and spatial scales in turbulent flow. In particular, we recall that for isotropic turbulence the range of spatial scales increases as $\mathrm{Re}^{3/4}$ in each direction, and that the range of temporal scales grows as $\sqrt{\mathrm{Re}}$. For sufficiently small Reynolds numbers Re, all the scales can be resolved. The term *direct numerical simulation* (DNS) is often used to describe computations of time-dependent flows in which all the scales are resolved.

For most fluid dynamical engineering applications turbulence effects are important and DNS of such turbulent flows is impractical. Hence, the Navier–Stokes equations are often augmented with turbulence models. *Reynolds-averaged Navier–Stokes* (RANS) turbulence models are the norm in engineering. (See the references in Sect. 1.4.2 for discussions of RANS models for incompressible flow and Wilcox (1993) for some particulars for compressible flow.) *Large-eddy simulation* (LES) models emerged in the 1970s as a research tool for turbulence flow physics research and began to make a limited impact in engineering applications in the 1990s. Various hybrid RANS/LES approaches have been developed. These use RANS models in part of the flow and LES models in other parts. One example is the *detached-eddy simulation* (DES) approach (see Spalart (2000)) that uses RANS models in attached boundary layers and LES models in regions of separated flow.

Sagaut (2006) provides extensive coverage of LES formulations and models, albeit only for incompressible flow. Pope (2000) discusses LES for compressible (and reacting) flows. Gatski, Hussaini and Lumley (1996) provide an introduction to DNS, LES and RANS modeling of turbulent flows. Spalart (2000) reviews RANS, LES and DES models.

The LES and RANS equations are derived from the Navier–Stokes equations by an averaging procedure. This involves filtering out the chaotic, fluctuating, small-scale, high-frequency motions, and modeling their effect on the slowly varying, smooth eddies for the dependent variables in the new governing equations. For example, for homogeneous flows a filter function $\bar{\mathbf{u}}(\mathbf{x}, t)$ is often defined as

$$\bar{\mathbf{u}}(\mathbf{x}, t) = \int \int G(\mathbf{x} - \mathbf{x}', t - t', \mathbf{x}, t)\, \mathbf{u}(\mathbf{x}', t')\, \mathrm{d}\mathbf{x}'\, \mathrm{d}t' \,, \tag{1.3.29}$$

where G is the weight or filter function in space and/or time. This has compact support (i. e., it vanishes outside a finite domain within which it assumes finite values) and satisfies

$$\int \int G(\mathbf{x} - \mathbf{x}', t - t', \mathbf{x}, t)\, \mathrm{d}\mathbf{x}'\, \mathrm{d}t' = 1 \,. \tag{1.3.30}$$

The filtering procedure needs to be properly adapted when one approaches the boundary; see, e. g., Sagaut (2006) for a discussion of appropriate LES filters near boundaries.

The key step in the derivation of the LES or RANS equations is averaging over small scales by the aforementioned procedures applied to the Navier–Stokes equations. (The RANS equations are often derived using an ensemble or temporal average; their equivalence to spatial averaging through some sort of ergodic hypothesis is not proven, but generally assumed.) What distinguishes LES from RANS is the definition of the small scales. LES assumes the small scales to be smaller than the mesh size Δ, and RANS assumes them to be smaller than the largest eddy scale L.

To illustrate the basic concept and the resulting closure problem, consider the simple scalar equation

$$\frac{\partial u}{\partial t} + \frac{\partial}{\partial x}(uu) = 0 \,, \tag{1.3.31}$$

and write

$$u = \bar{u} + u' \,, \tag{1.3.32}$$

where $\bar{u}$ is the filtered (or averaged) value of u, and u' is the fluctuating component. Applying the filtering operator (1.3.29) to (1.3.32) and assuming that G is chosen so that the filtering and differentiation operators commute, we have

$$\frac{\partial \bar{u}}{\partial t} + \frac{\partial}{\partial x}(\overline{uu}) = 0 \,. \tag{1.3.33}$$

After moving the nonlinear term to the right-hand side and adding $\partial(\bar{u}\bar{u})/\partial x$ to both sides, we obtain

$$\frac{\partial \bar{u}}{\partial t} + \frac{\partial}{\partial x}(\bar{u}\bar{u}) = -\frac{\partial}{\partial x}(\overline{uu} - \bar{u}\bar{u}) \,. \tag{1.3.34}$$

If the filter function G satisfies

$$G^2 = G \tag{1.3.35}$$

(such a filter is called a *Reynolds operator*), then

$$\bar{\bar{u}} = \bar{u} \,, \qquad \overline{u'} = 0 \,, \tag{1.3.36}$$

and $(\overline{uu} - \bar{u}\bar{u}) = \overline{u'u'}$. The filter functions used for RANS are time averages or ensemble averages; these RANS filters are typically Reynolds operators. The term $\overline{u'u'}$ is the *subgrid-scale stress* (also called the *residual stress*) for this simple equation (1.3.31), and it must be modeled. The filters used in LES usually do not satisfy (1.3.35). (An exception that does satisfy (1.3.36) is the sharp cutoff filter in Fourier space.)

We focus here on the LES equations since spectral methods have been used far more for LES computations than for RANS or DES computations. Indeed, classical spectral methods figured prominently in many of the early LES computations in the 1970s and 1980s. (As in CHQZ2, we distinguish between spectral methods in single domains, termed classical spectral methods, and the multidomain spectral methods appropriate for applications in complex geometries.) Applications of RANS (and DES) methods are almost invariably for problems in complex geometries, for which classical spectral methods are inappropriate. However, use of multidomain spectral methods for RANS of smooth flows in complex geometries would only be productive if the error due to the RANS modeling were less important than the discretization error. In most RANS applications for complex flows, the uncertainty due to the turbulence model overwhelms the uncertainty due to discretization error.

For compressible LES, as for RANS, the averaged equations simplify if a Favre (density-weighted) average, denoted by a tilde, is used for some of the variables. For example, the Favre-averaged velocity is given by

$$\tilde{\mathbf{u}} = \frac{\overline{\rho \mathbf{u}}}{\bar{\rho}} \,. \tag{1.3.37}$$

The LES equations for compressible flow that result from Favre averaging are the continuity equation

$$\frac{\partial \bar{\rho}}{\partial t} + \nabla \cdot (\bar{\rho}\tilde{\mathbf{u}}) = 0 \,, \tag{1.3.38}$$

the momentum equation (the meaning of the superscript c will be given below)

$$\frac{\partial (\bar{\rho}\tilde{\mathbf{u}})}{\partial t} + \nabla \cdot (\bar{\rho}\tilde{\mathbf{u}}\tilde{\mathbf{u}}^T) + \nabla \bar{p} - \nabla \cdot \underline{\boldsymbol{\tau}}^c = \nabla \cdot (\underline{\bar{\boldsymbol{\tau}}} - \underline{\boldsymbol{\tau}}^c) \,, \tag{1.3.39}$$

the energy equation

$$\frac{\partial(\bar{\rho}\tilde{E})}{\partial t} + \nabla \cdot (\bar{\rho}\tilde{\mathbf{u}}\tilde{E}) + \nabla \cdot (\bar{p}\tilde{\mathbf{u}}) + \nabla \cdot \mathbf{q}^c - \nabla \cdot (\underline{\tau}^c\tilde{\mathbf{u}}) \qquad (1.3.40)$$

$$= -\nabla \cdot \left[\bar{\rho}(\widetilde{\mathbf{u}E} - \tilde{\mathbf{u}}\tilde{E})\right] - \nabla \cdot (\bar{\mathbf{q}} - \mathbf{q}^c) - \nabla \cdot (\overline{p\mathbf{u}} - \bar{p}\tilde{\mathbf{u}}) + \nabla \cdot (\overline{\underline{\tau}\mathbf{u}} - \underline{\tau}^c\tilde{\mathbf{u}}) \,,$$

and the (ideal gas) equation of state

$$\bar{p} = \bar{\rho}\mathcal{R}\tilde{T} \,. \qquad (1.3.41)$$

Note that the total energy in (1.3.40) is given by

$$\bar{\rho}\tilde{E} = \frac{\bar{p}}{\gamma - 1} + \frac{1}{2}\bar{\rho}\widetilde{\mathbf{u} \cdot \mathbf{u}} \,. \qquad (1.3.42)$$

The resolved kinetic energy is computed from the trace of modeled subgrid-scale stresses, assuming that, unlike the typical situation for incompressible flows, the full subgrid-scale tensor (and not just its anisotropic part—see (1.4.11)) is modeled.

The energy equation (1.3.40) is only one of several possible formulations found in the literature. Alternate formulations include equations for the pressure, internal energy, enthalpy, and a modified total energy $\widehat{\rho E} = \bar{\rho}(\tilde{e} + \frac{1}{2}\tilde{\mathbf{u}}\cdot\tilde{\mathbf{u}})$.

Terms on the right-hand side of (1.3.39)–(1.3.40) must be modeled. The full descriptions of the compressible LES models are quite complicated. We refer the reader to Yoshizawa (1986) and to Speziale, Erlebacher, Zang and Hussaini (1988) for some early models, and to Pope (2000) for a more thorough coverage. A representative complete LES model for the simpler case of incompressible flow is provided below in Sect. 1.4.2.

A superscript c refers to nonlinear functions in which each variable is replaced by its resolved value (simple average or Favre average as the case may be) to permit an approximate evaluation. In particular, the spatial average of the viscous stresses are

$$\bar{\tau}_{ij} = \overline{2\mu S_{ij}} + \overline{\lambda \delta_{ij} S_{kk}} \,,$$

while a computable form of the viscous stress tensor is

$$\tau_{ij}^c = 2\mu^c \tilde{S}_{ij} + \lambda^c \delta_{ij} \tilde{S}_{kk} \,.$$

The viscosities are in general functions of temperature, and their computable forms are defined by $\mu^c = \mu(\tilde{T})$ and $\lambda^c = \lambda(\tilde{T})$. Note that the temperature appears naturally as a Favre-averaged quantity. In the energy equation, the heat fluxes are

$$\bar{\mathbf{q}} = -\overline{\kappa\nabla T} \,, \qquad \mathbf{q}^c = -\kappa^c\nabla\tilde{T} \,,$$

with a resolved-scale, temperature-dependent conductivity $\kappa^c = \kappa(\tilde{T})$.

1.3.4 Compressible Euler Equations

As noted in Sect. 1.3.2, various limiting cases of the nondimensionalized Navier–Stokes equations are of interest. If one lets Re $\to \infty$, i.e., $\mu \to 0$, in (1.3.23)–(1.3.25), then all the terms on the right-hand sides vanish, and one obtains the *Euler equations*. In dimensional form these are

$$\frac{\partial \rho}{\partial t} + \nabla \cdot (\rho \mathbf{u}) = 0 \, , \tag{1.3.43}$$

$$\frac{\partial (\rho \mathbf{u})}{\partial t} + \nabla \cdot (\rho \mathbf{u} \mathbf{u}^T) + \nabla p = \mathbf{0} \, , \tag{1.3.44}$$

$$\frac{\partial (\rho E)}{\partial t} + \nabla \cdot (\rho \mathbf{u} E) + \nabla \cdot (p \mathbf{u}) = 0 \, , \tag{1.3.45}$$

supplemented with an equation of state. (These equations coincide with the Navier–Stokes equations (1.3.11)–(1.3.13) if the right-hand side terms are disregarded.) Quite apart from their engineering use in obtaining acceptable physical solutions, the Euler equations are frequently used in the analysis and development of numerical algorithms that are eventually applied to the Navier–Stokes equations.

This system is hyperbolic. It admits weak or discontinuous solutions (Courant and Friedrichs (1976), Lax (1973)). The order of the Euler equations is reduced by one compared with the full Navier–Stokes equations, leading to the loss of the velocity no-slip boundary condition and the boundary condition on the temperature (or other thermodynamic variable). The boundary conditions at stationary solid walls are no-flux:

$$\mathbf{u} \cdot \mathbf{n} = 0 \, .$$

The conditions at inflow or outflow boundaries are dependent upon the characteristic directions there. Appropriate boundary conditions are discussed in Sect. 4.3.

The flow is called *subsonic* at a point $\mathbf{x}$ and a time t if $|\mathbf{u}| < c$ and *supersonic* if $|\mathbf{u}| > c$. The (local) *Mach number* M is given by

$$M = |\mathbf{u}|/c \, . \tag{1.3.46}$$

In order for the flow to change from a supersonic state to a subsonic one, it must (except in rare circumstances) pass through a shock wave. Here the flow variables are discontinuous and the differential equations themselves do not apply, although the more basic integral conservation laws still hold.

1.3.5 Compressible Potential Equation

Experience has shown that rather accurate predictions can be made of a number of compressible flows of practical importance under the assumption that the flow is irrotational, i. e., that the vorticity is zero (Lighthill (1989), Holst and Ballhaus (1979)). In this case, the velocity is derivable from a *velocity potential* ϕ, i. e.,

$$\mathbf{u} = \nabla \phi \,, \tag{1.3.47}$$

and (1.3.43) becomes the *compressible potential equation*

$$\frac{\partial \rho}{\partial t} + \nabla \cdot (\rho \nabla \phi) = 0 \,, \tag{1.3.48}$$

where ρ is related to $\nabla \phi$ by the isentropic relation

$$\rho = \rho_0 \left[1 + \frac{\gamma - 1}{2} \frac{|\nabla \phi|^2}{c^2} \right]^{1/(1-\gamma)} \,, \tag{1.3.49}$$

in which ρ_0 is the stagnation density, i. e., the density at a point at which the velocity is zero (a stagnation point). The momentum equation (1.3.44) is automatically satisfied. For steady flow, we have simply

$$\nabla \cdot (\rho \nabla \phi) = 0 \,. \tag{1.3.50}$$

The compressible potential equation is often referred to as the *full potential equation* to distinguish it from related but more severe approximations, such as the transonic small disturbance equation. (It is worth recalling that the transonic small disturbance equation was the setting for the type-dependent (elliptic/hyperbolic) difference scheme of Murman and Cole (1971), which was the breakthrough that ignited the intense developments in CFD for aerospace applications in the ensuing decades.)

Despite its seemingly restrictive assumption, the full potential equation has been a popular design tool in industry. As an irrotational flow is isentropic (the Crocco theorem—see, for example, Liepmann and Roshko (2001)), it is but consistent to replace a shock with a singular surface, across which all flow quantities are discontinuous except entropy. Since, as shown for example in Moore (1989), the entropy jump across a shock is proportional to the third power of the shock strength (measured by the jump in Mach number), the potential flow approximation has been found particularly useful in the formulation of transonic flow problems where the shock Mach number is less than 1.3. The full potential solution can be coupled with the boundary-layer solution to incorporate viscous effects. The relatively low computational requirements (such as low storage per cell) of the potential model (Flores et al. (1985)) and its innate preclusion of spurious entropy generation at stagnation points (often afflicting Euler and Navier–Stokes algorithms) makes it an attractive alternative for treating flows past complex geometries of industrial interest on highly dense grids. However, it must be pointed out that although the full potential equation may provide accurate fixed-point solutions in valid regimes, it may not accurately predict trends such as the lift-versus-slope curve (Salas and Gumbert (1985)).

The desire for efficient numerical solutions of (1.3.50) for use in aircraft design motivated considerable progress in computational fluid dynamics in the late 1970s and early 1980s; see, e. g., Jameson (1978) and Glowinski (1984).

1.3.6 Compressible Boundary-Layer Equations

Prandtl (1904) introduced the concept of a *boundary layer*—a thin layer adjacent to solid walls to which the effects of viscosity are often confined. The so-called *boundary-layer equations* are the companion inner equations to the (outer) Euler equations. These inner equations are obtained by a special limit process. For a two-dimensional flow, let the coordinate system (x^*, y^*) be such that the x^*-coordinate is along the boundary of the solid body in the direction of the free stream, and y^* is normal to the boundary. Introducing the scaled normal coordinate and normal velocity,

$$\bar{y}^* = \frac{y^*}{\sqrt{\mathrm{Re}}} \,, \quad \bar{v}^* = \frac{v^*}{\sqrt{\mathrm{Re}}} \,,$$

into (1.3.23)–(1.3.25) (see (1.5.14) for a particular choice of reference quantities) and letting $\mathrm{Re} \to \infty$, one obtains the compressible boundary-layer equations, which in nondimensional form are

$$\frac{\partial \rho^*}{\partial t^*} + \frac{\partial (\rho^* u^*)}{\partial x^*} + \frac{\partial (\rho^* \bar{v}^*)}{\partial \bar{y}^*} = 0 \,, \tag{1.3.51}$$

$$\frac{\partial (\rho^* u^*)}{\partial t^*} + \frac{\partial (\rho^* u^* u^*)}{\partial x^*} + \frac{\partial (\rho^* u^* \bar{v}^*)}{\partial \bar{y}^*} + \frac{\partial p^*}{\partial x^*} = \frac{\partial}{\partial \bar{y}^*} \left(\mu^* \frac{\partial u^*}{\partial \bar{y}^*} \right) \,, \tag{1.3.52}$$

$$\frac{\partial (\rho^* E^*)}{\partial t^*} + \frac{\partial (\rho^* u^* E^*)}{\partial x^*} + \frac{\partial (\rho^* \bar{v}^* E^*)}{\partial \bar{y}^*} + (\gamma - 1) \frac{\partial (p^* u^*)}{\partial x^*} + (\gamma - 1) \frac{\partial (p^* \bar{v}^*)}{\partial \bar{y}^*}$$
$$= \frac{\gamma}{\mathrm{Pr}} \frac{\partial}{\partial \bar{y}^*} \left(\kappa^* \frac{\partial T^*}{\partial \bar{y}^*} \right) + \gamma (\gamma - 1) \mathrm{M}_{\mathrm{ref}}^2 \left(\frac{\partial u^*}{\partial \bar{y}^*} \right)^2 \,. \tag{1.3.53}$$

The wall-normal momentum equation reduces to $\partial p^* / \partial \bar{y}^* = 0$, which means that the pressure is constant across the boundary layer.

The customary version of the compressible boundary-layer equations for steady, two-dimensional flow is, in dimensional form,

$$\frac{\partial (\rho u)}{\partial x} + \frac{\partial (\rho v)}{\partial y} = 0 \,, \tag{1.3.54}$$

$$\rho u \frac{\partial u}{\partial x} + \rho v \frac{\partial u}{\partial y} = -\frac{\partial p}{\partial x} + \frac{\partial}{\partial y} \left(\mu \frac{\partial u}{\partial y} \right) \,, \tag{1.3.55}$$

$$\rho C_p u \frac{\partial T}{\partial x} + \rho C_p v \frac{\partial T}{\partial y} = u \frac{\partial p}{\partial x} + \frac{\partial}{\partial y} \left(\kappa \frac{\partial T}{\partial y} \right) + \mu \left(\frac{\partial u}{\partial y} \right)^2 \,, \tag{1.3.56}$$

where the total energy E has been replaced with the temperature T. These are supplemented by an equation of state. The boundary conditions are (for fixed wall temperature)

$$
\begin{aligned}
u = v = 0, \quad & T = T_w \qquad \text{at } y = 0, \\
u = u_\infty, \quad & T = T_\infty \qquad \text{at } y = \infty,
\end{aligned}
\tag{1.3.57}
$$

plus appropriate inflow conditions at some $x = x_{\text{inflow}}$. The quantities u_∞ and T_∞, as well as the (y-independent) pressure (all potentially functions of x) come from the outer solution. Some alternative wall temperature conditions are an adiabatic wall or a specified wall heat flux.

An important measure of the size of the boundary layer for flows past a solid body is the *displacement thickness* δ^*, defined (in the case of flow past a flat plate) by

$$
\delta^* = \int_0^\infty \left(1 - \frac{\rho u}{\rho_\infty u_\infty} \right) dy .
\tag{1.3.58}
$$

It measures the deflection of the incident streamlines in the direction normal to the wall. A related measure is the *boundary-layer thickness* δ, commonly defined as the distance from the wall at which the velocity parallel to the wall has reached 99.5% (or sometimes 99.9%) of its free-stream value. In practice, this is a less precise measure than the displacement thickness.

1.3.7 Compressible Stokes Limit

In the limit Re $\rightarrow 0$, the viscous forces and the pressure forces balance each other. Hence, the appropriate scaling of pressure is

$$
p^+ = \left(\frac{L_{\text{ref}}}{U_{\text{ref}} \mu_{\text{ref}}} \right) p = \left(\frac{\text{Re}}{\gamma \text{M}_{\text{ref}}^2} \right) p^* .
\tag{1.3.59}
$$

Introducing this new pressure variable into (1.3.23)–(1.3.25) and taking Re $\rightarrow 0$, one obtains the compressible Stokes equations

$$
\frac{\partial \rho^*}{\partial t^*} + \nabla^* \cdot (\rho^* \mathbf{u}^*) = 0 ,
\tag{1.3.60}
$$

$$
\nabla^* p^+ = \nabla^* \cdot \underline{\tau}^* ,
\tag{1.3.61}
$$

$$
0 = \frac{1}{\text{Pr}} \nabla \cdot (\kappa^* \nabla T^*) + \gamma(\gamma - 1) \text{M}_{\text{ref}}^2 \nabla^* \cdot (\underline{\tau}^* \mathbf{u}^*) ,
\tag{1.3.62}
$$

usually accompanied by the calorically perfect gas equation of state

$$
\rho^* e^* = 1 ,
\tag{1.3.63}
$$

with e having the same scaling as E in (1.3.21).

1.3.8 Low Mach Number Compressible Limit

In the limit $M_{ref} \to 0$, the pressure term becomes singular in (1.3.24), thereby implying that the nondimensionalization of pressure with respect to the reference value in (1.3.21) is inappropriate. An appropriate nondimensionalized pressure for this limiting case is

$$\tilde{p} = \frac{p}{\rho_{ref} U_{ref}^2} = \frac{p^*}{\gamma M_{ref}^2} , \tag{1.3.64}$$

as pressure plays a dynamic role rather than a thermodynamic one.

Introducing this re-scaled pressure into (1.3.11)–(1.3.13), and then letting $M_{ref} \to 0$ yields the nondimensional compressible flow equations at zero Mach number:

$$\frac{\partial \rho^*}{\partial t^*} + \nabla^* \cdot (\rho^* \mathbf{u}^*) = 0 , \tag{1.3.65}$$

$$\frac{\partial (\rho^* \mathbf{u}^*)}{\partial t^*} + \nabla^* \cdot (\rho^* \mathbf{u}^* \mathbf{u}^{*T}) + \nabla^* \tilde{p} = \frac{1}{Re} \nabla^* \cdot \underline{\tau}^* , \tag{1.3.66}$$

$$\frac{\partial (\rho^* E^*)}{\partial t^*} + \nabla^* \cdot (\rho^* \mathbf{u}^* E^*) = \frac{1}{PrRe} \nabla^* \cdot (\kappa^* \nabla^* T^*) . \tag{1.3.67}$$

1.4 Incompressible Fluid Dynamics Equations

The equations described in the preceding section all have special forms in the incompressible limit. There are several different formulations of the incompressible Navier–Stokes equations. The most commonly used ones are the primitive variable (velocity and pressure), streamfunction-vorticity, and vorticity-velocity formulations. The primitive-variable formulation has been the one most extensively employed in large-scale three-dimensional calculations using spectral methods.

1.4.1 Incompressible Navier–Stokes Equations

The incompressible Navier–Stokes equations in primitive variable form can be derived from the low Mach number compressible limit if the additional assumption is made that the value of the nondimensional internal energy at the solid wall is $e_w^* = 1$. Then the equation of internal energy and the equation of state are identically satisfied by $\rho^* = 1$ and $e^* = 1$, and the above equations reduce to the incompressible flow equations. Thus a viscous fluid may be considered incompressible if the heat due to viscous dissipation and conduction is negligible, and the pressure variation causes negligible changes in internal energy or enthalpy.

In dimensional form the incompressible Navier–Stokes equations in a domain Ω are usually written as

$$\frac{\partial \mathbf{u}}{\partial t} + \mathbf{u} \cdot \nabla \mathbf{u} = -\nabla p + \frac{1}{2}\nabla \cdot \left[\nu \left(\nabla \mathbf{u} + \nabla \mathbf{u}^T\right)\right] + \mathbf{f} , \tag{1.4.1}$$

$$\nabla \cdot \mathbf{u} = 0 , \tag{1.4.2}$$

where $\mathbf{u}$ is the velocity vector, p is the pressure (divided by ρ), $\nu = \mu/\rho$ is the *kinematic viscosity*, $\mathbf{f}$ are the volumetric forces (divided by ρ), and ρ is constant (and taken to be unity). Although for constant ν the viscous term in (1.4.1) can be written as just $\nu \Delta \mathbf{u}$ (where Δ is the Laplacian operator), we use the present form because of its greater generality. Equation (1.4.1) is the momentum equation and (1.4.2) is the continuity constraint. The equations above are in nonconservation form, unlike those in, say, (1.3.12) and (1.3.11). Observe that, due to (1.4.2), $\mathbf{u} \cdot \nabla \mathbf{u} = \nabla \cdot (\mathbf{u}\mathbf{u}^T)$. (This identity is generally lost at the discrete level.)

One would also include the limiting form of the energy equation if there were a temperature-dependent body force in the momentum equation or if the viscosity depended upon the temperature. The former situation includes the important case of Rayleigh–Bénard flow, for which (1.4.1) includes a buoyancy force linear in temperature, and the limiting form of the energy equation is an advection-diffusion equation for the temperature. However, since our analyses and examples in this text do not include Rayleigh–Bénard flow, we stick with the simpler set of equations above.

The pressure in the incompressible Navier–Stokes equations is not a thermodynamic variable satisfying an equation of state. Rather, it is an implicit dynamic variable that adjusts itself instantaneously in a time-dependent flow to satisfy the incompressibility, or divergence-free, condition. From the mathematical point of view, it may be considered as a Lagrange multiplier that ensures the kinematical constraint of incompressibility (i. e., solenoidity of the velocity field). Hence, no initial or boundary conditions are required for the pressure. Various results on the existence and uniqueness of solutions to the Navier–Stokes equations are furnished in the treatises on the mathematical analysis of these equations (J.-L. Lions (1969), Ladyženskaya (1969), Temam (2001), Kreiss and Lorentz (1989), Chorin and Marsden (1993), Foias, Manley, Rosa and Temam (2001), Galdi (1994a, 1994b), P-L. Lions (1996), Majda and Bertozzi (2002)).

1.4.2 Incompressible Navier–Stokes Equations with Turbulence Models

The LES equations for compressible flow were given in (1.3.38)–(1.3.40). Their incompressible version in the absence of volumetric forces is

$$\frac{\partial \bar{\mathbf{u}}}{\partial t} + \nabla \cdot (\bar{\mathbf{u}}\bar{\mathbf{u}}^T) + \nabla \bar{p} - \nabla \cdot \underline{\bar{\tau}} = -\nabla \cdot \left(\overline{\mathbf{u}\mathbf{u}^T} - \bar{\mathbf{u}}\bar{\mathbf{u}}^T\right) , \tag{1.4.3}$$

$$\nabla \cdot \bar{\mathbf{u}} = 0 , \tag{1.4.4}$$

with the viscous stress given by

$$\bar{\tau} = -2\nu\bar{\mathbf{S}} ,$$

where the strain tensor $\bar{\mathbf{S}}$ is given in (1.3.3), the overbar denotes a filtering operation described by (1.3.30), and the viscosity has been assumed to be constant. Since the density is constant, the Favre averaging (1.3.37) reduces to the filtering operation (1.3.30).

The term on the right-hand-side of the momentum equation is the negative of the divergence of the total subgrid-scale stress

$$\tau_{ij}^{SGS} = \overline{\mathbf{u}\mathbf{u}^T} - \bar{\mathbf{u}}\bar{\mathbf{u}}^T , \tag{1.4.5}$$

and it requires a model. The remaining terms are all directly computable in terms of the average velocity and pressure, which are the dependent variables of the computation.

The Leonard (1974) decomposition of the subgrid-scale stress reads

$$\tau_{ij}^{SGS} = L_{ij} + C_{ij} + R_{ij} , \tag{1.4.6}$$

where

$$L_{ij} = (\overline{\bar{u}_i\bar{u}_j} - \overline{u_i u_j}) , \tag{1.4.7}$$

$$C_{ij} = (\overline{u_i'\bar{u}_j} + \overline{\bar{u}_i u_j'}) , \tag{1.4.8}$$

$$R_{ij} = \overline{u_i' u_j'} , \tag{1.4.9}$$

are, respectively, the *Leonard stress*, the *cross stress* and the *Reynolds stress*. The Leonard and cross stresses vanish for Reynolds filters.

Many models for the Reynolds subgrid-scale stress are based on the simple *Smagorinsky model* (1963), which assumes that (i) the subgrid-scale fluctuations are isotropic and homogeneous, (ii) there is a Kolmogorov inertial range, $E(k) = C_K\epsilon^{2/3}k^{-5/3}$, and (iii) the mean dissipation rate $\epsilon = 2\nu\bar{\Delta}\int_0^{1/\bar{\Delta}} k^2 E(k)dk$. In the preceding expressions, k is the spatial wavenumber, $E(k)$ is the energy density, C_K is the Kolmogoroff constant and $\bar{\Delta}$ is the filter width; see, e. g., Hinze (1975) for a discussion of isotropic turbulence and the Kolmogoroff theory. The above assumptions lead to the choice of *eddy viscosity* as

$$\nu_t = (C_s\bar{\Delta})^2\bar{S} \quad \text{with} \quad \bar{S} = \sqrt{2\sum_{i,j=1}^{3}\bar{S}_{ij}\bar{S}_{ij}} , \tag{1.4.10}$$

where $C_s^2 = (1/\pi^2)(1/(3C_K))^{3/2}$. The resulting Smagorinsky subgrid-scale model for the Reynolds stress is

$$\tau_{ij}^{Smag} = R_{ij} - \frac{1}{3}\delta_{ij}\sum_{k=1}^{3} R_{kk} = -2\nu_t\bar{S}_{ij} , \tag{1.4.11}$$

with ν_t given by (1.4.10). Typically,

$$\bar{\Delta} = 2(\Delta x \Delta y \Delta z)^{1/3} , \qquad (1.4.12)$$

with Δx, Δy and Δz the computational grid spacings in the three coordinate directions. The Smagorinsky constant is usually taken to be $C_s \approx 0.1$. (We caution the reader that several different conventions are used for this model; one will often not see a factor of 2 inside the square root in (1.4.10), and sometimes the Smagorinksy constant C_s is not squared.) Note that only the anisotropic part of the Reynolds stress is modeled. The isotropic part

$$\frac{1}{3}\delta_{ij} \sum_{k=1}^{3} R_{kk}$$

can be absorbed into the pressure for incompressible flow (but not, of course, for compressible flow).

One improvement to this model that is often employed is the *dynamic Smagorinsky model*, which was proposed by Germano, Piomelli, Moin and Cabot (1991) and refined by Lilly (1992). This uses filters with two different widths to make the "constant" C_s depend upon time and usually also upon space. In addition to the *grid filter* on the scale $\bar{\Delta}$, one also employs a larger *test filter* on the scale $\hat{\Delta}$ (usually $\hat{\Delta} = 2\bar{\Delta}$). One then defines the so-called resolved turbulent stresses $\boldsymbol{\tau}^{\text{SGS,resolved}}$ and the subtest stresses $\boldsymbol{\tau}^{\text{SGS,subtest}}$, given by

$$\tau_{ij}^{\text{SGS,resolved}} = \widehat{\bar{u}_i \bar{u}_j} - \hat{\bar{u}}_i \hat{\bar{u}}_j , \qquad \tau_{ij}^{\text{SGS,subtest}} = \widehat{\overline{u_i u_j}} - \hat{\bar{u}}_i \hat{\bar{u}}_j . \qquad (1.4.13)$$

Then, the modeling assumption

$$\tau_{ij}^{\text{SGS}} - \frac{1}{3}\delta_{ij} \sum_{k=1}^{3} \tau_{kk}^{\text{SGS}} = -2C_s \alpha_{ij} ,$$

$$\tau^{\text{SGS,subtest}} - \frac{1}{3}\delta_{ij} \sum_{k=1}^{3} \tau_{kk}^{\text{SGS,subtest}} = -2C_d \beta_{ij} , \qquad (1.4.14)$$

for some tensors $\boldsymbol{\alpha}$ and $\boldsymbol{\beta}$ is made. There are several choices for these tensors. One simple choice is

$$\alpha_{ij} = \hat{\Delta}^2 |\hat{\bar{\mathbf{S}}}| \hat{\bar{S}}_{ij} , \qquad \beta_{ij} = \bar{\Delta}^2 |\bar{\mathbf{S}}| \bar{S}_{ij} . \qquad (1.4.15)$$

Upon substitution of (1.4.14) into the *Germano identity* (Germano (1992)),

$$\tau_{ij}^{\text{SGS,resolved}} = \tau_{ij}^{\text{SGS,subtest}} - \hat{\tau}_{ij}^{\text{SGS}} , \qquad (1.4.16)$$

and invocation of a least-squares minimization process, the dynamic Smagorinsky "constant" C_d is then computed from

$$C_d = -\frac{1}{2} \frac{< \sum_{i,j=1}^{3} \tau_{ij}^{\text{SGS,resolved}} \gamma_{ij} >}{< \sum_{i,j=1}^{3} \gamma_{ij} \gamma_{ij} >} , \qquad (1.4.17)$$

where $\gamma_{ij} = \beta_{ij} - \hat{\alpha}_{ij}$, and $< \cdot >$ represents an appropriate spatial averaging process. For example, in the case of homogeneous turbulence this would be a full three-dimensional spatial average, whereas for more complex flows, the spatial averaging would be more localized. All the quantities in (1.4.17) depend either directly upon the dependent variables in the LES computation, i. e., the β_{ij}, or can be computed by application of the test filter to the dependent variables, i. e., $\tau_{ij}^{\text{SGS,resolved}}$ and α_{ij}. (See Germano et al. (1991), Lilly (1992), Piomelli (2004), or Sagaut (2006) for more details and refinements.)

Alternative approaches to turbulence modeling based on the use of separate equations for the large and small scales, with the modeling confined to terms in the equations for the small scales, have been taken by Temam and coworkers (see, e.g., Dubois, Jauberteau and Temam (1998)) with the so-called *nonlinear Galerkin* method, and by Hughes and coworkers (see, e.g., the review paper by Hughes, Scovazzi and Franca (2004)) with the so-called *variational multiscale* method. They differ in the models used for the unresolved terms. These models are intrinsically connected to the underlying discretization and cannot be directly described solely in terms of a PDE system, as can the models already described. We defer further description to Sect. 3.3.5, where discretization approaches are discussed.

We refer the reader to the review by Lesieur and Metais (1996) and the text by Sagaut (2006) for thorough discussions of the subtle issues and the various subgrid-scale models that have been utilized in large-eddy simulation of incompressible flows. Some mathematical aspects of LES are discussed in Berselli, Iliescu and Layton (2006). RANS models for incompressible flow are very well-developed. As spectral methods have rarely been applied to RANS, we simply refer to Speziale (1991), Wilcox (1993), Chen and Jaw (1998) and Bernard and Wallace (2002) as some standard references on the subject.

1.4.3 Vorticity–Streamfunction Equations

One of the most interesting characteristics of a flow is its vorticity. This is denoted by $\boldsymbol{\omega}$ and is given by

$$\boldsymbol{\omega} = \nabla \times \mathbf{u} . \qquad (1.4.18)$$

It represents (half) the local rotation rate of the fluid. A dynamical equation for the vorticity is derived by taking the curl of (1.4.1), which for the constant viscosity, unforced case yields

$$\frac{\partial \boldsymbol{\omega}}{\partial t} + \mathbf{u} \cdot \nabla \boldsymbol{\omega} = \boldsymbol{\omega} \cdot \nabla \mathbf{u} + \nu \Delta \boldsymbol{\omega} . \qquad (1.4.19)$$

This is an advection-diffusion equation with the additional term $\boldsymbol{\omega} \cdot \nabla \mathbf{u}$. This term represents the effects of vortex stretching. It is identically zero for two-dimensional flows and is responsible for many of the interesting aspects of three-dimensional flows.

The vorticity can be combined with the streamfunction ψ to yield a concise description of two-dimensional flows. By setting $\mathbf{u} = (u, v, 0)^T$, a streamfunction ψ is defined by the relations

$$u = \frac{\partial \psi}{\partial y} \,, \qquad v = -\frac{\partial \psi}{\partial x} \,; \tag{1.4.20}$$

the existence of such a function is guaranteed by the solenoidal property of $\mathbf{u}$. Taking the curl of the velocity and setting $\boldsymbol{\omega} = (0, 0, \omega)^T$, we obtain

$$\Delta \psi = -\omega \,. \tag{1.4.21}$$

Equation (1.4.19) reduces to

$$\frac{\partial \omega}{\partial t} + \frac{\partial \psi}{\partial y}\frac{\partial \omega}{\partial x} - \frac{\partial \psi}{\partial x}\frac{\partial \omega}{\partial y} = \nu \Delta \omega \,. \tag{1.4.22}$$

The flow is parallel to curves of constant ψ—the streamlines. In the case of steady, rigid walls, the boundary conditions that accompany (1.4.21)–(1.4.22) are

$$\psi = 0 \,, \qquad \frac{\partial \psi}{\partial n} = 0 \,, \tag{1.4.23}$$

where $(\partial \psi / \partial n)$ represents the partial derivative of ψ in the direction normal to the wall.

Equations (1.4.21)–(1.4.23) provide a complete description of a two-dimensional incompressible flow; the velocity is then recovered through (1.4.20). Note that the pressure is not needed. The subtlety of the streamfunction-vorticity formulation is that there are no physical boundary conditions on the vorticity, but two boundary conditions on the streamfunction.

The elimination of the vorticity leads to the pure streamfunction formulation

$$\frac{\partial}{\partial t}(\Delta \psi) + \frac{\partial \psi}{\partial y}\frac{\partial}{\partial x}(\Delta \psi) - \frac{\partial \psi}{\partial x}\frac{\partial}{\partial y}(\Delta \psi) = \nu \Delta^2 \psi \,. \tag{1.4.24}$$

The extension of these approaches to three-dimensional flows requires the introduction of a second streamfunction or the use of a vector potential. The appropriate equations for the former can be found in Murdock (1986) and those for the latter in Brosa and Grossman (2002).

1.4.4 Vorticity–Velocity Equations

Another approach to eliminating the pressure from the incompressible Navier–Stokes equations is to take the curl of the momentum equation. There are several versions of the resulting vorticity-velocity equations (see, e. g., Trujillo and Karniadakis (1999)). An example, for constant viscosity, is given by

$$\frac{\partial \omega}{\partial t} + \nabla \times (\omega \times \mathbf{u}) = -\nu \nabla \times (\nabla \times \omega) \,, \tag{1.4.25}$$

$$\Delta \mathbf{u} = -\nabla \times \omega \,, \tag{1.4.26}$$

$$\nabla \cdot \mathbf{u} = 0 \,. \tag{1.4.27}$$

The initial and boundary conditions on the velocity must be supplemented with initial and boundary conditions on the vorticity. The former are usually derived from the curl of the initial velocity field, and the latter from the boundary values of the curl of the instantaneous velocity field.

Although the vorticity-velocity equations have six dependent variables instead of the four associated with the primitive variable formulations, there are some circumstances in which they present computational advantages. One simple example will be given in Sect. 3.4.1.

1.4.5 Incompressible Boundary-Layer Equations

Prandtl's boundary-layer approximation for incompressible flow yields the following lowest-order terms from (1.4.1) and (1.4.2):

$$u\frac{\partial u}{\partial x} + v\frac{\partial u}{\partial y} = -\frac{\partial p}{\partial x} + \nu\frac{\partial^2 u}{\partial y^2} \,, \tag{1.4.28}$$

$$\frac{\partial u}{\partial x} + \frac{\partial v}{\partial y} = 0 \,. \tag{1.4.29}$$

The boundary conditions are

$$\begin{aligned} u = v = 0 & \quad \text{at } y = 0 \,, \\ u = u_\infty & \quad \text{at } y = \infty \,, \end{aligned} \tag{1.4.30}$$

plus appropriate inflow conditions at some $x = x_\infty$ and the prescribed pressure gradient.

1.5 Linear Stability of Parallel Flows

Even though a particular time-dependent Navier–Stokes problem may admit an equilibrium solution, i. e., a solution of the steady Navier–Stokes equations, that equilibrium solution may not be physically attainable due to instability (in time or in space) of the flow to small disturbances. The question of whether a given equilibrium solution is stable or unstable is crucial to many applications.

Rayleigh (1880) initiated the development of incompressible, inviscid linear stability theory. A subsequent series of papers by Rayleigh (see Mack

(1984)) established the theory of linear instability of inviscid flows as an eigenvalue problem governed by a second-order ordinary differential equation for the amplitude of the disturbance, with the disturbance wavenumber and frequency as parameters; this equation is now known as the Rayleigh equation. Apart from its importance in its own right, it provided two of the four independent fundamental solutions of the asymptotic viscous theory developed later. A key result of the inviscid linear theory is that the existence of an inflection point in the equilibrium velocity profile of the flow, i. e., a point at which the curvature of the profile vanishes, is a necessary condition for instability. In other words, there can be neither unstable nor neutral waves in a flow characterized by a velocity profile without an inflection point. As viscosity is supposed to have a diffusive effect, this led to the disturbing conclusion that flows with convex velocity profiles (e. g., boundary-layer flows) are stable, which conflicts with observations.

Although the formative ideas on the destabilizing influence of viscosity were propounded in Taylor (1915) and independently by Prandtl (1921), the genesis of an asymptotic viscous theory is ascribed to Tollmien (1929) and Schlichting (1933) (see Schlichting and Gersten (1999), Mack (1984)). The viscosity-induced instability waves of unidirectional flows are usually called Tollmien–Schlichting waves. The asymptotic viscous theory was put on a rather rigorous mathematical basis by Lin (1945) and Wasow (1948). Despite all these mathematical developments, the asymptotic viscous theory only attracted serious attention in the fluid mechanics community after its validation by the landmark experiments of Schubauer and Skramsdat (1947), where unstable waves just like those predicted by the theory were observed. Since the early 1960s, the asymptotic theories have been supplanted in practical applications by numerical solutions of the governing differential equations. See (Drazin and Reid (2004), Schmid and Henningson (2001), and Criminale, Jackson and Joslin (2003) for a thorough coverage of fluid dynamics stability.

In applications to flows past such vehicles as aircraft and submarines, a limitation of the linear theory is that although it can predict the critical value of the Reynolds number at which instability commences, it can predict neither where the laminar boundary layer will start to break down (transition onset) nor where the laminar-turbulent transition will be complete. However, linear stability theory has underpinned a semi-empirical criterion, known as the e^N method, for predicting the onset of transition in low disturbance environments. This was proposed by Smith and Gamberoni (1956) and Van Ingen (1956) and is still widely used in engineering applications (Malik (1989)). This method states that transition occurs roughly when linear theory predicts that an initial disturbance will have grown by a factor of e^N. The optimum choice of N is application-dependent, but it is usually in the range $N = 9$–11.

In this section we describe the mathematical formulation of the linear stability problem for *parallel flows*, which are flows for which the mean flow is a function of a single coordinate direction, and the mean velocity is perpendicular to that direction.

1.5.1 Incompressible Linear Stability

Let $\mathbf{u}_0$ and p_0 denote the velocity and pressure of an equilibrium solution (hereafter termed "mean flow") to the constant-viscosity, unforced, incompressible Navier–Stokes equations. Then write $\mathbf{u} = \mathbf{u}_0 + \mathbf{u}'$ and $p = p_0 + p'$, where $\mathbf{u}'$ and p' are perturbations to the mean flow. If $\mathbf{u}'$ and p' are presumed small, and quadratic terms are neglected, then (1.4.1)–(1.4.2) become

$$\frac{\partial \mathbf{u}}{\partial t} + \mathbf{u}_0 \cdot \nabla \mathbf{u} + \mathbf{u} \cdot \nabla \mathbf{u}_0 = -\nabla p + \nu \Delta \mathbf{u} , \qquad (1.5.1)$$

$$\nabla \cdot \mathbf{u} = 0 , \qquad (1.5.2)$$

where the primes have been dropped for notational simplicity.

A canonical stability problem of long-standing interest is flow in a plane channel (Fig. 1.2), also referred to as plane Poiseuille flow, which is confined in the wall-normal (y) direction between infinite plates. This idealized laminar flow is driven by a constant pressure gradient in the streamwise (x) direction, and there are no boundaries in either the streamwise or spanwise (z) directions. The nondimensional mean flow is parallel and is given by

$$\mathbf{u}_0(\mathbf{x}) = (u_0(y), 0, 0)^T , \quad u_0(y) = 1 - y^2 , \quad p_0(\mathbf{x}) = -\frac{2}{\mathrm{Re}} x , \qquad (1.5.3)$$

where distances have been scaled by the half-channel width h, velocities by the centerline velocity $u_c = u_0(0)$, and the Reynolds number Re is given by $\mathrm{Re} = u_c h / \nu$. (Just as we dropped the primes for notational convenience, we also drop the "*", heretofore used to denote nondimensional quantities.)

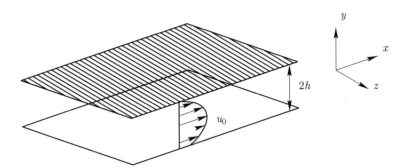

Fig. 1.2. Plane channel flow. The x, y and z directions are called the streamwise, normal and spanwise directions, respectively

The linear stability of this flow is assessed by studying perturbations of the form

$$\mathbf{u}(\mathbf{x}, t) = Re\left\{\hat{\mathbf{u}}(y)e^{i(\alpha x + \beta z) - i\omega t}\right\} ,$$
$$p(\mathbf{x}, t) = Re\left\{\hat{p}(y)e^{i(\alpha x + \beta z) - i\omega t}\right\} ,$$

(1.5.4)

where α, β and ω are complex constants, and Re denotes the real part of a complex quantity. (Note the slight difference between the symbols for the Reynolds number (Re) and the real part (Re).) Equations (1.5.1) and (1.5.2) become, in component form,

$$\{\mathcal{D}^2 - (\alpha^2 + \beta^2) - i\alpha Re\, u_0\}\hat{u} - Re\,(\mathcal{D}u_0)\hat{v} - i\alpha Re\,\hat{p} = -i\omega Re\,\hat{u} , \quad (1.5.5)$$

$$\{\mathcal{D}^2 - (\alpha^2 + \beta^2) - i\alpha Re\, u_0\}\hat{v} - Re\,\mathcal{D}\hat{p} = -i\omega Re\,\hat{v} , \quad (1.5.6)$$

$$\{\mathcal{D}^2 - (\alpha^2 + \beta^2) - i\alpha Re\, u_0\}\hat{w} - i\beta Re\,\hat{p} = -i\omega Re\,\hat{w} , \quad (1.5.7)$$

$$i\alpha\hat{u} + \mathcal{D}\hat{v} + i\beta\hat{w} = 0 , \quad (1.5.8)$$

where $\mathcal{D} = d/dy$. The boundary conditions are

$$\hat{u} = \hat{v} = \hat{w} = 0 \qquad \text{at } y = \pm 1 . \quad (1.5.9)$$

The system (1.5.5)–(1.5.9) describes a *dispersion relation* between α, β and ω with Re as a parameter. If four real quantities out of α, β and ω are prescribed, then the dispersion relation constitutes an eigenvalue problem for the remaining two real quantities. If α and β are fixed, real quantities, then ω is the complex eigenvalue. When approached in this manner, the problem is one of *temporal stability*. If $Im(\omega) > 0$, then the corresponding mode grows in time, and the mean flow will be disrupted. The equilibrium solution, then, is unstable if a growing mode exists for any real α and β. An alternative approach to this problem is one of *spatial stability*. Here, ω is real and fixed, and two relations are imposed upon α and β to complete the specification of the problem (Nayfeh (1980), Cebeci and Stewartson (1980)). If $Im(\alpha) > 0$ or $Im(\beta) > 0$, then the mode grows in space. If such growing modes exist for any real ω and for any orientations of the waves, then the flow is spatially unstable. Gaster (1962) has given a procedure for relating the results of temporal and spatial stability analyses. Huerre and Monkewitz (1990) provide a valuable discussion of the physical aspects of the spatial stability problem.

By manipulating (1.5.5)–(1.5.8) we arrive at

$$[\mathcal{D}^2 - (\alpha^2 + \beta^2)]^2\hat{v} - i\alpha Re\, u_0[\mathcal{D}^2 - (\alpha^2 + \beta^2)]\hat{v} + i\alpha Re\,(\mathcal{D}^2 u_0)\hat{v}$$
$$= -i\omega Re\,[\mathcal{D}^2 - (\alpha^2 + \beta^2)]\hat{v}$$

(1.5.10)

and

$$[\mathcal{D}^2 - (\alpha^2 + \beta^2)](\alpha\hat{w} - \beta\hat{u}) - i\alpha\mathrm{Re}\, u_0(\alpha\hat{w} - \beta\hat{u})$$
$$= -i\omega\mathrm{Re}\,(\alpha\hat{w} - \beta\hat{u}) - \beta\mathrm{Re}\,(\mathcal{D}u_0)\hat{v}\,. \tag{1.5.11}$$

The first of these is the celebrated *Orr–Sommerfeld equation*, and it is subjected to the boundary conditions

$$\hat{v} = \mathcal{D}\hat{v} = 0 \qquad \text{at } y = \pm 1\,. \tag{1.5.12}$$

(The condition on $\mathcal{D}\hat{v}$ follows from (1.5.8) and (1.5.9).) The quantity $\alpha\hat{w} - \beta\hat{u}$ is the normal component of the perturbation vorticity. It satisfies

$$\alpha\hat{w} - \beta\hat{u} = 0 \qquad \text{at } y = \pm 1\,. \tag{1.5.13}$$

For this reason (1.5.11) is often referred to as the *vertical vorticity equation* (although Herbert (1983b) called it the *Squire equation*). Hence, there are two distinct classes of solutions to the sixth-order system (1.5.5)–(1.5.9). The first class comprises the eigenmodes of (1.5.10) and (1.5.12), with (1.5.11) serving merely to determine the vertical vorticity of this mode. The second class has $\hat{v} \equiv 0$ and contains the eigenmodes of (1.5.11) and (1.5.13). Squire (1933) showed that all solutions of the second class are damped modes. Until the role of the vertical vorticity modes in the weakly nonlinear stage of transition was recognized in the 1980s (Herbert (1983b)), attention had been focused almost exclusively on the Orr–Sommerfeld solutions. Note that in the temporal stability problem the eigenvalue ω enters linearly, whereas in the spatial stability problem the eigenvalue α enters nonlinearly.

There are numerous other incompressible flows whose linear stability can be assessed by similar mathematical formulations, e. g., circular Poiseuille flow (pipe flow), Taylor–Couette flow (flow between rotating cylinders or rotating spheres), and free shear layers. Of course, the nondimensionalizations and coordinate systems may differ. Moreover, in many cases, e. g., Taylor–Couette flow, the set (1.5.5)–(1.5.8) of three second-order equations and one first-order equation cannot be reduced to the Orr–Sommerfeld and vertical-vorticity equations (1.5.12)–(1.5.13).

1.5.2 Compressible Linear Stability

The stability of compressible flows has not attracted nearly the amount of attention devoted to the stability of incompressible flows. Indeed, there has yet to appear a single text devoted to the subject, and it goes unmentioned in all but the most recent texts on hydrodynamic stability, such as Schmid and Henningson (2001) and Criminale, Jackson and Joslin (2003). The basic concepts and approach to the stability theory of compressible laminar boundary layers are similar to those of the incompressible counterpart. However, there are some fundamental differences, which will be discussed here following a brief historical overview.

Although Kuchemann (1938) must be credited with the first attempt to develop a compressible stability theory (which neglected the viscosity and the mean temperature gradient and was thus too restrictive), it was Lees and Lin (1946) who laid the foundation of an asymptotic theory analogous to that for the incompressible case. This asymptotic theory was further developed by Dunn (1953) and Dunn and Lin (1955). Mack (1969) provided comprehensive viscous and inviscid instability results for the flat-plate boundary layer for Mach numbers up to 10.

The first major difference from the incompressible linear stability theory is that from a mathematical point of view, the eigenvalue problem of the incompressible parallel flow is governed by a sixth-order system of ordinary differential equations (sometimes reducible to a fourth-order equation and a second-order equation), whereas the linear stability of compressible parallel flow is governed by an eighth-order system of ordinary differential equations. A key result is that the normal derivative of the mass-weighted streamwise velocity gradient $(\rho u')'$, where the prime stands for $\mathcal{D} = d/dy$, plays the same critical role as the curvature of the streamwise velocity (proportional to u'') in the incompressible theory. Consequently, the compressible flat-plate boundary layer is unstable to purely inviscid disturbances in contrast to the incompressible case where the instability, called the Tollmien–Schlichting instability, is of viscous origin. In supersonic boundary layers, these disturbances (known as the first modes) are most amplified when oblique. Wall cooling stabilizes these disturbances, as it tends to eliminate the generalized inflection point (where $(\rho u')' = 0$) within the boundary layer.

A second major difference between the compressible and incompressible theories arises in those circumstances in which the mean flow is supersonic relative to the phase velocity of the disturbance. Whenever the relative flow is supersonic over some portion of the boundary-layer profile, there are an infinite number of wavenumbers for the single phase velocity (Mack 1969). Associated with each of these so-called neutral disturbances is a family of unstable disturbances. The first of these modes is called Mack's second mode, and it is the dominant mode of instability in the hypersonic regime. This mode is destabilized by wall cooling, as that tends to increase the region of supersonic relative mean flow within the boundary layer.

Following Mack's pioneering numerical work, Malik (1982, 1990) developed efficient, high-order computational techniques for the solution of compressible stability problems, and demonstrated the use of the theory for analyzing supersonic and hypersonic transition experiments (Malik 1989). The theory has been extended to include real gas effects and the second mode disturbances were found to be relevant in boundary-layer transition over re-entry vehicles (Malik 2003).

Flow past solid boundaries, such as the flat plate illustrated in Fig. 1.3, have been the subject of many compressible linear stability studies. The cus-

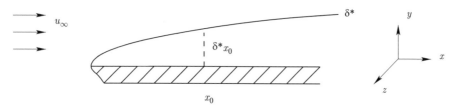

Fig. 1.3. Flow past a flat plate. The x, y and z directions are called the streamwise, normal and spanwise directions, respectively

tomary scaling for such flows is that distances are scaled by the boundary-layer displacement thickness δ^* (see (1.3.58)), while the velocity, density and temperature are scaled by their free-stream values at $y = \infty$, i.e., by u_∞, ρ_∞ and T_∞, respectively. We choose to scale the pressure by $p_\infty = \rho_\infty \mathcal{R} T_\infty$, where $\mathcal{R}$ is the gas constant. (An alternative scaling for the pressure uses $\rho_\infty u_\infty^2$.) We assume that the gas is thermally perfect and that the ratio, λ/μ, of the second coefficient of viscosity to the first coefficient of viscosity and the Prandtl number Pr are also constants. The Reynolds number Re is defined by

$$\mathrm{Re} = \rho_\infty U_\infty \delta^* / \mu_\infty \,, \qquad (1.5.14)$$

and the free-stream Mach number M_∞ is given by

$$\mathrm{M}_\infty = U_\infty / \sqrt{\gamma \mathcal{R} T_\infty} \,. \qquad (1.5.15)$$

As indicated by the figure, the boundary-layer displacement thickness increases in the streamwise direction. The flow is certainly nonparallel within the boundary layer, and the velocity in the wall-normal (y) direction is nonzero. However, $|v_0| \ll u_0$, and the x dependence is much weaker than that on y. The simplest approach to boundary-layer stability analysis is to make the *parallel flow approximation*. Here, one analyzes the stability in the vicinity of some point x_0 by supposing that $u_0(x, y)$ is given by $u_0(x_0, y)$ for all x (likewise for the temperature and pressure) and that v_0 is negligible. This approximation is illustrated in Fig. 1.4. Within this approximation it is customary to take for the mean flow the solution to the boundary-layer equations. It is assumed that the variation of the mean flow over the streamwise wavelength $2\pi/\alpha$ of the perturbation is negligible. This approximation has proven to capture many, but not all, of the qualitative features of boundary-layer stability, and often provides acceptable quantitative results. See Sect. 1.6 for more sophisticated approaches.

As in the case of the incompressible linear stability equations, let the mean flow quantities be denoted by the subscript 0, and take perturbations in the same form as (1.5.4). The equation of state (1.2.15) becomes

$$\hat{p} = \frac{\hat{\rho}}{\rho_0} + \frac{\hat{T}}{T_0} \,. \qquad (1.5.16)$$

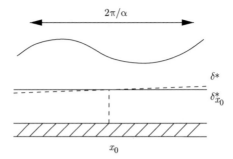

Fig. 1.4. The parallel boundary layer. The variation of the displacement thickness δ^* over one wave length $2\pi/\alpha$ of a perturbation is neglected

Following Malik and Orszag (1987), we use the disturbance variables in the combination

$$\hat{\mathbf{q}} = (\alpha\hat{u} + \beta\hat{w}, \hat{v}, \hat{p}, \hat{T}, \alpha\hat{w} - \beta\hat{u})^T \; . \tag{1.5.17}$$

(Note that $\hat{T}$ denotes the temperature perturbations, whereas the superscript T denotes the transpose.) Furthermore, let

$$\zeta = 2 + \frac{\lambda}{\mu} \; , \qquad \mathcal{G} = (\gamma - 1)\mathrm{M}_\infty^2 \mathrm{Pr} \; , \qquad \mathcal{U}_0 = \alpha u_0 + \beta w_0 \; ,$$
$$\mathcal{V}_0 = \alpha w_0 - \beta u_0 \; , \qquad \varpi^2 = \alpha^2 + \beta^2 \; , \qquad \varphi = \mathcal{U}_0 - \omega \; . \tag{1.5.18}$$

Starting from the continuity equation (1.3.11), the nonconservative form of the momentum equation (1.3.12) and the temperature equation (1.3.20), with the equation of state used to replace the density with the pressure, the compressible linear stability equations can be written as one first-order equation and four second-order equations:

$$A\,\mathcal{D}^2\hat{\mathbf{q}} + B\,\mathcal{D}\hat{\mathbf{q}} + C\hat{\mathbf{q}} = \mathbf{0} \; , \tag{1.5.19}$$

subject to the boundary conditions that the velocity and temperature disturbances vanish at both boundaries:

$$\begin{aligned}
\hat{v} &= 0 & &\text{at } y = 0, \infty \; , \\
\alpha\hat{u} + \beta\hat{w} &= 0 & &\text{at } y = 0, \infty \; , \\
\alpha\hat{w} - \beta\hat{u} &= 0 & &\text{at } y = 0, \infty \; , \\
\hat{T} &= 0 & &\text{at } y = 0, \infty \; .
\end{aligned} \tag{1.5.20}$$

There is no boundary condition on the disturbance pressure. Even if the mean flow satisfies an adiabatic wall temperature boundary condition, i.e., $\mathcal{D}T_0 = 0$, the disturbance temperature is assumed to vanish at the wall on the grounds that the time scale for the disturbance is too short for it to achieve an adiabatic state.

The coefficient matrices in (1.5.19) are given by

$$A = \begin{pmatrix} 1 & 0 & 0 & 0 & 0 \\ 0 & 1 & 0 & 0 & 0 \\ 0 & 0 & 0 & 0 & 0 \\ 0 & 0 & 0 & 1 & 0 \\ 0 & 0 & 0 & 0 & 1 \end{pmatrix}, \tag{1.5.21}$$

$$B = \begin{pmatrix} \dfrac{1}{\mu_0}\dfrac{\mathrm{d}\mu_0}{\mathrm{d}T_0}T_0' & i(\zeta-1)\varpi^2 & 0 & \dfrac{1}{\mu_0}\dfrac{\mathrm{d}\mu_0}{\mathrm{d}T_0}\mathcal{U}_0' & 0 \\[2ex] i\dfrac{\zeta-1}{\zeta} & \dfrac{1}{\mu_0}\dfrac{\mathrm{d}\mu_0}{\mathrm{d}T_0}T_0' & -\dfrac{\mathrm{Re}}{\mu_0\zeta} & 0 & 0 \\[2ex] 0 & 1 & 0 & 0 & 0 \\[2ex] \dfrac{2\mathcal{G}\mathcal{U}_0'}{\varpi^2} & 0 & 0 & \dfrac{2}{\mu_0}\dfrac{\mathrm{d}\mu_0}{\mathrm{d}T_0}T_0' & \dfrac{2\mathcal{G}\mathcal{V}_0'}{\varpi^2} \\[2ex] 0 & 0 & 0 & \dfrac{1}{\mu_0}\dfrac{\mathrm{d}\mu_0}{\mathrm{d}T_0}\mathcal{V}_0' & \dfrac{1}{\mu_0}\dfrac{\mathrm{d}\mu_0}{\mathrm{d}T_0}T_0' \end{pmatrix}, \tag{1.5.22}$$

and

$$C = \begin{pmatrix} -\dfrac{i\mathrm{Re}}{\mu_0 T_0}\varphi - \zeta\varpi^2 & C_{12} & -\dfrac{i\mathrm{Re}}{\mu_0}\varpi & C_{14} & 0 \\[2ex] i\dfrac{\zeta-1}{\zeta\mu_0}\dfrac{\mathrm{d}\mu_0}{\mathrm{d}T_0}T_0' & -\dfrac{i\mathrm{Re}}{\mu_0 T_0\zeta}\varphi - \dfrac{\varpi^2}{\zeta} & 0 & \dfrac{i}{\zeta\mu_0}\dfrac{\mathrm{d}\mu_0}{\mathrm{d}T_0}\mathcal{U}_0' & 0 \\[2ex] i & -\dfrac{T_0'}{T_0} & i\gamma\mathrm{M}_\infty^2\kappa & -\dfrac{i}{T_0}\varphi & 0 \\[2ex] 0 & C_{42} & \dfrac{i\mathrm{Re}\,\mathcal{G}}{\mu_0}\varphi & C_{44} & 0 \\[2ex] 0 & -\dfrac{\mathrm{Re}}{\mu_0 T_0}\mathcal{V}_0' & 0 & C_{54} & -\dfrac{i\mathrm{Re}}{\mu_0 T_0}\kappa - \varpi^2 \end{pmatrix}, \tag{1.5.23}$$

where a prime on a mean flow variable denotes the wall-normal derivative of the quantity, e. g., $T_0' = \mathcal{D}T_0$. The remaining matrix elements are

$$C_{12} = -\frac{\mathrm{Re}}{\mu_0 T_0}\mathcal{U}_0' + \frac{i}{\mu_0}\frac{\mathrm{d}\mu_0}{\mathrm{d}T_0}T_0'\varpi^2, \qquad C_{14} = \mathcal{U}_0'\frac{1}{\mu_0}\frac{\mathrm{d}^2\mu_0}{\mathrm{d}T_0^2}T_0' + \mathcal{U}_0''\frac{1}{\mu_0}\frac{\mathrm{d}\mu_0}{\mathrm{d}T_0},$$

$$C_{42} = -\frac{\mathrm{Re}\,\mathrm{Pr}}{\mu_0 T_0}T_0' + 2i\mathcal{G}\mathcal{U}_0',$$

$$C_{44} = -\frac{i\mathrm{Re}\,\mathrm{Pr}}{\mu_0 T_0}\varphi - \varpi^2 + \frac{\mathcal{G}}{\mu_0}\frac{\mathrm{d}\mu_0}{\mathrm{d}T_0}(U_0'^2 + W_0'^2) + \frac{1}{\mu_0}\frac{\mathrm{d}^2\mu_0}{\mathrm{d}T_0^2}(T_0')^2 + \frac{1}{\mu_0}\frac{\mathrm{d}\mu_0}{\mathrm{d}T_0}T_0'',$$

$$C_{54} = \frac{1}{\mu_0}\frac{\mathrm{d}^2\mu_0}{\mathrm{d}T_0^2}T_0'\mathcal{V}_0' + \frac{1}{\mu_0}\frac{\mathrm{d}\mu_0}{\mathrm{d}T_0}\mathcal{V}_0''. \tag{1.5.24}$$

This formulation allows reduction in order of the governing equations if viscous dissipation is neglected (i. e., $B_{25} = 0$), which has a small effect on the unstable eigenvalue for small Mach numbers. Other forms of the compressible stability equations are given in Malik (1990).

1.6 Stability Equations for Nonparallel Flows

Only since the late 1970s have more sophisticated approaches than the parallel-flow assumption been employed for the study of the stability of nonparallel flows. These approaches include matched asymptotic expansions, triple-deck theory and parabolized equations. We shall focus here on the latter, as spectral methods have rarely been used in the context of the other approaches.

The parabolized stability equations (PSE) approach was developed for incompressible flow by Herbert (1991). Further details and developments can be found in Bertolotti (1991) and Bertolotti et al. (1992) for incompressible flows, and in Chang et al. (1991), Bertolotti and Herbert (1991) for compressible flow and Malik (2003) for hypersonic flow including chemistry effects. Schmid and Henningson (2001) furnish detailed derivations and equations for the incompressible case. Herbert (1997) reviews applications of PSE to both linear and nonlinear stability problems, for incompressible as well as compressible flows. The PSE permit the stability analysis of many mean flows that depend upon both the streamwise and wall-normal directions. These include flow past a flat plate (see Fig. 1.3), as well as flows of engineering interest such as flow past an airplane or a submarine.

The parabolized stability equations are developed by representing the total flow as a mean flow plus a disturbance field and decomposing the disturbance into a rapidly varying wave-like part and a slowly varying shape function. As a complete description of the PSE would take up several pages, we confine ourselves here to just providing an outline of the approach and illustrating the mathematical structure of the equations, using two-dimensional incompressible flow as the setting. Let $\mathbf{q} = (u, v, p)^T$ represent the disturbance field and write

$$\mathbf{q}(x, y, t) = Re\left\{\hat{\mathbf{q}}(x, y)\chi(x, t)\right\} , \qquad (1.6.1)$$

with

$$\chi(x, t) = \exp\left(i \int_{x_0}^{x} \alpha(\psi) \, d\psi - \omega t\right) , \qquad (1.6.2)$$

where $\alpha(x)$ is a local complex wavenumber that captures the rapidly varying part, ω is the frequency, and $\hat{\mathbf{q}}$ is the slowly varying (in x) shape function. Constraints of the type

$$\int_{0}^{\infty} \left(\bar{\hat{u}}\frac{\partial \hat{u}}{\partial x} + \bar{\hat{v}}\frac{\partial \hat{v}}{\partial x}\right) \, dy = 0 , \qquad (1.6.3)$$

where the overbar denotes the complex conjugate, impose a condition on $\alpha(x)$ such that most of the waviness and growth of the disturbance are absorbed into the exponential function χ in (1.6.2), making the shape function $\hat{q}$ slowly varying with x. The ellipticity is retained for the wave part while parabolization is applied to the shape function. Hence, the $\hat{q}_{xx}$ terms in the governing equations are dropped, and we arrive at a set of equations in which the only second-order derivatives are those with respect to y.

The linear parabolized stability equations are derived from the Navier–Stokes equations with the usual linearization procedure plus the parabolization approach discussed above. The linear PSE may be written as

$$\frac{\partial \hat{q}}{\partial x} = L_0 \hat{q} + L_1 \frac{\partial \hat{q}}{\partial y} + L_2 \frac{\partial^2 \hat{q}}{\partial y^2} . \tag{1.6.4}$$

The boundary conditions are

$$\hat{u} = 0 , \quad \hat{v} = 0 \quad \text{at} \quad y = 0, \infty , \tag{1.6.5}$$

supplemented with appropriate inflow conditions at $x = x_0$.

The matrices in (1.6.4) are given by

$$L_0 = \begin{pmatrix} -i\alpha & 0 & 0 \\ 0 & -i\alpha + \frac{i\omega}{u_0} - \frac{\alpha^2}{u_0 \text{Re}} - \frac{1}{u_0} \frac{\partial v_0}{\partial y} & 0 \\ i\omega - \frac{\alpha^2}{\text{Re}} - \frac{\partial u_0}{\partial x} & -\frac{\partial u_0}{\partial y} & -i\alpha \end{pmatrix} , \tag{1.6.6}$$

$$L_1 = \begin{pmatrix} 0 & -1 & 0 \\ 0 & -\frac{v_0}{u_0} & -\frac{1}{u_0} \\ -v_0 & u_0 & 0 \end{pmatrix} , \quad L_2 = \begin{pmatrix} 0 & 0 & 0 \\ 0 & \frac{1}{\text{Re}\, u_0} & 0 \\ \frac{1}{\text{Re}} & 0 & 0 \end{pmatrix} .$$

Note that α is a complex function of x, u_0 and v_0 are real functions of x and y, and Re and ω are real constants.

For the nonlinear PSE approximation, the disturbance function can be expressed by the following truncated Fourier series

$$q(x, y, t) = \sum_{m=-M}^{M} \hat{q}_m(x, y) \chi_m(x, t) + \text{complex conjugate} , \tag{1.6.7}$$

where

$$\chi_m(x, t) = \exp\left(i \int_{x_0}^{x} \alpha_m(\psi) d\psi - m\omega t \right) . \tag{1.6.8}$$

Following a similar parabolization strategy, one obtains for each Fourier mode $\hat{q}_m(x, y)$ an equation that is structurally the same as (1.6.4) with the addition of a forcing term that accounts for the nonlinear interactions.

For further details, see Bertolotti, Herbert and Spalart (1992) for the full, nonlinear PSE description of incompressible flow, and see Chang et al. (1991) and Bertolotti and Herbert (1991) for the compressible formulation.

Finally, we conclude this survey with some brief remarks on triple deck theory. One line of approach has employed triple deck theory, (e. g., Smith (1979) and Hall (1983)), to derive a set of equations based on formal asymptotic expansions to account consistently for the spatial evolution of the mean flow and the instability. As the triple deck approaches have invariably employed finite-difference methods, we shall not discuss them further. (There appears to be no fundamental obstacle to the use of spectral methods for the triple deck approach.)

2. Single-Domain Algorithms and Applications for Stability Analysis

2.1 Introduction

Classical (single-domain) spectral methods in fluid dynamics have been applied primarily to investigations of the basic flow physics phenomena associated with stability, transition, and turbulence. Although spectral methods have received great visibility for their role in direct numerical simulation and large-eddy simulation of fluid flows, as illustrated by the examples in Chaps. 3 and 4, they play an important role in the less computationally intensive tools used to study linear and nonlinear stability in fluid dynamics. Figure 2.1 illustrates the components of each basic stage of a stability and transition investigation—the physical phenomenon, its mathematical model, and the typical output most appropriate for each stage.

The starting point is the determination of the mean, or base, flow. This refers to the unperturbed laminar flow. In some cases, such as that of the plane channel flow problem discussed in Sect. 1.5.1, this can be found in analytical form. At the other extreme, say, for analyzing the flow over a particular aircraft wing accounting for the fuselage and engines, the mean flow has to be computed by numerical solution of the full Navier–Stokes equations. In many cases of practical interest the two-dimensional boundary-layer equations furnish an acceptable description of the mean flow. Spectral methods have not been used for mean flow computations with the full Navier–Stokes equations (at least using just a single domain), but they have been used for solution of the boundary-layer equations. Spectral methods for mean flow computations are discussed in Sect. 2.2.

The second step is the determination of the linear stability properties of the mean flow. Here a linearized form of the Navier–Stokes equations is appropriate. Spectral methods have been particularly favored for this stage of transition physics analysis. Broad, multidimensional parametric studies are conducted to cover a range of physical parameters, such as the Reynolds number, as well as a range of wave parameters, such as temporal frequencies and spatial wavenumbers, to determine the linearly unstable regimes of the mean flow. Section 2.3 is devoted to spectral methods for linear stability for incompressible parallel flows, while Sect. 2.4 covers the compressible case. More precisely, in those two sections we focus on the use of spectral methods for approximating the eigenvalues of the linearized Navier–Stokes equations, as eigenvalue analysis was the dominant approach to linear stability in fluid dynamics until roughly 1990. We do, however, make a few comments in Sect. 2.6 about applications of spectral methods to transient growth analysis, which now appears to be more appropriate in some applications. Spectral linear stability algorithms for the more physically relevant case of nonparallel flow are treated in Sect. 2.5.

The next stage in a systematic study of transition is to examine the effects of weak nonlinearities. Once again, parametric studies are conducted to ascertain which nonlinear interactions are the most likely candidates to trigger transition to turbulence. Spectral methods have been employed for this type of study, although finite-difference methods have often been used in some coordinate directions, especially for compressible flows. Methods for this type of analysis are discussed in Sect. 2.7.

The final stage in a study of transition to turbulence is to perform unsteady, 3D simulations using the Navier–Stokes equations to examine the strongly nonlinear stages of transition and perhaps even the complete transition to turbulent flow. Needless to say, this can only be done for a small number of cases. Ideally, the information gleaned from the linear and weakly nonlinear stability analyses have led to the choice of the most physically relevant cases. Spectral methods have played a very prominent role in these sorts of simulations. Spectral methods for this type of problem are discussed in Chap. 3 for incompressible flows and in Chap. 4 for compressible flows.

Our discussion of spectral approximations to particular fluid dynamics applications in Chaps. 2–4, and even more so the subsequent discussion of multidomain spectral methods in Chaps. 5–7, presumes that the reader is already familiar with the fundamentals of spectral methods in single domains. In this category we include such essentials as the nodal and modal basis functions and the interpolation, projection, and transform procedures for Fourier and Jacobi (especially Chebyshev and Legendre) polynomials, as well as the basic spectral discretization approaches using collocation, tau, Galerkin, and Galerkin with numerical integration (G-NI) methods. Naturally, we would recommend CHQZ2 for a thorough treatment of these fundamentals. However, Chap. 8 of the present text contains a primer on this material, and we recommend that the newcomer to spectral methods at least peruse Chap. 8 prior to reading this and the intervening chapters. This review is especially important because it turns out that the spectral algorithms used for the boundary-layer flows, which are logically the starting point for stability investigations, as illustrated in Fig. 2.1, employ the most non-standard methods discussed in this chapter.

In most cases the spectral methods applied to the type of problems discussed in this chapter have been of collocation or tau type. Curiously, linear stability eigenvalue problems appear to have attracted little theoretical attention from the numerical analysis community.

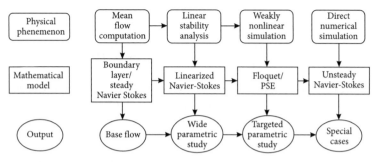

Fig. 2.1. Schematic of the physical phenomena, the mathematical models and the outputs from the various stages of stability and transition investigations

Numerous numerical examples are included in Chaps. 2–4 to illustrate the applications of classical spectral methods to fluid dynamics. For the most part, we have drawn heavily upon examples generated by ourselves and by our colleagues. We trust that the reader will appreciate that this was primarily for convenience, as these data and codes were readily available to us. These examples are meant to be *representative* of spectral methods applications. In no way do we mean to suggest that the examples that we have included are the definitive applications of spectral methods.

2.2 Boundary-Layer Flows

The compressible, two-dimensional boundary-layer equations are presented in Sect. 1.3.6, and the incompressible versions are given in Sect. 1.4.5. One of the simplest boundary layers is the one sketched in Fig. 1.3 for two-dimensional flow over a flat plate. The displacement thickness δ^*, given in dimensional terms by (1.3.58), measures the deflection of the incident streamlines in the direction normal to the wall. The displacement thickness is typically three times the less precise boundary-layer thickness δ (see Sect. 1.3.6) at low free-stream Mach numbers, and is even larger at supersonic Mach numbers.

2.2.1 Incompressible Boundary-Layer Flows

For the special case of incompressible, constant-viscosity flow over a flat plate subjected to a pressure gradient, a similarity solution to (1.4.28)–(1.4.29) exists. The transformation

$$\eta = y\sqrt{\frac{m+1}{2}\frac{u_\infty}{\nu x}}\,, \qquad \psi = \sqrt{\frac{2}{m+1}}\,\nu u_\infty x f(\eta)\,, \qquad (2.2.1)$$

where

$$u_\infty \propto x^m \, , \qquad u = \frac{\partial \psi}{\partial y} = u_\infty f'(\eta) \, ,$$

$$v = -\frac{\partial \psi}{\partial x} = \frac{1}{2}\sqrt{\nu u_\infty/x}\,[\eta f'(\eta) - f(\eta)] \, , \tag{2.2.2}$$

and the prime denotes differentiation with respect to η, leads to the *Falkner-Skan similarity equation* (Schlichting and Gersten (1999))

$$f''' + ff'' + \beta(1 - f'^2) = 0 \, , \tag{2.2.3}$$

where $\beta = 2m/(m+1)$ is the pressure gradient parameter. The boundary conditions are

$$f = f' = 0 \quad \text{at} \quad \eta = 0 \, ,$$
$$f' \longrightarrow 1 \quad \text{as} \quad \eta \longrightarrow \infty \, . \tag{2.2.4}$$

The *Blasius equation* is the special case of the Falkner–Skan equation with zero pressure gradient ($\beta = 0$).

The application of spectral methods to the Falkner–Skan equation is not entirely straightforward. For one thing, the function $f(\eta)$ is unbounded as $\eta \to \infty$; indeed, it grows linearly with η. If one desires an expansion over the full infinite interval, e. g., in terms of Laguerre functions or in terms of Chebyshev or Legendre polynomials combined with a mapping of the reference interval $[-1,1]$ into the semi-infinite interval $[0, \infty]$ (see, e. g., Sects. 8.6 and 8.8.2 or Boyd (2001), Chap. 17), then one should write $f(\eta) = \eta + g(\eta)$, and transform the Falkner–Skan equation into an equation for $g(\eta)$; the solutions will be well-behaved since $g(\eta)$ decays faster than exponentially as $\eta \to \infty$; indeed, g decays as an $(1/\eta^2)\,e^{-\eta^2}$ (see Lagerstram (1989)). Alternatively, one can truncate the semi-infinite domain $[0, \infty)$ to a finite domain $[0, \eta_{\max}]$ and perhaps add an additional mapping to cluster the grid points where most needed (see Sect. 8.8). This would avoid the relatively minor nuisance of shifting from $f(\eta)$ to $g(\eta)$. Another complication of the Falkner–Skan equation is the double boundary condition at $\eta = 0$. Collocation methods do not handle such double boundary conditions gracefully. The most obvious option is to enforce $f(0) = 0$ and $\mathrm{d}f/\mathrm{d}\eta(0) = 0$ and drop the differential equation at the collocation point closest to the wall. (For a G-NI method the Neumann condition would be readily enforced in a weak form.) However, as discussed in Sect. 2.3, such a collocation approximation would not converge as rapidly as the G-NI method. Finally, (2.2.3) is nonlinear, and its discrete approximations require iterative solution schemes.

Zebib (1984) solved the Blasius equation by combining domain truncation (using the simple affine transformation (8.8.12)) with an integral Chebyshev tau method (see Sect. 2.3.3 and CHQZ2, Sect. 4.1.2) and utilized Newton's method to solve the implicit equations. Streett, Zang and Hussaini (1984) also used domain truncation in their spectral collocation solution of the general Falkner–Skan equation, but they eschewed Newton's method in favor of a preconditioned iterative scheme (with a local linearization at each iteration) for solution of the implicit equations.

We focus on the latter method henceforth, as it has been generalized both to nonsimilar and also to compressible boundary layers; the latter generalization is due to Pruett and Streett (1991), who provide a detailed explanation of the spectral algorithm. A low-order finite-difference approximation of the linearized equations was originally used by Streett et al. as a preconditioner due to the memory limitations of the computers of that era. As discussed, for example, in CHQZ2, Sect. 4.4.2, while considering first-order equations, the subtle part of applying low-order (finite-difference or finite-element) preconditioning schemes to a third-order equation is the preconditioning of the odd-order derivatives. Ordinarily, this would require that the preconditioning grid have, say, 50% more points than the spectral grid. Streett et al. avoided both this complication and the complication due to the double boundary condition at $\eta = 0$ by abandoning the use of the single equation (2.2.3) description of the boundary-layer problem in favor of a system consisting of a second-order differential equation and an integral equation. This system arises most naturally as a special case of the general boundary-layer equations for a nonsimilar problem. In this context the Görtler variables (Görtler (1957); see also Cebeci and Bradshaw (1977)) are convenient. In the case of constant viscosity, the independent variables are

$$\xi = \int_0^x u_\infty(x')\, \mathrm{d}x' , \qquad \eta = u_\infty(x)y/\sqrt{2\xi\nu} , \qquad (2.2.5)$$

and the dependent ones are

$$\bar{u} = u/u_\infty(x) , \qquad \bar{v} = \frac{\sqrt{2\xi/\nu}}{u_\infty(x)}v + \frac{2\xi u}{\sqrt{\nu}\, u_\infty(x)}\frac{\partial \eta}{\partial x} . \qquad (2.2.6)$$

The corresponding boundary-layer equations are

$$\frac{\partial^2 \bar{u}}{\partial \eta^2} - \bar{v}\frac{\partial \bar{u}}{\partial \eta} + \beta(1 - \bar{u}^2) - 2\xi\bar{u}\frac{\partial \bar{u}}{\partial \xi} = 0 , \qquad (2.2.7)$$

$$\frac{\partial \bar{v}}{\partial \eta} + \bar{u} + 2\xi\frac{\partial \bar{u}}{\partial \xi} = 0 , \qquad (2.2.8)$$

the boundary conditions are

$$\bar{u} = \bar{v} = 0 \quad \text{at } \eta = 0 ,$$

$$\bar{u} \longrightarrow 1 \quad \text{as } \eta \longrightarrow \infty , \qquad (2.2.9)$$

and suitable inflow conditions are applied at some $\xi = \xi_0$. The pressure gradient parameter β is equal to $(2\xi/u_\infty(x))(\partial u_\infty/\partial \xi)$. For the case of a truncated domain in the η coordinate, (2.2.9) is replaced with

$$\bar{u} = \bar{v} = 0 \quad \text{at } \eta = 0 \,,$$

$$\bar{u} = 1 \qquad \text{at } \eta = \eta_{\max} \,. \tag{2.2.10}$$

These equations reduce to an alternative form of the similar boundary-layer equations when $\partial \bar{u} / \partial \xi = 0$:

$$\frac{\partial^2 \bar{u}}{\partial \eta^2} - \bar{v} \frac{\partial \bar{u}}{\partial \eta} + \beta(1 - \bar{u}^2) = 0 \,, \tag{2.2.11}$$

$$\frac{\partial \bar{v}}{\partial \eta} + \bar{u} = 0 \,, \tag{2.2.12}$$

with boundary conditions again given solely by (2.2.9) or (2.2.10). (Note that (2.2.11) and (2.2.12) combine into a single equation for $\bar{v}$ that is analogous to (2.2.3).)

Streett, Zang and Hussaini (1984) and Pruett and Streett (1991) map the computational coordinate $\zeta \in [-1, 1]$ into the physical coordinate $\eta \in [0, \eta_{\max}]$ via (8.8.15), or, in terms of the current variables,

$$\eta = \frac{\eta_{\max}}{2} \frac{[1 - \tanh(\sigma)](1 + \zeta)}{1 - \tanh[\sigma(1 + \zeta)/2]} \,, \tag{2.2.13}$$

where $\eta_{\max}$ is the location of the domain truncation, and $\sigma \geq 0$ is a stretching parameter. This mapping clusters the collocation points near $\eta = 0$, as illustrated in the left part of Fig. 2.2 for $\eta_{\max} = 20$ and $\sigma = 0, 0.7$ and 1.2. This figure shows the collocation points in both the computational coordinate (ζ) and the physical coordinate (η). In the limit $\sigma \to 0$, the mapping reduces to

$$\eta = \frac{\eta_{\max}}{2}(1 + \zeta) \,,$$

which is just an affine mapping. As σ increases, the clustering near the solid boundary at $\eta = 0$ increases. The value $\sigma = 0.7$ is typical for computations

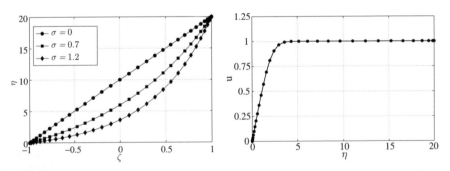

Fig. 2.2. Stretching (*left*) and mean flow (*right*) for the incompressible boundary layer using $N = 32$

of flows with M $\approx$ 5 (Pruett and Streett (1991)), whereas $\sigma = 1.2$ was recommended by Streett, Zang and Hussaini (1984) for incompressible (M = 0) flows.

Let ζ_j, for $j = 0, 1, \ldots, N$, denote the standard Gauss–Lobatto collocation points in the computational coordinate, and denote their images in the physical coordinate through the mapping (2.2.13) by η_j. Then the spectral method of Streett et al. for the incompressible boundary-layer equations is based on a collocation approximation to

$$\frac{\partial^2 \bar{u}}{\partial \eta^2} - \bar{v}\frac{\partial \bar{u}}{\partial \eta} + \beta\left(1 - \bar{u}^2\right)\bigg|_{\eta=\eta_j} = 0 \, , \qquad j = 1, \ldots, N-1 \, , \qquad (2.2.14)$$

$$\bar{v}_j - \bar{v}_{j-1} + \int_{\eta_{j-1}}^{\eta_j} \bar{u}(\eta)\mathrm{d}\eta = 0 \, , \qquad j = 1, \ldots, N \, , \qquad (2.2.15)$$

$$\bar{u}_0 = \bar{v}_0 = 0, \qquad \bar{u}_N = 1 \, . \qquad (2.2.16)$$

Note that (2.2.14) is collocated in standard form, whereas (2.2.15) is collocated in integral form.

In terms of spectral differentiation and integration matrices (which act on the values of the discrete solution at the Gauss–Lobatto points), this system is

$$R_N \left(\tilde{D}_N^{(2)}\mathbf{u} - \mathbf{v}^T\tilde{D}_N\mathbf{u} + \beta(1 - \mathbf{u} \boxtimes \mathbf{u})\right) = \mathbf{0} \, , \qquad (2.2.17)$$

$$\bar{v}_j - \bar{v}_{j-1} + \tilde{E}_N^j\mathbf{u} = 0 \, , \qquad j = 1, \ldots, N \, , \qquad (2.2.18)$$

$$\bar{u}_N = \bar{v}_N = 0, \qquad \bar{u}_0 = 1 \, , \qquad (2.2.19)$$

where $\mathbf{u} = (\bar{u}_0, \ldots, \bar{u}_N)^T$, $\mathbf{v} = (\bar{v}_0, \ldots, \bar{v}_N)^T$, R_N is the rectangular matrix that selects the interior $N-1$ rows, $\tilde{D}_N$ is the first-derivative interpolation matrix (see Sect. 8.2; it is sometimes just called the first-derivative matrix), $\tilde{D}_N^2$ is the second-derivative interpolation matrix, $\mathbf{u} \boxtimes \mathbf{v}$ is the componentwise product of two vectors $\mathbf{u}$ and $\mathbf{v}$, i.e., $\mathbf{u} \boxtimes \mathbf{v} = (\bar{u}_0\bar{v}_0, \ldots, \bar{u}_N\bar{v}_N)^T$, and $\tilde{E}_N^j$ is the matrix representing the integral in (2.2.15); the tilde on the last three matrices indicates that these are the matrices with respect to the mapped coordinate η.

The differentiation and integration matrices in the physical coordinate η are obtained by appropriate multiplications of the corresponding matrices in the standard interval $(-1, 1)$ by the derivative of the mapping, which can be represented by the diagonal matrix S with the entries

$$S_{jl} = \frac{1}{(\mathrm{d}\eta/\mathrm{d}\zeta)|_j}\delta_{jl} \, , \qquad (2.2.20)$$

where δ_{jl} is the Kronecker delta symbol. For example, $\tilde{D}_N = SD_N$, where D_N is the first-derivative interpolation matrix in the reference interval $(-1, 1)$ (see Sect. 8.2 in general and (8.2.35) in particular; see also CHQZ2, Chap. 2). The second-derivative interpolation matrix is $\tilde{D}_N^{(2)} = \tilde{D}_N \tilde{D}_N$. Many practitioners prefer to compute the $(d\eta/d\zeta)|_j$ term by always applying the first-derivative interpolation matrix D_N to the mapping formula (2.2.13) rather than using an analytical expression for $(d\eta/d\zeta)|_j$ (when possible).

The first-derivative matrix is given explicitly in (8.3.27) for Chebyshev expansions and in (8.4.19) for Legendre expansions. Note that formula (8.3.27) for the Chebyshev first-derivative interpolation matrix in the reference domain assumes that the collocation points there are given by $\zeta_j = \cos(\pi j/N)$. This is natural for implementations that use fast transform methods (see Appendix A.10 for the fast Chebyshev transform) for computing interpolation derivatives. In this boundary-layer application, we adopt the more universal left-to-right ordering of the Gauss–Lobatto points, i. e., $\zeta_j = -\cos(\pi j/N)$. Hence, one must add a negative sign to the right-hand sides of the formulas in (8.3.27).

Pruett and Streett (1991) provide the details of a spectrally accurate approximation to integrals such as those that appear in (2.2.15). We have that

$$\int_{\eta_{j-1}}^{\eta_j} \overline{u}(\eta)\, d\eta = \int_{\zeta_{j-1}}^{\zeta_j} \overline{u}(\zeta)(d\eta/d\zeta)\, d\zeta \ .$$

Write the integrand on the right-hand side as $\tilde{u}(\zeta) = \overline{u}(\zeta)(d\eta/d\zeta)$ and approximate $\tilde{u}$ by its interpolating polynomial $I_N(\tilde{u}) = \sum_{k=0}^{N} \tilde{u}_k \phi_k$, where ϕ_k is the k-th orthogonal polynomial of the chosen family, and the discrete expansion coefficients are given in (8.2.24). Then

$$\int_{\zeta_{j-1}}^{\zeta_j} \tilde{u}(\zeta)\, d\zeta \approx \sum_{k=0}^{N} \tilde{u}_k \int_{\zeta_{j-1}}^{\zeta_j} \phi_k(\zeta)\, d\zeta \ .$$

In the case of a Chebyshev expansion, we substitute $\zeta = -\cos\theta$ and obtain

$$Q_{jk} = \int_{\zeta_{j-1}}^{\zeta_j} T_k(\zeta)d\zeta = \int_{(j-1)\pi/N}^{j\pi/N} \cos(k\theta)\sin\theta d\theta \ , \qquad (2.2.21)$$

with T_k denoting the Chebyshev polynomial of degree k; this expression can readily be evaluated analytically. Therefore, the matrix for which $\tilde{E}_N^j$ comprises the rows is the product of Q, C and S, where Q is given by (2.2.21), C is the matrix representing the discrete orthogonal transform (8.2.24) and S is given by (2.2.20). For Chebyshev expansions, C is given in analytical form by (8.3.17) (but with the multiplicative term $(-1)^k$ inserted on the right-hand side when the $\zeta_j = -\cos(\pi j/N)$ ordering is used).

Equation (2.2.14) can be preconditioned effectively by a standard second-order finite-difference scheme and the trapezoidal rule is an obvious approximation to the integral in (2.2.15). (See CHQZ2, Sects. 4.4 and 4.5, and Appendix A.10 for a generic discussion of this preconditioned iteration approach to linear systems resulting from spectral discretizations.) Streett et al. obtained efficient Chebyshev and Legendre collocation solutions of the similar boundary-layer equations by this method. Figure 2.3 illustrates the convergence for both spectral solutions of the Blasius equation as well as for a fourth-order finite-difference method (Wornom (1978)). The convergence illustration is provided for the displacement thickness δ^*, which is given by (1.3.58), with $\rho(y) = \rho_\infty$ in the incompressible case; $\delta^* \approx 1.7208$ for this example. The rapid decay of the error in the Chebyshev and Legendre solutions is indica-

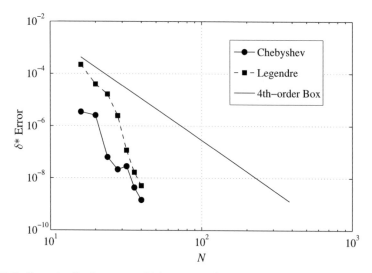

Fig. 2.3. Error in displacement thickness as a function of resolution for Chebyshev spectral, Legendre spectral and fourth-order finite-difference solutions to the similar boundary-layer equations. [From data in Streett, Zang and Hussaini (1984)]

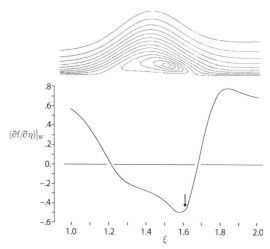

Fig. 2.4. Streamlines (*top*) and skin friction (*bottom*) from a Chebyshev solution to the nonsimilar boundary-layer equations. The arrow marks the region of the flow that is most sensitive to the numerical resolution. [Courtesy of C. L. Streett]

tive of spectral accuracy. (See Sect. 8.1 for the meaning of the terms *spectral accuracy*, *infinite-order convergence*, and *exponential convergence*.) Notice the extremely high accuracy achievable from the Chebyshev spectral method with a mere thirty collocation points.

Streett et al. (1984) solved the nonsimilar boundary-layer equations (2.2.7)–(2.2.9) in an analogous fashion. The preconditioning chosen for the streamwise direction ξ was a 3-point (second-order) upwind scheme. (This is a standard discretization in a finite-difference marching scheme.)

Since the spectral discretization is global in ξ, it is capable of handling separated flow. Figure 2.4 illustrates such a flow calculated by this spectral method. Even for this challenging case, four-digit accuracy in skin friction ($\nu(\partial u/\partial y)$) was obtained with a 26×40 grid, as opposed to the 240×200 grid (and an order of magnitude greater computer time) required by a standard second-order finite-difference scheme.

2.2.2 Compressible Boundary-Layer Flows

Pruett and Streett (1991) extended the method discussed in the previous subsection to the more challenging case of compressible flow, in particular to nonsimilar, compressible boundary layers on a flat plate or a cylinder. Various subsequent adaptations (most unpublished) were made for additional applications, such as to two-dimensional or axisymmetric boundary layers with free-stream conditions that depend upon x, and Pruett (1994) eventually generalized this to the case of infinite swept wings. The standard form of the compressible boundary-layer equations is given in (1.3.54)–(1.3.57). Pruett and Streett (1991) and Pruett (1994, 1996) applied different transformations to facilitate the numerical solution of these equations. The particular transformation used in the numerical examples of this subsection is

$$\xi = x/L \,, \qquad \eta = \sqrt{\frac{u_\infty}{x\rho_\infty\mu_\infty}} \int_0^y \rho(y')\,\mathrm{d}y' \,, \qquad (2.2.22)$$

where

$$L = \sqrt{\mu_\infty x/\rho_\infty u_\infty} \qquad (2.2.23)$$

is the boundary-layer length scale. The equations that result from applying this transformation to (1.3.54)–(1.3.57) are more complex than the transformed incompressible boundary-layer equations (2.2.7)–(2.2.8), and they include an equation for the temperature; see Pruett and Streett (1991) and Pruett (1994) for the details. Nevertheless, the basic approach of their marching-in-ξ/spectral collocation-in-η solution of the compressible boundary-layer equations is quite similar to the approach described in the previous subsection for the incompressible boundary-layer equations. The discretization in the wall-normal direction is Chebyshev collocation, and a marching scheme of up to fifth-order is used in the streamwise direction. (Third-order proved to be sufficient in most cases.) However, the preconditioned iterative scheme for solving the implicit equations was supplemented with a Newton iteration scheme, for by then computer memory limitations were not an issue. The latter was used except when a good initial guess was unavailable, as was often the case for the first station or two in the streamwise marching. The subtleties involved in extracting the wall-normal velocity are discussed by Pruett (1993).

We provide two numerical examples using the Pruett–Streett code for the compressible boundary-layer equations. The first example is for nearly incompressible flow over a flat plate under zero pressure gradient. Some of the key parameters of this example are Mach number $M = 0.1$, edge pressure $p_\infty = 102,607\,\text{Pa}$, and edge temperature $T_\infty = 309\,°\text{K}$; Sutherland's viscosity law (1.3.9), constant Prandtl number $Pr = 0.72$, adiabatic constant $\gamma = 1.4$, and adiabatic wall boundary conditions are used. At this low Mach number, the streamwise velocity profile is indistinguishable from that shown in Fig. 2.2, and the nondimensional temperature profile is nearly flat, departing from $T = 1$ by no more than 0.04%.

Figure 2.5 illustrates the accuracy of the resulting approximation to the displacement thickness δ^* (see (1.3.58)), and the first and second derivatives of the velocity at the wall, $du/d\eta$ and $d^2u/d\eta^2$. (These quantities have the values 1.7219, 0.57174, and -3.6737×10^{-3}, respectively.) This problem is rather sensitive to round-off errors due to a combination of the nonlinear algebraic system, the poor conditioning of the spectral approximation for large N, and the loss of precision in the compressible equations for small M (costing about four digits as judged by the temperature profile). Therefore, all the computations for the boundary-layer equations that are reported here were performed in 128-bit arithmetic. The "exact" values used for comparison were taken from computations made with $N = 128$ and $\eta_{max} = 20$, with the implicit equation residuals reduced to less than 10^{-25}. The left half of the figure illustrates the convergence for fixed $N = 96$ as a function of the outer boundary η_{max}. The error decays faster than exponentially in η_{max}; indeed, it

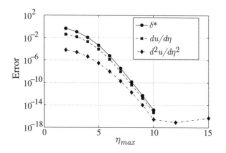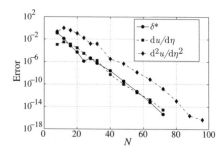

Fig. 2.5. Convergence of the Chebyshev collocation approximation to the compressible, similar boundary-layer equations for $M = 0.1$ flow past a flat plate. Errors as a function of the domain truncation η_{max} (*left*) and as a function of N (*right*)

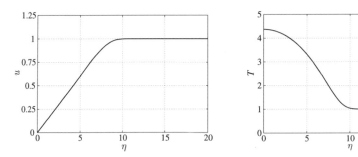

Fig. 2.6. Velocity (*left*) and temperature (*right*) for M = 4.5 boundary-layer flow past a flat plate in terms of the transformed coordinate η

is nearly linear in a semi-log plot with abscissa $\eta_{\max}^2$ (not shown here). This is expected since the dependent variables in the compressible boundary-layer problem decay as a Gaussian in η (and also y). The right half of the figure demonstrates the convergence for fixed $\eta_{\max} = 15$ as a function of N. This convergence is exponential in N. As discussed by Boyd (2001), Chap. 17, for problems on infinite intervals, infinite-order convergence, i. e., convergence faster than any finite power of $1/N$, requires that the domain truncation increase with N.

The second compressible boundary-layer example is for similar, supersonic flow past a flat plate under zero pressure gradient. Some of the key parameters of this example are Mach number M = 4.5, edge pressure $p_\infty = 958\,\mathrm{Pa}$, and edge temperature $T_\infty = 61\,^\circ\mathrm{K}$. The mean flow profiles are shown in Fig. 2.6.

Figure 2.7 illustrates the accuracy of the resulting approximation to the displacement thickness δ^*, and the first derivative of the velocity at the wall, $\mathrm{d}u/\mathrm{d}\eta$, the temperature T at the wall and the second derivative of the temperature at the wall $\mathrm{d}^2T/\mathrm{d}\eta^2$. (These quantities have the values 7.1228, 0.84050, 4.3718 and -4.0055, respectively.) For this example, no appreciable

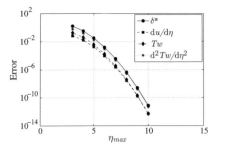 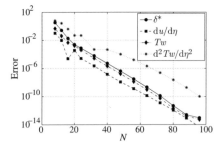

Fig. 2.7. Convergence of the Chebyshev collocation approximation to the compressible, similar boundary-layer equations for M = 4.5 flow past a flat plate. Errors as a function of the domain truncation $\eta_{\max}$ (*left*) and as a function of N (*right*)

loss of digits occurred, unlike the previous example in which the low Mach number led to significant round-off errors. Although the algebraic system cannot be solved to full machine precision, very accurate solutions (even for second derivatives) are obtainable for modest size grids. As discussed, for example, in Drazin and Reid (2004), the stability properties of a flow are very sensitive to the second derivative of the mean flow. This means that very accurate computations of the mean flow are essential. For example, in the case of supersonic boundary-layer flow, Pruett and Zang (1992) reported that errors in the fourth decimal place in the mean flow profiles produced detectable discrepancies in integrated growth rates. This consideration makes spectral solutions of the boundary-layer equations attractive.

Finally, in Fig. 2.8 we record the results of mean flow computations at M = 0, 1.6 and 4.5 using the Pruett–Streett spectral compressible boundary-layer code. Results such as these were the initial phase (the mean flow computation column in Fig. 2.1) of several compressible boundary-layer transition studies performed in the early 1990s. Results of the spectral computations used for the subsequent second and third phases (the linear stability analysis and weakly nonlinear simulation columns in Fig. 2.1) are covered in Sects. 2.4 and 2.7.2, respectively). Results from the fourth phase (direct numerical simulation) are discussed in Chap. 4 (Sect. 4.6.2). At this point we have turned to the particular format for these sorts of plots that is more customary for flow physics studies (the wall-normal coordinate is now the abscissa rather than the ordinate), and we have shifted from the transformed variable η (see (2.2.22)) used for the earlier illustrations of numerical convergence to the physical coordinate y, scaled by the boundary-layer displacement thickness δ^*. In all cases, a mere 65 collocation points sufficed for six-digit accuracy on the sensitive second derivatives of the mean flow.

As the Mach number increases, the displacement thickness grows with respect to the boundary-layer length scale L, given in (2.2.23). (This is because

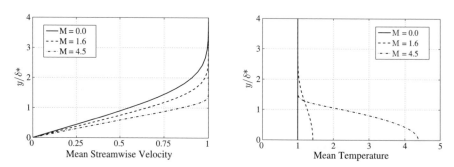

Fig. 2.8. Velocity (*left*) and temperature (*right*) for boundary-layer flow past a flat plate in terms of the physical coordinate y nondimensionalized by the displacement thickness δ^*

as $y \to 0$ the density ρ, which appears in (1.3.58), decreases as the temperature T grows.) For the cases illustrated in Fig. 2.8, we have that $\delta^* = 1.72\,L$ for M = 0, $\delta^* = 2.72\,L$ for M = 1.6, and $\delta^* = 10.5\,L$ for M = 4.5. Hence, the edge of the boundary layer as measured by the displacement thickness moves increasing further from the wall as M increases.

2.3 Linear Stability of Incompressible Parallel Flows

Spectral methods have proven to be a useful technique for supplying accurate, efficient answers to linear stability problems for both incompressible and compressible flow. The mathematical formulation of such linear stability problems as eigenvalue problems is covered in Sect. 1.5. In this section we focus on spectral approximations to linear stability analyses of incompressible parallel flows. (Recall from Sect. 1.5 that parallel flows are flows for which the mean flow is a function of a single coordinate direction, and the mean velocity is perpendicular to that direction.) A convenient aspect of this class of flows is that periodicity may be assumed in all directions except the direction normal to the mean flow; see, e. g., the expansions in (1.5.4). The following section discusses solution of the eigenvalue problem for compressible flows, and in Sect. 2.5 we discuss methods for linear stability analysis of nonparallel flows. Some brief comments on the analysis of transients are provided in Sect. 2.6. In Sect. E.1 of Appendix E we summarize some standard numerical techniques for solving the algebraic eigenvalue problems produced by these spectral discretizations of linear stability problems.

Collocation, tau and Galerkin spectral methods have all been applied to linear stability problems. The first such application to fluid dynamic stability was made by Orszag (1971b), who used a Chebyshev tau method to compute the eigenvalues of plane channel flow with unprecedented accuracy. This paper and the Orszag (1971a) and Patterson and Orszag (1971) papers on spectral methods for isotropic turbulence, which appeared at nearly the same time, catalyzed the interest of the fluid dynamics community in spectral methods. Use of the Cheyshev tau method for eigenvalue problems goes back at least to Chaves and Ortiz (1968), who applied it to a second-order eigenvalue problem with polynomial coefficients.

2.3.1 Spectral Approximations for Plane Poiseuille Flow

Chebyshev tau approximations to the Orr-Sommerfeld equation (1.5.10) for plane Poiseuille flow are obtained as follows. The dependent variable $\hat{v}$ is expanded in terms of Chebyshev polynomials:

$$\hat{v}^N(y) = \sum_{k=0}^{N} \tilde{v}_k T_k(y) \,, \tag{2.3.1}$$

and the mean flow is written as

$$u_0(y) = \sum_{k=0}^{N} \tilde{b}_k T_k(y) \, . \tag{2.3.2}$$

For plane channel flow, $u_0(y) = 1 - y^2$ (see (1.5.3)); whence,

$$\tilde{b}_0 = \tfrac{1}{2} \, , \qquad \tilde{b}_2 = \tfrac{1}{2} \, , \qquad \tilde{b}_k = 0 \quad \text{otherwise} \, .$$

For the tau approximation, (1.5.10) is enforced in Chebyshev transform space for $k = 0, 1, \ldots, N-4$. Four equations for the boundary conditions are used in lieu of the Chebyshev representation of (1.5.10) for $k = N-3, N-2, N-1, N$.

The Chebyshev recursion relation (8.3.23) yields strictly upper triangular representations for derivatives in transform space, e. g., the Chebyshev coefficients of the second derivative are given by (8.3.25). The transform space representation of the terms in (1.5.10) with constant coefficients in y are readily derived. See Orszag (1971b) for a detailed description of how to write the terms involving u_0 and $D^2 u_0$ for a general mean flow profile. The boundary conditions (1.5.12) yield

$$\sum_{k=0}^{N} \hat{v}_k = 0 \, , \qquad \sum_{k=0}^{N} (-1)^k \hat{v}_k = 0 \, ,$$
$$\sum_{k=0}^{N} k^2 \hat{v}_k = 0 \, , \qquad \sum_{k=0}^{N} (-1)^k k^2 \hat{v}_k = 0 \, . \tag{2.3.3}$$

The first row of the above equations follows from (8.3.7), and the second row from (8.3.9).

For a symmetric (in y) mean flow, as in the plane channel problem, the even and odd Chebyshev coefficients in the discrete equations arising from the differential equation are decoupled. The even and odd terms are also decoupled in the boundary conditions; they may be rewritten as

$$\sum_{\substack{k=0 \\ k \text{ even}}}^{N} \tilde{v}_k = \sum_{\substack{k=1 \\ k \text{ odd}}}^{N} \tilde{v}_k = \sum_{\substack{k=2 \\ k \text{ even}}}^{N} k^2 \tilde{v}_k = \sum_{\substack{k=1 \\ k \text{ odd}}}^{N} k^2 \tilde{v}_k = 0 \, . \tag{2.3.4}$$

The end result of the Chebyshev tau discretization of the Orr–Sommerfeld problem is a generalized eigenvalue problem of the form

$$A\tilde{\mathbf{v}} = \omega B\tilde{\mathbf{v}} \, , \tag{2.3.5}$$

where $\tilde{\mathbf{v}} = (\tilde{v}_0, \ldots, \tilde{v}_N)^T$, and A and B are the resulting matrices. (Solution methods for the generalized eigenvalue problem are surveyed in Sect. E.1 of Appendix E.)

Herbert (1977a) observed that one must be careful to avoid round-off error in the computation of the matrix elements. This application of the Chebyshev tau method requires an expression for the Chebyshev coefficients of $d^4 \hat{v}/dy^4$:

$$\hat{v}_k^{(4)} = \frac{1}{c_k} \sum_{\substack{p=k+4 \\ p+k \text{ even}}}^{N} p[p^2(p^2-4)^2 - 3k^2 p^4 + 3k^4 p^2 - k^2(k^2-4)^2]\tilde{v}_p \ . \quad (2.3.6)$$

For large k and p, there can be substantial loss of significance in the bracketed term. Herbert suggested a straightforward re-arrangement of the terms within the bracket using $p = q + k$:

$$[q^4(q^2-8) + kq^3(6q^2-32) + (12k^2q^2 + 8k^3q)(q^2-4) + 16q^2 + 32kq] \ .$$

The round-off errors for this expression are much lower.

Collocation approximations to the Orr–Sommerfeld equation are more subtle because of the double boundary condition at the walls. Suppose that the standard set of Chebyshev Gauss–Lobatto (CGL) collocation points is employed:

$$y_j = \cos(\pi j/N) \ , \qquad j = 0, \ldots, N \ . \quad (2.3.7)$$

If one enforces (1.5.10) for $j = 1, \ldots, N-1$ and (1.5.12) for $j = 0$ and N, then the problem is overdetermined. One way to remove the overdeterminacy is to enforce (1.5.10) for $j = 2, \ldots, N-2$ only, dropping the differential equation conditions at $j = 1$ and $j = N-1$, the interior points nearest the boundary. The matrices A and B can be constructed readily from the differentiation matrix D_N (see (8.3.27)), its powers, and diagonal matrices representing the constant terms, u_0 and $D^2 u_0$. Schmid and Henningson (2001) adopt this approach, although they represent the solution by its Chebyshev coefficients rather than by its physical space values. Their method amounts to postmultiplying all the matrices by the matrix C^{-1} given in (8.3.19), which represents the transformation from transform space to physical space.

Herbert (1977a) has devised another collocation method. He replaces the collocation points (2.3.7) with

$$y_j = \cos \frac{\pi j}{N-4} \ , \qquad j = 0, 1, \ldots, N-4 \ . \quad (2.3.8)$$

The $N+1$ coefficients $\tilde{v}_k$ in the expansion (2.3.1) are determined by enforcing (1.5.10) at $j = 0, 1, \ldots, N-4$, along with the conditions (2.3.4). The result is a linear system similar to (2.3.1) in which $\tilde{v}_k$ are the unknowns, as in the collocation method described above.

The Galerkin (or the G-NI) discretization starts from the weak formulation of (1.5.10) and (1.5.12) and requires that the approximate solution $\hat{v}^N$, sought in the space X_N of the polynomials of degree $\leq N$ obeying the four boundary conditions (1.5.12), satisfies

$$a(\hat{v}^N, \phi_k) = \omega b(\hat{v}^N, \phi_k) , \qquad k = 1, \ldots, N - 3 , \qquad (2.3.9)$$

for suitable basis functions ϕ_k in X_N. The bilinear forms are given by

$$a(\psi, \phi) = (\mathscr{D}\psi, \mathscr{D}\phi) - i\alpha\mathrm{Re}(u_0\mathscr{D}\psi, \phi) + i\alpha\mathrm{Re}((\mathcal{D}^2 u_0)\psi, \phi) \qquad (2.3.10)$$

and

$$b(\psi, \phi) = -i\alpha\mathrm{Re}(\mathscr{D}\psi, \phi) , \qquad (2.3.11)$$

with $\mathscr{D} = \mathcal{D}^2 - (\alpha^2 + \beta^2)I$. (Recall from Sect. 1.5 that Re is the Reynolds number, α and β are the streamwise and spanwise wavenumbers, respectively, ω is the temporal frequency, $u_0(y)$ is the streamwise mean flow, and $\mathcal{D} = d/dy$.)

Melenk, Kirchner and Schwab (2000) and Kirchner (2000) utilized a Legendre Galerkin approximation with the modal basis functions

$$\phi_k = \sqrt{\frac{2k+3}{2}} \left(\frac{L_{k+3} - L_{k+1}}{(2k+3)(2k+5)} - \frac{L_{k+1} - L_{k-1}}{(2k+1)(2k-1)} \right) , \qquad (2.3.12)$$
$$k = 1, \ldots, N - 3 ,$$

which correspond to

$$\mathcal{D}\phi_k = \frac{1}{\sqrt{2(2k+3)}} (L_{k+2} - L_k) , \qquad (2.3.13)$$

$$\mathcal{D}^2\phi_k = \sqrt{\frac{2k+3}{2}} L_{k+1} . \qquad (2.3.14)$$

The basis functions (2.3.12) are obtained by integrating (2.3.14) twice and incorporating the boundary conditions. The approximate solution is then expressed as

$$\hat{v}^N = \sum_{k=1}^{N-3} \tilde{v}_k \phi_k . \qquad (2.3.15)$$

The matrices A and B are assembled from the linear combination of inner products, such as (ϕ_h, ϕ_k), $(\mathcal{D}^2\phi_h, \phi_k)$, and $(u_0\phi_h, \phi_k)$. For the channel flow problem, where $u_0(y) = 1 - y^2$, all the integrals can be performed analytically. Moreover, the odd and even modes are uncoupled, and both matrices have small bandwidth. In particular, A has only seven nonzero diagonals and B only five. Kirchner (2000) has provided closed-form expressions for all the matrix elements. The tau and collocation methods, in contrast, produce full matrices. This enables the use of far more efficient eigenvalue computation procedures for the Galerkin method, at least on the plane channel problem.

The Galerkin with numerical integration (G-NI) version of (2.3.9) is obtained by evaluating the integrals numerically, e.g., using $N + 1$ Legendre Gauss–Lobatto (LGL) points. It reads

$$a_N(\hat{v}^N, \phi) = \omega b_N(\hat{v}^N, \phi) , \qquad \text{for all } \phi \in X_N , \qquad (2.3.16)$$

where the index N appended to the bilinear forms stands for the replacement of exact integration with numerical quadrature (see (8.2.20)). For the present example, only a few of the matrix elements for the G-NI method disagree with those of the Galerkin method if the modal basis (2.3.12) is used, e. g., (ϕ_{N-3}, ϕ_{N-3}) is the integration of a polynomial of degree $2N$, which exceeds the precision of the LGL quadrature rule. For the channel flow problem, the quadrature errors are removed by using $N + 3$ LGL points. For more general mean flows, such as the flat-plate boundary layer, for which u_0 is more complicated (indeed, often only available numerically), then the pure Galerkin method must be supplanted by the G-NI method, and quadrature errors are unavoidable.

The G-NI method cannot be interpreted as a collocation method in this case. (We refer to Sect. 8.9.4 for an example of equivalence between GNI and collocation at the internal nodes.) To get the correct interpretation of the method, let us use the exactness of the discrete inner product to integrate back by parts twice in the first term of $a_N(\hat{v}^N, \phi)$. We obtain

$$(r^N, \phi)_N = 0 , \qquad \text{for all } \phi \in X_N ,$$

where

$$r^N = r^N(\hat{v}^N, \omega) = \mathscr{D}^2 \hat{v}^N - i\alpha \operatorname{Re} u_0 \mathscr{D} \hat{v}^N + i\alpha \operatorname{Re}(D^2 u_0)\hat{v}^N + i\omega\alpha \operatorname{Re} \mathscr{D} \hat{v}^N$$

denotes the "residual" of the Orr–Sommerfeld equation. The maximal localization is achieved by choosing as test functions ϕ the polynomials $\tilde{\psi}_j$, with $2 \le j \le N - 2$, defined by the conditions $\tilde{\psi}_j(x_k) = \delta_{jk}$ for $2 \le k \le N - 2$, as well as $\tilde{\psi}_j(\pm 1) = \tilde{\psi}'_j(\pm 1) = 0$ (note that these polynomials do not vanish, although they are small, at the first and last internal nodes x_1 and x_{N-1}). With such a choice, the G-NI method is equivalent to the set of relations, for $2 \le j \le N - 2$,

$$\tilde{\psi}_j(x_1)w_1 r^N_{x_1} + w_j r^N_{x_j} + \tilde{\psi}_j(x_{N-1})w_{N-1} r^N_{x_{N-1}} = 0$$

(where w_j are the weights of the Legendre Gauss–Lobatto quadrature formula), i. e., to $N - 3$ particular 3-point linear combinations of the Orr–Sommerfeld equations.

Rigorous numerical analyses of collocation and tau methods for the Orr–Sommerfeld eigenvalue problem are not available (to our knowledge). Melenk, Kirchner and Schwab (2000) provide numerous theoretical results for their Galerkin method, most of which carry over to the G-NI method. For example, they prove that no spurious eigenvalues exist (see Sect. 2.3.2 for a discussion of spurious modes), provide bounds on the Euclidean norms of the matrices A and B, such as

$$\|A\|_2 \le C \operatorname{Re} , \qquad \|B\|_2 \le C \operatorname{Re} , \qquad \|B^{-1}\|_2 \le C\frac{N^4}{\operatorname{Re}} , \qquad (2.3.17)$$

(where C depends only on α, β and u_0), and derive convergence estimates for the eigenvalues and eigenfunctions. In particular, they prove that the convergence is exponential in N if $u_0(y)$ is analytic, and demonstrate that proper resolution in approximating an eigensolution $(\hat{v}, \omega)$ requires a condition of the form

$$\frac{\mathrm{Re}(1 + |\omega|)}{N^2} \leq C$$

to be satisfied. Except for the presence of $|\omega|$, this estimate is the same as that discussed in CHQZ2, Sect. 7.2 for the advection-diffusion equation. Here again, we observe that spectral methods require that the number of modes be inversely proportional to the boundary-layer thickness (which scales as $1/\sqrt{\mathrm{Re}}$).

2.3.2 Numerical Examples for Plane Poiseuille Flow

We now provide some numerical examples of various spectral methods on a channel flow linear stability problem with $\mathrm{Re} = 7500$, $\alpha = 1.0$, and $\beta = 0.25$. The structure of the eigenfunction of the least stable mode of the Orr–Sommerfeld equation is illustrated in the left part of Fig. 2.9. (Its eigenfrequency is $\omega = 0.2546468 + 0.00168654i$.) This particular mode is a so-called *wall mode*; its structure is concentrated next to the wall, where the Chebyshev and Legendre methods naturally cluster the grid points. The phase speed of this mode is $Re\{\omega\}/\alpha = 0.2546468$. This matches the local mean flow velocity at $y_{\mathrm{crit}} \approx \pm 0.863$. The region near these points are the *critical layers*, where most of the eigenfunction structure is concentrated. Plane channel flow also admits *center modes*, i. e., modes whose structure is located near the center of the channel. (These center modes are always linearly stable; see Zang and Krist (1989) for a further discussion.) Figure 2.9 (right) illustrates the least

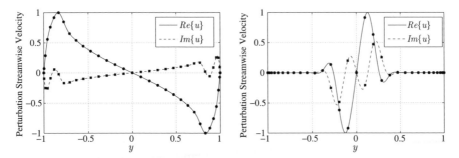

Fig. 2.9. Streamwise velocity eigenfunctions of the most unstable wall mode (*left*) and the least stable center mode (*right*) of the Orr–Sommerfeld equation for plane channel flow at $\mathrm{Re} = 7500$. The circles denote the Chebyshev collocation points for $N = 32$

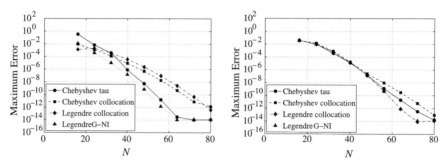

Fig. 2.10. Absolute value of errors in the least stable Orr–Sommerfeld (*left*) and vertical-vorticity (*right*) wall modes for Re = 7500, $\alpha = 1.0, \beta = 0.25$ from various spectral methods

stable center mode with the same parameters as the wall mode. (Its eigenfrequency is $\omega = 0.959155 - 0.405607i$, corresponding to critical layers at $y_{\text{crit}} \approx \pm 0.202$.) Note that the center mode structure is most complex in the region in which the Chebyshev and Legendre grids are coarsest. The vertical-vorticity equation also admits wall and center modes. The eigenfrequency of the least stable wall mode for it is $\omega = 0.161816 - 0.090663i$.

A comparison of the accuracy from some of these spectral methods for the wall modes of the Orr–Sommerfeld equation for plane channel flow is furnished in the left half of Fig. 2.10. (The collocation computations were performed with the MATLAB© code in Appendix A of Schmid and Henningson (2001).) For all methods the error decays exponentially fast. However, the collocation methods exhibit a distinctly slower (but still spectral) rate of convergence than the tau and G-NI methods. Furthermore, we observe that for a given N (until the round-off error floor is reached), the Legendre G-NI method is about one order of magnitude more accurate than the Chebyshev tau method, and that the Chebyshev collocation method exhibits a similar advantage over the Legendre collocation method.

The right half of Fig. 2.10 shows corresponding results of these same spectral methods for the vertical-vorticity eigenvalue problem—(1.5.11) with $\hat{v} = 0$ plus (1.5.13). For this problem the collocation method enforces (1.5.11) at all the interior points, whereas the collocation equations for two of the interior points were dropped for the higher order Orr–Sommerfeld problem. The Legendre G-NI methods use the modal basis functions given in (2.3.12), for which the function but not the first derivative vanishes at both boundaries. For this second-order problem, the collocation methods converge at the same rate as the tau and G-NI methods. (It might be difficult to see in these plots, but the symbols for the two methods overlap.) Indeed, the Legendre G-NI method is algebraically equivalent to the Legendre collocation method. The Legendre G-NI results are also the more accurate ones here as well, but only distinctly so for the larger values of N.

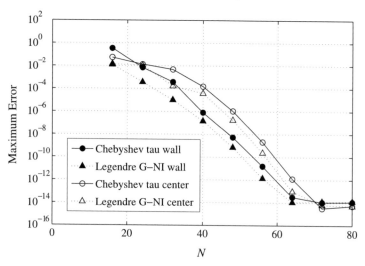

Fig. 2.11. Absolute value of errors in the least stable Orr–Sommerfeld wall and center modes for $\mathrm{Re} = 7500, \alpha = 1.0, \beta = 0.25$ from Chebyshev tau and Legendre G-NI spectral methods

Finally, Fig. 2.11 compares the accuracy of the Chebyshev tau and Legendre G-NI methods on the wall and center modes. As one would expect, the wall mode is resolved better than the center mode for a given N, in this case with about two orders of magnitude smaller error for a given value of N.

As noted in Orszag's (1971b) paper, the original version of the Chebyshev tau method for the Orr–Sommerfeld problem produces a small number of eigenvalues with very large, positive imaginary parts. The full spectra of the numerical eigenvalues for the above plane Poiseuille flow numerical example are displayed in Fig. 2.12. The domains for the displays in the left column encompass all the numerical eigenvalues for the particular numerical method, whereas the domains for the displays in the right column focus on the physically relevant eigenvalues. The numerical eigenvalues for the Chebyshev tau method that are located in the upper left of the top left frame of Fig. 2.12 are numerical artifacts that have no physical counterpart; these artificial eigenvalues with large positive imaginary parts are commonly referred to as *spurious eigenvalues*. The corresponding collocation and G-NI results, also illustrated in Fig. 2.12, do not exhibit any spurious eigenvalues with large imaginary parts, although there are some purely numerical eigenvalues with large negative imaginary parts. For problems whose stability properties are well-understood, such as plane channel flow, these spurious eigenvalues are not particularly troublesome. However, spurious eigenvalues are a potentially dangerous aspect of the standard tau method, as they may inhibit proper interpretation of the stability properties of less well-understood flows. Various explanations (and some remedies) have been offered for the origin of these spurious eigenvalues from the tau method, including the singularity

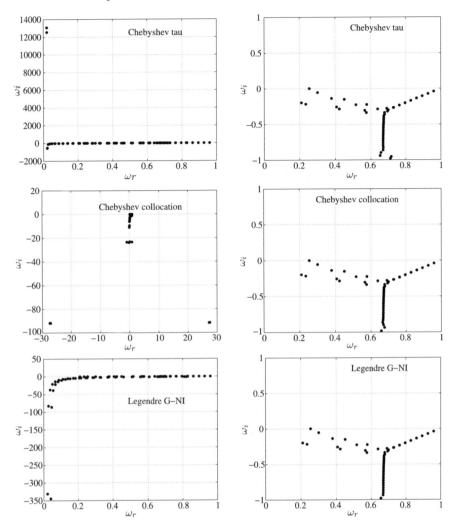

Fig. 2.12. Spectra from Chebyshev tau, Chebyshev collocation and Legendre G-NI spectral methods for the Orr–Sommerfeld problem for plane channel flow with $Re = 7500, \alpha = 1.0, \beta = 0.25$ using $N = 100$. The plotting domains in the left column encompass all the numerical eigenvalues, whereas the domains in the right column are confined to the physically relevant eigenvalues

of the matrix B. See Dongarra, Straughan and Walker (1996) for a fairly extensive discussion of spurious eigenvalues produced by the tau method. They advocate converting the fourth-order Orr–Sommerfeld problem into two second-order problems, noting that this removes spurious eigenvalues.

For the particular numerical examples in Figs. 2.10 and 2.11, the computations of the least stable eigenvalue converge to nearly machine precision

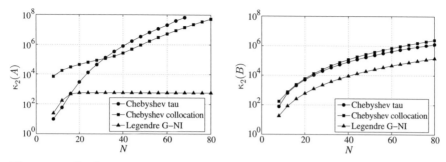

Fig. 2.13. Condition numbers of the Chebyshev tau, Chebyshev collocation and Legendre G-NI matrices for the Orr-Sommerfeld problem with for $Re = 7500, \alpha = 1.0, \beta = 0.25$. L_2-condition numbers of A (*left*) and B (*right*)

with $N < 100$. However, if one is interested in the accurate computation of many of the eigenvalues or in problems with higher Reynolds numbers, both of which require larger values of the discretization parameter N, then round-off errors become a concern. As a rule of thumb, the condition numbers of the derivative matrices in collocation and tau methods scale as N^{2q} where q is the order of the highest derivative in the problem; see CHQZ2, Sects 4.3 and 7.3. Dongarra, Straughan and Walker (1996) were also motivated to convert the fourth-order Orr–Sommerfeld problem into two second-order problems because of round-off error concerns; see their paper for numerical examples. They view the dual benefits of reduced round-off errors and elimination of spurious modes as being worth the extra expense in the eigenvalue computations produced by the larger matrices that ensue.

The matrix A, which contains the fourth-derivative operator, is much better behaved for the Galerkin and G-NI methods than for the tau and collocation methods. This is illustrated in Fig. 2.13, where the condition numbers of the matrices A and B in the 2-norm are plotted for the tau, collocation and G-NI methods. For the tau and collocation cases, we observe the expected growth in the condition number of A as N^8 and that of B as N^4, whereas for the G-NI method we see that the condition number of A is relatively well-behaved while that of B also scales as N^4, in conformance with the estimate (2.3.17). These results indicate that round-off errors are less of a concern with the Galerkin and G-NI formulations than with tau and collocation ones.

2.3.3 Some Other Incompressible Linear Stability Problems

So far in this section we have limited our discussion of the applications of spectral methods to linear stability problems to the simple case of plane channel flow. The tau method has been used for studying the stability of many fluid dynamics problems. Dongarra, Straughan and Walker(1996) furnish a summary of many of the more complex flow problems to which the

Chebyshev tau method has been applied. Here we comment briefly on some other applications using variations of tau, collocation and Galerkin methods.

Zebib (1984) developed a tau method for the integral formulation of the linear stability problem. (This was discussed in CHQZ2, Sect. 4.1.2 in the context of a simple second-order equation.) For solving the Orr–Sommerfeld problem for the plane channel flow, he expands the highest-order derivative in Chebyshev polynomials, e. g.,

$$\hat{v}^{(4)}(y) = \sum_{k=0}^{N} \tilde{v}_k^{(4)} T_k(y) \ . \tag{2.3.18}$$

He then integrates this four times to obtain a polynomial of degree $N + 4$ for $\hat{v}$ with four constants of integration. These constants are determined in terms of the $\tilde{v}_k^{(4)}$ by enforcing the four boundary conditions. The expansion coefficients $\tilde{v}_k^{(4)}$ are then determined by applying the usual tau procedure to (1.5.10) and (1.5.12). Bridges and Morris (1984) developed a similar integral variant of the tau method.

A different spectral Galerkin procedure originated with Leonard and Wray (1982). Their approach was to choose the expansion functions to satisfy not only the boundary conditions but also the incompressibility constraint. Leonard and Wray describe this approach for the pipe flow problem using a particular choice of shifted Jacobi polynomials. Spalart, Moser and Rogers (1991) discuss some appropriate choices for the basis functions (based on Jacobi polynomials) for the flat plate and free-shear-layer problems. These methods were actually developed primarily for use in direct numerical simulations. Their application to the linear stability problem was a secondary motivation. See Sect. 3.4.2 for a further discussion of these expansions in the context of plane channel flow. Meseguer and Trefethen (2003) developed a Galerkin method based on Chebyshev polynomials for incompressible pipe flow stability analysis. These Galerkin stability methods have produced extremely accurate solutions with a very small number of terms in the expansion, significantly fewer than those required by the typical collocation method. Of course, the expansion functions must be tuned to the specific problem, in contrast to the greater generality of collocation methods.

Khorrami, Malik and Ash (1989) developed a Chebyshev collocation method that worked directly with (1.5.5)–(1.5.9) rather than with the Orr–Sommerfeld and vertical-vorticity equations. (In their application, the mean flow did not permit the clean separation into these two equations that exists in plane channel flow.) In addition to reducing the largest derivative from fourth-order to second-order, there is only the need for one condition on each of the velocity components at each boundary; this is easily done in the usual collocation manner. This problem is well-posed mathematically without any boundary conditions on the pressure. However, if the Gauss–Lobatto points are used for constructing the interpolants of the pressure and for enforcing the collocation condition on (1.5.5), there is a nonzero pressure mode whose gradient is zero at all the Gauss–Lobatto points, and which therefore can have arbitrary amplitude. This subject is discussed at much greater length in Sect. 3.5. Here we simply note that Khorrami, Malik and Ash considered two different Neumann boundary conditions on the pressure—a homogeneous condition and an inhomogeneous condition derived from taking the component of the momentum equation normal to the boundary. Their method was developed to study stability problems in cylindrical geometries, including pipe flow and flow in a cylindrical annulus. Khorrami, Malik and Ash observed no significant difference (to at least seven digits) between the results of their homogeneous and inhomogeneous pressure boundary conditions, nor with other published results on plane channel flow, pipe flow and rotating pipe flow. They did note that use of a staggered grid for the pressure would alleviate all concerns. (Again, see Sect. 3.5 for a further discussion.) The analog of this for the tau method applied to (1.5.5)–(1.5.9) would be to restrict the expansion of the pressure to a polynomial of degree $\leq N - 1$ while the velocity components are represented by polynomials of degree $\leq N$.

Linear stability investigations of channel flow, pipe flow and annular flow are fairly tractable because the mean flow is strictly parallel, i.e., $\mathbf{u}_0$ is parallel to the wall and is independent of x. Even such a simple flow as that over a flat plate is nonparallel, as was discussed in Sect. 1.5.2. As noted there, a reasonable approximation to boundary-layer stability is to make the parallel flow approximation (see Fig. 1.4). Within this approximation, it is customary to take for $u_0(y)$ the solution to the similar boundary-layer equations. This stability problem is also governed by (1.5.10)–(1.5.13) except that the boundary conditions are imposed at $y = 0$ and $y = \infty$. The same strategies that were discussed in Sect. 2.2 for the Chebyshev spectral solution of the mean flow are also available here—a suitable mapping in y (bearing in mind that the solution to the stability problem decays exponentially in y in contrast to the y^2 exponential decay of the boundary-layer solution) and either the standard Chebyshev expansion over $[-1, 1)$ in the computational coordinate or else an expansion over $[0, 1)$. Bridges and Morris (1987) applied their tau method to the flat-plate boundary layer.

Fischer (1993) applied Spalart's (1984) Galerkin method, which combines the exponential mapping (8.8.11) with an expansion in a fairly general set of Jacobi polynomials to the Orr–Sommerfeld problem for a flat-plate boundary layer. He conducted a rigorous theoretical analysis of the convergence properties in terms of the mapping parameter L and the discretization parameter N. He proved that for fixed L, the approximations converge with finite order, whereas if L depends upon N in proportion to N^σ for $0 < \sigma < 1$ arbitrary, then convergence is infinite order.

Von Kerczek (1982) has applied a standard Chebyshev tau method to the linear stability problem of oscillatory channel flow. In this flow the pressure gradient is given not by (1.5.3) but rather by

$$p_0(x, t) = -\frac{2x}{\text{Re}} \left[1 + \Lambda \cos \omega t\right] , \qquad (2.3.19)$$

and the mean velocity $u_0(y, t)$ is also periodic in time. (Here, the frequency ω is real.) The appropriate Orr–Sommerfeld equation includes a time dependence. Floquet theory guarantees that the perturbations may be written as

$$\mathbf{u}(\mathbf{x}, t) = \hat{\mathbf{u}}(y, t)e^{i(\alpha x + \beta z) + \lambda t} , \tag{2.3.20}$$

where $\hat{\mathbf{u}}(y, t)$ is periodic in t with period $2\pi/\omega$. The Floquet exponent λ determines the stability of infinitesimal perturbations (Davis (1976)). Von Kerczek discusses several techniques for computing these Floquet exponents in Chebyshev approximations.

The parallel, spatial stability problem can also be attacked by Chebyshev methods. Bridges and Morris (1984) have given an extensive discussion of this problem in the context of spectral methods. The principal algorithmic difficulty is the solution of the eigenvalue problem: ω is fixed and α is the eigenvalue (if $\beta = 0$). This enters nonlinearly in (1.5.10). (Section E.1 of Appendix E also covers solution methods for the spatial eigenvalue problem.) Khorrami, Malik and Ash (1989) applied their collocation approach to spatial stability problems as well. They noted that the computational savings realized for spatial stability problems by using the Orr–Sommerfeld and Squire equations instead of the usual continuity and momentum equations were much smaller than the savings realized for temporal stability analyses.

2.4 Linear Stability of Compressible Parallel Flows

Most compressible stability analyses, such as those of Mack (1969, 1984) and Malik (1989), have utilized finite-difference methods. Part of the reason for this is that much of the seminal work in the subject predated the use of spectral methods for such applications. Numerous algorithmic developments for compressible, linear stability analysis took place during the late 1980s and the 1990s, driven by the interest in understanding and predicting transition for supersonic and hypersonic flows past aerospace vehicles. While most of these developments were in the context of traditional finite-difference methods, there were several applications of spectral methods to compressible stability problems. However, compressible stability has proven to be a particular challenge to spectral methods. Here we survey those few applications of spectral methods to this topic and discuss the challenges.

All of the spectral compressible stability methods (of which we are aware) have utilized Chebyshev collocation. Macaraeg, Streett and Hussaini (1988) (see also Streett and Macaraeg (1989) and Macaraeg and Streett (1991)), Malik (1990) and Herbert (1990) (see also Duck, Erlebacher and Hussaini (1994)) developed slightly different spectral methods for temporal and spatial compressible stability problems using (1.5.19)–1.5.20). The Macaraeg, Streett and Hussaini collocation scheme uses a staggered grid in which the velocity components and the temperature are defined at the Gauss–Lobatto nodes whereas the pressure is defined at the Gauss nodes. (Recall that on the interval $[-1, 1]$, the Chebyshev Gauss–Lobatto nodes are given by $x_j = \cos(\pi j/N)$, $j = 0, 1, \ldots, N$, and the Chebyshev Gauss nodes by $x_j = \cos(\pi(j - 1/2)/N)$, $j = 0, 1, \ldots, N$. Further discussion on staggered grids is provided in Sects. 3.4.3 and 4.6.1.) The momentum equation and the energy equation are collocated at the Gauss–Lobatto nodes, and the continuity equation is collocated at the Gauss nodes. Malik's Chebyshev scheme uses a nonstaggered grid and has the same type of Neumann boundary condition for the pressure as the Khorrami, Malik and Ash (1989) scheme for incompressible stability. Both Macaraeg, Streett and Hussaini (1988) and Malik (1990) implemented spectral multidomain algorithms as well. Other implementations of Malik's Chebyshev nonstaggered-grid approach to compressible linear stability have been made, e. g., by Hanifi, Schmid and Henningson (1996) and by Ma and Zhong (2003). Our convergence and application illustrations, though, are confined to the staggered-grid method.

The early applications of the Macaraeg, Streett and Hussaini method were made to the temporal and spatial stability of free shear layers and to boundary layers on flat plates and cones. They used an algebraic mapping, with clustering near the internal critical layers, to handle the infinite (free-shear-layer) and semi-infinite (flat-plate and cone boundary-layer) domains. The mean flow for their computations relied upon the spectral compressible boundary-layer code discussed in Sect. 2.2.2.

We first provide a convergence illustration for the linear stability analysis of an ideal gas, compressible free shear layer, such as that illustrated in Fig. 2.14. The flow is assumed to be parallel with streamwise velocity given by the hyperbolic tangent profile $u_0 = \tanh(y)$ for $M = 0.5$ flow at $y = \pm\infty$ with a Reynolds number (based on the displacement thickness) of $Re = 100$. (The hyperbolic tangent profile is not an exact solution for the mean flow. Although it is a simple test case for the numerical method, one must imagine that there is an additional forcing term in the mean flow equations to produce this profile.) Sutherland's formula (1.3.8) was used for the viscosity, the constant Prandtl number $Pr = 0.7$ was taken, the ratio of specific heats of $\gamma = 1.4$ was used, and the far-field temperature was $T_\infty = 180\,°K$. Figure 2.15 illustrates the mean flow profile along with the streamwise velocity perturbation of the most unstable mode ($\omega = 0.1274549\,i$) for perturbations with $\alpha = 0.4$ and $\beta = 0$. The circles denote the locations in physical space of the collocation points generated by the mapping

$$y = \frac{y_{\max}\,(C_2 - 1)^{C_1}(1 + \xi)}{(C_2 - (1 + \xi)^2)^{C_1}}.$$ (2.4.1)

The computational coordinate ξ varies in $[-1, 1]$, the domain in y is truncated to $[-y_{\max}, y_{\max}]$, and the values of the constants for this example are $C_1 = 4$, $C_2 = 1.3$. (See Macaraeg, Streett and Hussaini (1988) or Streett and Macaraeg (1989) for more details.) In this free-shear-layer example the critical

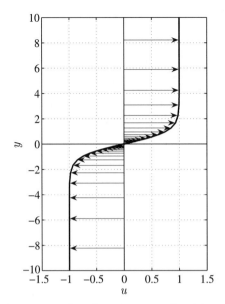

Fig. 2.14. Hyperbolic tangent free-shear-layer profile. The x, y and z directions are called the streamwise, normal and spanwise directions, respectively

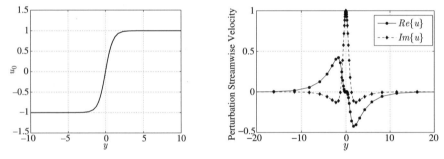

Fig. 2.15. Compressible free-shear-layer mean flow (*left*) and streamwise velocity eigenfunction (*right*) for $M = 0.5, Re = 100, \alpha = 0.4, \beta = 0.0$

layer, i. e., the region in which the phase speed of the wave matches that of the mean velocity, is at $y = 0$, and the mapping (2.4.1) appropriately clusters the collocation points in physical space around the critical layer.

Figure 2.16 provides the evidence of spectral convergence for this problem. The "exact" solution was taken from a computation with $y_{max} = 150$ and $N = 160$. In both cases the other parameter was taken sufficiently large so that it did not impact the accuracy. Note that the error decays exponentially fast with respect to both parameters. (Recall, though, that both parameters must increase together to achieve this.) However, the exponential decay is only linear in y_{max} rather than quadratic in y_{max} as it was for the

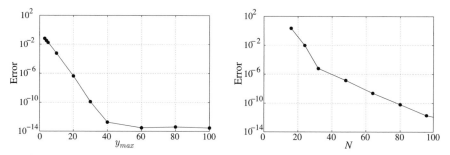

Fig. 2.16. Convergence of the Chebyshev collocation approximations to the linear stability of a compressible free shear layer at M = 0.5. Errors are given as a function of the domain truncation y_{max} (*left*) and as a functions of N (*right*)

compressible boundary-layer solution (see Fig. 2.5). Indeed, asymptotic analysis shows that the solution to the compressible boundary-layer approaches 1 as $e^{-\sigma y^2}$ for some $\sigma > 0$, whereas the solution to the linear stability problem approaches 0 as $e^{-\alpha y}$ (see Moore (1989)). Macaraeg, Streett and Hussaini (1988) demonstrated an order of magnitude reduction in the number of grid points needed for three-digit accuracy on the growth rates compared with a contemporary finite-difference algorithm.

Most compressible linear stability problems are more challenging numerically than the free-shear-layer example above. If one considers a free shear layer in which the velocity is 0 at $y = -\infty$ and 1 at $y = +\infty$, as in the case of gas injected into a quiescent fluid, and the Mach number (at $y = +\infty$) is larger than 1, then there are some linear modes whose phase speeds relative to the mean flow are supersonic. These *supersonic modes* are highly oscillatory in y with relatively slow decay as $y \to \infty$; they are difficult to capture with a global discretization (as opposed to an adaptive shooting method). An illustration of the perturbation pressure for such a supersonic mode is provided in Fig. 2.17 from Macaraeg, Streett and Hussaini (1988). This is for a self-similar free shear layer with M = 4 at $y = +\infty$ and Re = 10,000. The disturbance wavenumbers are $\alpha = 0.30416$ and $\beta = 0.2017$. The temperature perturbation has a much finer structure near $y = 0$ than the pressure perturbation. Such a mode cannot be resolved efficiently by a classical,

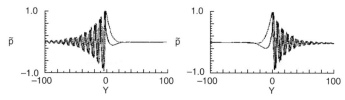

Fig. 2.17. Perturbation pressure for a representative supersonic mode of a compressible free shear layer at M = 4. [From Macaraeg, Streett, Hussaini (1988)]

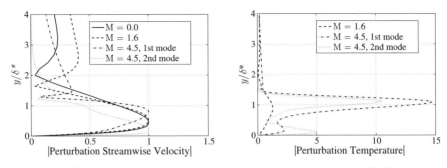

Fig. 2.18. Compressible flat plate perturbation streamwise velocity (*left*) and perturbation temperature (*right*) for linear instabilities. [Adapted from Ng, Erlebacher (1992) using data supplied by Erlebacher]

Table 2.1. Parameters for compressible stability computations

M	Re	T_∞	α	β	A	h	β_2
0	1043	520 °R	0.35	0.0	1%	0	0.34
1.6	1675	520 °R	0.24	0.0	1.5%	0	0.4
4.5	10,000	110 °R	0.6	0.0	—	—	—
4.5	10,000	110 °R	2.52	0.0	6%	1	2.1

single-domain spectral method. Indeed, this particular result is taken from a spectral multidomain computation.

The perturbations of interest for supersonic flows past solid walls are not as challenging as the supersonic modes for the free shear layer. Nevertheless, they are more difficult to handle with a single-domain method than are incompressible stability analyses. Figure 2.18 illustrates the streamwise velocity and temperature perturbations for typical unstable modes for flow past a flat plate. Some of the key parameters for these cases are listed in Table 2.1. (The last three columns are not relevant here; they are discussed in Sect. 2.7.2.) The mean flows are plotted in Fig. 2.8. As the Mach number increases from zero, the nature of the dominant instability initially evolves gradually from the incompressible case, although it becomes oblique ($\beta \neq 0$) beyond M = 1. Beyond M $\approx$ 2.2 a qualitatively different type of unstable mode appears, although it does not become the dominant instability until M $\approx$ 3.7. (This new mode is referred to as the second mode, and the other mode is called the first mode.) As is evident in Fig. 2.18, the second mode has a much different structure. It is far more concentrated near the critical layer (close to $y/\delta^* = 1$), decays far more rapidly as $y \to \infty$, but yet has an important, pronounced structure in the near-wall region. This makes the second mode more difficult to resolve with a single-domain method than the first mode. The compressible stability solutions illustrated in this figure were computed by Ng and Erlebacher (1992) using the staggered-grid method with no more than 65 Chebyshev collocation points. Their convergence studies demonstrated that at least five-digit accuracy was obtained for the eigenvalue. (Such precision was desired in their application because the linear eigenfunction was an input to the subsequent secondary instability computation—see Sect. 2.7.2.) This accuracy could only be obtained with such few points with the use of a mapping that clustered the points around $y/\delta^* = 1$. They utilized the hyperbolic tangent clustering (8.8.7) together with the truncated algebraic mapping (8.8.13).

2.5 Nonparallel Linear Stability

Classical, single-domain spectral methods have seen relatively little use on nonparallel linear stability problems. Here we survey their applications to the parabolized stability equations and to global stability analysis.

2.5.1 Linear Parabolized Stability Equations

The parabolized stability equations (PSE) are described in Sect. 1.6. PSE can be solved by marching along the parabolic streamwise direction. Herbert (1991) and Bertolotti et al. (1992) discretized (1.6.4) with a spectral collocation method in the wall-normal direction and a backwards Euler finite-difference method in the streamwise direction. (Since the parabolized stability equations can be solved by a marching method in x, spectral methods would be a poor choice for the discretization in that coordinate direction.) The solution at $x = x_0 + \Delta x$ is computed from (1.6.4)–(1.6.5) with $\chi(x_0 + \Delta x) = \chi(x_0)$ as a first approximation, then a new χ is calculated using (1.6.3). Equations (1.6.4)–(1.6.5) are solved again with the new value of χ. This process continues until the solution converges. The marching is then carried out to the next x-station. The one-step difference in the marching direction is very special in that the error is proportional to the product of the step-size and the second derivative, i.e., $\frac{\partial^2 \hat{q}}{\partial x^2}\Delta x$. (Recall that the vector of dependent variables is $\mathbf{q} = (u, v, p)^T$ in the two-dimensional, incompressible case and that the hat denotes the slowly varying shape function.) If the one-step difference were to cause large error for reasonable step-sizes, it must mean that the second derivative is large—a violation of the PSE assumption. One must then carefully examine the validity of solutions obtained in such cases. On the other hand, if one-step difference causes negligible changes in the solution for a range of sufficiently large step-sizes, it must imply that the PSE assumption holds.

Strictly speaking, the parabolized stability equations do not constitute a well-posed problem, and it has been shown that the solution eventually blows up when the marching step size is reduced. For the incompressible primitive variable formulation, Li and Malik (1996, 1997) show that a stable solution can only be obtained if the step size satisfies the condition

$$\Delta x > \frac{1}{|\alpha_r|} , \qquad (2.5.1)$$

where α_r is the real part of the spatial wavefunction α in (1.6.2). The step-size limitation is relaxed somewhat for the stream function-vorticity formulation of the incompressible PSE equations (Li and Malik (1996)) and for the PSE equations for compressible flow. For 3D disturbances in compressible flow, $\hat{\phi} = (\hat{p}, \hat{u}, \hat{v}, \hat{w}, \hat{T})^T$ and the matrices $L_1 - L_4$ in the PSE system (1.6.4) are of order 5.

The PSE equations have produced impressive quantitative agreement with direct numerical simulations of linear instabilities in spatially evolving flows. See Bertolotti (1991), Bertolotti et al. (1992), Joslin, Streett and Chang (1993) for incompressible flow, and Pruett, Zang, Chang and Carpenter (1995) for compressible flow. Figure 2.19, taken from Bertolotti et al. (1992), compares incompressible PSE results with the much more expensive direct numerical simulation results (using the spectral fringe-region method of Spalart that is discussed in Sect. 3.6.1). (The Reynolds number Re in the abscissa of the figure is based on the distance from the leading edge of the flat plate.) The agreement is excellent. The figure also illustrates the nonparallel effects that are missed by the parallel flow approximation. This type of evidence has led to their acceptance as a trusted engineering tool for predicting fluid dynamical stability in aeronautical applications.

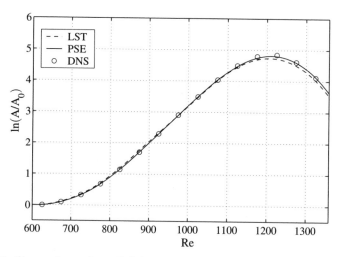

Fig. 2.19. Comparison of parallel, linear stability theory (LST), parabolized stability equations (PSE) and direct numerical simulation (DNS) solutions for instability growth in a flat-plate boundary layer. [Adapted with permission from F.P. Bertolotti, Th. Herbert, P.R. Spalart (1992); ©1992, Cambridge University Press]

The original work on incompressible PSE by Herbert (1991) and Bertolotti et al. (1992) utilized Chebyshev collocation for the discretization in the wall-normal direction, using on the order of 50 points. Applications to compressible flow problems have tended to use high-order finite-difference methods (or, in the case of Bertolotti and Herbert (1991), multidomain spectral methods), in part because of their greater flexibility in resolving the internal critical layers that dominate the physics of transition. We comment briefly on the multidomain PSE applications in Sect. 5.13. Some other applications of linear PSE to incompressible flows using single-domain spectral methods can be found, for example, in Herbert (1997), Koch et al. (2000), and Pralits, Hanifi and Henningson (2002).

2.5.2 Two-Dimensional Global Stability Analysis

The linear stability methods discussed in Sects. 2.3 and 2.4 reduce to the problem of solving one-dimensional eigenvalue problems. These methods require very modest computing resources, and thorough parameter studies have been feasible for several decades. The PSE method discussed in Sect. 2.5.1 requires the solution of a one-dimensional boundary-value problem at each marching step (because of the implicit marching method). Computing resources were not a barrier to parameter studies with PSE even when they were first introduced.

Spectral collocation and Galerkin methods have also been applied to linear stability problems with more than one implicit coordinate direction. The term *global stability analysis* is often used to describe this class of linear stability problem.

A relatively simple example is the stability analysis performed by Reddy et al. (1998). They used a Fourier Galerkin–Chebyshev collocation method to study the stability of plane Couette flow with a finite-amplitude, steady spanwise disturbance for which the assumed mean velocity profile is given by

$$u(\mathbf{x}) = u_0(y) + \sum_{m=-M}^{M} \hat{u}_m(y)e^{i\beta mz} . \tag{2.5.2}$$

(This models the streamwise streaks that are prominent in some boundary-layer transition scenarios.) In their case $\hat{u}_m$ decay fairly rapidly as m increases, and so the resulting eigenvalue problem is quite manageable.

A more complex example is the flow in a rectangular duct. The mean velocity $\mathbf{u}_0(\mathbf{x}) = (u_0(y, z), 0, 0)^T$ is a function of two coordinate directions and the mean pressure is linear in x. (An exact solution for $u_0(y, z)$ was given by Saint-Venant (1855) in terms of an infinite series. Tatsumi and Yoshimura (1990) computed the mean flow using Legendre expansions.) The following form for the perturbations is assumed in the stability analysis:

$$\mathbf{u}(\mathbf{x}, t) = \hat{\mathbf{u}}(y, z)e^{i\alpha x - i\omega t} , \tag{2.5.3}$$

where α and ω are complex constants. Equations similar to (1.5.5)–(1.5.9) are readily derived. Tatsumi and Yoshimura (1990) eliminated $\hat{u}$ and $\hat{p}$ to obtain a system of two higher-order equations for $\hat{v}$ and $\hat{w}$. They then utilized a two-dimensional Legendre–Galerkin expansion, based on suitable combinations of Legendre polynomials that satisfied the boundary conditions, to discretize the linear stability eigenvalue problem.

For the temporal stability problem one obtains a generalized eigenvalue problem of the form

$$A\mathbf{q} = \omega B\mathbf{q} , \qquad (2.5.4)$$

where $\mathbf{q}$ represents the unknowns in, say, lexicographic ordering, and the matrices A and B represent the two-dimensional discretizations in y and z. The size of this generalized eigenvalue problem for two-dimensional global stability analyses is roughly the square of the size of the one-dimensional problems considered in Sect. 2.3. (Again, see Sect. E.1 of Appendix E for appropriate algorithms for this algebraic eigenvalue problem.)

As noted by Theofilis (2003) in his extensive review of methods and physical results for multidimensional linear stability problems, computation of the complete eigenvalue spectrum is typically not performed due to the much larger size of these eigenvalue problems. Iterative methods for finding some or all of the eigenvalues and eigenvalues, such as the Arnoldi and Jacobi-Davidson algorithms mentioned in Appendix D.3 are the preferred approach. We refer the reader to Theofilis' (2003) review article for numerous examples of global stability analysis, many using spectral methods.

2.6 Transient Growth Analysis

All the applications discussed above have been in the context of the traditional eigenvalue-based approach to linear stability. However, convincing evidence had emerged by the 1990s that in many cases the transition from laminar to turbulent flows was associated with the growth of very strong transients rather than with the growth of unstable linear eigenmodes; see Schmid (2007). Linear analysis can still provide insight into this phenomena. However, instead of looking just at the eigenvalues of the operator defined by the relevant form of the linearized Navier–Stokes equations, one also looks at the maximum possible transient growth.

Figure 2.20, taken from the simple example at the very start of the Trefethen and Embree (2005) text, illustrates the origin and potential impact of transients. The curves in this plot represent the norm of the matrix exponential e^{At} corresponding to the solution operator for a simple 2×2 ODE system

$$\frac{dy}{dt} = Ay , \qquad A = \begin{pmatrix} -1 & 5 \\ 0 & -2 \end{pmatrix} . \qquad (2.6.1)$$

The dashed and dotted curves in the figure represent the evolution of the norms of solutions whose initial conditions are normalized eigenfunctions

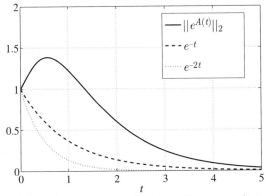

Fig. 2.20. Norm of the solution operator for the ODE system (2.6.1) for the matrix A given in (2.6.1) compared with the evolution of its two normalized eigenmodes

corresponding to the eigenvalues -1 and -2, respectively. They exhibit pure exponential decay. The solid curve traces the norm of the matrix exponential e^{At}, which is the solution operator for the ODE system. This measures the maximum solution norm over all normalized initial conditions for the ODE. Such a discrepancy between the transient growth of the solution operator and the pure decay of the eigenmodes exists because the matrix A does not have an orthogonal set of eigenvectors, i. e., A is a nonnormal matrix. (A *normal matrix* is one which commutes with its transpose; equivalently, a normal matrix has an orthogonal basis of eigenvectors.) The notion of nonnormality extends to operators such as those associated with the linear stability equations. These are often nonnormal, as in the case of plane channel flow. In the simple matrix example the transient growth is fairly modest—less than a factor of 1.4. As we shall see, there are examples in fluid dynamics stability problems in which transient growth can be several orders of magnitude. This cannot be deduced from eigenvalue analysis alone.

The transient growth can be examined by several means, and spectral approximations to the linear stability operators have proven useful in this regard. Reddy and Henningson (1993) introduced the use of the matrix exponential corresponding to the solution operator for the linear stability equations (1.5.10)–(1.5.11). Details of how the maximum growth is computed from the discretized system can be found in Schmid and Henningson (2001).

Figure 2.21, taken from Reddy and Henningson (1993), illustrates the results for plane Poiseuille flow. The figure contains three curves, all for two-dimensional disturbances with $\alpha = 1$ and $\beta = 0$. The "modal" curve represents the growth of the most unstable eigenvalue for Re $= 8000$, the "stable" curve is the maximum growth permitted for the linearly stable case of Re $= 5000$, and the "unstable" curve gives the maximum growth for the linearly unstable case of Re $= 8000$. These results indicate that even though the linearly stable case ultimately decays, as indicated by its eigenvalues, it nevertheless admits a transient growth by a factor of 30.

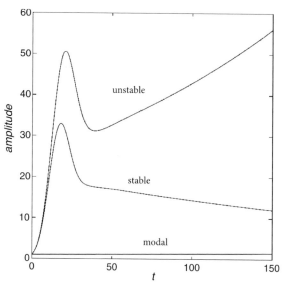

Fig. 2.21. Growth curves for plane Poiseuille flow with $\alpha = 1$, $\beta = 0$. The bottom curve is the linear eigenmode for Re = 8000, while the middle curve and top curves are the maximum growth curves at Re = 5000 and Re = 8000, respectively. [Reprinted with permission from S.C. Reddy, D.S. Henningson (1993); © 1993, Cambridge University Press]

An alternative approach uses the *pseudospectra* of the linear operator to the same end. See Trefethen (1997) for a review of origins of the study of pseudospectra and Trefethen and Embree (2005) for both a comprehensive coverage of pseudospectra and a review (in their Chap. 20) of some of the key evidence for transient growth in fluid dynamics transition.

Rather than deal with the technicalities of the pseudospectra of operators, we shall give a simplified description of the pseudospectra of a matrix. For any $\epsilon > 0$ the ϵ-*pseudospectrum* of a matrix L is defined to be that subset of the complex plane defined by

$$\Lambda_\epsilon(L) = \{z \in \mathcal{C} : z \in \Lambda(L + E) \text{ for some } E \text{ with } \|E\| \leq \epsilon\}. \qquad (2.6.2)$$

The usual spectrum is produced for $\epsilon = 0$. Loosely speaking, for $\epsilon > 0$, the ϵ-pseudospectrum is the set of points that are elements of the spectrum of some matrix that differs from L (in norm) by no more than ϵ. For a normal matrix, $\Lambda_\epsilon(L)$ is the same as the set $\{z \in \mathcal{C} : \|z - \Lambda(L)\| \leq \epsilon\}$, where $\Lambda(L)$ is the set of the eigenvalues of L. However, if L is not normal, $\Lambda_\epsilon(L)$ can be a much larger set. In CHQZ2, Sect. 4.3.2 we discussed the pseudospectra of the first-derivative matrices arising from spectral discretizations, and how the nonnormality of these matrices (except for the Fourier case) leads to significant round-off errors in the computations of the matrix eigenvalues. Here, in Fig. 2.22 we illustrate the pseudospectra of the Orr–Sommerfeld operator for plane Poiseuille flow under the same conditions as those of the previous examples for this flow (Re = 7500, $\alpha = 1$ and $\beta = 0.25$). (The

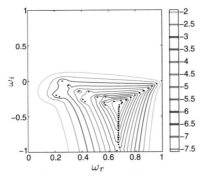

Fig. 2.22. Pseudospectra of the Orr–Sommerfeld operator for plane Poiseuille flow for Re = 7500, $\alpha = 1.0, \beta = 0.25$. The dots represent the eigenvalues and the curves are the ϵ-pseudospectra. The outermost isoline is for $\epsilon = -2.0$ and the innermost one is for $\epsilon = -7.5$

computations in the figure utilized the Eigtool software of Wright; see Wright and Trefethen (2001) for the details of the pseudospectra computation.) The dots in this figure represent the numerically computed eigenvalues—they are, of course, identical to those shown in the right frame in the middle row of Fig. 2.12. The solid lines are the ϵ-pseudospectra for values of ϵ ranging from 10^{-7} to 10^{-2}. The excursions of the ϵ-pseudospectra curves into the (unstable) right-half plane imply in this case a purely linear growth of at least an order of magnitude for the associated transients. The Orr–Sommerfeld discretization was the Chebyshev collocation method for the Orr–Sommerfeld equation discussed in Sect. 2.3.1. This computation used the code given in Schmid and Henningson (2001).

Most applications of transient growth analysis have been to incompressible flows. One exception is the work of Hanifi, Schmid and Henningson (1996), who examined transient growth in compressible boundary layers using a spectral collocation method based on that of Malik (1990).

Trefethen and Embree (2005) discuss applications of pseudospectra analysis to fluid dynamics stability in some detail. This text also has several chapters devoted to numerical algorithms for the efficient and accurate computation of pseudospectra. See Trefethen and Embree (2005) and Schmid and Henningson (2001) for extensive discussions of the underlying mathematics of the analysis of transient growth of operators, and for the details of how the maximum growth can be computed.

2.7 Nonlinear Stability

2.7.1 Quasi-Steady Finite-Amplitude Solutions

Nonlinear stability analyses are a relatively efficient tool for conducting parameter studies of the early nonlinear stages of transition. Direct numerical simulation techniques—discussed at length in Chap. 3 and especially in Sect. 3.4—are a far more expensive approach to such parameter studies.

For some problems quasi-steady, finite-amplitude solutions are of interest. For the particular case of the plane Poiseuille flow problem considered in Sect. 2.3, these are nontrivial solutions to the incompressible Navier–Stokes equations that are periodic in x and/or z, but neither decay nor grow in time. Herbert (1977b, 1983a) and Orszag and Patera (1983) used Fourier–Chebyshev methods to determine finite-amplitude solutions to channel flow perturbations. Their approach is to write the total solution for the velocity as

$$\mathbf{u}(\mathbf{x}, t) = \mathbf{u}_0(y) + A \sum_{m=-\infty}^{\infty} \hat{\mathbf{u}}_m(y) e^{im\alpha(x-ct)} , \qquad (2.7.1)$$

where c is a real phase speed and A is an amplitude. Here they make a two-dimensional approximation—Fourier Galerkin in x and Chebyshev tau or collocation in y. The result is a neutral stability *surface* defined by a characteristic equation involving Re, A and α. Spectral methods permit this surface, and the corresponding finite-amplitude solutions, to be computed with great precision. An example for channel flow, due to Herbert (1983a), is provided in Fig. 2.23, where E denotes the perturbation kinetic energy divided by the mean kinetic energy. To compute just a single point on this surface via a direct numerical simulation requires a very long time integration (hundreds or more of temporal periods) for the transients to subside.

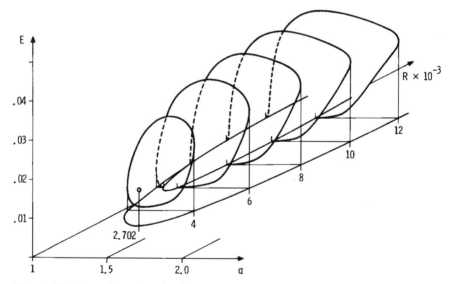

Fig. 2.23. Neutral surface for finite-amplitude perturbations in plane Poiseuille flow. [From Th. Herbert (1983a)]

2.7.2 Secondary Instability Theory

The nonlinear interaction of waves is of considerable interest in investigations of fluid dynamic stability and transition. The most commonly studied problem is the interaction of a finite-amplitude wave (the *primary wave*) with various other small amplitude waves. One seeks to identify the most important *secondary instability* induced by the interaction of an infinitesimally small wave with the mean flow and the primary wave. For plane Poiseuille flow this secondary instability may be assessed by performing a linear stability analysis of the mean flow plus a finite-amplitude two-dimensional wave such as represented by (2.7.1).

However, most investigations using *secondary instability theory* (SIT) have not used an exact finite-amplitude primary disturbances such as those described above. Instead, the usual practice has been to use as the new mean flow (termed the *modulated base flow*) just the undisturbed mean flow plus a small multiple of a linear mode, i. e.,

$$\mathbf{u}_b(x, y, t) = \mathbf{u}_0(y) + Re\left\{A \ \mathbf{u}_1(y)e^{i(\alpha x - \omega_r t)}\right\} , \qquad (2.7.2)$$

in the case of a two-dimensional primary disturbance. The magnitude of the imposed primary wave is denoted by A and it measures the maximum streamwise velocity disturbance relative to the free-stream velocity. The subscripts 0, b, and 1 refer, respectively, to the laminar mean flow, the modulated base flow, and the primary disturbance. The streamwise wave number and the temporal frequency of the primary disturbance are denoted by α and ω, respectively. In the temporal formulation, the temporal growth of the primary disturbance is neglected; hence only the real part of ω is used in (2.7.2). In the spatial formulation, ω is necessarily real, and now only the real part of α is used in (2.7.2).

For a two-dimensional primary disturbance the total flow variable $\mathbf{q} = (u, v, p)^T$ can be written in the truncated Floquet form

$$\mathbf{q}(x, y, t) = \mathbf{q}_b(x, y, t) + \epsilon e^{\sigma t + \gamma x' + i\beta_2 z + ih\alpha x'/2} \sum_{m=-M_-}^{M_+} \mathbf{q}_{2,m}(y)e^{im\alpha x'} ,$$

$$(2.7.3)$$

where $x' = x - c_r t$, and $c_r = Re\{\omega/\alpha\}$ is the phase velocity of the primary wave. The subscript 2 refers to the secondary disturbance, its spanwise wavenumber is denoted by β_2, $h \in [0, 1]$ relates the streamwise wavelength of the secondary wave to that of the primary wave. Secondary waves for which the streamwise wavenumber of the secondary wave is half that of the primary are called *subharmonic modes*; they are described by $h = 1$. The *fundamental modes* are those for which the streamwise wavenumber of the secondary is the same as that of the primary, and they correspond to $h = 0$. The secondary waves for which $h \in (0, 1)$ are referred to as *detuned modes*. The subharmonic modes can be described by (2.7.3) with $M_- = 1$ and $M_+ = 0$, whereas the fundamental modes require at least $M_- = 1$ and $M_+ = 1$.

The secondary disturbance amplitude ϵ is assumed sufficiently small to permit linearization of the Navier–Stokes equations about the modulated base flow. This produces a coupled set of one-dimensional equations (in y) for the Fourier harmonics $\mathbf{q}_{2,m}(y)$. For the temporal secondary instability problem, $\gamma = 0$, and $\sigma \neq 0$ is the complex eigenvalue to be determined. For the spatial problem, one writes $\sigma = \gamma c_r$ in (2.7.3) and solves for γ as the eigenvalue. The temporal and spatial growth rates are given by the real parts of σ and γ, respectively.

Methods similar to those applied to the purely linear problem may be used here. (Of course, the algebraic problem is more complex because of the coupling in x, which makes the problem of Floquet type.) In most applications the series (2.7.3) is truncated to have no more than three Fourier modes. Secondary instability theory results for channel flow were obtained by Orszag and Patera (1983) using a Chebyshev tau method. Numerous pioneering investigations of the temporal and spatial secondary instability of channel flow and flat-plate boundary-layer flow under the parallel flow approximation were conducted by Herbert (1983a, 1983b, 1988) with a Chebyshev collocation scheme.

Several groups applied secondary instability theory (SIT) to compressible transition problems in the early 1990s. Most used low-order methods. However, Ng and Erlebacher (1992), Pruett, Ng and Erlebacher (1991) and Ng and Zang (1993) applied Chebyshev collocation to the compressible extension of secondary instability theory to supersonic flow past flat plates, cylinders and sharp cones under the parallel flow approximation. The temporal secondary instability problem was solved in the first two references, whereas the third reference solved the spatial secondary instability problem. These analyses required the numerical solution of the compressible boundary-layer equations (for the mean flow), the compressible linear stability equations (for the primary disturbance), and the secondary instability equations (for the secondary disturbance). All three types of problems were solved with spectral accuracy. See Sect. 2.2.2 for the spectral algorithm used for the boundary layer solution and Sect. 2.4 for the spectral compressible linear stability method. The complicated differential equations for the secondary instability problem were generated with the aid of a symbolic manipulation language and then discretized with a standard Chebyshev collocation approximation. The same mapping and collocation points were used in the unbounded direction y for the secondary instability equations as were used for the linear stability computations described in Sect. 2.4. (The linear stability results appear in the coefficients of the secondary instability equations.) They used the Gauss–Lobatto nodes for all the equations and obtained four-digit accuracy on the eigenmodes of the secondary instability problem with at most 3 Fourier modes in x and 65 Chebyshev points in y.

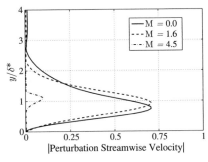 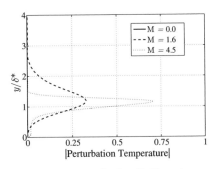

Fig. 2.24. Compressible flat plate perturbation streamwise velocity (*left*) and perturbation temperature (*right*) for subharmonic secondary instabilities. [Adapted from Ng and Erlebacher (1992) using data supplied by Erlebacher]

Figure 2.24 illustrates some of the Ng and Erlebacher results for the eigenfunctions of the secondary instabilities. The primary wave was a first mode in the M = 0 and M = 1.6 cases, and a second mode for M = 4.5. (See Table 2.1 for the exact parameters of these cases, and compare them with Fig. 2.8 for the mean flow and Fig. 2.18 for the structure of the primary waves. Do note from Fig. 2.18 that for supersonic flows the temperature perturbation is much larger than the velocity perturbation. Hence, even a 1% velocity perturbation at M = 4.5 corresponds to a nearly 15% perturbation in temperature relative to its free-stream value.) For the results at M = 4.5 with a second-mode primary wave, there is the same concentration of the eigenfunction near the critical layer for the secondary mode as there is for the primary. Also note how much larger the temperature perturbations are than the velocity perturbations in the case of the second-mode secondary instability.

An illustration of the utility of the secondary instability approach is given in Fig. 2.25, which illustrates the type of parameter study that can be conducted rapidly using the secondary instability approach. The particular case examined was the subharmonic secondary instability of the second mode at M = 4.5 under the conditions given in the fourth row of Table 2.1. A single low-resolution direct numerical simulation run would be far more expensive and would generate but a single point on this plot.

Finally, we close this section by illustrating in Fig. 2.26 the cross-verification of SIT and DNS codes. The frame on the left, taken from Pruett and Zang (1992), compares temporal SIT with temporal DNS for Mach 4.5 flow past a cylinder. (More discussion about this particular case can be found in Sect. 4.6.2.) The initial condition for the DNS consisted of the mean flow, a 2D second-mode primary wave, and a subharmonic secondary mode. The curves are taken from the DNS results, with the solid curve labeled E_{10} denoting the energy contained in the 2D primary wave, and the dashed curve labeled $E_{\frac{1}{2}1}$ the energy in the 3D subharmonic wave. The

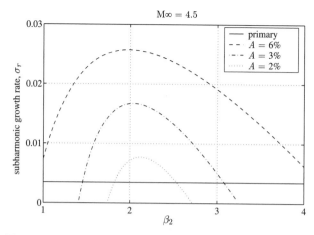

Fig. 2.25. Typical parametric study of the growth rate of a temporal secondary instability as a function of the spanwise wavenumber β_2 and primary wave amplitude A [Reprinted with permission from L.L. Ng, G. Erlebacher (1992); © 1992, American Institute of Physics]

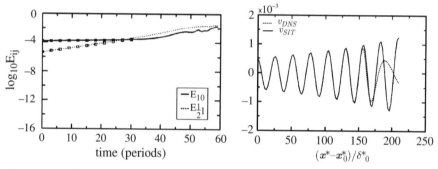

Fig. 2.26. Comparison of compressible secondary instability with direct numerical simulation for a boundary layer. The left frame compares temporal SIT and DNS for a subharmonic instability of Mach 4.5 flow past a cylinder. [Reprinted with permission from C.P. Pruett, T.A. Zang (1992); © 1992, American Institute of Physics]. The right frame compares spatial SIT vs. DNS for a fundamental instability in Mach 1.6 flow past a flat plate. [Reprinted with permission from L.L. Ng, T.A. Zang (1993); © 1993, American Institute of Physics]

symbols are taken from compressible linear stability theory and compressible secondary instability theory for the respective modes. The particular case shown in the figure used relatively large amplitudes for the primary and secondary waves because this simulation aimed to illustrate the breakdown of the high-speed flow. However, for other simulations with substantially lower amplitudes, Pruett and Zang were able to reproduce in the DNS the growth rates to six significant digits. As they noted: "Six-digit accuracy has led us to uncover errors in the formulation of minor viscous terms, which typically affect growth rates only in the fourth significant digit for Re $\approx 10^5$". The right frame in the figure shows a comparison between spatial SIT and spatial DNS for the wall-normal (v) velocity component of the secondary instability. This case (due to Ng and Zang (1993)) is for Mach 1.6 flow past a flat plate with a 2D first-mode primary and a fundamental secondary mode. The discrepancy at larger values of x is due to the outflow boundary condition in the DNS. (Ng and Zang (1993) described the compressible version of Streett's buffer domain treatment of the outflow boundary (see Sect. 3.6.2): the equations are parabolized by smoothly decreasing the viscous terms to zero, the wall velocity is gradually increased to its free-stream value, and a source term is added to balance the changes in the mean flow.) Otherwise, the curves are indistinguishable. The high accuracy of spectral methods is especially useful for this type of code verification study because low-order methods are unlikely to catch errors in the viscous terms.

2.7.3 Nonlinear Parabolized Stability Equations

The parabolized stability equations have proven to be a more versatile tool than SIT methods for studying weakly nonlinear phenomena, as they can handle nonparallel effects. Nonlinear PSE computations can be carried out in a similar fashion as for linear disturbances. Use of single-domain spectral methods for nonlinear PSE has been confined to incompressible flow. As one example of a nonlinear PSE computation using spectral methods, Fig. 2.27 compares PSE results and DNS results for the early nonlinear stages of transition in a flat-plate boundary layer with the experimental results of Kachanov and Levchenko (1984). (As in Fig. 2.19, the Reynolds number Re in the abscissa of figure is based on the distance from the leading edge of the flat plate.) The spatial DNS results are due to Fasel et al. (1990), and the PSE results are from Bertolotti (1991). The figure, based on data in Bertolotti (1991), is taken from Herbert (1997). The inflow conditions consisted of a primary wave, a 2D disturbance, labeled $(2,0)$ in the figure, and a 3D subharmonic secondary mode, denoted by $(1,1)$, with amplitudes and wavenumbers chosen to match the forcing in the experiment. The PSE results, which are orders of magnitude cheaper than the DNS results, agree very well with the DNS and with the experiment.

In Fig. 2.26 we illustrated the use of spatial secondary instability theory in the verification of DNS codes (and vice versa). Of course, this type of comparison can only be made to great precision if the DNS uses a parallel-flow approximation. Comparisons of DNS codes with PSE results do not suffer from this limitation. However, there do not appear to be any comparisons of nonlinear PSE results with DNS results, where both results were generated with spectral methods. We refer the reader to Joslin, Streett and Chang

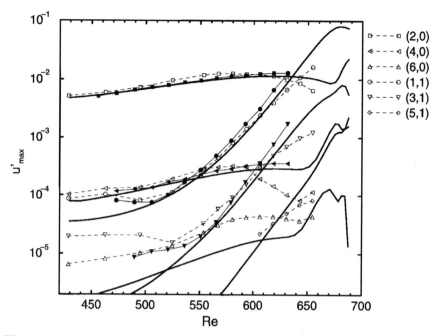

Fig. 2.27. Nonlinear parabolized stability equations (PSE) results for subharmonic secondary instability in an incompressible flat-plate boundary layer. The PSE results (*solid lines*) are compared with DNS results (*solid symbols*) and experimental data (*open symbols*). The modes are labeled by the streamwise and spanwise harmonic number. [From Th. Herbert (1997) based on data in Bertolotti (1991)]

(1993), Pruett, Zang, Chang and Carpenter (1995)) for examples of meticulous verifications of direct numerical simulation codes against nonlinear PSE results, albeit with PSE codes and DNS codes using nonspectral methods.

3. Single-Domain Algorithms and Applications for Incompressible Flows

3.1 Introduction

There are essentially four different formulations of the incompressible Navier–Stokes equations—the primitive-variable (velocity and pressure), vorticity–streamfunction, streamfunction and vorticity–velocity formulations. The equations for these formulations are given in Sect. 1.4. The primitive-variable formulation has been the one most extensively employed in three-dimensional spectral calculations; spectral methods based on the primitive variable formulation in Cartesian coordinates will be the focus of this chapter. In particular, we focus on the details of spectral methods for fully periodic (homogeneous) flows and for flows with two periodic and one nonperiodic direction. A special application of spectral methods using the vorticity–velocity equations is described in Sect. 3.4.1. Discussions of spectral methods for the streamfunction–vorticity formulation can be found in Quartapelle (1993) and Peyret (2002). Bernardi, Dauge and Maday (1999) discuss spectral methods for axisymmetric flows in both the primitive-variable and streamfunction–vorticity formulations.

For ease of exposition in this chapter, we will generally simplify the Navier–Stokes equations, given for arbitrary kinematic viscosity ν in Sect. 1.4, by taking the viscosity to be constant in space and time. In this case, the Navier–Stokes equations on a domain Ω are usually written as (see (1.4.1) and (1.4.2))

$$\frac{\partial \mathbf{u}}{\partial t} + \mathbf{u} \cdot \nabla \mathbf{u} = -\nabla p + \nu \Delta \mathbf{u} + \mathbf{f} \quad \text{in } \Omega, \quad t > 0, \tag{3.1.1}$$

$$\nabla \cdot \mathbf{u} = 0 \qquad \text{in } \Omega, \quad t > 0, \tag{3.1.2}$$

where $\mathbf{u}$ is the velocity vector, p is the pressure and $\mathbf{f}$ are the volumetric forces. (Both p and $\mathbf{f}$ have been divided by the constant density.) Equation (3.1.1) is the momentum equation and (3.1.2) is the continuity equation. Boundary conditions have to be enforced over all the boundary $\partial \Omega$. Usually they are conditions of periodicity, or the assignment of the velocity or the normal component of the stress tensor; different conditions may be prescribed on different portions of the boundary. Finally, the initial conditions are

$$\mathbf{u} = \mathbf{u}_0 \quad \text{in } \Omega \quad \text{at } t = 0. \tag{3.1.3}$$

The initial velocity field must be divergence-free; otherwise, the continuous problem itself fails to have a classical solution (see Heywood and Rannacher (1986)). Moreover, an initial velocity field, even though divergence-free, must satisfy appropriate compatibility conditions with the boundary data in order to produce smooth solutions. Otherwise, transients that overwhelm delicate time-dependent stability mechanisms (see Deville, Kleiser and Montigny-Rannou (1984)) may be induced. The review article by Gresho (1991) discusses a wide range of compatibility issues for incompressible flow simulations. Boyd and Flyer (1999) focus on space–time compatibility issues in the particular context of spectral methods, and discuss a variety of approaches to adjusting the initial conditions to satisfy better the compatibility conditions.

The simplest problem formulations are those for which there are in all coordinate directions only periodic boundary conditions and/or solid bodies with no-slip boundary conditions. In many problems, however, the physical domain Ω is unbounded in one or more coordinate directions. For some of these problems there are spectrally accurate expansion functions that can readily resolve the salient features of the solution. In other cases a mapping onto a bounded computational domain is effective (see Sect. 8.8). As a last resort, artificial boundaries have to be used. Boundary conditions need to be placed on the primitive variables at these artificial boundaries. Rigorous results with regard to the implications of these boundary conditions for the existence and uniqueness of the Navier–Stokes solutions for some but not all cases are available; see, e. g., Galdi (1994b).

For any discretization of the incompressible Navier–Stokes equations, the principal conceptual difficulty is the treatment of the pressure. Unlike the velocity, there is no evolution equation for the pressure. Instead, it is determined by the constraint (3.1.2). Some numerical methods, known as penalty methods or artificial-compressibility methods, have circumvented this problem by introducing a time derivative of the pressure into the continuity equation (see Temam (2001)). Since the artificial compressibility method has seen little use in this context of spectral methods, we shall not discuss it further in this chapter. Furthermore, we still confine ourselves in this and the next chapter to classical spectral methods as applied on a single, tensor-product domain. The full perspective on the use of spectral methods for incompressible flow problems is deferred until the discussion of spectral methods in complex domains in Chap. 5. A general framework regarding solution algorithms for fully nonperiodic flows is provided in Chap. 7.

This chapter contains detailed discussions of a number of complete single-domain spectral algorithms for fully homogeneous flows, and for flows with

Fig. 3.1. The roadmap of Chap. 3. The upper and middle blocks quote full algorithms, the lower block just lists some of the choices for velocity-pressure polynomial spaces on a reference domain

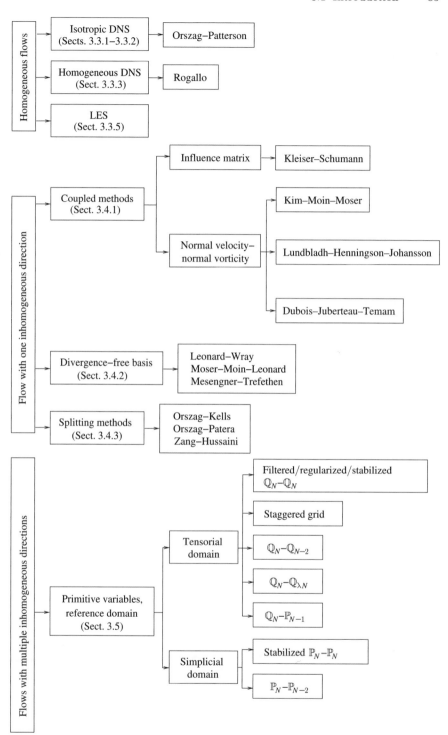

one inhomogeneous direction. For flows with multiple inhomogeneous directions, we confine ourselves to the basic principles underlying the design of complete Navier–Stokes algorithms. Figure 3.1 provides an outline of the subcategories of these major categories and a listing of the principal complete algorithms that we discuss.

3.2 Conservation Properties and Time-Discretization

The topics of the conservation properties of the incompressible Navier–Stokes equations and the choice of time-discretization are hardly unique to spatial discretization by spectral methods. However, since the choice of the spectral discretization is strongly influenced by these aspects, we review some of the essential properties here.

3.2.1 Conservation Properties

As a rule, the algebraic structure of the discretized Navier–Stokes equations should mimic as many of the key properties of the continuous system as possible. These include conservation and symmetry properties as well as consistency between the divergence, gradient and Laplacian operators. The inviscid part of the Navier–Stokes equations conserves linear and angular momentum, and kinetic energy. Ideally, one should ensure that the semi-discrete (continuous in time, discrete in space) system and even the fully discrete system conserves all these quantities. Often it is not possible to satisfy all of these properties simultaneously. Then, one must assign priority to those properties that are most important to the physics of the problem under consideration.

The conservation properties of discrete approximations to the Navier–Stokes equations depend upon the precise form of the momentum equation in Ω that is used. There are four such forms:

The Convection Form.

$$\frac{\partial \mathbf{u}}{\partial t} + \mathbf{u} \cdot \nabla \mathbf{u} = -\nabla p + \nu \Delta \mathbf{u} + \mathbf{f} \ . \tag{3.2.1}$$

The Divergence Form.

$$\frac{\partial \mathbf{u}}{\partial t} + \nabla \cdot (\mathbf{u}\,\mathbf{u}^T) = -\nabla p + \nu \Delta \mathbf{u} + \mathbf{f} \ . \tag{3.2.2}$$

The Rotation Form.

$$\frac{\partial \mathbf{u}}{\partial t} + \boldsymbol{\omega} \times \mathbf{u} = -\nabla q + \nu \Delta \mathbf{u} + \mathbf{f} \ . \tag{3.2.3}$$

The Skew-Symmetric Form.

$$\frac{\partial \mathbf{u}}{\partial t} + \frac{1}{2}\mathbf{u} \cdot \nabla \mathbf{u} + \frac{1}{2}\nabla \cdot (\mathbf{u}\,\mathbf{u}^T) = -\nabla p + \nu \Delta \mathbf{u} + \mathbf{f} \ . \tag{3.2.4}$$

In (3.2.3), $\boldsymbol{\omega} = \nabla \times \mathbf{u} = (\omega_1, \omega_2, \omega_3)^T$ is the vorticity and $q = p + (1/2)|\mathbf{u}|^2$ is the dynamic pressure. (The symbol P is more commonly used for the dynamic pressure than the symbol q. We use the latter in (3.2.3) to maintain consistency with the convention used in this book that upper case symbols refer to matrices.) These partial differential equations are all equivalent analytically, due to the condition $\nabla \cdot \mathbf{u} = 0$ in Ω, but this is not usually the case with their discrete counterparts.

Consider now the conservation properties of the differential system in the absence of viscosity and body forces. Any form of the momentum equation can be used in the discussion, due to the equivalence mentioned above. We choose the form (3.2.2), which we write as

$$\frac{\partial \mathbf{u}}{\partial t} + \nabla \cdot \boldsymbol{\mathcal{F}} = \mathbf{0} \quad \text{in } \Omega \ , \tag{3.2.5}$$

where the flux tensor $\boldsymbol{\mathcal{F}}$ is defined as $\boldsymbol{\mathcal{F}}(\mathbf{u}, p) = \mathbf{u}\,\mathbf{u}^T + p\boldsymbol{\mathcal{I}}$, where $\boldsymbol{\mathcal{I}}$ is the identity tensor. By integrating (3.2.2) over Ω and applying Gauss' Theorem, we have

$$\frac{\mathrm{d}}{\mathrm{d}t}\int_{\Omega}\mathbf{u} = -\int_{\partial\Omega}\boldsymbol{\mathcal{F}}\cdot\mathbf{n} \ . \tag{3.2.6}$$

Hence, the only integral changes in $\mathbf{u}$ (the linear momentum) are those due to fluxes through the boundaries.

Now take the inner product of $\mathbf{u}$ with (3.2.2), obtaining

$$\frac{1}{2}\frac{\mathrm{d}}{\mathrm{d}t}(\mathbf{u}, \mathbf{u}) + (\nabla \cdot \boldsymbol{\mathcal{F}}, \mathbf{u}) = 0 \ . \tag{3.2.7}$$

However, under the assumption of periodic or no-slip ($\mathbf{u} \cdot \mathbf{n} = 0$) boundary conditions on $\partial\Omega$, one has

$$(\nabla \cdot \boldsymbol{\mathcal{F}}, \mathbf{u}) = -\left(p - \frac{1}{2}\mathbf{u} \cdot \mathbf{u}, \nabla \cdot \mathbf{u}\right) = 0 \ ;$$

hence, the quadratic conservation law

$$\frac{\mathrm{d}}{\mathrm{d}t}(\mathbf{u}, \mathbf{u}) = \frac{\mathrm{d}}{\mathrm{d}t}\|\mathbf{u}\|^2 = 0 \tag{3.2.8}$$

holds.

As discussed in CHQZ2, Sect. 3.6, the analysis of the conservation properties of semi-discrete spectral methods (discrete in space, continuous in time) relies on the symmetry or skew-symmetry properties of the differentiation matrices (which depend upon which form of the momentum equation is used in their derivation) and on whether the discrete inner products have sufficient precision to be equivalent to the continuous inner products for all test and trial functions.

Analytically, the divergence operator is the adjoint of the (negative) gradient operator, and the Laplacian is the composition of the divergence and the gradient. Both properties are automatic for the discrete operators generated by Fourier methods and Legendre Galerkin methods, as is the latter for Chebyshev methods. The former property is violated by most Chebyshev methods, since their first derivative operators are not skew-symmetric. Moreover, the physical conservation properties are for inner products with unit weight, which is satisfied by Fourier and Legendre methods, but not by Chebyshev methods.

Let us now illustrate some discrete conservation properties for the unforced, inviscid Navier–Stokes equations with periodic boundary conditions.

The Rotation Form. Consider first the Galerkin approximation, which can be written as

$$\left(\frac{\partial \mathbf{u}^N}{\partial t} + \boldsymbol{\omega}^N \times \mathbf{u}^N + \nabla q^N, \mathbf{v}\right) = 0 \qquad \text{for all } \mathbf{v} \in S_N \,, \qquad (3.2.9)$$

where S_N is the space of all velocities whose components are trigonometric polynomials of degree $\leq N/2$ in each variable and $(\cdot, \cdot)$ is the L^2-inner product. (See Sect. 8.9 for the fundamentals of spectral discretizations for differential equations.) The conservation of the α-component ($\alpha = 1, 2, 3$) of linear momentum can be demonstrated by choosing $\mathbf{v} = \hat{\mathbf{e}}_\alpha$, where $\hat{\mathbf{e}}_\alpha$ is the constant velocity whose α-component is 1 and whose other components are 0. Then we obtain

$$\frac{\mathrm{d}}{\mathrm{d}t} \int_\Omega u_\alpha^N + \left(\boldsymbol{\omega}^N \times \mathbf{u}^N, \hat{\mathbf{e}}_\alpha\right) + \left(\nabla q^N, \hat{\mathbf{e}}_\alpha\right) = 0 \,.$$

The last term on the left-hand side is readily seen to vanish since $\left(\nabla q^N, \hat{\mathbf{e}}_\alpha\right) = -\left(q^N, \nabla \cdot \hat{\mathbf{e}}_\alpha\right)$ and the divergence of a constant is zero. The middle term also vanishes, but demonstrating this is more involved. Consider the $\alpha = 1$ case. For simplicity, set $\mathbf{u}^N = (u^N, v^N, w^N)^T$. Then,

$$\left(\boldsymbol{\omega}^N \times \mathbf{u}^N, \hat{\mathbf{e}}_1\right) = \int_\Omega \left(w^N \omega_2^N - v^N \omega_3^N\right)$$

$$= \int_\Omega \left(-v^N \frac{\partial v^N}{\partial x} - w^N \frac{\partial w^N}{\partial x} + w^N \frac{\partial u^N}{\partial z} + v^N \frac{\partial u^N}{\partial y}\right)$$

$$= \int_\Omega \left(-v^N \frac{\partial v^N}{\partial x} - w^N \frac{\partial w^N}{\partial x} - u^N \frac{\partial w^N}{\partial z} - u^N \frac{\partial v^N}{\partial y}\right)$$

$$= \int_\Omega \left(-v^N \frac{\partial v^N}{\partial x} - w^N \frac{\partial w^N}{\partial x} + u^N \frac{\partial u^N}{\partial x}\right)$$

$$= \frac{1}{2} \int_\Omega \left(-\frac{\partial (v^N)^2}{\partial x} - \frac{\partial (w^N)^2}{\partial x} + \frac{\partial (u^N)^2}{\partial x}\right) = 0 \,.$$

$$(3.2.10)$$

The third line follows by integrating by parts the third term with respect to z and the fourth term with respect to y and noting that the boundary terms vanish because of periodicity. The fourth line follows from the continuity equation. Thus,

$$\frac{\mathrm{d}}{\mathrm{d}t} \int_{\Omega} \mathbf{u}^N = \mathbf{0} \,. \qquad (3.2.11)$$

Conservation of kinetic energy is derived by choosing $\mathbf{v} = \mathbf{u}^N$:

$$\frac{1}{2}\frac{\mathrm{d}}{\mathrm{d}t}\left(\mathbf{u}^N, \mathbf{u}^N\right) + \left(\boldsymbol{\omega}^N \times \mathbf{u}^N, \mathbf{u}^N\right) + \left(\nabla q^N, \mathbf{u}^N\right) = 0 \,.$$

However, $(\boldsymbol{\omega}^N \times \mathbf{u}^N, \mathbf{u}^N) = 0$ because a vector is orthogonal to its cross product with another vector, and $\left(\nabla q^N, \mathbf{u}^N\right) = -\left(q^N, \nabla \cdot \mathbf{u}^N\right) = 0$. Therefore, we have the quadratic conservation law

$$\frac{\mathrm{d}}{\mathrm{d}t}\left(\mathbf{u}^N, \mathbf{u}^N\right) = \frac{\mathrm{d}}{\mathrm{d}t}\|\mathbf{u}^N\|^2 = 0 \,. \qquad (3.2.12)$$

To examine the conservation properties of collocation approximations we introduce some additional notation. Let

$$\mathbf{x_j} = \frac{2\pi}{N}(j_1, j_2, j_3), \quad j_\alpha = 0, 1, \ldots, N-1 \,, \qquad (3.2.13)$$

be the collocation nodes; let $I_N v$ denote the multidimensional trigonometric interpolant of a continuous function v at these nodes (see (8.1.24) and (8.1.25)). Finally, let

$$(\varphi, \psi)_N = \left(\frac{2\pi}{N}\right)^3 \sum_{\mathbf{j}} \varphi(\mathbf{x_j})\psi(\mathbf{x_j}) \qquad (3.2.14)$$

be the associated discrete inner product, with summation extended over all nodes. First, the *interpolation gradient operator* $\mathbb{G}_N$ (which maps a (continuous) function into its interpolation partial derivatives) is defined by

$$\mathbb{G}_N\, q = (\mathcal{D}_N^x q, \mathcal{D}_N^y q, \mathcal{D}_N^z q)^T \,, \qquad (3.2.15)$$

with the one-dimensional interpolation differentiation operators $\mathcal{D}_N^x$, $\mathcal{D}_N^y$ and $\mathcal{D}_N^z$ (also indicated as $\mathcal{D}_N^{(1)}$, $\mathcal{D}_N^{(2)}$ and $\mathcal{D}_N^{(3)}$ in the sequel) representing differentiation of the interpolant $I_N q$, with respect to the x, y and z-coordinates, respectively (see (8.1.39)). Likewise, the *interpolation divergence operator* $\mathbb{D}_N$ is defined by

$$\mathbb{D}_N\, \mathbf{u} = \mathcal{D}_N^x u_1 + \mathcal{D}_N^y u_2 + \mathcal{D}_N^z u_3 \,. \qquad (3.2.16)$$

The one-dimensional interpolation differentiation operators are all skew-symmetric (see CHQZ2, Sect. 2.1.3), e. g., $(\mathcal{D}_N^x u, v)_N = -(u, \mathcal{D}_N^x v)_N$ for all trigonometric polynomials u, v of degree $\leq N/2$ in one variable. For the vector operators, this means that

$$(\mathbb{D}_N\, b u, q)_N = -(\mathbf{u}, \mathbb{G}_N q)_N \,. \qquad (3.2.17)$$

Returning now to the examination of conservation properties, the collocation approximation to (3.2.3) may be expressed as

$$\left(\frac{\partial \mathbf{u}^N}{\partial t} + \boldsymbol{\omega}^N \times \mathbf{u}^N + \mathbb{G}_N q^N, \mathbf{v}\right)_N = 0 \qquad \text{for all } \mathbf{v} \in S_N . \tag{3.2.18}$$

Again, by selecting $\mathbf{v} = \hat{\mathbf{e}}_\alpha$ we can demonstrate conservation of linear momentum, for we have that

$$\frac{d}{dt}\left(\left(\frac{2\pi}{N}\right)^3 \sum_j u_\alpha(\mathbf{x_j})\right) + \left(\boldsymbol{\omega}^N \times \mathbf{u}^N, \hat{\mathbf{e}}_\alpha\right)_N + \left(\mathbb{G}_N q^N, \hat{\mathbf{e}}_\alpha\right)_N = 0 .$$

For the third term, we have that $\left(\mathbb{G}_N q^N, \hat{\mathbf{e}}_\alpha\right)_N = -\left(q^N, \mathbb{D}_N \hat{\mathbf{e}}_\alpha\right)_N = 0$. A derivation similar to that of (3.2.11), except that the discrete inner product is used, shows that the second term vanishes. (The skew-symmetry of the interpolation differentiation operators permits the same integration-by-parts operations.) Hence,

$$\frac{d}{dt}\left(\left(\frac{2\pi}{N}\right)^3 \sum_j u_\alpha(\mathbf{x_j})\right) = \frac{d}{dt}\left(u_\alpha^N, 1\right)_N = 0 . \tag{3.2.19}$$

Choosing $\mathbf{v} = \mathbf{u}^N$ as for the Galerkin case, we have

$$\frac{1}{2}\frac{d}{dt}\left(\mathbf{u}^N, \mathbf{u}^N\right)_N + \left(\boldsymbol{\omega}^N \times \mathbf{u}^N, \mathbf{u}^N\right)_N + \left(\mathbb{G}_N q^N, \mathbf{u}^N\right)_N = 0 .$$

The second inner product on the left-hand side of the above equations is identically zero as before, and the third one is equal to $-(q^N, \mathbb{D}_N \mathbf{u}^N)_N$, which vanishes. Hence, as for the Galerkin case,

$$\frac{d}{dt}\left(\mathbf{u}^N, \mathbf{u}^N\right)_N = \frac{d}{dt}\|\mathbf{u}^N\|_N^2 = 0 , \tag{3.2.20}$$

where now we use the discrete inner product and norm. Note that in the Fourier case, indeed $\|\mathbf{u}^N\|_N = \|\mathbf{u}^N\|$ by the exactness of the quadrature rule (see (8.1.30)).

The same conservation properties apply to Legendre Galerkin and Legendre collocation (or G-NI) approximations to problems with no-slip boundary conditions.

The Skew-Symmetric Form. Now consider the inviscid Navier–Stokes equations using the skew-symmetric form. The Galerkin approximation is equivalent to

$$\left(\frac{\partial \mathbf{u}^N}{\partial t} + \frac{1}{2}\mathbf{u}^N \cdot \nabla \mathbf{u}^N + \frac{1}{2}\nabla \cdot \left(\mathbf{u}^N (\mathbf{u}^N)^T\right) + \nabla p^N, \mathbf{v}\right) = 0 \qquad \forall \mathbf{v} \in S_N . \tag{3.2.21}$$

Take $\mathbf{v} = \mathbf{u}^N$. Then, (3.2.21) becomes

$$\frac{1}{2}\frac{d}{dt}(\mathbf{u}^N, \mathbf{u}^N) + \frac{1}{2}\left(\mathbf{u}^N \cdot \nabla\mathbf{u}^N, \mathbf{u}^N\right) + \frac{1}{2}\left(\nabla \cdot (\mathbf{u}^N (\mathbf{u}^N)^T), \mathbf{u}^N\right) + \left(\nabla q^N, \mathbf{u}^N\right) = 0\,.$$

However, as before, $\left(\nabla q^N, \mathbf{u}^N\right) = -\left(q^N, \nabla \cdot \mathbf{u}^N\right) = 0$.

On the other hand,

$$\left(\nabla \cdot \left(\mathbf{u}^N (\mathbf{u}^N)^T\right), \mathbf{u}^N\right) = -\left(\mathbf{u}^N \cdot \nabla\mathbf{u}^N, \mathbf{u}^N\right)\,.$$

Hence, we have (3.2.12).

Similarly, the collocation approximation is

$$\left(\frac{\partial\mathbf{u}^N}{\partial t}, \mathbf{v}\right)_N + \frac{1}{2}\sum_{\alpha=1}^{3}\left(\mathcal{D}_N^{(\alpha)}(u_\alpha^N\mathbf{u}^N), \mathbf{v}\right)_N$$

$$+ \frac{1}{2}\sum_{\alpha=1}^{3}\left(u_\alpha^N\mathcal{D}_N^{(\alpha)}\mathbf{u}^N, \mathbf{v}\right)_N + \left(\mathbb{G}p^N, \mathbf{v}\right)_N = 0 \qquad \forall\mathbf{v} \in S_N\,. \tag{3.2.22}$$

Using $\mathbf{v} = \mathbf{u}^N$, we have

$$\frac{1}{2}\frac{d}{dt}\left(\mathbf{u}^N, \mathbf{u}^N\right)_N + \frac{1}{2}\sum_{\alpha=1}^{3}\left(\mathcal{D}_N^{(\alpha)}(u_\alpha^N\mathbf{u}^N), \mathbf{u}^N\right)_N$$

$$+ \frac{1}{2}\sum_{\alpha=1}^{3}\left(u_\alpha^N\mathcal{D}_N^{(\alpha)}\mathbf{u}^N, \mathbf{u}^N\right)_N + \left(\mathbb{G}_N p^N, \mathbf{u}^N\right)_N = 0\,.$$

The last term vanishes for the usual reason. Exploiting the skew-symmetry of $\mathcal{D}_N^{(\alpha)}$ and the properties of the inner product, we have

$$\left(\mathcal{D}_N^{(\alpha)}(u_\alpha^N\mathbf{u}^N), \mathbf{u}^N\right)_N = -\left(u_\alpha^N\mathbf{u}^N, \mathcal{D}_N^{(\alpha)}\mathbf{u}^N\right)_N$$

$$= -\left(\mathbf{u}^N, u_\alpha^N\mathcal{D}_N^{(\alpha)}\mathbf{u}^N\right)_N = -\left(u_\alpha^N\mathcal{D}_N^{(\alpha)}\mathbf{u}^N, \mathbf{u}^N\right)_N\,.$$

Hence, we have (3.2.20).

To demonstrate discrete conservation of linear momentum for the collocation approximation, we choose $\mathbf{v} = \hat{\mathbf{e}}_\beta$ in (3.2.22), yielding

$$\frac{d}{dt}\left(u_\beta^N, 1\right)_N + \frac{1}{2}\sum_{\alpha=1}^{3}\left(\mathcal{D}_N^{(\alpha)}(u_\alpha^N u_\beta), 1\right)_N$$

$$+ \frac{1}{2}\sum_{\alpha=1}^{3}\left(u_\alpha^N\mathcal{D}_N^{(\alpha)}u_\beta, 1\right)_N + \left(\mathbb{G}_N p^N, \hat{\mathbf{e}}_\beta\right)_N = 0\,.$$

The last term vanishes for the usual reasons, and integrating by parts the second and third terms gives

$$-\frac{1}{2}\sum_{\alpha=1}^{3}\left(u_\alpha^N u_\beta, \mathcal{D}_N^{(\alpha)}(1)\right)_N - \frac{1}{2}\sum_{\alpha=1}^{3}\left(u_\beta, \mathcal{D}_N^{(\alpha)}u_\alpha^N\right)_N\,.$$

The first of these obviously vanishes, and the second term does so as well because the collocation approximation enforces $\sum_{\alpha=1}^{3} \mathcal{D}_N^{(\alpha)} u_\alpha^N = 0$. Hence, we have the discrete conservation of linear momentum (3.2.19). Conservation of linear momentum for the Galerkin approximation follows from similar arguments, but using the continuous inner product instead.

Again, the same conservation properties apply to Legendre Galerkin and Legendre collocation (or G-NI) approximations to problems with no-slip boundary conditions.

Convection and Divergence Forms. It is easily seen that the Fourier Galerkin approximations to the incompressible Navier–Stokes equations with all four forms, (3.2.1)–(3.2.4), of the momentum equation are equivalent. Hence, the convection and divergence forms also yield quadratic conservation for Galerkin approximations.

However, the four forms (3.2.1)–(3.2.4) all yield different solutions for Fourier collocation and G-NI approximations. The manipulations necessary to demonstrate conservation for the convection and divergence forms are blocked by the lack of the necessary precision of the discrete inner products. In these cases, the integrands are polynomials of degree $3N$ whereas the quadrature rule only has degree $2N$.

In practice, approximations to fully periodic problems (homogeneous flows) must use the rotation or skew-symmetric forms to avoid temporal instabilities unless the viscosity is very large. See Sect. 3.3.6 for some numerical examples. However, this guideline does not necessarily apply to problems with multiple inhomogeneous directions, as there are additional considerations (see Sect. 3.5.1).

3.2.2 General Guidelines for Time-Discretization

There is a rich variety of strategies for discretizing the Navier–Stokes equations in time and space. The choice of spatial discretization is strongly driven by the specific problem at hand, in particular, by the geometry of the domain and by the given boundary conditions. The spatial discretization will be covered in the remaining sections of this chapter, as well as in some sections of Chap. 5. We devote this section to an overview of the most commonly used time-discretization strategies. Some of them are splitting techniques, which allow the decoupled updating of pressure and velocity at different stages within each time-step. Other techniques lead to a so-called Stokes-like system in which velocity and pressure are coupled; however, their computation can still be decoupled while solving the system by elimination techniques.

For additional ease of exposition of the spectral algorithms, we will use first-order time-discretizations—forward Euler for explicit terms and backward Euler for implicit terms. In practice higher order time-discretizations are invariably used. In the late 1970s, second-order Adams–Bashforth for explicit terms and second-order Crank–Nicolson for implicit terms were a common choice. Low-storage Runge–Kutta (third-order and fourth-order) for explicit terms became popular in the 1980s. In the 1990s the third-order backward-difference scheme came into use for implicit terms. These and other standard time-discretizations are summarized in Appendix D (see also Chap. 7).

The forthcoming discussion will be confined to the primitive variable formulation (3.1.1)–(3.1.2), using the convection form of the momentum equation for simplicity. Some of the algorithms we describe can be rephrased for other basic formulations, such as the vorticity–velocity formulation, as well. For simplicity, we assume that no-slip conditions are enforced on the part $\partial \Omega_0$ of the boundary where periodicity is not imposed.

3.2.3 Coupled Methods

We first address those strategies that at each time-step lead to a problem in which velocity and pressure are still coupled. In some literature, these are called *monolithic methods*. Most strategies rely on the use of finite differences to discretize the acceleration term.

Fully Implicit Schemes. If all terms of the equations are dealt with implicitly, one obtains a nonlinear, fully coupled problem to be solved at each time-step; this can be accomplished by a Newton–Krylov iteration (with a suitable preconditioner). The computational cost per time-step of this approach may be very high, although the strategy may be more economical on balance in those cases for which the time-step is not constrained for accuracy considerations, such as, e. g., when using a time-dependent problem to approximate a steady-state solution.

Semi-Implicit Schemes. In other cases, it may be convenient to reserve the implicit treatment for some of the terms of the equations. This is invariably the case for the terms that account for the flow incompressibility; indeed, the continuity equation must be enforced at the new time-level; this is possible only if the pressure is free to adjust itself at the new time-level as well.

Explicit Convection and Diffusion. If both the diffusion and inertial terms are treated explicitly, then at each step one is left with a *linear Darcy problem*, which for a Euler time-advancing scheme reads

$$\frac{1}{\Delta t}\mathbf{u}^{n+1} + \nabla p^{n+1} = \mathbf{f}^{n+1} + \frac{1}{\Delta t}\mathbf{u}^n + \nu \Delta \mathbf{u}^n - \mathbf{u}^n \cdot \nabla \mathbf{u}^n \quad \text{in } \Omega,$$
$$\nabla \cdot \mathbf{u}^{n+1} = 0 \quad \text{in } \Omega,$$

(3.2.23)

where $\mathbf{f}^{n+1}$ is shorthand notation for $\mathbf{f}(t^{n+1})$, with $t^{n+1} = (n+1)\Delta t$, while $\mathbf{u}^n$ and p^n denote approximations to $\mathbf{u}(t^n)$ and $p(t^n)$, respectively. This has to be accompanied by the boundary condition

$$\mathbf{u}^{n+1} \cdot \mathbf{n} = 0 \quad \text{on } \partial \Omega_0.$$

A simple, albeit non-unique, way to solve this problem is to take the divergence of the first equation, getting a Poisson equation for the pressure. It is supplemented by a Neumann boundary condition which is obtained by taking the normal component of the first equation on $\partial\Omega_0$. A drawback of this approach is that it does not allow a clean and straightforward enforcement of the tangential boundary condition of the velocity. A further drawback comes from the severe stability restriction on the time-step (after a space discretization is introduced), related to the generous presence of explicit terms in the scheme.

Explicit Convection, Implicit Diffusion. The first difficulty mentioned above does not appear, while the second one is partially cured, if the viscous term is also treated implicitly. In this case, at each time-step we end up with the *generalized Stokes problem*

$$\frac{1}{\Delta t}\mathbf{u}^{n+1} - \nu\Delta\mathbf{u}^{n+1} + \nabla p^{n+1} = \mathbf{f}^{n+1} + \frac{1}{\Delta t}\mathbf{u}^n - \mathbf{u}^n\cdot\nabla\mathbf{u}^n \qquad \text{in } \Omega\,,$$

$$\nabla\cdot\mathbf{u}^{n+1} = 0 \qquad \text{in } \Omega\,,$$

$$\mathbf{u}^{n+1} = \mathbf{0} \qquad \text{on } \partial\Omega_0\,.$$

$$(3.2.24)$$

(The term "generalized" refers to the presence of the zero-order terms in the momentum equation.) The algebraic equivalent of this system, obtained after introducing a spatial discretization, can be solved by different strategies, as we will see in detail in Sect. 7.3. The first strategy consists of solving the full system by a preconditioned iterative method (e. g., a conjugate gradient in the symmetric case or GMRES otherwise). A second one aims at eliminating the velocity field using the momentum equation, yielding an independent reduced system for the pressure alone; the resulting equation is often called the *pressure Schur complement* equation, although in some quarters it is referred to as the *Uzawa* equation. For its solution we refer to Sect. 7.3.2. Finally, we mention the *influence-matrix* technique, which is particularly effective for flows with only one inhomogeneous direction (see Sect. 3.4 for details).

Example of a Higher-order Time Discretization. The Euler method is just a pedagogical example. In practical computations, one uses schemes that yield higher accuracy under milder stability restrictions. An example is given by the third-order scheme proposed by Karniadakis, Israeli and Orszag (1991):

$$\frac{11}{6\Delta t}\mathbf{u}^{n+1} - \nu\Delta\mathbf{u}^{n+1} + \nabla p^{n+1} = \mathbf{f}^{n+1} + \frac{1}{6\Delta t}(18\mathbf{u}^n - 9\mathbf{u}^{n-1} + 2\mathbf{u}^{n-2})$$

$$- (3\mathbf{u}^n\cdot\nabla\mathbf{u}^n - 3\mathbf{u}^{n-1}\cdot\nabla\mathbf{u}^{n-1} + \mathbf{u}^{n-2}\cdot\nabla\mathbf{u}^{n-2}) \qquad \text{in } \Omega\,,$$

$$\nabla\cdot\mathbf{u}^{n+1} = 0 \qquad \text{in } \Omega\,,$$

$$\mathbf{u}^{n+1} = \mathbf{0} \qquad \text{on } \partial\Omega_0\,.$$

$$(3.2.25)$$

This combines the backward-difference scheme BDF3 with extrapolation of the convection term. (See also Sect. 7.2 and in particular Table 7.1.) The problem (3.2.25) can also be regarded as a generalized Stokes problem. When the convection term is treated semi-implicitly (by means of a linearization procedure), then instead of a generalized Stokes problem we obtain an Oseen problem. In any case, after spatial discretization both the generalized Stokes and Oseen problems give rise to a blockwise system (see (7.3.1)), whose solution will be discussed in Sect. 7.4.

3.2.4 Splitting Methods

Now we turn to schemes that, at each time-step, produce a cascade of subproblems involving the velocity field and the pressure separately. However, this is achieved at the expense of introducing an error depending on the timestep, which is called the splitting error. These methods are commonly called *splitting methods*. They are also referred to as *fractional-step methods*, or, in some literature, as *segregated methods*. In fact, fractional-step methods can be used whenever the differential problem is governed by an operator that can be split into the sum of two or more operators of simpler form. Then, a time-step advancement is achieved through sub-steps involving only a single operator (with appropriate boundary conditions) at each substep. The pioneering work on fractional-step methods is due to Yanenko (1971) and Marchuk (1975).

The progenitor of splitting methods in fluid dynamics is the Chorin–Temam method, proposed by Chorin (1968) and Temam (1969) in the late 1960s. It has two stages. In the first one, an intermediate velocity field (which is not divergence-free) is computed by discretizing in time the momentum equation, in which the pressure term has been dropped or treated explicitly, and applying the prescribed boundary conditions. In the second stage, the computed velocity is projected upon the space of divergence-free vector fields (this motivates the term *projection method*, which is often associated with the Chorin–Temam method); the projection step amounts to solving a Darcy problem, which in turn can be reduced, as has already been mentioned above, to the solution of a Poisson problem for the pressure followed by the update of the velocity. The result is taken as the velocity field at the new time-level.

Below is a possible realization of this method.

i) *First step: preliminary velocity*

$$\frac{1}{\Delta t}(\mathbf{u}^{n+1/2} - \mathbf{u}^n) - \nu\triangle\mathbf{u}^{n+1/2} = \mathbf{f}^{n+1} - C(\mathbf{u}^n) \qquad \text{in } \Omega \ ,$$

$$\mathbf{u}^{n+1/2} = \mathbf{0} \qquad \text{on } \partial\Omega_0 \ ,$$
$$\tag{3.2.26}$$

where $C(\mathbf{u}^n)$ accounts for the convection term, which is often treated explicitly, e. g., $C(\mathbf{u}^n) = \mathbf{u}^n \cdot \nabla\mathbf{u}^n$.

ii) *Second step: projection*

$$\frac{1}{\Delta t}(\mathbf{u}^{n+1} - \mathbf{u}^{n+1/2}) + \nabla p^{n+1} = \mathbf{0} \qquad \text{in } \Omega ,$$
$$\nabla \cdot \mathbf{u}^{n+1} = 0 \qquad \text{in } \Omega , \qquad (3.2.27)$$
$$\mathbf{u}^{n+1} \cdot \mathbf{n} = 0 \qquad \text{on } \partial \Omega_0 .$$

This step, which yields a Darcy problem for the unknowns $\mathbf{u}^{n+1}$ and p^{n+1}, is in fact reducible to a Poisson problem for the pressure:

$$\triangle p^{n+1} = \frac{1}{\Delta t} \nabla \cdot \mathbf{u}^{n+1/2} \qquad \text{in } \Omega ,$$
$$\frac{\partial p}{\partial n}^{n+1} = 0 \qquad \text{on } \partial \Omega_0 , \qquad (3.2.28)$$

followed by an update of the velocity field:

$$\mathbf{u}^{n+1} = \mathbf{u}^{n+1/2} - \frac{1}{\Delta t} \nabla p^{n+1} . \qquad (3.2.29)$$

The latter equality, which follows from the Helmholtz decomposition principle, can in fact be regarded as a correction (by a pressure gradient) of $\mathbf{u}^{n+1/2}$ in order to ensure that the new velocity $\mathbf{u}^{n+1}$, also called the end-of-step velocity, is divergence-free.

Although the Chorin–Temam method is very simple to use, it suffers from quite low accuracy in time. In particular, the no-slip boundary condition is not exactly satisfied at the new time-level, due to the presence of an $O(\Delta t)$ splitting error in the tangential component; this generates an unphysical pressure boundary layer of thickness $O(\nu\sqrt{\Delta t})$ (see Rannacher (1992)).

Several improvements of the Chorin–Temam method have been proposed for the purpose of reducing the inherent splitting error. Overviews are given, e. g., in Prohl (1997), Gresho and Sani (1998), Guermond, Minev and Shen (2006). Section 7.2 will be devoted to the discussion of some high-order splitting schemes, and Sect. 7.4 will cover the alternative algebraic factorization methods.

3.2.5 Other Integration Methods

Several other time-advancing techniques that do not directly fit into the previous families have been proposed.

Operator Integration Factors. This technique is based on decomposing the spatial terms of the momentum equation into a component that is advanced using the integrating-factor approach, and a component that is advanced by a conventional time-discretization method. Assume that the momentum equation is written as

$$\mathbf{u}_t + A(\mathbf{u}) + B(\mathbf{u}) = \mathbf{0} . \qquad (3.2.30)$$

An appropriate integrating factor $\Phi = \Phi(t; t_n, \Delta t)$ is a time-dependent family of operators (i. e., matrices, after the spatial discretization has been introduced), which satisfy the differential equation

$$\Phi_t = \Phi A \ , \tag{3.2.31}$$

plus the condition that $\Phi = I$ at $t = t_n$ or at $t = t_{n+1}$. Applying Φ to (3.2.30), one gets

$$(\Phi \mathbf{u})_t + \Phi B(\mathbf{u}) = \mathbf{0} \ , \tag{3.2.32}$$

which is then treated by the preferred time-discretization technique. Several tricks may be applied to compute the action of Φ in an efficient manner.

In this framework we have the following methods.

The Rogallo Method. In some special cases involving constant viscosity and periodic boundary conditions in three or two directions, the pressure variable and the continuity constraint can be eliminated from the discrete system (see Sects. 3.3.1 and 3.4.2, respectively). In these cases the integrating factor may be applied to those spatial derivatives in the viscous terms that correspond to the periodic directions, and the solution to (3.2.31) obtained analytically. What remains for the $B(\mathbf{u})$ term in (3.2.32) in these special cases are the convection terms and possibly the derivative in the viscous terms corresponding to a single nonperiodic direction. A time-discretization that is explicit for convection can be implemented very efficiently, since even if there is a remaining viscous term that is treated implicitly, this requires only a one-dimensional implicit solution. This approach was developed by Rogallo (1977) for simulations of homogeneous turbulence. The details are given in Sects. 3.3.1 and 3.4.2.

The OIFS Method. In the general case, of course, the pressure variable and the continuity constraint cannot be eliminated, and all time integrations must be performed numerically. The *operator implicit factor splitting (OIFS)* method, proposed by Maday, Patera and Rønquist (1990), solves the necessarily implicit Stokes problem using a large time-step, while using smaller time-steps to advance the convection terms explicitly. Hence, the operator $A(\mathbf{u})$ in (3.2.30) pertains to the convection terms and the operator $B(\mathbf{u})$ to the Stokes problem. The overall interval of integration $[t_n, t_{n+1}]$ is chosen to satisfy the accuracy constraints for the Stokes problem. The integrating factor Φ is computed numerically by explicit time integration of (3.2.31) using a time-step that is sufficiently small to satisfy both temporal stability and the desired temporal accuracy. The generalized Stokes problem represented by (3.2.32) is then solved implicitly. The details of the method can be found in Deville, Fischer and Mund (2002) and in Karniadakis and Sherwin (2005).

Characteristics Methods. A similar principle inspires the characteristics methods, in which the convection equation is integrated by an ODE technique moving on the characteristic curves. Then, a Stokes-like problem is left to be solved. In this respect, these methods can be regarded as monolithic methods. We refer, e. g., to Mohammadi and Pironneau (2001) or Quarteroni and Valli (1994) for further details.

3.3 Homogeneous Flows

Von Neumann (1949) and Emmons (1949) proposed numerical simulation of turbulence as early as the late 1940s. Nevertheless, nearly two decades elapsed before this vision became reality. It materialized at the National Center for Atmospheric Research in Boulder, Colorado in the late 1960s, due in large part to the availability of the Control Data Corporation 6600 computer. Deardorff (1970) combined a finite-difference method with a subgrid-scale model to compute turbulent channel flow on a 24 × 20 × 14 grid. Orszag and Patterson (1972) performed the first direct simulation of homogeneous, isotropic turbulence on a 32^3 grid using a spectral Galerkin method. To this day this algorithm remains the workhorse for numerical simulations of homogeneous turbulence. Several refinements of their basic algorithm have been developed, notably those by Rogallo (1977, 1981) and Basdevant (1983), and by Brachet et al. (1983) for flows with special symmetries.

Homogeneous flows are flows that are translation invariant. For such flows, periodic boundary conditions in all three directions are justified. Equations (3.1.1)–(3.1.3) plus the periodic boundary conditions become

$$\frac{\partial \mathbf{u}}{\partial t} + \mathbf{u} \cdot \nabla \mathbf{u} = -\nabla p + \nu \Delta \mathbf{u}, \tag{3.3.1}$$

$$\nabla \cdot \mathbf{u} = 0 \,, \tag{3.3.2}$$

$$\mathbf{u} \quad 2\pi\text{-periodic with respect to } \mathbf{x} \,, \tag{3.3.3}$$

$$\mathbf{u} = \mathbf{u}_0 \text{ at } t = 0 \,. \tag{3.3.4}$$

We have assumed for simplicity that the periodicity lengths in all three directions are 2π. Rigorous theories for this problem can be found in Temam (1995).

3.3.1 Fourier Galerkin Approximation for Isotropic Turbulence

The solution to the problem has the Fourier series representation

$$\mathbf{u}(\mathbf{x}, t) = \sum_{\mathbf{k}} \hat{\mathbf{u}}_{\mathbf{k}}(t) e^{i\mathbf{k} \cdot \mathbf{x}} \,, \tag{3.3.5}$$

$$p(\mathbf{x}, t) = \sum_{\mathbf{k}} \hat{p}_{\mathbf{k}}(t) e^{i\mathbf{k} \cdot \mathbf{x}} \,. \tag{3.3.6}$$

In Fourier space (3.3.1) and (3.3.2) become

$$\left(\frac{d}{dt} + \nu |\mathbf{k}|^2 \right) \hat{\mathbf{u}}_{\mathbf{k}} = -i\mathbf{k}\hat{p}_{\mathbf{k}} + \hat{\mathbf{c}}_{\mathbf{k}} \,, \tag{3.3.7}$$

$$i\mathbf{k} \cdot \hat{\mathbf{u}}_{\mathbf{k}} = 0 \,. \tag{3.3.8}$$

The term

$$\hat{\mathbf{c}}_{\mathbf{k}} = -\widehat{(\mathbf{u} \cdot \nabla \mathbf{u})}_{\mathbf{k}} \qquad (3.3.9)$$

is nonlinear and it is responsible for most of the algorithmic complexities of the problem. Its numerical treatment is discussed extensively below. The pressure may be eliminated by taking $i\mathbf{k}$ dotted with (3.3.7) and using (3.3.8). Hence,

$$\hat{p}_{\mathbf{k}} = -\frac{1}{|\mathbf{k}|^2} i\mathbf{k} \cdot \hat{\mathbf{c}}_{\mathbf{k}} , \qquad (3.3.10)$$

$$\left(\frac{\mathrm{d}}{\mathrm{d}t} + \nu|\mathbf{k}|^2\right)\hat{u}_k = \hat{\mathbf{c}}_{\mathbf{k}} - \mathbf{k}\frac{\mathbf{k} \cdot \hat{\mathbf{c}}_{\mathbf{k}}}{|\mathbf{k}|^2} . \qquad (3.3.11)$$

The Fourier Galerkin approximation consists of truncating the sums at $|k_1|, |k_2|, |k_3|, < N/2$. (For the reasons given in CHQZ2, Sect. 3.3.1, the modes for which any $k_\alpha = -N/2, \alpha = 1, 2$, or 3 are not retained.) Note that in (3.3.7) and (3.3.8) the negative gradient operator is the adjoint of the divergence operator, and the Laplacian operator is the composition of the divergence and the gradient.

The original Orszag–Patterson algorithm (Orszag (1969), Patterson and Orszag (1971), Orszag (1971a), Orszag and Patterson (1972)) employed leap frog time-differencing for the nonlinear term and Crank–Nicolson time-differencing for the viscous term. (However, in most applications, an explicit scheme for the viscous term will suffice, since the stability restriction arising from the convection terms can be more severe than the viscous stability limit.) Other implementations, for example, that by Rogallo (1977), have used fourth-order Runge–Kutta formulas for the nonlinear terms and an integrating-factor technique on diffusion. Precisely, (3.3.11) is multiplied by the integrating factor $e^{-\nu|\mathbf{k}|^2(t-t_n)}$ to yield

$$\frac{\mathrm{d}}{\mathrm{d}t}\left(e^{-\nu|\mathbf{k}|^2(t-t_n)}\hat{u}_k\right) = e^{-\nu|\mathbf{k}|^2(t-t_n)}\left(\hat{\mathbf{c}}_{\mathbf{k}} - \mathbf{k}\frac{\mathbf{k} \cdot \hat{\mathbf{c}}_{\mathbf{k}}}{|\mathbf{k}|^2}\right) .$$

The integration over one time-step is performed analytically for the left-hand side and numerically for the right-hand side. In this way, the only time-step limitation arises from the terms on the right-hand side.

3.3.2 De-aliasing Using Transform Methods

Let $\mathbf{u} = (u, v, w) = (u_1, u_2, u_3)$. In component form the nonlinear term is

$$(\hat{\mathbf{c}}_{\mathbf{k}})_\alpha = -ik_\beta \sum_{\mathbf{m}+\mathbf{n}=\mathbf{k}} \hat{u}_{\beta,\mathbf{m}}\hat{u}_{\alpha,\mathbf{n}} \qquad (3.3.12)$$

(where repeated indices—in this case β—imply summation). This convolution sum is the standard Galerkin approximation to the nonlinear term—$\mathbf{u} \cdot \nabla \mathbf{u}$.

Let us focus on a typical addend of (3.3.12), namely,

$$\hat{s}_{\mathbf{k}} = \sum_{\mathbf{m}+\mathbf{n}=\mathbf{k}} \hat{u}_{\mathbf{m}} \hat{v}_{\mathbf{n}} \ , \tag{3.3.13}$$

which is a triple convolution sum. Efficient transform methods for the one-dimensional version of (3.3.13) are discussed in CHQZ2, Sect. 3.4. We recall that in one dimension (3.3.13) reads

$$\hat{s}_k = \sum_{m+n=k} \hat{u}_m \hat{v}_n \ .$$

The pseudospectral transform method consists of transforming $\hat{u}_n$ and $\hat{v}_n$, $n = -N/2, \ldots, N/2-1$, to physical space using the discrete Fourier transform with N points, forming the physical space products $u_i v_i$, $i = 0, \ldots, N - 1$, and then transforming these products back to Fourier space. The result is

$$\tilde{s}_k = \hat{s}_k + \sum_{m+n=k \pm N} \hat{u}_m \hat{v}_n$$

(see CHQZ2, formula (3.4.9)). The three-dimensional analog is

$$\tilde{s}_{\mathbf{k}} = \hat{s}_{\mathbf{k}} + \sum_{\mathbf{m}+\mathbf{n}=\mathbf{k}\pm N\mathbf{e}_1} \hat{u}_{\mathbf{m}} \hat{v}_{\mathbf{n}} + \sum_{\mathbf{m}+\mathbf{n}\mathbf{k}\pm N\mathbf{e}_2} \hat{u}_{\mathbf{m}} \hat{v}_{\mathbf{n}}$$
$$+ \sum_{\mathbf{m}+\mathbf{n}=\mathbf{k}\pm N\mathbf{e}_3} \hat{u}_{\mathbf{m}} \hat{v}_{\mathbf{n}} + \sum_{\mathbf{m}+\mathbf{n}=\mathbf{k}\pm N\mathbf{e}_1 \pm N\mathbf{e}_2} \hat{u}_{\mathbf{m}} \hat{v}_{\mathbf{n}} + \sum_{\mathbf{m}+\mathbf{n}=\mathbf{k}\pm N\mathbf{e}_1 \pm N\mathbf{e}_3} \hat{u}_{\mathbf{m}} \hat{v}_{\mathbf{n}}$$
$$+ \sum_{\mathbf{m}+\mathbf{n}=\mathbf{k}\pm N\mathbf{e}_2 \pm N\mathbf{e}_3} \hat{u}_{\mathbf{m}} \hat{v}_{\mathbf{n}} + \sum_{\mathbf{m}+\mathbf{n}=\mathbf{k}\pm N\mathbf{e}_1 \pm N\mathbf{e}_2 \pm N\mathbf{e}_3} \hat{u}_{\mathbf{m}} \hat{v}_{\mathbf{n}} \ , \tag{3.3.14}$$

where $\mathbf{e}_j$ is a unit vector in the j-th direction. The second, third, and fourth terms on the right-hand side are the single-aliased contributions; the fifth, sixth and seventh are the double-aliased ones; and the last term is the triple-aliased contribution. The basic techniques for de-aliasing are truncation and phase shifts. The 3/2-*rule*, in which the discrete transforms are evaluated with $3N/2$ points, extends to three dimensions in a straightforward manner. The phase-shift technique, however, becomes much more involved. Aliasing removal by phase shifts thus requires eight separate evaluations of the convolution terms. (See Rogallo (1981) for details.)

Another option is combining truncation with just two shifted grids, the standard Fourier collocation grid in three dimensions:

$$\mathbf{x_j} = \frac{2\pi}{N} (j_1, j_2, j_3) \ , \quad j_\alpha = 0, 1, \ldots, N - 1 \ , \tag{3.3.15}$$

and the grid offset from this by a half-cell in each direction:

$$\mathbf{x}_j' = \mathbf{x}_j + \left(\frac{\pi}{N}, \frac{\pi}{N}, \frac{\pi}{N} \right) \,. \tag{3.3.16}$$

This eliminates both the single-aliased and triple-aliased contributions.

The double-aliased and triple-aliased terms vanish if the spherical truncation (Patterson and Orszag (1971))

$$k_1^2 + k_2^2 + k_3^2 \leq \left(\frac{\sqrt{2}}{3} N \right)^2 \,, \tag{3.3.17}$$

or the polyhedral truncation (Orszag (1971a))

$$\begin{aligned} |k_\alpha| &\leq N/2 \,, & \alpha &= 1,2,3 \,, \\ |k_\alpha \pm k_\beta| &\leq 2N/3 \,, & \alpha, \beta &= 1,2,3 \,, \ \alpha \neq \beta \,, \end{aligned} \tag{3.3.18}$$

is used. The latter truncation has the advantage of retaining 27% more modes than the former one.

A pseudospectral method employing random shifts at each step was discussed at the end of Sect. 3.2.3 in CHQZ2. This technique was in fact developed by Rogallo (1977) for the present application of homogeneous turbulence simulations. The spherical truncation (3.3.17) removes the double-aliasing and triple-aliasing errors. At the first stage of a second-order Runge–Kutta method, the convolution sum is evaluated by a pseudospectral transform method on the grid $\mathbf{x}_j$ given by (3.3.15), except that on the right-hand side each j_α is replaced by $j_\alpha + \Delta_\alpha$, where the Δ_α are random numbers in $(0, 2\pi/N)$. At the second stage the grid (3.3.16) is used in the pseudospectral transform method. This reduces the single-aliased errors to $O(\Delta t^2)$ times their ordinary pseudospectral values. The random shifts at each time-step inhibit these remaining aliasing errors from accumulating over many time-steps. This strategy can be adapted to most time-discretization schemes.

The truncations described by (3.3.17) and (3.3.18) are both isotropic (in addition to eliminating the double-aliased and triple-aliased errors). Since homogeneous turbulence appears to be isotropic on the smallest scales, one may wish to incorporate the isotropic truncation

$$k_1^2 + k_2^2 + k_3^2 < (N/2)^2 \tag{3.3.19}$$

in those methods that do not employ a truncation for de-aliasing purposes.

Table 3.1 summarizes the various transform methods available for three-dimensional convolution sums. The operation counts are for the evaluation of one triple convolution sum and they presume that a single full three-dimensional FFT on an N^3 grid takes $(15/2)N^3 \log_2 N$ operations and that the three-dimensional FFT used for the 3/2-rule methods takes only $(235/16)N^3 \log_2 N$ operations because of the large number of zeros in the

Table 3.1. Transform methods for a triple convolution sum

Methods	Operations	Active Modes	Operations/ Mode	Aliasing Errors
pseudospectral	$22.5N^3 \log_2 N$	N^3	$22.5 \log_2 N$	single, double and triple
$\frac{3}{2}$-rule	$44N^3 \log_2(\frac{3}{2}N)$	N^3	$44 \log_2(\frac{3}{2}N)$	none
8 shifted grids	$180N^3 \log_2 N$	N^3	$180 \log_2 N$	none
$\frac{3}{2}$-rule + (3.3.19)	$44N^3 \log_2(\frac{3}{2}N)$	$\frac{\pi}{6}N^3$	$84 \log_2(\frac{3}{2}N)$	none
random pseudospectral + (3.3.17)	$22.5N^3 \log_2 N$	$\frac{8\pi\sqrt{2}}{81}N^3$	$51 \log_2 N$	$O(\Delta t^2)$ single
2 shifted grids + (3.3.17)	$45N^3 \log_2 N$	$\frac{8\pi\sqrt{2}}{81}N^3$	$103 \log_2 N$	none
2 shifted grids + (3.3.18)	$45N^3 \log_2 N$	$\frac{5}{9}N^3$	$81 \log_2 N$	none

data. (Addition and multiplication count as separate operations and the simple radix-2 transform is assumed; more general and more efficient transforms are discussed in Appendix B.) The active modes are those that survive the truncation. The last column indicates which, if any, aliasing errors are present.

There are several refinements that can profitably be employed to reduce the storage requirements, CPU time, or I/O cost for an out-of-core simulation. Rogallo (1981) and Basdevant (1983) provide extended discussions. Basdevant, for example, shows how isotropic simulations may be performed (with the 3/2-rule) at a cost of roughly eight three-dimensional FFT's per step. The generation of divergence-free initial conditions has been discussed by Rogallo (1981) and Schumann (1985). The paper by Brachet et al. (1983) describes how symmetries in special flows such as the Taylor–Green vortex may be exploited to provide a wider dynamic range.

An alternative, but less efficient, method of performing de-aliasing is with the 2/3-*rule*. In this case the pseudospectral transform method is applied on an N^3 grid, but all Fourier coefficients with some $k_i \geq N/3$, $i = 1, 2, 3$, are set to zero after the transform back to Fourier space.

Some three decades after its introduction, the Orszag–Patterson algorithm continues to be used for numerical investigations of isotropic turbulence at higher and higher Reynolds numbers. For example, Kaneda and Ishihara (2006) exploited 512 nodes of the Earth Simulator (then the world's fastest computer) to perform isotropic turbulence simulations using a very similar, Fourier spectral algorithm on grids as large as 4096^3. (The sustained speed

Fig. 3.2. Direct numerical simulation of incompressible isotropic turbulence by Kaneda and Ishihara (2006) on a 2048^3 grid. The regions of intense vorticity in a subdomain with $1/4$ the length in each coordinate direction of the full domain are shown [Courtesy of Y. Kaneda, T. Ishihara (2006)]

was as fast as 16 TFlop.) Figure 3.2 illustrates the regions of intense vorticity in $1/64$ of the volume of their 2048^3 simulation for Re $= 16,135$. The macroscopic scale L is approximately 80% the size of one edge of the figure, and the microscopic scale η is 0.06% of the edge length. Among the many results obtained from their high-resolution simulations was convincing evidence that the scaled energy spectrum (where the wavenumber is scaled by the inverse of the Kolmogorov length scale $\eta = (\nu^3/\bar{\epsilon})^{1/4}$, with ν the viscosity and $\bar{\epsilon}$ the average dissipation rate) is not the classical Kolmogorov result of $k^{-5/3}$, but rather k^{-m} with $m \simeq 5/3 - 0.10$.

3.3.3 Pseudospectral and Collocation Methods

The pseudospectral version of the Orszag–Patterson algorithm uses (3.3.9) and (3.3.11), but approximates the nonlinear terms $\hat{\mathbf{c}}_{\mathbf{k}}$ pseudospectrally, as discussed in CHQZ2, Sect. 3.4.1. In this approach the fully aliased convolution sum (3.3.14) is used in place of the correct sum (3.3.13). The operation count is clearly $(45/2)N^3 \log_2 N$. This is one half the cost of a true Galerkin method implemented via the 3/2-rule. If, for physical or esthetical reasons, an isotropic truncation is desired, then the effective cost per mode of a pseudospectral method rises to 60% of the Galerkin cost.

Collocation algorithms for homogeneous flows are based on the Navier–Stokes equations in physical space. Although we focus the following discussion around the nonconservative form of the equations (3.1.1), in practice either the rotation form or the skew-symmetric form must be used to ensure numerical stability, since, as noted in Sect. 3.2.1, only these satisfy a discrete quadratic conservation property. The semi-discrete collocation approximation enforces (3.1.1)–(3.1.2) at the points (3.3.15). In order to cast the scheme in matrix form, let us introduce some notation. Let $\boldsymbol{u}^N$ and $\mathbf{p}^N$ be vectors that collect the discrete values of velocity and pressure at the collocation points; note the notationally minor but logically major difference between $\mathbf{u}^N$ (a vector-valued function defined in $\mathbb{R}^3$) and $\boldsymbol{u}^N$ (an algebraic vector with $3N^3$ components). Let G_N and D_N be rectangular matrices corresponding to the discrete operators $\mathbb{G}_N$ and $\mathbb{D}_N$ defined in (3.2.15) and (3.2.16). For instance, G_N acts on a vector $\mathbf{q}$ of values at the collocation points as

$$G_N\, \mathbf{q} = (D_N^x \mathbf{q}, D_N^y \mathbf{q}, D_N^z \mathbf{q})^T \;, \tag{3.3.20}$$

where D_N^x, D_N^y, D_N^z are the matrices realizing the first-order partial derivatives at the collocation points (see (8.7.2)). Let

$$L_N = D_N G_N \tag{3.3.21}$$

be the matrix realizing the discrete Laplace operator. Finally, let $\boldsymbol{u}^N \cdot G_N$ denote the matrix

$$\boldsymbol{u}^N \cdot G_N = \operatorname{diag}(\boldsymbol{u}_x^N) D_N^x + \operatorname{diag}(\boldsymbol{u}_y^N) D_N^y + \operatorname{diag}(\boldsymbol{u}_z^N) D_N^z \;, \tag{3.3.22}$$

where $\boldsymbol{u}_x^N$, $\boldsymbol{u}_y^N$ and $\boldsymbol{u}_z^N$ are the block components of $\boldsymbol{u}^N$. Then, the collocation scheme can be written as

$$\frac{d\boldsymbol{u}^N}{dt} + \boldsymbol{u}^N \cdot G_N \boldsymbol{u}^N = -G_N \mathbf{p}^N + \nu L_N \boldsymbol{u}^N \;, \tag{3.3.23}$$

$$D_N \boldsymbol{u}^N = \mathbf{0} \;. \tag{3.3.24}$$

The fully discrete, first-order temporal approximation to (3.1.1) and (3.1.2) with explicit treatment of the convection and diffusion terms can be written as

$$\boldsymbol{u}^{n+1} = \boldsymbol{u}^n - \Delta t\, \boldsymbol{z}^n + \nu \Delta t\, L_N \boldsymbol{u}^n - \Delta t\, G_N \mathbf{p}^{n+1} \;, \tag{3.3.25}$$
$$D_N \boldsymbol{u}^{n+1} = \mathbf{0} \;, \tag{3.3.26}$$

where $\boldsymbol{z}^n = \boldsymbol{u}^n \cdot G_N \boldsymbol{u}^n$, and we have dropped the superscript N (denoting the degree of the Fourier approximation) to make room for the superscript n (denoting the time-step).

Taking the discrete divergence of (3.3.25) and using (3.3.26), which applies to $\boldsymbol{u}^n$ as well, plus the property that D_N and L_N commute for Fourier methods, we obtain

$$L_N \mathbf{p}^{n+1} = D_N \boldsymbol{z}^n \;. \tag{3.3.27}$$

In Fourier space, (3.3.25)–(3.3.27) become

$$\hat{\mathbf{u}}_{\mathbf{k}}^{n+1} = \hat{\mathbf{u}}_{\mathbf{k}}^n - \nu\Delta t\, k^2\hat{\mathbf{u}}_{\mathbf{k}}^n - \Delta t\, \hat{\mathbf{z}}_k^n + \Delta t\, \frac{\mathbf{k}}{|\mathbf{k}|^2}(\mathbf{k}\cdot\hat{\mathbf{z}}_{\mathbf{k}}^n)\,, \qquad (3.3.28)$$

where $\hat{\mathbf{z}}^n$ is $\mathbf{z}^n$ in transform space, i.e., the discrete Fourier transform of $\mathbf{u}^n\cdot\nabla\mathbf{u}^n$. (Note that the viscous term may be treated implicitly at essentially no extra cost.) The collocation method may be identified with a particular pseudospectral method. Unlike Galerkin methods, pseudospectral methods, which write the nonlinear term in different (but equivalent) forms for the continuous problem, will not be equivalent. Thus, the algorithms based on the rotation and skew-symmetric forms, which use $\boldsymbol{\omega}^n\times\mathbf{u}^n$ and $\frac{1}{2}\mathbf{u}^n\cdot\nabla\mathbf{u}^n + \frac{1}{2}\nabla(\mathbf{u}^n(\mathbf{u}^n)^T)$, respectively, in place of $\mathbf{u}^n\cdot\nabla\mathbf{u}^n$ in (3.3.28) will yield a different result.

In the event that the fluid has variable transport properties (as in computations that include a turbulence model), the viscosity will depend on $\mathbf{x}$ and t. The following version of the Chorin–Temam splitting method (see Sect. 3.2.4) is then appropriate:

$$\mathbf{u}^{n+1/2} = \mathbf{u}^n - \Delta t\, \mathbf{z}^n + \Delta t\, D_N(\boldsymbol{\nu}G_N\mathbf{u}^n)\,, \qquad (3.3.29)$$

$$\mathbf{u}^{n+1} = \mathbf{u}^{n+1/2} - \Delta t\, G_N\mathbf{p}^{n+1}\,,$$
$$D_N\mathbf{u}^{n+1} = \mathbf{0}\,, \qquad (3.3.30)$$

where $\boldsymbol{\nu}$ is the diagonal matrix representing the values of the viscosity at the grid points.

The first, intermediate velocity step, given by (3.3.29), is straightforward. A full implicit treatment of the viscous term with a spatially or temporally varying viscosity requires the use of an iterative method, which as discussed in CHQZ2, Chap. 4, are still fairly expensive. A more attractive approach is to use a semi-implicit method that splits the viscous term in (3.3.29) into a part with just the spatial–temporal average viscosity (treated implicitly) and a part with the fluctuating part (treated explicitly). For the second, correction step, (3.3.30) implies that the pressure satisfies

$$L_N\mathbf{p}^{n+1} = \frac{1}{\Delta t}D_N\mathbf{u}^{n+1/2}\,. \qquad (3.3.31)$$

This equation is readily solved in Fourier space.

The principal difference between the collocation and Galerkin methods is the inclusion of aliasing interactions in the former. One can also perform a dealiased collocation computation by performing the computations in the usual collocation manner, but applying the 2/3-rule to remove the upper-third of the modes in each coordinate direction during the solution of (3.3.27) in Fourier space. Some numerical examples and a summary of broader numerical experience are provided in Sect. 3.3.6.

3.3.4 Rogallo Transformation for Homogeneous Turbulence

The original Orszag–Patterson algorithm was suitable only for isotropic turbulent flows. Rogallo (1977, 1981) extended it to all homogeneous turbulent flows. These have the form $\mathbf{u} = \bar{\mathbf{u}} + \mathbf{u}'$ and $p = \bar{p} + p'$ with the mean flow

$$\bar{\mathbf{u}} = A\,\mathbf{x}\,, \qquad \bar{p} = \mathbf{x}^T B\mathbf{x}\,, \tag{3.3.32}$$

where the tensors A and B depend only on t (and B is symmetric). (Any more general form of the mean flow necessarily leads to inhomogeneous turbulence.) The computations are done in a moving coordinate system $\mathbf{x}'$ described by

$$\mathbf{x}' = C\mathbf{x}\,, \tag{3.3.33}$$

where C also depends only on t. The tensor A may be written

$$A = A_s + A_a\,, \tag{3.3.34}$$

where A_s and A_a are the symmetric and antisymmetric parts and represent the strain rate and vorticity of the mean flow. Rogallo (1981) shows that, subject to

$$\sum_{i=1}^{3}(A_s)_{ii} = 0\,, \tag{3.3.35}$$

A_s is an arbitrary function of time and the other tensors are related by

$$\frac{\mathrm{d}A_a}{\mathrm{d}t} + A_s A_a + A_a A_s = 0\,, \tag{3.3.36}$$

$$\frac{\mathrm{d}A_s}{\mathrm{d}t} + A_s^2 + A_a^2 + 2B = 0\,, \tag{3.3.37}$$

$$\frac{\mathrm{d}C}{\mathrm{d}t} + CA = 0\,. \tag{3.3.38}$$

The equation governing the fluctuating part of the flow is, with tensor notation employed,

$$\frac{\partial u_i'}{\partial t} + A_{ij}u_j' + C_{kj}(u_i'u_j')_{,k} = -C_{ji}p_{,j}' + \nu C_{kj}C_{lj}u_{i,kl}'\,, \tag{3.3.39}$$

$$C_{ji}u_{i,j}' = 0\,, \tag{3.3.40}$$

where the derivatives are taken with respect to $\mathbf{x}'$ and derivatives with respect to x_k' are denoted by $,k$. Since the coefficients of (3.3.39) and (3.3.40) are independent of $\mathbf{x}'$, Fourier approximations are suitable.

In some cases it is advisable to periodically (in time) perform a re-gridding operation. We illustrate this here for the frequently studied case of homogeneous turbulence in uniform shear flow. The top left frame of Fig. 3.3 illustrates the mean velocity for the particular case of a mean flow in the

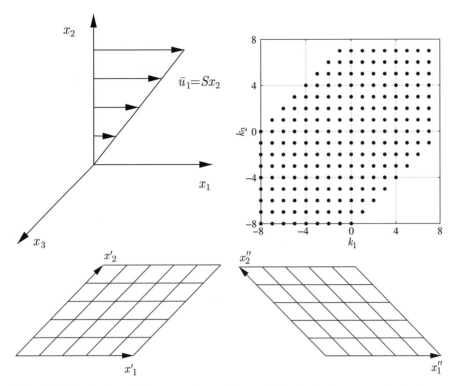

Fig. 3.3. Illustration of the mean flow and re-gridding for homogeneous turbulence in uniform shear flow. Mean velocity for uniform shear flow (*top left*), the distorted grids before (*bottom left*) and after (*bottom right*) re-gridding, and illustration of modes that are not affected by aliasing during the re-gridding process for uniform shear flow (*top right*)

x-direction with a uniform shear rate S in the x_2 direction, i.e., the mean velocity has the gradient $\partial \bar{u}_1/\partial x_2 = S$, i.e., $\bar{\mathbf{u}} = (Sx_2, 0, 0)$. In this case the general transformation (3.3.33) specializes to the transformation

$$x_1' = x_1 - (St)x_2 , \quad x_2' = x_2 , \quad x_3' = x_3 . \tag{3.3.41}$$

Assume that the computation starts at $t = 0$ on a standard cubic grid in physical space. Then, by $St = 0.5$, the grid in the reference rectangular grid in (x_1', x_2') coordinates corresponds to the skewed grid in physical space shown in the bottom left frame of Fig. 3.3. This grid skewness becomes increasingly more severe as time increases. To alleviate the deleterious effects of this upon the computational results, at $St = 0.5$ the grid is transformed to the grid skewed in the opposite direction, as illustrated in the bottom right frame of Fig. 3.3. Since the grid points of the two grids in the physical domain (x_1, x_2) coincide (incorporating periodicity), one can readily perform this in physical space. However, this produces aliasing effects. Delorme's

(1985) remedy for this was to simply de-alias in the x_2' direction all the modes that would be subjected to aliasing both before and after re-gridding. In particular, ignoring the x_3 direction and selecting a general re-gridding time t_R, if a variable, say u, has the Fourier expansion in (x_1', x_2') prior to re-gridding:

$$u(x_1', x_2', t) = \sum_{k_1,k_2=-N/2}^{N/2-1} \hat{u}_{k_1,k_2}(t) e^{i\mathbf{k}\cdot\mathbf{x}'} , \qquad (3.3.42)$$

then, after re-gridding, with $(x_1'', x_2'') = (x_1' - (2St_R)x_2', x_2')$,

$$\begin{aligned} u(x_1'', x_2'', t) &= \sum_{k_1,k_2=-N/2}^{N/2-1} \hat{u}_{k_1,k_2}(t) e^{i\mathbf{k}\cdot\mathbf{x}''} \\ &= \sum_{k_1,k_2=-N/2}^{N/2-1} \hat{u}_{k_1,k_2}(t) e^{i[k_1 x_1' + (k_2 - (2St_R)k_1)x_2']} . \end{aligned} \qquad (3.3.43)$$

Hence, when $|k_2 - (2St_R)k_1| > N/2$, aliasing occurs. The cure in this case is to zero out, both before and after re-gridding, all modes $\hat{u}_{k_1,k_2}(t)$ such that $|k_2 - (2St_R)k_1| > N/2$. If $St_R = 0.5$, then this condition becomes $|k_2 - k_1| > N/2$. Clearly, the more frequently one re-grids, the fewer modes are discarded. The top right frame of Fig. 3.3 illustrates by dots those modes in (k_1, k_2) space that are not affected by aliasing during re-gridding. The particular case shown there is for a 16^2 grid in the $x_1 - x_2$ plane with re-gridding performed when $S_R t = 0.5$. The modes in the upper left and lower right are affected by aliasing, and it is customary to zero out these modes during the re-gridding process. Similar re-gridding considerations apply to other instances of the Rogallo transformation.

3.3.5 Large-Eddy Simulation of Isotropic Turbulence

The basic formulation of large-eddy simulation models was presented in Sects. 1.3.3 and 1.4.2. Sagaut (2006) provides extensive coverage of various LES models. Here we illustrate the Smagorinsky model and the multiscale model of Hughes, Mazzei and Jansen (2000) for the simple case of isotropic turbulence. Traditional LES models, such as the Smagorinsky model and the dynamic Smagorinsky model, when discretized via a spectral method, typically use a collocation method (perhaps with the 3/2-rule used for de-aliasing), whereas the multilevel and multiscale models use a Galerkin (or G-NI) formulation. However, we will illustrate all three types of models here using a Galerkin discretization with N modes in each coordinate direction in the domain $(0, 2\pi)^3$.

For a direct numerical simulation the differential equations are given by (3.3.7)–(3.3.9). The filtered variables in large-eddy simulations of incompressible turbulence satisfy (1.4.3)–(1.4.4). LES of isotropic turbulence using Fourier spectral methods on these equations have typically used for the filter function G (see (1.3.29)) either a Gaussian filter with half-width 2Δ in physical space or a sharp cut-off filter in Fourier space that keeps all modes with $|\mathbf{k}| \leq \bar{N}/2$ with $\bar{N} \leq N$ and discards the rest.

For the simple Smagorinsky model the momentum and continuity equations for the filtered variables (denoted by bars) are

$$\left(\frac{d}{dt} + \nu|\mathbf{k}|^2\right)\hat{\bar{\mathbf{u}}}_{\mathbf{k}} = -i\mathbf{k}\hat{\bar{p}}_{\mathbf{k}} + \hat{\bar{\mathbf{c}}}_{\mathbf{k}} + i\mathbf{k}\cdot\hat{\bar{\boldsymbol{\tau}}}_{\mathbf{k}}^{\mathrm{Smag}} , \qquad (3.3.44)$$

$$i\mathbf{k}\cdot\hat{\bar{\mathbf{u}}}_{\mathbf{k}} = 0 , \qquad (3.3.45)$$

where $\hat{\bar{\boldsymbol{\tau}}}_{\mathbf{k}}^{\mathrm{Smag}}$ is the $\mathbf{k}$-th Fourier coefficient of the subgrid-scale stress $\hat{\bar{\boldsymbol{\tau}}}^{\mathrm{Smag}}$, given by (1.4.10)–(1.4.12), with $\Delta x = \Delta y = \Delta z = 2\pi/N$. Spectral methods for these isotropic LES problems have typically treated the last two terms on the right-hand side of (3.3.44) explicitly and the remaining terms implicitly.

The multiscale approach uses a sharp cutoff filter in Fourier space yielding the following decomposition

$$\mathbf{u} = \mathbf{u}^l + \mathbf{u}^s , \qquad (3.3.46)$$

with

$$\mathbf{u}^l(\mathbf{x}, t) = \sum_{|\mathbf{k}|<\bar{N}/2} \hat{\mathbf{u}}_{\mathbf{k}}(t)e^{i\mathbf{k}\cdot\mathbf{x}} , \qquad (3.3.47)$$

$$\mathbf{u}^s(\mathbf{x}, t) = \sum_{\bar{N}/2\leq|\mathbf{k}|<N/2} \hat{\mathbf{u}}_{\mathbf{k}}(t)e^{i\mathbf{k}\cdot\mathbf{x}} , \qquad (3.3.48)$$

where $\bar{N} < N$ divides the large and small scales. Both components are required to satisfy the continuity equation. The momentum equation for the large scales is left unchanged and a model is used only for the small scales, namely,

$$\left(\frac{d}{dt} + \nu|\mathbf{k}|^2\right)\hat{\mathbf{u}}_{\mathbf{k}}^l = -i\mathbf{k}\hat{p}_{\mathbf{k}}^l + \hat{\mathbf{c}}_{\mathbf{k}} , \qquad |\mathbf{k}| < \bar{N}/2 , \qquad (3.3.49)$$

$$i\mathbf{k}\cdot\hat{\mathbf{u}}_{\mathbf{k}}^l = 0 ,$$

and

$$\left(\frac{d}{dt} + \nu|\mathbf{k}|^2\right)\hat{\mathbf{u}}_{\mathbf{k}}^s = -i\mathbf{k}\hat{p}_{\mathbf{k}}^s + \hat{\mathbf{c}}_{\mathbf{k}} + i\mathbf{k}\cdot\hat{\boldsymbol{\tau}}_{\mathbf{k}}^{MS} , \qquad \bar{N}/2 \leq |\mathbf{k}| < N/2 ,$$

$$i\mathbf{k}\cdot\hat{\mathbf{u}}_{\mathbf{k}}^s = 0 ,$$

$$(3.3.50)$$

where $\boldsymbol{\tau}^{MS} = 2\nu_t^{MS}\underline{\bar{S}}^s$ is given by a formula similar to (1.4.11), except that $\underline{\bar{S}}$ is replaced by the small-scale strain tensor $\underline{\bar{S}}^s$, which in turn is computed from (1.3.3) using only the small-scale velocity component. The large and small scales in (3.3.49)–(3.3.50) are coupled through the nonlinear term $\hat{\mathbf{c}}$.

The interested reader should consult Kang, Chester and Meneveau (2003) and Hughes, Mazzei, Oberai and Wray (2001) for some representative computations using these and other subgrid-scale models for isotropic turbulence. A more complex approach based on multilevel decompositions is described in Dubois, Jauberteau and Temam (1998).

3.3.6 The Taylor–Green Vortex Example: Stability, Accuracy and Aliasing

In this section we demonstrate the accuracy of Fourier spectral methods on a particular, well-studied isotropic flow, the Taylor–Green vortex. The objective is to illustrate the convergence and conservation properties of the various spectral methods discussed earlier in this section. We also present comparisons with finite-order methods (second-order central differences, fourth-order central differences and sixth-order compact differences).

Taylor and Green (1937) used an asymptotic expansion in time to study the time evolution of a three-dimensional, incompressible flow with a special, (triply) periodic initial condition. This asymptotic expansion, however, diverges for $t \gtrsim 3$. Orszag (1974) performed the initial numerical computations for the Taylor–Green vortex. Brachet et al. studied the generalized Taylor–Green vortex with the family of initial conditions (in θ) on the domain $\Omega = (0, 2\pi)^3$:

$$u(\mathbf{x}, 0) = \frac{2}{\sqrt{3}} \sin\left(\theta + \frac{2\pi}{3}\right) \sin x \, \cos y \, \cos z \, ,$$

$$v(\mathbf{x}, 0) = \frac{2}{\sqrt{3}} \sin\left(\theta - \frac{2\pi}{3}\right) \cos x \, \sin y \, \sin z \, , \qquad (3.3.51)$$

$$w(\mathbf{x}, 0) = \frac{2}{\sqrt{3}} \sin(\theta) \cos x \, \cos y \, \sin z \, ,$$

by means of both asymptotic expansions in t for the inviscid flow and numerical computations for the viscous case with $\theta = 0$.

As discussed in CHQZ2, Sect. 1.3, the Brachet et al. computations, which were performed with a special Fourier spectral method on an effective 256^3 grid, de-aliased with the 2/3-rule (see CHQZ2, Sect. 3.4.2), were the first numerical computations of isotropic turbulence to produce an identifiable inertial range. (Because of the symmetries of the generalized Taylor–Green vortex, spectral computations that exploit the symmetries required a factor of 64 times fewer independent variables than the straightforward spectral methods discussed in this section. See Brachet et al. (1983) for the details.) Subsequently, Brachet (1991) performed 864^3 computations at even larger Reynolds numbers.

In two dimensions, the Taylor–Green vortex has the exact solution

$$u(x, y, t) = -e^{-2\nu t} \cos x \, \sin y \, ,$$

$$v(x, y, t) = e^{-2\nu t} \sin x \, \cos y \, , \qquad (3.3.52)$$

$$p(x, y, t) = -\frac{1}{4} e^{-4\nu t} (\cos(2x) + \cos(2y)) \, .$$

This is a useful exact solution for code verification and convergence assessment for Navier–Stokes algorithms in two dimensions; see, e. g., Rabenold (2006) and Sect. 3.7.2.

Brachet et al. reported three-dimensional computations from $t = 0$ to $t = 16$ with Reynolds numbers ranging from Re $= 100$ to Re $= 3000$. Here we use their Re $= 100$ and Re $= 800$ cases to compare spectral methods with finite-difference methods and to compare different types of spectral methods with each other. The time of maximum dissipation, which we use as the point of comparison, is approximately $t = 5$ for the former case and $t = 9$ for the latter case. The finite-difference methods illustrated are second-order central differences (FD2), fourth-order central differences (FD4), and sixth-order compact differences (CP6). The second-order and fourth-order finite-difference methods used here are the standard central-difference methods with 3-point and 5-point explicit stencils, respectively. The sixth-order compact method employs the classical 3-point Padé approximations (see, for example, Collatz (1966) and Lele (1992)); these are

$$u'_{j-1} + 3u'_j + u'_{j+1} = \frac{7}{3\Delta x}(u_{j+1} - u_{j-1}) + \frac{1}{12\Delta x}(u_{j+2} - u_{j-2}) \, , \quad (3.3.53)$$

and

$$\frac{2}{11}u''_{i-1} + u''_i + \frac{2}{11}u''_{i+1} = \frac{12}{11(\Delta x)^2}(u_{i-1} - 2u_i + u_{i+1})$$

$$+ \frac{3}{44(\Delta x)^2}(u_{i-2} - 2u_i + u_{i+2}) \, , \qquad (3.3.54)$$

for the first and second derivatives, respectively, where Δx is the grid spacing, and u'_j and u''_j denote the approximation to the first and second derivatives at $x_j = j\Delta x$. (Of course, when nonperiodic boundary conditions are present, special stencils are needed for points at, and sometimes also adjacent to, the boundary; see Sect. 4.2.)

The spectral methods illustrated all use Fourier series, and include de-aliased computations with the 3/2-rule (Galerkin), along with aliased computations using the rotation form (3.2.3), the skew-symmetric form (3.2.4), the convection form (3.2.1) and the divergence form (3.2.2). (Recall that all forms of the momentum equation are algebraically equivalent for de-aliased Fourier computations, but not for aliased ones.) The time-discretization is performed with the Carpenter–Kennedy fourth-order, low-storage Runge–Kutta method (see Appendix D), with $\Delta t = 0.005$. This time-step suffices to keep the time-discretization error below 10^{-8} for all the examples below.

Figure 3.4 illustrates one velocity and one vorticity component for the initial conditions ($t = 0$) and at the approximate times of maximum energy dissipation rate ($t = 5$ for Re $= 100$ and $t = 9$ for Re $= 800$) of the two cases to follow. The figure emphasizes the much greater small-scale structure present in the higher Reynolds number case.

The integral quantities that we use to evaluate the various methods are the energy

$$E = \frac{1}{2} \int_\Omega \mathbf{u} \cdot \mathbf{u} \,, \tag{3.3.55}$$

the dissipation rate (which happens to be 2ν times the enstrophy, which is the mean-square vorticity)

$$\epsilon = \nu \int_\Omega (\nabla \mathbf{u}) : (\nabla \mathbf{u}) \,, \tag{3.3.56}$$

the (derivative) skewness, for $i, j = 1, 2, 3$,

$$(Sk)_{ij} = \int_\Omega (\partial u_i / \partial x_j)^3 \left/ \left(\int_\Omega (\partial u_i / \partial x_j)^2 \right)^{3/2} \right. , \tag{3.3.57}$$

and the (derivative) flatness, for $i, j = 1, 2, 3$,

$$(Fl)_{ij} = \int_\Omega (\partial u_i / \partial x_j)^4 \left/ \left(\int_\Omega (\partial u_i / \partial x_j)^2 \right)^2 \right. . \tag{3.3.58}$$

(Note that there is no summation over repeated indices in (3.3.57)–(3.3.58).) With the exception of the factor of ν in (3.3.56), these are the lowest-order statistical moments of interest in the isotropic turbulence problem. (The dissipation rate is used here for ready comparison with some of the results in Brachet et al. (1983).) For the Taylor–Green vortex with $\theta = 0$, the (derivative) skewness is diagonal, and $Sk_{11} = Sk_{22}$; hence, its only independent entries are Sk_{11} and Sk_{33}. The (derivative) flatness is symmetric, and $Fl_{11} = Fl_{22}$, leaving it with 5 independent entries.

The Re $= 100$ case is readily resolved, as illustrated in Fig. 3.5 for the time history of the dissipation rate. The 24^3 spectral Galerkin result is graphically indistinguishable from the exact solution, whereas this level of accuracy requires 32^3 for CP6, 48^3 for FD4 and 128^3 for FD2. The spectral collocation computations using the skew-symmetric form are virtually indistinguishable from the Galerkin (de-aliased) computations on the same grid, and are also clearly more accurate than the collocation method based on the rotation form.

A more quantitative measure of the various methods is furnished in Fig. 3.6, which displays the relative errors on the energy, dissipation rate, and the derivative skewness and flatness at $t = 5$, which is close to the

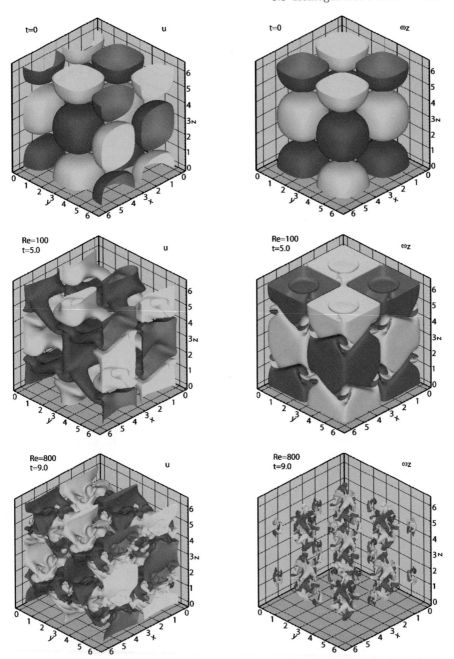

Fig. 3.4. Velocity and vorticity visualization for the Taylor–Green vortex problem. The *left column* illustrates the *u*-component of velocity with contour levels of −0.25 and +0.25. The *right column* illustrates the *z*-component of the vorticity, with contour levels of approximately ±25 percent of the instantaneous peak vorticity. The *darker contours* are the ones for the negative levels

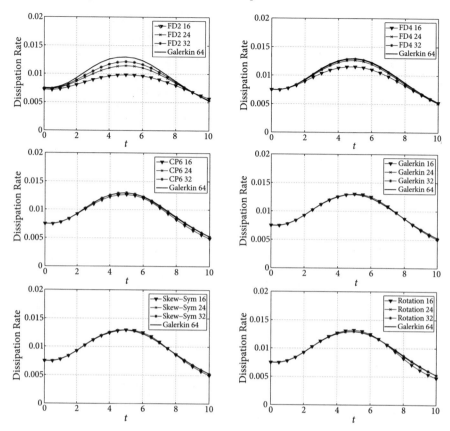

Fig. 3.5. Convergence of the dissipation rate for the Re = 100 Taylor–Green vortex problem for several finite-difference and spectral methods

time of maximum dissipation. The results shown for the skewness and flatness are the root-mean-square (rms) errors for the independent components (two for skewness and five for flatness). (The "exact" values are taken from a Galerkin computation on a 192^3 grid, for which the time discretization errors dominate the spatial discretization errors.) The top two rows compare the three finite-difference methods with the spectral Galerkin method on a log–log plot to emphasize the finite rate of convergence of the lower order methods.

From these two rows we conclude that (1) the spectral method is always at least two orders of magnitude more accurate than the second-order method and one order more accurate than the fourth-order method; the spectral method becomes orders of magnitude more accurate than the sixth-order method at high accuracy levels (errors less than 10^{-3}). The bottom two rows compare various spectral approximations on a semi-log plot to highlight the exponential convergence of the spectral methods. In general, the Galerkin

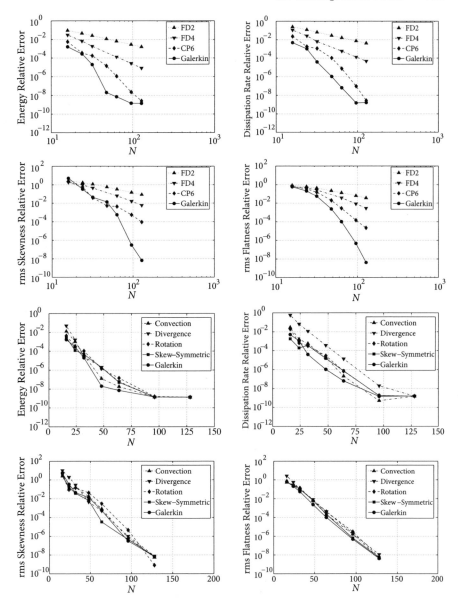

Fig. 3.6. Convergence of the integral quantities at $t = 5$ for the Re = 100 Taylor–Green vortex problem for several finite-difference methods and several spectral methods

and skew-symmetric collocation methods are the most accurate, whereas the divergence and rotation form collocation methods are the least accurate. For this case the physical viscosity is evidently sufficiently large to suppress the temporal instability for the spectral collocation method using the convection and divergence forms for inviscid, incompressible flow (see Sect. 3.1). The Galerkin method is clearly more accurate on the energy and dissipation rate, whereas the collocation method in skew-symmetric form appears to have the edge for the skewness and flatness quantities.

For this case, a spectral (Galerkin or skew-symmetric collocation) method with $N = 64$ achieves an accuracy of 10^{-8} on energy, 10^{-6} on the dissipation rate, 10^{-4} on skewness and 10^{-2} on flatness. As expected, the accuracy diminishes as the order of the turbulence quantity increases. An accuracy of 10^{-3} on skewness requires $N = 64$ for the spectral method, $N = 96$ for CP6, $N = 192$ for FD4 and $N = 1024$ for FD2. The spatial resolution requirements provided in CHQZ2, Sect. 1.3—$N \approx c_1 \sqrt{\text{Re}}$, with $c_1 = 2$, 6 and 24 for spectral, FD4 and FD2, respectively—are remarkably close to these actual results.

The Reynolds number in the second example (Re = 800) requires significantly more resolution—by roughly a factor of 5 in each direction according to the $N \approx c_1 \sqrt{\text{Re}}$ scaling—than the Re = 100 case. This is slightly beyond the Re = 500 threshold at which Brachet et al. (1983) report a qualitative change in the flow. The spherically averaged energy spectrum $E(k,t)$, given by

$$E(k,t) = \frac{1}{2} \sum_{k-\frac{1}{2}<|\mathbf{k}|<k+\frac{1}{2}} |\hat{u}_{\mathbf{k}}(t)|^2 , \qquad (3.3.59)$$

where $\hat{u}_{\mathbf{k}}(t)$ is the $\mathbf{k}$-th Fourier coefficient of the velocity as in (3.3.5), is plotted for the present Re = 800 case at $t = 9$ as well as the previous Re = 100 case at $t = 5$ in Fig. 3.7. For the larger Reynolds number case, there is a small inertial range, for which $E(k,t)$ is close to the expected Kolmogorov $k^{-5/3}$ law. No such range exists for the low Reynolds number case. Moreover, the viscosity is sufficiently small for Re = 800 so that the collocation computations using the convection and divergence forms are temporally unstable, as illustrated in Fig. 3.8. The rotation and skew-symmetric form collocation

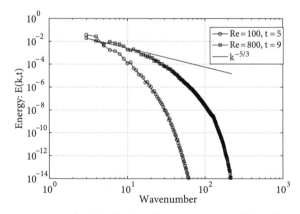

Fig. 3.7. Energy spectra for the Taylor–Green vortex problem: Re = 100 at $t = 5$ and Re = 800 at $t = 9$. The Kolmogoroff law is shown by the *straight line*

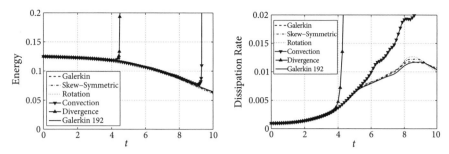

Fig. 3.8. Energy (*left*) and dissipation rate (*right*) history for the Re = 800 Taylor–Green vortex problem for several spectral methods

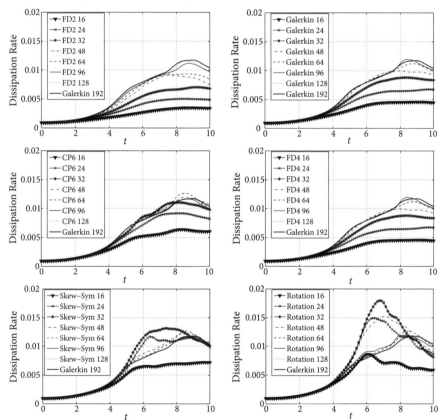

Fig. 3.9. Convergence of the dissipation rate for the Re = 800 Taylor–Green vortex problem for three finite-difference methods and three spectral methods

results are temporally stable, as expected from their inviscid conservation properties.

Figure 3.9 illustrates the time histories of the dissipation rate at various resolutions. Here the coarser grid results are much less accurate than

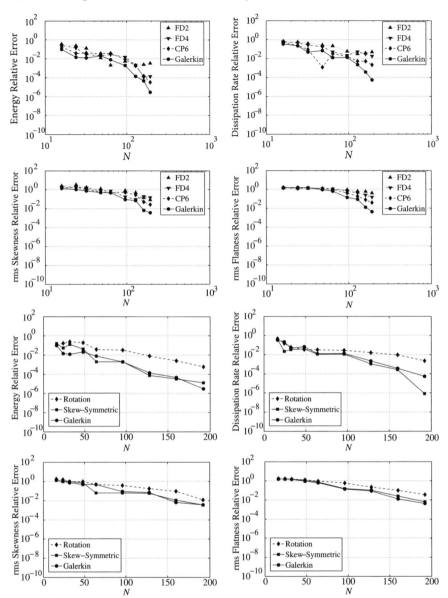

Fig. 3.10. Convergence of the integral quantities at $t = 9$ for the Re $= 800$ Taylor–Green vortex problem for several finite-difference methods and several spectral methods

those displayed in Fig. 3.5 for the Re $= 100$ case. Here the spectral Galerkin and skew-symmetric collocation methods require $N = 128$ to be graphically indistinguishable from the baseline result. The CP6 result is slightly distinguishable from the baseline result and is noticeably superior to the spectral rotation form collocation result. Notice, though, that the FD4, CP6 and spectral Galerkin and skew-symmetric results are all very inaccurate for $N \leq 96$. This behavior is typical of spectral methods; namely, they are no better than low-order methods until the basic flow features are resolved.

More quantitative convergence results are presented in Fig. 3.10, in which errors are computed with respect to a skew-symmetric collocation computation on a 256^3 grid. From the top two rows, which compare the low-order methods with the spectral Galerkin method, we see that there is no clear pattern for $N \leq 96$, but that for larger values of N, the expected asymptotic convergence rates of the methods emerge. The spectral Galerkin method achieves the conventionally acceptable accuracy (for flow physics simulations, the principal domain of classical spectral methods) of 10^{-2} on flatness for $N \approx 160$, whereas even the CP6 method requires $N \approx 256$ for this accuracy. The main conclusions from the bottom two rows are that the rotation form collocation results are markedly inferior, whereas the Galerkin and skew-symmetric collocation results are nearly comparable.

An aliasing effect that is very common in DNS and LES computations is the artificial accumulation of energy at high wavenumbers balanced by a decrease in energy at low wavenumbers. We illustrate this effect here in Fig. 3.11, which shows the one-dimensional spectra in the x-direction of the

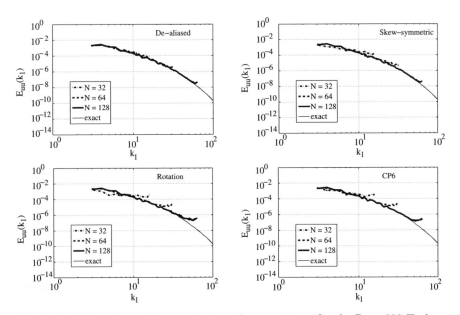

Fig. 3.11. 1D spectra of the u-velocity with respect to x for the $Re = 800$ Taylor–Green vortex problem at $t = 9$. These are shown for a Fourier de-aliased computation (*top left*), for Fourier aliased computations using the skew-symmetric form (*top right*) and the rotation form (*bottom left*), and for a sixth-order compact scheme using the skew-symmetric form (*bottom right*)

u-component of velocity. This quantity is defined by

$$E_{uu}(k_1)(t) = \frac{1}{2\pi}\overline{\left| \int_0^{2\pi} u(x,y,z,t)e^{-ik_1x}\mathrm{d}x \right|^2} , \qquad (3.3.60)$$

where the overbar denotes the average over the y and z-directions. (The "exact" spectra in these figures is taken from the 256^3 collocation computation using the skew-symmetric form.) As can be seen from Fig. 3.9, all of these computations are significantly under-resolved for $N = 32$ and $N = 48$, and the Galerkin and skew-symmetric collocation computations are reasonably well-resolved for the velocity components themselves for $N = 128$. There is but a minor accumulation of energy at high wavenumbers apparent in Fig. 3.11 for the Galerkin method, but such an accumulation is very obvious for the aliased collocation computation with the rotation form and even for the aliased sixth-order compact scheme computation with the skew-symmetric form. On the other hand, the high-wavenumber energy pile-up is only marginally worse for the aliased skew-symmetric collocation computation than for the de-aliased Fourier computation. We refer the reader to Blaisdell, Spryopolous and Qin (1996) for an analytical explanation of the difference in the aliasing effects resulting from the different forms of the convection terms in collocation computations of the Navier–Stokes equations. They showed that aliasing had a much greater impact upon computations using the rotation form than those using the skew-symmetric form.

A long-standing, controversial issue has been whether, in practical calculations, it is necessary to remove the aliasing terms in order to obtain a reliable result. As shown in Sect. 3.7 we do have the theoretical consolation that de-aliased and aliased computations of the Navier–Stokes equations have the same asymptotic convergence rate. This result does not address the problem-dependent issue of what the difference is likely to be for poorly resolved or even marginally resolved computations. These unsteady, Taylor–Green vortex computations are far from definitive, as they are perhaps the simplest possible turbulence computations because there are neither boundary effects nor forcing terms. We do observe that, provided a conservative form (rotation or skew-symmetric) is utilized, the aliased computations exhibit the same asymptotic convergence rate as the de-aliased computations. (See Sect. 3.7.2 for a summary of the numerical analysis for this problem.) However, for poorly resolved flows, the aliased computations in rotation form can be considerably worse than the de-aliased results, whereas aliased computations using the skew-symmetric form are nearly as good as de-aliased computations.

Nevertheless, there is the possibility that in more challenging flows, aliased computations using the skew-symmetric form may lead to misleading results if the flow is only marginally resolved. Kravchenko and Moin (1997) presented an alternative analytical perspective of the effect of aliasing on the various forms of the incompressible Navier–Stokes equations, reaching the same general conclusions as Blaisdell, Spryopoulous and Qin (1996). Their numerical example was for turbulent flow in a channel, using the normal velocity-normal vorticity algorithm discussed in Sect. 3.4.1. They found an instance in which the computed solution incorrectly laminarized for a marginally resolved aliased computation using the skew-symmetric form, but not for a de-aliased computation. We must then advise a practitioner performing an aliased computation to include an assessment of aliasing effects in the solution verification activities.

3.4 Flows with One Inhomogeneous Direction

The inhomogeneous flows that have been fruitfully investigated by three-dimensional spectral solutions of the incompressible Navier–Stokes equations include channel flow, pipe flow, the parallel boundary layer, Taylor–Couette flow, the Rayleigh–Bénard problem, and free shear layers. In each of these applications the boundary conditions are periodic in two directions, and non-periodic in one. In all but the last case, no-slip conditions are required at one or more walls. Problems with two or more nonperiodic directions are more complex and will be addressed in Sect. 3.5.

The channel flow problem illustrates the salient features of all the wall-bounded algorithms cited above, even in a two-dimensional setting, and this is used here to describe the different spectral numerical methods. Then, the unique features of spectral algorithms for the other problems are discussed.

We consider plane channel flow, also called Poiseuille flow (see Fig. 1.2 and the discussions in Sects. 1.5.1 and 2.3), within the framework of the two-dimensional Navier–Stokes equations with constant viscosity ν. The flow is driven by a mean pressure gradient, $\bar{p}_x = -2\nu$, in the x-direction. We write (3.1.1)–(3.1.2) as

$$\frac{\partial \mathbf{u}}{\partial t} + \mathbf{u} \cdot \nabla \mathbf{u} = -\nabla p - \bar{p}_x \mathbf{e}_1 + \nu \Delta \mathbf{u} \quad \text{in } \Omega , \tag{3.4.1}$$

$$\nabla \cdot \mathbf{u} = 0 \quad \text{in } \Omega , \tag{3.4.2}$$

where $\mathbf{u} = (u, v)^T$, with u and v denoting the velocity components in the x and y directions, p denotes the actual pressure minus the contribution from the mean pressure gradient, and $\mathbf{e}_1$ is the unit vector in the x-direction (thus, the true pressure is $p + \bar{p}_x x$). (In the remainder of this discussion on plane Poiseuille flow we shall refer to p as just the pressure.) We adopt the same nondimensionalization as in Sect. 1.5.1 and hereafter use ν in lieu of $1/\mathrm{Re}$. The domain Ω is $(0, L_x) \times (-1, 1)$. Periodic boundary conditions in x and no-slip boundary conditions in y:

$$\mathbf{u} = \mathbf{0} \quad \text{on} \quad y = \pm 1 \tag{3.4.3}$$

are assumed, as are appropriate initial conditions. A Fourier representation in x and a Chebyshev representation in y are customary, but the latter could be equally well replaced by any Jacobi expansion (in particular the Legendre one), although the nonlinear terms would be handled less efficiently for large N_y due to the lack of a fast transform.

Three classes of spectral methods will be discussed here in detail: (i) coupled (or monolithic) methods in which a single system combining the continuity equation with the implicit contributions of the pressure gradient and viscous terms of the momentum equations is inverted at each time-step (see Sect. 3.4.1); (ii) Galerkin methods with velocity trial functions that are divergence-free in addition to complying with the appropriate boundary conditions (see Sect. 3.4.2); and (iii) splitting (or fractional-step) methods that decouple the viscous and nonlinear terms from the pressure gradient and the divergence-free constraint (see Sect. 3.4.3). All of the algorithms described in the next three subsections have been applied to three-dimensional flows. They use a Fourier representation in z. We are focusing on the two-dimensional versions solely to simplify the discussion.

For all the coupled methods and splitting methods that we discuss, the dependent variables have discrete Fourier representations in x of the form

$$\mathbf{u}(x, y, t) = \sum_{k=-N_x/2}^{N_x/2-1} \hat{\mathbf{u}}_k(y, t)e^{2\pi ikx/L_x} , \qquad (3.4.4)$$

$$p(x, y, t) = \sum_{k=-N_x/2}^{N_x/2-1} \hat{p}_k(y, t)e^{2\pi ikx/L_x} , \qquad (3.4.5)$$

and Chebyshev polynomial representations in y of the form:

$$\hat{\mathbf{u}}_k(y, t) = \sum_{m=0}^{N_y} \tilde{\mathbf{u}}_{k,m}(t)T_m(y) , \qquad (3.4.6)$$

$$\hat{p}_k(y, t) = \sum_{m=0}^{N_y} \tilde{p}_{k,m}(t)T_m(y) . \qquad (3.4.7)$$

(This section is an exception to our usual practice of using ^'s for exact coefficients and ~'s for discrete ones.)

For most of the collocation methods that we discuss, the collocation points in x and y are

$$x_j = jL_x/N_x , \qquad j = 0, 1, \ldots, N_x - 1 , \qquad (3.4.8)$$

$$y_j = \cos\frac{\pi j}{N_y} , \qquad j = 0, 1, \ldots, N_y , \qquad (3.4.9)$$

respectively. (The exception is the particular splitting method of Zang and Hussaini (1986) discussed in Sect. 3.4.3.) We will use the same symbols to denote the continuous variables $\mathbf{u}$ and p, and their discrete physical space approximations.

Since the pressure adjusts itself instantaneously to changes in the velocity according to the divergence-free constraint, an implicit treatment of the pressure gradient term is imperative. (In fact, since the pressure acts as an advection term with infinite wave speed, its time-discretization must be A-stable. As seen in Sect. 3.2.2, an explicit time-discretization of the viscous term is possible, although rarely practical for realistic viscosities and grids. The explicit viscous stability limit scales as $1/N_x^2$ in x and $1/N_y^4$ in y (see CHQZ2, Chap. 4), where N_x and N_y are the number of modes in the x and y-directions, respectively. The latter restriction is the most severe and virtually mandates an implicit treatment of the normal diffusion term. The principal algorithmic complexity for these problems arises from the need to invert the operators which represent the implicit treatment of the pressure and normal diffusion terms.

For most of the algorithms discussed in this section, efficient direct solution methods are available. Moreover, for many flows there are symmetry properties that can be exploited to reduce the cost of the computation. For example, in the channel flow transition problem the linear and secondary instability modes often used as initial conditions satisfy spanwise symmetry conditions that can be handled, as, for example, by Gilbert and Kleiser (1990), by using sine or cosine expansions in the spanwise direction in lieu of Fourier expansions. However, especially for the case of turbulent flows, with the use of these symmetries comes the risk of overlooking some important physical mechanisms. So the symmetry properties must be exploited with care.

3.4.1 Coupled Methods

The temporal discretization of choice for this channel flow problem has been explicit for convection and implicit for diffusion (and pressure). The first-order time-discretization of (3.4.1)–(3.4.2) becomes, after a discrete Fourier transform in x, but with y still continuous,

$$\frac{1}{\Delta t}\hat{\mathbf{u}}_k^{n+1} + \nu\hat{k}^2\hat{\mathbf{u}}_k^{n+1} - \nu\frac{\partial^2 \hat{\mathbf{u}}_k^{n+1}}{\partial y^2} + \widehat{\nabla}\hat{p}_k^{n+1} = \frac{1}{\Delta t}\hat{\mathbf{u}}_k^n + \hat{\mathbf{h}}_k^n \,, \qquad (3.4.10)$$

$$\widehat{\nabla}\cdot\hat{\mathbf{u}}_k^{n+1} = 0 \,, \qquad (3.4.11)$$

the boundary conditions reduce to

$$\hat{\mathbf{u}}_k^{n+1}(\pm 1) = \mathbf{0} \,, \qquad (3.4.12)$$

where $\hat{\mathbf{u}}_k^{n+1} = (\hat{u}_k^{n+1}, \hat{v}_k^{n+1})^T$, $\hat{k} = 2\pi k/L_x$, and

$$\mathbf{h} = -\mathbf{u}\cdot\nabla\mathbf{u} - \bar{p}_x\mathbf{e}_1 \,. \qquad (3.4.13)$$

The superscript represents the time-level. The x-discrete, y-continuous operators are

$$\widehat{\nabla}\hat{p}_k^{n+1} = \left(i\hat{k}\hat{p}_k^{n+1}, \frac{\partial\hat{p}_k^{n+1}}{\partial y}\right), \tag{3.4.14}$$

$$\widehat{\nabla}\cdot\hat{\mathbf{u}}_k^{n+1} = i\hat{k}\hat{u}_k^{n+1} + \frac{\partial\hat{v}_k^{n+1}}{\partial y}. \tag{3.4.15}$$

After a Chebyshev spectral discretization method in the y-direction has been applied to the system (3.4.10)–(3.4.12) (for each wave number k), we are left with an algebraic system of the form

$$L\hat{\mathbf{w}} = \hat{\mathbf{f}}, \tag{3.4.16}$$

where $\hat{\mathbf{w}}$ is a suitably ordered vector collecting the unknowns for the two velocity components and the pressure, whereas $\hat{\mathbf{f}}$ represents the known right-hand side. The matrix L is a full $M \times M$ matrix where $M \cong 3N_y$. Barring some special structure of the matrix L, a direct solution of (3.4.16) by Gauss elimination methods requires $O(M^2)$ storage and $O(M^3)$ arithmetic operations. An iterative solution, on the other hand, requires only $O(M)$ storage and $O(M\log_2 M)$ operations per iteration.

Kleiser–Schumann Algorithm. Equations (3.4.10)–(3.4.12) can be expressed, after dropping the subscript k for simplicity, and changing the sign on the momentum equation to conform to the notation of Kleiser and Schumann, as

$$\nu\hat{\mathbf{u}}'' - \lambda\hat{\mathbf{u}} - \widehat{\nabla}\hat{p} = -\hat{\mathbf{r}}, \tag{3.4.17}$$

$$\widehat{\nabla}\cdot\hat{\mathbf{u}} = 0, \tag{3.4.18}$$

$$\hat{\mathbf{u}}(\pm 1) = \mathbf{0}, \tag{3.4.19}$$

where $\lambda = (1/\Delta t) + \nu\hat{k}^2$, $\hat{\mathbf{r}} = (1/\Delta t)\hat{\mathbf{u}}^n + \hat{\mathbf{h}}^n$, primes denote derivatives with respect to y, and the superscripts denoting time-level on the left-hand side are omitted for convenience. Moin and Kim (1980) and Kleiser and Schumann (1980) have devised particular methods for solving the equations resulting from a Chebyshev tau discretization for the y-derivative terms in (3.4.17) and (3.4.18). The resulting system of equations is nearly block-tridiagonal. Moin and Kim used a direct inversion method. We shall focus here on the Kleiser–Schumann algorithm, both because it has been used much more extensively than the Moin–Kim algorithm and also because there are some theoretical results available for it (see Sect. 3.7).

Let us first consider the continuous (in y) version of the problem (3.4.17)–(3.4.19). Taking the divergence of (3.4.17) yields the equation for pressure

$$\hat{p}'' - \hat{k}^2\hat{p} = \widehat{\nabla}\cdot\hat{\mathbf{r}}, \tag{3.4.20}$$

and the boundary condition is

$$\widehat{\nabla} \cdot \hat{\mathbf{u}}(\pm 1) = 0 , \quad \text{i. e., } \hat{v}'(\pm 1) = 0 . \tag{3.4.21}$$

The equation for $\hat{v}$ is

$$\nu \hat{v}'' - \lambda \hat{v} - \hat{p}' = -\hat{r}_2 , \qquad \hat{v}(\pm 1) = 0 , \tag{3.4.22}$$

and the equation for $\hat{u}$ reads

$$\nu \hat{u}'' - \lambda \hat{u} - i\hat{k}\hat{p} = -\hat{r}_1 , \qquad \hat{u}(\pm 1) = 0 . \tag{3.4.23}$$

Equations (3.4.20)–(3.4.22) form a complete set for $\hat{p}$ and $\hat{v}$. Setting $\mathcal{D} = d/dy$, let

$$\mathcal{L} = \begin{pmatrix} \mathcal{D}^2 - \hat{k}^2 & 0 \\ -\mathcal{D} & \nu \mathcal{D}^2 - \lambda \end{pmatrix} , \qquad \hat{f} = \begin{pmatrix} \widehat{\nabla} \cdot \hat{\mathbf{r}} \\ -\hat{r}_2 \end{pmatrix} , \tag{3.4.24}$$

and write these equations as

$$\mathcal{L} \begin{pmatrix} \hat{p} \\ \hat{v} \end{pmatrix} = \hat{f} , \tag{3.4.25}$$
$$\hat{v}(\pm 1) = \hat{v}'(\pm 1) = 0 .$$

Once $\hat{p}$ is available, $\hat{u}$ can be obtained either from the u-momentum equation (3.4.23) or from the continuity equation (3.4.11).

Kleiser and Schumann called (3.4.25) the *A-problem*. It consists of two coupled Helmholtz equations. Kleiser and Schumann obtained its solution by considering the related *B-problem*:

$$\mathcal{L} \begin{pmatrix} \hat{p} \\ \hat{v} \end{pmatrix} = \hat{f} , \tag{3.4.26}$$
$$\hat{p}(\pm 1) = \hat{p}_{b\pm} \quad \text{and} \quad \hat{v}(\pm 1) = 0 .$$

The pressure $\hat{p}_{b\pm}$ at the walls is unknown a priori, but it is required to be consistent with the conditions $\hat{v}'(\pm 1) = 0$. However, the B-problem consists of two uncoupled scalar Helmholtz equations (for each wavenumber k) that can be solved by the techniques discussed in CHQZ2, Sect. 4.1. In the case of the tau approximation the solution cost is $O(N_y)$ per wavenumber.

The solution to the A-problem can be expressed as a linear combination of the solution to three versions of the B-problem: (1) the inhomogeneous differential equation with homogeneous boundary conditions:

$$\mathcal{L} \begin{pmatrix} \hat{p}_p \\ \hat{v}_p \end{pmatrix} = \hat{f} , \tag{3.4.27}$$
$$\hat{p}_p(\pm 1) = \hat{v}_p(\pm 1) = 0 ,$$

(2) the homogeneous differential equation with unit pressure at $y = +1$:

$$\mathcal{L}\begin{pmatrix} \hat{p}_+ \\ \hat{v}_+ \end{pmatrix} = 0 \,,$$

$$\hat{p}_+(+1) = 1 \,, \quad \hat{p}_+(-1) = \hat{v}_+(\pm 1) = 0 \,, \tag{3.4.28}$$

and (3) the homogeneous differential equation with unit pressure at $y = -1$:

$$\mathcal{L}\begin{pmatrix} \hat{p}_- \\ \hat{v}_- \end{pmatrix} = 0 \,,$$

$$\hat{p}_-(-1) = 1 \,, \quad \hat{p}_-(+1) = \hat{v}_-(\pm 1) = 0 \,. \tag{3.4.29}$$

In particular, the A-problem can be expanded as

$$\begin{pmatrix} \hat{p} \\ \hat{v} \end{pmatrix} = \begin{pmatrix} \hat{p}_p \\ \hat{v}_p \end{pmatrix} + \delta_+ \begin{pmatrix} \hat{p}_+ \\ \hat{v}_+ \end{pmatrix} + \delta_- \begin{pmatrix} \hat{p}_- \\ \hat{v}_- \end{pmatrix} \,. \tag{3.4.30}$$

The boundary conditions of the A-problem require

$$\begin{pmatrix} \hat{v}'_+(+1) & \hat{v}'_-(+1) \\ \hat{v}'_+(-1) & \hat{v}'_-(-1) \end{pmatrix} \begin{pmatrix} \delta_+ \\ \delta_- \end{pmatrix} = - \begin{pmatrix} \hat{v}'_p(+1) \\ \hat{v}'_p(-1) \end{pmatrix} \,. \tag{3.4.31}$$

This determines δ_+ and δ_-, and hence the correct pressure boundary condition

$$\hat{p}(\pm 1) = \hat{p}_{b\pm} = \delta_\pm \,. \tag{3.4.32}$$

The 2×2 coefficient matrix in (3.4.31) is called the *influence matrix*, and is calculated once initially for each k and stored. In summary, for each time-step (1) the B-problem (3.4.27) is solved with homogeneous pressure boundary conditions, (2) the correct pressure boundary conditions $\delta_\pm$ are defined from (3.4.31)–(3.4.32), and then (3) the solution to the A-problem is obtained from (3.4.30). Having now the final $\hat{p}$ and $\hat{v}$, then $\hat{u}$ can be obtained from (3.4.23). (Alternatively, one could use the continuity equation (3.4.19) to determine $\hat{u}$).

One must keep in mind that the derivation of (3.4.20) from (3.4.17)–(3.4.18) was based on the properties of the continuous differentiation operators. Had this derivation instead started from the fully discretized version of (3.4.17)–(3.4.19), there would have been additional terms in the right-hand side of (3.4.20). As a result, the discrete solution to (3.4.20)–(3.4.23) does not yield a vanishing discrete divergence. On the other hand if the continuity equation is used to solve for $\hat{u}$, then the discrete u-momentum equation (3.4.23) is not satisfied. This applies for any discretization method in y and not just to the tau approximation used by Kleiser and Schumann. Kleiser and Schumann actually included an additional step in their algorithm which accounted for the discretization effects and thereby produced a solution that fully satisfied the tau approximation to (3.4.10)–(3.4.12). Section E.2 of

Appendix E furnishes the details of the *tau correction* step of the Kleiser–Schumann algorithm.

An important question is whether one need bother with the tau correction. Kleiser and Schumann (1980) estimate the magnitude of the error that would be made in the interior divergence as

$$\frac{N_y}{\nu \Delta t} \tilde{u}_{k,m} , \quad m = N_y - 1, \, N_y . \tag{3.4.33}$$

Since $\tilde{u}_{N_y-1}$ and $\tilde{u}_{N_y}$ decrease rapidly with N_y, this error should be small. A rigorous proof of the convergence of this method (for the steady problem) is available and is summarized in Sect. 3.7.5. The theory states that spectral accuracy is achieved for both the corrected and uncorrected versions. However, unpublished results of Kleiser indicate that lower stability limits on the time-step arose in some of his time-dependent numerical experiments when the tau correction was ignored.

The influence-matrix approach has been used in many spectral algorithms, including ones with multiple inhomogeneous directions. Tuckerman (1989a) has provided an elegant, comprehensive formulation in algebraic terms of the influence-matrix method for solution of Stokes problems. Her treatment covers both collocation and tau methods, discretization corrections (such as the tau correction discussed above), various coordinate systems, and multiple inhomogeneous directions.

We shall see in Sects. 7.3.2 and 7.4 that some solution schemes for the Stokes problem on complex domains using spectral methods use what amounts to an influence-matrix approach. In these applications it is better known as the Schur complement method.

This algorithm was later used by Gilbert and Kleiser (1990) for the first simulation of the complete transition to turbulence process in a wall-bounded flow using a 128^3 grid. Figure 3.12 illustrates the evolution of one of the principal diagnostics of a transitional flow—the wall-normal shear of the streamwise velocity $\partial u/\partial y$. The ordinate in the top part of the figure is the Reynolds number based on the wall shear velocity; it is given by $\text{Re}_\tau = \sqrt{\frac{1}{\nu} \frac{\partial \bar{u}}{\partial y}} \, h$, where h is the channel half-width and $\bar{u}(y,t)$ is the average over x and z of the streamwise velocity. The bottom part of the figure illustrates the evolution of the vertical shear at the spanwise station containing the peak shear. These detailed results compared very favorably with the vibrating ribbon experiments of Nishioka, Asai and Iida (1980). The $t = 136$ frame was already computed by Kleiser and Schumann (1984) at lower resolution.

Normal Velocity–Normal Vorticity Algorithms. An alternative coupled approach to plane channel flow uses equations for the wall-normal velocity and the wall-normal vorticity in lieu of equations for the primitive variables. Kim, Moin and Moser (1987) developed a channel flow algorithm in terms of these vorticity–velocity variables, as originally suggested by Orszag and Patera (1980).

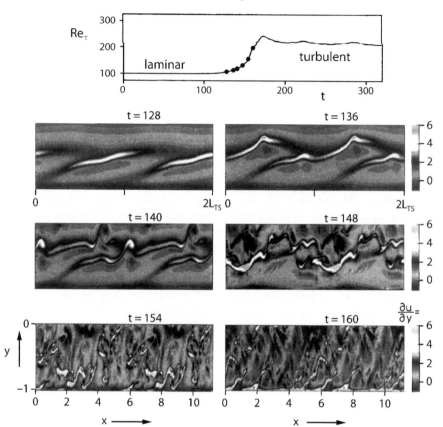

Fig. 3.12. DNS of transition to turbulence in plane channel flow by Gilbert and Kleiser (1990). The *top figure* illustrates the evolution in time of the Reynolds number based on wall friction velocity. The *remaining frames* illustrate the shear, $\partial u/\partial y$, in the bottom half of the channel in a two-dimensional slice at the spanwise (z) location containing the maximum shear [Reprinted with permission from N. Gilbert, L. Kleiser (1990); © 1990, Taylor and Francis Group]

Application of the Laplacian to (3.4.1) yields for the full three-dimensional equations

$$\frac{\partial}{\partial t}\Delta\mathbf{u} = -\nabla\Delta p + \Delta\mathbf{h} + \nu\Delta^2\mathbf{u} , \qquad (3.4.34)$$

where $\mathbf{h}$ is again given by (3.4.13). Taking the divergence of (3.4.1) and using the equation of continuity, we obtain

$$\Delta p = \nabla\cdot\mathbf{h} . \qquad (3.4.35)$$

Substituting this in (3.4.34) yields

$$\frac{\partial}{\partial t}\Delta\mathbf{u} = -\nabla(\nabla\cdot\mathbf{h}) + \Delta\mathbf{h} + \nu\Delta^2\mathbf{u} , \qquad (3.4.36)$$

while taking the curl of (3.4.1) produces

$$\frac{\partial \boldsymbol{\omega}}{\partial t} = -\nabla \times \mathbf{h} + \nu \Delta \boldsymbol{\omega} . \tag{3.4.37}$$

Using (3.4.36) the equations for the wall-normal component of velocity v can be written as the fourth-order system

$$\frac{\partial \phi}{\partial t} = h_v + \nu \Delta \phi ,$$
$$\Delta v = \phi , \tag{3.4.38}$$
$$v = \frac{\partial v}{\partial y} = 0 \qquad \text{at} \qquad y = \pm 1 .$$

Likewise the equations for the wall-normal vorticity ω_2 are

$$\frac{\partial \omega_2}{\partial t} = h_\omega + \nu \Delta \omega_2 , \tag{3.4.39}$$
$$\omega_2 = 0 \qquad \text{at} \quad y = \pm 1 .$$

In the above equations,

$$h_v = \left(\frac{\partial^2}{\partial x^2} + \frac{\partial^2}{\partial z^2} \right) h_2 - \frac{\partial}{\partial y} \left(\frac{\partial h_1}{\partial x} + \frac{\partial h_3}{\partial z} \right) \tag{3.4.40}$$

and

$$h_\omega = \frac{\partial h_1}{\partial z} - \frac{\partial h_3}{\partial x} . \tag{3.4.41}$$

Once the equations for v and ω_2 are solved, the other components of velocity can be obtained from the equation of continuity along with the expression for ω_2:

$$\frac{\partial u}{\partial x} + \frac{\partial w}{\partial z} = -\frac{\partial v}{\partial y} ,$$
$$\frac{\partial u}{\partial z} - \frac{\partial w}{\partial x} = -\omega_2 , \tag{3.4.42}$$
$$u = w = 0 \qquad \text{at} \quad y = \pm 1 .$$

In the two-dimensional case (3.4.39) is not needed, and (3.4.42) reduces to

$$\frac{\partial u}{\partial x} = -\frac{\partial v}{\partial y} , \tag{3.4.43}$$
$$u = 0 \qquad \text{at} \quad y = \pm 1 .$$

After a Fourier transform in x and application of backward Euler for the viscous terms and forward Euler for the remaining terms, the two-dimensional version of (3.4.38) can be written as

$$\nu\hat{\phi}'' - \lambda\hat{\phi} = -\hat{r} , \tag{3.4.44}$$

$$\nu\hat{v}'' - k^2\hat{v} = \hat{\phi} , \tag{3.4.45}$$

$$\hat{v}(\pm 1) = \hat{v}'(\pm 1) = 0 , \tag{3.4.46}$$

where $\lambda = (1/\Delta t) + \nu\hat{k}^2$ as before, but now $\hat{r} = (1/\Delta t)\hat{\phi}^n + \hat{h}_v^n$. These equations can be written as the system

$$\mathcal{L}\begin{pmatrix} \hat{v} \\ \hat{\phi} \end{pmatrix} = \hat{f} , \tag{3.4.47}$$

$$\hat{v}(\pm 1) = \hat{v}'(\pm 1) = 0 ,$$

where

$$\mathcal{L} = \begin{pmatrix} \mathcal{D}^2 - \hat{k}^2 & -1 \\ 0 & \nu\mathcal{D}^2 - \lambda \end{pmatrix} , \qquad \hat{f} = \begin{pmatrix} 0 \\ -\hat{r} \end{pmatrix} , \tag{3.4.48}$$

again with $\mathcal{D} = d/dy$.

This system is very similar to the Kleiser–Schumann A-problem (3.4.25) and can be solved by the same type of influence-matrix method. The Kim, Moin and Moser algorithm used a Chebyshev tau approximation in y. They do not need to use a *tau-correction*-like step to ensure that the computed solution satisfies both the incompressibility constraint and the momentum equation because the same degree of approximation is used for all variables in all stages of the algorithm.

The method developed by Lundbladh, Henningson and Johansson (1992) is nearly identical except that they use the integral Chebyshev tau method (see (8.5.2) and CHQZ2, Sect. 4.1.2) of solving the integrated forms of the differential equations, and they exploit the channel flow symmetry properties in y to reduce the influence-matrix process to a particular solution and just one homogeneous solution. They apply an *integration correction* to ensure that the incompressibility constraint is satisfied exactly by the discrete solution. They note that failure to employ this correction can lead to an instability for low spatial resolutions as the time step is decreased. However, they report that in practice failure to apply this correction does not lead to observable difficulties.

Dubois, Jauberteau and Temam (1999, Chap. 3) take yet another approach to the spectral solution of the vorticity–velocity formulation for channel flow. They utilize the Babuška–Shen Legendre basis (see (8.5.2) and CHQZ2, Sects. 2.3.3 and 4.1.3) for their Legendre Galerkin discretization of the implicit equations. This yields an efficient direct solution to (3.4.44)–(3.4.46) with less effect of round-off errors than for the tau method. Dubois and Maranville (1994) provide a mathematical analysis of a set of PDEs that are closely related to those above in (3.4.38)–(3.4.42).

The derivation of the normal velocity-normal vorticity method, based on the application of the Laplacian to the momentum equation makes the mathematical context rather unusual. Its rigorous analysis is challenging. This approach appears to be quite efficient in practice and exploits the particular structure of the equations and the geometry. This has emerged as perhaps the most widely used of the various algorithms for plane channel flow, and many significant physical results have been obtained, as illustrated for the adaptation to free shear-layer problems in Fig. 3.16 in Sect. 3.4.5, and for the adaptation to boundary-layer flow in Fig. 3.26 in Sect. 3.6.1.

3.4.2 Galerkin Methods Using Divergence-Free Bases

A class of spectral Galerkin methods for wall-bounded flows based on Jacobi–Fourier polynomials was proposed by Leonard. He suggested the use of velocity trial functions that are divergence-free and also satisfy the viscous boundary conditions. This was first applied to pipe flow by Leonard and Wray (1982). Moser, Moin and Leonard (1983) developed the method for both plane and curved channels. In the spirit of this section we shall present a two-dimensional version of their plane channel algorithm.

First, the velocity field is expanded as

$$\mathbf{u}(x, y, t) = \sum_{j=0}^{N_y} \sum_{k=-N_x/2}^{N_x/2-1} \alpha_{jk} \hat{\mathbf{u}}_j(k, y) e^{2\pi i k x / L_x} . \qquad (3.4.49)$$

The trial functions are vector functions satisfying the divergence-free constraint and the no-slip boundary conditions:

$$\nabla \cdot \left[\hat{\mathbf{u}}_j(k, y) e^{2\pi i k x / L_x} \right] = 0 , \qquad (3.4.50)$$

$$\hat{\mathbf{u}}_j(k, y) = \mathbf{0} \quad \text{at } y = \pm 1 . \qquad (3.4.51)$$

The test functions $\boldsymbol{\Phi}_{jk}$ are chosen as

$$\boldsymbol{\Phi}_{jk}(x, y) = \hat{\boldsymbol{\xi}}_j(k, y) e^{2\pi i k x / L_x} \qquad (3.4.52)$$

to satisfy the divergence-free constraint and the no-slip condition at the boundaries:

$$\nabla \cdot \boldsymbol{\Phi}_{jk} = 0 , \qquad (3.4.53)$$

$$\hat{\boldsymbol{\xi}}_j(k, y) = \mathbf{0} \quad \text{at } y = \pm 1 . \qquad (3.4.54)$$

We define an inner product of L^2-type as

$$(\boldsymbol{\Phi}, \boldsymbol{\Psi}) = \int_0^{L_x} dx \int_{-1}^1 \boldsymbol{\Phi}^H \boldsymbol{\Psi} \, dy . \qquad (3.4.55)$$

Substituting (3.4.49) into (3.4.1) with constant viscosity, and taking the inner product with the test functions (3.4.52), the following set of equations are obtained:

$$\left(\frac{\partial \mathbf{u}}{\partial t}, \boldsymbol{\Phi}_{jk}\right) = -\nu(\Delta \mathbf{u}, \boldsymbol{\Phi}_{jk}) - (\mathbf{u} \cdot \nabla \mathbf{u}, \boldsymbol{\Phi}_{jk}) . \tag{3.4.56}$$

Notice that $(\nabla p, \boldsymbol{\Phi}_{jk}) = -(p, \nabla \cdot \boldsymbol{\Phi}_{jk}) = 0$, and thus, the pressure is effectively eliminated from the problem. This relationship requires only that $\boldsymbol{\Phi}_{jk} \cdot \mathbf{n}$ vanish at $y = \pm 1$, and is guaranteed by (3.4.54). In their explanation of this method, Moser et al. state that they require only this "inviscid" boundary condition on their test functions. However, the functions they actually do choose (see (3.4.59)–(3.4.60)) satisfy (3.4.54). From a mathematical point of view, the test functions ought to satisfy the full no-slip conditions (Pasquarelli, Quarteroni and Sacchi-Landriani (1987)). Notice that in this discretization the divergence operator is indeed the adjoint of the (negative) gradient operator.

The resulting equations are uncoupled in k. Each set can be written in the compact form

$$A \frac{d\boldsymbol{\alpha}}{dt} = \nu B \boldsymbol{\alpha} + \mathbf{F} , \tag{3.4.57}$$

where A and B are $(N_y + 1) \times (N_y + 1)$ matrices with elements

$$A_{ij} = (\hat{\mathbf{u}}_i, \hat{\boldsymbol{\xi}}_j) , \qquad B_{ij} = \left(\frac{d^2 \hat{\mathbf{u}}_i}{dy^2} - \left(\frac{2\pi k}{L_x}\right)^2 \hat{\mathbf{u}}_i, \hat{\boldsymbol{\xi}}_j\right) , \tag{3.4.58}$$

and $\mathbf{F}$ represents a similar contribution from the nonlinear term.

The freedom in the choice of the vectors $\hat{\mathbf{u}}_j$ and $\hat{\boldsymbol{\xi}}_j$ is exercised in favor of those that yield matrices A and B with small bandwidths. A convenient choice is

$$\hat{\mathbf{u}}_j = \begin{pmatrix} i f_j' \\ \frac{2\pi k}{L_x} f_j \end{pmatrix} , \qquad f_j(\pm 1) = f_j'(\pm 1) = 0 ,$$

and

$$\hat{\boldsymbol{\xi}}_j = \begin{pmatrix} i g_j' \\ \frac{2\pi k}{L_x} g_j \end{pmatrix} , \qquad g_j(\pm 1) = 0 .$$

(The case $k = 0$ is treated separately.) Equation (3.4.57) is a set of ordinary differential equations that can be solved by any standard explicit or implicit numerical scheme. Note that explicit schemes for the viscous term have no computational advantage unless A is much sparser than B.

A simple choice for the quasi-orthogonal functions f_j and g_j is:

$$f_j(y) = (1 - y^2)^2 T_j(y) , \tag{3.4.59}$$

$$g_j(y) = \left(\frac{T_{j+2}(y)}{j(j+1)} - \frac{2T_j(y)}{(j+1)(j-1)} + \frac{T_{j-2}(y)}{j(j-1)}\right) \Big/ 4(1 - y^2)^{1/2} . \tag{3.4.60}$$

Thus, the trial functions are polynomials, whereas the test functions are polynomials divided by $\sqrt{1-y^2}$. Notice that $f_j'(\pm 1) = g_j'(\pm 1) = 0$. Although the trial and test functions satisfy the same conditions, they are chosen from different spaces of functions. Hence, this method is properly referred to as a Petrov-Galerkin method. The factor $(1-y^2)^{-1/2}$ is included in the test functions, so that the inner products in (3.4.56) involve integrals of Chebyshev polynomials, low degree polynomials, and the Chebyshev weight. As a result, the test and trial functions are quasi-orthogonal—the inner products in (3.4.56) are nonzero only for a small separation between the order of the test and trial functions.

This Galerkin method may be interpreted in another way. The test functions $\boldsymbol{\varPhi}_{jk}$ are now taken to be given by (3.4.52), (3.4.59) and (3.4.60), *without* the $(1-y^2)^{-1/2}$ factor in the last equation, and the inner product (3.4.55) is replaced by one that includes the Chebyshev weight. In the new inner product, $(\nabla p, \boldsymbol{\varPhi}_{jk})$ still vanishes, because the integration-by-parts operation yields

$$- \int_0^{L_x} \mathrm{d}x \int_{-1}^1 p \, \nabla \cdot (\boldsymbol{\varPhi}_{jk} w) \, \mathrm{d}y \; ,$$

where $w(y) = (1-y^2)^{-1/2}$, and the divergence term is identically zero. From this point of view both the trial and test functions are polynomials which satisfy no-slip conditions. However, the trial functions are themselves divergence-free, whereas the test functions are not.

3.4.3 Splitting Methods

In the decades since the original Chorin (1968) and Temam (1969) papers on splitting methods for the incompressible Navier–Stokes equations, there have been many refinements of the splitting technique aimed at improving the temporal accuracy. Until the paper by Perot (1993), splitting formulations were usually made in the context of the differential equations. Perot proposed instead an algebraic perspective that has proven advantageous; much of the subsequent work has shifted to the algebraic factorization approach. A more complete survey of this topic is provided in Chap. 7.

In keeping with our use in the present chapter of low-order time discretizations to simplify the presentation of the spectral spatial discretizations, we consider splitting methods of differential type for the simple channel flow problem that is the focus of this section. A variety of splitting methods have been employed in plane Poiseuille flow simulations with spectral methods, starting with Orszag and Kells (1978) and Orszag and Patera (1983). Here we will describe the Zang and Hussaini (1986) version, not just for parochial reasons, but also to illustrate a staggered grid in the wall-normal direction.

Chebyshev Staggered Grid. The Chebyshev staggered grid for channel flow was introduced by Malik, Zang and Hussaini (1985) and is illustrated in

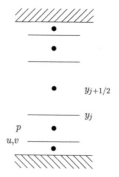

Fig. 3.13. The staggered grid for the normal (y)-direction in channel flow

Fig. 3.13. The collocation points in the y-direction are

$$y_j = \cos \frac{\pi j}{N_y} , \qquad j = 0, 1, \ldots, N_y , \qquad (3.4.61)$$

$$y_{j+1/2} = \cos \frac{\pi \left(j + \frac{1}{2}\right)}{N_y} , \qquad j = 0, 1, \ldots, N_y - 1 , \qquad (3.4.62)$$

which are the Chebyshev Gauss–Lobatto points for the velocity and the Chebyshev Gauss points for the pressure, respectively.

The discrete Fourier representation in x is the same as given in (3.4.4), the discrete representation in y for the velocity is the same as (3.4.6), but the pressure representation in y is given here by

$$\hat{p}_k(y, t) = \sum_{m=0}^{N_y - 1} \tilde{p}_{k,m}(t) T_m(y) . \qquad (3.4.63)$$

The momentum equation is enforced at the Gauss–Lobatto points (3.4.61), whereas the continuity equations is enforced at the Gauss points (3.4.62). Note that the maximum order of the trial functions for the pressure is one less than that for the velocity (see Sect. 3.7). If the summation (3.4.63) had included the $m = N_y$ term, and the same collocation points were used for the pressure and the continuity equation as for the velocity and momentum equation, then the linear system for the discrete dependent variables would have been underdetermined. In particular, the pressure mode $\tilde{p}_{0,N_y}(t) T_{N_y}(y)$ has no effect upon the velocity. The reason for this is that the pressure affects the velocity only through the momentum equation. At the interior Gauss–Lobatto points the gradient of this pressure mode vanishes. For this reason it is called a spurious (or parasitic) mode. The mode $\tilde{P}_{0,0}(t)$ also has no effect upon the velocity. Since it represents the mean value of the pressure and has a counterpart in the continuous formulation of the problem, it is not usually called spurious. We observe that the coupled methods described in the previous subsection do not suffer from the presence of spurious pressure modes.

Indeed, their pressure is determined as the solution of a Poisson equation, obtained by taking the divergence of the momentum equation, supplemented by appropriate boundary conditions; this provides a natural obstruction to the onset of spurious modes. The issue of spurious pressure modes is addressed again in Sect. 3.5 for problems with more than one nonperiodic direction and also in Sect. 3.7 from the theoretical point of view.

Chebyshev interpolation is the natural process for transferring variables between the grids of (3.4.61) and (3.4.62). Let a_j, for $j = 0, 1, \ldots, N_y$, denote the values of a function $a(y)$ at the points (3.4.61). The Chebyshev coefficients are given by the usual quadrature rule

$$
\begin{aligned}
\tilde{a}_m &= \frac{2}{N_y \bar{c}_m} \sum_{j=0}^{N_y} \bar{c}_j^{-1} a_j T_m(y_j) \\
&= \frac{2}{N_y \bar{c}_m} \sum_{j=0}^{N_y} \bar{c}_j^{-1} a_j \cos \frac{m \pi j}{N_y}, \qquad m = 0, 1, \ldots, N_y .
\end{aligned}
\tag{3.4.64}
$$

The interpolated values of a are

$$
\begin{aligned}
a_{j+1/2} &= \sum_{m=0}^{N_y-1} \tilde{a}_m T_m \left(y_{j+1/2} \right) \\
&= \sum_{m=0}^{N_y-1} \tilde{a}_m \cos \frac{(j+1/2)\pi m}{N_y}, \qquad j = 0, 1, \ldots, N_y - 1 .
\end{aligned}
\tag{3.4.65}
$$

(Note that $T_{N_y}(y_{j+1/2}) = 0$ for $j = 0, 1, \ldots, N_y - 1$.) The fast Fourier transform may be used to evaluate both sums (3.4.64) and (3.4.65). The less familiar sum in (3.4.65) may be handled by the technique presented in Appendix B.

Zang–Hussaini Algorithm. The splitting scheme used by Zang and Hussaini (1986) consists of a velocity step

$$
\begin{aligned}
\frac{1}{\Delta t}(\mathbf{u}^{n+1/2} - \mathbf{u}^n) - \nu \Delta \mathbf{u}^{n+1/2} &= \mathbf{h}^n & &\text{in } \Omega , \\
\mathbf{u}^{n+1/2} &= \mathbf{g}^{n+1/2} & &\text{at } y = \pm 1 ,
\end{aligned}
\tag{3.4.66}
$$

followed by a pressure step

$$
\begin{aligned}
\frac{1}{\Delta t}(\mathbf{u}^{n+1} - \mathbf{u}^{n+1/2}) + \nabla p^{n+1} &= 0 & &\text{in } \Omega , \\
\nabla \cdot \mathbf{u}^{n+1} &= 0 & &\text{in } \Omega , \\
v^{n+1} &= 0 & &\text{at } y = \pm 1 ,
\end{aligned}
\tag{3.4.67}
$$

where, as usual, we have assumed an Euler time-discretization for simplicity. The velocity boundary condition on the first fractional step is denoted by $\mathbf{g}$, the choice of which is discussed below. The pressure step may be viewed as a projection of the velocity field onto a divergence-free space. Note that in the velocity step all of the velocity components are specified at the boundary, whereas in the pressure step only the normal velocity component is specified on the boundary, and the differential equation itself is used for the tangential velocity component there.

The boundary conditions for each step are determined by the condition that the individual step be well-posed. At the end of a full step the flow will be divergence-free, but there will be a slip velocity at the boundary. If the boundary conditions on the intermediate velocity step are

$$\mathbf{g}^{n+1/2} = \mathbf{0} , \tag{3.4.68}$$

then heuristic analyses, e. g., Orszag, Israeli and Deville (1986), and numerical experience indicates that the slip velocity at the end of the full step will be $O(\Delta t)$. This can be reduced by resorting to intermediate boundary conditions of the type suggested by Fortin, Peyret and Temam (1971):

$$g_x^{n+1/2} = \Delta t \frac{\partial p^n}{\partial x}, \quad g_y^{n+1/2} = 0 . \tag{3.4.69}$$

This reduces the final slip velocity to $O(\Delta t^2)$. Higher-order corrections are also possible, for example,

$$g_x^{n+1/2} = \Delta t \frac{\partial p^n}{\partial x} + (\Delta t)^2 \frac{\partial^2 p^n}{\partial t \partial x} \cong \Delta t \left(2 \frac{\partial p^n}{\partial x} - \frac{\partial p^{n-1}}{\partial x} \right) , \tag{3.4.70}$$

which reduces the slip velocity to $O(\Delta t^3)$. As noted by Zang and Hussaini (1986), replacing the right-hand side of (3.4.69) with $\Delta t\, \partial p^n/\partial y$ leads to explosive numerical instability, whereas replacing it (in the constant viscosity case) with $\Delta t\, \nu\, \partial^2 v^n/\partial y^2$ leads to an accurate, well-behaved solution.

As has already been mentioned, Chap. 7 discusses later approaches to reducing the splitting error from both the differential and algebraic perspectives. Here, however, we concentrate on the spectral discretization aspects of the Zang–Hussaini algorithm. Taking a Fourier transform in x, and dropping the subscript k, (3.4.66) becomes

$$\nu(\hat{\mathbf{u}}^{n+1/2})'' - \lambda \hat{\mathbf{u}}^{n+1/2} = -\hat{\mathbf{r}} , \tag{3.4.71}$$

$$\hat{\mathbf{u}}^{n+1/2}(\pm 1) = \hat{\mathbf{g}}^{n+1/2}(\pm 1) , \tag{3.4.72}$$

where, as in the discussion of the Kleiser–Schumann algorithm, $\lambda = (1/\Delta t) + \nu \hat{k}^2$, $\hat{\mathbf{r}} = (1/\Delta t)\hat{\mathbf{u}}^n + \hat{\mathbf{h}}^n$, primes denote derivatives with respect to y, $\hat{k} = 2\pi k/L$, and $\mathbf{h}$ is given by (3.4.13). This system is readily solved by a Chebyshev tau method to yield the intermediate velocity field $\hat{\mathbf{u}}^{n+1/2}$ at the Gauss–Lobatto points.

The pressure correction step (3.4.66) must be performed in a manner consistent with the fully discrete equations (3.4.24) and account for the use of the staggered grid. This complication prevents the use of an efficient ad hoc solution scheme such as those available for the coupled algorithms discussed in Sect. 3.4.1. Zang and Hussaini used Chebyshev collocation in y. Let $\hat{\boldsymbol{u}}$ denote the vector collecting the grid values of $\hat{\boldsymbol{u}}$ at the collocation points; let $\hat{\boldsymbol{u}} = (\hat{\boldsymbol{u}}_x, \hat{\boldsymbol{u}}_y)$ be its splitting along the x- and y-components. Let C_0 and C_+ be the matrices which represent the operations of computing Chebyshev coefficients from function values at the velocity nodes (3.4.61) and the pressure nodes (3.4.62), respectively. (The matrix C_0 represents the operations in (3.4.64) and C_+^{-1} corresponds to (3.4.65).) Let D represent the differentiation operator in terms of the Chebyshev coefficients.

The matrix representing the two-dimensional discrete divergence operator is

$$D_N \hat{\boldsymbol{u}} = \left(C_+^{-1} C_0\right) \left[ik\hat{\boldsymbol{u}}_x + C_0^{-1} D C_0 \hat{\boldsymbol{u}}_y\right] . \tag{3.4.73}$$

(The matrix $C_+^{-1} C_0$ represents interpolation from the cell edges to the cell centers.) The two-dimensional discrete gradient matrix is

$$G_N \hat{\boldsymbol{p}} = \left(C_0^{-1} C_+\right) \left(ik\hat{\boldsymbol{p}}, C_+^{-1} D C_+ \hat{\boldsymbol{p}}\right) . \tag{3.4.74}$$

Letting $\hat{\boldsymbol{q}} = (\Delta t)\,\hat{\boldsymbol{p}}$, the fully discrete version of (3.4.67) is

$$\hat{\boldsymbol{u}}^{n+1} = \hat{\boldsymbol{u}}^{n+1/2} - G_N \hat{\boldsymbol{q}} \qquad \text{at interior Gauss–Lobatto nodes}, \tag{3.4.75}$$

$$D_N \hat{\boldsymbol{u}}^{n+1} = \boldsymbol{0} \qquad \text{at Gauss nodes}, \tag{3.4.76}$$

$$\hat{\boldsymbol{u}}_x^{n+1} = \hat{\boldsymbol{u}}_x^{n+1/2} - (G_N \hat{\boldsymbol{q}})_x \qquad \text{at } y = \pm 1 , \tag{3.4.77}$$

$$\hat{\boldsymbol{u}}_y^{n+1} = \boldsymbol{0} \qquad \text{at } y = \pm 1 . \tag{3.4.78}$$

Equations (3.4.75) and (3.4.78) combine to yield

$$\hat{\boldsymbol{u}}^{n+1} = Z_N \left(\hat{\boldsymbol{u}}^{n+1/2} - G_N \hat{\boldsymbol{q}}\right) , \tag{3.4.79}$$

where Z_N represents the operation of setting the boundary values of the y component to zero. Applying (3.4.76) to (3.4.79) we arrive at

$$L\hat{\boldsymbol{q}} = \boldsymbol{f} , \tag{3.4.80}$$

with

$$L = D_N Z_N G_N \quad \text{and} \quad \boldsymbol{f} = D_N Z_N \hat{\boldsymbol{u}}^{n+1/2} . \tag{3.4.81}$$

Zang and Hussaini used the minimum residual method to compute the pressure from (3.4.80), and the velocities were then adjusted via (3.4.75)–(3.4.78). The equation for the pressure is singular for $k = 0$, but this merely reflects the indeterminacy of the pressure to within an additive constant. Note that no pressure boundary condition is included in (3.4.80). The right-hand side, however, contains the desired boundary conditions on the normal velocity. (The variable p which appears in this splitting scheme differs from the true pressure by a quantity of order $\nu\,\Delta t$ (Kim and Moin (1985)).

This particular splitting algorithm has been used for several large-scale DNS of channel flow transition (e. g., Zang and Krist (1989)) and for LES of turbulent channel flow. (See Sects. 1.3.3 and 1.4.2 for a summary of LES and

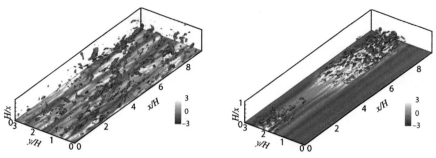

Fig. 3.14. Turbulent fluctuations near the bottom wall in incompressible pulsating channel flow from the LES computations of Scotti and Piomelli (2001). The left frame is near the end of the acceleration phase and the right frame is at the middle of the deceleration phase of the cycle [Reprinted with permission from A. Scotti, U. Piomelli (2001); © 2001, American Institute of Physics]

Sagaut (2006) for a thorough discussion of the subject.) For example, Scotti and Piomelli (2001) performed LES of pulsating channel flow with this splitting algorithm. Figure 3.14 illustrates the flow structures at a fully turbulent phase of the oscillation (left half of the figure) and at a relaminarization phase (right half). The solid surface is a contour of the fluctuating streamwise velocity. These 64^3 LES by Scotti and Piomelli (2001) used the dynamic Smagorinsky subgrid-scale model. The small-scale surfaces are contours of a measure of the coherent vorticity due to rotational motions. Note that the grid used for this large-eddy simulation was significantly coarser than that used in many of the examples above for transitional and turbulent flows. This illustrates a major attraction of the LES approach. The smaller grid permits wide parameter studies to be performed as opposed to the one-of-a-kind simulations typical of direct numerical simulations for such flows. Scotti and Piomelli did parametric studies using LES to characterize the detailed physics of such pulsating flows.

Adaptations of the algorithm have been used for DNS of flat-plate boundary-layer transition (see Fig. 3.15 in Sect. 3.4.5) and for LES of transition (e. g., Piomelli, Zang, Speziale and Hussaini (1990)).

3.4.4 Other Confined Flows

Efficient spectral algorithms have been applied to a variety of flows with one inhomogeneous direction. In this and the following subsection we mention but a few of the many adaptations of the channel flow algorithms described above to other applications in bounded and unbounded domains. Whether or not spectral methods are computationally efficient for other applications is determined largely by whether an efficient scheme is found for solution of the implicit equations. Tuckerman (1989b) has provided a systematic approach to transform the linear systems arising from expansions in orthogonal polynomials into equivalent linear systems with banded matrices. Most of the methods below used more ad hoc approaches to accomplish this.

For confined flows, the simplest adaptation is to plane Couette flow, which requires but a change in the wall boundary conditions to any of the algorithms discussed above; see Lundbladh and Johansson (1991) for use of the normal velocity–normal vorticity method. Another ready adaption is to the Rayleigh–Bénard problem (see Chandrasekhar (1961)) in a channel. The conventional description of this convection problem uses the Boussinesq equations, which consist of the incompressible Navier–Stokes equations, with a term linear in the temperature added to the normal momentum equation, plus an additional equation for the temperature. See McLaughlin and Orszag (1982) for the use of the Orszag–Kells splitting method in their three-dimensional simulations of Rayleigh–Bénard transition.

The algorithms described above can also be applied in cylindrical co-ordinates to flow in a curved channel. The use of cylindrical coordinates introduces geometric factors (inverse powers of r) into the equations. They pose no difficulty for the explicitly treated terms in spectral algorithms. But they may affect the efficiency of the solution of the implicit terms. The direct solution methods used in the Kleiser–Schumann algorithm become more expensive. In cylindrical geometry the tau method is still applicable, although the bandwidth of the matrix increases and the cost roughly triples. The matrix-diagonalization technique is an attractive alternative to the tau method for the direct solution of the implicit equations. However, it does increase the asymptotic operation count of the entire algorithm from $O(N_x N_y N_z \log_2 N)$ to $O(N_x N_y^2 N_z)$, where $N = \max(N_x, N_y, N_z)$. Suitable combinations of Chebyshev polynomials were devised for the curved channel problem by Moser, Moin and Leonard (1983). Nevertheless, due to the increased bandwidth, the implicit step of this Galerkin curved channel algorithm is three times as expensive as the corresponding plane channel one. This algorithm has been used by Moser and Moin (1987) in an extensive study of turbulent curved channel flow.

Provided that the flow is assumed to be periodic in the axial direction, flow between concentric, rotating cylinders (cylindrical Taylor–Couette flow) can be handled relatively efficiently by spectral methods. Marcus (1984a, 1984b) used a matrix-diagonalization technique for the solution of the implicit equations in his influence-matrix method for this flow. Marcus and Tuckerman (1987a, 1987b) used a similar method for axisymmetric flow between concentric rotating spheres (spherical Taylor–Couette flow).

The earliest spectral pipe flow algorithms were devised by Leonard and Wray (1982) and Orszag and Patera (1983). The former used a Galerkin method with a special class of shifted Jacobi polynomials. They handle the coordinate singularity at the origin automatically. Although the implicit equations have small bandwidth, this method has an asymptotic operation count of $O(N_x N_y^2 N_z)$ due to the lack of a fast Jacobi transform for evaluating the nonlinear terms. Meseguer and Trefethen (2003) developed an alternative Galerkin method using divergence-free basis functions built from shifted Chebyshev polynomials that recovers the use of fast transforms. Orszag and

Patera (1983) used a splitting method for their simulations of pipe flow transition. They dealt with the coordinate singularity by using an expansion that incorporates the radial behavior of each mode (see CHQZ2, Sect. 3.9). The implicit equations were solved by matrix diagonalization. Priymak and Miyazaki (1998) solve the pipe flow problem using a Fourier–Chebyshev influence-matrix method combined with a change of variables that enables the pipe flow solution to be obtained at nearly the same efficiency as the channel flow solution.

3.4.5 Unbounded Flows

For flows in which the nonperiodic, transverse (y) direction is unbounded, one cannot directly utilize expansions in orthogonal polynomials on $(-1, 1)$, as we saw in Chap. 2 for the case of the boundary-layer equations and for linear stability analyses of unbounded flows. In this section we focus on algorithms for full Navier–Stokes computations of flat-plate boundary layer flows (see Fig. 1.3) and free shear layer flows (see Fig. 2.14). In general, mappings of the type discussion in Sect. 8.8 need to be utilized.

Flat-Plate Boundary-Layer Flows. Calculations for true boundary-layer flows require a method that can handle two inhomogeneous directions. Methods for this problem are discussed in Sect. 3.5. A simpler and yet useful problem with but one inhomogeneous direction is the so-called parallel boundary layer (Fig. 1.4). Here the boundary-layer profile at some interesting streamwise location is singled out. It becomes an exact solution to the steady Navier–Stokes equations with an appropriate forcing term. Most numerical simulations of this time-dependent problem have approximated some of the spatial effects by choosing the forcing function so that the unperturbed boundary layer thickens in time (Spalart and Yang (1987)).

Spalart (1986a) applied a spectral Galerkin method within Leray's (1933) weak formulation of the incompressible Navier–Stokes equations to the parallel boundary-layer problem. (The description of this method is more readily available in Spalart (1984) and Spalart, Moser and Rogers (1991).) In this case, (3.4.56) is replaced by

$$\left(\frac{\partial \mathbf{u}}{\partial t}, \boldsymbol{\Phi}_{jk}\right) = -\nu(\nabla \mathbf{u}, \nabla \boldsymbol{\Phi}_{jk}) - (\mathbf{u} \cdot \nabla \mathbf{u}, \boldsymbol{\Phi}_{jk}), \qquad (3.4.82)$$

and the test and trial functions are identical. Spalart used both the even and odd polynomials on $(0, 1)$ (see Sect. 8.8.2) in addition to a special basis function (for each Fourier wavenumber) which is essentially $e^{-|k|y}$. This represents the exponentially decaying irrotational component. The remaining, rotational part of the flow also decays exponentially as y tends to infinity, but at a much faster rate. Spalart uses this decomposition into rotational and irrotational components to improve the resolution of the rotational component, which is much more confined to a thin boundary layer than the irrotational component. Thus, an exponential mapping yields finite, but high order accuracy for the rotational part of the flow. If the exponential mapping is combined with a special set of Jacobi polynomials, then a small bandwidth results for the implicit terms. Although the evaluation of the explicit terms requires $O(N_x N_y^2 N_z)$ operations in 3D since a fast transform is not available in y, the number of expansion functions needed in y is typically much smaller than are used in x and z. Spalart has applied this method to a variety of boundary-layer flows, including transitional flow (Spalart and Yang (1987)) and turbulent flow (Spalart (1988b)) under the parallel boundary layer approximation, and also, after some additional approximations, to sink flow boundary layers (Spalart (1986b)).

Both the Galerkin method for the boundary layer and the more conventional collocation algorithms of Orszag and Patera (1983), using splitting, and Laurien (1986), using an influence matrix, apply, in effect, zero-perturbation boundary conditions as y tends to infinity through their use of full semi-infinite mappings. An alternative is to truncate the domain at a finite distance $y_{\max}$ and apply asymptotic boundary conditions there. Malik, Zang and Hussaini (1985) used the boundary conditions

$$\frac{d\hat{\mathbf{u}}_k}{dy} + |k|\hat{\mathbf{u}}_k = 0 \ , \tag{3.4.83}$$

which follow from the behavior of the inviscid linearized problem. For large $y_{\max}$, both asymptotic and zero-perturbation boundary conditions perform comparably, but the asymptotic boundary conditions are clearly better for smaller $y_{\max}$ (Malik, Zang and Hussaini (1985)). Lundbladh et al. (1994) generalized these conditions to

$$\frac{d\hat{\mathbf{u}}_k}{dy} + |k|\hat{\mathbf{u}}_k = \frac{d\hat{\mathbf{v}}_k}{dy} + |k|\hat{\mathbf{v}}_k \ . \tag{3.4.84}$$

Lundbladh et al. report that using (3.4.84) permitted the top boundary to be located at half the distance required for pure Neumann conditions on the Fourier coefficients of the velocity components.

Figure 3.15 illustrates results from the flat-plate adaptation of the Zang and Hussaini (1986) splitting method applied to a direct numerical simulation of transition in flow past a flat plate using the parallel boundary layer approximation. The left half of the figure is taken from the experiments of Hama and Nutant (1963) who used a hydrogen bubble flow visualization technique to illustrate the strongly nonlinear stage of transition. The right half of the figure, from Zang, Hussaini and Erlebacher (see Zang, Krist, Erlebacher and Hussaini (1987) and Zang and Hussaini (1987)), shows how well the roll-up of the bubbles was reproduced in the numerical computations using a $128 \times 144 \times 288$ grid. These authors demonstrated that the fine details of the vortex roll-ups were not present in the streamwise symmetry plane but only appeared in a streamwise plane displaced by a small fraction of the spanwise

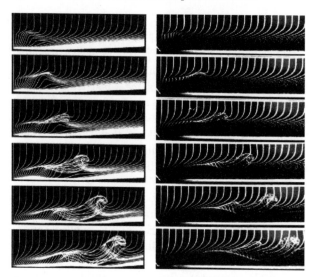

Fig. 3.15. Comparison of hydrogen bubble flow visualizations (*left*) of incompressible flat plate boundary-layer transition with DNS results of Zang, Hussaini and Erlebacher (*right*) [Reprinted with permission from T.A. Zang, M.Y. Hussaini (1987); © 1987 ASME]

wavelength from the symmetry plane. The semi-infinite domain was handled with a composition of the truncated algebraic mapping (8.8.13) with a local clustering (8.8.6) near the wall-normal region containing the roll-up features. Moreover, the asymptotic boundary conditions (3.4.83) were employed.

Free-Shear-Layer Flows. Free-shear-layer problems are posed on the infinite domain $y \in (-\infty, \infty)$ with quiescent boundary conditions at $y = \pm\infty$. In such problems, splitting errors are inconsequential. The free mixing layer has been simulated numerically by Cain, Reynolds and Ferziger (1984) and by Metcalfe et al. (1987) under the assumption of periodicity in x (and z). Cain et al. used the mapping (8.8.22) in y together with a Fourier expansion in ξ. (The mapping (8.8.21) is inappropriate because the mean flow has different values at $y = \pm\infty$.) The Poisson and Helmholtz equations can be solved in $O(N_x N_y N_z \log_2 N)$ operations by the direct method described in CHQZ2, Sect. 4.1.1. Metcalfe et al. have used both algebraic and exponential mappings (see (8.8.18) and (8.8.19)) combined with a Chebyshev polynomial expansion in ξ. They used the matrix-diagonalization technique for the solution of the implicit equations.

Spalart, Moser and Rogers (1991) presented an adaptation of the normal velocity–normal vorticity coupled method for mixing layers and wakes that starts with the solution of the full three-dimensional equation (1.4.19) for the vorticity. Unlike the case for wall-bounded flows, for these applications all components of the vorticity are known to vanish at the boundaries $y = \pm\infty$. Spalart et al. combine the exponential mapping (8.8.19) in y with an expansion for the vorticity based on the Jacobi polynomials $P_n^{(1,1)}$. The resulting implicit equations for the fully discrete vorticity equation with the viscous term treated implicitly are tridiagonal and therefore inexpensive to solve. The components of the velocity parallel to the wavevector (k_x, k_y) can be obtained directly from the vorticity, although the $(k_x, k_y) = (0, 0)$ case must be treated separately. A second-order equation still needs to be solved for the wall-normal velocity component. This requires the use of some additional expansion functions. For the details, see Spalart, Moser and Rogers (1991).

Figure 3.16 shows results from Rogers and Moser (1992) using this algorithm for DNS of transition in incompressible, free shear layers. This figure, based on computations on a $64 \times 128 \times 64$ grid, illustrates several aspects of the vorticity from a simulation that is representative of experiments on vortex roll-up in mixing layers. The thin, shaded surfaces correspond to the rib vortices (large component of vorticity normal to the spanwise direction), the cross-hatched surfaces denote the "cups" (regions of strong spanwise vorticity) that are critical to free-shear-layer transition, and the lines are vortex lines that comprise the rib vortices. See Moser, Rogers and Ewing (1998) for applications to wakes.

A numerical method for a different model of a free shear layer was used by Riley and Metcalfe (1980) and by Metcalfe et al. (1987). The simulation is conducted on a finite domain in y, say, $y \in (0, \pi)$, with no mapping. Free-slip boundary conditions are applied at $y = 0$ and $y = \pi$. This is achieved by using a cosine expansion in y for u (and w) and a sine expansion for v. Consequently, there is an infinite array of image mixing layers stacked in the

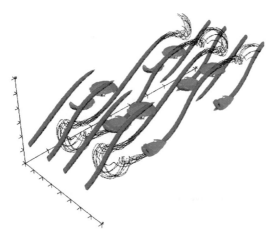

Fig. 3.16. DNS of vortex rollup in an incompressible free shear layer by Rogers and Moser (1992). The surfaces denote two types of regions of strong vorticity and the lines are vortex lines [Reprinted with permission from M.M. Rogers, R.D. Moser (1992); © 1992, Cambridge University Press]

y-direction. Curry et al. (1984) have used a similar expansion for a Rayleigh–Bénard computation.

3.4.6 A Numerical Example: Accuracy

The numerical example in this section is focused on demonstrating the rapid convergence of spectral approximations to plane channel flow. We recall from Sects. 1.5 and 2.3.1 that the solution to the linear stability problem for plane channel flow has the form

$$\mathbf{u}(x, y, z, t) = (1 - y^2)\hat{\mathbf{e}}_x + \varepsilon Re\{\hat{\mathbf{u}}(y)e^{i(\alpha x + i\beta z) - i\omega t}\} \qquad (3.4.85)$$

for the velocity. Here α and β are specified real constants, ε is a specified (small positive) amplitude parameter, while the function $\hat{\mathbf{u}}(y)$ and the complex temporal frequency ω come from solutions to (1.5.5)–(1.5.9). For the present example, we solve the nonlinear problem (3.4.1)–(3.4.3) for plane channel flow with initial condition given by (3.4.85) for $t = 0$. For small ε, this nonlinear solution should be closely approximated by (3.4.85).

The particular problem chosen for this illustration has the same parameters as the illustration in Sect. 2.3.2 for the spectral solution of the linear stability problem for plane channel flow, namely, $\nu = (7500)^{-1}$, $\alpha = 1$ and $\beta = 0.25$; we recall that the eigenfrequency of the only growing mode is $\omega = \omega_r + i\omega_i$, where $\omega_r = 0.2546468$ and $\omega_i = 0.00168654$. See Fig. 2.9 for the spatial structure of the streamwise velocity of this wall mode. (The most unstable mode for this plane channel problem is always a two-dimensional mode. We are using this particular mode just to illustrate the comparative accuracy of various numerical schemes.) The streamwise velocity component of the eigenfunction $\hat{\mathbf{u}}(y)$ for this mode is shown in Fig. 2.9. The amplitude parameter for the present numerical examples is $\varepsilon = 0.00001$. The natural choice of the streamwise and spanwise periodicity lengths are $L_x = 2\pi/\alpha = 2\pi$ and $L_z = 2\pi/\beta = 8\pi$. Four discretizations in x and z were used: (1) Fourier collocation (FC), (2) second-order finite differences (FD2), (3) fourth-order finite differences (FD4), and (4) sixth-order compact differences (CP6). Only Chebyshev collocation was considered in y. For descriptions of these well-known finite-difference discretization methods, see Sect. 3.3.6. All grids were uniform in x and z and used the Chebyshev Gauss–Lobatto points in y. The time-discretization was a second-order semi-implicit scheme (explicit convection, implicit diffusion—see Sect. 3.2.3) with the time-step chosen so small that the spatial errors predominated. All runs were made from $t = 0$ to $t = 4\pi/\omega_r$. This is twice the length of time $T (= 2\pi/\omega_r)$ that it takes for the perturbation wave to propagate through the streamwise computational domain.

The scalar quantity from the numerical solution that we use to measure accuracy is the maximum streamwise velocity perturbation

$$u'_{max}(t) = \max_{x,y,z} \left(u(x, y, z, t) - (1 - y^2) \right) . \qquad (3.4.86)$$

For sufficiently small ε this ought to exhibit the exponential behavior

$$u'_{max}(t) = u'_{max}(0)e^{-i\omega t} , \qquad (3.4.87)$$

where $u'_{max}(0)$ is the maximum streamwise velocity perturbation at $t = 0$. Since the initial perturbation is normalized so that $u'_{max}(0) = 1$, we expect any deviations from (3.4.87) due to nonlinear effects to be on the order of 10^{-10} for $t \in [0, 2T]$. As noted above, the time-step was sufficiently small that time-discretization errors were no more than 10^{-10}. The spectral numerical results that follow with this example were computed with a coupled method (Malik, Zang and Hussaini (1985)). Hence, there are no splitting errors. Similar results would be expected from any of the coupled methods discussed in Sect. 3.4.1.

The performance of the Chebyshev collocation discretization in y of this problem is illustrated in Figs. 3.17 and 3.18. Since the perturbation

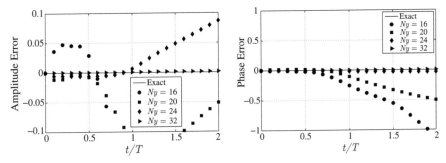

Fig. 3.17. Computed absolute amplitude and phase errors of the maximum streamwise perturbation velocity for 4-point Fourier spectral discretizations in x and z and a Chebyshev spectral discretization in y with various resolutions

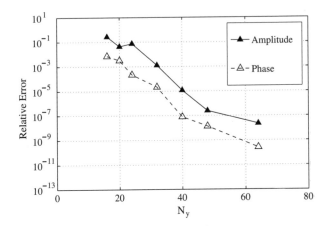

Fig. 3.18. Relative amplitude and phase errors of the maximum streamwise perturbation velocity for a Fourier–Chebyshev spectral method

is a trigonometric function, four Fourier collocation points in the streamwise and spanwise directions suffice for essentially perfect resolution in those directions. The first figure illustrates the time history of the absolute errors in the amplitudes and phases of the quantity $u'_{max}(t)/u'_{max}(0)$. At the end of two periods ($t = 2T$) the exponential factor in (3.4.87) corresponds to an amplitude ratio of 1.086789 and a phase change of 4π with respect to $t = 0$. The second figure indicates that $N_y = 32$ suffices for less than a 0.1% error in both amplitude and phase over this interval, where the errors are substantially larger for $N_y = 24$. As one can tell from Fig. 2.9, 32 points in y are needed to resolve the rapid oscillation of the velocity perturbation near the walls. Figure 3.18 illustrates the spectral convergence on this example. Note

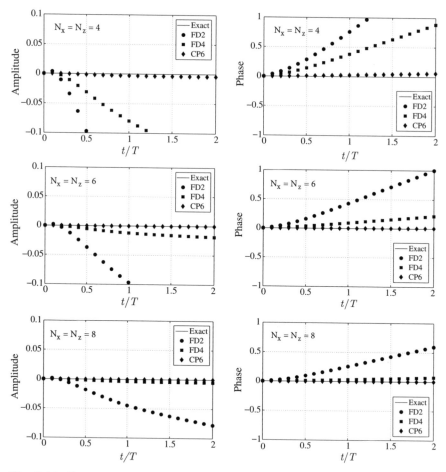

Fig. 3.19. Computed absolute amplitude and phase errors in the maximum streamwise perturbation velocity for finite-difference approximations in x and z and a Chebyshev spectral method in y. The number of grid points in x and z increases from 4 to 6 to 8 from the top to the bottom rows

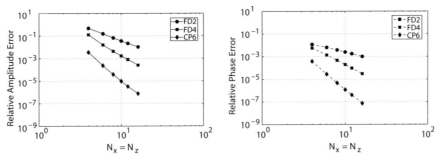

Fig. 3.20. Relative amplitude (*left*) and phase (*right*) errors of the maximum streamwise perturbation velocity for finite-difference-Chebyshev methods [Courtesy of Y. Kenada, T. Ishihara (2006)]

that the phase error is consistently lower than the amplitude error. This is characteristic of spectral methods for wave propagation problems.

As noted above, $N_x = N_z = 4$ suffices for the horizontal resolution with a Fourier discretization in those directions. Figure 3.19 illustrates the evolution of the absolute errors for FD2, FD4 and CP6 discretizations in the horizontal directions; the vertical discretization is Chebyshev collocation with $N_y = 64$, which produces a relative error of less than 10^{-7} from the vertical approximation (see Fig. 3.18). Figure 3.20 illustrates the amplitude and phase errors from the finite-difference approximations in the horizontal directions. These results suggest that in order to achieve 0.1% relative accuracy in both amplitude and phase, one needs 6 points (per wavelength) with the sixth-order compact scheme, 12 with the fourth-order central scheme, and more than 48 points per wavelength with the second-order method. (The conclusions for FD4 and FD6 follow from picking the minimum $N_x = N_z$ at which both curves (actually the amplitude curve, since it is always the highest) are below a relative error of 10^{-3}. The conclusion for FD2 requires an extrapolation of the curves.) One should bear in mind that most high-order finite-difference schemes have lower order at boundaries for efficiency and/or numerical stability reasons. (See, for example, the discussion in Sect. 4.2.1 and in CHQZ2, Sect. 3.7.) Hence, for nonperiodic problems high-order finite-difference schemes may require more points per wavelength than suggested by these results.

3.5 Flows with Multiple Inhomogeneous Directions

Flows with but a single inhomogeneous direction, such as plane channel flow, the parallel boundary layer, and the classical Taylor–Couette flow between infinite, rotating cylinders are only idealizations of the flows which occur in nature. The streamwise direction in channel and boundary-layer flow is, in fact, inhomogeneous (especially for the boundary layer, for which even

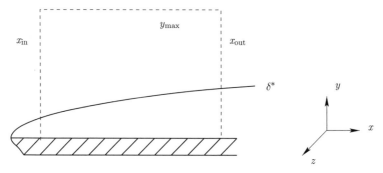

Fig. 3.21. Computational domain for a spatially developing boundary layer. There is an inflow boundary at $x = x_{\mathrm{in}}$, and an outflow boundary at $x = x_{\mathrm{out}}$. Domain truncation in y is applied at $y = y_{\mathrm{max}}$

the mean flow depends on x), and rotating cylinders do have finite length. Spectral algorithms for these problems must have at least two nonperiodic directions. The spatially developing boundary layer (Fig. 3.21) is a problem of considerable physical interest. The semi-infinite domain in y can be handled by mapping, as discussed in Sect. 3.4.5. However, inflow/outflow boundary conditions are required at the two streamwise boundaries. See Sect. 3.6 for a further discussion on these boundary conditions.

The numerical discretization of the incompressible Navier–Stokes equations with multiple inhomogeneous directions requires three main ingredients: (i) the choice of a spatial discretization for velocity and pressure, (ii) the choice of a time-discretization technique, and (iii) the choice of a solution technique for the algebraic system obtained after spatial and temporal discretizations are applied. The latter ingredient is dictated by the fact that pressure is defined by an implicit equation, and most often the viscous term is time advanced implicitly for stability reasons. In the case of multiple inhomogeneous directions, the stage for solution of the algebraic system becomes dramatically more cumbersome than in the case of fully homogeneous flows (for which it is trivial), or flows with one inhomogeneous direction (for which one can reduce the problem to a cascade of simple one-dimensional systems). Indeed, in the situation considered in this section, large algebraic systems are obtained, in which the unknowns in all directions are coupled; except for the smallest problems their efficient solution usually requires sophisticated numerical linear algebra tools, such as iterative methods with suitable forms of preconditionings. In these cases, the level of conceptual complexity depends rather insignificantly both on the particulars of the boundary conditions, as well as whether a single domain or multiple domains are used. For these and other reasons, we prefer to devote an entire chapter, Chap. 7, to discussing time-discretization strategies and algebraic solution techniques which are especially suitable for the incompressible Navier–Stokes equations under general boundary conditions and in general geometries. We devote the

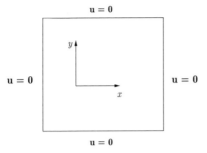

Fig. 3.22. Two-dimensional cavity flow geometry

rest of this section to discussing the issue of spatial discretization in a single domain, focusing on the interplay between the velocity and pressure discrete representations. We will continue this discussion in Chap. 5, in the framework of multidomain spectral methods.

For the sake of simplicity, we choose the (physically uninteresting) problem of flow in a two-dimensional cavity (Fig. 3.22) to illustrate the essentials of the various spatial discretization techniques. The extension to a (physically more interesting) three-dimensional cavity or duct is straightforward, using the techniques of the previous section if the flow may justifiably be taken to be periodic in the third direction, and extending the methods discussed here if it must be treated as fully inhomogeneous. In this discussion we revert to considering the general case of the incompressible Navier–Stokes equations with a forcing function **f**. (The Rayleigh–Bénard problem is a case in which there is a forcing function, namely, **f** is proportional to the temperature.) The basic equations, then, are again given by (3.1.1) and (3.1.2) on the domain $\Omega = (-1,1)^2$ with the boundary conditions

$$\mathbf{u} = \mathbf{0} \quad \text{on} \quad \partial\Omega. \tag{3.5.1}$$

3.5.1 The Choice of Spatial Discretization in a Cavity

The spatial discretization of the Navier–Stokes equations (3.1.1)–(3.1.2) with initial conditions (3.1.3) and boundary conditions (3.5.1) can be conducted with one of the standard strategies used in spectral methods, namely Galerkin, G-NI, collocation or tau (see Chap. 8, or CHQZ2). The peculiar issue related to the discretization of the primitive variable formulation is the possible onset of spurious pressure modes that are not controlled by the discretization scheme. To avoid their appearance, the discretizations of pressure and velocity must be carefully related to each other. Alternative strategies consist in systematically filtering out the spurious modes as soon as they are generated, or in modifying the discrete equations to prevent their onset.

At the outset let us introduce the notation for the polynomial spaces that will be used in the rest of the chapter. We will denote by

$$\mathbb{Q}_N = \mathrm{span}\{x^{j_1}y^{j_2} \mid 0 \le j_1, j_2 \le N\} \tag{3.5.2}$$

the space of all polynomials of degree $\le N$ in each variable, i.e., $\mathbb{Q}_N = \mathbb{P}_N^{(x)} \otimes \mathbb{P}_N^{(y)}$, if $\mathbb{P}_N^{(s)}$ is the space of all polynomials of degree $\le N$ in the variable s. We will also use the smaller space

$$\mathbb{P}_N = \mathrm{span}\{x^{j_1}y^{j_2} \mid 0 \le j_1, j_2 , \; j_1 + j_2 \le N\} \tag{3.5.3}$$

of all polynomials of total degree $\le N$. The extension of these definitions to the 3D case is obvious. We remark that in CHQZ1 and CHQZ2 the space $\mathbb{Q}_N$ above was actually denoted by $\mathbb{P}_N$, according to the notation historically used in the spectral methods community. In CHQZ2, Sect. 2.9, the space $\mathbb{P}_N$ above was indicated by $\mathcal{P}_N$. Here we conform to the conventional notation in the finite-element literature, as that notation has been adopted for multidomain spectral methods.

In order to explain the nature of spurious modes, let us examine the spatially discretized Navier–Stokes equations. Suppose that the discrete velocity and pressure are such that $\mathbf{u}^N \in (\mathbb{Q}_N)^2$ and $p^N \in \mathbb{Q}_N$, and that, in addition, $\mathbf{u}^N$ vanishes on $\partial\Omega$. Let us expand both variables along (nodal or modal) bases, and let us indicate by $\boldsymbol{u}$ and $\mathbf{p}$ the vectors of the velocity and pressure unknowns corresponding to these bases. Let G_N, D_N and M_N be the matrices that realize the gradient, divergence and mass operators according to these bases, and let L_N be the matrix associated with the discretization of the viscous term $-\nu\triangle$. The discrete problem reads

$$M_N \frac{\partial \boldsymbol{u}}{\partial t} + C_N(\boldsymbol{u}) + L_N \boldsymbol{u} + G_N \mathbf{p} = \boldsymbol{f} , \tag{3.5.4}$$

$$D_N \boldsymbol{u} = \mathbf{0} , \tag{3.5.5}$$

where $\boldsymbol{f}$ is a vector accounting for the volumetric forces, whereas $C_N(\boldsymbol{u})$ convection for the nonlinear terms. For instance, if the convection form of the momentum equation is used, and $\mathbf{u}^N$ is expanded along a nodal basis, we have $C_N(\boldsymbol{u}) = \boldsymbol{u} \cdot G_N \boldsymbol{u}$, with the same notation as in (3.3.22).

By definition, any discrete pressure $\mathbf{q}$ that is nonconstant in Ω and satisfies

$$G_N \mathbf{q} = \mathbf{0} \tag{3.5.6}$$

is termed a *spurious pressure mode* (also called a parasitic mode). Since pressure enters the Navier–Stokes equations only through its gradient, and no boundary conditions are applied on it, we have that any couple $(\boldsymbol{u}, \mathbf{p} + \mathbf{q})$ solves the equations provided that $(\boldsymbol{u}, \mathbf{p})$ does. Thus, the existence of spurious pressure modes leads to nonuniqueness of the solution of the Navier–Stokes equations.

Condition (3.5.6) for a nonconstant pressure $\mathbf{q}$ is obviously equivalent to the matrix $G_N^T G_N$ having a zero eigenvalue with multiplicity greater than 1. Equivalently, the matrix G_N has the zero singular value with multiplicity greater than 1. (Recall that the singular values of a matrix A are the square roots of the eigenvalues of the matrix $A^T A$, and that the number of nonzero singular values is the rank of A.)

A slightly more sophisticated concept is that of *pseudo-spurious pressure modes*. Instead of looking at the eigenvalues of the matrix $G_N^T G_N$, let us consider the generalized eigenvalue problem

$$G_N^T L_N^{-1} G_N \mathbf{q} = \mu M_N^p \mathbf{q} \,, \tag{3.5.7}$$

where M_N^p indicates the pressure mass matrix. Again, spurious pressure modes are associated with the multiplicity > 1 of the generalized eigenvalue $\mu = 0$. Let us denote by β_N^* the square root of the smallest nonzero generalized eigenvalue μ. If this quantity is not bounded from below by a strictly positive constant independent of N, i.e., if the sequence β_N^* tends to 0 as $N \to \infty$, we say that the numerical method has pseudo-spurious pressure modes. Precisely, the latter are the pressure eigenmodes associated with eigenvalues μ which tend to 0 as $N \to \infty$. The quantity β_N^* is very important, since it influences the quality of the approximation of the pressure and sometimes also that of the velocity (see Sect. 3.7.5 for more details). Furthermore, it influences the condition number of the *pressure Schur complement matrix* (or *Uzawa operator*), which will be introduced in Sect. 7.3.1 as a method to solve the discrete Stokes problem.

Our discussion now focuses on the spurious pressure modes associated with various kinds of spectral discretizations. A thorough investigation of their impact on the theoretical properties of the spectral schemes will be carried out in Sect. 3.7.

If we consider simple collocation methods, spurious pressure modes can be related to the employed grid(s) in an intuitive way. (Note that, for Legendre approximations, the collocation scheme may coincide with a G-NI scheme, due to the no-slip boundary conditions chosen in this section.) Three grid possibilities are illustrated in Fig. 3.23. The standard, or nonstaggered, grid uses the Gauss–Lobatto points (in both x and y) as the nodes for identifying all the unknowns. Thus, discrete velocities and pressures are both polynomials of degree N in each variable. For this reason, the resulting method will be termed the $\mathbb{Q}_N - \mathbb{Q}_N$ method (it is traditionally referred to as the $\mathbb{P}_N - \mathbb{P}_N$ method in the spectral methods community). The half-staggered grid uses the Gauss–Lobatto points for the velocity and the Gauss points for the pressure. In this case, discrete velocities are polynomials (in both variables) of degree N, whereas discrete pressures are polynomials of degree $N - 1$; we obtain a $\mathbb{Q}_N - \mathbb{Q}_{N-1}$ method. The fully staggered grid uses different nodes for each primitive variable. The u component of velocity is defined at the Gauss–Lobatto points in x and the Gauss points (plus the boundary points)

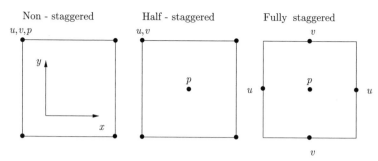

Fig. 3.23. Alternative velocity and pressure nodes for flows with two inhomogeneous directions

in y; hence, it is a polynomial of degree N in the x variable and degree $N+1$ in the y variable. The v component is handled in reverse fashion. Finally, the pressure is defined at the Gauss points in both directions; hence, it is a polynomial of degree $N-1$ in each variable. (For hyperbolic problems, yet another type of staggered grid is sometimes used; see Sect. 4.6.1.)

(a) *The nonstaggered* $(\mathbb{Q}_N - \mathbb{Q}_N)$ *method*

The collocation scheme based on the standard Gauss–Lobatto grid is the most straightforward one to implement. The momentum equations are enforced at the interior nodes, whereas the continuity equation is collocated at all nodes. Since the divergence of a discrete velocity is a polynomial of degree N at most, it turns out that the discrete solution is divergence-free all over the domain. However, the method suffers from the presence of seven spurious pressure modes; they are characterized as those nonconstant pressures that have vanishing gradients at all the interior nodes, where the momentum equations are collocated. Therefore, such pressures have no effect whatsoever upon the velocity. The specific spurious modes for the standard Chebyshev grid are the line mode $q = T_N(x)$, the column mode $q = T_N(y)$, the checkerboard mode $q = T_N(x)T_N(y)$, and four corner modes $q = T'_N(x)(1 \pm x)\,T'_N(y)(1 \pm y)$. A corner mode is one that vanishes at all the nodes (including those on the boundary) save at one of the corners. Its gradient vanishes at all the interior nodes and on the nodes on the two sides that do not form the corner. If a Legendre grid is used instead, the modification is obvious (see Sect. 3.7.1).

The presence of spurious modes need not disqualify a discretization for practical applications. The simplicity of using the same grid and polynomial degree for all variables may warrant the use of supplementary techniques such as *filtering, regularization* and *stabilization*.

Filtering is perhaps the most obvious approach. Since the spurious pressure modes do not affect the velocity, all that needs to be done with them is to filter them from the pressure before making use of the computed solution. Huberson and Morchoisne (1983) and Métivet (1987) discuss some filtering techniques based on orthogonal projections.

In her general formulation of the influence-matrix method for solving the Stokes problem, Tuckerman (1989a) uses a regularization procedure to solve the resulting underdetermined linear system. This appears to automatically remove the spurious modes. Phillips and Roberts (1993) apply a singular-value decomposition (SVD) to the Stokes matrix expressing the collocation equations.

Finite-element preconditioning (see CHQZ2, Sect. 4.4 for a thorough discussion) may also produce a spurious mode control. Demaret and Deville (1989) suggest the use of the $\mathbb{Q}_2 - \mathbb{Q}_1$ element (which is uniformly inf-sup stable, see e. g., Brezzi and Fortin (1992)) on the quadrilateral mesh defined by the Gauss–Lobatto nodes. If the resulting finite-element matrix is used as a preconditioner for the full Stokes matrix, no spurious modes are generated during the iterative process.

The stabilization approach (inspired by a similar approach in use in the h-FEM community, see e. g. Brooks and Hughes (1982) and Franca and Frey (1992)) consists of modifying the discrete continuity equation (in the framework of a Galerkin or G-NI method) by adding a term depending on the spectral residual of the momentum equation; in this way, consistency is preserved and spectral convergence is possible. The added term provides an extra control on the discrete pressure, which prohibits the presence of spurious modes. Various forms of the stabilizing term have been proposed. Canuto and Van Kemenade (1996) and Canuto, Russo and Van Kemenade (1998), following a strategy already applied for stabilizing advection-diffusion equations (see CHQZ2, Sect. 7.2.1), first enrich the space of discrete velocities by "bubble" functions associated with the cells of the quadrilateral mesh defined by the Gauss–Lobatto nodes. Next, bubbles are eliminated by static condensation, with the result of appending a local, cell-by-cell projection of the spectral residual to the continuity equation. Gervasio and Saleri (1998) provide an SUPG-type (streamline upwind Petrov Galerkin) stabilization in the framework of a spectral element discretization (see Sect. 5.6), thus adding a global form of the spectral residual within each subdomain. Both approaches are supplemented with recipes to automatically select the stabilization parameters. Among the advantages of the stabilization strategy are the positive effects on the convection instabilities, in the case of convection-dominated flows, and the immediate availability of a natural finite-element preconditioner for solving the resulting algebraic system.

(b) *The half-staggered grid* $(\mathbb{Q}_N - \mathbb{Q}_{N-1})$ *method*

The implementation of a collocation method based on the half-staggered grid is slightly more involved. The same type of interpolation used in the Zang–Hussaini algorithm discussed in Sect. 3.4.3 applies. Now the momentum equations are enforced at the interior Gauss–Lobatto nodes, whereas the continuity equation is enforced at all Gauss nodes. Since discrete pressures are one degree less than discrete divergences, the computed velocity is not divergence-free all over Ω. This discretization contains a single spurious mode, which for a Chebyshev grid is $q = T'_N(x)T'_N(y)$. At the interior velocity nodes this has the form

$$q(x_j, y_k) = (-1)^{j+k} \bigg/ \left[\sin\left(\frac{(2j+1)\pi}{2N}\right) \sin\left(\frac{(2k+1)\pi}{2N}\right) \right], \qquad (3.5.8)$$

which is close to a pure checkerboard pattern. In addition to possessing fewer spurious modes, the method also avoids the need for a pressure boundary condition if the resulting generalized Stokes problem is solved by a projection method. However, several pseudo-spurious modes are present. Montigny-Rannou and Morchoisne (1987) describe an algorithm which uses the half-staggered grid.

(c) The fully staggered grid method

The collocation method based on the fully staggered grid was proposed by Bernardi and Maday (1988b). It is similar to the celebrated finite-difference marker-and-cell scheme (Harlow and Welsh (1965)) and is obviously the most cumbersome one to use. Each component of the momentum equation is enforced at those internal nodes which identify the corresponding velocity component. The calculation of the interpolation derivatives of u in the x-direction is straightforward. The computation of interpolation derivatives with respect to y can be accomplished by writing $u(x, y) = (1 - y^2)u_b(x, y)$ and using the Leibnitz rule with $u_b(x, y)$ differentiated via interpolation. In a Chebyshev method, since $u_b(x, y)$ is a polynomial in y of degree $\leq N - 1$, its Chebyshev coefficients can be computed exactly from its values at the Gauss points in y. As discussed in Appendix B, this, as well as the inverse operation, can be performed via the FFT. The continuity equation is enforced at all Gauss nodes, leading again to a solution that is not divergence-free. However, unlike the previous schemes, the present method possesses no spurious modes. There still remain pseudo-spurious modes; indeed the constant β_N^* decays like $N^{-1/2}$ (and like N^{-1} in the three-dimensional version of the method).

Of the three schemes presented above, only the first one is amenable to an efficient implementation and therefore to practical use; for the two others, the trade-off between accuracy and cost is negative. Yet they serve to highlight the crucial aspect of the issue we are discussing in the section. Spurious pressure modes are high-frequency modes (this is true also for pseudo-spurious modes). Hence, they can be reduced in number and even eliminated by introducing a large enough gap between the polynomial degrees of velocities and pressures. The last example indicates that a gap of two is the minimal one to ensure spurious-free pressures. On the other hand, too large a gap weakens the constraint on the divergence of the velocity, leading to a poor enforcement of the continuity equation.

(d) *The $\mathbb{Q}_N - \mathbb{Q}_{N-2}$ method*

These considerations induced Maday, Patera and Rønquist (1987) to propose the so-called $\mathbb{Q}_N - \mathbb{Q}_{N-2}$ method (under the name of the $\mathbb{P}_N - \mathbb{P}_{N-2}$ method), which soon became very popular and widely used to solve the incompressible Navier–Stokes equations. One should actually speak of two variants of the method that share the feature that the discrete velocities are polynomials of degree N and the pressures are polynomials of degree $N-2$, in each variable. In other words, discrete velocities belong to the space

$$V_N = \left\{ \mathbf{v} \in (\mathbb{Q}_N)^2 : \mathbf{v} = \mathbf{0} \text{ on } \partial\Omega \right\},$$

whereas discrete pressures belong to the space

$$Q_N = \left\{ q \in \mathbb{Q}_{N-2} : \int_\Omega q = 0 \right\}.$$

Velocities are invariably identified through their values at the traditional N-degree Gauss–Lobatto grid, i. e., they are expanded along the nodal basis associated with these points. Pressures may be identified through their values either at the internal nodes of this grid (option (a)), or at all nodes of the $(N-2)$-degree Gauss grid (option (b)). Momentum and continuity equations are enforced by the G-NI approach (thus in the framework of Legendre rather than Chebyshev approximations). The method applied to the Stokes problem consists of seeking $\mathbf{u}^N \in V_N$ and $p^N \in Q_N$ such that

$$\begin{aligned}
(\nabla \mathbf{u}^N, \nabla \mathbf{v})_N - (\nabla \cdot \mathbf{v}, p^N)_N &= (\mathbf{f}, \mathbf{v})_N && \text{for all } \mathbf{v} \in V_N, \\
(\nabla \cdot \mathbf{u}^N, q)_N &= 0 && \text{for all } q \in Q_N.
\end{aligned}$$

In the term accounting for the Laplacian (after integration by parts) and in the right-hand side term, the integrals are approximated by the N-degree Gauss–Lobatto quadrature formula. In the pressure and divergence terms, one can use either this formula (consistent with choice (a) for the pressure degrees of freedom), or the $(N-2)$-degree Gauss formula (consistent with choice (b)). Note however that both formulas yield the exact value of the integrals, since the product $(\nabla \cdot \mathbf{v})\, q$ is a polynomial of degree $2N-2$, whereas both formulas integrate all polynomials of degree $2N-1$ exactly.

Using the Gauss–Lobatto formula in the pressure term and taking as velocity test functions the characteristic Lagrange polynomials (discrete delta-functions—see Sect. 8.2) at the internal nodes (these are the nodal basis functions), one immediately sees that the momentum equation is collocated at these nodes. Conversely, using the Gauss formula in the divergence term and taking as pressure test functions the characteristic Lagrange polynomials at all nodes, one sees that the continuity equation is collocated at the Gauss points. The alternative use of the Gauss–Lobatto formula in the divergence term shows that a certain linear combination of nodal divergence values is required to vanish at each node of such grids. In this sense, the method is not classified as of standard collocation type. The discrete velocity defined by the scheme is not fully divergence-free in Ω.

As observed by Wilhelm and Kleiser (2000), when the method is applied to the full Navier–Stokes equations the divergence error at the Gauss–Lobatto grid may lead to temporal instabilities. They are not caused by nonlinear effects, but rather by the discretization of the particular form of the convection term (see Sect. 3.2.1). The divergence form, the skew-symmetric form and the rotation form (3.2.3) are unstable, whereas the convection form and the rotation form which includes the term $-\frac{1}{2}\nabla|\mathbf{u}|^2$ are stable.

Both choices of pressure degrees of freedom are reported to be equivalent in terms of computational costs. Option (a) involves a single grid, but requires pressures to be extrapolated at the Gauss–Lobatto boundary nodes; option (b) requires re-interpolation from the Gauss grid to the Gauss–Lobatto grid.

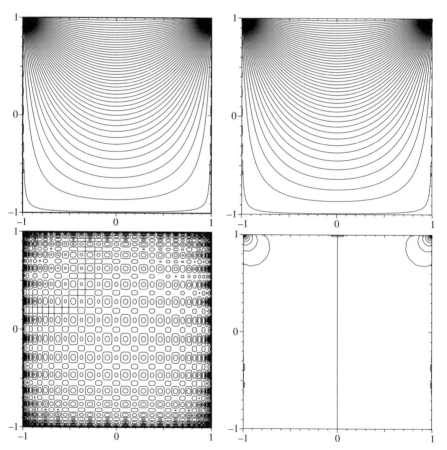

Fig. 3.24. Streamfunctions (*top row*) and pressure contours (*bottom row*) obtained by the $\mathbb{Q}_N - \mathbb{Q}_N$ method (*left column*) and the $\mathbb{Q}_N - \mathbb{Q}_{N-2}$ method (*right column*)

The $\mathbb{Q}_N - \mathbb{Q}_{N-2}$ method is free of spurious pressure modes. The constant β_N^* still decays like $N^{-1/2}$ (and like N^{-1} in the three-dimensional version of the method), thus indicating the presence of pseudo-spurious modes. These features are similar to those of the fully staggered-grid method described above. However, the efficiency of the implementation is drastically improved, as well as the accuracy in the approximation of both velocity and pressure (this aspect will be detailed in Sect. 3.7.2).

Figure 3.24 illustrates the nature of the spurious modes in two dimensions for regularized driven-cavity flow. The discretization uses $N = 32$; the boundary conditions are homogeneous Dirichlet except for the top side, where $\mathbf{u}(x, 1) = (\arctan(50(1 - |x|)), 0)^T$. The spurious modes dramatically impact the computed pressure, but the velocity is entirely unaffected. Indeed, the velocity is essentially the same in the $\mathbb{Q}_N - \mathbb{Q}_N$ and $\mathbb{Q}_N - \mathbb{Q}_{N-2}$ methods.

(e) *The $\mathbb{Q}_N - \mathbb{Q}_{\lambda N}$ method*

The possibility of getting rid of the pseudo-spurious pressure modes as well is related to the asymptotic widening of the gap between the velocity and pressure polynomial degrees. Indeed, Bernardi and Maday (1999) have considered the $\mathbb{Q}_N - \mathbb{Q}_{\lambda N}$ method, which uses velocities of degree N and pressures of degree M equal to the largest integer $\leq \lambda N$, where $\lambda \in (0, 1)$ is a fixed real parameter. Under the condition that $\lambda N \leq N - 2$, the method is free of both spurious and pseudo-spurious pressure modes; actually, the constant β_N^* satisfies the uniform (in N) lower bound $\beta_N^* \geq c(1 - \lambda)^{1/2}$ (or $\beta_N^* \geq c(1 - \lambda)$ in the three-dimensional version of the method). As a consequence, the discretization error in both velocity and pressure is optimal with respect to N, i. e., it decays with the same asymptotics as the best approximation error allowed by the choice of the polynomial spaces. From the computational point of view, the discrete pressures are identified by their values at the nodes of an M-degree Gauss grid, and then they are re-interpolated at the nodes of the standard N-degree Gauss–Lobatto grid.

(f) *The $\mathbb{Q}_N - \mathbb{P}_{N-1}$ method*

An alternative choice, which again guarantees the absence of spurious and pseudo-spurious pressure modes, is the $\mathbb{Q}_N - \mathbb{P}_{N-1}$ method; it is a classical method in the h-version of finite elements on quadrilateral meshes. Bernardi and Maday (1999) prove the bound $\beta_N^* \geq \beta > 0$ to be independent of N. This method can no longer be given a collocation-like interpretation. Unlike the $\mathbb{Q}_N - \mathbb{Q}_{\lambda N}$ method, tensorization is no longer possible, but the approximation of the pressure is more accurate, i. e., the decay rate of the pressure error is faster as N increases.

3.5.2 The Choice of Spatial Discretization on a Reference Domain

So far, we have considered a flow confined inside a square cavity, i. e., we have imposed homogeneous Dirichlet boundary conditions on the velocity field. In view of the treatment of multidomain spectral discretizations in Chap. 5, we need velocity spaces on a reference domain $\hat{\Omega}$ whose elements need not vanish on $\partial\hat{\Omega}$ (as well as pressure spaces on $\hat{\Omega}$ whose elements need not have zero average therein). Let us denote such spaces by $V_N(\hat{\Omega})$ and $Q_N(\hat{\Omega})$, respectively. (Note that hereafter N will play the role of a parameter, which may not coincide with the maximal degree of the polynomials contained in these spaces.) They will be used to generate the local spaces of velocities and pressures on each element via appropriate mappings.

Here, we provide a list of admissible pairs $V_N(\hat{\Omega})$ and $Q_N(\hat{\Omega})$. Some of them have been already considered above, whereas others have been proposed in the hp FEM literature (see, e.g., Stenberg and Suri (1996), Schwab and Suri (1999), Chilton and Suri (2001)).

Let $\hat{\Omega}$ be the reference square $\hat{\Omega}_C^2 = (-1,1)^2$ or the reference cube $\hat{\Omega}_C^3 = (-1,1)^3$. We consider:

$$V_N(\hat{\Omega}) = (\mathbb{Q}_N)^d , \qquad\qquad Q_N(\hat{\Omega}) = \mathbb{Q}_N ; \qquad\qquad (3.5.9)$$

$$V_N(\hat{\Omega}) = (\mathbb{Q}_N)^d , \qquad\qquad Q_N(\hat{\Omega}) = \mathbb{Q}_{N-2} ; \qquad\qquad (3.5.10)$$

$$V_N(\hat{\Omega}) = (J_N \oplus E_{N-1})^d , \qquad Q_N(\hat{\Omega}) = \mathbb{Q}_{N-2} ; \qquad\qquad (3.5.11)$$

$$V_N(\hat{\Omega}) = (\mathbb{Q}_N)^d , \qquad\qquad Q_N(\hat{\Omega}) = \mathbb{Q}_{\lambda N} , \quad 0 < \lambda < 1; \quad (3.5.12)$$

$$V_N(\hat{\Omega}) = (\mathbb{Q}_N)^d , \qquad\qquad Q_N(\hat{\Omega}) = \mathbb{P}_{N-1} ; \qquad\qquad (3.5.13)$$

$$V_N(\hat{\Omega}) = (\mathbb{Q}_N \cap \mathbb{P}_{N+2})^d , \qquad Q_N(\hat{\Omega}) = \mathbb{P}_{N-1} ; \qquad\qquad (3.5.14)$$

$$V_N(\hat{\Omega}) = (\mathbb{Q}_N)^d , \qquad\qquad Q_N(\hat{\Omega}) = \mathbb{Q}_{N-2} \cup \mathbb{P}_{N-1} . \qquad (3.5.15)$$

In (3.5.11), $J_N = \mathbb{Q}_N \cap H_0^1(\hat{\Omega}) = \{bv : v \in \mathbb{Q}_{N-2}\}$ and b is the bubble $b(\hat{\mathbf{x}}) = \prod_{i=1}^d (1 - \hat{x}_i^2)$, whereas $E_{N-1} = (\mathbb{P}_1^{(x)} \otimes \mathbb{P}_{N-1}^{(y)}) \oplus (\mathbb{P}_{N-1}^{(x)} \otimes \mathbb{P}_1^{(y)})$ for $d = 2$ or $E_{N-1} = (\mathbb{P}_1^{(x)} \otimes \mathbb{Q}_{N-1}^{(y,z)}) \oplus (\mathbb{P}_1^{(y)} \otimes \mathbb{Q}_{N-1}^{(x,z)}) \oplus (\mathbb{P}_1^{(z)} \otimes \mathbb{Q}_{N-1}^{(x,y)})$ for $d = 3$.

The pair (3.5.9) admits spurious modes; all others have pseudo-spurious modes, with the exception of (3.5.12) and (3.5.13). The space $V_N(\hat{\Omega})$ in (3.5.11) has a slightly smaller dimension than $(\mathbb{Q}_N)^d$, with similar approximation properties; so does $V_N(\hat{\Omega})$ in (3.5.14). Finally, (3.5.15) provides the maximal space of non-spurious pressures associated with the choice $V_N(\hat{\Omega}) = (\mathbb{Q}_N)^d$ for velocities.

If $\hat{\Omega} = \hat{\Omega}_S^d = \{(\hat{x}_1, \ldots, \hat{x}_d) \in \mathbb{R}^d : -1 \le \hat{x}_i \text{ for all } i, \hat{x}_1 + \cdots + \hat{x}_d \le 0\}$ is the reference triangle ($d = 2$) or tetrahedron ($d = 3$), the admissible pairs of spaces are

$$V_N(\hat{\Omega}) = (\mathbb{P}_N)^d , \qquad\qquad Q_N(\hat{\Omega}) = \mathbb{P}_N ; \qquad\qquad (3.5.16)$$

$$V_N(\hat{\Omega}) = (\mathbb{P}_N)^d , \qquad\qquad Q_N(\hat{\Omega}) = \mathbb{P}_{N-2} ; \qquad (3.5.17)$$

$$V_N(\hat{\Omega}) = (\mathbb{P}_{N-1} \oplus (\mathbb{P}_N \cap H_0^1(\hat{\Omega})))^d , \qquad Q_N(\hat{\Omega}) = \mathbb{P}_{N-2} . \qquad (3.5.18)$$

The first pair obviously admits spurious modes (the divergence of a velocity in $(\mathbb{P}_N)^d$ belongs to $\mathbb{P}_{N-1}$). Neither of the two other pairs exhibit spurious modes, but have pseudo-spurious modes. The latter pair has a slightly smaller dimension than the former. We refer to Sect. 3.7.3, Part c), for further properties of these spaces.

3.6 Outflow Boundary Conditions

As noted in the opening section of this chapter, problems requiring an artificial boundary condition in the numerical solution require very careful treatment to ensure that any spurious wave reflections at an artificial boundary do not contaminate the flow in the region of interest. The inflow boundary conditions are usually well-known for transition problems. Lund, Wu and Squires (1998) have provided a procedure for specifying consistent inflow conditions for turbulence simulations. Treatment of far-field boundary conditions was discussed in Sect. 3.4.5. By far the most subtle artificial boundary conditions are those at the outflow boundary for spatially developing flows. Except in the near-wall region the advection terms ensure that all but the strongest disturbances are convected downstream. The streamwise viscous terms propagate information upstream through diffusion. The major challenge for incompressible flow problems is that pressure disturbances are instantaneously propagated throughout the flow because the speed of sound is effectively infinite. In this section we summarize two approaches to treating an unbounded domain in the direction of the mean flow, both described in the particular context of the true, spatially developing flat-plate boundary layer illustrated in Fig. 3.21.

3.6.1 Fringe Regions

Spalart (1988a, 1988b) developed two methods for approximating spatially developing flows that permit the use of a Fourier expansion in the x-direction: the fringe-region method and a multiscale method. We focus here on the former due to its fairly widespread use. Figure 3.25 illustrates the basic concept of the *fringe-region* method for the classical case of the boundary layer on a flat plate. The left half of the figure illustrates the boundary-layer thickness in the computational domain $(x_{\text{start}}, x_{\text{out}})$ under the assumption of periodicity in x with periodicity length $L = x_{\text{out}} - x_{\text{start}}$. The region of physical interest in the streamwise direction is $(x_{\text{in}}, x_{\text{out}})$.

The Navier–Stokes equations are modified in such a way that the flow in the domain of interest is a close approximation to the real flow in a spatially developing boundary layer, but in the *fringe region* $(x_{\text{start}}, x_{\text{in}})$ the disturbances which leave the periodic domain at $x = x_{\text{out}}$ are gradually damped so that they are insignificant by the time they have propagated from $x = x_{\text{start}}$ (due to periodicity) to x_{in}. The modified Navier–Stokes equations include an x-dependent forcing function added to the momentum equation to enforce the streamwise periodicity, i.e., (3.1.1) becomes

$$\frac{\partial \mathbf{u}}{\partial t} + \mathbf{u} \cdot \nabla \mathbf{u} = -\nabla p + \nu \Delta \mathbf{u} + \lambda(x)(\mathbf{u}_r - \mathbf{u}) \quad \text{in } \Omega , \qquad (3.6.1)$$

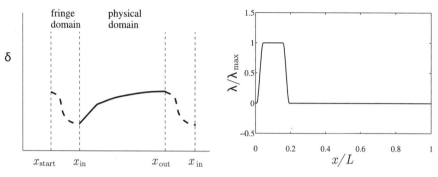

Fig. 3.25. Illustration of the computational domain for the fringe-region method (*left*) and the damping function λ (*right*)

where $\Omega = (x_{\text{start}}, x_{\text{out}}) \times (0, \infty)$ in the case of the boundary layer. The right half of the figure illustrates the desirable shape of the fringe function λ, i. e., vanishing in the physical domain, flat in most of the interior of the fringe region, while decaying smoothly to 0 at the boundaries of the fringe region. A suitable such function is

$$\lambda(x) = \lambda_{\max} \left[S \left(\frac{x - x_{\text{start}}}{d_{\text{rise}}} \right) - S \left(\frac{x - x_{\text{in}}}{d_{\text{fall}}} \right) + 1 \right] , \qquad (3.6.2)$$

where

$$S(x) = \begin{cases} 0 & x \leq 0 , \\ 1 / \left[1 + e^{\frac{1}{x-1} + \frac{1}{x}} \right] & 0 < x < 1 , \\ 1 & x \geq 1 , \end{cases} \qquad (3.6.3)$$

and d_{rise} and d_{fall} are parameters that control the widths of the thin increasing and decreasing regions of $\lambda(x)$, respectively. The vector $\mathbf{u}_r$ is a prescribed function that mimics the desired solution $\mathbf{u}$ at the inflow x_{in}. There is a trade-off between the relative length of the fringe region and the magnitude, $\lambda_{\max}$ of the damping function. Large $\lambda_{\max}$ produces more rapid damping (in x), but eventually restricts the size of the allowable time-step since this term is treated explicitly. A larger fringe region allows for more damping, but increases the cost of the computation due to the larger number of streamwise grid points. In practice, the fringe region typically takes 10–20% of the streamwise computational domain.

The specific description given above follows the version of the fringe method used by Lundblad et al. (1994), which is a slight adaptation of Spalart's version. Nordström, Nordin and Henningson (1999) have furnished a detailed analysis of this particular version of the fringe method for a linearized model of the incompressible Navier–Stokes equations. Any of the various spectral algorithms developed for temporally developing flows can be readily modified for application of the fringe-region method. The reader interested in further details should consult papers such as the two mentioned directly above, as well as the papers by Bertolotti, Herbert and Spalart (1992)

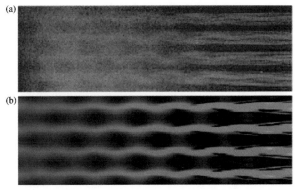

Fig. 3.26. Streamwise velocity flow visualizations of incompressible boundary-layer transition by Berlin, Wiegel and Henningson (1999): experiment (**a**) and spatial computation (**b**) [Reprinted with permission from S. Berlin, M. Wiegel, D.S. Hennigson (1999); © 1999, Cambridge University Press]

for boundary-layer stability simulations and by Spalart and Watmuff (1993) for turbulent flow simulations. (The direct numerical simulation results in Fig. 2.19 used the fringe-region method.) Although the fringe-region method solves an adulterated set of equations, and the resulting discrepancy between the fringe-region method solution and the true spatially developing solution is difficult to bound, numerous impressive studies have been performed with this technique.

The spatial simulation of oblique transition in a boundary layer on a $1200 \times 64 \times 96$ grid by Berlin, Wiegel and Henningson (1999) is a prime example of a high-resolution DNS using the fringe-region method with a Fourier–Chebyshev algorithm that is an adaptation to boundary-layer flow of the normal velocity–normal vorticity method discussed in Sect. 3.4.1. Figure 3.26 illustrates a comparison of their numerical results with flow visualizations of their experiment on transition in a boundary layer. The transition mechanism in this example is one of transient growth (see Sect. 2.6) rather than linear or secondary instability.

3.6.2 Buffer Domains

Streett and Macaraeg (1989) took a less drastic approach to handling the outflow boundary condition of a spatially developing flow. They retain genuine inflow and outflow boundaries, but augment the original physical domain of interest with a *buffer domain* tacked onto the downstream edge, as illustrated in Fig. 3.27. The Navier–Stokes equations are unchanged in the physical domain, but three modifications are applied in the buffer domain. First, the viscosity which appears in all the streamwise derivatives of all the viscous terms is multiplied by an attenuation function $\lambda(x)$ which smoothly diminishes from 1 to 0 in the streamwise direction. Second, this attenuation function multiplies the source terms in the pressure Poisson equation in

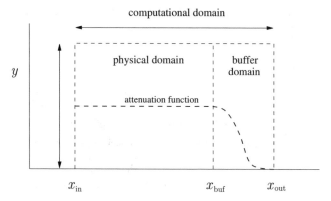

Fig. 3.27. Illustration of the computational domain for the buffer-domain method with the attenuation function overlaid

their splitting method. Third, the advection terms are linearized about the mean flow. The first two modifications reduce the upstream influence from the viscous terms and the pressure waves. The third modification ensures that the outflow boundary has only outgoing convected waves. The attenuation function chosen by Streett and Macaraeg was based upon a hyperbolic tangent function. Virtually all uses of the buffer-domain technique have been in the context of spectral methods for complex domains or for schemes which relied upon finite-difference methods in the flow direction (and often spectral methods in the other directions). The experience of the early tests of this buffer-domain approach by Street and Macaraeg (1989), Danabasoglu, Biringen and Streett (1990) and Joslin, Streett and Chang (1993), covering cylindrical Taylor–Couette flow, plane Poiseuille flow and flat-plate boundary flow, respectively, indicate that a buffer domain length of three wavelengths of the primary wave is sufficient. (The spatial discretization in the first case was fully spectral collocation, but with a multidomain Chebyshev discretization using patching (see Sect. 5.13) in the streamwise direction. In the second case a fourth-order finite-difference method was used in all directions. In the third case, Chebyshev collocation was used in y, Fourier collocation in z and fourth-order finite differences in x.) As noted by Rempfer (2003), the buffer-domain method has become the standard method for handling outflow boundary conditions in bonafide, spatially developing incompressible flow simulations, regardless of the type of spatial discretization.

3.7 Analysis of Spectral Methods for Incompressible Flows

This section is devoted to the mathematical analysis of spectral methods for the incompressible Navier–Stokes equations. Here we draw upon the general framework for the theoretical analysis of spectral methods that was presented in CHQZ2, Chaps. 5 and 7. Our investigation will be confined to time-independent problems, as the spectral schemes are invariably used to handle the spatial discretization only. The main purpose of the mathematical analysis is to prove the convergence of the approximate polynomial solution to the exact solution of the Navier–Stokes equations, as the degree of the polynomials tends to infinity. An isolated (or nonsingular) solution is considered in most cases, but the analysis can be extended to cover turning points and bifurcations as well. Error estimates in Sobolev norms are obtained, and they show that the error decays spectrally for infinitely smooth solutions.

Since the focus of this chapter has been on algorithms for the primitive variable formulation of incompressible flows, we confine the details of the theory to this formulation. Single-domain results for the pure streamfunction formulation, which requires the discretization of a fourth-order problem, can be found in Maday and Métivet (1987), and Bernardi and Maday (1988a) (see also Bernardi and Maday (1992)). The vorticity-streamfunction formulation, which leads to a saddle-point problem of the type considered in Sect. 3.7.2, is analyzed, e. g., in Bernardi and Maday (1997), Sect. 28.

In the primitive variable formulation, the velocity and the pressure cannot be approximated independently. A compatibility condition must be satisfied by the finite-dimensional spaces in which they are sought in order to have a solvable system. This is a well-known issue in the finite-element method (see, e. g., Girault and Raviart (1986) and Brezzi and Fortin (1992)), and it has been already mentioned in Sect. 3.5.1. In those spectral methods for which the continuity equation is discretized directly, the pressure can be affected by spurious components (parasitic modes) that deteriorate the accuracy of the method (see the following Sect. 3.7.1). Parasitic modes can be completely characterized mathematically, and the theoretical analysis can even indicate a way to filter them out (see Sect. 3.7.2). Other spectral methods, like the Kleiser–Schumann method described in Sect. 3.4.1 or the Zang–Hussaini method described in Sect. 3.4.3, replace the continuity equation by an equation for the pressure. In these cases, parasitic modes may be implicitly filtered out by the solution process.

We consider the steady Navier–Stokes equations in primitive variables in a square domain $\Omega \subset \mathbb{R}^2$ of the plane under periodic and/or homogeneous Dirichlet boundary conditions. Thus, Ω is either $(0, 2\pi)^2$ (if fully periodic boundary conditions are imposed) or $(0, 2\pi) \times (-1, 1)$ (periodic/Dirichlet boundary conditions) or $(-1, 1)^2$ (fully Dirichlet boundary conditions). There is no difficulty in extending our discussion to the three-dimensional case. The results in the literature are actually given for three-dimensional problems, but we choose here the two-dimensional geometry for ease of exposition.

The velocity $\mathbf{u}$ belongs to the subspace V of $(H^1(\Omega))^2$ of the vector fields that satisfy the prescribed boundary conditions. The space $H^1(\Omega)$ is defined in Sect. A.8(a). V is a closed subspace under the norm of $(H^1(\Omega))^2$, which we denote by $\|\mathbf{v}\|_V$. The pressure p belongs to the closed subspace Q of $L^2(\Omega)$ of the functions with zero average in Ω. The norm $\|q\|_Q$ is the $L^2(\Omega)$ norm.

We have in mind to approximate the problem of seeking $\mathbf{u} \in V$ and $p \in Q$ satisfying

$$-\nu \Delta \mathbf{u} + \nabla p + \mathbf{u} \cdot \nabla \mathbf{u} = \mathbf{f} \quad \text{in } \Omega ,$$
$$\nabla \cdot \mathbf{u} = 0 \quad \text{in } \Omega . \tag{3.7.1}$$

The external force $\mathbf{f}$ is supposed to be square integrable in Ω, or even continuous in $\overline{\Omega}$ (more generally $\mathbf{f}$ can be an element in the dual space V' of V (see Sect. A.1(c)).

First, we observe that the analysis of an approximation to the full Navier–Stokes problem can be reduced to the analysis of the corresponding approximation to the Stokes problem, by resorting to an implicit function theorem or to more sophisticated functional analysis results. More precisely, consider the problem of seeking $\mathbf{u} \in V$ and $p \in Q$ satisfying

$$-\nu \Delta \mathbf{u} + \nabla p = \mathbf{g} \quad \text{in } \Omega ,$$
$$\nabla \cdot \mathbf{u} = 0 \quad \text{in } \Omega , \tag{3.7.2}$$

and define the operator $T : V' \to V \times Q$ as $T\mathbf{g} = (\mathbf{u}, p)$. Then define the nonlinear operator $G : V \times Q \to V'$ as $G(\mathbf{u}, p) = \mathbf{f} - \mathbf{u} \cdot \nabla \mathbf{u}$. Setting $\mathbf{w} = (\mathbf{u}, p)$ and $F(\mathbf{w}) = \mathbf{w} - TG(\mathbf{w})$, the solution $\mathbf{w}$ of (3.7.1) can be characterized by the equation

$$F(\mathbf{w}) = \mathbf{0} . \tag{3.7.3}$$

Consider now a spectral approximation of (3.7.1), in which the velocity $\mathbf{u}^N$ is sought in a subspace V_N of V, the pressure p^N in a subspace Q_N of Q, the Stokes operator T is approximated by an operator $T_N : V' \to V_N \times Q_N$ and the nonlinear operator G by an operator $G_N : V_N \times Q_N \to V'$. Set $\mathbf{w}^N = (\mathbf{u}^N, p^N)$ and $F_N(\mathbf{w}^N) = \mathbf{w}^N - T_N G_N(\mathbf{w}^N)$. As before, the approximate solution is defined by the equation

$$F_N\left(\mathbf{w}^N\right) = \mathbf{0} . \tag{3.7.4}$$

Abstract theorems concerning the approximation of nonlinear equations like (3.7.3) by a family of finite-dimensional equations like (3.7.4) can be invoked at this point (see, e.g., Brezzi, Rappaz and Raviart (1980, 1981a, 1981b), and Caloz and Rappaz (1997)). One can even take into account in the formulation the dependence of the solution upon the Reynolds number $\lambda = \nu^{-1}$, i.e., replace $F(\mathbf{w})$ by $F(\lambda, \mathbf{w})$ in (3.7.3) and $F_N(\mathbf{w}^N)$ by $F_N(\lambda, \mathbf{w}^N)$ in (3.7.4). An example of such abstract results, in the case of isolated solutions, is Theorem 7.3 given in CHQZ2, Sect. 7.8. There it is shown how to use it in the analysis of spectral approximations to the Burgers equation. The analysis for the Navier–Stokes equations follows the same guidelines, although it is technically more involved (see, e.g., Bernardi and Maday (1997),

Sect. 27). In order to check the assumptions, one exploits the approximation properties of T_N to T, using the stability results sketched hereafter and the consistency results given in CHQZ2, Chap. 5. Since G_N is typically the G-NI or collocation realization of G, estimates on the error between G_N and G can easily be obtained from the approximation results for the Fourier or polynomial interpolation, given again in CHQZ2, Chap. 5. Thus, the crucial point is the analysis of the error in the Stokes approximation. Hence, from now on we will deal with the Stokes problem (3.7.2) only.

3.7.1 Compatibility Conditions Between Velocity and Pressure

In most of the spectral algorithms set in a reference Cartesian-product domain, the space V_N of the discrete velocities is the subspace of V of the polynomials of degree $\leq N$ in each variable. Here by polynomial we mean trigonometric polynomial in the direction(s) of periodicity, and algebraic polynomial in the remaining direction(s). From now on, we assume that V_N is defined in this way. A few exceptions exist in the literature, e. g., the method by Moser, Moin and Leonard (1983) (see Sect. 3.4.2), where the discrete velocities individually satisfy the continuity equation; hence they span a proper subspace of our V_N.

With our choice of the subspace V_N for the velocities, the most natural candidate for the space Q_N of the pressures is the subspace $\mathcal{Q}_N$ of Q of all the polynomials of degree $\leq N$ in each variable. However, such a space may be "too large" compared to the space V_N, in the sense that the approximation scheme may fail to define a unique pressure p_N in $\mathcal{Q}_N$ (the corresponding algebraic system being underspecified). This negative phenomenon has been already mentioned in Sect. 3.4.3. When it occurs, Q_N has to be restricted to a proper subspace of $\mathcal{Q}_N$, or in other words, one has to satisfy a *compatibility condition* between the spaces of velocity and pressure.

In order to investigate the phenomenon, let us assume that the momentum equation in (3.7.2) is discretized by a projection method of Galerkin, G-NI, collocation or tau type. All the existing algorithms treat the momentum equation by one of these methods. They differ primarily in the approximation of the continuity equation. For the sake of simplicity, we start by considering a Galerkin projection method with respect to the $L^2(\Omega)$ inner product $(\mathbf{u}, \mathbf{v})$; the G-NI and collocation approaches will be addressed later on. Thus, here we couple a Fourier Galerkin method in the direction(s) of periodicity with a Legendre Galerkin method in the other direction(s). The discretization of the momentum equation reads as follows: $\mathbf{u}^N \in V_N$ and $p^N \in \mathcal{Q}_N$ satisfy

$$-\nu(\Delta \mathbf{u}^N, \mathbf{v}) + (\nabla p^N, \mathbf{v}) = (\mathbf{g}, \mathbf{v}) \qquad \text{for all } \mathbf{v} \in V_N . \tag{3.7.5}$$

(The equivalent expression $\nu(\nabla \mathbf{u}^N, \nabla \mathbf{v}) - (p^N, \nabla \cdot \mathbf{v}) = (\mathbf{g}, \mathbf{v})$, which stems from integration by parts using the boundary conditions, can be used instead.) Now assume that there exist nonzero $p \in \mathcal{Q}_N$ such that

$$(\nabla p, \mathbf{v}) = 0 \quad \text{for all } \mathbf{v} \in V_N . \tag{3.7.6}$$

Then each couple $(\mathbf{u}^N, p^N + p)$ is also a solution of (3.7.5). Thus, the pressure is not uniquely determined by (3.7.5). The pressures $p \in \mathcal{Q}_N$ that satisfy (3.7.6) are called *spurious modes* or *parasitic modes*. They form a linear subspace $X_N \subset \mathcal{Q}_N$. In order to characterize X_N, we have to distinguish among different combinations of boundary conditions.

(a) *Fully periodic problem*

The space of velocities V_N is $[S_N \cap Q]^2$, where S_N denotes the space of the trigonometric polynomials of degree $\leq N$ in each variable and Q is, as before, the space of the square integrable functions with zero average in Ω. (We choose this normalization for the velocities since their zero mode is not affected by (3.7.5).) The space of pressures $\mathcal{Q}_N$ is $S_N \cap Q$. If $p \in \mathcal{Q}_N$, then $\nabla p \in V_N$; hence choosing $\mathbf{v} = \nabla p$ in (3.7.6), one obtains $\nabla p \equiv 0$; thus, $p = 0$. It follows that for pure Fourier methods there are no spurious modes, i.e., $X_N = \{0\}$. The spaces V_N, $\mathcal{Q}_N$ are precisely those used in the Orszag–Patterson method.

(b) *Mixed periodic–nonperiodic problem*

V_N is the space $[S_N \otimes \mathbb{P}_N^0]^2$, where S_N is now the space of the trigonometric polynomials of degree $\leq N$ in one variable, and $\mathbb{P}_N^0$ is the subspace of $\mathbb{P}_N$ of the algebraic polynomials in one variable that vanish at $y = \pm 1$. The space $\mathcal{Q}_N$ is $[S_N \otimes \mathbb{P}_N] \cap Q$.

Let us represent $p \in \mathcal{Q}_N$ as $p(x, y) = \sum_{|k|, n \leq N} \hat{p}_{kn} e^{ikx} L_n(y)$, where $L_n(y)$ denotes the n-th Legendre polynomial. Equation (3.7.6) is equivalent to

$$\left(p, \frac{\partial v}{\partial y} \right) = 0 \quad \text{for all } v \in S_N \otimes \mathbb{P}_N^0 , \tag{3.7.7}$$

$$\left(p, \frac{\partial v}{\partial x} \right) = 0 \quad \text{for all } v \in S_N \otimes \mathbb{P}_N^0 . \tag{3.7.8}$$

A basis in $S_N \otimes \mathbb{P}_N^0$ is given by

$$\left\{ e^{ikx}(1 - y^2) L_m'(y) \colon |k| \leq N , \ 1 \leq m \leq N - 1 \right\} . \tag{3.7.9}$$

Using this basis in (3.7.7) and taking into account the differential equation (8.4.1) satisfied by the Legendre polynomials, one obtains the set of equations

$$\hat{p}_{km} = 0 \quad \text{for } |k| \leq N , \ 1 \leq m \leq N - 1 , \tag{3.7.10}$$

which are equivalent to (3.7.7). Another basis for $S_N \otimes \mathbb{P}_N^0$ is given by

$$\left\{ e^{ikx}[L_m(y) - L_\alpha(y)] : |k| \leq N , \ 2 \leq m \leq N \right\} , \qquad (3.7.11)$$

where $\alpha = m(\mathrm{mod}\,2)$. Using this basis in (3.7.8) yields the set of equations

$$(L_m, L_m)\hat{p}_{km} = (L_\alpha, L_\alpha)\hat{p}_{k\alpha} \quad 0 < |k| \leq N , \ 2 \leq m \leq N . \qquad (3.7.12)$$

By (3.7.10) and (3.7.12), it follows that

$$\hat{p}_{km} = 0 \quad \text{for all } k, m, \text{ except } \hat{p}_{00} \text{ and } \hat{p}_{0N} . \qquad (3.7.13)$$

On the other hand, $\hat{p}_{00} = 0$ since $p \in \mathcal{Q}_N \subset Q$. We conclude that there exists one spurious mode; precisely we have $X_N = \mathrm{span}\{L_N(y)\}$.

(c) *Fully nonperiodic problem*
V_N is the space $(\mathbb{Q}_N)^2 \cap V = [\mathbb{P}_N^0 \otimes \mathbb{P}_N^0]^2$, and $\mathcal{Q}_N = \mathbb{Q}_N \cap Q$. We proceed as in the previous case, using now as test function v the product of a basis function $(1 - s^2)L_m'(s)$ in the direction of differentiation, and a basis function $L_m(s) - L_\alpha(s)$ in the other direction. It is easily seen that the non-trivial solutions of (3.7.6) are spanned by the four modes

$$L_i(x)L_j(y) , \quad i, j = 0 \text{ or } N , \qquad (3.7.14)$$

and by four other functions $\overline{p}_{ij}$ $(i, j = 0 \text{ or } 1)$ that are suitable combinations of the remaining modes with the same parity:

$$\overline{p}_{ij} = \sum_{\substack{k=i(\mathrm{mod}\,2) \\ m=j(\mathrm{mod}\,2)}} \overline{c}_{ijkm} L_k(x)L_m(y) . \qquad (3.7.15)$$

Thus, we have seven spurious modes, i.e., $\dim X_N = 7$. Spurious pressure modes were first characterized mathematically by Bernardi, Maday and Métivet (1986).

When a G-NI collocation method is used with the tensor product of the Gauss–Lobatto nodes x_j $(j = 0, \ldots, N)$ as quadrature or collocation points, then the spurious modes are spanned by the seven functions

$$\begin{cases} L_i(x)L_j(y) , & (i, j) = (N, 0), (0, N), (N, N) , \text{ and} \\ L_N'(x)L_N'(y)(1 \pm x)(1 \pm y) . \end{cases} \qquad (3.7.16)$$

The last four modes are called *corner modes*, as they vanish in all but one corner of the domain. It is worth noticing that in space dimension $d > 2$, the number of spurious modes actually depends on N, for instance $\dim X_N = 12N + 3$ when $d = 3$ (Bernardi and Maday (1992)). In the latter case, there are seven modes of the type $L_i(x)L_j(y)L_k(z)$ where i, j, k take the values 0 or N and are not all equal to 0. The remaining modes are obtained by taking the tensor product of any corner mode as defined above with any polynomial in $\mathbb{P}_N$, in all possible combinations of space directions.

If a Chebyshev method is used instead, the previous results hold after replacing each L_l by T_l, the l-th Chebyshev polynomial.

3.7.2 Direct Discretization of the Continuity Equation: The Inf-sup Condition

The most direct way of enforcing the divergence-free condition on the numerical flow consists of approximating the continuity equation by one of the projection methods (Galerkin, G-NI, collocation or tau) considered in this book. The Galerkin method is conceptually the simplest one, and it leads to the following approximation of the Stokes problem (3.7.2): one seeks $\mathbf{u}^N \in V_N$ and $p^N \in Q_N$ satisfying

$$
\begin{aligned}
-\nu(\Delta \mathbf{u}^N, \mathbf{v}) + (\nabla p^N, \mathbf{v}) = (\mathbf{g}, \mathbf{v}) \quad &\text{for all } \mathbf{v} \in V_N , \\
(\nabla \cdot \mathbf{u}^N, q) = 0 \qquad &\text{for all } q \in Q_N .
\end{aligned}
\tag{3.7.17}
$$

An equivalent formulation of the equations, obtained after integration by parts in the momentum equation, taking into account the (periodic and/or no-slip) boundary conditions, is as follows:

$$
\begin{aligned}
\nu(\nabla \mathbf{u}^N, \nabla \mathbf{v}) - (\nabla \cdot \mathbf{v}, p^N) = (\mathbf{g}, \mathbf{v}) \quad &\text{for all } \mathbf{v} \in V_N , \\
-(\nabla \cdot \mathbf{u}^N, q) = 0 \qquad &\text{for all } q \in Q_N
\end{aligned}
\tag{3.7.18}
$$

(the change of sign in the continuity equation is convenient for the subsequent analysis).

We recall that V_N is the space of the polynomial fields of degree $\leq N$ in each variable that satisfy the boundary conditions; Q_N denotes a suitable subspace of the space $\mathcal{Q}_N$ of the polynomials of degree $\leq N$ in each variable that have zero average over the domain Ω; finally, $(\mathbf{u}, \mathbf{v})$ denotes the $L^2(\Omega)$ inner product. The Galerkin method used in practice for fully periodic problems (the Orszag–Patterson method described in Sect. 3.3) can actually be written as (3.7.17). For nonperiodic problems, the tau method can be written in the form (3.7.17), except that the test functions $\mathbf{v}$ for the momentum equation are polynomials of degree $N - 2$ in the nonperiodic directions, and they need not satisfy the boundary conditions. The method by Moin and Kim (1980) but with a conventional tau discretization of the momentum equation (see Sect. 3.4.1) is precisely of this type.

The G-NI approach is a variant of formulation (3.7.18), in which the L^2-inner products are approximated by Gaussian quadratures, possibly of different type for the diffusion term and the pressure terms. Denoting the resulting inner products by $((\,.\,,\,.\,))_N$ and $(\,.\,,\,.\,)_N$ respectively, we obtain the G-NI scheme, which consists of looking for $\mathbf{u}^N \in V_N$ and $p^N \in Q_N$ satisfying

$$
\begin{aligned}
\nu((\nabla \mathbf{u}^N, \nabla \mathbf{v}))_N - (\nabla \cdot \mathbf{v}, p^N)_N = ((\mathbf{g}, \mathbf{v}))_N \quad &\text{for all } \mathbf{v} \in V_N , \\
-(\nabla \cdot \mathbf{u}^N, q)_N = 0 \qquad &\text{for all } q \in Q_N .
\end{aligned}
\tag{3.7.19}
$$

The G-NI approach is used within the spectral element method (Patera (1984)); see Sect. 5.6 for more details. If all quadrature formulas are based on the N-degree Legendre Gauss–Lobatto quadrature nodes, then the G-NI scheme coincides with the collocation scheme at these points. Other collocation schemes use different grids for the two sets of equations (see e. g., Métivet (1987), Bernardi and Maday (1986)); this results precisely in two different discrete inner products to be inserted in (3.7.18).

General Theory. The Galerkin and G-NI methods, as well as certain collocation methods, can be cast in the following abstract form, often referred to as a *saddle-point* formulation. One seeks $\mathbf{u}^N \in V_N$ and $p^N \in Q_N$, which satisfy

$$
\begin{aligned}
a_N(\mathbf{u}^N, \mathbf{v}) + b_N(\mathbf{v}, p^N) &= ((\mathbf{g}, \mathbf{v}))_N \quad && \text{for all } \mathbf{v} \in V_N \ , \\
b_N(\mathbf{u}^N, q) &= 0 && \text{for all } q \in Q_N \ ,
\end{aligned}
\tag{3.7.20}
$$

where $a_N : V_N \times V_N \to \mathbb{R}$ is a suitable approximation of the form $a(\mathbf{u}, \mathbf{v}) = \nu(\nabla\mathbf{u}, \nabla\mathbf{v})$, whereas $b_N : V_N \times Q_N \to \mathbb{R}$ is an approximation of $b(\mathbf{v}, q) = -(\nabla \cdot \mathbf{v}, q)$ and $((\mathbf{g}, \mathbf{v}))_N$ is an approximation of $(\mathbf{g}, \mathbf{v})$ (obviously, for the pure Galerkin method, each form actually coincides with the corresponding exact one). We observe that the bilinear forms $a(\mathbf{u}, \mathbf{v})$ and $b(\mathbf{v}, q)$ are continuous, in the sense that there exist two constants $\gamma > 0$ and $\delta > 0$ such that

$$
|a(\mathbf{u}, \mathbf{v})| \leq \gamma \|\mathbf{u}\|_V \|\mathbf{v}\|_V \quad \text{for all } \mathbf{u}, \mathbf{v} \in V \ , \tag{3.7.21}
$$
$$
|b(\mathbf{v}, q)| \leq \delta \|\mathbf{v}\|_V \|q\|_Q \quad \text{for all } \mathbf{v} \in V, \text{ for all } q \in Q \ . \tag{3.7.22}
$$

Since V_N and Q_N are finite-dimensional spaces, there exist constants $\gamma_N > 0$ and $\delta_N > 0$ such that estimates similar to the previous ones hold for the discrete forms as well; precisely,

$$
|a_N(\mathbf{u}, \mathbf{v})| \leq \gamma_N \|\mathbf{u}\|_V \|\mathbf{v}\|_V \quad \text{for all } \mathbf{u}, \mathbf{v} \in V_N \ , \tag{3.7.23}
$$
$$
|b_N(\mathbf{v}, q)| \leq \delta_N \|\mathbf{v}\|_V \|q\|_Q \quad \text{for all } \mathbf{v} \in V_N, \text{ for all } q \in Q_N \ . \tag{3.7.24}
$$

Concerning the right-hand side of the discrete momentum equation in (3.7.20), we suppose that there exists a constant $C(\mathbf{g}) \geq 0$ depending on $\mathbf{g}$ such that

$$
|((\mathbf{g}, \mathbf{v}))_N| \leq C(\mathbf{g}) \|\mathbf{v}\|_V \quad \text{for all } \mathbf{v} \in V_N \ . \tag{3.7.25}
$$

This condition holds, e. g., if $\mathbf{g} \in (C^0(\bar{\Omega}))^2$ and the discrete L^2-norm $\|\mathbf{v}\|_N$ is uniformly equivalent to the exact L^2-norm of $\mathbf{v}$ in V_N.

In order to discuss the stability and convergence properties of the approximation (3.7.20), we make use of a general result of the approximation of saddle-point problems such as (3.7.20) (see Brezzi (1974)), based on the so-called *inf-sup condition*. According to Brezzi's theorem, any such problem has a unique solution if the following two conditions are satisfied:

(i) *Setting*

$$
Z_N = \{ \mathbf{v} \in V_N \,|\, b_N(\mathbf{v}, q) = 0 \text{ for all } q \in Q_N \} \ , \tag{3.7.26}
$$

there exists a constant $\alpha_N > 0$ such that

$$a_N(\mathbf{v}, \mathbf{v}) \geq \alpha_N \|\mathbf{v}\|_V^2 \quad \text{for all } \mathbf{v} \in Z_N ; \tag{3.7.27}$$

(ii) *there exists a constant $\beta_N > 0$ such that*

$$\sup_{\mathbf{v} \in V_N} \frac{b_N(\mathbf{v}, q)}{\|\mathbf{v}\|_V} \geq \beta_N \|q\|_Q \quad \text{for all } q \in Q_N . \tag{3.7.28}$$

The latter condition, equivalently written as

$$\inf_{q \in Q_N} \sup_{\mathbf{v} \in V_N} \frac{b_N(\mathbf{v}, q)}{\|\mathbf{v}\|_V} \geq \beta_N > 0 ,$$

is precisely the inf-sup condition, which plays a crucial role in the subsequent analysis. The constant β_N will be referred to as the *inf-sup constant*.

Let us represent the system of equations (3.7.20) in matrix form with respect to (modal or nodal) bases in V_N and Q_N as

$$\begin{pmatrix} A & B \\ B^T & 0 \end{pmatrix} \begin{pmatrix} \mathbf{u} \\ \mathbf{p} \end{pmatrix} = \begin{pmatrix} \mathbf{g} \\ \mathbf{0} \end{pmatrix} \tag{3.7.29}$$

(i.e., $\mathbf{u}$ is the vector of the expansion coefficients of $\mathbf{u}^N$ with respect to the chosen basis in V_N, and similarly for $\mathbf{p}$). Then, condition (3.7.27) says that A is positive-definite on the space $\mathcal{Z}$ of the vectors $\mathbf{v}$ satisfying $B^T \mathbf{v} = \mathbf{0}$, whereas condition (3.7.28) is equivalent to the fact that B has maximal rank. A solution procedure for this linear system, based on the Uzawa algorithm, is discussed in Sect. 7.3.2 and analyzed in Sect. 3.7.4.

Condition (3.7.27) can be weakened to require that A be just nonsingular on $\mathcal{Z}$ (see Brezzi (1974), or the subsequent condition (3.7.50)). The weaker condition together with (3.7.28) is not only sufficient but also necessary for the well-posedness of problem (3.7.20).

Under the assumptions (3.7.27)–(3.7.28) and (3.7.25), the following estimate on the solution of (3.7.20) holds:

$$\|\mathbf{u}^N\|_V \leq \frac{1}{\alpha_N} C(\mathbf{g}) , \tag{3.7.30}$$

$$\|p^N\|_Q \leq \frac{1}{\beta_N} \left(1 + \frac{\gamma_N}{\alpha_N}\right) C(\mathbf{g}) . \tag{3.7.31}$$

It follows that the approximation (3.7.20) is stable if the constants α_N and β_N introduced in (3.7.27)–(3.7.28), as well as the constant γ_N defined in (3.7.23), are independent of N.

Let us now discuss convergence. The simplest situation occurs for the Galerkin method, for which all discrete forms coincide with the corresponding exact ones. The following abstract error estimates hold (see Brezzi and Fortin (1992)):

$$\|\mathbf{u} - \mathbf{u}^N\|_V \leq \left(1 + \frac{\gamma}{\alpha_N}\right) \inf_{\mathbf{w} \in Z_N} \|\mathbf{u} - \mathbf{w}\|_V + \frac{\delta}{\alpha_N} \inf_{q \in Q_N} \|p - q\|_Q \quad (3.7.32)$$

(where $(\mathbf{u}, p)$ is the solution of (3.7.2) and Z_N is defined in (3.7.26) with $b_N = b$), and

$$\begin{aligned}
\|p - p^N\|_Q &\leq \frac{\gamma}{\beta_N}\left(1 + \frac{\gamma}{\alpha_N}\right) \inf_{\mathbf{w} \in Z_N} \|\mathbf{u} - \mathbf{w}\|_V \\
&+ \left(1 + \frac{\delta}{\beta_N}\left(1 + \frac{\gamma}{\alpha_N}\right)\right) \inf_{q \in Q_N} \|p - q\|_Q \, .
\end{aligned} \quad (3.7.33)$$

In the G-NI or collocation case, the error estimates also account for the approximation of the bilinear and linear forms. Precisely, let us define the approximation errors

$$\begin{aligned}
\mathcal{E}_N(\mathbf{u}) &= \inf_{\mathbf{v} \in V_N} \left((\gamma + \gamma_N)\|\mathbf{u} - \mathbf{v}\|_V + \sup_{\mathbf{z} \in V_N} \frac{a(\mathbf{v}, \mathbf{z}) - a_N(\mathbf{v}, \mathbf{z})}{\|\mathbf{z}\|_V}\right) , \\
\mathcal{E}_N(p) &= \inf_{q \in Q_N} \left(\delta\|p - q\|_Q + \sup_{\mathbf{z} \in V_N} \frac{b(\mathbf{z}, q) - b_N(\mathbf{z}, q)}{\|\mathbf{z}\|_V}\right) , \\
\mathcal{E}_N(\mathbf{g}) &= \sup_{\mathbf{z} \in V_N} \frac{(\mathbf{g}, \mathbf{z}) - ((\mathbf{g}, \mathbf{z}))_N}{\|\mathbf{z}\|_V} .
\end{aligned} \quad (3.7.34)$$

Then, the velocity error is bounded as

$$\begin{aligned}
\|\mathbf{u} - \mathbf{u}^N\|_V &\leq \left(1 + \frac{\gamma_N}{\alpha_N}\right) \inf_{\mathbf{w} \in Z_N} \|\mathbf{u} - \mathbf{w}\|_V \\
&+ \frac{1}{\alpha_N} \left(\mathcal{E}_N(\mathbf{u}) + \mathcal{E}_N(p) + \mathcal{E}_N(\mathbf{g})\right) ,
\end{aligned} \quad (3.7.35)$$

whereas the pressure error is bounded as

$$\|p - p^N\|_Q \leq \frac{1}{\beta_N}\left(\|\mathbf{u} - \mathbf{u}^N\|_V + \mathcal{E}_N(\mathbf{u}) + \tilde{\mathcal{E}}_N(p) + \mathcal{E}_N(\mathbf{g})\right) , \quad (3.7.36)$$

where $\tilde{\mathcal{E}}_N(p)$ is defined as $\mathcal{E}_N(p)$ except that δ is replaced by $\delta + \beta_N$.

The first infimum on the right-hand side of (3.7.32) or (3.7.35) can be bounded in terms of the best approximation error in the whole V_N; precisely, one has

$$\inf_{\mathbf{w} \in Z_N} \|\mathbf{u} - \mathbf{w}\|_V \leq \left(1 + \frac{\delta}{\beta_N}\right) \inf_{\mathbf{v} \in V_N} \left\{\|\mathbf{u} - \mathbf{v}\|_V + \sup_{q \in Q_N} \frac{b(\mathbf{v}, q) - b_N(\mathbf{v}, q)}{\|q\|_Q}\right\} \quad (3.7.37)$$

(see, e. g., Brezzi and Fortin (1992)). However, this bound might not be optimal, due to the presence of the constant β_N (possibly decaying to 0 for increasing N) at the denominator. Often, a direct construction of a particular approximation $\mathbf{w} \in Z_N$ leads to a better error estimate, as shown later on.

3.7.3 Specific Applications

We now apply the previous general results to the spectral approximation (3.7.20) of the Stokes problem. It is clear from (3.7.30)–(3.7.36) that
the discrete spaces V_N and Q_N for velocities and pressures, as well as the
bilinear forms $a_N(\mathbf{u}, \mathbf{v})$ and $b_N(\mathbf{v}, q)$, should be chosen in such a way that
the constants α_N, β_N and γ_N are independent of N. If this is not the case,
α_N and β_N should decay with N, and γ_N should blow up with N at the
lowest possible rate. In the latter situation, the estimates mentioned above
will still imply convergence, provided that the infima on the right-hand side
decay to zero sufficiently fast to compensate for the growth of the constants.

In most situations of interest, α_N is indeed bounded from below, and γ_N is
bounded from above, uniformly in N. This occurs whenever $(a_N(\mathbf{v}, \mathbf{v}))^{1/2}$ is
a discrete norm of the gradient of $\mathbf{v}$, which turns out to be uniformly equiva-
lent to $(a(\mathbf{v}, \mathbf{v}))^{1/2} = \left(\nu \int_\Omega |\nabla \mathbf{v}|^2 \, dx \, dy\right)^{1/2}$. The last expression is equivalent
to the standard $H^1(\Omega)$ norm of V, thanks to the Poincaré inequality (A.10).
Note however that there are cases in which the uniform behavior of α_N does
not occur (see below), although the definition of V_N is different from the one
assumed in the present discussion.

So we concentrate on condition (3.7.28), which represents the true compatibility condition between V_N and Q_N. We recall (see above) that we have
chosen as V_N the subspace of V of the polynomial vectors of degree $\leq N$ in
each variable. Then, it is clear that Q_N cannot contain any element in the
space of spurious modes X_N, for otherwise the left-hand side of (3.7.28) would
be zero on these elements, while the right-hand side would not. Therefore, an
admissible choice for Q_N is any supplementary space of X_N in $\mathcal{Q}_N$, i. e., any
subspace of $\mathcal{Q}_N$ such that $\dim Q_N + \dim X_N = \dim \mathcal{Q}_N$ and $Q_N \cap X_N = \{0\}$.
More flexibility can be obtained by choosing as Q_N a proper subspace of
a supplementary space of X_N in $\mathcal{Q}_N$. Both situations are expressed by the
conditions

$$Q_N \subseteq \mathcal{Q}_N , \quad Q_N \cap X_N = \{0\} . \tag{3.7.38}$$

Any space Q_N satisfying these conditions will also satisfy (3.7.28) for a suitable constant β_N. Actually, if the left-hand side of (3.7.28) is zero, then
$q \in X_N$; hence, by (3.7.38), $q = 0$. Thus, the left-hand side of (3.7.28) is
a norm on Q_N, and therefore it is equivalent to any other norm on it, due to
the finite dimension of Q_N. We conclude that for each Q_N satisfying (3.7.38),
the Stokes approximation (3.7.17) has a unique solution $(\mathbf{u}^N, p^N)$.

Among the spaces satisfying (3.7.38), a natural candidate is the orthogonal space of X_N in $\mathcal{Q}_N$, with respect to the chosen (continuous or discrete)
L^2-inner product in $\mathcal{Q}_N$. However, this choice of Q_N may lead to a poor
approximation of the pressure; hence, it is usually discarded. Conversely, the
choice of a proper subspace of a supplementary space of X_N in $\mathcal{Q}_N$ may
have the positive effect of ruling out certain pseudo-spurious pressure modes
(see Sect. 3.5.1), yielding a larger constant β_N, even independent of N. Ob-

viously, too small a space Q_N would compromise again the accuracy of the approximation.

Finally, we mention that some algorithms actually use the whole of Q_N to represent the pressure. The linear systems that arise from these algorithms for the Stokes problem are singular because of the presence of spurious modes. The solution procedures, whether direct or iterative, select one particular pressure in Q_N, which can be represented as the sum of the "correct" pressure (lying in some supplementary space Q_N) and a certain linear combination of the spurious modes (in X_N). A filtering procedure is then applied to remove the spurious modes.

Let us now consider in some detail the three boundary conditions for the Stokes problem.

(a) *Fully periodic problem*

We consider the Galerkin scheme. Since $X_N = \{0\}$, we can choose $Q_N = \mathcal{Q}_N$. Then (3.7.28) is fulfilled with an inf-sup constant $\beta_N = \beta > 0$ independent of N (see Bernardi, Maday, and Métivet (1987b)). It is easily seen that the space Z_N defined in (3.7.26) is the subspace of V_N of the divergence-free velocities over all Ω. If $P_N : L^2(\Omega) \to S_N$ denotes the truncation of the Fourier series (see CHQZ2, formula (5.8.3)), then $\nabla \cdot (P_N \mathbf{u}) = P_N \nabla \cdot \mathbf{u} = 0$ if $\mathbf{u}$ is divergence-free. Thus, we can choose $\mathbf{w} = P_N \mathbf{u} \in Z_N$ and $q = P_N p$ in (3.7.32)–(3.7.33). Using the approximation properties for the operator P_N (see CHQZ2, Sect. 5.8.1), one obtains the following error estimate for $s \geq 1$:

$$\|\mathbf{u} - \mathbf{u}^N\|_{H^1(\Omega)} + \|p - p^N\|_{L^2(\Omega)} \leq C N^{1-s}(|u|_{H_p^s(\Omega)} + |p|_{H_p^{s-1}(\Omega)}) , \quad (3.7.39)$$

where $C > 0$ is a constant independent of N, $H_p^s(\Omega)$ are Sobolev spaces of the periodic functions as defined in Sect. A.8(c), and the seminorms on the right-hand side measure the derivatives of the highest order only.

This kind of result was first obtained by Maday and Quarteroni (1982) in their analysis of the Orszag–Patterson method without using explicitly an inf-sup condition. They consider a Fourier Galerkin method, where the nonlinear convective term is de-aliased, and a collocation method (which can actually be interpreted as a G-NI method) in which it is not. The analysis shows that in terms of rate of convergence the collocation method is just as good as the Galerkin method. We have here a rigorous demonstration for a complicated nonlinear system that the aliasing errors are of the same order as the truncation error. We recall that this feature was already observed in CHQZ2, Sects. 2.1.2 and 3.10, and numerical examples were furnished in Sect. 3.3.6. The same conclusions have been obtained by Bernardi, Maday, and Métivet (1987b) using the approach described above.

(b) *Periodic-nonperiodic problem*

Let us start again with the Galerkin scheme. Bernardi, Maday and Métivet (1987a) have shown that with the choice $Q_N = S_N \otimes \mathbb{P}_{N-1}$, the best inf-sup constant β_N in (3.7.28) is bounded from below by CN^{-1}. On the other hand, for all divergence-free vectors $\mathbf{u}$, one can construct an element $\mathbf{w} \in V_N$ such that $\nabla \cdot \mathbf{w} \equiv 0$ in Ω and the estimate $\|\mathbf{u} - \mathbf{w}\|_V \leq CN^{1-s}\|\mathbf{u}\|_{H_p^s(\Omega)}$ holds (where now $H_p^s(\Omega)$ is the Sobolev space of order s of the periodic functions in the x variable). This is accomplished by taking the H^1-projection upon $\mathbb{P}_{N-1}^0(-1, 1)$ of each Fourier coefficient of order $\leq N$ of the second component of $\mathbf{u}$, then modifying it to have zero average on $(-1, 1)$ without changing the boundary values, and finally defining the Fourier coefficient of the first component in order to satisfy the divergence-free condition. Obviously, $\mathbf{w}$ belongs to the space Z_N defined in (3.7.26). By the previous estimate and the error estimate for the best approximation of p in Q_N (see CHQZ2, estimate (5.8.39)), we obtain the following convergence results from (3.7.32)–(3.7.33):

$$\|\mathbf{u} - \mathbf{u}^N\|_{H^1(\Omega)} + N^{-1}\|p - p^N\|_{L^2(\Omega)}$$
$$\leq CN^{1-s}\{|\mathbf{u}|_{H_p^{s;N}(\Omega)} + |p|_{H_p^{s-1;N}(\Omega)}\}, \tag{3.7.40}$$

for $s \geq 1$ and a constant $C > 0$ independent of N; the seminorms on the right-hand side are defined analogously to CHQZ2, formula (5.4.10).

The same conclusions hold if one considers a collocation method. For instance, Bernardi, Maday, and Métivet (1987a, 1987b) have studied a collocation method in which a staggered grid in the direction of nonperiodicity is used; the families of Gauss–Lobatto points and of Gauss points with respect to the Legendre weight (see (8.4.14) and (8.4.12)) serve, respectively, as the collocation grids for the momentum and continuity equations. Recall that a similar approach is followed in many of the algorithms for inhomogeneous flows described in Sect. 3.4.

(c) Fully nonperiodic problem

In order to be more consistent with the actual implementation of the algorithms, here we carry on our discussion in the framework of a G-NI method. Furthermore, while we still refer explicitly to the two-dimensional case, we will give the asymptotic behavior of constants and errors in the general d-dimensional case.

We will invariably use the tensorized N-degree Legendre Gauss–Lobatto grid for discretizing the integral in the diffusion term $a(\mathbf{u}, \mathbf{v})$ and in the right-hand side $(\mathbf{g}, \mathbf{v})$. With such a choice, it is easily seen that the constants α_N and γ_N satisfy $\alpha_N \geq \alpha_* > 0$ and $\gamma_N \leq \gamma_*$ for suitable α_* and γ_* independent of N. All the results stated below hold for the Galerkin discretizations as well. Obviously, the spurious and pseudo-spurious pressure modes are different, but the structure is similar, and the asymptotic behavior of the best inf-sup constants β_N and the discretization errors are the same.

We discuss several choices of Q_N. If we use the same Legendre Gauss–Lobatto quadrature rule to discretize the pressure term $b(\mathbf{v}, q)$, then the space of spurious pressures X_N is spanned by the modes indicated in (3.7.16). The orthogonal space to X_N in $\mathcal{Q}_N$ is precisely the space spanned by the divergences $\nabla \cdot \mathbf{v}$ for $\mathbf{v}$ ranging over V_N. With the choice of this space as Q_N, the best possible inf-sup constant β_N decays as N^{-1} in any space dimension (Bernardi and Maday (1990)). However, due to the no-slip boundary conditions imposed on $\mathbf{v}$, each polynomial $\nabla \cdot \mathbf{v}$ vanishes at the four corners of Ω. Thus, the approximation properties of Q_N are very poor near these points for a general pressure, and spectral accuracy is lost. Furthermore, low-degree polynomials are not orthogonal to the corner spurious pressure modes $L'_N(x)L'_N(y)(1 \pm x)(1 \pm y)$ (recall (8.2.33)–(8.4.17)); hence, they do not belong to Q_N. For these reasons, Bernardi, Maday and Métivet (1987b) proposed as Q_N a supplementary space defined in such a way that the orthogonality conditions to the corner spurious modes are replaced by the orthogonality to the leading terms in their Legendre expansions. The resulting Q_N contains all the polynomials of degree $\leq \lambda N$ for a positive $\lambda < 1$, thus assuring the possibility of an asymptotically optimal approximation of the pressure. In addition, β_N retains the same asymptotic behavior as before.

As far as the approximation of the velocity is concerned, we observe that for each divergence-free vector $\mathbf{u}$ it is possible to find a divergence-free vector $\mathbf{w} \in V_N$ such that $\|\mathbf{u}-\mathbf{w}\|_{H^1(\Omega)} \leq CN^{1-s}\|\mathbf{u}\|_{H^s(\Omega)}$. Actually, since $\mathbf{u} = \nabla \times \phi$ for a suitable $\phi \in H_0^2(\Omega)$ (see Sect. A.8(b)), we can set $\mathbf{w} = \nabla \times \phi_N$, where ϕ_N is the projection of ϕ defined in CHQZ2, formula (5.8.16) with $m = l = 2$. (General results on the approximation of divergence-free vector fields by divergence-free polynomial fields have been given by Sacchi-Landriani and Vandeven (1987).) Using (3.7.35)–(3.7.36), we obtain the following error estimate:

$$
\|\mathbf{u} - \mathbf{u}^N\|_{H^1(\Omega)} + N^{-1}\|p - p^N\|_{L^2(\Omega)}
$$
$$
\leq CN^{1-s}\{|\mathbf{u}|_{H^{s;N}(\Omega)} + |p|_{H^{s-1;N}(\Omega)}\} + C'N^{-\sigma}|\mathbf{g}|_{H^{\sigma;N}(\Omega)} \tag{3.7.41}
$$

for $s \geq 1$, $\sigma > d/2$ and constants $C, C' > 0$ independent of N. Note that since Q_N is a supplementary space of X_N in $\mathcal{Q}_N$, the discrete continuity equation in (3.7.20) is actually fulfilled by all $q \in \mathcal{Q}_N$. We conclude that the G-NI scheme considered here can be interpreted as a collocation scheme in which the momentum equation is enforced at the internal Legendre Gauss–Lobatto nodes, whereas the continuity equation is enforced at the internal as well as the boundary nodes. The latter conditions are indeed equivalent to fulfilling the continuity equation in the whole of $\bar{\Omega}$.

The second choice of Q_N that we discuss is $Q_N = \mathbb{Q}_{N-2}(\Omega) \cap Q$. It is motivated by the fact that $M = N-2$ is the largest integer such that $\mathbb{Q}_M(\Omega)$ does not contain spurious pressure modes. The resulting method, the $\mathbb{Q}_N - \mathbb{Q}_{N-2}$ method proposed by Maday, Patera and Rønquist (1987), is very popular as it is often incorporated in the spectral element discretization of the Navier–Stokes equations in general domains (see Sect. 5.6). Obviously, Q_N is smaller than a supplementary space of X_N in $\mathcal{Q}_N$, thus implying that the continuity equation cannot be satisfied exactly all over $\bar{\Omega}$. Yet, such a drawback is surely compensated for by the ease of handling Q_N (we recall that with this choice the pressure term $b(\mathbf{v}, q)$ can be exactly evaluated by quadratures, even using two different sets of quadrature points, see Sect. 3.5.1). The best possible inf-sup constant β_N for the present choice of Q_N decays as $N^{(1-d)/2}$ (see Fig. 3.29 for the numerical values), which shows that pseudo-spurious pressure modes are still present. The corresponding error estimate is as follows:

$$
\begin{aligned}
&\|\mathbf{u} - \mathbf{u}^N\|_{H^1(\Omega)} + N^{(1-d)/2}\|p - p^N\|_{L^2(\Omega)} \\
&\leq CN^{1-s}\{|\mathbf{u}|_{H^{s;N}(\Omega)} + |p|_{H^{s-1;N}(\Omega)}\} + C'N^{-\sigma}|\mathbf{g}|_{H^{\sigma;N}(\Omega)} .
\end{aligned}
\tag{3.7.42}
$$

(We refer to Bernardi and Maday (1997) for other estimates in the case of inhomogeneous boundary conditions).

In order to rule out the pseudo-spurious pressure modes as well, one needs to further reduce the maximal polynomial degree allowed for the discrete pressures (while retaining the asymptotic approximation property for the pressure). Bernardi and Maday (1999) advocate two choices: either $Q_N = \mathbb{Q}_M(\Omega) \cap Q$, where M is the largest integer satisfying $M \leq \min(\lambda N, N - 2)$ for a fixed parameter $\lambda \in (0, 1)$, or $Q_N = \mathbb{P}_{N-1}(\Omega) \cap Q$. In both cases, the associated best possible constant β_N is uniformly bounded from below independently of N (precisely, with the former choice one has $\beta_N \geq c(1 - \lambda)^{(d-1)/2}$). This proves the absence of pseudo-spurious modes. The resulting error estimates are asymptotically optimal with respect to N, in the sense that the overall discretization error for velocity and pressure decays proportionally to the sum of the best approximation errors for $\mathbf{u}, p$ and $\mathbf{g}$. One has indeed

$$
\begin{aligned}
&\|\mathbf{u} - \mathbf{u}^N\|_{H^1(\Omega)} + \|p - p^N\|_{L^2(\Omega)} \\
&\leq CN^{1-s}\{|\mathbf{u}|_{H^{s;N}(\Omega)} + |p|_{H^{s-1;N}(\Omega)}\} + C'N^{-\sigma}|\mathbf{g}|_{H^{\sigma;N}(\Omega)} ,
\end{aligned}
\tag{3.7.43}
$$

where C actually depends on λ in the first choice.

Finally, we provide bounds for the inf-sup constants β_N of other methods discussed in Sect. 3.5.

The fully staggered collocation method by Bernardi and Maday (1988b) described in Sect. 3.5.1 fits into the abstract scheme (3.7.20) with the choice $V_N = (\mathbb{P}^0_N \otimes \mathbb{P}^0_{N+1}) \times (\mathbb{P}^0_{N+1} \otimes \mathbb{P}^0_N)$ and $Q_N = \mathbb{Q}_{N-1}(\Omega) \cap Q$. As anticipated in Sect. 3.5.1, the method is free of spurious pressure modes. However, the quality of the approximation is lower that for the previous methods. Indeed, not only does one have $\beta_N \geq N^{(1-d)/2}$, but also the constant α_N decays while N increases, precisely according to the law $\alpha_N \sim cN^{1-d}$.

Let us finally consider the methods listed in Sect. 3.5.2 and not yet discussed above. Assume that they are applied to the discretization of the Stokes problem in the reference domain $\hat{\Omega}$, with homogeneous Dirichlet boundary conditions on the velocity. Obviously, (3.5.11) coincides with the $\mathbb{Q}_N - \mathbb{Q}_{N-2}$ method for such boundary conditions. Choice (3.5.11) also yields an inf-sup constant that asymptotically behaves as that of the $\mathbb{Q}_N - \mathbb{Q}_{N-2}$ method; indeed, there are no pseudo-spurious modes in $\mathbb{P}_{N-1}$, as mentioned above.

Choice (3.5.14) again yields an inf-sup constant that is bounded from below by $CN^{(1-d)/2}$. Finally, choice (3.5.17) in the reference triangle yields an inf-sup constant β_N, which can be proven to be bounded from below by CN^{-3} (Schwab and Suri (1999)); numerical experiments actually indicate the asymptotic behavior $\beta_N \sim cN^{-1}$.

Numerical Results. We now show a few numerical results obtained by solving the Stokes equations by the G-NI method on a single domain.

The *Kovasznay solution* to the steady Stokes equations provides the two-dimensional flow field behind a periodic array of cylinders (see Kovasznay (1948)). It has the following expression:

$$u(x, y) = 1 - e^{\lambda x} \cos(2\pi y) , \quad v(x, y) = \frac{\lambda}{2\pi} e^{\lambda x} \sin(2\pi y) , \qquad (3.7.44)$$

$$p(x, y) = -e^{2\lambda x}/2 , \quad \lambda = \frac{1}{2\nu} - \sqrt{\frac{1}{4\nu^2} + 4\pi^2} . \qquad (3.7.45)$$

For this numerical example, we use $\nu = 0.025$ and the computational domain $\Omega = (-1, 3) \times (0.5, 2.5)$. On the boundary of Ω we impose a Dirichlet condition. The numerical solution is obtained by the $\mathbb{Q}_N - \mathbb{Q}_{N-2}$ method in its G-NI implementation, which uses staggered grids. The forcing term for the momentum equation is $\mathbf{f} = (e^{\lambda x}((\lambda^2 - 4\pi^2)\nu \cos(2\pi y) - \lambda e^{\lambda x}), e^{\lambda x} \nu \frac{\lambda}{2\pi} \sin(2\pi y)(-\lambda^2 + 4\pi^2))^T$. The associated error curves are reported in Fig. 3.28 (left).

Next, we consider a forced, time-dependent Stokes problem in $(0, 2\pi)^2$ with Dirichlet boundary conditions, whose exact solution is given by (3.3.52). The numerical solution is obtained by the G-NI approach using $\mathbb{Q}_N - \mathbb{Q}_{N-2}$ polynomials in a single domain on staggered grids, and a BDF difference scheme of order 4 in time, with time-step $\Delta t = 10^{-5}$. For $N \geq 28$ the discretization error in time dominates the discretization error in space. The associated error curves at time $t = 0.01$ are reported in Fig. 3.28 (right).

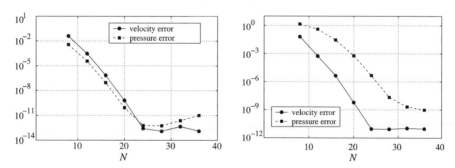

Fig. 3.28. Absolute errors in the H^1-norm for the velocity and the L^2-norm for the pressure on the Kovasznay solution of the steady Stokes problem (*left*) and the Taylor–Green solution of the time-dependent Stokes problem at $t = 0.01$ (*right*)

Extensions. So far, we have considered spectral approximations to the Stokes problem (3.7.2) that use Legendre polynomials in the directions of nonperiodicity. Whenever Chebyshev polynomials are preferred for their close relationship with Fourier polynomials, the standard L^2-inner product (u, v) is replaced by the weighted L^2-inner product $(u, v)_w$ in a Galerkin formulation (w denotes, as usual, the Chebyshev weight in each direction of nonperiodicity). A similar replacement takes place in the discrete inner products associated with collocation methods. Confining ourselves to Galerkin schemes for simplicity, the Chebyshev analog of (3.7.17) consists of seeking $\mathbf{u}^N \in V_N$ and $p^N \in Q_N$ satisfying

$$\nu(-\Delta \mathbf{u}^N, \mathbf{v})_w + (\mathbf{v}, \nabla p^N)_w = (\mathbf{g}, \mathbf{v})_w \quad \text{for all } \mathbf{v} \in V_N \ ,$$
$$(\nabla \cdot \mathbf{u}^N, q)_w = 0 \qquad \text{for all } q \in Q_N \ , \tag{3.7.46}$$

where V_N (Q_N, resp.) is a finite-dimensional subspace of a suitably weighted space V_w (Q_w, resp.). The new problem cannot be cast into the abstract form (3.7.20). Actually, due to the Chebyshev weight, we have

$$(\nabla \cdot \mathbf{v}, p)_w \neq -(\mathbf{v}, \nabla p)_w \ , \tag{3.7.47}$$

i.e., the negative gradient and the divergence operators are not adjoint to each other in the weighted Chebyshev inner product. Thus, one is led to extend the abstract formulation (3.7.20) by allowing two different bilinear forms $b_{1,N}(\mathbf{v}, q)$ and $b_{2,N}(\mathbf{v}, q)$ to be used. Another interesting situation that calls for such an extension is the solution of the Stokes problem in axisymmetric domains (see Bernardi, Dauge and Maday (1999)). Even with Legendre discretizations, the quite natural use of polar or cylindrical coordinates introduces a weight for which again (3.7.47) holds.

We consider the abstract formulation, which consists of seeking $\mathbf{u}^N \in V_N$ and $p^N \in Q_N$ such that

$$a_N(\mathbf{u}^N, \mathbf{v}) + b_{1,N}(\mathbf{v}, p^N) = ((\mathbf{g}, \mathbf{v}))_N \quad \text{for all } \mathbf{v} \in V_N \ ,$$
$$b_{2,N}(\mathbf{u}^N, q) = 0 \qquad \text{for all } q \in Q_N \ . \tag{3.7.48}$$

The solvability of (3.7.48) can be completely characterized in terms of inf-sup conditions on the forms a_N and $b_{i,N}$, $i = 1, 2$. These conditions generalize those given in Brezzi (1974) for problem (3.7.20). In order to state them, let us define

$$Z_N^{(i)} = \{\mathbf{v} \in V_N \mid b_{i,N}(\mathbf{v}, q) = 0, \text{ for all } q \in Q_N\}, \quad i = 1, 2 \ . \tag{3.7.49}$$

In Bernardi, Canuto and Maday (1988) (see also Nicolaides (1982)), it is proven that problem (3.7.48) has a unique solution if the following assumptions are fulfilled:

(i) *There exists a constant $\alpha_N > 0$ such that*

$$\sup_{\mathbf{v} \in Z_N^{(1)}} \frac{a_N(\mathbf{u}, \mathbf{v})}{\|\mathbf{v}\|_{V_w}} \geq \alpha_N \|\mathbf{u}\|_{V_w} \quad \text{for all } \mathbf{u} \in Z_N^{(2)} ; \tag{3.7.50}$$

(ii)

$$\dim Z_N^{(1)} = \dim Z_N^{(2)} ;$$

(iii) *there exist two constants $\beta_N^{(i)} > 0$, $(i = 1, 2)$ such that*

$$\sup_{\mathbf{v} \in V_N} \frac{b_{i,N}(\mathbf{v}, q)}{\|\mathbf{v}\|_{V_w}} \geq \beta_N^{(i)} \|q\|_{Q_w} \quad \text{for all } q \in Q_N . \tag{3.7.51}$$

These conditions are also necessary. Under the previous assumptions, one can obtain stability estimates similar to (3.7.30)–(3.7.31) and convergence estimates similar to (3.7.35)–(3.7.36).

For Chebyshev discretizations, the application of this abstract tool of analysis is quite technical (see, e. g., Bernardi, Canuto and Maday (1988)), yet the conclusions are essentially comparable with the results for the Legendre case discussed above. For spectral discretizations of axisymmetric problems, we refer to Bernardi, Dauge and Maday (1999) and the references therein.

Finally, we mention that the Stokes equations can be supplemented with boundary conditions other than the value of the velocity field, as considered so far. For instance, on a portion Γ of the boundary, one can prescribe the normal component of the stress tensor, i. e.,

$$\nu(\mathbf{n} \cdot \nabla)\mathbf{u} - p\mathbf{n} = \boldsymbol{\phi} \quad \text{on } \Gamma ,$$

or the pressure and the tangential component of the velocity,

$$p = \bar{p} \quad \text{and} \quad \mathbf{u} \times \mathbf{n} = \boldsymbol{\psi} \quad \text{on } \Gamma .$$

In the former case, if the boundary condition is enforced weakly within a Galerkin or G-NI scheme, the corresponding inf-sup condition is implied by that for the pure Dirichlet problem. In the latter case, we refer to Bernardi, Canuto and Maday (1991) for the analysis of several discretization schemes.

Boundary conditions that involve the vorticity can be handled by a vorticity–velocity-pressure formulation, whose spectral discretization is studied in Bernardi and Chorfi (2006).

3.7.4 The Inf-sup Condition and the Pressure Operator

In this section, we highlight the relationship between the constants α_N, β_N, γ_N and δ_N associated with the bilinear forms $a_N(\mathbf{u}, \mathbf{v})$ and $b_N(\mathbf{v}, q)$ (see

definitions (3.7.27), (3.7.28), (3.7.23) and (3.7.24)) and the spectral properties of the pressure, or pseudo-Laplacian, operator $G_N^T L_N^{-1} G_N$, considered in
Sects. 3.5.1 and 7.3.2. We recall that this operator originates from the Stokes
system by Gaussian elimination of the velocity unknowns. The resulting reduced system, involving the pressure unknowns only, can be solved by an
iterative method.

At first, we note that within the abstract setting of the previous section
(see in particular (3.7.29)), the pressure operator is written as $B^T A^{-1} B$.
Indeed, by elimination of the velocity, (3.7.29) is equivalent to

$$B^T A^{-1} B\mathbf{p} = B^T A^{-1}\mathbf{g} . \tag{3.7.52}$$

Let us denote by M the (pressure) mass matrix associated with a (continuous
or discrete) inner product in Q_N, which we represent as $(p, q)_N$. In other
words, let $(p, q)_N = \mathbf{q}^T M\mathbf{p}$ for all $p, q \in Q_N$, where $\mathbf{p}$, $\mathbf{q}$ denote as in the
previous section the vectors of coefficients of the pressures p, $q \in Q_N$ with
respect to a chosen basis. We assume that the induced norm $\|q\|_N = (q, q)_N^{1/2}$
is uniformly equivalent in Q_N to the standard L^2-norm, i.e., there exists
constants $C_1, C_2 > 0$ such that

$$C_1\|q\|_Q \le \|q\|_N \le C_2\|q\|_Q \qquad \text{for all } q \in Q_N . \tag{3.7.53}$$

Then, as discussed in Sect. 7.3.2, the solution to (3.7.52) can be obtained
by an iterative procedure (such as the conjugate gradient method, see Appendix C) applied to the preconditioned system

$$M^{-1}B^T A^{-1} B\mathbf{p} = M^{-1}B^T A^{-1}\mathbf{g} . \tag{3.7.54}$$

This indicates the relevance of the generalized eigenvalue problem

$$B^T A^{-1} B\mathbf{p} = \mu M\mathbf{p} , \tag{3.7.55}$$

already mentioned in (3.5.7).

Let $(\mu_l, \mathbf{p}_l)$ for $l = 1, \ldots, L$ be the eigenpairs of this problem; we suppose
that $0 \le \mu_1 \le \cdots \le \mu_L$ and that $\mathbf{p}_m^T M\mathbf{p}_l = 0$ for $l \ne m$. Going back to the
variational setting and denoting by $p_l \in Q_N$ the pressure associated with the
eigenvector $\mathbf{p}_l$ and by $\mathbf{u}_l \in V_N$ the velocity associated with $\mathbf{u}_l = A^{-1}\mathbf{p}_l$, we
equivalently have, for $l = 1, \ldots, L$,

$$a_N(\mathbf{u}_l, \mathbf{v}) + b_N(\mathbf{v}, p_l) = 0 \qquad \text{for all } \mathbf{v} \in V_N ,$$
$$b_N(\mathbf{u}_l, q) = -\mu_l(p_l, q)_N \quad \text{for all } q \in Q_N .$$

We note that $\mu_1 > 0$ if and only if $\beta_N > 0$, which is precisely the inf-sup condition (3.7.28). Furthermore, we have $(p_l, p_m)_N = 0$ for $l \ne m$; on the other
hand, $a_N(\mathbf{u}_l, \mathbf{u}_m) = -b_N(\mathbf{u}_m, p_l) = \mu_m(p_m, p_l)_N$; hence, $a_N(\mathbf{u}_l, \mathbf{u}_m) = 0$
for $l \ne m$, and $a_N(\mathbf{u}_l, \mathbf{u}_l) = \mu_l(p_l, p_l)_N$ for all l. The eigenpressures
$\{p_l \mid l = 1, \ldots, L\}$ form an orthogonal basis in Q_N for the inner product

$(p, q)_N$, whereas the eigenvelocities $\{\mathbf{u}_l \mid l = 1, \ldots, L\}$ form an orthogonal system in V_N for the inner product $a_N(\mathbf{u}, \mathbf{v})$. They span a subspace that we denote by E_N.

From now on, we assume that the coercivity condition (3.7.27) indeed holds for all $\mathbf{v} \in V_N$. Consequently, the quantity $\|\mathbf{v}\|_{N,V} = (a_N(\mathbf{v}, \mathbf{v}))^{1/2}$ is a discrete norm in V_N, which satisfies $\alpha_N \|\mathbf{v}\|_V^2 \leq \|\mathbf{v}\|_{N,V}^2 \leq \gamma_N \|\mathbf{v}\|_V^2$ for all $\mathbf{v} \in V_N$. In addition, we assume that β_N and δ_N denote the best possible constants for which (3.7.28) and (3.7.24) hold true, i.e., we assume that

$$\beta_N = \inf_{q \in Q_N} \sup_{\mathbf{v} \in V_N} \frac{b_N(\mathbf{v}, q)}{\|\mathbf{v}\|_V}$$

and

$$\delta_N = \sup_{q \in Q_N} \sup_{\mathbf{v} \in V_N} \frac{b_N(\mathbf{v}, q)}{\|\mathbf{v}\|_V} .$$

Then, we shall prove that the following inequalities hold:

$$C_1 \alpha_N \mu_1 \leq \beta_N^2 \leq C_2 \gamma_N \mu_1 \tag{3.7.56}$$

and

$$C_1 \alpha_N \mu_L \leq \delta_N^2 \leq C_2 \gamma_N \mu_L , \tag{3.7.57}$$

where the constants C_1 and C_2 are defined in (3.7.53). In the most common situation in which α_N and γ_N can be bounded independently of N, these inequalities show that β_N^2 (δ_N^2, resp.) behaves asymptotically as the smallest (the largest, resp.) eigenvalue of problem (3.7.55). Furthermore, they allow us to estimate the condition number (in the Euclidean norm)

$$\mathcal{K}_N = \mathrm{cond}_2\left((M^{-1/2})^T B^T A^{-1} B M^{-1/2}\right) = \frac{\mu_L}{\mu_1}$$

of the iterative matrix in (3.7.54). Indeed, we have

$$\frac{C_1 \alpha_N}{C_2 \gamma_N} \frac{\delta_N^2}{\beta_N^2} \leq \mathcal{K}_N \leq \frac{C_2 \gamma_N}{C_1 \alpha_N} \frac{\delta_N^2}{\beta_N^2} . \tag{3.7.58}$$

Thus, *in the relevant and common situation in which α_N, γ_N and δ_N can be bounded independently of N, the condition number $\mathcal{K}_N$ behaves asymptotically as β_N^{-2}.* This is a further element that highlights the importance of the inf-sup condition (3.7.28).

Numerical evidence is given in Fig. 3.29, where we report the value of β_N versus N for the $\mathbb{Q}_N - \mathbb{Q}_{N-2}$ method applied to the Stokes problem in the square $\Omega = (-1, 1)^2$, with homogeneous Dirichlet boundary conditions on the velocity field.

The rest of the subsection is devoted to the proof of (3.7.56) and (3.7.57). Let us start by observing that, given

$$\mathbf{v} = \sum_{l=1}^{L} c_l \mathbf{u}_l \in E_N \qquad \text{and} \qquad q = \sum_{l=1}^{L} d_l p_l \in Q_N ,$$

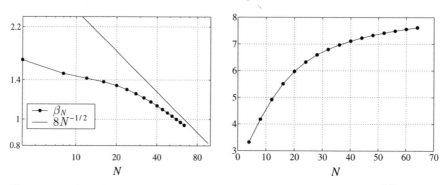

Fig. 3.29. The inf-sup constant β_N versus N (*left*) and the ratio $\beta_N/N^{-1/2}$ (*right*), for the $\mathbb{Q}_N - \mathbb{Q}_{N-2}$ method

we have

$$b_N(\mathbf{v}, q) = \sum_{l=1}^{L} c_l d_l \mu_l(p_l, p_l)_N \quad \text{and} \quad \|\mathbf{v}\|_{N,V} = \left(\sum_{l=1}^{L} c_l^2 \mu_l(p_l, p_l)_N \right)^{1/2}.$$

Next, given any $\mathbf{v} \in V_N$, we define $\mathbf{v}_E \in E_N$ as its orthogonal projection on E_N, i.e., it satisfies

$$a_N(\mathbf{v}_E, \mathbf{w}) = a_N(\mathbf{v}, \mathbf{w}) \quad \text{for all } \mathbf{w} \in E_N .$$

Then, it is easily seen that

$$b_N(\mathbf{v}, q) = b_N(\mathbf{v}_E, q) \quad \text{for all } \mathbf{v} \in V_N \text{ and } q \in Q_N .$$

Next, we observe that, given any $q \in Q_N$, we have

$$\sup_{\mathbf{v} \in V_N} \frac{b_N(\mathbf{v}, q)}{\|\mathbf{v}\|_{N,V}} = \sup_{\mathbf{v} \in E_N} \frac{b_N(\mathbf{v}, q)}{\|\mathbf{v}\|_{N,V}}$$

$$= \sup_{\mathbf{c} \in \mathbb{R}^L} \frac{\sum_{l=1}^{L} c_l d_l \mu_l(p_l, p_l)_N}{\left(\sum_{l=1}^{L} c_l^2 \mu_l(p_l, p_l)_N \right)^{1/2}} = \left(\sum_{l=1}^{L} d_l^2 \mu_l(p_l, p_l)_N \right)^{1/2} . \tag{3.7.59}$$

The first equality is obtained by observing that on the one side $E_N \subseteq V_N$, whereas on the other side $\|\mathbf{v}\|_{N,V}^2 = \|\mathbf{v}_E\|_{N,V}^2 + \|\mathbf{v} - \mathbf{v}_E\|_{N,V}^2$ for all $\mathbf{v} \in V_N$ by definition of orthogonal projection. The remaining steps are standard.

Inequalities (3.7.56) and (3.7.57) easily follow from (3.7.59). Indeed, for all $q \in Q_N$ we have

$$\sup_{\mathbf{v} \in V_N} \frac{b_N(\mathbf{v}, q)}{\|\mathbf{v}\|_V} \geq \sqrt{\alpha_N} \sup_{\mathbf{v} \in V_N} \frac{b_N(\mathbf{v}, q)}{\|\mathbf{v}\|_{N,V}} = \sqrt{\alpha_N} \left(\sum_{l=1}^{L} d_l^2 \mu_l(p_l, p_l)_N \right)^{1/2}$$

$$\geq \sqrt{\alpha_N \mu_1} \|q\|_N \geq C_1 \sqrt{\alpha_N \mu_1} \|q\|_Q .$$

Taking the infimum over all $q \in Q_N$ we obtain the inequality on the left-hand side of (3.7.56). On the other hand, taking $q = p_1$ in (3.7.59), we obtain

$$\sup_{\mathbf{v} \in V_N} \frac{b_N(\mathbf{v}, p_1)}{\|\mathbf{v}\|_V} \leq \sqrt{\gamma_N} \sup_{\mathbf{v} \in V_N} \frac{b_N(\mathbf{v}, p_1)}{\|\mathbf{v}\|_{N,V}} = \sqrt{\gamma_N \mu_1} \|p_1\|_N \leq C_2 \sqrt{\gamma_N \mu_1} \|p_1\|_Q ,$$

which immediately yields the inequality on the right-hand side of (3.7.56). The proof of (3.7.57) follows the same guidelines, now taking $q = p_L$ to obtain the inequality on the left-hand side.

3.7.5 Discretizations of the Continuity Equation by an Influence-Matrix Technique: The Kleiser–Schumann Method

An alternative approach to the direct discretization of the continuity equation consists of discretizing a Poisson equation for the pressure. This equation is obtained by taking the divergence of the momentum equation in (3.7.2) and using the continuity equation to obtain

$$\Delta p = \nabla \cdot \mathbf{g} \quad \text{in } \Omega . \tag{3.7.60}$$

A proper boundary condition to be associated with this equation is

$$\nabla \cdot \mathbf{u} = 0 \quad \text{on } \partial\Omega . \tag{3.7.61}$$

As a matter of fact, if $(\mathbf{u}, p)$ is a (sufficiently smooth) solution of the Stokes system (3.7.2), then (3.7.60) and (3.7.61) hold. Conversely, if $(\mathbf{u}, p)$ is a (sufficiently smooth) solution of the momentum equation and of (3.7.60) and (3.7.61), then it is easily seen that the function $\eta = \nabla \cdot \mathbf{u}$ satisfies

$$\begin{aligned} \Delta\eta &= 0 \quad \text{in } \Omega , \\ \eta &= 0 \quad \text{on } \partial\Omega . \end{aligned} \tag{3.7.62}$$

Thus, $\mathbf{u}$ is divergence-free in Ω, i.e., $(\mathbf{u}, p)$ is also a solution of (3.7.2).

The boundary condition (3.7.61) is of non-standard type, since a boundary condition for the pressure is usually associated with (3.7.60). However, Kleiser and Schumann (1980) suggested an efficient method for correctly imposing the boundary condition on the pressure in order to satisfy the continuity equation at the walls. Their method is described in Sect. 3.4.1.

The Kleiser–Schumann method has been theoretically investigated by Canuto and Sacchi-Landriani (1986), who considered a Fourier–Legendre tau method for the Stokes problem, and by Sacchi-Landriani (1986) who extended the analysis to the Fourier–Chebyshev case for the full Navier–Stokes equations. Here is a brief account of their analysis.

In this section we maintain the same notation of Sect. 3.4.1. Thus, the basic set of equations to be approximated is (3.4.17)–(3.4.19). Since the time-discretization does not affect the treatment of the continuity equation, we consider a time-independent version of the scheme, i. e., we set $\Delta t = \infty$ in the definition of λ. Furthermore, we assume for simplicity that $\nu = 1$, so that $\lambda = k^2$.

Canuto and Sacchi-Landriani consider a Legendre tau discretization in the y-variable of the boundary-value problems (3.4.20)–(3.4.23). First, they prove that the influence matrix M_N, which appears in (3.4.31) and which corresponds to the tau discretization of the "B-problem", is nonsingular for all wavenumbers $k \neq 0$ and for all degrees N of the polynomials.

Next, some estimates on the solutions of the approximate "B-problems" are obtained. Hereafter, we denote by $\|v\|$ the norm of the space $L^2(-1, 1)$. For the solution $(\hat{v}_p^N, \hat{p}_p^N)$ of the tau approximation to the "B-problem" with homogeneous Dirichlet boundary conditions for the pressure, the estimates

$$\|(\hat{v}_p^N)''\|^2 + \lambda\|(\hat{v}_p^N)'\|^2 + \lambda^2\|\hat{v}_p^N\|^2 \leq C\|\hat{\mathbf{r}}\|^2 \,, \qquad (3.7.63)$$

$$\|(\hat{p}_p^N)'\| + \lambda\|\hat{p}_p^N\|^2 \leq C\|\hat{\mathbf{r}}\|^2 \qquad (3.7.64)$$

hold with a constant C independent of N. These estimates are proven by a suitable choice of test functions in the tau equations for $\hat{v}_p^N$ and $\hat{p}_p^N$. As a by-product of these estimates we get an estimate for the right-hand side of the algebraic system (3.4.31). More precisely,

$$|(\hat{v}_p^N)'(\pm 1)| \leq C\lambda^{-1/4}\|\hat{\mathbf{r}}\| \,. \qquad (3.7.65)$$

On the other hand, let $(\hat{v}_\pm^N, \hat{p}_\pm^N)$ be the tau solutions of the homogeneous B-problems with boundary conditions $\hat{p}_-^N(-1) = 1$, $\hat{p}_-^N(1) = 0$ and $\hat{p}_+^N(-1) = 0$, $\hat{p}_+^N(1) = 1$, respectively. Then we have the following estimates:

$$\|(\hat{v}_\pm^N)''\|^2 + \lambda\|(\hat{v}_\pm^N)'\|^2 + \lambda^2\|\hat{v}_\pm^N\|^2 \leq C\lambda^{1/2} \,, \qquad (3.7.66)$$

$$\|(\hat{p}_\pm^N)'\|^2 + \lambda\|\hat{p}_\pm^N\|^2 \leq C\lambda^{1/2} \,. \qquad (3.7.67)$$

The estimates are a consequence of an analysis of the behavior of the Legendre coefficients of $(\hat{v}_\pm^N, \hat{p}_\pm^N)$, using a sort of maximum principle in frequency space (Canuto (1988); see CHQZ2, Sect. 7.2 for similar results in the Chebyshev case). As a by-product of (3.7.66), one proves that the Euclidean norm of the inverse of the influence matrix M_N is bounded independently of λ and N, i. e.,

$$\|M_N^{-1}\|_2 \leq C \,. \qquad (3.7.68)$$

From this estimate and (3.7.65) one obtains an estimate on the solution of the algebraic system (3.4.31):

$$|\delta_\pm| \leq C\lambda^{-1/4}\|\hat{\mathbf{r}}\| \,. \qquad (3.7.69)$$

Recalling the decomposition (3.4.30), we obtain from (3.7.63)–(3.7.64), (3.7.66)–(3.7.67) an estimate for the approximate solution $(\hat{v}^N, \hat{p}^N)$ of the

"*A*-problem", i. e.,

$$\|(\hat{v}^N)''\|^2 + \lambda\|(\hat{v}^N)'\|^2 + \lambda^2\|\hat{v}^N\|^2 \le C\|\hat{\mathbf{r}}\|^2 , \tag{3.7.70}$$

$$\|(\hat{p}^N)'\| + \lambda\|\hat{p}^N\|^2 \le C\|\hat{\mathbf{r}}\|^2 . \tag{3.7.71}$$

The previous estimates concern one fixed mode (of wavenumber k) in the Fourier expansion. Since the dependence on k is explicit, it is possible to obtain from them a stability estimate on the complete solution produced by the Kleiser–Schumann method, which we denote by $(\mathbf{u}^N, p^N)$. Precisely, denoting by $\mathbf{r}$ the function defined in $\Omega = (0, 2\pi) \times (-1, 1)$ whose k-th Fourier coefficient in the x-direction is $\hat{\mathbf{r}}$, we have

$$\|\mathbf{u}^N\|_{H^2(\Omega)} + \|p^N\|_{H^1(\Omega)/\mathbb{R}} \le C\|\mathbf{r}\|_{L^2(\Omega)} , \tag{3.7.72}$$

for a constant C independent of N.

The convergence analysis is carried out by comparing the approximate solution to a suitable orthogonal projection of the exact solution and by using the approximation results of CHQZ2, Sects. 5.1.2 and 5.4.2 for checking consistency. Denoting here the exact solution by $(\mathbf{u}, p)$, one can prove the following convergence estimate:

$$\begin{aligned}
\|\mathbf{u} - \mathbf{u}^N\|_{H^2(\Omega)} + \|p - p^N\|_{H^1(\Omega)/\mathbb{R}} \\
\le CN^{2-m}\{\|u\|_{H_p^m(\Omega)} + \|p\|_{H_p^{m-1}(\Omega)}\} ,
\end{aligned} \tag{3.7.73}$$

for all $m \ge 2$ for which the right-hand side is finite. Here $H_p^m(\Omega)$ denotes the Sobolev space of order m of the functions periodic in the first variable.

The same estimate holds if we take into account the tau correction proposed by Kleiser and Schumann in order to satisfy exactly the divergence-free condition all over the domain. Thus, spectral accuracy is guaranteed for the velocity field in the Kleiser–Schumann method for the periodic channel whether or not the tau correction is applied. Furthermore, the method selects one pressure among all the approximate solutions of the momentum equation (see (3.7.5) and (3.7.6)). Since by (3.7.73) the exact pressure is approximated with spectral accuracy, we conclude that the computed pressure does not contain any component along the spurious modes exhibited in Sect. 3.7.1.

4. Single-Domain Algorithms
and Applications for Compressible Flows

4.1 Introduction

Spectral methods have been used far less frequently for compressible flow calculations than for incompressible ones. A principal reason is that few compressible flows possess the high degree of regularity in the primitive variables that is ideal for spectral approximations. Incompressible flows may possess singularities, but primarily for geometric reasons, such as sharp edges or corners. Compressible flows, however, are also subject to singularities arising from nonlinear wave propagation. (Singularities already present a difficulty for spectral methods for linear hyperbolic problems if the initial data or boundary conditions are discontinuous.) Another delicate issue is the enforcement of boundary conditions in hyperbolic systems—since the effects of an incorrect treatment propagate to the interior (as opposed to being confined to the vicinity of the boundary as is the case for elliptic problems).

The purpose of this chapter is to highlight the key ingredients of the application of high-order methods in a single domain to hyperbolic problems and compressible flows. The particular issues that are discussed here include boundary conditions, high-frequency control, Euler equations, applications to smooth, homogeneous flows, applications to smooth, inhomogeneous flows, shock fitting and shock capturing. Many of the more problematic issues for high-order methods for compressible flows are better addressed in a multidomain context, in particular, in the framework of discontinuous Galerkin methods; see Chap. 5.

4.2 Boundary Treatment for Hyperbolic Systems

The proper treatment of boundary conditions is important to obtain accurate and stable solutions. Section 3.7 of CHQZ2 discussed several alternative boundary treatments for scalar hyperbolic problems—strong enforcement, weak enforcement using the variational formulation, and weak enforcement through explicit penalty terms. The present section focuses on additional issues that arise for hyperbolic systems. These include the use of characteristic compatibility conditions at boundaries, both for linear hyperbolic systems and for the nonlinear Euler equations that describe inviscid, compressible flow. Theoretical results for both accuracy and stability are reviewed. The principles covered here for proper treatment of physical boundary conditions in hyperbolic systems are also germane to interface conditions for the particular type of spectral multidomain methods discussed in the latter half of Chap. 5.

4.2.1 Characteristic Compatibility Conditions

An Example of Unstable Treatment. An instructive example of sensitivity to boundary conditions for a hyperbolic system was provided by Gottlieb, Gunzburger and Turkel (1982). Consider the following linear system for $\mathbf{u} = (u, v)^T$:

$$\frac{\partial \mathbf{u}}{\partial t} + A \frac{\partial \mathbf{u}}{\partial x} = \mathbf{0} , \quad -1 < x < 1 , \quad t > 0 , \tag{4.2.1}$$

where A is the constant matrix

$$A = \begin{pmatrix} -1/2 & -1 \\ -1 & -1/2 \end{pmatrix} ; \tag{4.2.2}$$

the boundary conditions are

$$\begin{aligned} u(-1, t) &= \sin(-2 + 3t) + \cos(-2 - t) , \\ u(+1, t) &= \sin(2 + 3t) + \cos(2 - t) , \end{aligned} \tag{4.2.3}$$

and the initial conditions are

$$\begin{aligned} u(x, 0) &= \sin(2x) + \cos(2x) , \\ v(x, 0) &= \sin(2x) - \cos(2x) . \end{aligned} \tag{4.2.4}$$

This a well-posed problem. Here we use a Chebyshev collocation scheme based on the Gauss–Lobatto points (8.3.15) together with an explicit time-discretization to demonstrate the importance of appropriate numerical boundary conditions; the lessons drawn here extend to other types of spectral algorithms; see the discussion below (Sect. 4.2.2) on weak approximations. The most tempting boundary treatment would be to update both u and v in the interior according to (4.2.1), to fix u at the boundaries via (4.2.3), and to update v at the boundaries according to (4.2.1). Note that the x-derivatives of u and v that are required in (4.2.1) are readily available at the boundaries from the usual Chebyshev first-derivative interpolation formula. Computed solutions to this problem are strongly unstable, as illustrated in the left frame of Fig. 4.1 at $t = 3$ for $N = 16$. Indeed, this computed solution grows without bound as t increases. The reason is that the use of (4.2.1) for v at the boundaries is an incorrect extrapolation of the PDE to the boundary.

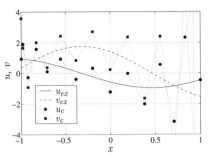 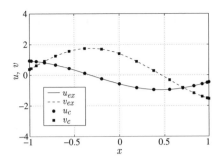

Fig. 4.1. Illustration of noncharacteristic (*left*) and characteristic (*right*) boundary conditions for the two-dimensional hyperbolic system at $t = 3$. The solid lines are the exact solution, and the symbols represent the computed solution

The Characteristic Compatibility Method (CCM). The eigenvalues of A are $-3/2$ and $1/2$, corresponding to the (right) eigenvectors $(1/2, 1/2)^T$ and $(1/2, -1/2)^T$, respectively. Letting

$$W = \begin{pmatrix} 1/2 & 1/2 \\ 1/2 & -1/2 \end{pmatrix} , \tag{4.2.5}$$

we have that

$$W^{-1} = \begin{pmatrix} 1 & 1 \\ 1 & -1 \end{pmatrix} , \tag{4.2.6}$$

and

$$\Lambda = W^{-1} A\, W = \begin{pmatrix} -3/2 & 0 \\ 0 & 1/2 \end{pmatrix} . \tag{4.2.7}$$

Equation (4.2.1) can be transformed to read

$$\frac{\partial \mathbf{z}}{\partial t} + \Lambda \frac{\partial \mathbf{z}}{\partial x} = \mathbf{0} , \tag{4.2.8}$$

where $\mathbf{z} = W^{-1}\mathbf{u}$ are the *characteristic variables*, or

$$\frac{\partial}{\partial t} \begin{pmatrix} u + v \\ u - v \end{pmatrix} + \begin{pmatrix} -3/2 & 0 \\ 0 & 1/2 \end{pmatrix} \frac{\partial}{\partial x} \begin{pmatrix} u + v \\ u - v \end{pmatrix} = 0 . \tag{4.2.9}$$

The characteristic variable $(u+v)$ is propagated to the left with speed $3/2$ and $(u - v)$ moves to the right with speed $1/2$. Hence, the Chebyshev collocation method for the scalar case (see CHQZ2, Sect. 3.7.1) suggests that the partial differential equation for $u+v$ be collocated at $x = -1$, and the partial differential equation for $u - v$ at $x = 1$. Since the time-advancing scheme is explicit, this scheme is equivalently accomplished by advancing both the physical unknowns at each boundary using (4.2.1), and then retaining only the linear combination corresponding to the outgoing characteristic variable. Specifically, let u_{PDE} denote the value of u at a boundary derived from the partial differential equation and use a similar notation for v_{PDE}. Then, for all $t > 0$, the final values of u and v at the boundaries satisfy the conditions

$$
\begin{aligned}
u(-1) &= \sin(-2 + 3t) + \cos(-2 - t) \, , \\
u(+1) &= \sin(2 + 3t) + \cos(2 - t) \, , \\
u(-1) + v(-1) &= u_{\text{PDE}}(-1) + v_{\text{PDE}}(-1) \, , \\
u(+1) - v(+1) &= u_{\text{PDE}}(+1) - v_{\text{PDE}}(+1) \, .
\end{aligned}
\tag{4.2.10}
$$

The first two equations came from (4.2.3) and the last two from the discrete solution of (4.2.9). An alternative way of writing (4.2.10) is

$$
u(-1) = \sin(-2 + 3t) + \cos(-2 - t) \, , \tag{4.2.11a}
$$
$$
\{u_t + v_t\} - (3/2)\{u_x + v_x\} = 0 \quad \text{at} \quad x = -1 \, ,
$$

$$
u(+1) = \sin(2 + 3t) + \cos(2 - t) \, , \tag{4.2.11b}
$$
$$
\{u_t - v_t\} + (1/2)\{u_x - v_x\} = 0 \quad \text{at} \quad x = +1 \, .
$$

We note that the second equations in (4.2.11) entail those particular linear combinations of the scalar differential equations that reproduce the equations satisfied by the outgoing characteristic variables. This way of enforcing the boundary conditions is termed the *characteristic compatibility* method, or, in short, CCM; it will be extended to cover the most general linear case later in this subsection.

Computed solutions for these boundary conditions exhibit no problems, and indeed, spectral accuracy is achieved as the discretization is refined. The right frame of Fig. 4.1 illustrates the results at $t = 3$ using these characteristic boundary conditions for computations using Chebyshev collocation in space (with $N = 16$) and fourth-order Runge–Kutta (see (D.2.17)) in time. Recall that for a spectral method the approximation of the derivatives at the boundary is taken from the same polynomial approximation as that used for the derivatives at interior points. For finite-order differencing schemes, a special approximation is invariably used for boundary points and sometimes near-boundary points as well.

Figure 4.2 provides convergence and temporal stability illustrations for the hyperbolic system (4.2.1)–(4.2.4) using the characteristic compatibility boundary conditions (4.2.10). The Chebyshev collocation solution is temporally stable (see Sect. D.1) and exhibits spectral accuracy. Also shown for comparison are results for those fourth-order and sixth-order compact difference schemes that are temporally stable for the scalar problem. The notation and stencils at and near the boundary for the compact schemes can be found in Carpenter, Gottlieb and Abarbanel (1993); $(3-4-3)$ denotes the classical fourth-order stencil

$$
u'_{j-1} + 4u'_j + u'_{j+1} = \frac{3}{\Delta x}(u_{j+1} - u_{j-1}) \tag{4.2.12}
$$

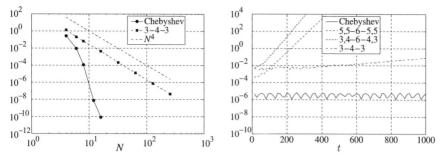

Fig. 4.2. Maximum error at $t = 8$ for Chebyshev collocation and a fourth-order compact scheme (*left*) and maximum error for $N = 16$ as a function of t (*right*) for Chebyshev collocation and several compact schemes for a hyperbolic system

in the interior with third-order stencils at both boundaries; $(3, 4 - 6 - 4, 3)$ denotes the classical sixth-order compact stencil (3.3.53) in the interior with third-order stencils at a boundary point and fourth-order stencils at a point adjacent to the boundary; $(5, 5 - 6 - 5, 5)$ denotes the classical sixth-order compact stencil in the interior with the particular stencils given by Carpenter, Gottlieb and Abarbanel (1993) on p. 293. As proven by Carpenter, Gottlieb and Abarbanel and illustrated by CHQZ2 in Sect. 3.7, all of these compact schemes are temporally stable for the scalar hyperbolic problem. However, as also proven by Carpenter, Gottlieb and Abarbanel and illustrated here in Fig. 4.2, these are all nevertheless temporally unstable for a system, although the growth for the $(3-4-3)$ scheme is very slow. We should note that Carpenter, Gottlieb and Abarbanel (1994) have derived some globally fourth-order compact schemes that are both stable and temporally stable for hyperbolic systems; however, globally sixth-order accurate compact schemes that are temporally stable for a hyperbolic system are not known. See Sect. D.1 for the definitions of stability and temporal stability. Here we are using the definition of stability in its PDE context (where it is sometimes called Lax stability), for which in the bound (D.1.5) the constant C is independent of Δt, ε and the spatial discretization parameter N, the norm is independent of N, but σ will in general be a function of N.

The boundary treatment described above advances the differential system at all the Gauss–Lobatto points ignoring the boundary conditions, and then applies the characteristic boundary conditions. However, this recipe may fail if an implicit time-advancing scheme is used. For instance, if the differential system (4.2.1) is advanced by one step of the Crank–Nicolson method (see (D.2.8)), and then the characteristic boundary corrections are applied, the resulting stability limit is roughly three-quarters of the stability limit for the modified Euler method (see (D.2.15)), as shown by Canuto and Quarteroni (1987).

The remedy to this consists, of course, of also treating the boundary conditions implicitly. At each boundary point two equations have to be satisfied:

one physical boundary condition from (4.2.3) and the partial differential equation (4.2.9) for the outgoing characteristic variable. Both of these equations have to be included in the implicit linear system for each time-step. This is precisely the implementation of the characteristic compatibility method within implicit time-stepping schemes. As observed by Canuto and Quarteroni (1987), the matrix corresponding to the spatial part of the differential system has eigenvalues with negative real parts. It follows that if the time-derivative is discretized by an implicit A-stable method (see Sect. D.1), such as the Crank–Nicolson scheme, no stability restriction on the time-step will occur. This is particularly appealing in the simulation of slow transients or in the convergence to steady state.

CCM for a General 1D System. The previous considerations extend in a straightforward manner to the case of a general one-dimensional, linear, constant-coefficient hyperbolic system. Consider

$$\frac{\partial \mathbf{u}}{\partial t} + A\, \frac{\partial \mathbf{u}}{\partial x} = \mathbf{f}\,, \quad -1 < x < 1\,, \quad t > 0\,, \tag{4.2.13}$$

where $\mathbf{u} = \mathbf{u}(x,t) \in \mathbb{R}^q$, A is a real, nonsingular and diagonalizable matrix of order q with real eigenvalues, and $\mathbf{f} = \mathbf{f}(x,t)$ is a given data. Let Λ be the real diagonal matrix of the eigenvalues of A, and let W be the square matrix whose columns are the (right) eigenvectors of A. Then $\Lambda = VAW$, where $V = W^{-1}$ in general, and $V = W^T$ when A is symmetric (in which case the eigenvectors can be chosen so that they are orthogonal). The characteristic variables are defined as $\mathbf{z} = V\mathbf{u}$ and satisfy the diagonal system

$$\frac{\partial \mathbf{z}}{\partial t} + \Lambda \frac{\partial \mathbf{z}}{\partial x} = V\mathbf{f}\,, \quad -1 < x < 1\,, \quad t > 0\,. \tag{4.2.14}$$

We split Λ as $\Lambda = \mathrm{diag}(\Lambda^+, \Lambda^-)$, where Λ^+ is a diagonal, positive-definite matrix of order p, $0 \le p \le q$, and Λ^- is a diagonal, negative-definite matrix of order $q - p$. Consequently, we split $\mathbf{z}$ as $\mathbf{z} = (\mathbf{z}^+, \mathbf{z}^-)^T$; $\mathbf{z}^+$ $(\mathbf{z}^-)$ are the p $(q - p)$ characteristic variables that are constant on the characteristic lines with positive (negative) slope, i. e., they are traveling rightward (leftward). At the right boundary point $x = 1$, $\mathbf{z}^+$ are the outgoing characteristic variables, whereas $\mathbf{z}^-$ are the incoming ones; the opposite association applies at the left boundary point $x = -1$.

So, we have the two characteristic compatibility conditions

$$\frac{\partial \mathbf{z}^+}{\partial t} + \Lambda^+ \frac{\partial \mathbf{z}^+}{\partial x} = \mathbf{h}^+ \tag{4.2.15}$$

and

$$\frac{\partial \mathbf{z}^-}{\partial t} + \Lambda^- \frac{\partial \mathbf{z}^-}{\partial x} = \mathbf{h}^-\,, \tag{4.2.16}$$

where $\mathbf{h} = V\mathbf{f}$, and the associated splitting is $\mathbf{h} = (\mathbf{h}^+, \mathbf{h}^-)^T$, with $\mathbf{h}^+$ the first p components of $\mathbf{h}$, and $\mathbf{h}^-$ the last $q - p$ components of $\mathbf{h}$.

A trivial situation occurs if we prescribe the values of the incoming characteristic variables at each boundary point, since (4.2.13) yields q independent scalar boundary-value problems. Usually, at each boundary point one prescribes linear combinations $B\mathbf{u} = \mathbf{g}$ of physical variables that correspond to linear combinations $C\mathbf{z} = \mathbf{g}$, with $C = BW$, of characteristic variables. None of the outgoing characteristic variables should be determined by these combinations, as the resulting values would, in general, be incompatible with those propagated from the interior by the hyperbolic system. Instead, the boundary conditions should allow the incoming characteristic variables to be determined in terms of the outgoing ones and the data. To be precise, let

$$B_L\mathbf{u}(-1,t) = \mathbf{g}_L(t) \,, \qquad B_R\mathbf{u}(1,t) = \mathbf{g}_R(t) \,, \qquad t > 0 \,, \qquad (4.2.17)$$

be the prescribed boundary conditions. Let us focus on the left boundary point $x = -1$. Since p (assumed > 0) characteristic variables are entering therein, B_L has to be a $p \times q$ matrix. Setting $C_L = B_L W$ and using the splitting $\mathbf{z} = (\mathbf{z}^+, \mathbf{z}^-)^T$, with the associated splitting

$$W = (W^+ \; W^-), \qquad V = W^{-1} = \begin{pmatrix} V^+ \\ V^- \end{pmatrix} \qquad (4.2.18)$$

of the eigenvector matrix and its inverse, we get

$$C_L\mathbf{z}(-1,t) = C_L^+\mathbf{z}^+(-1,t) + C_L^-\mathbf{z}^-(-1,t) = \mathbf{g}_L(t) \,, \qquad (4.2.19)$$

where $C_L^+ = B_L W^+$ is a $p \times p$ matrix and $C_L^- = B_L W^-$ is a $p \times (q-p)$ matrix. We require C_L^+ to be nonsingular, so that the incoming characteristic variables can be prescribed as

$$\mathbf{z}^+(-1,t) = S_L\mathbf{z}^-(-1,t) + \mathbf{z}_L(t) \,, \qquad (4.2.20)$$

where $S_L = -(C_L^+)^{-1}C_L^-$ is a $p \times (q-p)$ matrix and $\mathbf{z}_L(t) = (C_L^+)^{-1}\mathbf{g}_L(t)$.

Similarly, at the right boundary point $x = 1$, we arrive at prescribing the incoming characteristic variables as

$$\mathbf{z}^-(1,t) = S_R\mathbf{z}^+(1,t) + \mathbf{z}_R(t) \,, \qquad (4.2.21)$$

where S_R is a $(q-p) \times p$ matrix. The matrices S_L and S_R are called *reflection matrices*. The hyperbolic system (4.2.13) is therefore supplemented by the boundary conditions (4.2.17), which are equivalently reformulated as (4.2.20) and (4.2.21), and by the initial condition $\mathbf{u}(x,0) = \mathbf{u}_0(x)$, $-1 \le x \le 1$.

CCM for the Collocation Method. For a collocation method, the proper discretization is summarized as follows. At the interior Gauss–Lobatto nodes, enforce the PDE (4.2.13). At the left boundary, enforce (4.2.16) for the left-moving characteristics, along with (4.2.20) for the incoming boundary data.

At the right boundary, enforce (4.2.15) for the right-moving characteristics, along with (4.2.21) for the incoming boundary data.

In more detail for the case of an explicit time-advancing scheme, the collocation approach using Gauss–Lobatto points starts with collocating (4.2.13) at both the interior and boundary points. This yields a set of preliminary values derived from the PDE, $\{\mathbf{u}^{n+1}_{\mathrm{PDE},j}\,, j = 0,\ldots,N\}$. The internal preliminary values are retained, i. e., $\{\mathbf{u}^{n+1}_j = \mathbf{u}^{n+1}_{\mathrm{PDE},j}\,, j = 1,\ldots,N-1\}$, whereas the preliminary values at the boundaries are only used to calculate the outgoing characteristic variables, that is,

$$(\mathbf{z}^-)^{n+1}_0 = V^- \mathbf{u}^{n+1}_{\mathrm{PDE},0} \quad \text{at } x = -1$$

and

$$(\mathbf{z}^+)^{n+1}_N = V^+ \mathbf{u}^{n+1}_{\mathrm{PDE},N} \quad \text{at } x = 1 .$$

These equations for the outgoing characteristics are supplemented with those in (4.2.20)–(4.2.21) for the incoming characteristic variables. Hence, at the boundary we have

$$\begin{aligned}
(\mathbf{z}^+)^{n+1}_0 &= S_L(\mathbf{z}^-)^{n+1}_0 + \mathbf{z}^{n+1}_L \,, \\
(\mathbf{z}^-)^{n+1}_0 &= V^- \mathbf{u}^{n+1}_{\mathrm{PDE},0} \,,
\end{aligned} \tag{4.2.22}$$

and

$$\begin{aligned}
(\mathbf{z}^+)^{n+1}_N &= V^+ \mathbf{u}^{n+1}_{\mathrm{PDE},N} \,, \\
(\mathbf{z}^-)^{n+1}_N &= S_R(\mathbf{z}^+)^{n+1}_N + \mathbf{z}^{n+1}_R \,.
\end{aligned} \tag{4.2.23}$$

The final boundary values of the original variables $\mathbf{u}$ are extracted from

$$\mathbf{u}^{n+1} = W\mathbf{z}^{n+1} = W\left((\mathbf{z}^+)^{n+1}, (\mathbf{z}^-)^{n+1}\right)^T$$

at the boundary. For implicit time-discretizations, the modifications follow the principles discussed earlier.

Equivalently, the boundary conditions can be written as general characteristic compatibility conditions, à la (4.2.11) as

$$\begin{aligned}
B_L \mathbf{u}(-1,t) &= \mathbf{g}_L(t) \,, \\
V^- \frac{\partial \mathbf{u}}{\partial t} + V^- A \frac{\partial \mathbf{u}}{\partial x} &= V^- \mathbf{f} \quad \text{at} \quad x = -1 \,,
\end{aligned} \tag{4.2.24}$$

$$\begin{aligned}
B_R \mathbf{u}(1,t) &= \mathbf{g}_R(t) \,, \\
V^+ \frac{\partial \mathbf{u}}{\partial t} + V^+ A \frac{\partial \mathbf{u}}{\partial x} &= V^+ \mathbf{f} \quad \text{at} \quad x = +1 \,.
\end{aligned} \tag{4.2.25}$$

For the problem (4.2.1)–(4.2.2) considered earlier in this subsection, the general notation yields

$$W = \begin{pmatrix} 1/\sqrt{2} & 1/\sqrt{2} \\ 1/\sqrt{2} & -1/\sqrt{2} \end{pmatrix}, \qquad \Lambda = \begin{pmatrix} 1/2 & 0 \\ 0 & -3/2 \end{pmatrix},$$

$$\mathbf{z} = \begin{pmatrix} z^+ \\ z^- \end{pmatrix} = \begin{pmatrix} (u-v)/\sqrt{2} \\ (u+v)/\sqrt{2} \end{pmatrix}, \qquad B_L = B_R = (1 \quad 0),$$

$$g_L = \sin(-2+3t) + \cos(-2-t), \qquad g_R = \sin(2+3t) + \cos(2-t),$$

$$C_L = C_R = (1/\sqrt{2} \quad 1/\sqrt{2}).$$

$$(4.2.26)$$

CCM for a General Multidimensional System. Consider now a general linear, hyperbolic system of q (≥ 2) equations on a domain $\Omega \subset \mathbb{R}^q$:

$$\frac{\partial \mathbf{u}}{\partial t} + \sum_{j=1}^{d} A_j \frac{\partial \mathbf{u}}{\partial x_j} + C\mathbf{u} = \mathbf{g}, \qquad \mathbf{x} \in \Omega, t > 0, \qquad (4.2.27)$$

with appropriate initial and boundary conditions, where A_j and C are real matrices. Let $\mathbf{n}$ be the outward (unit) normal at any point on the boundary $\partial\Omega$, and let

$$A(\mathbf{n}) = \sum_{j=1}^{d} n_j A_j. \qquad (4.2.28)$$

Since the system is hyperbolic, at each point on the boundary the matrix $A(\mathbf{n})$ has real eigenvalues and can be diagonalized; denote by $\Lambda(\mathbf{n})$ the resulting diagonal matrix. Then using the splitting $\Lambda(\mathbf{n}) = (\Lambda^+ \Lambda^-)$ between positive and negative eigenvalues, we can apply the CCM at that point. This called a *locally one-dimensional* CCM treatment.

Again, the recommended strategy consists of collocating at the boundary points the linear combinations of the physical equations that correspond to the outgoing locally one-dimensional characteristic variables.

The previous discussion can be easily extended to the case in which the matrix has null eigenvalues. The corresponding characteristic variables have to be invariably treated as are the outgoing characteristic variables, i. e., they are advanced by the differential equation at the boundary point. In other words, the diagonal block Λ^+ contains the nonnegative eigenvalues of $A(\mathbf{n})$, whereas Λ^- contains the strictly negative ones.

Some authors have recommended the use of the more complicated, multidimensional characteristics (see Butler (1960), Zannetti and Colasurdo (1981)). (In two dimensions, these are called "bi-characteristics".) In the context of spectral methods on a single domain, experience such as that cited below in Sects. 4.6.1 and 4.7 indicates that, except for points in the corner of the domain, the locally one-dimensional treatment is quite sufficient, i. e., no solution filtering (see Sect. 4.4) is needed to suppress spurious oscillations, spectrally accurate solutions are obtained, and fully converged solution are obtained efficiently for time-marching solutions to steady-state problems. The difficulty with points in the corner of a domain is that they are simultaneously part of two boundaries, and therefore there is ambiguity about which boundary condition to apply.

As a simple example, consider a two-dimensional problem of the form

$$\frac{\partial \mathbf{u}}{\partial t} + A\frac{\partial \mathbf{u}}{\partial x} + B\frac{\partial \mathbf{u}}{\partial y} + C\mathbf{u} = \mathbf{g}, \quad -1 < x, y < 1, \quad t > 0, \quad (4.2.29)$$

where both A and B are real, nonsingular and diagonalizable matrices, with real eigenvalues. For purposes of assigning boundary conditions at any boundary point $(-1, y)$ or $(1, y)$, we diagonalize with respect to the x-derivative term and use

$$\mathbf{f} = \mathbf{g} - B\frac{\partial \mathbf{u}}{\partial y} - C\mathbf{u}$$

in the preceding formulas, obtaining

$$\frac{\partial \mathbf{z}}{\partial t} + \Lambda\frac{\partial \mathbf{z}}{\partial x} = V\mathbf{g} - VBW\frac{\partial \mathbf{z}}{\partial y} - VCW\mathbf{z} \quad (4.2.30)$$

in place of (4.2.14). We then use the same splitting into $\mathbf{z}^+$ and $\mathbf{z}^-$ to determine the locally one-dimensional boundary conditions at $(-1, y)$ and $(1, y)$ analogous to (4.2.22)–(4.2.23). Likewise, the locally 1D characteristic compatibility conditions analogous to (4.2.11) are based on

$$V^-\frac{\partial \mathbf{u}}{\partial t} + V^- A\frac{\partial \mathbf{u}}{\partial x} = V^-\mathbf{g} - V^- B\frac{\partial \mathbf{u}}{\partial y} - V^- C\mathbf{u},$$

$$V^+\frac{\partial \mathbf{u}}{\partial t} + V^+ A\frac{\partial \mathbf{u}}{\partial x} = V^+\mathbf{g} - V^+ B\frac{\partial \mathbf{u}}{\partial y} - V^+ C\mathbf{u}.$$

References and Outlook. The importance of basing the boundary conditions of finite-difference schemes upon the characteristic variables has been stressed for many years by Moretti (1968). Gustafsson, Kreiss and Sundström (1972) and Osher (1969) have developed a rigorous mathematical framework for ascertaining the stability of interior and boundary finite-difference schemes for initial-boundary-value problems. Their discussions implicitly suggest the use of characteristic variables in the boundary conditions, since the stability criterion is given for the equations in characteristic form. This mathematical theory is rather technically sophisticated. However, Trefethen (1983) has provided a useful physical interpretation of the technical stability criterion in terms of group velocity. The particular boundary condition formula given in (4.2.10) has been stressed by Gottlieb and Turkel (1985), who simplified the stability criterion for systems by showing that it could be reduced to the criterion for a scalar equation.

The preceding discussion has been in the context of appropriate numerical boundary conditions for problems on finite domains with well-posed boundary conditions and known boundary data. A long-standing problem has been the formulation and numerical approximation of boundary conditions at artificial boundaries imposed to truncate unbounded domains. The usual goal is to permit outgoing waves to pass through the artificial boundary without generating spurious reflections of incoming waves. Tsynkov (1998) and Hagstrom (1999) provide reviews of various numerical treatments of artificial boundaries.

4.2.2 Boundary Treatment for Linear Systems in Weak Formulations

The weak imposition of boundary conditions, discussed in CHQZ2, Sect. 3.7.1, for scalar equations, can be extended to systems by resorting to the characteristic variables, as done in the Chebyshev collocation example illustrated in the previous subsection. In particular, the *weak* enforcement of the boundary conditions (4.2.17) relies on the integral formulation of the homogeneous version of (4.2.13) and integration by parts; we have

$$\int_{-1}^{1} \mathbf{v}^T \frac{\partial \mathbf{u}}{\partial t}\, dx - \int_{-1}^{1} \left(\frac{\partial \mathbf{v}}{\partial x}\right)^T A\, \mathbf{u}\, dx + [\mathbf{v}^T A \mathbf{u}]_{-1}^{1} = 0$$

for all $t > 0$ and all differentiable test functions $\mathbf{v}$. Setting $V\mathbf{v} = \mathbf{y} = (\mathbf{y}^+, \mathbf{y}^-)^T$, we express each boundary term as

$$\mathbf{v}^T A \mathbf{u} = \mathbf{y}^T \Lambda \mathbf{z} = (\mathbf{y}^+)^T \Lambda^+ \mathbf{z}^+ + (\mathbf{y}^-)^T \Lambda^- \mathbf{z}^- . \tag{4.2.31}$$

Using (4.2.20) and (4.2.21) to enforce the boundary conditions, we obtain

$$\int_{-1}^{1} \mathbf{v}^T \frac{\partial \mathbf{u}}{\partial t}(t)\, dx - \int_{-1}^{1} \left(\frac{\partial \mathbf{v}}{\partial x}\right)^T A\, \mathbf{u}(t)\, dx$$
$$- (\mathbf{y}^+)^T(-1,t)\Lambda^+ S_L \mathbf{z}^-(-1,t) - (\mathbf{y}^-)^T(-1,t)\Lambda^- \mathbf{z}^-(-1,t) \tag{4.2.32}$$
$$+ (\mathbf{y}^+)^T(1,t)\Lambda^+ \mathbf{z}^+(1,t) + (\mathbf{y}^-)^T(1,t)\Lambda^- S_R \mathbf{z}^+(1,t)$$
$$= (\mathbf{y}^+)^T(-1,t)\Lambda^+ \mathbf{z}_L(t) - (\mathbf{y}^-)^T(1,t)\Lambda^- \mathbf{z}_R(t) .$$

A G-NI scheme is obtained by replacing integrals by Legendre Gauss–Lobatto quadrature formulas in this relation, and by requiring that all components of the G-NI solution $\mathbf{u}^N(t)$ and of the test functions $\mathbf{v}$ be polynomials of degree $\leq N$, i.e., $\mathbf{u}^N(t), \mathbf{v} \in (\mathbb{P}_N(-1,1))^q$. An equivalent formulation is obtained by counter-integrating by parts in (4.2.32) and again using (4.2.31). We obtain

$$\int_{-1}^{1} \mathbf{v}^T \frac{\partial \mathbf{u}}{\partial t}(t)\, dx + \int_{-1}^{1} \mathbf{v}^T A\, \frac{\partial \mathbf{u}}{\partial x}(t)\, dx$$
$$+ (\mathbf{y}^+)^T(-1,t)\Lambda^+ \left(\mathbf{z}^+(-1,t) - S_L \mathbf{z}^-(-1,t)\right) \tag{4.2.33}$$
$$- (\mathbf{y}^-)^T(1,t)\Lambda^- \left(\mathbf{z}^-(1,t) - S_R \mathbf{z}^+(1,t)\right)$$
$$= (\mathbf{y}^+)^T(-1,t)\Lambda^+ \mathbf{z}_L(t) - (\mathbf{y}^-)^T(1,t)\Lambda^- \mathbf{z}_R(t) .$$

As above, from (4.2.33) we obtain the G-NI scheme by discretizing the integrals via the Legendre Gauss–Lobatto quadrature rule and by restricting the trial and test functions to polynomial vectors. With obvious notation, the G-NI solution $\mathbf{u}^N(t) \in (\mathbb{P}_N(-1,1))^q$ satisfies, for all $t > 0$ and all

$\mathbf{v} \in (\mathbb{P}_N(-1,1))^d$,

$$
\begin{aligned}
(\mathbf{u}_t^N(t) &+ A\mathbf{u}_x^N(t), \mathbf{v})_N \\
&+ (\mathbf{y}^+)^T(-1,t)\Lambda^+\big((\mathbf{z}^N)^+(-1,t) - S_L(\mathbf{z}^N)^-(-1,t)\big) \\
&- (\mathbf{y}^-)^T(1,t)\Lambda^-\big((\mathbf{z}^N)^-(1,t) - S_R(\mathbf{z}^N)^+(1,t)\big) \\
&= (\mathbf{y}^+)^T(-1,t)\Lambda^+\mathbf{z}_L(t) - (\mathbf{y}^-)^T(1,t)\Lambda^-\mathbf{z}_R(t) ,
\end{aligned}
\tag{4.2.34}
$$

with $\mathbf{z}^N = V\mathbf{u}^N$ and $\mathbf{y} = V\mathbf{v}$.

If A is symmetric, the L^2-stability of this scheme is assured provided the conditions

$$
\Lambda^- + S_L^T \Lambda^+ S_L < 0 , \qquad \Lambda^+ + S_R^T \Lambda^- S_R > 0 ,
\tag{4.2.35}
$$

are satisfied, i. e., if the indicated matrices are negative-definite and positive-definite, respectively. A different condition guaranteeing L^2-stability is the *dissipativity* of the reflection matrices, expressed as

$$
\|S_L\| \, \|S_R\| < 1
\tag{4.2.36}
$$

(where $\|S\|$ is the Euclidean norm of the rectangular matrix S, i. e., the square root of the largest eigenvalue of $S^T S$). The detailed arguments leading to these conditions are reported in Sect. 4.2.4.

No general proof of stability is available for the nonsymmetric case. However, one can sometimes perform a transformation that produces a symmetric matrix. This can be done for the Euler equations, as noted in Sect. 4.3.

As usual, a pointwise interpretation of the G-NI scheme (4.2.34) is obtained by picking as $\mathbf{v}$ the polynomial vectors that, at one quadrature node, coincide with a column of the identity matrix, while vanishing at all other nodes. It is easily seen that the hyperbolic system (4.2.13) is collocated at the internal nodes, whereas at the boundary nodes one has

$$
\begin{aligned}
\Big(\frac{\partial \mathbf{u}^N}{\partial t} &+ A\,\frac{\partial \mathbf{u}^N}{\partial x}\Big)(-1,t) \\
&+ \frac{1}{w_0} W^+ \Lambda^+\big((\mathbf{z}^N)^+(-1,t) - S_L(\mathbf{z}^N)^-(-1,t) - \mathbf{z}_L(t)\big) = \mathbf{0}
\end{aligned}
\tag{4.2.37}
$$

and

$$
\begin{aligned}
\Big(\frac{\partial \mathbf{u}^N}{\partial t} &+ A\,\frac{\partial \mathbf{u}^N}{\partial x}\Big)(1,t) \\
&- \frac{1}{w_N} W^- \Lambda^-\big((\mathbf{z}^N)^-(1,t) - S_R(\mathbf{z}^N)^+(1,t) - \mathbf{z}_R(t)\big) = \mathbf{0} ,
\end{aligned}
\tag{4.2.38}
$$

where $w_0 = w_N = 2/N(N+1)$ are the boundary weights of the Legendre Gauss-Lobatto formula of order N. As for the scalar case, this shows that the boundary conditions (4.2.20) and (4.2.21) are enforced by a penalty approach. Funaro and Gottlieb (1991) investigate the L^2-stability of penalty methods for hyperbolic systems (see also Sect. 4.2.4). Hesthaven and Gottlieb (1996) and Hesthaven (1997, 1999) discuss the application of penalty methods to the spectral solution of the compressible Navier–Stokes equations.

4.2.3 Spectral Accuracy and Conservation

From the mathematical point of view, it is possible to keep track of the effect of the boundary conditions upon the overall accuracy of the scheme. Consider, for instance, an explicit time-advancing/Chebyshev collocation approximation to the system of the conservation laws

$$\frac{\partial \mathbf{u}}{\partial t} + \frac{\partial \mathbf{f}(\mathbf{u})}{\partial x} = \mathbf{0} \,, \tag{4.2.39}$$

where $\mathbf{u}$ is a vector in $\mathbb{R}^q$ and $\mathbf{f}(\mathbf{u}) = A\mathbf{u}$ as above, or, more generally, is a nonlinear function of $\mathbf{u}$. Let $\tilde{\mathbf{u}}^{n+1}$ be the preliminary values of the numerical solution (which is a polynomial of degree N) obtained by applying in a straightforward way a stable time-advancing scheme of order r (such as the fourth-order Runge–Kutta scheme) at all the points. Let $\mathbf{u}^{n+1}$ denote the values equal to $\tilde{\mathbf{u}}^{n+1}$ at the interior nodes and modified at the boundary nodes according to the CCM approach described in Sect. 4.2.1. The error equation for this approximation (see CHQZ2, Sect. 6.6) has the structure

$$\begin{aligned} \mathbf{u}^{n+1} = \mathbf{u}^n - \Delta t \frac{\partial}{\partial x} \mathbf{F}_N(\mathbf{u}^n, \Delta t) \\ + \frac{1}{2} \{ \boldsymbol{\tau}_+^{n+1}(1+x) + \boldsymbol{\tau}_-^{n+1}(1-x) \} T_N'(x) \,. \end{aligned} \tag{4.2.40}$$

Here, $(\partial/\partial x)\mathbf{F}_N(\mathbf{u}^n, \Delta t)$ denotes the interpolation derivative of $\mathbf{F}(\mathbf{u}^n, \Delta t)$, where the latter quantity is the approximation of $\mathbf{f}(\mathbf{u})$ generated by one step of the time-advancing scheme starting from $\mathbf{u}^n$; the error coefficients τ are defined as

$$\begin{aligned} \boldsymbol{\tau}_+^{n+1} &= \frac{1}{N^2} \frac{\mathbf{u}_0^{n+1} - \tilde{\mathbf{u}}_0^{n+1}}{\Delta t} \,, \\ \boldsymbol{\tau}_-^{n+1} &= \frac{1}{N^2} \frac{\mathbf{u}_N^{n+1} - \tilde{\mathbf{u}}_N^{n+1}}{\Delta t} \,. \end{aligned} \tag{4.2.41}$$

To examine the conservation properties of the scheme, we integrate (4.2.40) from -1 to 1, obtaining

$$\begin{aligned} \int_{-1}^{1} \mathbf{u}^{n+1} \mathrm{d}x = \int_{-1}^{1} \mathbf{u}^n \mathrm{d}x - \Delta t [\mathbf{F}(\mathbf{u}^n, \Delta t)]_{-1}^{+1} \\ + \frac{\delta_N}{N^2} \left\{ \frac{\mathbf{u}_0^{n+1} - \tilde{\mathbf{u}}_0^{n+1}}{\Delta t} - \frac{\mathbf{u}_N^{n+1} - \tilde{\mathbf{u}}_N^{n+1}}{\Delta t} \right\} \,, \end{aligned} \tag{4.2.42}$$

with $\delta_N = 2(1 + 1/(N^2 - 1))$. Thus, the scheme is globally conservative up to an error that decays as $\Delta t \to 0$ and $N \to \infty$ and that depends on the boundary conditions. Moreover, using (4.2.40) again, it is possible to prove that the consistency error of this method is of order r in time and is infinite-order in space. Thus, the boundary treatment does not destroy the spectral accuracy of the Chebyshev method.

Implicit in the preceding discussion has been the condition that the differential initial-boundary-value problem is well-posed. This places a certain restriction on the allowable boundary conditions. Numerical experience for linear problems indicates that if a spectral method is used with well-posed boundary conditions implemented by the CCM, then the boundary treatment will produce no instabilities. Sufficient conditions for the well-posedness of the problem are provided in next section. Elementary discussions of how to determine whether the initial-boundary-value problem is well-posed are given by Oliger and Sundström (1978) and by Kreiss and Lorenz (1989).

4.2.4 Analysis of Spectral Methods for Symmetric Hyperbolic Systems

Various spectral discretizations of an initial-boundary-value problem for the linear hyperbolic system

$$\frac{\partial \mathbf{u}}{\partial t} + A\,\frac{\partial \mathbf{u}}{\partial x} = \mathbf{0}\,,\quad -1 < x < 1\,,\qquad t > 0$$

(where $\mathbf{u} = \mathbf{u}(x,t) \in \mathbb{R}^q$ and A is a real, nonsingular matrix of order q with real eigenvalues) have been considered in Sect. 4.2.2. We have documented that the enforcement of the boundary conditions requires particular care; this aspect is also reflected by the intrinsic difficulty of the stability analysis for certain schemes of practical use. The analysis here makes the additional assumption that A is symmetric.

The G-NI scheme (4.2.34) incorporating the boundary conditions (4.2.17) (or, equivalently, (4.2.20)–(4.2.21)) in a weak manner not only works well in practice but allows a natural and simple stability analysis in the L^2-sense. Hereafter, we will prove that *this scheme is L^2-stable provided either one of the assumptions (4.2.35) or (4.2.36) is satisfied.*

The first, classical strategy of analysis consists of choosing $\mathbf{v} = \mathbf{u}^N$ in (4.2.34). The temporal term $(\mathbf{u}_t^N, \mathbf{u}^N)_N$ equals $\frac{1}{2}\frac{d}{dt}\|\mathbf{u}^N\|_N^2$; here, $\|\mathbf{v}\|_N$ denotes the discrete L^2-norm of the vector function $\mathbf{v}$ at the Legendre Gauss–Lobatto nodes, which is uniformly equivalent to the exact L^2-norm

$$\|\mathbf{v}\|_{L^2(-1,1)} \;=\; \left(\int_{-1}^1 \|\mathbf{v}(x)\|_{\mathbb{R}^q}^2\,dx\right)^{1/2} \quad \text{if } \mathbf{v} \in (\mathbb{P}_N(-1,1))^q \text{ (see CHQZ2,}$$

Sect. 5.3). The spatial term $(A\mathbf{u}_x^N, \mathbf{u}^N)_N$ coincides with the L^2-inner product $(A\mathbf{u}_x^N, \mathbf{u}^N)$ due to the exactness of the quadrature formula; furthermore, the integration-by-parts formula for symmetric matrices, $(A\mathbf{u}_x, \mathbf{v}) = -(A\mathbf{v}_x, \mathbf{u}) + [\mathbf{v}^T A\,\mathbf{u}]_{-1}^1$, yields $(A\mathbf{u}_x^N, \mathbf{u}^N) = \frac{1}{2}[(\mathbf{u}^N)^T A\,\mathbf{u}^N]_{-1}^1$. (When A depends upon x, one should resort to the skew-symmetric form of the spatial term.) We now substitute into (4.2.34) and use the decomposition (4.2.31) of the boundary terms; writing $\mathbf{z}^N = (\mathbf{z}^+, \mathbf{z}^-)^T$ and neglecting the time argument for the sake of simplicity, we get

$$
\frac{1}{2}\frac{d}{dt}\|\mathbf{u}^N\|_N^2
$$
$$
+ \left(\tfrac{1}{2}(\mathbf{z}^+)^T \Lambda^+ \mathbf{z}^+ - (\mathbf{z}^+)^T \Lambda^+ S_L \mathbf{z}^- - \tfrac{1}{2}(\mathbf{z}^-)^T \Lambda^- \mathbf{z}^-\right)_{|x=-1}
$$
$$
+ \left(-\tfrac{1}{2}(\mathbf{z}^-)^T \Lambda^- \mathbf{z}^- + (\mathbf{z}^-)^T \Lambda^- S_R \mathbf{z}^+ + \tfrac{1}{2}(\mathbf{z}^+)^T \Lambda^+ \mathbf{z}^+\right)_{|x=1}
$$
$$
= (\mathbf{z}^+)^T \Lambda^+ \mathbf{z}_{L\,|x=-1} - (\mathbf{z}^-)^T \Lambda^- \mathbf{z}_{R\,|x=1}\ .
$$

(Recall that $\Lambda = W^T A\, W$.) We now make repeated use of the inequality $|\mathbf{x}^T \Lambda \mathbf{y}| \le \frac{\alpha}{2}\mathbf{x}^T|\Lambda|\mathbf{x} + \frac{1}{2\alpha}\mathbf{y}^T|\Lambda|\mathbf{y}$, where $\mathbf{x}, \mathbf{y}$ are arbitrary vectors, Λ is Λ^+ or Λ^-, $|\Lambda|$ is the diagonal matrix whose entries are the absolute values of the entries of Λ, and $\alpha > 0$ is an arbitrary scalar. We obtain, for all α and $\beta > 0$,

$$
\frac{1}{2}\frac{d}{dt}\|\mathbf{u}^N\|_N^2 + \left(\tfrac{1}{2}(\mathbf{z}^+)^T \Lambda^+ \mathbf{z}^+ - \tfrac{\alpha}{2}(\mathbf{z}^+)^T \Lambda^+ \mathbf{z}^+\right.
$$
$$
\left. - \tfrac{1}{2\alpha}(\mathbf{z}^-)^T S_L^T \Lambda^+ S_L \mathbf{z}^- - \tfrac{1}{2}(\mathbf{z}^-)^T \Lambda^- \mathbf{z}^-\right)_{|x=-1}
$$
$$
+ \left(-\tfrac{1}{2}(\mathbf{z}^-)^T \Lambda^- \mathbf{z}^- + \tfrac{\alpha}{2}(\mathbf{z}^-)^T \Lambda^- \mathbf{z}^-\right.
$$
$$
\left. + \tfrac{1}{2\alpha}(\mathbf{z}^+)^T S_R^T \Lambda^- S_R \mathbf{z}^+ + \tfrac{1}{2}(\mathbf{z}^+)^T \Lambda^+ \mathbf{z}^+\right)_{|x=1}
$$
$$
\le \left(\tfrac{\beta}{2}(\mathbf{z}^+)^T \Lambda^+ \mathbf{z}^+ + \tfrac{1}{2\beta}\mathbf{z}_L^T \Lambda^+ \mathbf{z}_L\right)_{|x=-1}
$$
$$
+ \left(\tfrac{\beta}{2}(\mathbf{z}^-)^T |\Lambda^-| \mathbf{z}^- + \tfrac{1}{2\beta}\mathbf{z}_R^T |\Lambda^-| \mathbf{z}_R\right)_{|x=1}\ ,
$$

i.e.,

$$
\frac{d}{dt}\|\mathbf{u}^N\|_N^2 + \left((1 - \alpha - \beta)(\mathbf{z}^+)^T \Lambda^+ \mathbf{z}^+ - \tfrac{1}{\alpha}(\mathbf{z}^-)^T(\alpha\Lambda^- + S_L^T \Lambda^+ S_L)\mathbf{z}^-\right)_{|x=-1}
$$
$$
+ \left((1 - \alpha - \beta)(\mathbf{z}^-)^T |\Lambda^-| \mathbf{z}^- + \tfrac{1}{\alpha}(\mathbf{z}^+)^T(\alpha\Lambda^+ + S_R^T \Lambda^- S_R)\mathbf{z}^+\right)_{|x=1}
$$
$$
\le \tfrac{1}{\beta}\left(\mathbf{z}_L^T \Lambda^+ \mathbf{z}_{L\,|x=-1} + \mathbf{z}_R^T |\Lambda^-| \mathbf{z}_{R\,|x=1}\right)\ .
$$

Now suppose that (4.2.35) holds; consequently, there exists γ, $0 < \gamma < 1$, such that

$$
(1 - \gamma)\Lambda^- + S_L^T \Lambda^+ S_L \le 0\ , \qquad (1 - \gamma)\Lambda^+ + S_R^T \Lambda^- S_R \ge 0\ .
$$

Choosing $\alpha = 1 - \gamma$ and $\beta = \gamma$, we finally obtain the L^2-stability estimate

$$
\frac{d}{dt}\|\mathbf{u}^N(t)\|_{L^2(-1,1)}^2 \le C\left(\|\mathbf{z}_L(t)\|_{\mathbb{R}^p}^2 + \|\mathbf{z}_R(t)\|_{\mathbb{R}^{q-p}}^2\right)\ , \qquad t > 0\ , \qquad (4.2.43)
$$

for some constant C depending on the spectral radius of A but independent of N and t.

Let us now show that L^2-stability can also follow from condition (4.2.36). Observe that the term $(\mathbf{u}_t^N + A\mathbf{u}_x^N, \mathbf{v})_N$ appearing in (4.2.34) can be equivalently written as $(\mathbf{z}_t^N + \Lambda\mathbf{z}_x^N, \mathbf{y})_N$, since the eigenvector matrix W is orthogonal. Dropping again the index N, let us choose $\mathbf{y} = (\mathbf{y}^+, \mathbf{y}^-)^T$ as test function, with $\mathbf{y}^{\pm} = c^{\pm}|\Lambda^{\pm}|^{-1}\mathbf{z}^{\pm}$, where $c^{\pm}$ are positive constants to be determined later on. Then, we have

$$
\frac{c^+}{2}\frac{\mathrm{d}}{\mathrm{d}t}((\Lambda^+)^{-1}\mathbf{z}^+,\mathbf{z}^+)_N + \frac{c^-}{2}\frac{\mathrm{d}}{\mathrm{d}t}(|\Lambda^-|^{-1}\mathbf{z}^-,\mathbf{z}^-)_N
$$
$$
+ c^+(\mathbf{z}_x^+,\mathbf{z}^+) - c^-(\mathbf{z}_x^-,\mathbf{z}^-)
$$
$$
+ c^+\left(\|\mathbf{z}^+\|^2 - (\mathbf{z}^+)^T S_L \mathbf{z}^-\right)_{|x=-1} + c^-\left(\|\mathbf{z}^-\|^2 - (\mathbf{z}^-)^T S_R \mathbf{z}^+\right)_{|x=1}
$$
$$
= (\mathbf{z}^+)^T \mathbf{z}_{L\,|x=-1} + (\mathbf{z}^-)^T \mathbf{z}_{R\,|x=1} .
$$

We now integrate by parts and apply the bounds

$$
|(\mathbf{z}^+)^T S_L \mathbf{z}^-| \le \|S_L\|\left(\frac{\alpha^-}{2}\|\mathbf{z}^+\|^2 + \frac{1}{2\alpha^-}\|\mathbf{z}^-\|^2\right)
$$

and

$$
|(\mathbf{z}^+)^T \mathbf{z}_L| \le \frac{\beta c^+}{2}\|\mathbf{z}^+\|^2 + \frac{1}{2\beta c^+}\|\mathbf{z}_L\|^2
$$

at $x = -1$, and similar bounds for the terms at $x = 1$. Here, $\alpha^{\pm}$ and β are arbitrary positive constants. We obtain

$$
\frac{\mathrm{d}}{\mathrm{d}t}\left(c^+((\Lambda^+)^{-1}\mathbf{z}^+,\mathbf{z}^+)_N + c^-(|\Lambda^-|^{-1}\mathbf{z}^-,\mathbf{z}^-)_N\right)
$$
$$
+ \left(c^+ - c^-\|S_R\|\frac{1}{\alpha^+}\right)\|\mathbf{z}^+(1,t)\|^2 + c^+(1 - \alpha^-\|S_L\| - \beta)\|\mathbf{z}^+(-1,t)\|^2
$$
$$
+ c^-(1 - \alpha^+\|S_R\| - \beta)\|\mathbf{z}^-(1,t)\|^2 + \left(c^- - c^+\|S_L\|\frac{1}{\alpha^-}\right)\|\mathbf{z}^-(-1,t)\|^2
$$
$$
\le \frac{1}{\beta c^+}\|\mathbf{z}_L(t)\|^2 + \frac{1}{\beta c^-}\|\mathbf{z}_R(t)\|^2 .
$$

We now assume that (4.2.36) holds, and we choose $\alpha^+ = \|S_L\|$, $\alpha^- = \|S_R\|$, $c^+ = 1$, $c^- = \|S_L\|/\|S_R\|$ and $\beta = 1 - \|S_L\|\|S_R\| > 0$. Observing that the term differentiated in time above is a norm uniformly equivalent to the L^2-norm of $\mathbf{u}^N(t)$, we again obtain the L^2-stability bound (4.2.43), possibly with a different constant C.

As usual, stability together with the consistency of the G-NI procedure implies the convergence of the approximation as $N \to \infty$, as well as the related error estimates.

We close this section by briefly mentioning theoretical results concerning Chebyshev discretizations.

At first, we note that even for hyperbolic systems the initial-boundary-value problem need not be well-posed in the standard Chebyshev norm. The same negative results for scalar equations (see CHQZ2, (7.6.24)–(7.6.25)) persist for systems, as an effect of the propagation and reflection of waves at the boundary points, where the weight w becomes unbounded.

Unfortunately, the choice of modified Chebyshev weights vanishing at the outflow boundary points—which proves successful in the analysis of scalar problems—does not work for systems of hyperbolic equations, due to the coupling of the unknowns at the boundary. To illustrate the phenomenon,

Gottlieb and Turkel (1985) consider the system

$$u_t - u_x = 0 \,,$$
$$v_t + v_x = 0 \,,$$

(4.2.44)

with boundary conditions

$$u(1,t) = \alpha v(1,t) \,, \quad v(-1,t) = \beta u(-1,t) \,, \qquad \alpha\beta \neq 0 \,,$$

and initial conditions $v(x,0) \equiv 0$,

$$u(x,0) = u_0^{\varepsilon}(x) = \begin{cases} 1 - \dfrac{|x|}{\varepsilon} & \text{if } |x| \leq \varepsilon \,, \\ 0 & \text{if } |x| > \varepsilon \,. \end{cases}$$

It is easily seen that at time $t = 1 + \varepsilon$ one has

$$\int_{-1}^{1} v^2(x, 1+\varepsilon)\tilde{w}(x)dx =\sim c\varepsilon^{1/2} \,, \quad \text{whereas} \int_{-1}^{1} u^2(x,0)\tilde{w}(x)\mathrm{d} \sim c\varepsilon \,,$$

where $\tilde{w}$ is either the Chebyshev weight or the modified weight $\tilde{w}(x) = (1-x)w(x)$.

These difficulties led Reyna (1982) to propose a Chebyshev collocation method with a smoothing consisting of a high-mode cut-off in Legendre transform space rather than in Chebyshev transform space. The extra cost compared with a more conventional Chebyshev smoothing is the multiplication by a matrix that transforms between Chebyshev expansions and Legendre expansions. The size of this full matrix depends upon the amount of smoothing. The resulting numerical scheme is stable and convergent in the L^2-norm, and the stability estimate corresponds to the estimate for the continuous problem.

A stability and convergence analysis based on the use of the error equation and the Fourier–Laplace transform was carried out by Gottlieb, Lustman and Tadmor (1987a, 1987b). One remarkable aspect of the analysis is that (algebraic) stability for the system (in a suitable weighted L^2-norm) can be proven as a consequence of a sufficiency criterion that deals exclusively with the properties of the scalar equation. Thus, the difficulty inherent in the coupling of the scalar equations through the boundary conditions is avoided. The criterion includes the (algebraic) stability (see CHQZ2, Sect. 6.4.2) of the scheme applied to the scalar equation; it is fulfilled, e. g., by a Chebyshev or Legendre collocation method that uses as collocation points the zeros of T'_{N+1} or L'_{N+1}. For these methods, the weight function in the spatial L^2-norm is given by $(1 \pm x)w(x)$, where $w(x)$ is either the Chebyshev or the Legendre weight, and the sign is chosen in order to annihilate the weight at the outflow boundary.

4.3 Boundary Treatment for the Euler Equations

An examination of the Euler equations (see Sect. 1.3.4) is instructive. The one-dimensional version of (1.3.43)–(1.3.45) produces the Euler equations in conservation form:

$$\frac{\partial \rho}{\partial t} + \frac{\partial (\rho u)}{\partial x} = 0 \ ,$$

$$\frac{\partial (\rho u)}{\partial t} + \frac{\partial (\rho u u)}{\partial x} + \frac{\partial p}{\partial x} = 0 \ , \tag{4.3.1}$$

$$\frac{\partial (\rho E)}{\partial t} + \frac{\partial (\rho u E)}{\partial x} + \frac{\partial (pu)}{\partial x} = 0 \ .$$

An alternative version that is useful for both analysis and some applications (at least for applications to smooth flows) writes the density and momentum equations in nonconservative form and uses the pressure in lieu of the total energy (see (1.3.20)):

$$\frac{\partial \rho}{\partial t} + u \frac{\partial \rho}{\partial x} + \rho \frac{\partial u}{\partial x} = 0 \ ,$$

$$\frac{\partial u}{\partial t} + u \frac{\partial u}{\partial x} + \frac{1}{\rho} \frac{\partial p}{\partial x} = 0 \ , \tag{4.3.2}$$

$$\frac{\partial p}{\partial t} + u \frac{\partial p}{\partial x} + \gamma p \frac{\partial u}{\partial x} = 0 \ .$$

In compact form, (4.3.2) reads as

$$\frac{\partial \mathbf{q}}{\partial t} + A(\mathbf{q}) \frac{\partial \mathbf{q}}{\partial x} = 0 \ , \tag{4.3.3}$$

where

$$\mathbf{q} = (\rho, u, p)^T \tag{4.3.4}$$

and

$$A(\mathbf{q}) = \begin{pmatrix} u & \rho & 0 \\ 0 & u & 1/\rho \\ 0 & \gamma p & u \end{pmatrix} \ . \tag{4.3.5}$$

The eigenvalues of A are $u + c$, u and $u - c$, where the sound speed c is given by (1.2.21). The character of the flow depends upon the relative magnitudes of u and c. Recall from (1.3.46) that the Mach number is the ratio $M = |u|/c$. If $M > 1$, then all the eigenvalues have the same sign. The flow is supersonic, and all information propagates downstream. If $M < 1$, then the eigenvalue $u - c$ has the opposite sign to u and $u + c$ (assuming $u > 0$). The flow is subsonic, and information is able to propagate upstream.

Supersonic flow usually adjusts to downstream obstacles by undergoing a discontinuous change referred to as a shock. The Euler equations themselves do not apply at a shock. The conditions there must be derived from the physical conservation laws, which are given by the integral version of (4.3.1). These produce the Rankine–Hugoniot conditions

$$[\rho u] = 0 \ ,$$

$$[\rho u^2 + p] = 0 \ , \tag{4.3.6}$$

$$[\rho u E + pu] = 0 \ ,$$

together with the condition that the entropy does not decrease across the shock (which follows from the fact that the Euler equations are the limit of the compressible Navier–Stokes equations for vanishingly small dissipation). The brackets in (4.3.6) denote the jump across the shock *in the frame moving with the shock*. Letting $v = u - v_{sh}$, where v_{sh} is the velocity of the shock, and invoking (1.2.4), conditions (4.3.6) are equivalent to

$$[\rho v] = 0 \,,$$
$$[\rho v^2 + p] = 0 \,,$$
$$\left[\rho v (e + \frac{1}{2} u^2) + pu\right] = 0 \,.$$

(4.3.7)

Of course, if the flow is subsonic everywhere, or if it is smooth even though supersonic, then there is no a priori reason to be skeptical of applying spectral methods. However, spectral methods do appear to be ill-suited to problems with discontinuities in the solution or even in some of its low-order derivatives. The standard integration-by-parts estimate of the size of the spectral coefficients of such nonsmooth functions implies that they decay in a slow, algebraic manner. Left as is, this behavior produces the familiar Gibbs oscillations in the solution together with slow, global convergence of the numerical results to the true solution. Nevertheless, techniques that seek to extract useful and accurate information from spectral solutions to some discontinuous problems have been developed; see Sect. 4.8 for remarks on fluids applications and CHQZ2, Sect. 7.6, for a general discussion.

For nonlinear problems, such as those that arise in gas dynamics, one often resorts to the use of the *linearized characteristic correction method* for applying the boundary conditions. The simplest point of linearization is the most recent time-level. In explicit spectral methods the major source of error is usually the time discretization. The linearization error is often far smaller.

Consider the case of the one-dimensional Euler equations in the form (4.3.3) for x in $(-1, 1)$ and $t > 0$. The matrix A in (4.3.5) is nonsymmetric; accordingly, the eigenvectors are not orthogonal. The matrix of eigenvalues is

$$\Lambda = \begin{pmatrix} u + c & 0 & 0 \\ 0 & u & 0 \\ 0 & 0 & u - c \end{pmatrix} \,,$$

(4.3.8)

and the matrix of right eigenvectors (in the columns) corresponding to the eigenvalues $u + c$, u and $u - c$ is

$$W = \frac{1}{2} \begin{pmatrix} \dfrac{1}{c^2} & -\dfrac{2}{c^2} & \dfrac{1}{c^2} \\ \dfrac{1}{\rho c} & 0 & -\dfrac{1}{\rho c} \\ 1 & 0 & 1 \end{pmatrix} \,,$$

(4.3.9)

the inverse matrix (with left eigenvectors on the rows) is

$$V = W^{-1} = \begin{pmatrix} 0 & \rho c & 1 \\ -c^2 & 0 & 1 \\ 0 & -\rho c & 1 \end{pmatrix} , \tag{4.3.10}$$

and the characteristic combinations of the dependent variables are

$$\mathbf{z} = V \begin{pmatrix} \rho \\ u \\ p \end{pmatrix} = \begin{pmatrix} p + (\rho c)u \\ p - c^2\rho \\ p - (\rho c)u \end{pmatrix} . \tag{4.3.11}$$

We assume that the boundary conditions have the form $B_L \mathbf{q} = \mathbf{g}_L$ and $B_R \mathbf{q} = \mathbf{g}_R$ for suitable matrices B_L and B_R.

Consider the left boundary at $x = -1$ at some time t_n, and let $\bar{A}$ be the matrix given in (4.3.5) evaluated at the state $\bar{\mathbf{q}} = \mathbf{q}(-1, t_n)$. Assume subsonic inflow, i.e., $u < c$. Then, in terms of the discussion of Sect. 4.2.1, we have

$$\mathbf{z}^+ = (p + (\bar{\rho}\bar{c})u, p - \bar{c}^2\rho)^T , \tag{4.3.12}$$

$$\mathbf{z}^- = (p - (\bar{\rho}\bar{c})u) . \tag{4.3.13}$$

The quantity $\mathbf{z}^-$ is determined from the interior solution, whereas $\mathbf{z}^+$ is determined by the boundary data $B_L \mathbf{q} = \mathbf{g}_L$. Application of (4.2.20), and then transformation from $\mathbf{z}$ to $\mathbf{q}$ provides the two remaining conditions at the left boundary.

For example, suppose that $\rho = \rho_L$ and $u = u_L$ are the prescribed data at the inflow boundary. Then

$$B_L = \begin{pmatrix} 1 & 0 & 0 \\ 0 & 1 & 0 \end{pmatrix} , \quad C_L^+ = \begin{pmatrix} \dfrac{1}{2\bar{c}^2} & -\dfrac{1}{\bar{c}^2} \\ \dfrac{1}{2\bar{\rho}\bar{c}} & 0 \end{pmatrix} , \quad C_L^- = \begin{pmatrix} \dfrac{1}{2\bar{c}^2} \\ -\dfrac{1}{2\bar{\rho}\bar{c}} \end{pmatrix} , \quad S_L = \begin{pmatrix} 1 \\ 1 \end{pmatrix} ,$$

and (4.2.20) yields

$$\rho_0 = \rho_L ,$$
$$u_0 = u_L , \tag{4.3.14}$$
$$p_0 = p_{\mathrm{PDE}} + (\bar{\rho}\bar{c})[u_L - u_{\mathrm{PDE}}] = [p_{\mathrm{PDE}} - (\bar{\rho}\bar{c})\,u_{\mathrm{PDE}}] + (\bar{\rho}\bar{c})u_L ,$$

for the values at the left boundary at the new time-step. (Recall that the subscript PDE denotes values at the left boundary obtained from advancing the PDE.)

Another derivation of the proper boundary conditions at $x = -1$, is to note that the relevant characteristic combination from the interior is $p - (\bar{\rho}\bar{c})u$, and to just take from (4.3.2) the combination $(\partial p/\partial t) - (\bar{\rho}\bar{c})(\partial u/\partial t)$, which yields

$$\frac{\partial p}{\partial t} - (\bar{\rho}\bar{c})\frac{\partial u}{\partial t} = -\bar{u}\frac{\partial p}{\partial x} - \gamma\bar{p}\frac{\partial u}{\partial x} + (\bar{\rho}\bar{c})u\frac{\partial u}{\partial x} + \bar{c}\frac{\partial p}{\partial x} \qquad \text{at} \qquad x = -1 \ . \quad (4.3.15)$$

The time derivatives on the left-hand side and the spatial derivatives on the right-hand side are evaluated from the same schemes used to update the interior points. The resulting condition that relates p and u at the new time-level is combined with the prescribed boundary conditions to completely determine the variables at the new time-level. The result (4.3.15) is equivalent to (4.3.14).

The Euler equations in three dimensions in nonconservative form can be written as

$$\frac{\partial \mathbf{q}}{\partial t} + \sum_{j=1}^{3} A_j(\mathbf{q})\frac{\partial \mathbf{q}}{\partial x_j} = 0 \ , \qquad (4.3.16)$$

where $\mathbf{q} = (\rho, u, v, w, p)^T$. We refer the reader to Warming, Beam and Hyett (1975) for a transformation that simultaneously symmetrizes the individual matrices $A_j(\mathbf{q})$ and also diagonalizes the matrix (4.2.28) used in linearized, locally one-dimensional characteristic compatibility methods. This paper also contains a transformation that accomplishes the same for the Jacobian matrices of the conservation form of the Euler equations.

As noted earlier, the proper specification of numerical boundary conditions at artificial (nonphysical) boundaries is challenging. For the one-dimensional case, suppose that the right boundary is such an artificial boundary, and that the flow there is subsonic. Again using the incoming linearized characteristic variable, i.e., (4.3.13), a common practice is to impose the condition

$$\frac{\partial p}{\partial t} - (\bar{\rho}\bar{c})\frac{\partial u}{\partial t} = 0 \qquad (4.3.17)$$

along with the linearized characteristic compatibility conditions for the other two characteristic variables:

$$\begin{aligned} \frac{\partial p}{\partial t} + (\bar{\rho}\bar{c})\frac{\partial u}{\partial t} &= -\bar{u}\frac{\partial p}{\partial x} - \gamma\bar{p}\frac{\partial u}{\partial x} - (\bar{\rho}\bar{c})u\frac{\partial u}{\partial x} - \bar{c}\frac{\partial p}{\partial x} \ , \\ \frac{\partial p}{\partial t} - \bar{c}^2\frac{\partial \rho}{\partial t} &= -\bar{u}\frac{\partial p}{\partial x} - \gamma\bar{p}\frac{\partial u}{\partial x} - \bar{c}^2(u\frac{\partial \rho}{\partial x} + \bar{\rho}\frac{\partial u}{\partial x}) \ . \end{aligned} \qquad (4.3.18)$$

For more discussion on artificial boundary conditions for the Euler equations, see the articles by Thompson (1987) and Giles (1990), which contain detailed discussions of this linearized characteristics approach to inflow/outflow boundary conditions, albeit just for inviscid flow and for low-order methods. Poinsot and Lele (1992) discuss the generalization to viscous flow.

In some approaches for handling artificial boundaries, the underlying PDE is modified in a "sponge layer" or "perfectly matched layer" that is interposed between the domain of interest and the artificial boundary. The buffer domain approach for incompressible flow discussed in Sect. 3.5 is a particular example of this approach. (A reference to the compressible extension of this is provided in Sect. 2.7.2.) Hu (2001) describes a perfectly matched layer approach to the linearized Euler equations. Rigorous theoretical results on the suitability of numerical implementations of these artificial boundary conditions are rare in general and virtually nonexistent for spectral methods.

4.4 High-Frequency Control

A common challenge for the application of spectral methods to all but the simplest hyperbolic problems (linear, constant-coefficient problems are the exception) is coping with the tendency for the high-frequency content of the real solution to grow in time. Both variable-coefficient linear problems and nonlinear problems will have this feature. These can lead to temporal instability in finite-dimensional approximations. Since the spatial discretization by spectral methods introduces no (periodic boundary conditions) or very little (nonperiodic boundary conditions) numerical dissipation, there is no a priori feature of the numerical algorithm to control the growth of high-frequency content. The same tendency afflicts central-difference schemes, but the rate of high-frequency growth is lower in those methods. Various techniques for controlling high-frequency growth in spectral methods have been employed: de-aliasing, solution filtering, derivative filtering, and hyperviscosity.

De-aliasing procedures are only strictly applicable for polynomial nonlinearities, e. g., the quadratic nonlinearities for the incompressible Navier–Stokes equations (see Sect. 3.3.2). They become increasingly expensive as the order of the nonlinearity increases. However, for some problems the PDE can be rewritten in a form that reduces aliasing effects; see, for example, the discussions of skew-symmetric forms in Sect. 3.2.1 and 3.3.6 for incompressible flow and in Sect. 4.5 for compressible flow.

Solution filtering consists of applying a low-pass filter to the entire solution. Consider the case of a scalar variable $u(x, t)$ subject to periodic boundary conditions, say, on $(0, 2\pi)$. At the end of time-step n the unfiltered numerical solution $u^n(x)$ has the discrete Fourier series

$$u^n(x_j) = \sum_{k=-N/2}^{N/2-1} \tilde{u}_k e^{ikx_j} , \qquad j = 0, \dots, N-1 . \tag{4.4.1}$$

The filtered solution $u^{n,f}$ is then computed from

$$u^{n,f}(x_j) = \sum_{k=-N/2}^{N/2-1} \sigma_k \tilde{u}_k e^{ikx_j} , \qquad j = 0, \dots, N-1 , \tag{4.4.2}$$

where

$$\sigma_k = \sigma(2k\pi/N) , \qquad k = -N/2, \ldots, N/2 , \qquad (4.4.3)$$

and $\sigma = \sigma(\theta)$, $\theta \in [\pi, \pi]$ is a *filtering function*, or simply a *filter*, with the properties given at the end of Sect. 8.1. Following the application of the filter, the time advancement continues with $u^{n,f}$ rather than u^n.

Detailed discussions of filtering functions for spectral methods can be found, e. g., in CHQZ2, Sects. 2.1.4 and 7.6.3. The most widely-used filters for Fourier approximations are the *exponential filter*

$$\sigma(\theta) = e^{-\alpha\theta^p} , \qquad \alpha > 0 \qquad (4.4.4)$$

and the *Vandeven filter* (1991)

$$\sigma(\theta) = 1 - \frac{(2p-1)!}{(p-1)!} \int_0^{\theta/\pi} [t(1-t)]^{p-1} \, dt , \qquad (4.4.5)$$

where p is the order of the filter (see Sect. 8.1). The exponential filter does not strictly vanish as $|\theta| \to \pi$; however, choosing α so that $\sigma(\pi)$ is machine zero effectively satisfies this condition.

Figure 4.3 illustrates these two families of filters. The low-frequency components are relatively untouched, whereas the high-frequency components are increasingly suppressed as $|\theta| \to \pi$. Solution filtering reduces the accuracy of a spectral approximation to at most order p, and it rather drastically affects more than half the modes (in each coordinate direction). In practice, solution filtering is typically applied periodically in time (and not at every time-step); the frequency of filtering is usually based on trial-and-error, seeking the best compromise between accuracy and temporal stability.

Solution filtering for nonperiodic problems using strong enforcement of the boundary conditions is not as straightforward because of the need to respect the boundary conditions during the filtering. This can be done for

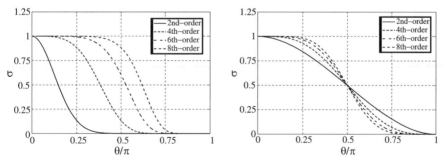

Fig. 4.3. Exponential (*left*) and Vandeven (*right*) filtering functions of various orders. For the exponential filter, the coefficient α in (4.4.4) has been chosen so that $e^{-\alpha} = 10^{-14}$

simple boundary conditions, such as Dirichlet boundary conditions, by applying the filter to an expansion in a suitable combination of the polynomials that satisfies the boundary conditions. However, there does not appear to be an approach that works for strong enforcement of general boundary conditions, particularly for hyperbolic problems when characteristic correction methods are employed. This difficulty is not present for weak enforcement of the boundary conditions.

An alternative, less drastic, filtering approach is that of *derivative filtering* (suggested by Majda, McDonough and Osher (1978) and Kreiss and Oliger (1979)). Here the solution itself is not touched, but the derivative approximation is given by

$$\frac{du^{n,f}}{dx}(x_j) \approx \sum_{k=-N/2}^{N/2-1} (ik)\sigma_k \tilde{u}_k e^{ikx_j} , \qquad j = 0, \ldots, N-1 . \qquad (4.4.6)$$

For a collocation method, derivative filtering can be implemented at no additional expense; if derivatives are computed by a transform method, then in the transform space stage, one simply multiplies the discrete Fourier coefficient by a stored variable that contains $ik\sigma_k$; likewise, if matrix multiplies are used for differentiation, then one builds the filtering function into the stored matrix. See CHQZ2, Sect. 7.6 for further details and theoretical analysis.

For expansions in Jacobi polynomials, derivative filtering can be implemented in analogous fashion to the Fourier case. The only difference is that the function $\sigma(\theta)$ used for defining the discrete factors σ_k is now defined on $[0, 1]$ and $\sigma_k = \sigma(k/N)$.

The final strategy, termed *hyperviscosity*, is the addition of an explicit high-order spatial operator (often nonlinear) to the PDE. For example, the hyperviscosity version of the Euler equations (4.2.3) is

$$\frac{\partial \mathbf{q}}{\partial t} + A\,\frac{\partial \mathbf{q}}{\partial x} = \mathbf{h}_\epsilon(\mathbf{q}) , \qquad (4.4.7)$$

where the term on the right-hand side is the hyperviscosity; such terms typically depend on a small parameter ϵ so that the original PDE is obtained as $\epsilon \to 0$.

Tadmor (1989) developed a form of hyperviscosity that was spectrally small in smooth regions, the so-called *spectral vanishing viscosity* method. For the scalar version of the nonlinear hyperbolic equation (4.2.39) with nonperiodic boundary conditions, this involves the addition of a term to the right-hand side:

$$\frac{\partial u_N}{\partial t} + \frac{\partial f(u_N)}{\partial x} = \epsilon(-1)^{s+1} \tilde{u}_N , \qquad (4.4.8)$$

where Q is a first-order differential operator and $\tilde{u}_N$ contains only the higher-frequency components of u_N, with $\epsilon \approx N^{1-2s}$ and $s \approx \log N$. Note that the order of the artificial viscosity increases (as $2 \log N$) with N, as opposed to the low, fixed order of traditional artificial viscosity terms. This approach has undergone a series of improvements (and extensions to expansions in orthogonal polynomials). See CHQZ2, Sect. 7.7 for further details and theoretical analysis.

4.5 Homogeneous Turbulence

The Fourier spectral algorithms for incompressible isotropic turbulence discussed in Sect. 3.3 can be extended to the compressible case. Provided that the velocity fluctuations are sufficiently small that no shocks develop, the solution will be smooth. Of course, the flow field will contain features on scales as small as the dissipation scales, and these must be resolved.

However, there are several additional complications in the compressible case. The inviscid nonlinearities of the compressible Navier–Stokes equations are not merely quadratic, as they are for incompressible flow. Moreover, for simulations of real fluids the viscous terms contain transcendental nonlinearities arising from appropriate empirical laws for the transport coefficients, such as Sutherland's formula (1.3.8). The pressure is a dynamic variable rather than a constraint; since now the pressure cannot be obtained analytically, the time-step restrictions on explicit schemes become severe as $M \to 0$. The first two complications make a strict Fourier Galerkin approximation rather unwieldy, and most simulations of isotropic turbulence have been made with a collocation approach. The last complication has prompted either the use of semi-implicit schemes or the avoidance of cases at low Mach number.

4.5.1 Algorithmic Considerations

The basic elements of Fourier collocation-in-space/explicit discretization-in-time approximations for homogeneous, compressible Navier–Stokes flows are straightforward. (See., e. g., the discussion in Sect. 3.3.3—apart from the discretization of the pressure constraint equation—for the spatial approximation, and the discussions in Sects. 3.2.2 and Appendix C.3 for various time-discretizations.) In this subsection we focus on the unique algorithmic and physical considerations for the compressible case. In particular, we discuss the extension of the Rogallo transformation (see Sect. 3.3.4), conservation properties, semi-implicit schemes, and strong density fluctuations. In the following subsection we survey representative DNS and LES applications.

Rogallo's transformation for treating homogeneous turbulence with fully periodic boundary conditions can be extended to a limited class of compressible flows. Strictly speaking this requires the first and second coefficients of viscosity, μ and λ, and the thermal conductivity κ to be constant in space and time. (One also needs that μ be constant in the incompressible case, but for most fluids this is a more drastic assumption in the compressible regime.)

Even then the general transformation (3.3.32)–(3.3.33) leaves residual terms in the Navier–Stokes equations that prohibit use of periodic boundary conditions in all coordinate directions except in a few special cases, namely, mean flows with uniform shear or isotropic strain. Compressible, homogeneous shear flow was first studied by Feiereisen, Reynolds and Ferziger (1981) and Delorme (1985). As these reports may not be readily available, see, e. g., Sarkar, Erlebacher and Hussaini (1991) and Blaisdell (1993) for more discussion of the homogeneous shear case. Blaisdell (1991) derived the most general form of the Rogallo transformation that applies to compressible flow. See Blaisdell, Mansour and Reynolds (1993) for more discussion of the homogeneous shear case, and see Blaisdell, Coleman and Mansour (1996) for details on the isotropic strain case. Some other homogeneous flows have been simulated with Fourier approximations using the Rogallo transformation, but only subject to approximations; e. g., Cambon, Coleman and Mansour (1993) assumed isentropic flow and simulated cases such as non-isotropic compression.

We focus on the uniform shear case here, following the approach of Feiereisen, Reynolds and Ferziger (1981). The top left frame of Fig. 3.3 illustrates the mean velocity for the particular case of a mean flow in the x-direction with a uniform shear rate S in the x_2-direction, i. e., the mean velocity has the gradient $\partial \bar{u}_1/\partial x_2 = S$, i. e., $\bar{\mathbf{u}} = (Sx_2, 0, 0)^T$. (We assume a periodicity length of 2π in each coordinate direction, omitting the details of generalizing this to an arbitrary periodicity length in each coordinate direction.) The mean density, pressure (and temperature) are uniform in space. The total velocity $\mathbf{u}$ can be written as a mean velocity plus a fluctuating component (denoted by primes):

$$u_i = Sx_z\delta_{i1} + u_i' , \qquad i = 1, 2, 3 . \tag{4.5.1}$$

The spatial coordinates are transformed as given in (3.3.41). Hence,

$$\frac{\partial}{\partial x_1} = \frac{\partial}{\partial x_1'} , \qquad \frac{\partial}{\partial x_2} = \frac{\partial}{\partial x_2'} - (St)\frac{\partial}{\partial x_1'} , \qquad \frac{\partial}{\partial x_3} = \frac{\partial}{\partial x_3'} . \tag{4.5.2}$$

The following set of equations on $(0, 2\pi)^3$ result for the fluctuating components:

$$\rho_{,t} + (\rho u_i')_{,i} - (St)(\rho u_2')_{,1} = 0 , \tag{4.5.3}$$

$$(\rho u_i')_{,t} + \frac{1}{2}\left[(\rho u_i' u_j')_{,j} + \rho u_j' u_{i.j}' + u_i'(\rho u_j')_{,j}\right] + p_{,i}$$
$$= \tau_{ij,j}' - S\rho u_2'\delta_{i1} + (St)(\rho u_2' u_i')_{,1} + (St)p_{,1}\delta_{i2} - (St)\tau_{i2,1}' , \tag{4.5.4}$$

$$p_{,t} + u_j' p_{,j} + \gamma p u_{j,j}' = (St)u_2' p_{,1} + \gamma(St)pu_{2,1}'$$
$$+ (\gamma - 1)\kappa\left[T_{,jj} - 2(St)T_{,12} + (St)^2 T_{,11}\right] \tag{4.5.5}$$
$$+ (\gamma - 1)\Phi ,$$

with

$$p = \rho RT \ , \qquad \tau_{ij} = \mu(u_{i,j} + u_{j,i}) - \frac{2}{3}\mu(u_{k,k})\delta_{ij} \ , \qquad \Phi = \tau_{ij}u_{i,j} \ . \qquad (4.5.6)$$

Derivatives in these equations are denoted by $,t$ for derivatives with respect to time, and by $,j$ for those with respect to x'_j. Also, in both sets of equations, we have used the summation convention, i. e., a sum (from 1 to 3) is performed over all indices that are repeated in individual terms. In (4.5.3)–(4.5.5) the velocity (with primes) does not include the mean velocity, whereas in (4.5.6) the velocity (without primes) includes the mean velocity and the derivatives are with respect to x_j.

Two comments about the form of (4.5.3)–(4.5.5) are in order. First, the equation for the pressure (4.5.5) was used in place of the more customary equation for the total energy because no residual terms appear in the former. Second, the momentum equation (4.5.4) differs from the standard form (1.3.12) in that the identity

$$\nabla \cdot (\rho \mathbf{u}\mathbf{u}^T) = \rho \mathbf{u} \cdot \nabla \mathbf{u} + \mathbf{u}\nabla \cdot (\rho \mathbf{u}) \qquad (4.5.7)$$

has been employed. Feiereisen et al. demonstrated that a Fourier collocation method applied to the $S = 0$, inviscid version of (4.5.3)–(4.5.5) conserves mass—$\sum \rho_{l,m,n}$, momentum—$\sum (\rho \mathbf{u})_{l,m,n}$, and total energy—$\sum(1/(\gamma-1)p+ \frac{1}{2}\rho|\mathbf{u}|^2)_{l,m,n}$, in the absence of time-discretization errors. Blaisdell, Mansour and Reynolds (1993) replaced the pressure equation above with an equation for what they termed the "pseudo total energy", namely, $p/(\gamma-1)+\frac{1}{2}\rho\,\mathbf{u}'\cdot\mathbf{u}'$. They observed that this alternative takes three fewer derivatives than the Feiereisen et al. version. They also discuss the Rogallo transformation for other uniform shear cases. Blaisdell, Coleman and Mansour (1996) applied the Rogallo transformation to isotropic mean strain in a compressible flow.

A Fourier collocation approximation coupled with an explicit time discretization to (4.5.3)–(4.5.6) is quite straightforward except for the need to periodically (in time) re-grid in the $x_1 - x_2$-plane. See Sect. 3.3.4 for the details of the re-gridding. Many spectral simulations of homogeneous, compressible turbulence employed no de-aliasing procedure (apart from the re-gridding process), some used the 2/3-rule (de-aliases only the quadratic products), and others the 1/2-rule (de-aliases quadratic and cubic products but not transcendental terms). Most simulations of compressible, isotropic turbulence have at least employed an isotropic truncation of the Fourier modes after each time-step. Blaisdell, Mansour and Reynolds (1993) noted that the skew-symmetric terms in the pressure equation reduce the effects of aliasing errors. A more detailed analysis for the latter point can be found in Blaisdell, Spryopoulos and Qin (1996). Actually, the latter work used a slightly different skew-symmetric form for the convection terms in the momentum equation than those in (4.5.4) and also uses the internal energy equation, with a skew-symmetric form for its convection terms, instead of the pressure equation.

Virtually all spectral computations of homogeneous, compressible turbulence have used explicit time-stepping procedures. The time-step limitation is usually imposed by the acoustic terms at low Mach numbers and by the advection and viscous terms otherwise. However, Delorme (1984) developed a class of schemes employing a semi-implicit treatment of the advection and/or diffusion terms, whereas Erlebacher, Hussaini, Speziale and Zang (1992) utilized a semi-implicit procedure that exploits an analytical integration of the acoustic waves. This has some advantages at low Mach numbers when the details of the small-scale sound waves are not of interest.

4.5.2 Representative Applications

We now turn to some of the applications of these methods and begin by defining two key parameters for homogeneous, compressible turbulence. The fluctuating velocity field is customarily split as $\mathbf{u}' = \mathbf{u}^{I'} + \mathbf{u}^{C'}$, where the solenoidal (or incompressible) component satisfies $\nabla \cdot \mathbf{u}^{I'} = 0$ and the irrotational (or compressible) component satisfies $\nabla \times \mathbf{u}^{C'} = \mathbf{0}$. The overall strength of the velocity fluctuations is measured by the root-mean-square velocity fluctuation, $u_{\mathrm{rms}} = \sqrt{\overline{(u_1')^2 + (u_2')^2 + (u_3')^2}}$, where the overbar denotes a spatial average. The root-mean-square fluctuations of the compressible and incompressible components are defined similarly. The fluctuating, or turbulent, Mach number is defined as

$$\mathrm{M}_t = \sqrt{\overline{(u_1')^2 + (u_2')^2 + (u_3')^2}}/c \,, \tag{4.5.8}$$

and the fraction of the initial velocity field that is compressible is measured by the parameter

$$\chi = \frac{u_{\mathrm{rms}}^C}{u_{\mathrm{rms}}} \,. \tag{4.5.9}$$

For sufficiently large M_t, the flow will have supersonic regions. These produce the so-called "eddy shocklets", which are small-scale shock waves (more precisely, extremely thin internal layers) where the flow compresses to subsonic. Passot and Pouquet (1987) performed 256^2 simulations of isotropic turbulence at various levels of M_t and found that for $\mathrm{M}_t \geq 0.3$, eddy shocklets appeared, as has been observed experimentally. See the left frame of Fig. 4.4 for an example taken from their work. Relatively high resolution is needed to capture these eddy shocklets. (The oscillations resulting from poorly resolved flow structures prove fatal at sufficiently large amplitudes in the compressible case because they lead to negative densities. This difficulty does not arise for incompressible simulations.) Blaisdell, Mansour and Reynolds (1993) performed 192^3 computations of compressible turbulence in uniform shear flow. A 2D slice from one of their computations, given in the right frame of Fig. 4.4, illustrates the eddy shocklets in 3D flow. They are elongated in the x_1-direction, of course, by the mean shear.

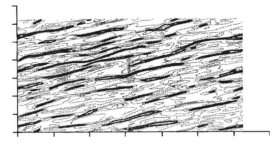

Fig. 4.4. Density contours from compressible turbulence DNS at large fluctuating Mach number. A 2D DNS of isotropic turbulence (*left*) [Reprinted with permission from T. Passot, A. Pouquet (1987); © 1987, Cambridge University Press] and a slice from a 3D DNS of homogeneous turbulence in uniform shear flow (*right*) [Reprinted with permission from G.A. Blaisdell, N.N. Mansour and W.C. Reynolds (1993); © 1993, Cambridge University Press]

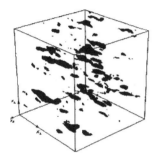

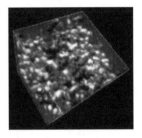

Fig. 4.5. Visualizations of 3D compressible turbulence. Regions of supersonic flow in a 3D DNS of compressible turbulence in uniform shear flow (*left*) [From S. Sarkar, S. Erlebacher, M.Y. Hussaini (1991)]. Contours of the incompressible (*middle*) and solenoidal (*right*) components of velocity from an LES of isotropic, compressible turbulence [Reprinted with permission from G. Erlebacher, M.Y. Hussaini, C.G. Speziale, T.A.Zang (1992); © 1992, Cambridge University Press]

For an illustration of the supersonic regions in a 3D DNS of compressible turbulence in uniform shear flow, see the left frame of Fig. 4.5, taken from a 128^3 DNS by Sarkar, Erlebacher and Hussaini (1991) for computations with $M_t = 0.3$ and $\chi = 0.09$. Some visualizations of other aspects of compressible turbulence are also provided in Fig. 4.5, from Erlebacher, Hussaini, Speziale and Zang (1992). These illustrate the incompressible (middle) and compressible (right) velocity components from a 128^3 simulation of compressible isotropic turbulence. Figure 4.6, from Cambon, Coleman and Mansour (1993), shows the turbulent Mach number before (left) and after (right) compressible turbulence is subjected to uniform compression. (As noted above, the case required the additional approximation that the flow is isentropic in order for the Rogallo transformation to be applied.)

Passot and Pouquet (1988) derived a class of hyperviscosities for use in numerical simulations of compressible turbulence as an approach to deal with

Fig. 4.6. Contours of turbulent Mach number in a 3D DNS of compressible turbulence under uniform axial compression [Reprinted with permission from C. Cambon, G.N. Coleman, N.N. Mansour (1993); © 1993, Cambridge University Press]

the severe resolution requirements posed by eddy shocklets at all but the lowest Reynolds numbers. Their hyperviscosity was a nonlinear version of a fourth-order differential operator that ensured that the artificial dissipation was everywhere positive. However, it was nonzero everywhere for nonconstant functions.

Apart from their use in exhibiting the relatively striking feature of eddy shocklets, for which there is no incompressible counterpart, spectral DNS of compressible turbulence have been used in a number of studies that have elucidated some fundamental aspects of compressible turbulence and also for calibration and refinement of LES models. For example, Erlebacher et al. (1990) used DNS to confirm the asymptotic theory (due to Kreiss, Lorenz and Naughton (1991) and Erlebacher et al. (1990)) of the properties of compressible turbulence. The key result of the asymptotic theory applied to the initial-value problem of compressible two-dimensional isotropic turbulence is that the various types of the low turbulent Mach number regimes have been classified with respect to the initial conditions. Specifically, the theory provides the initial conditions whose evolution precludes the occurrence of eddy shocklets, and it also provides the necessary (but not the sufficient) conditions for the occurrence of eddy shocklets. The application of the theory to homogeneous compressible turbulence by Sarkar et al. (1991) led to the appropriate modeling of the dilatational terms—the pressure dilatation and the compressible dissipation—in the Reynolds stress transport equations, which predicts the significant reduction in growth rate of the compressible mixing layer.

The early large-eddy simulations of homogeneous, compressible turbulence were routinely performed with Fourier collocation methods. The LES equations for compressible flow were presented in Sect. 1.3.3; see (1.3.38)–

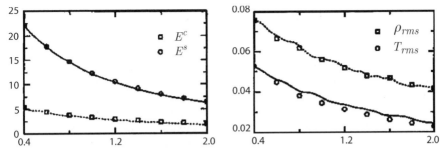

Fig. 4.7. LES of isotropic, compressible turbulence with $M_t = 0.3$ and $\chi = 0.2$ on a 32^3 grid. The LES results are the solid lines and the results form the corresponding 96^3 DNS are given by the symbols [Reprinted with permission from T.A. Zang, R.B. Dahlburg, J.P. Dahlburg (1992); © 1992, American Institute of Physics]

(1.3.41). These add additional complications to the Navier–Stokes equations that make Galerkin methods even less attractive than they are for DNS. An example of a LES used for calibration and refinement of LES models is provided in Fig. 4.7 from Zang, Dahlburg and Dahlburg (1992). These 32^3 LES and 96^3 DNS spectral computations of isotropic, compressible turbulence demonstrated that the LES model of Speziale, Erlebacher, Zang and Hussaini (1988) did a reasonable job for an admittedly extremely simple flow. Some representative references to other work of groups who have used spectral DNS of homogeneous compressible turbulence for developing and assessing turbulence models are Kida and Orszag (1990), Simone, Coleman and Cambon (1997), Hamba (1999) and Carati, Winckelmans and Jeanmart (2001).

The aforementioned applications notwithstanding, spectral methods have been less favored for homogeneous, compressible turbulence simulations, particularly large-eddy simulations, than they have been for incompressible ones. The difficulties that spectral methods have with handling thin shock-like structures (except for the shock-fitting methods discussed in Sect. 4.7) have led many groups to choose traditional shock-capturing schemes. Even for marginally resolved simulations where the oscillations arising from poorly resolved structures are sufficiently small to avoid negative densities, concern about the effects of aliasing have led some to prefer schemes such as sixth-order compact differences because of their inherent dissipation of the least well-resolved modes (see Lele (1992)). These sixth-order schemes are often used in conjunction with additional explicit solution filtering procedures to moderate the build-up of high-frequency oscillations.

Solution filtering has rarely been used in conjunction with spectral simulations of compressible turbulence, because these filters, even the ones discussed in Sect. 4.4, lead to finite-order results.

4.6 Smooth, Inhomogeneous Flows

For flows that are inhomogeneous yet smooth, i. e., with no discontinuities in the solution or low-order derivatives, computations with single-domain spectral methods have been confined to a few demonstration problems and even fewer serious applications. For inviscid, and even more so for viscous problems, the severe time-step restrictions accompanying spectral methods with nonperiodic boundary conditions and explicit time-discretization have certainly been a barrier to wider use. As we shall discuss in Chap. 5, the use of multidomain spectral methods not only opens up a much wider class of applications, but also offers some relief from the time-step limitation. In this section, we provide a short survey of the use of single-domain spectral methods for the Euler and Navier–Stokes equations.

4.6.1 Euler Equations

Kopriva et al. (1984) provided an early demonstration that spectral accuracy can be obtained for two-dimensional, nonperiodic, smooth Euler flow. The particular problem they solved is for a class of steady, isentropic flows between curved walls that admits an analytic solution derived by Ringleb (1940). Their computations were performed with the 2D Euler equations in nonconservative form. Since the entropy is constant, there is only the need to solve the equations for the density and the two velocity components. (Actually, Kopriva et al. used the calorically perfect equation of state to eliminate the density in lieu of the pressure in the continuity equation.) The computational domain of one of the cases presented by Kopriva et al. (1984) is illustrated in the left frame of Fig. 4.8. The boundaries $\mathbf{a} - \mathbf{b}$ and $\mathbf{c} - \mathbf{d}$ have been chosen to match the streamlines of the analytic solution; hence, the boundary conditions there are for zero normal velocity. The flow at the inflow boundary $\mathbf{c} - \mathbf{a}$ is subsonic. Hence only two boundary conditions are needed; a characteristics-based method similar to the CCM was used for the third condition. The outflow boundary $\mathbf{d} - \mathbf{b}$ is supersonic, and therefore no boundary condition is required there. However, a bi-characteristic treatment was required at the corner points. The transformation used to map the reference square $[-1, 1]^2$ into the curved domain in physical coordinates utilized the potential and streamfunction of the exact solution.

Kopriva et al. (1984) used a straightforward Chebyshev collocation method combined with an explicit time-discretization. The system was integrated forward in time until the computed solution was steady to better than 1 part in 10^{-10}. Two cases were computed: a transonic case with subsonic inflow and supersonic outflow and a purely supersonic case. The middle frame of Fig. 4.8 illustrates the Mach number of the computed transonic solution. It is clearly very smooth. In particular, there are no deleterious effects at the sonic points, where the flow transitions from subsonic to supersonic. The convergence of the computed solution as a function of the number of grid-points N_y

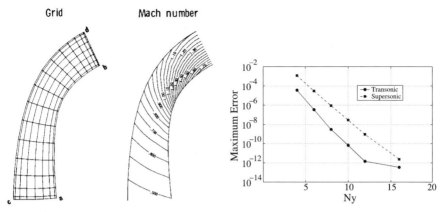

Fig. 4.8. Spectral solution for Ringleb flow. The *left frame* illustrates the Chebyshev grid use for the transonic case, the middle frame illustrates the computed Mach number contours, and the *right frame* shows the maximum error in the computed solution as a function of the number of polynomials used in the narrow direction

in the direction normal to the side boundaries is given in the right frame of the figure, not only for this transonic case but also for the purely supersonic case. (In the former, $N_x = 2N_y$, and in the latter $N_x = N_y$.) The solution does indeed exhibit spectral accuracy.

For problems in this type of geometry—two solid walls and two inflow/outflow boundaries—the treatment of the corner points in the domain usually requires the use of boundary conditions based on bi-characteristics. See Kopriva (1991) for an extended discussion of the use of bi-characteristic boundary conditions in the context of spectral methods.

If one switches from the conventional, nonstaggered grid used in the Ringleb example, and shown in the left frame of Fig. 4.9 for the reference domain, to a staggered grid, shown in the right frame of the figure, then there are obviously no corner points to worry about. Aspects of one-dimensional staggered grids are discussed in CHQZ2, Sect. 3.7—discretization for a scalar hyperbolic equation—and in Sect. 3.4.3 of the present text—processes for interpolating between grids using fast Chebyshev transforms. Consider now a two-dimensional hyperbolic problem, e. g.,

$$\frac{\partial \mathbf{q}}{\partial t} + \frac{\partial \boldsymbol{\mathcal{F}}_1}{\partial x} + \frac{\partial \boldsymbol{\mathcal{F}}_2}{\partial y} = \mathbf{0} \,, \qquad -1 < x < 1, -1 < y < 1, \quad t > 0$$

$$\mathbf{q}(x, y, 0) = \mathbf{q}_0(x, y) \,,$$

(4.6.1)

where, in the case of the two-dimensional generalization of the nonconservative Euler equations (4.3.2), $\mathbf{q} = (\rho, u, v, p)^T$. The Gauss/Gauss points are where the solution $\mathbf{q}$ is defined, the Gauss–Lobatto/Gauss points are where the x-component of the flux $\boldsymbol{\mathcal{F}}_1$ is defined, and the Gauss/Gauss–Lobatto points are where the y-component of the flux $\boldsymbol{\mathcal{F}}_2$ is defined. Note that this

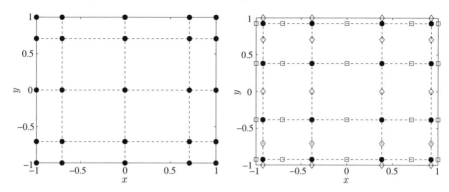

Fig. 4.9. Chebyshev nonstaggered (*left*) and staggered (*right*) grids in two dimensions for $N_x = N_y = 4$. For the nonstaggered grid the filled circles are the Gauss–Lobatto/Gauss–Lobatto points. For the staggered grid the filled circles are the Gauss–Gauss points, the open diamonds are the Gauss/Gauss–Lobatto points, and the open squares are the Gauss–Lobatto/Gauss points

staggered grid for the Euler equations is neither the half-staggered grid nor the fully staggered grid used for incompressible flow (see Sect. 3.5.1). The use of the Gauss/Gauss points to define all the components of the solution greatly simplifies the implementation compared with the use of the fully staggered grid. Indeed, in the present case all the interpolations between solutions at the Gauss/Gauss points and the fluxes (at the Gauss–Lobatto/Gauss or the Gauss/Gauss–Lobatto points) are strictly one-dimensional.

Hence, to advance one time-step using an explicit method on (4.6.1) one treats the $\partial \boldsymbol{\mathcal{F}}_1/\partial x$ term as follows. First, transform the solution $\mathbf{q}$ from the Gauss/Gauss points to the Gauss–Lobatto/Gauss points. Second, evaluate the flux $\boldsymbol{\mathcal{F}}_1(\mathbf{q})$ and then the derivative $\partial \boldsymbol{\mathcal{F}}_1/\partial x$ at the Gauss–Lobatto/Gauss points. Finally, transform this result back to the Gauss/Gauss points for the purpose of the time advancement of the dependent variables. Boundary conditions are imposed weakly. Full details on the use of staggered grids for hyperbolic problems, including the Euler equations, can be found in Kopriva and Kolias (1996). Important details include the imposition of solid wall and locally one-dimensional CCM boundary conditions, as well as the proper treatment of metric terms on the staggered grid. For use in three-dimensional problems, the solution is now defined at the Gauss/Gauss/Gauss points, and the x-component of the flux, say, is defined at the Gauss–Lobatto/Gauss/Gauss points.

The staggered grid method has seen little use in classical, single domain spectral methods. Kopriva and Kolias' (1996) development of the spectral staggered grid method for hyperbolic systems was focused on spectral domain decomposition methods. See Sect. 5.13 for further discussion and some examples in the multidomain context.

4.6.2 Navier–Stokes Equations

Very few simulations of transition and turbulence for inhomogeneous, shock-free compressible flows in simple geometries have been performed with spectral discretizations in all three directions. For wall-bounded flows, such as the flat-plate and channel flows discussed in Sect. 3.4, Chebyshev discretizations in the wall-normal direction produce a severe time-step limitation for an explicit time-stepping scheme, with the time-step limit scaling as $1/N^4$ due to the viscous terms. Not only are the viscous terms nonlinear in the dependent variables, but in most cases it is important to include the temperature dependence of the various transport coefficients, at least for wall-bounded flows. This makes the development of efficient implicit schemes for global discretizations quite challenging. Effective implicit schemes for handling the viscous terms are only available for flows with nearly constant temperature. In this subsection, we discuss those few fully spectral algorithms that have been employed for this application and then mention some work that has used spectral methods in two of the directions.

Some of the earliest numerical simulations of transition in compressible flow, by Erlebacher and Hussaini (1990, 1991), were performed with a fully spectral method. They simulated the early stages of transition for Mach 4.5 flow past a flat plate, using the simplest version of the parallel-flow approximation in which the wall-normal mean velocity is neglected and forcing terms are added to the equations to ensure that the initial mean flow is maintained undisturbed in the absence of perturbations. Their three-dimensional simulations used the Navier–Stokes equations in nonconservative form with the pressure equation in place of the energy equation. The wall boundary conditions were no-slip on velocity and zero perturbation for the temperature, which for the mean flow satisfied an adiabatic wall boundary condition. (In these types of problems the mean flow is normally assumed to satisfy an adiabatic wall temperature condition, whereas the perturbations are taken to be isothermal.) The semi-infinite domain in y was truncated at a finite distance, at which zero-perturbation boundary conditions were applied on the velocity and temperature. The density at both the wall and the free-stream were determined from the PDE, and the pressure was then computed from the equation of state.

Erlebacher and Hussaini employed Chebyshev collocation in the wall-normal direction and Fourier collocation in the streamwise and spanwise directions. In the wall-normal direction they utilized the hyperbolic tangent clustering (8.8.7) together with the truncated algebraic mapping (8.8.13) to resolve the detailed structure of the instability modes, which is concentrated at the boundary-layer edge (see Fig. 2.18). The time-stepping was purely explicit (third-order Runge–Kutta). The PDEs require 27 partial derivatives to be computed at each stage of the time-stepping. This work explored the weakly nonlinear stages of supersonic transition for both first and second modes, and needed grids no finer than $24 \times 65 \times 24$. (These studies took place at a time preceding the development of the more efficient secondary instability theory (SIT) and parabolized stability equations (PSE) tools.)

Implicit schemes for handling the viscous terms are non-trivial, especially since the temperature varies strongly in a boundary layer with an adiabatic wall condition (see Fig. 2.8). Consequently, DNS and LES tools for the strongly nonlinear regime of transition were superseded with methods with high, but finite, order (such as sixth-order compact schemes) in the wall-normal direction, while the fully spectral DNS codes became more of a verification tool for the more efficient SIT methods for the weakly nonlinear regime. One example of this use, due to Pruett, Ng and Erlebacher (1991), is provided in Fig. 4.10, which compares the evolution of various modes as predicted by the spectral SIT algorithm (discussed in Sect. 2.7.2) and the Erlebacher–Hussaini spectral DNS code. The particular case shown in the figure is the one given in the last row of Table 2.1, i. e., a subharmonic secondary instability for Mach 4.5 flow past a flat plate. Recall from Sect. 2.7.2 that secondary instability theory assumes that the amplitude of the primary wave is constant. The primary wave in this simulation is unstable. The SIT predictions can be adjusted to account for the increasing amplitude of the primary wave. The figure illustrates that this produces a much better comparison with DNS results.

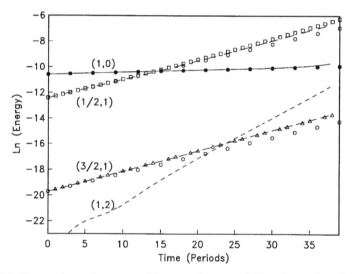

Fig. 4.10. Comparison of compressible secondary instability theory (*symbols*) with direct numerical simulation (*lines*) for various modes in the early nonlinear stage of transition for Mach 4.5 flow past a flat-plate boundary layer. For the SIT predictions of the subharmonic mode, the circles correspond to neglecting the slow growth of the primary wave, whereas the squares are for results that include the effect of the growth of the primary wave [Reprinted with permission from C.D. Pruett, L.L. Ng, G. Erlebacher (1991); © 1991, American Institute of Physics]

Fig. 4.11. Normal vorticity in a near-wall $x - z$-plane from a DNS of incompressible (*left*) and Mach 3.0 compressible (*right*) channel flow [Reprinted with permission from G.N. Coleman, J. Kim, R.D. Moser (1995); ©1995, Cambridge University Press]

The above examples for transitional flows have been for rather small-scale simulations. There has been at least one set of large-scale, purely spectral computations of compressible turbulent flow, which is due to Coleman, Kim and Moser (1995). They simulated compressible flow in a channel at Mach numbers of 1.5 and 3.0 with isothermal wall boundary conditions. They used a semi-implicit Fourier–Legendre method (due to Buell (1990)) on $144 \times 118 \times 80$ grids. Buell adds and subtracts a constant temperature term from the Navier–Stokes equations and then treats the constant temperature term implicitly, and the difference between the real viscous term and its constant temperature approximation explicitly. Because of the isothermal boundary condition, the portion of the viscous term treated explicitly is relatively small. This permits much larger time-steps to be used than has been possible for the adiabatic wall boundary that is appropriate for flow past flat plates, cylinders and cones. Figure 4.11 from the Coleman et al. work, illustrates an important effect of compressibility. This figure contains results from an incompressible channel flow, due to Kim, Moin and Moser (1987) using the normal velocity-normal vorticity algorithm discussed in Sect. 3.4.1 and from Mach 3 compressible channel flow. The contours of normal vorticity are a useful indicator of the streamwise "streaks" that are prominent in the near-wall regions of turbulent flow. This figure indicates that these streaks become more coherent as the Mach number increases.

Apart from special applications such as this, large-scale, fully spectral simulations have not been performed for wall-bounded compressible transitional and turbulent flows. However, numerous temporal simulations have been performed with Fourier spectral approximations in the two horizontal directions. Typically, sixth-order compact schemes have been employed in the wall-normal direction, permitting, as reported by Pruett and Zang (1992) for their $96 \times 144 \times 144$ DNS of Mach 4.5 flow past a hollow cylinder, an increase in the time-step by at least a factor of 10. El-Hady, Zang and Piomelli (1994) used the same method in their large-eddy simulations of transition in Mach 4.5 flow past a cylinder. Sandham and Reynolds (1991) and Adams and Kleiser (1996) used a very similar numerical approach—Fourier collocation in two directions and sixth-order compact differencing in the transverse direction—for their $96 \times 100 \times 96$ free-shear-layer and $128 \times 191 \times 128$ flat-plate boundary-layer simulations, respectively. Both used the total energy equation rather than the pressure equation and employed the Thompson (1987) version of nonreflecting boundary conditions at the artificial, free-stream boundary rather than zero-disturbance boundary conditions there. The transformation (8.8.22) could have been used in the former case for the compressible free-shear-layer simulations without the severe time-step restriction that accompanies Chebyshev discretization for the flat-plate boundary layer. Sandham and Reynolds opted for the sixth-order compact scheme with an artificial free-stream boundary out of concern for adverse effects from free-stream sound waves which would be left underresolved in the farfield. Direct numerical simulations of turbulent, supersonic wall-bounded flows have been performed with two periodic directions on several occasions. See Guarini et al. (2000) and Maeder, Adams and Kleiser (2001) for simulations of turbulent supersonic boundary layers.

The experience of the 1990s indicates that single-domain spectral methods (in all spatial directions) are not well-suited to simulations of smooth compressible, wall-bounded flows because of the severe time-step restriction, and no general semi-implicit scheme has yet proved sufficiently efficient. Although compact differencing methods have been favored for use in inhomogeneous directions, there are concerns about their temporal stability on systems of equations (see Sect. 4.2.1). Solution filtering is typically used with these methods, but a high-order scheme that did not need to resort to this device would seem preferable. Spectral domain decomposition methods, discussed in Chaps. 5–7, may offer a more promising alternative to achievement of high-order accuracy with an acceptable time-step restriction for compressible wall-bounded flows.

4.6.3 Numerical Example

We close this discussion with a numerical example comparing spectral and high-order methods for a viscous, compressible flow problem. The most commonly used finite-difference methods are fourth-order explicit central differences, fourth-order or sixth-order compact differences and third or higher order essentially non-oscillatory (ENO) schemes (Shu and Osher (1988, 1989)). The central-difference schemes are preferred for problems without sharp discontinuities such as unresolved shock waves, whereas the ENO schemes are attractive for compressible flows with shocks. The trade-offs between spectral methods and alternative finite, but high-order methods for compressible simulations are very problem-dependent.

Shu et al. (1992) made comparisons between several schemes for the temporal simulation of a compressible free shear layer. Their accuracy comparisons were made for the evolution of a small amplitude linear instability wave (computed to high accuracy by the spectral collocation method of Macaraeg, Streett and Hussaini (1988)) for a free shear layer with a hyperbolic tangent mean velocity profile (see Fig. 2.14). The mean flow has a Mach number of 0.5 in the free stream and a Reynolds number of 100. The growth rate

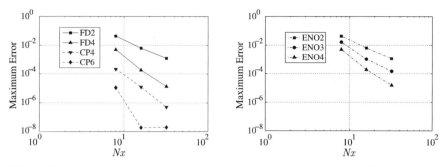

Fig. 4.12. Error in growth rate after ten time-steps for the compressible free-shear-layer linear instability. [From data in Shu et al. (1992)]

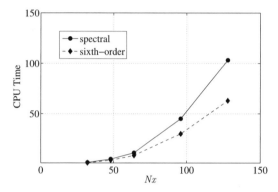

Fig. 4.13. CPU time for one time-step for the free shear layer problem on an N_x^3 grid. [From data in Zang (1990)]

for the linear instability is 0.127454941, and its eigenfunction is displayed in Fig. 2.15. The transformation (8.8.22) was used to permit a cosine expansion in y. All computations used 128 points in y, which effectively removed the y-discretization errors from the comparison. Figure 4.12 reports the errors in the growth rate determined from these simulations after 10 time-steps for several alternative methods. (Four points in x are sufficient to capture 8-digit accuracy with a spectral method.) As has been amply demonstrated in other contexts, the price that ENO methods pay for their ability to handle strong oscillations is significantly lower accuracy than the central-differencing schemes. With respect to the sixth-order scheme, the results of this example are quite consistent with those of similar examples for incompressible flow presented in Sects. 3.3.6 and 3.4.6. This and other numerical evidence in the literature suggests that a spectral method has comparable resolution to a sixth-order compact-difference method with between 25% and 50% more grid-points in each coordinate direction.

Figure 4.13, taken from Zang (1990), provides a direct comparison of computational times (on a single processor of a Cray YMP) between the sixth-order method and the spectral method. On problems such as this, where the spectral method does not suffer the severe time-step restriction that it does for wall-bounded flows, the two approaches are roughly comparable in cost for equivalent accuracy.

4.7 Shock Fitting

There are essentially three types of numerical methods for treating conservation laws with discontinuous solutions—shock capturing, shock tracking and shock fitting. A shock-capturing method, as the name implies, captures the shock automatically, but it requires explicit or implicit numerical dissipation to do so. Woodward and Colella (1984) provide an excellent review of shock-capturing methods. In a shock-tracking method, the shock is "captured" and identified with a level-set function, which is then tracked (Glimm, Li, Liu and Zhao (2001)). The next section discusses the status of shock-capturing using spectral methods. The only spectral shock-tracking methods have been multidomain ones; these are covered in Sect. 5.12. Spectral shock-fitting methods are discussed in this section in the single-domain context.

In a shock-fitting method, the shock is treated as a boundary, for which a separate shock evolution equation is derived and solved. The general approach was pioneered by Moretti (see Moretti (1987, 2002)). Provided that there are no additional shocks or slip surfaces behind the primary shock, the flow within the computational domain is smooth. Therefore, a carefully designed spectral shock-fitted solution will be highly accurate. Although shock-fitted spectral solutions to the Euler equations were first presented by Salas, Zang and Hussaini (1982) (see also Hussaini et al. (1985b)), Kopriva, Zang and Hussaini (1991) made some critical improvements to the solid-wall boundary conditions and the shock-fitting conditions. Kopriva (1992) derived the general two-dimensional shock-fitting equations for time-dependent flows. Subsequently, Kopriva (1993) extended this to viscous problems. The shock-fitting approach for viscous problems assumes that the shock is sufficiently sharp as to be treated as a discontinuity. See Moretti and Salas (1969) for the range in Mach number-Reynolds number space for which this assumption is reasonable. Our discussion here concentrates on the inviscid, axisymmetric case and uses much of the notation of Brooks and Powers (2004), which is closer to the notation of this text than that of Kopriva, Zang and Hussaini (1991).

The blunt body problem, illustrated in Fig. 4.14, has been the focus of most of the spectral shock-fitting work. The objective is the determination of the shock location and the computation of the flow in the downstream, high-pressure region 2, between the shock and the surface of the blunt body. The flow in the upstream, low-pressure region 1 is supersonic, and can be taken as the prescribed inflow condition at the shock. The computation is performed

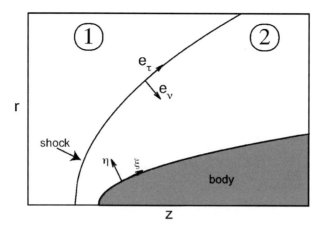

Fig. 4.14. Geometry of the blunt body problem for supersonic flow. The blunt body (hatched), shock (solid line), the shock tangent ($\mathbf{e}_\tau$) and the shock normal ($\mathbf{e}_\nu$), pointing into the high pressure region) are shown. The upstream flow in region 1 is supersonic. The downstream flow in region 2, along with the shock position, is computed by the numerical method

in a reference domain $(\xi(r, z, t), \eta(r, z, t)) \in [0, 1]^2$, which is based on a body-fitted grid, for which the coordinates at the body are tangential and normal to the body. The coordinates of the (stationary) body surface are given by $(r_b(\xi), z_b(\xi))$. The coordinate η is normal to the body but oblique to the shock. (The time dependence arises because one boundary of the coordinate system is the shock, which moves during the computation. We omit the details of the mapping between (r, z) and (ξ, η); these mapping details can be found in Brooks and Powers (2004).

We record here the formulas for the several unit vectors needed in the sequel. In the physical domain the unit vectors in the ξ and η directions are, respectively,

$$\mathbf{e}_\xi = \frac{\dfrac{dr_b}{d\xi}\mathbf{e}_r + \dfrac{dz_b}{d\xi}\mathbf{e}_z}{\sqrt{\left(\dfrac{dr_b}{d\xi}\right)^2 + \left(\dfrac{dz_b}{d\xi}\right)^2}} \,, \qquad \mathbf{e}_\eta = \frac{\dfrac{dz_b}{d\xi}\mathbf{e}_r - \dfrac{dr_b}{d\xi}\mathbf{e}_z}{\sqrt{\left(\dfrac{dr_b}{d\xi}\right)^2 + \left(\dfrac{dz_b}{d\xi}\right)^2}} \,, \qquad (4.7.1)$$

where $\mathbf{e}_r$ and $\mathbf{e}_z$ are the unit vectors in the r and z directions, respectively. The unit vectors tangential and normal to the shock (at $\eta = 1$) are denoted by $\mathbf{e}_\tau$ and $\mathbf{e}_\nu$, respectively (see Fig. 4.14); they are

$$\mathbf{e}_\tau = \frac{-\eta_z\mathbf{e}_r + \eta_r\mathbf{e}_z}{\sqrt{\eta_r^2 + \eta_z^2}} \,, \qquad \mathbf{e}_\nu = \frac{\eta_r\mathbf{e}_r + \eta_z\mathbf{e}_z}{\sqrt{\eta_r^2 + \eta_z^2}} \,. \qquad (4.7.2)$$

In addition, we have the following relationships between the metrics for the coordinate transformation and its inverse:

$$\xi_z = \frac{1}{J} r_\eta , \quad \eta_z = -\frac{1}{J} r_\xi , \quad \xi_r = -\frac{1}{J} z_\eta , \quad \eta_r = \frac{1}{J} z_\xi , \quad J = r_\eta z_\xi - r_\xi z_\eta .$$
$$(4.7.3)$$

The fluid motion is modeled by the axisymmetric Euler equations in non-conservative form. Denote the density, velocity and pressure by ρ, $\mathbf{u} = (u, w)^T$ and p, as usual. In terms of the computational coordinates the Euler equations are

$$\frac{\partial \mathbf{q}}{\partial t} + A \frac{\partial \mathbf{q}}{\partial \xi} + B \frac{\partial \mathbf{q}}{\partial \eta} + \mathbf{h} = \mathbf{0} , \qquad (4.7.4)$$

where

$$\mathbf{q} = (\rho, u, w, p)^T , \qquad \mathbf{h} = (\rho u/r, 0, 0, \gamma p u/r)^T ,$$

$$A = \begin{pmatrix} u^c & \rho \xi_r & \rho \xi_z & 0 \\ 0 & u^c & 0 & \dfrac{\xi_r}{\rho} \\ 0 & 0 & u^c & \dfrac{\xi_z}{\rho} \\ 0 & \gamma p \xi_r & \gamma p \xi_z & u^c \end{pmatrix} , \qquad B = \begin{pmatrix} w^c & \rho \eta_r & \rho \eta_z & 0 \\ 0 & w^c & 0 & \dfrac{\eta_r}{\rho} \\ 0 & 0 & w^c & \dfrac{\eta_z}{\rho} \\ 0 & \gamma p \eta_r & \gamma p \eta_z & w^c \end{pmatrix} , \qquad (4.7.5)$$

and the contravariant velocity components are given by

$$u^c = \xi_t + u \xi_r + w \xi_z , \qquad w^c = \eta_t + u \eta_r + w \eta_z . \qquad (4.7.6)$$

In the spectral shock-fitting method, (4.7.4) is discretized in space with a Chebyshev collocation method and advanced in time by an explicit time-discretization. The above PDE, of course, needs to be supplemented with appropriate boundary conditions. The boundary conditions at the solid body ($\eta = 0$), at the symmetry plane ($\xi = 0$) and at the outflow line ($\xi = 1$) are straightforward (albeit tedious in places), whereas the boundary conditions at the shock ($\eta = 1$) are quite non-standard.

At the (presumed stationary) solid boundary, zero normal flow is imposed, i.e., $\mathbf{u} \cdot \mathbf{n} = 0$, where $\mathbf{n}$ is the outward unit normal at the body boundary, and $\mathbf{n} = -\mathbf{e}_\eta$. At the body, we thus have that $\mathbf{u} \cdot \mathbf{e}_\eta = u \dfrac{dz_b}{d\xi} - w \dfrac{dr_b}{d\xi} = J(u \eta_r + w \eta_z) = 0$, so that the zero normal flow condition can be written $w^c = u \eta_r + w \eta_z = 0$. (Since $\eta = 0$ at the body, we have that $\eta_t = 0$ there.)

The full set of body boundary conditions comes from application of the linearized, locally one-dimensional, characteristic compatibility method discussed in Sects. 4.2.1 and 4.3. (See (4.3.15) for an example of the linearization.) For the solid boundary at $\eta = 0$, the normal to the boundary is antiparallel to the η-direction. The locally one-dimensional CCM then uses the matrix $-B$, which has the eigenvalues $-w^c - c\sqrt{\eta_r^2 + \eta_z^2}$, $-w^c$, $-w^c$ and

$-w^c + c\sqrt{\eta_r^2 + \eta_z^2}$. The corresponding linearized characteristic variables are

$$\mathbf{z} = \begin{pmatrix} \dfrac{\bar{\rho}\bar{c}\eta_r}{\sqrt{\eta_r^2 + \eta_z^2}}\, u + \dfrac{\bar{\rho}\bar{c}\eta_z}{\sqrt{\eta_r^2 + \eta_z^2}}\, w + p \\[3mm] \eta_z\, u - \eta_r\, w \\[2mm] p - \bar{c}^2\, \rho \\[2mm] -\dfrac{\bar{\rho}\bar{c}\eta_r}{\sqrt{\eta_r^2 + \eta_z^2}}\, u - \dfrac{\bar{\rho}\bar{c}\eta_z}{\sqrt{\eta_r^2 + \eta_z^2}}\, w + p \end{pmatrix}, \tag{4.7.7}$$

where the overbars denote flow variables evaluated at the current time-step. The first characteristic variable represents information carried into the flow from the direction of the solid body, and therefore it is discarded in favor of the zero normal flow condition. The second and third eigenvalues are zero at the wall. As noted in Sect. 4.2.1, the corresponding characteristics should be treated as outflow characteristics, along with the fourth characteristic, whose eigenvalue is strictly positive. The last three characteristic variables, then, represent information carried from the PDE, and they are the basis of the differential equations applied at the boundary. Therefore, the solid body boundary conditions are

$$\eta_r u + \eta_z w = 0\,,$$

$$\eta_z \frac{\partial u}{\partial t} - \eta_r \frac{\partial w}{\partial t} = u^c\left(\eta_r \frac{\partial w}{\partial \xi} - \eta_z \frac{\partial u}{\partial \xi}\right) + \frac{1}{\rho}(\eta_r \xi_z - \eta_z \xi_r)\frac{\partial p}{\partial \xi}\,,$$

$$\frac{\partial p}{\partial t} - c^2 \frac{\partial \rho}{\partial t} = -u^c \frac{\partial p}{\partial \xi} + c^2 u^c \frac{\partial \rho}{\partial \xi}\,,$$

$$\frac{\partial p}{\partial t} = c\sqrt{\eta_r^2 + \eta_z^2}\,\frac{\partial p}{\partial \eta} + \frac{c(\eta_r \xi_r + \eta_z \xi_z)}{\sqrt{\eta_r^2 + \eta_z^2}}\frac{\partial p}{\partial \xi} + \frac{\rho c u^c}{\sqrt{\eta_r^2 + \eta_z^2}}\left(\eta_r \frac{\partial u}{\partial \xi} + \eta_z \frac{\partial w}{\partial \xi}\right)$$

$$\qquad - u^c \frac{\partial p}{\partial \xi} - \rho c^2\left(\xi_r \frac{\partial u}{\partial \xi} + \xi_z \frac{\partial w}{\partial \xi} + \eta_r \frac{\partial u}{\partial \eta} + \eta_z \frac{\partial w}{\partial \eta}\right) - \frac{\rho c^2 u}{r}\,. \tag{4.7.8}$$

At the symmetry plane $\xi = 0$, one applies three symmetry conditions along with the condition that the entropy on the body equals the entropy on the symmetry plane just behind the shock (denoted by s_0):

$$\frac{\partial w}{\partial \xi} = \frac{\partial p}{\partial \xi} = u = 0\,, \qquad s = s_0\,. \tag{4.7.9}$$

When the flow is supersonic across the outflow plane, no boundary conditions are needed. When all or part of the outflow boundary is subsonic, then the linearized, locally one-dimensional CCM should be applied (as in (4.3.17)–(4.3.18)), although for the boundary at $\xi = 1$, the normal is not necessarily parallel to the ξ-direction; hence, one needs to examine the appropriate combination of the matrices A and B per (4.2.28). In virtually all applications of spectral shock-fitting to inviscid blunt body problems, the outflow boundary is indeed chosen to be in a region in which the flow is supersonic all along the boundary, and this extra complication is avoided.

The shock-fitting procedure uses the Rankine-Hugoniot conditions along with the CCM to determine the shock motion and the solution at the shock front. Let $v_{sh,\nu}$ and $v_{sh,\eta}$ denote the speed of the shock in the directions $\mathbf{e}_\nu$ and $\mathbf{e}_\eta$, respectively. The normal component of the fluid velocity, relative to the shock velocity, on either side of the shock is

$$\delta_i = \mathbf{u}_i \cdot \mathbf{e}_\nu - v_{sh,\nu} = \mathbf{u}_i \cdot \mathbf{e}_\nu + v_{sh,\eta}(\mathbf{e}_\eta \cdot \mathbf{e}_\nu) , \qquad i = 1,2 . \qquad (4.7.10)$$

(Recall that the subscripts 1 and 2 denote, respectively, the low-pressure and high-pressures sides of the shock, that $\mathbf{e}_\nu$, given in (4.7.2), is the unit normal to the shock, and that $\mathbf{e}_\eta$, given in (4.7.1), is the unit normal to the body, with both normals pointing into the downstream flow region 2.)

The Rankine-Hugoniot conditions are invoked to determine the jump in the solution across the shock. The two-dimensional version of the Rankine-Hugoniot relations (4.3.7), employed in a form that accounts for the shock velocity and shape, yields

$$\delta_2 = \frac{\gamma-1}{\gamma+1}\delta_1 + \frac{2\gamma}{\gamma+1}\frac{p_1}{\rho_1\delta_1} ,$$

$$p_2 = \frac{2}{\gamma+1}\left(\rho_1\delta_1^2 - \frac{\gamma-1}{2}p_1\right) ,$$

$$\frac{\rho_2}{\rho_1} = \frac{\delta_1}{\delta_2} ,$$

$$\mathbf{u}_2 \cdot \mathbf{e}_\tau = \mathbf{u}_1 \cdot \mathbf{e}_\tau . \qquad (4.7.11)$$

We need to determine the equation for the shock acceleration, which we integrate to get the shock velocity, and then integrate again to update the shock position. The time derivatives of the first two equations of (4.7.11) yield

$$\frac{\partial\delta_2}{\partial t} = a\frac{\partial\delta_1}{\partial t} + b\frac{\partial p_1}{\partial t} + d\frac{\partial\rho_1}{\partial t} ,$$

$$\frac{\partial p_2}{\partial t} = e\frac{\partial\delta_1}{\partial t} + f\frac{\partial p_1}{\partial t} + g\frac{\partial\rho_1}{\partial t} . \qquad (4.7.12)$$

The coefficients are given by

$$a = \frac{\gamma-1}{\gamma+1} - \frac{2\gamma}{\gamma+1}\frac{p_1}{\rho_1\delta_1^2} , \qquad b = \frac{2\gamma}{(\gamma+1)\rho_1\delta_1} , \qquad d = -\frac{bp_1}{\rho_1} ,$$

$$e = \frac{4\rho_1\delta_1}{\gamma+1} , \qquad f = -\frac{\gamma-1}{\gamma+1} , \qquad g = \frac{2\delta_1^2}{\gamma+1} . \qquad (4.7.13)$$

From (4.7.10) we also have, for $i = 1, 2$, that

$$\frac{\partial\delta_i}{\partial t} = \frac{\partial\mathbf{u}_i}{\partial t} \cdot \mathbf{e}_\nu + \mathbf{u}_i \cdot \frac{\partial\mathbf{e}_\nu}{\partial t} + (\mathbf{e}_\eta \cdot \mathbf{e}_\nu)\frac{\partial v_{sh,\eta}}{\partial t} + v_{sh,\eta}\left(\mathbf{e}_\eta \cdot \frac{\partial\mathbf{e}_\nu}{\partial t}\right) . \qquad (4.7.14)$$

The final equation comes from the observation that since the velocity normal to the shock on the downstream (high-pressure) side is necessarily subsonic, there is one characteristic variable which carries information into the shock from the downstream side, namely, $p_2 - (\rho_2 c_2)\mathbf{u}_2 \cdot \mathbf{e}_\nu$. (See the discussion of a subsonic inflow boundary condition in Sect. 4.3). Application of the CCM to (4.7.4) then yields

$$\frac{\partial p_2}{\partial t} - (\rho_2 c_2)\frac{\partial \mathbf{u}_2}{\partial t} \cdot \mathbf{e}_\nu = q \,, \tag{4.7.15}$$

where

$$q = -u^c \frac{\partial p}{\partial \xi} - w^c \frac{\partial p}{\partial \eta} - \gamma p \left(\xi_r \frac{\partial u}{\partial \xi} + \xi_z \frac{\partial w}{\partial \xi} + \eta_r \frac{\partial u}{\partial \eta} + \eta_z \frac{\partial w}{\partial \eta} \right) - \frac{\gamma p u}{r}$$
$$- \frac{\rho c}{\sqrt{\eta_r^2 + \eta_z^2}} \left[\eta_r \left(u^c \frac{\partial u}{\partial \xi} + w^c \frac{\partial u}{\partial \eta} + \frac{1}{\rho} \left(\xi_r \frac{\partial p}{\partial \xi} + \eta_r \frac{\partial p}{\partial \eta} \right) \right) \right. \tag{4.7.16}$$
$$\left. + \eta_z \left(u^c \frac{\partial w}{\partial \xi} + w^c \frac{\partial w}{\partial \eta} + \frac{1}{\rho} \left(\xi_z \frac{\partial p}{\partial \xi} + \eta_z \frac{\partial p}{\partial \eta} \right) \right) \right] \,.$$

We also note that

$$\frac{\partial \mathbf{e}_\nu}{\partial t} = \left(\eta_z \frac{\partial \eta_r}{\partial t} - \eta_r \frac{\partial \eta_z}{\partial t} \right) \frac{(\eta_z \mathbf{e}_r - \eta_r \mathbf{e}_z)}{(\eta_r^2 + \eta_z^2)^{3/2}} \,. \tag{4.7.17}$$

Combining (4.7.12), (4.7.14) and (4.7.15), we obtain the following equation for $\partial v_{sh,\eta}/\partial t$ in terms of the known upstream quantities ρ_1, $\mathbf{u}_1$ and p_1:

$$\frac{\partial v_{sh,\eta}}{\partial t} = \frac{(-e + \rho_2 c_2 a)(\mathbf{u}_1 + v_{sh,\eta}\mathbf{e}_\eta) \cdot \dfrac{\partial \mathbf{e}_\nu}{\partial t}}{(\mathbf{e}_\eta \cdot \mathbf{e}_\nu)[-e + \rho_2 c_2(a-1)]}$$
$$+ \frac{\rho_2 c_2 (\mathbf{u}_2 + v_{sh,\eta}\mathbf{e}_\eta) \cdot \dfrac{\partial \mathbf{e}_\nu}{\partial t} - q}{(\mathbf{e}_\eta \cdot \mathbf{e}_\nu)[-e + \rho_2 c_2(a-1)]} \tag{4.7.18}$$
$$+ \frac{(e - \rho_2 c_2 a)\dfrac{\partial \mathbf{u}_1}{\partial t} \cdot \mathbf{e}_\nu + (f - \rho_2 c_2 b)\dfrac{\partial p_1}{\partial t} + (g - \rho_2 c_2 d)\dfrac{\partial \rho_1}{\partial t}}{(\mathbf{e}_\eta \cdot \mathbf{e}_\nu)[-e + \rho_2 c_2(a-1)]} \,.$$

(The third line of this equation vanishes if the upstream flow is steady.)

Finally, the shock distance $r_{sh,\eta}$ in the body normal direction is obtained from integrating

$$\frac{\partial r_{sh,\eta}}{\partial t} = v_{sh,\eta} \,. \tag{4.7.19}$$

The basic elements of a single time-step for the spectral shock-fitting method are (1) update the shock velocity using (4.7.18); (2) update the solution at all interior points and at the supersonic outflow line via (4.7.4); (3)

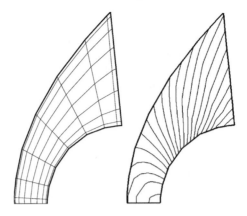

Fig. 4.15. Grid (*left*) and pressure (*right*) for the shock-fitted, Chebyshev collocation solution for Mach 4 flow past a circular cylinder [Adapted from Kopriva, Zang
and Hussaini (1991)]

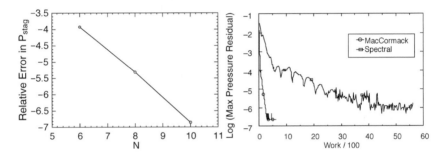

Fig. 4.16. Error in the stagnation pressure as a function of the number of grid
points, N, in each direction for the shock-fitted, Chebyshev collocation solution for
Mach 4 flow past a circular cylinder (*left*). Decay of the maximum residual (*right*)
in the pressure as a function of the time step (iteration) for the Chebyshev collocation and second-order finite-difference solutions [Adapted from Kopriva, Zang and
Hussaini (1991)]

apply the boundary conditions (4.7.8) at the wall and (4.7.9) at the symmetry
plane; (4) update the shock position with (4.7.19); and (5) update the solu-
tion at the shock front using the given upstream values of ρ_1, $\mathbf{u}_1$ and p_1 along
with (4.7.11) and the characteristic combination (4.7.15) from the computed
interior solution. All spatial derivatives, including the various metric terms,
are computed by spectral collocation.

Most demonstrations of spectral shock-fitting methods have been to
steady flow past blunt bodies. Kopriva, Zang and Hussaini (1991) reported
demonstrations of spectral accuracy on shock-fitted Chebyshev collocation
solutions to several blunt body problems. Figure 4.15 illustrates the computational grid and the spectral solution for the pressure on a 10×10 grid for

Mach 4 flow past a circular cylinder. The stagnation pressure (in this case, the pressure at the flow symmetry line on the cylinder) is known exactly for this problem. The left frame of Fig. 4.16 demonstrates spectral accuracy for the numerical solution of the stagnation pressure, whereas the right frame shows not only that the spectral solution converges to machine zero (these were 32-bit computations), but also that the spectral solution is faster (by 6 times) than second-order MacCormack finite-difference solutions of comparable accuracy (6×6 and 21×21 grid, respectively). (The time-step count has been normalized to equivalent CPU time for the 2 methods.) No filtering of any kind was needed to produce these oscillation-free spectral results.

4.8 Shock Capturing

The challenge of using spectral methods to compute solutions with discontinuities, particularly flows with shock waves and/or contact discontinuities, has been tackled since the early 1980s. As discussed, e. g., in CHQZ2, Sect. 2.1.4, spectral approximations of discontinuous functions are afflicted with the Gibbs phenomenon, which manifests itself in oscillatory solutions. For nonlinear problems these oscillations can lead to actual breakdown of the computation, e. g., by producing negative densities in compressible flow computations. A usable method requires a stable solution; moreover, the ideal would be to produce spectral accuracy for both the location of discontinuities and for the smooth part of the flow field.

The early work on spectral shock-capturing methods (Gottlieb, Lustman and Orszag (1981), Taylor, Myers and Albert (1981), Zang and Hussaini (1981), Cornille (1982), Sakell (1984), Hussaini et al. (1985a), Streett, Zang and Hussaini (1985)) typically utilized some combination of explicit filtering (see Sect. 4.5) and explicit artificial viscosity. Solution filtering was applied periodically, and sometimes derivative filtering was also employed at every time-step. In that era, the Lanczos, raised cosine, sharpened raised cosine and exponential filters (see CHQZ2, Sects. 2.1.4 and 7.6.3) were all in use. The artificial viscosities were patterned after the second-order and fourth-order artificial viscosities then in vogue for finite-difference methods. The Zang and Hussaini (1981) results for the shocked flow in a quasi-one-dimensional nozzle utilized linear hyperviscosity for stabilization, whereas the Gottlieb, Lustman and Orszag (1981) approach to the one-dimensional shock-tube problem relied on solution filtering. The work of Gottlieb, Lustman and Orszag was unique among these in the inclusion of singularity detection and reconstruction algorithms.

Figure 4.17, from Streett, Zang and Hussaini (1985), is one such early result of spectral shock capturing. The problem that was solved was two-dimensional transonic potential flow past a lifting airfoil, described by (1.3.50) with appropriate boundary conditions on the airfoil surface and in the free stream. The approximation was Chebyshev collocation in radial distance from

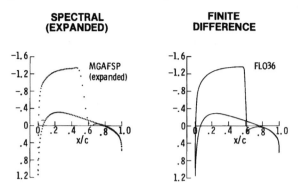

Fig. 4.17. Surface pressures for expanded spectral (*left*) and finite-difference (*right*) solutions to a transonic, lifting airfoil [From Streett, Zang and Hussaini (1985)]

the airfoil surface and Fourier collocation in circumferential angle. Nonlinear hyperviscosity was added via the artificial density approach of Hafez, South and Murman (1979), and solution filtering was included to reduce the oscillations. The implicit equations were solved with a spectral multigrid method (CHQZ2, Sect. 4.6.2) utilizing approximate factorization of a second-order finite-difference preconditioner (CHQZ2, Sect. 4.4.3). The spectral solution on a 18×64 grid was demonstrated to be more accurate than a contemporary state-of-the-art second-order finite-difference method on a 32×192 grid. (The figure displays the spectral result interpolated to a 32×192 grid.) Nevertheless, the spectral solution does exhibit low-amplitude, high-frequency oscillations throughout the flow (only visible in the figure near the shock), and spectral accuracy of the airfoil lift was not demonstrated for the flow with a shock (although at least sixth-order accuracy was demonstrated by Streett (1983) for subsonic cases).

While these early spectral shock-capturing methods were able to produce stable numerical solutions to simple one- and two-dimensional flows with shocks, all of the solutions exhibited oscillations (albeit much milder oscillations than from naive spectral solutions), and none exhibited true spectral accuracy (either for the shock location or the smooth solution). Moreover, the computational times of most of these methods were substantially longer than those of contemporary finite-difference and finite-volume methods with a comparable global accuracy due to both the greater cost of evaluating derivatives and the greater stiffness of the equations (which led to many more time-steps/iterations to reach a solution). Later developments in spectral shock-capturing methods focused on high-order filters (eighth-order and above) and spectral vanishing viscosity. See Sect. 4.4 for some brief comments on both high-order filters and spectral viscosity. See also CHQZ2, Sect. 7.7 for a more extensive discussion, and see, e. g., Chap. VI in Bernardi and Maday (1997) for some representative calculations with these methods.

The singularity detection and reconstruction techniques due to Gottlieb and coworkers are reviewed, e. g., in Gelb and Tanner (2006) and in CHQZ2, Sect. 7.6.3. For some representative results of these more modern shock-capturing methods, see Don (1994) and Shu et al. (2005). This body of work has demonstrated that two-dimensional flows with complex patterns of discontinuities can be successfully computed with spectral methods, albeit with low-amplitude oscillations. However, a demonstration of spectral accuracy in the results is still lacking.

The decades-old result shown in Fig. 4.17 is one of the few published numerical results (and may be the only result) of single-domain spectral shock capturing in two dimensions that demonstrates a lower computational time to achieve an accuracy equal to or better than the accuracy that would be acceptable in an engineering application from a contemporary low-order method. As noted above, even this example was not spectrally accurate. To our knowledge there are no three-dimensional results of single-domain spectral shock capturing.

5. Discretization Strategies
for Spectral Methods in Complex Domains

5.1 Introduction

Barely a decade after Orszag's (1969) and Kreiss and Oliger's (1972) pioneering work on (global) spectral methods, the first heuristic steps were taken towards combining spectral methods with domain decomposition approaches. Orszag (1980) and Morchoisne (1983) proposed spectral domain decomposition approaches for simple elliptic problems based on the strong form (patching) of the equations. Both employed standard spectral collocation techniques within the interiors of the subdomains and on the full domain boundary. However, the former used non-overlapping domains and applied conditions at each collocation node on subdomain interfaces to enforce the continuity of the solution and its normal derivative, whereas the latter used overlapping subdomains (the Schwarz method) and enforced continuity of the solution at the internal subdomain boundaries. Patera (1984), on the other hand, employed a weak (variational) formulation as the discretization principle for his spectral element approach to spectral domain decomposition.

The principal motivations for the development of spectral domain decomposition methods were to extend spectral methods to domains that cannot be mapped into a single reference domain, to enable local grid refinements on even simple domains, to take advantage of the different behavior of the solution in different parts of the domain, and perhaps to get some relief from the severe time-step restriction of single-domain spectral methods using explicit time discretizations. A similar perspective underlies the development of *hp* finite-element methods (see, e.g., Babuška and Suri (1994), Schwab (1998), Babuška and Strouboulis (2001), Melenk (2003)). For spectral approximations the domain decomposition approach allows the exploitation of local tensor-product bases.

In this chapter we provide a general introduction to discretization methods and theoretical analysis of spectral methods in complex geometries. The next chapter will be devoted to solution algorithms, in particular those that are suitable for parallel implementation. Our scope is mostly confined to model problems that illustrate the fundamentals and establish the relationship between classical spectral methods and their domain decomposition progeny. Domain decomposition methods based on the weak formulation have a broader user base and a wider and more complex application history. Consequently, we put considerably more attention on these methods than on those based on the strong form of the equation. Nonetheless we will point out the analogies between the weak and strong approaches.

Detailed expositions of domain decomposition methods can be found in the texts by Smith et al. (1996), Quarteroni and Valli (1999), Toselli and Widlund (2005), and Wohlmuth (2001). Particular attention to spectral element methods is given in the texts by Karniadakis and Sherwin (1999, 2005) and Deville, Fischer and Mund (2002). Readers are encouraged to rely on texts such as these for a more thorough treatment of many aspects of the subject.

In this chapter we explain how high-order spectral methods can be adapted to the approximation of differential problems that are set in a computational domain of complex form. More specifically, we will consider a domain $\Omega \subset \mathbb{R}^d$, $d = 1, 2, 3$, that can be represented as the union of subdomains Ω_m, $m = 1, \ldots, M$ (for a suitable integer $M \geq 2$). Each of these subdomains can be obtained through a mapping from a reference domain (also called a parent domain or a master domain) $\hat{\Omega}$, as described in Sect. 8.8.4.

The partition of Ω can be either *geometrically conforming*, when neighboring subdomains share either a vertex or a complete edge or face, or *geometrically nonconforming* when interfaces do not match completely. In the former case we will introduce the spectral element method (SEM) (together with its extension SEM-NI that makes use of numerical integration). The continuity of the solution will be implied by the choice of trial functions, whereas the continuity of its flux will be automatically accounted for by the weak (integral) formulation, as in finite-elements. We will also address the case of spectral discontinuous Galerkin methods (SDGM), in which solutions are looked for independently in each subdomain, and jump terms are introduced at subdomain interfaces in order to ensure that the limit solution (when the polynomial degree tends to infinity) is globally continuous. SDGM can be used for geometrically conforming as well as geometrically nonconforming approaches. The same flexibility is enjoyed by the spectral mortar element method (MEM). In addition, both SDGM and MEM can accommodate nonuniform polynomial degrees in the different subdomains far more easily than

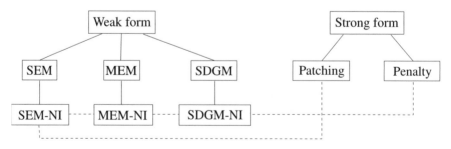

Fig. 5.1. Multidomain discretization strategies. Dashed lines indicate the existence of analogies at the algebraic level

SEM does. In particular, the continuity requirements at subdomain interfaces in MEM methods are satisfied through integral relations between the
solutions of adjacent elements along edges or faces.

At the obvious risk of oversimplifying the situation, in Fig. 5.1 we provide
a rough classification of various types of multidomain discretization strategies, that is, the way to extend spectral methods to (complex) domains partitioned into subdomains of simple shape. This will serve as a roadmap for our
discussion. The discretization strategies that we will consider in this chapter are the spectral element method, the spectral mortar element method,
the spectral discontinuous Galerkin method, and the patching approach. See
Hesthaven (2000) and Gottlieb and Hesthaven (2001) for discussions of spectral multidomain methods using the penalty approach.

5.2 The Spectral Element Method (SEM) in 1D

SEM represents a special case of Galerkin methods in which the finite-
dimensional space of trial/test functions is made of continuous piecewise (images of) algebraic polynomials of high degree on each element of a partition
of the computational domain. We focus first on the one-dimensional problem
for ease of exposition, then we address the extension to multiple dimensions.

5.2.1 SEM Formulation

Consider the one-dimensional advection-diffusion-reaction problem

$$\mathcal{L}u = -(\alpha u' - \beta u)' + \gamma u = f\,, \qquad a < x < b\,,$$
$$u(a) = u(b) = 0\,,$$

$$(5.2.1)$$

where the prime denotes the derivative with respect to x, the given coefficients
α, β, γ satisfy $0 < \alpha_0 \le \alpha \le \alpha_1$, $|\beta| \le \beta_1$, $|\gamma| \le \gamma_1$ for suitable constants
$\alpha_0, \alpha_1, \beta_1, \gamma_1$, whereas $f \in L^2(a, b)$. As can be seen in Sect. 8.9, or in CHQZ2,
Sect. 1.2.3, the weak formulation of (5.2.1) consists of finding a function u in
a suitable space V that satisfies

$$\int_a^b (\alpha u' - \beta u)v' + \gamma uv = \int_a^b fv \qquad \forall v \in V\,.$$

$$(5.2.2)$$

This set of equations is obtained after multiplication of the equation $\mathcal{L}u = f$
by a test function v vanishing at the endpoints of the interval and applying
integration by parts. The integral form above can be written equivalently in
abstract form as

$$u \in V: \qquad a(u, v) = (f, v) \qquad \forall v \in V\,,$$

$$(5.2.3)$$

where

$$V = H_0^1(a, b) = \{v \in H^1(a, b) \,|\, v(a) = v(b) = 0\}$$

is the Sobolev space for trial and test functions,

$$a(u, v) = \int_a^b (\alpha u' - \beta u)v' + \gamma uv \qquad (5.2.4)$$

is the bilinear form associated with $\mathcal{L}$, and $(f, v) = \int_a^b fv$ is the scalar product in $L^2(a, b)$.

We split $\Omega = (a, b)$ into a set of $M \geq 2$ disjoint subintervals (the "elements") $\Omega_m = (\overline{x}_{m-1}, \overline{x}_m)$, $m = 1, \ldots, M$, with $a = \overline{x}_0 < \overline{x}_1 < \ldots < \overline{x}_{M-1} < \overline{x}_M = b$. Let $h_m = \overline{x}_m - \overline{x}_{m-1}$ be the length of the m-th element and let $h = \max_m h_m$. We approximate (5.2.1) by projecting the differential equation onto a space of functions that are piecewise polynomials of degree $N_m \geq 1$ on the element Ω_m, continuous at the internal endpoints and null at the two boundary points. Precisely, we set

$$V_\delta = \{v \in C^0(\overline{\Omega}) \,|\, v_{|\Omega_m} \in \mathbb{P}_{N_m}, \, m = 1, \ldots, M, \quad v(a) = v(b) = 0\}.$$

Then, the approximation of (5.2.1) by SEM reads

$$u_\delta \in V_\delta : \quad a(u_\delta, v) = (f, v) \qquad \forall v \in V_\delta. \qquad (5.2.5)$$

We stress the fact that at element interfaces the solution u_δ is merely continuous. If we define $N = \min_m N_m$, the continuity of the flux $\alpha u'_\delta - \beta u_\delta$ of u_δ is ensured only in the limit, when $N \to \infty$, by the weak form (5.2.3). In spite of this minimal smoothness, u_δ will converge with spectral accuracy (with respect to N) to the exact solution when the latter is smooth.

Equivalently, (5.2.5) can be restated as

$$u_\delta \in V_\delta : \quad \sum_{m=1}^M a_{\Omega_m}(u_\delta, v) = \sum_{m=1}^M (f, v)_{\Omega_m} \qquad \forall v \in V_\delta, \qquad (5.2.6)$$

where $(v, w)_{\Omega_m} = \int_{\overline{x}_{m-1}}^{\overline{x}_m} vw$ is the scalar product of $L^2(\Omega_m)$, while

$$a_{\Omega_m}(u, v) = \int_{\overline{x}_{m-1}}^{\overline{x}_m} (\alpha u' - \beta u)v' + \gamma uv \qquad (5.2.7)$$

is the restriction of $a(\cdot, \cdot)$ upon Ω_m.

SEM was first introduced by Patera (1984) for Chebyshev expansions, then generalized to the Legendre case by Maday and Patera (1989). The SEM discretization depends on both the geometric partition and the polynomial degrees N_m for $m = 1, \ldots, M$. If the partition into subintervals is frozen, while the local polynomial degrees are increased to improve accuracy, the method shares the same structure as the *p-version* of the finite-element method (FEM) (where the local polynomial degrees are usually denoted by p_m). A more flexible strategy, in the same spirit as the *hp-version* of the finite-element method, consists of simultaneously refining the partition and increasing the local polynomial degrees (usually according to available information on the local behavior of the solution). In general, the latter strategy is the most efficient one, and it will be the primary focus of this chapter.

From the algorithmic point of view, both approaches (SEM and the p- or hp-version of FEM) make use of a parent element, say $\hat{\Omega} = (-1, 1)$, on which the basis functions are constructed. From the implementation perspective, the main difference arises from the way the basis functions are chosen (and therefore on the structure of the corresponding stiffness matrix); indeed, in its most classical version, SEM uses a nodal approach whereas the p-version is generally based on a modal approach.

5.2.2 Construction of SEM Basis Functions

To generate the nodal basis functions in Ω_m, $m = 1, \ldots, M$, we proceed as follows.

On the parent element $\hat{\Omega}$, we consider the N_m+1 Legendre Gauss–Lobatto (LGL for short) nodes $\hat{x}_i$, $i = 0, \ldots, N_m$, introduced in Sect. 8.4 and the corresponding $N_m + 1$ characteristic Lagrange polynomials of degree N_m (the discrete delta-functions), which satisfy $\hat{\psi}_i(\hat{x}_j) = \delta_{ij}$, for $i, j = 0, \ldots, N_m$ (see (8.2.30)).

Now, we take as basis functions $\psi_i^{(m)}(x)$ in Ω_m those functions obtained by mapping $\hat{\psi}_i(\hat{x})$ from $\hat{\Omega}$ to Ω_m, that is, for all $x \in \Omega_m$, we set

$$\psi_i^{(m)}(x) = \hat{\psi}_i(F_m^{-1}(x)), \quad i = 0, \ldots, N_m, \tag{5.2.8}$$

where

$$x = F_m(\hat{x}) = \frac{h_m}{2}\hat{x} + \frac{\overline{x}_m + \overline{x}_{m-1}}{2} \tag{5.2.9}$$

is the affine function that maps $(-1, 1)$ into Ω_m. The functions (5.2.8) form a boundary-adapted basis in each element (see Sect. 8.5, or CHQZ2, Sect. 2.3.3).

On each element Ω_m the SEM solution u_δ is therefore represented in terms of the basis functions (5.2.8) as follows

$$u_\delta(x) = \sum_{i=0}^{N_m} u_i^{(m)} \psi_i^{(m)}(x), \quad x \in \Omega_m, \tag{5.2.10}$$

where $u_i^{(m)}$ represents the unknown nodal value $u_\delta(x_i^{(m)})$, and

$$x_i^{(m)} = F_m(\hat{x}_i) = \frac{h_m}{2}\hat{x}_i + \frac{\overline{x}_m + \overline{x}_{m-1}}{2} \tag{5.2.11}$$

is the i-th LGL point in Ω_m.

A global representation for u_δ can be obtained by generating a global basis for V_δ as follows. For every $m = 1, \ldots, M$ and every $i = 1, \ldots, N_m - 1$, extend $\psi_i^{(m)}(x)$ by zero outside Ω_m, that is, consider the function

$$\tilde{\psi}_i^{(m)}(x) = \begin{cases} \psi_i^{(m)}(x) & \text{for } x \in \overline{\Omega}_m, \\ 0 & \text{otherwise.} \end{cases} \tag{5.2.12}$$

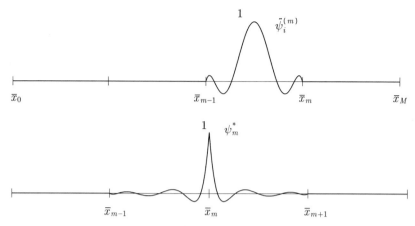

Fig. 5.2. SEM basis functions

The function $\tilde{\psi}_i^{(m)}$ is referred to as a *characteristic Lagrange function*. Moreover, for every interelement node $\overline{x}_m$, $m = 1, \ldots, M - 1$, define the function ψ_m^* obtained by "glueing" $\psi_{N_m}^{(m)}$ and $\psi_0^{(m+1)}$, precisely

$$\psi_m^*(x) = \begin{cases} \psi_{N_m}^{(m)}(x) & \text{for } x \in \overline{\Omega}_m, \\ \psi_0^{(m+1)}(x) & \text{for } x \in \overline{\Omega}_{m+1}, \\ 0 & \text{otherwise.} \end{cases} \qquad (5.2.13)$$

See Fig. 5.2 for an example. This is another characteristic Lagrange function. (We term it a "function" and not a "polynomial" as in the single-domain case because in the global interval (a, b) it is indeed a patch of distinct polynomials.)

In such a way we obtain $n_{\text{int}} = \sum_{m=1}^{M}(N_m - 1)$ "internal" basis functions plus $n_\Gamma = M - 1$ "interface" basis functions, yielding $n = n_{\text{int}} + n_\Gamma$ basis functions for the space V_δ. Correspondingly, for every $x \in (a, b)$, the SEM solution can be represented as

$$u_\delta(x) = \sum_{m=1}^{M-1} u_m^{(\Gamma)} \psi_m^*(x) + \sum_{m=1}^{M} \sum_{i=1}^{N_m-1} u_i^{(m)} \tilde{\psi}_i^{(m)}(x)\,, \qquad (5.2.14)$$

with $u_m^{(\Gamma)}$ indicating the (unknown) value of u_δ at the interface point $\overline{x}_m$, $m = 1, \ldots, M - 1$, and $u_i^{(m)}$ the value of u_δ at the i-th internal LGL point $x_i^{(m)}$ in Ω_m. Note that since every function used in expansion (5.2.14) vanishes at the boundary points, so does u_δ. After a suitable reordering we can write

$$u_\delta(x) = \sum_{k=1}^{n} u_k \psi_k(x), \qquad x \in (a, b)\,,$$

where we have indicated by u_k the unknown value of u at the k-th node x_k and by ψ_k the basis function associated with that node (ψ_k coincides with one

Fig. 5.3. An example of ordering $\{x_1, \ldots, x_n\}$ of the nodes of the SEM discretization

of the ψ_m^*'s or one of the $\tilde{\psi}_i^{(m)}$'s depending upon whether x_k is an interface node or an internal node; see, e.g., Fig. 5.3).

Exactly the same construction can be used to generate a basis in V_δ starting from any boundary-adapted basis on the parent element $\hat{\Omega}$. In particular, a modal basis in V_δ can be constructed starting from the modal basis (8.5.2).

5.2.3 SEM-NI and its Collocation Interpretation

For many algebraic realizations of the SEM, LGL quadrature formulas are used for the evaluation of the entries of the stiffness and other matrices (see Sect. 8.9.3) and the right-hand side. In this case, the SEM problem (5.2.5) is replaced by the more flexible SEM-NI version:

$$u_\delta \in V_\delta: \qquad \sum_{m=1}^{M} a_{N_m, \Omega_m}(u_\delta, v) = \sum_{m=1}^{M} (f, v)_{N_m, \Omega_m} \qquad \forall v \in V_\delta, \quad (5.2.15)$$

where

$$(u, v)_{N_m, \Omega_m} = \sum_{j=0}^{N_m} u(x_j^{(m)}) v(x_j^{(m)}) w_j^m \qquad (5.2.16)$$

is the analog in Ω_m of the Legendre Gauss–Lobatto inner product; thus, $w_j^m = w_j \frac{h_m}{2}$, while $x_j^{(m)}$ is the j-th LGL point in Ω_m. Moreover, $a_{N_m, \Omega_m}(u, v)$ is the approximation of the elemental bilinear form $a_{\Omega_m}(u, v)$ where LGL numerical integration with $N_m + 1$ nodes has been used. Precisely, still considering the case of the boundary-value problem (5.2.1) as an instance, the approximate form reads as

$$a_{N_m, \Omega_m}(u, v) = (\alpha u' - \beta u, v')_{N_m, \Omega_m} + (\gamma u, v)_{N_m, \Omega_m},$$

by analogy with (5.2.7).

The nodal basis constructed in the previous section is the natural choice within SEM-NI. The pattern of the associated stiffness matrix, for two different orderings of the basis functions, is shown in Fig. 5.4. Four spectral elements are used, with polynomials of degree four in each element. Node numbering is either done sequentially from left to right (matrix on the left) or else by gathering first the internal nodes and then the nodes lying on the boundaries of the elements (matrix on the right). The unknown on the left boundary of the domain has been eliminated via the Dirichlet condition imposed therein.

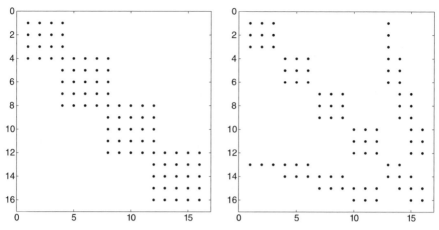

Fig. 5.4. Stiffness matrices associated with the SEM-NI problem (5.2.15) where u_δ satisfies a Dirichlet condition on the left-hand side and a flux condition on the right-hand side

We should note that many users of the spectral element method consider numerical quadrature to be an inherent part of the method because numerical quadrature was already used by Patera (1984) in his original description of the SEM; they often do not mention explicitly that they use numerical integration.

The SEM-NI formulation can be given an equivalent interpretation as a *generalized collocation* method. For any element Ω_m, $m = 1, \ldots, M$, and any function v defined in $\overline{\Omega}_m$, let us set

$$\mathcal{L}_{N_m}^{(m)} v = -\mathcal{D}_{N_m}^{(m)}(\alpha v' - \beta v) + \gamma v = -\left(I_{N_m}^{(m)}(\alpha v' - \beta v)\right)' + \gamma v \,,$$

where $\mathcal{D}_{N_m}^{(m)}(w) = \left(I_{N_m}^{(m)}(w)\right)'$ is the interpolation derivative in Ω_m (see (8.2.34)), i.e., $I_{N_m}^{(m)}(w)$ denotes the polynomial of degree N_m that interpolates the continuous function w at the $N_m + 1$ LGL points (5.2.11). For any internal LGL node $x_j^{(m)}$ in Ω_m, let us take as test function v in (5.2.15) the characteristic Lagrange function $\tilde{\psi}_j^{(m)}$ defined in (5.2.12). Counterintegrating by parts, we easily find that

$$\mathcal{L}_{N_m}^{(m)} u_\delta^{(m)} = f \qquad \text{at } x_j^{(m)} \,, \tag{5.2.17}$$

where $u_\delta^{(m)}$ denotes the restriction upon Ω_m of u_δ. Moreover, at every point $\overline{x}_m = x_{N_m}^{(m)} = x_0^{(m+1)}$, $1 \le m \le M - 1$, standing at the interface between the two neighboring elements Ω_m and Ω_{m+1}, the following two matching conditions are satisfied:

$$u_\delta^{(m)}(\overline{x}_m) = u_\delta^{(m+1)}(\overline{x}_m) \tag{5.2.18}$$

and

$$w_{N_m}^{(m)} \left(\mathcal{L}_{N_m}^{(m)} u_\delta^{(m)} - f \right) |_{x=\overline{x}_m} + w_0^{(m+1)} \left(\mathcal{L}_{N_m}^{(m+1)}, u_\delta^{(m+1)} - f \right) |_{x=\overline{x}_m}$$

$$= - \left(\alpha(u_\delta^{(m)})' - \beta u_\delta^{(m)} \right) |_{x=\overline{x}_m} + \left(\alpha(u_\delta^{(m+1)})' - \beta u_\delta^{(m+1)} \right) |_{x=\overline{x}_m} . \quad (5.2.19)$$

The latter condition follows by taking as a test function the characteristic Lagrange function ψ_m^* defined in (5.2.13). Finally, $u_\delta^{(1)}(a) = u_\delta^{(M)}(b) = 0$. This approach extends the weak treatment of Neumann conditions, which is typical of the G-NI scheme, to the SEM approach (see Sect. 8.9.5 for more details).

5.3 SEM for Multidimensional Problems

We now generalize the SEM approach to multiple dimensions and describe its essential constituents.

5.3.1 Construction of SEM Function Spaces

To define SEM on a multidimensional domain $\Omega \subset \mathbb{R}^d$ ($d = 2, 3$), we introduce a partition $\mathcal{T} = \{\Omega_m\}$ of Ω. Each element Ω_m is obtained by a transformation F_m from a reference (or parent) element $\hat{\Omega}$, which is either the reference d-cube

$$\hat{\Omega}_C^d = \{\hat{\mathbf{x}} = (\hat{x}_1, \ldots, \hat{x}_d) : -1 < \hat{x}_i < 1, \ i = 1, \ldots, d\} = (-1, 1)^d ,$$

or the reference d-simplex (a triangle in 2D or a tetrahedron in 3D):

$$\hat{\Omega}_S^d = \{\hat{\mathbf{x}} = (\hat{x}_1, \ldots, \hat{x}_d) : -1 < \hat{x}_1, \ldots, \hat{x}_d, \ \hat{x}_1 + \ldots + \hat{x}_d < 0\} .$$

The case of 3D elements obtained as transformations of prisms or pyramids can be treated similarly.

The transformation F_m is bijective and differentiable with its inverse. If Ω_m is a quadrilateral element with straight edges, $F_m : \hat{\Omega}_C^2 \to \Omega_m$ is a bilinear affine map, whereas if Ω_m is a parallelepipedal element with plane faces, $F_m : \hat{\Omega}_C^3 \to \Omega_m$ is a trilinear affine transformation.

In the case of curvilinear elements, the transformation F_m can be constructed using the Gordon–Hall map described in Sect. 8.8.4, where edges and faces are parameterized with polynomials of the same degree as those that are used to construct the SEM solution. If Ω_m is a simplex (a triangle or a tetrahedron) with straight edges or faces, then $F_m : \hat{\Omega}_S^d \to \Omega_m$ is the affine map

$$\mathbf{x} = F_m(\hat{\mathbf{x}}) = B_m \hat{\mathbf{x}} + \mathbf{b}_m ,$$

where $\mathbf{b}_m$ is a vector with d components, whereas B_m is an invertible $d \times d$ matrix.

The case in which Ω_m has curvilinear edges or faces can be treated as indicated in Sect. 8.8.4. For instance, if Ω_m is a triangle with curved edges, then

$$\mathbf{x} = F_m^{\mathrm{curv}}(\hat{\mathbf{x}}) = F_m(\hat{\mathbf{x}}) + \sum_{i=1}^{3} \varphi_{m,i}(\hat{\mathbf{x}}) \,,$$

where $F_m(\hat{\mathbf{x}})$ is the affine transformation defined above in the case of straight edges, while $\varphi_{m,i}$ is a suitable correction that accounts for the fact that the i-th edge is curved. The same approach can be pursued for tetrahedra and other 3D elements (see also Karniadakis and Sherwin (2005)).

On the reference element we introduce a polynomial space that we generically indicate by $\hat{P}_N$, for $N \geq 1$. More specifically,

$$\hat{P}_N = \begin{cases} \hat{\mathbb{Q}}_N = \mathrm{span}\{\hat{x}_1^{j_1} \cdots \hat{x}_d^{j_d} : 0 \leq j_1, \ldots, j_d \leq N\} & \text{if } \hat{\Omega} = \hat{\Omega}_C^d \,, \\ \hat{\mathbb{P}}_N = \mathrm{span}\{\hat{x}_1^{j_1} \cdots \hat{x}_d^{j_d} : 0 \leq j_1, \ldots, j_d \,, \\ \qquad\qquad\qquad\qquad j_1 + \ldots + j_d \leq N\} & \text{if } \hat{\Omega} = \hat{\Omega}_S^d \,, \end{cases} \quad (5.3.1)$$

that is, $\hat{\mathbb{Q}}_N$ denotes the space of polynomials of degree $\leq N$ in each coordinate variable $\hat{x}_i$, whereas $\hat{\mathbb{P}}_N$ that of polynomials of degree $\leq N$ globally.

Suitable modal bases for $\hat{\mathbb{Q}}_N$ and $\hat{\mathbb{P}}_N$ are constructed in Sect 8.7 and in CHQZ2, Sects. 2.8 and 2.9.

Nodal bases are constructed as characteristic Lagrange bases with respect to interpolation points. As indicated in CHQZ2, Sects. 2.8 and 2.9, the latter are tensor products of LGL (Legendre Gauss–Lobatto) nodes in $\hat{\Omega}_C^d$, while they can be obtained according to different optimality criteria in the case of the simplex $\hat{\Omega}_S^d$. In either case, we will indicate (for simplicity of notation) by $\hat{\varphi}_i$ and $\hat{\mathbf{x}}_i$, $i = 1, \ldots, \hat{N}$, the set of basis functions and nodes in $\hat{\Omega}$. Here $\hat{N}$ indicates the dimension of the space $\hat{P}_N$, which is

$$\hat{N} = \dim\hat{P}_N = \begin{cases} (N+1)^d & \text{if } \hat{P}_N = \hat{\mathbb{Q}}_N \,, \\ \displaystyle\prod_{k=1}^{d}(N+k)/d! & \text{if } \hat{P}_N = \hat{\mathbb{P}}_N \,. \end{cases} \quad (5.3.2)$$

We assume that $\mathcal{T}$ is a *geometrically conforming partition* (or decomposition) of Ω. This means that $\overline{\Omega} = \bigcup_m \overline{\Omega}_m$, and that for $m \neq k$, $\overline{\Omega}_m \cap \overline{\Omega}_k$

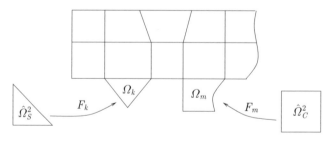

Fig. 5.5. Domain partition and maps from reference elements

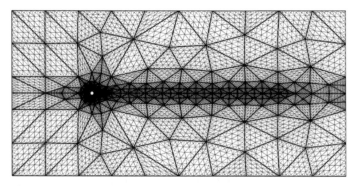

Fig. 5.6. A partition made of 490 triangular spectral elements. Within each element the small triangles have vertices made of the electrostatic points corresponding to a polynomial of degree $N = 9$ [Courtesy of T. Warburton, L.F. Pavarino and J.S. Hesthaven]

is either empty, or a vertex or a complete edge or a complete face for both elements. See Fig. 5.5 for an illustrative example in 2D, and Fig. 5.6 for a partition of Ω into triangular spectral elements with nodal points given by the electrostatic nodes (see CHQZ2, Sect. 2.9.2).

Edges and faces are images of edges and faces in $\hat{\Omega}$ under the action of F_m. If Ω_m and Ω_k share a common edge or a face, then F_m^{-1} and F_k^{-1} must agree there (up to the composition with an affine mapping).

We will set $h_m = \operatorname{diam} \Omega_m$. For every Ω_m, let $N_m \geq 1$ be the polynomial degree that we want to use in Ω_m. Then, we define

$$V_{N_m}(\Omega_m) = \{v \,:\, v = \hat{v} \circ F_m^{-1} \text{ for some } \hat{v} \in \hat{P}_N\}, \qquad (5.3.3)$$

i.e., the functions of $V_{N_m}(\Omega_m)$ are images through F_m of polynomial functions $\hat{v} \in \hat{P}_N$ in $\hat{\Omega}$. We recall that $v = \hat{v} \circ F_m^{-1}$ means $v(\mathbf{x}) = \hat{v}(F_m^{-1}(\mathbf{x}))$ for all $\mathbf{x} \in \Omega_m$.

Finally, we define the SEM space as

$$X_\delta = \{v \in C^0(\overline{\Omega}) \,:\, v_{|\Omega_m} \in V_{N_m}(\Omega_m), \forall \Omega_m \in \mathcal{T}\}, \qquad (5.3.4)$$

where the subindex δ is an abridged notation for "discrete" that accounts for the local geometric sizes $\{h_m\}$ and the local polynomial degrees $\{N_m\}$.

The actual space V_δ that will be used is a subspace of X_δ whose functions vanish on the Dirichlet portion (if any) of the boundary (see (5.3.10)).

5.3.2 Construction of SEM Basis Functions

For every $\Omega_m \in \mathcal{T}$, a basis $\{\varphi_i^{(m)}\}$ for $V_{N_m}(\Omega_m)$ is obtained as the image of a suitably chosen boundary-adapted basis $\{\hat{\varphi}_i\}$ of $\hat{P}_N$, that is,

$$\varphi_i^{(m)} = \hat{\varphi}_i \circ F_m^{-1}, \quad \text{or } \varphi_i^{(m)}(\mathbf{x}) = \hat{\varphi}_i(\hat{\mathbf{x}}), \quad \text{with } \mathbf{x} = F_m(\hat{\mathbf{x}}). \qquad (5.3.5)$$

A basis for the whole space X_δ is then obtained by "glueing" together the elemental basis functions on each element Ω_m in order to guarantee global continuity. Each bubble basis function within one element immediately generates a global basis function by extending it by zero outside the element. Vertex, edge and (in 3D) face basis functions living in contiguous elements are properly matched to generate global basis functions.

The matching is fairly trivial if, as in most SEM or SEM-NI realizations, the polynomial degree on both contiguous domains is the same. To fix ideas, suppose that Ω_m and Ω_k share a common edge (in 2D) or a common face (in 3D), which we denote by $\Gamma_{km} = \partial\Omega_m \cap \partial\Omega_k$. If the nodal basis (5.3.5) is used in each element, then the nodes on Γ_{km} are the same for both elements. Consequently, the two basis functions (characteristic Lagrange polynomials) associated with the same node $\mathbf{x}$ on Γ_{km}, say $\psi_i^{(m)}$ living in $\overline{\Omega}_m$ and $\psi_j^{(k)}$ living in $\overline{\Omega}_k$, coincide on Γ_{km}; hence, they give rise to a continuous function across Γ_{km}. If $\mathbf{x}$ is internal to Γ_{km}, the function

$$\psi = \begin{cases} \psi_i^{(m)} & \text{in } \overline{\Omega}_m\,, \\ \psi_j^{(k)} & \text{in } \overline{\Omega}_k\,, \\ 0 & \text{elsewhere}\,, \end{cases} \tag{5.3.6}$$

is the global basis function of nodal type associated with the node $\mathbf{x}$. If $\mathbf{x}$ is a vertex or lies on an edge (in 3D), the global basis function is obtained by glueing together all the local basis functions associated with $\mathbf{x}$ and extending the resulting function by zero outside the patch of elements containing $\mathbf{x}$.

The construction of the global basis is fairly similar if the local bases are of modal type (8.7.4). The only subtlety is that edge or face basis functions with the same wavenumber may have opposite sign, due to the different local orientation of the interfaces; hence, an adjustment of sign may be required before glueing the local functions (see Karniadakis and Sherwin (2005), Chap. 4, for more details).

The matching procedure just described applies as well if in each element one allows for a different polynomial degree along each direction, provided the polynomial degrees along an interface between two elements agree on both sides. A simple example of such a situation is given in Fig. 5.7.

Less trivial is the situation in which we allow for totally different polynomial degrees in contiguous elements (unless only higher order polynomial bubbles are added in certain subdomains). Continuity across an interface Γ_{km} between two subdomains Ω_k and Ω_m, with (say) $N_m < N_k$, requires that the restriction to Γ_{km} of any function in $V_{N_k}(\Omega_k)$ be indeed (the image of) a polynomial of degree N_m. If the modal bases are used in each element, this requirement is easily satisfied: only the edge basis functions in Ω_m associated with Γ_{km} contribute to the global basis functions; they are extended to the contiguous domain Ω_k as described above. In other words, one discards the contribution from the edge basis functions in Ω_k associated with Γ_{km},

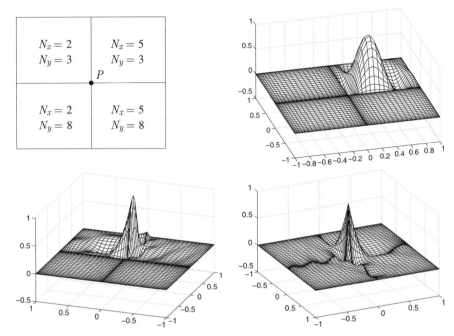

Fig. 5.7. An example of spectral element discretization with different polynomial degrees in the different elements (*top left*) and three basis functions associated with an internal node (*top right*), an edge node (*bottom left*) and the cross-point P (*bottom right*)

having wavenumber higher than N_m. In algebraic terms, the corresponding rows and columns of the local stiffness matrix of Ω_k are deleted.

On the other hand, if the nodal bases are used instead, one proceeds as follows. Let $\psi_i^{(m)}$, $i \in I$, be the basis functions in $\overline{\Omega}_m$ associated with the nodes on Γ_{km}; let $\psi_j^{(k)}$, $j \in J$, be defined similarly for the other domain. Since the restriction of $\psi_i^{(m)}$ to Γ_{km} is (the image of) a polynomial of degree $\leq N_m$, there exist coefficients r_{ij} such that

$$\psi_i^{(m)} = \sum_{j \in J} r_{ij}\psi_j^{(k)} \qquad \text{on } \Gamma_{km} \qquad (5.3.7)$$

(precisely, $r_{ij} = \psi_i^{(m)}(\mathbf{x}_j^{(k)})$, where $\mathbf{x}_j^{(k)}$ is the node on $\Gamma_{km} \subset \overline{\Omega}_k$ associated with $\psi_j^{(k)}$). Thus, $\psi_i^{(m)}$ is glued across Γ_{km} with the particular linear combination of edge basis functions in Ω_k given by the right-hand side of (5.3.7). At the algebraic level, this amounts to condensing properly the stiffness matrix of Ω_k: the set of rows with indices $j \in J$ is replaced by their linear combinations with coefficients r_{ij}, for all $i \in I$; a similar transformation is applied to the set of columns with indices $j \in J$.

We observe that the matching procedure described above is just a special case of the mortar matching that will be discussed in detail in Sect. 5.7. The general requirement is that two functions $v_m \in V_{N_m}(\Omega_m)$ and $v_k \in V_{N_k}(\Omega_k)$ agree in the sense that

$$\int_{\Gamma_{km}} (v_m - v_k)\varphi d\gamma = 0 \qquad \text{for all } \varphi \in Y_{km} , \qquad (5.3.8)$$

where Y_{km} is a suitable function space on Γ_{km}. If Y_{km} is the space of restrictions of $V_{N_k}(\Omega_k)$ on Γ_{km}, then v_k is forced to coincide with v_m on Γ_{km} (yielding that indeed the degree of v_k reduces to that of v_m therein). Should instead Y_{km} coincide with the space of restriction of $V_{N_m}(\Omega_m)$ on Γ_{km} (or even with a smaller space), condition (5.3.8) would only imply continuity in the least-square sense. The latter is indeed the most common type of matching within the mortar method.

5.3.3 SEM and SEM-NI Formulations

At this stage, all ingredients are in place to set up the spectral element approximation of a boundary-value problem. For the sake of illustration, let us assume that the boundary-value problem (5.2.1) is replaced by its multidimensional advection-diffusion-reaction analog

$$\begin{aligned} \mathcal{L}u = -\nabla \cdot (\alpha\nabla u - \beta u) + \gamma u = f & \qquad \text{in } \Omega \subset \mathbb{R}^d \ (d = 2, 3), \\ u = 0 & \qquad \text{on } \Gamma_D , \\ (\alpha\nabla u - \beta u) \cdot \mathbf{n} = \psi & \qquad \text{on } \Gamma_N , \end{aligned}$$

$$(5.3.9)$$

where β is now a vector field, α is a positive scalar function with $\alpha(\mathbf{x}) \geq \alpha_0 > 0$ for all $\mathbf{x} \in \Omega$, and f and ψ are given functions. The boundary $\partial\Omega$ is the union of two non-intersecting parts Γ_D and Γ_N, that is $\Gamma_D \cup \Gamma_N = \partial\Omega$, $\text{int}(\Gamma_D) \cap \text{int}(\Gamma_N) = \emptyset$. Either Γ_D or Γ_N may be empty. Suitable assumptions on the coefficients that guarantee the well-posedness of the problem are given in Sect. 5.4.2. Let $H^1_{\Gamma_D}(\Omega) = \{v \in H^1(\Omega) \ : \ v_{|\Gamma_D} = 0\}$ be the Sobolev space of the functions vanishing on Γ_D.

The weak formulation of (5.3.9) reads as follows:

$$u \in H^1_{\Gamma_D}(\Omega) : \int_{\Omega} (\alpha\nabla u - \beta u) \cdot \nabla v + \gamma u v = \int_{\Omega} f v + \int_{\Gamma_N} \psi v \quad \forall v \in H^1_{\Gamma_D}(\Omega) .$$

If we define

$$V_\delta = \{v \in X_\delta \ : \ v_{|\Gamma_D} = 0\} \subset H^1_{\Gamma_D}(\Omega) , \qquad (5.3.10)$$

the spectral element approximation of (5.3.9) consists in finding $u_\delta \in V_\delta$ such that

$$\sum_m a_{\Omega_m}(u_\delta, v) = \sum_m \left\{ (f, v)_{\Omega_m} + \int_{\partial\Omega_m \cap \Gamma_N} \psi v \right\} \quad \forall v \in V_\delta , \qquad (5.3.11)$$

where now, for every m,

$$a_{\Omega_m}(u,v) = \int_{\Omega_m} (\alpha\nabla u - \boldsymbol{\beta}u)\nabla v + \gamma uv, \qquad (f,v)_{\Omega_m} = \int_{\Omega_m} fv. \quad (5.3.12)$$

A limitation intrinsic to the definition of V_δ is that if Γ_m is an edge in 2D or a face in 3D lying on the boundary $\partial\Omega$, then Γ_m bears either a Dirichlet condition or a Neumann condition, that is, either $\Gamma_m \subset \Gamma_D$ or $\Gamma_m \subset \Gamma_N$.

Just as was done for the one-dimensional problem, in the multidimensional case we also can consider the SEM-NI variant of SEM, which consists of finding $u_\delta \in V_\delta$ such that

$$\sum_m a_{N_m,\Omega_m}(u_\delta,v) = \sum_m (f,v)_{N_m,\Omega_m} - \sum_n (\psi,v)_{\Gamma_n} \qquad \forall v \in V_\delta. \quad (5.3.13)$$

The notation $a_{N_m,\Omega_m}(\cdot,\cdot)$ and $(\cdot,\cdot)_{N_m,\Omega_m}$ indicate that the integrals on Ω_m are computed using suitable numerical integration formulas on the corresponding reference element $\hat{\Omega}$. These are invariably the LGL nodes if $\hat{\Omega} = \hat{\Omega}_C^d$, while on $\hat{\Omega}_S^d$ the choice of nodes is less standard: several variants are available, such as Fekete or electrostatic points, as discussed in CHQZ2, Sect. 2.9.

The last term in (5.3.13) refers to the sum on all edges (in 2D) or faces (in 3D) Γ_n, lying on the boundary Γ_N where the Neumann boundary conditions are enforced. On each Γ_n the integrals are approximated by using the numerical integration formula on the corresponding edge (or face) of $\hat{\Omega}$, which is induced by the above integration formula on the whole $\hat{\Omega}$.

As for 1D problems, the SEM-NI method can be interpreted as a *generalized collocation* method. Indeed, if Ω_m is the image of $\hat{\Omega}_C^d$, from (5.3.13) we can deduce a collocation equation at each internal node, by choosing as test function v the (image of the) characteristic Lagrange function at that node, φ_i^m, which extends by zero outside Ω_m. This can be shown by proceeding as done in Sect. 5.2.1 for the one-dimensional case.

Similarly, we can obtain that at a common nodal point on Γ_{km}, the interface between two elements Ω_k and Ω_m, a pointwise equation is satisfied stating that the jump of the normal fluxes is equal to a linear combination of the two residuals at that point, with weights that vanish as N_k and N_m tend to infinity.

At a cross-point P (where several elements merge), the equation corresponding to the choice of the characteristic Lagrange function at P yields that a linear combination of jumps of the "normal" derivatives across the edges (in 2D, faces in 3D) merging at P must equate a linear combination of elemental residuals.

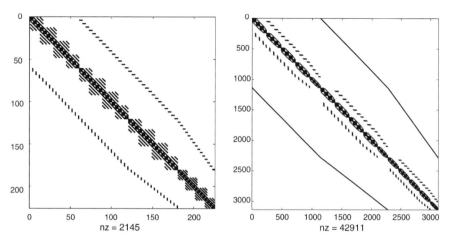

Fig. 5.8. At *left* the pattern of a SEM-NI matrix for a 2D Poisson problem, with $\Omega = (0, 1)^2$. The polynomial degrees are $N_x = N_y = 4$, while the number of elements is $M = 4 \times 4$. On the *right* the pattern of a SEM-NI matrix for a 3D elliptic problem, with $\Omega = (0, 1)^3$. The polynomial degrees are $N_x = N_y = N_z = 4$, while the number of elements is $M = 3 \times 4 \times 5$

5.3.4 Algebraic Aspects of SEM and SEM-NI

SEM (SEM-NI, resp.) gives rise to an algebraic system whose stiffness matrix A has entries

$$A_{ij} = \sum_m a_{\Omega_m}(\varphi_j^m, \varphi_i^m) \qquad \left(A_{ij} = \sum_m a_{N_m, \Omega_m}(\varphi_j^m, \varphi_i^m), \text{ resp.} \right) .$$

The matrix A can be built up by assembling the local stiffness matrices related to every element Ω_m.

Let us consider the special case in which Ω is a square or a cube partitioned into squares or cubes Ω_m, $m = 1, \ldots, M$, of equal size, and that the same polynomial degree N is used on each of them. This means that $V_N(\Omega_m) = \mathbb{Q}_N$ is used for every m. The patterns of A in 2D and 3D are plotted in Fig. 5.8 for a particular choice of N and M; the usual lexicographical ordering of nodes has been used.

The numerical linear algebra properties of a matrix are strongly tied to its condition number (see (C.1.11)). For a symmetric and positive-definite matrix, the spectral condition number $\mathcal{K}$, i.e., the ratio of the largest to the smallest eigenvalue (see (C.1.12) and (C.1.13)), is particularly relevant. Chapter 4 in CHQZ2 illustrates the behavior of the spectral condition number of various stiffness matrices arising from single-domain spectral discretizations. For the Poisson problem in 2D with Dirichlet boundary conditions, the condition number of the SEM-NI stiffness matrix, which uses in each element the tensor-product nodal basis (8.7.3) on quadrilateral elements of equal linear size h and polynomial degree N, can be bounded by a constant (independent

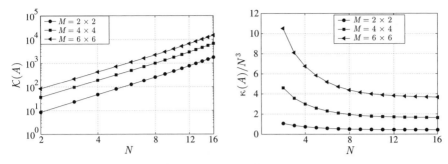

Fig. 5.9. The spectral condition number $\mathcal{K}(A)$ where A is the SEM-NI matrix of the Laplacian in 2D (*left*) and the ratio $\mathcal{K}(A)/N^3$ (*right*)

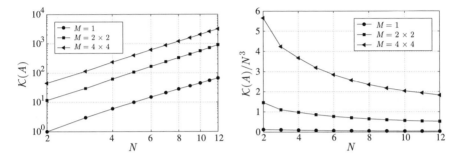

Fig. 5.10. The spectral condition number $\mathcal{K}(A)$ where A is the SEM-NI matrix of the Laplacian in 3D (*left*) and the ratio $\mathcal{K}(A)/N^3$ (*right*)

of the discretization) times $h^{-2}N^3$ (Melenk (2002)). See Figs. 5.9 and 5.10, where the condition number of the SEM-NI matrix is plotted versus N for several subdomain partitions. If the vertex functions in each direction are replaced by linear polynomials, the condition number scales less favorably with respect to N, namely it is proportional to $N^4 \log N$ (Maitre and Pourquier (1996)). If the tensor-product modal basis (8.7.4) is used in each element within SEM, the resulting condition number scales as N^4 (Hu, Guo and Katz (1998)) and, after appropriate scaling, this becomes $h^{-2}N^4$.

5.3.5 Finite-Element Preconditioning of SEM-NI Matrices

Finite-element matrices constructed on piecewise linear (in 1D), bilinear (in 2D) or trilinear (in 3D) shape functions centered at LGL nodes of the reference cube $\hat{\Omega}_C^d$ provide optimal preconditioners for spectral G-NI matrices built on the single-domain $\hat{\Omega}_C^d$. (See Fig. 5.11, left, for an example in 2D.) This was illustrated in CHQZ2, Sect. 4.4.2 for the case $d = 1$, and Sect. 4.4.3 for the case $d = 2$. Indeed, we have shown in CHQZ2, Table 4.8, that there are several possible ways to construct a finite-element preconditioner and, correspondingly, several forms of the preconditioned system.

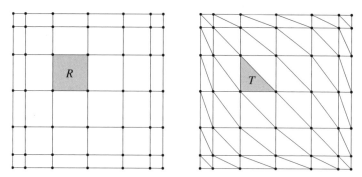

Fig. 5.11. Finite elements induced by LGL nodes on the reference square $\hat{\Omega}_C^2 = (-1, 1)^2$: rectangular (*left*), triangular (*right*)

Among them, two emerge for their overall efficiency: the preconditioned matrices $P_2 = (M_{FE}^{-1} K_{FE})^{-1} (M_{GNI}^{-1} K_{GNI})$ and $P_3 = K_{FE,app}^{-1} K_{GNI}$. The former, called *strong* preconditioning, amounts to preconditioning the collocation matrix $L_{coll} = M_{GNI}^{-1} K_{GNI}$ by the corresponding strong form of the finite-element matrix $M_{FE}^{-1} K_{FE}$. (We recall that K and M refer, respectively, to the stiffness and mass matrices, and the subscript qualifies either the finite-element or the G-NI approximation.) The latter, called *weak* preconditioning, is obtained by applying to the G-NI stiffness matrix the finite-element stiffness matrix, in an approximate form; this is achieved by computing numerically all the integrals by replacing every integrand with its bilinear (or trilinear) interpolant at the LGL nodes. See CHQZ2, Table 4.9 and the conclusions drawn in its discussion.

These preconditioned matrices feature a spectral condition number that is uniformly bounded with respect to the polynomial degree N. This property is often referred to as *FEM–G-NI equivalence*.

A similar kind of construction holds if the finite-elements are simplicial, i. e., triangles in 2D or tetrahedra in 3D, still with vertices at the LGL nodes. See Fig. 5.11, right, for a 2D case. To distinguish these kinds of preconditioners from the previous types, let us denote them with a subscript S (S stands for simplices): $P_{2,S}$ and $P_{3,S}$.

Finite-element preconditioners for the SEM-NI stiffness matrix can be built up using the same principle, by assembling all local preconditioners described above on every single spectral element. The corresponding preconditioned matrices will still be denoted by the same names, P_2 and P_3 in the case of rectangular finite-elements, $P_{2,S}$ and $P_{3,S}$ in that of triangular finite-elements.

A simple Rayleigh quotient argument, reported in Sect. 6.3.5, shows that the spectral condition number is optimal, that is, uniformly bounded with respect to both the polynomial degrees N_m and the spectral element sizes h_m in the case of quadrilateral (or parallelepipedal) spectral elements. This is known as the *FEM-SEM equivalence*.

Table 5.1. Iterative condition number of the preconditioned matrices P_2, $P_{2,S}$ and P_3 associated with the problem $-\Delta u = f$ in $\Omega = (-1,1)^2$ with homogeneous Dirichlet boundary conditions

N	$M = 2$			$M = 4$			$M = 8$		
	P_2	$P_{2,S}$	P_3	P_2	$P_{2,S}$	P_3	P_2	$P_{2,S}$	P_3
2	1.714	2.058	2.348	1.907	2.680	2.571	1.975	2.905	2.641
3	1.656	2.460	2.531	1.685	2.759	2.552	1.695	2.849	2.559
4	1.592	2.588	2.687	1.592	2.763	2.695	1.592	2.827	2.697
5	1.535	2.622	2.815	1.535	2.751	2.820	1.535	2.789	2.822
6	1.472	2.677	2.919	1.472	2.772	2.922	1.472	2.795	2.922
7	1.382	2.719	3.003	1.382	2.784	3.005	1.382	2.802	3.005
8	1.384	2.740	3.073	1.384	2.786	3.074	1.384	2.799	3.074
9	1.383	2.757	3.131	1.383	2.788	3.132	1.383	2.799	3.132
10	1.377	2.768	3.180	1.377	2.786	3.181	1.377	2.796	3.181
11	1.371	2.774	3.223	1.371	2.783	3.223	1.371	2.793	3.223
12	1.369	2.778	3.259	1.369	2.783	3.260	1.369	2.788	3.260

In Table 5.1 the iterative condition number of the preconditioned SEM-NI stiffness matrix corresponding to the Laplace operator with Dirichlet boundary conditions is shown. The domain Ω is split into $M \times M$ squared spectral elements of equal size, and the same polynomial degree is used in each spectral element. Three kinds of finite-element preconditioners are used, yielding the preconditioned matrices called P_2, $P_{2,S}$ and P_3, respectively.

Figure 5.12 (taken from Warburton, Pavarino and Hesthaven (2000)) illustrates that similar conclusions hold for the preconditioned matrix $P_{3,S}$ constructed on non-uniform triangular grids on each rectangular spectral element. Conversely, Figs. 5.13 and 5.14 (taken from the same paper) show that the finite-element preconditioners are not optimal on triangles when constructed on Fekete points or on electrostatic points (see CHQZ2, Sect. 2.9.2).

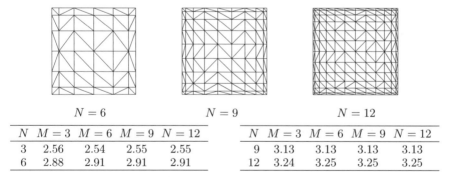

| | $N = 6$ | | $N = 9$ | | $N = 12$ | |

N	$M = 3$	$M = 6$	$M = 9$	$N = 12$		N	$M = 3$	$M = 6$	$M = 9$	$N = 12$
3	2.56	2.54	2.55	2.55		9	3.13	3.13	3.13	3.13
6	2.88	2.91	2.91	2.91		12	3.24	3.25	3.25	3.25

Fig. 5.12. *Top*: example of finite-element meshes built on the LGL nodes in each quadrilateral element for $N = 6, 9, 12$. *Bottom*: spectral condition numbers $\mathcal{K}(P_{3,S})$. The FEM-SEM spectral equivalence holds [Courtesy of T. Warburton, L.F. Pavarino and J.S. Hesthaven]

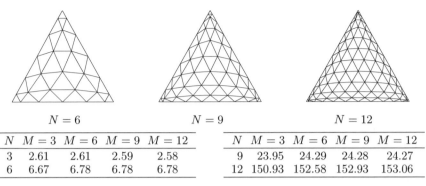

N	M = 3	M = 6	M = 9	M = 12
3	2.61	2.61	2.59	2.58
6	6.67	6.78	6.78	6.78

N	M = 3	M = 6	M = 9	M = 12
9	23.95	24.29	24.28	24.27
12	150.93	152.58	152.93	153.06

Fig. 5.13. *Top*: example of finite-element meshes built on the electrostatic nodes in each triangular element for $N = 6, 9, 12$. *Bottom*: spectral condition numbers $\mathcal{K}(P_{3,S})$. There is no FEM-SEM spectral equivalence [Courtesy of T. Warburton, L.F. Pavarino and J.S. Hesthaven]

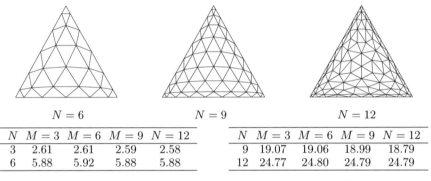

N	M = 3	M = 6	M = 9	N = 12
3	2.61	2.61	2.59	2.58
6	5.88	5.92	5.88	5.88

N	M = 3	M = 6	M = 9	N = 12
9	19.07	19.06	18.99	18.79
12	24.77	24.80	24.79	24.79

Fig. 5.14. *Top*: example of finite-element meshes built on the Fekete nodes in each triangular element for $N = 6, 9, 12$. *Bottom*: spectral condition numbers $\mathcal{K}(P_{3,S})$. There is no FEM-SEM spectral equivalence [Courtesy of T. Warburton, L.F. Pavarino and J.S. Hesthaven]

The FEM-SEM equivalence holds also for operators with nonconstant coefficients. Table 5.2 reports the iterative condition numbers of P_2 and $P_{2,S}$ for the Poisson problem, $-\mathrm{div}(\nu \nabla u) = f$ in $\Omega = (-1, 1)^2$, with $\nu = 3 + x + y$ and homogeneous Dirichlet boundary conditions.

More results on FEM-SEM equivalence, also for other kinds of FEM-preconditioners and non-Cartesian computational domains can be found in Canuto, Gervasio and Quarteroni (2007).

The construction of local matrices has been addressed in CHQZ2, Sects. 3.8.1 and 4.2.1. Efficient matrix assembly techniques are those typically used in finite-element-based domain decomposition methods (see, e. g., Smith et al. (1996)), following the additive structure of its entries A_{ij}. Specific approaches suitable for spectral elements on quadrilaterals are described in Deville, Fischer and Mund (2002), Sect. 4.5.1, and in Karniadakis and Sherwin (2005), Sect. 4.2.

Table 5.2. Iterative condition numbers of P_2 and $P_{2,S}$ associated with the problem $-\mathrm{div}(\nu\nabla u) = f$ in $\Omega = (-1,1)^2$ with homogenous Dirichlet boundary conditions. $\nu = 3 + x + y$

N	$M = 2$		$M = 4$		$M = 8$	
	P_2	$P_{2,S}$	P_2	$P_{2,S}$	P_2	$P_{2,S}$
2	1.278	2.038	1.908	2.658	1.976	2.903
3	1.501	2.461	1.679	2.759	1.694	2.848
4	1.508	2.586	1.593	2.764	1.593	2.827
5	1.532	2.611	1.535	2.752	1.535	2.789
6	1.477	2.626	1.475	2.768	1.473	2.794
7	1.446	2.677	1.429	2.783	1.412	2.801
8	1.424	2.709	1.400	2.785	1.384	2.798
9	1.391	2.728	1.382	2.786	1.382	2.798
10	1.379	2.739	1.377	2.785	1.377	2.795
11	1.376	2.746	1.370	2.780	1.371	2.793
12	1.374	2.748	1.369	2.782	1.369	2.787

5.4 Analysis of SEM and SEM-NI Approximations

We provide the details of the a priori and a posteriori error analysis in the one-dimensional case, while we confine ourselves to stating the main results in the multidimensional case.

5.4.1 One-Dimensional Analysis

The SEM method (5.2.5) is a standard Galerkin approximation of the boundary-value problem (5.2.1). As such, its analysis can be carried out by resorting to the Lax–Milgram Theorem (see Sect. A.3) and using approximation results for piecewise Lagrange interpolation, according to the general theory presented in CHQZ2, Sect. 6.4.1.

Let us assume that the coefficients β and γ of the operator $\mathcal{L}$ satisfy the inequality $\frac{1}{2}\beta' + \gamma \geq 0$ in (a, b). Then, it is easily seen by partial integration that the bilinear form $a(\cdot, \cdot)$ is coercive, that is,

$$a(v, v) \geq \alpha_0 \|v'\|_{L^2(a,b)}^2 \qquad \text{for all } v \in H_0^1(a, b) .$$

Using the Poincaré inequality $\|v\|_{L^2(a,b)} \leq C_p \|v'\|_{L^2(a,b)}$, for all $v \in H_0^1(a, b)$, where C_p denotes the Poincaré constant (see Sect. A.10), we actually have

$$a(v, v) \geq \alpha_* \|v\|_{H^1(a,b)}^2 \qquad \text{for all } v \in H_0^1(a, b) ,$$

with $\alpha_* = \alpha_0 (1 + C_p^2)^{-1/2}$. Furthermore, the bilinear form is continuous, i. e.,

$$|a(v, w)| \leq A_0 \|v\|_{H^1(a,b)} \|w\|_{H^1(a,b)} \qquad \text{for all } v, w \in H^1(a, b) ,$$

where $A_0 = \alpha_1 + \beta_1 + \gamma_1$ is the continuity constant. Then, (5.2.5) has a unique solution (owing to the Lax–Milgram Theorem), which is uniformly stable.

Precisely,

$$\|u_\delta\|_{H^1(a,b)} \leq \frac{1}{\alpha_*}\|f\|_{L^2(a,b)} \ .$$

A Priori Error Analysis for SEM. We focus on the convergence properties of SEM. Subtracting (5.2.5) from (5.2.3), we obtain the so-called *Galerkin
orthogonality* relations

$$a(u - u_\delta, v_\delta) = 0 \qquad \text{for all } \ v_\delta \in V_\delta \ ,$$

which immediately imply

$$a(u - u_\delta, u - u_\delta) = a(u - u_\delta, u - v_\delta) \qquad \text{for all } \ v_\delta \in V_\delta \ . \qquad (5.4.1)$$

Using the coercivity (on the left-hand side) and the continuity (on the right-
hand side) of the bilinear form a, we easily get the bound

$$\|u - u_\delta\|_{H^1(a,b)} \leq \frac{A_0}{\alpha_*} \inf_{v_\delta \in V_\delta} \|u - v_\delta\|_{H^1(a,b)} \ , \qquad (5.4.2)$$

known as *Cea's Lemma* (see CHQZ2, Sect. 6.4.1). This result shows that the
error $u - u_\delta$ in the energy norm is proportional to the best approximation
error of u in V_δ in the same norm.

At this point, we are left with the problem of finding a particular element
$v_\delta \in V_\delta$ for which the norm of $u - v_\delta$ behaves asymptotically as the infimum
above. A natural choice is the SEM interpolant $I_\delta u$; precisely, $I_\delta u$ is the
function of V_δ that, when restricted to the interval Ω_m, interpolates the
function u at the $N_m + 1$ LGL points of Ω_m. It can be proven that the
interpolation error satisfies

$$\|u - I_\delta u\|_{H^k(a,b)}$$
$$\leq \left\{ \sum_{m=1}^{M} C(s_m) h_m^{2(\min(N_m+1,s_m)-k)} N_m^{2(k-s_m)} |u|^2_{H^{s_m};N_m(\Omega_m)} \right\}^{1/2} \qquad (5.4.3)$$

for $k = 0, 1$, where $H^{s_m}(\Omega_m)$ measures the local smoothness of u in Ω_m;
the constants $C(s_m) > 0$ are independent of h_m and N_m. This inequality
can be deduced by applying the local interpolation error estimate (5.4.42)
of CHQZ2 on each subinterval Ω_m (see the complete proof in Sect. 5.4.3).
Then, using (5.4.2), we arrive at the final a priori error bound

$$\|u - u_\delta\|_{H^1(a,b)}$$
$$\leq \frac{A_0}{\alpha_*} \left\{ \sum_{m=1}^{M} C(s_m) h_m^{2(\min(N_m+1,s_m)-1)} N_m^{2(1-s_m)} |u|^2_{H^{s_m};N_m(\Omega_m)} \right\}^{1/2} \ . \qquad (5.4.4)$$

If the polynomial degrees are the same for all Ω_m, say $N_m = N$, and the element sizes satisfy $ch \leq h_m \leq c'h$ for all m, we obtain the error estimate

$$\|u - u_\delta\|_{H^1(a,b)} \leq C_1(s) h^{\min(N+1,s)-1} N^{1-s} |u|_{H^{s;N}(a,b)} ,$$

provided $u \in H^s(a,b) \cap H_0^1(a,b)$, for a suitable $s \geq 1$. This estimate can be obtained from (5.4.4) by summing up on m and noting that $|u|_{H^{s;N}(a,b)} = (\sum_{m=1}^M |u|_{H^{s;N}(\Omega_m)}^2)^{1/2}$.

If u is very smooth (i.e., s is very large), it is advantageous to improve the quality of the approximation by keeping h fixed and increasing N. A simple argument supports this guideline. Assume that the domain is divided into M elements of equal size $h = \frac{b-a}{M}$ and that polynomials of degree up to N are used in each element. Then the dimension D of V_δ, i.e., the total number of degrees of freedom, satisfies $D \sim MN$. If $N \geq s - 1$, the previous bound yields

$$\|u - u_\delta\|_{H^1(a,b)} \leq C_2(s) \frac{1}{M^{s-1}} \frac{1}{N^{s-1}} \sim C_2(s) \frac{1}{D^{s-1}} ,$$

i.e., the predicted rate of decay of the error is essentially insensitive to whether D is obtained from few elements carrying a high polynomial degree or from more elements carrying a lower polynomial degree. Computational efficiency considerations may suggest the latter choice. On the contrary, if $N < s - 1$, one has

$$\|u - u_\delta\|_{H^1(a,b)} \leq C_3(s) \frac{1}{M^{N-1}} \frac{1}{N^{s-1}} \sim C_3(s) M^{s-N} \frac{1}{D^{s-1}} ,$$

and the predicted rate of decay of the error is maximal for $M = 1$, i.e., when a single element is used.

Obviously, such analysis presumes the Sobolev norm of order s of u to be "uniformly distributed" throughout the interval (a,b). Often, the scenario is different, with isolated points of non-smoothness or concentration of high-order Sobolev norms in limited portions of the domain. In these cases, a judicious combination of (non-uniform) mesh refinement and polynomial degree increment may still guarantee a fast decay of the error, even of exponential type, in terms of the number D of total degrees of freedom. Schwab (1998) (see also Melenk (2003)) discusses several situations of this kind: i) the case of a piecewise-analytic function; ii) the case of an infinitely smooth function but with a strong transition (a layer) in a small portion of the domain; iii) the case of an infinitely smooth function except for a point, where an algebraic singularity occurs.

If u is a continuous function composed of M pieces that are analytic in the closure of each interval Ω_m, $m = 1, \ldots, M$, then the error decays exponentially fast, i.e., an estimate of the form

$$\|u - u_\delta\|_{H^1(a,b)} \leq C B^{-D} \tag{5.4.5}$$

holds for a suitable $B > 1$ depending on the regions of analyticity in the complex plane of the smooth pieces of u. In particular, if u is analytic in

$[a, b]$, exponential decay of the error occurs with one element. The estimate follows from approximation results similar to estimate (5.5.27) of CHQZ2, which also holds in the Legendre setting.

If u is the boundary-layer function defined in $[a, b]$ as $u(x) = 1 - e^{-(x-a)/d}$ for some small $d > 0$, then an estimate of the form (5.4.5) holds as well for a suitable $B > 1$ uniformly in d and in the polynomial degrees, provided the discretization is chosen as follows. Until the product dN is smaller than a fixed constant, two elements suffice: $\Omega_1 = (a, c)$ with $h_1 = c - a$ proportional to dN and $N_1 = N$, and $\Omega_2 = (c, b)$ with $N_2 = 1$; as soon as dN exceeds the constant, one element suffices: $\Omega_1 = (a, b)$ with $N_1 = N$. Note that the use of a single element for all regimes would lead to a uniform-in-d decay of the error proportional to $N^{-1} \log \sqrt{N}$ before the exponential convergence takes place. Functions having a local behavior of the type just considered occur as solutions of singularly perturbed boundary-value problems, such as reaction-diffusion or convection-diffusion problems; see CHQZ2, Sect. 7.2.

At last, if u behaves as $(x - a)^\alpha$ near a, for some non-integer $\alpha > 0$, and is smooth elsewhere, then an error decay of the form

$$\|u - u_\delta\|_{H^1(a,b)} \leq C B^{-\sqrt{D}},$$

with $B > 1$, can be established provided the discretization is chosen as follows. The mesh is geometrically graded towards $x = a$, i.e., the endpoints of the elements are defined as $\overline{x}_m = a + \sigma^{M-m}(b - a)$ for $m = 1, \ldots, M$, with $0 < \sigma < 1$ fixed; furthermore, $N_1 = 1$, $N_m = \max(2, [\mu m])$ for $m = 2, \ldots, M$ and a suitable constant $\mu > 0$. Guo and Babuška (1986) suggest the choice $\sigma = (\sqrt{2} - 1)^2$ for the grading factor.

A Priori Analysis for SEM-NI. A stability and convergence analysis, similar to the one given above for the SEM discretization, can be performed for the SEM-NI version (5.2.15), which represents an instance of a *generalized Galerkin* method. The latter is an *internal* (or *conforming*) approximation of the original problem, in the sense that the numerical solution is sought in a space V_δ, which is a subspace of the functional space $H_0^1(a, b)$ to which the exact solution u belongs. However, the bilinear and linear forms in (5.2.15) do not coincide with those of the original problem.

The analysis of generalized Galerkin methods can be carried out by resorting to the abstract theory presented in CHQZ2, Sect. 6.4.3, which is based on the well-known first Strang Lemma (see, e.g., Quarteroni and Valli (1994), Theorem 5.5.1). The bilinear form $a_\delta(u, v) = \sum_{m=1}^{M} a_{N_m, \Omega_m}(u, v)$, which appears on the left-hand side of (5.2.15), is uniformly coercive provided the following inequality holds for the lower order coefficients β and γ on each element Ω_m:

$$\frac{1}{2}(I_{N_m}\beta)' + \gamma \geq 0.$$

Indeed, using the exactness of the LGL quadrature formula, one has, as for the SEM case,

$$a_\delta(v,v) \geq \alpha_* \|v\|^2_{H^1(a,b)} \qquad \text{for all } v \in V_\delta .$$

On the other hand, using the uniform (in N_m and m) equivalence between the continuous norm $\|v\|_{L^2(\Omega_m)}$ and the discrete norm $\|v\|_{N_m,\Omega_m} = \sqrt{(v,v)_{N_m,\Omega_m}}$ for all functions v that are polynomials of degree N_m on Ω_m (this result stems from estimate (5.3.2) in CHQZ2), we can bound the right-hand side of (5.2.15) as

$$\sum_{m=1}^{M} (f,v)_{N_m,\Omega_m} \leq \left(\sum_{m=1}^{M} \|f\|^2_{N,\Omega_m} \right)^{1/2} \left(\sum_{m=1}^{M} \|v\|^2_{N,\Omega_m} \right)^{1/2}$$

$$\leq C_2 \left(\sum_{m=1}^{M} \|f\|^2_{N,\Omega_m} \right)^{1/2} \|v\|_{L^2(a,b)} ,$$

and the discrete norm of f can be bounded by the norm $\|f\|_{C^0([a,b])}$. This proves that the SEM-NI problem has a unique solution, which is uniformly stable as it satisfies the bound

$$\|u_\delta\|_{H^1(a,b)} \leq \frac{C_2}{\alpha_*} \|f\|_{C^0([a,b])} .$$

Furthermore, the difference between the original bilinear form $a(u,v)$ and the approximate one $\sum_m a_{N_m,\Omega_m}(u,v)$ can be bounded in terms of the SEM-interpolation error $u - I_\delta u$, and so is the difference between the right-hand side $\int_\Omega fv$ and $\sum_m (f,v)_{N_m,\Omega_m}$ for all $v \in V_\delta$. Hence, one can prove that, if f is smooth enough, u_δ converges to u with the same order as in the SEM case.

A Posteriori Error Analysis. We now aim at deriving sharp estimates of the error $u - u_\delta$ in terms of known and computable quantities, such as u_δ itself. We will confine ourselves to the so-called *residual-based* estimates; other approaches are possible, such as estimates based on the solution of suitably defined local problems (see, e. g., Ainsworth and Oden (1992)).

We assume that u_δ is the SEM approximation of problem (5.2.1), i.e., it is the solution of (5.2.5). For the sake of simplicity, we assume that the coefficients α, β, γ are constant (while f may be a variable function). Denote, as above, by $u_\delta^{(m)}$ the restriction of u_δ to the m-th element Ω_m; let $f^{(m)}$ be defined similarly. In order to define a local error indicator η_m in Ω_m, let us introduce the L^2-orthogonal projection of $f^{(m)}$ upon the space $\mathbb{P}_{N_m}(\Omega_m)$, i.e., the truncation of the (shifted) Legendre series of $f^{(m)}$, which we denote by $f_{N_m}^{(m)}$. Next, let us introduce the *element residual* in Ω_m

$$r_m = f_{N_m}^{(m)} - L u_\delta^{(m)} = f_{N_m}^{(m)} + (\alpha(u_\delta^{(m)})')' - \beta u_\delta^{(m)})' - \gamma u_\delta^{(m)} , \qquad (5.4.6)$$

which is a polynomial of degree $\leq N_m$. Finally, let us introduce the weight function

$$w_m(x) = (\overline{x}_m - x)(x - \overline{x}_{m-1}) \tag{5.4.7}$$

vanishing at the endpoints of the interval, and the associated weighted L^2-norm $\|g\|_{L^2_{w_m}(\Omega_m)} = \left(\int_{\Omega_m} g^2(x) \, w_m(x) \, dx \right)^{1/2}$.

The *local error indicator* is defined as

$$\eta_m^2 = \frac{1}{N_m(N_m + 1)} \|r_m\|^2_{L^2_{w_m}(\Omega_m)} , \tag{5.4.8}$$

whereas a *global error estimator* is defined as

$$\eta^2 = \sum_{m=1}^{M} \left(\eta_m^2 + \frac{1}{N_m(N_m + 1)} \|f^{(m)} - f_{N_m}^{(m)}\|^2_{L^2_{w_m}(\Omega_m)} \right) . \tag{5.4.9}$$

The main a posteriori results, due to Bernardi (1996), are as follows. The global H^1-norm of the error $u - u_\delta$ can be estimated from above as

$$\|u - u_\delta\|_{H^1(a,b)} \leq \frac{\eta}{\alpha_*} ; \tag{5.4.10}$$

this shows that η is a *reliable* error estimator, i. e., if the computed solution u_δ is such that $\eta \leq \alpha_* \mathrm{TOL}$, then we are guaranteed that $\|u - u_\delta\|_{H^1(a,b)} \leq \mathrm{TOL}$. On the other hand, there exists a constant $C > 0$ independent of δ such that for $m = 1, \ldots, M$ the local H^1-norm of the error can be estimated from below as

$$\eta_m^2 \leq C\|u - u_\delta\|^2_{H^1(\Omega_m)} + \frac{2}{N_m(N_m + 1)} \|f^{(m)} - f_{N_m}^{(m)}\|^2_{L^2_{w_m}(\Omega_m)} . \tag{5.4.11}$$

This leads to the lower bound for the global error

$$\eta^2 \leq C\|u - u_\delta\|^2_{H^1(a,b)} + \sum_{m=1}^{M} \frac{2}{N_m(N_m + 1)} \|f^{(m)} - f_{N_m}^{(m)}\|^2_{L^2_{w_m}(\Omega_m)} , \tag{5.4.12}$$

which shows that η is an *efficient* error estimator, i. e., (5.4.10) is never an overly pessimistic bound.

We observe that, since $0 \leq w_m(x) \leq \frac{1}{4}h_m^2$, we have

$$\eta_m^2 \leq \frac{h_m^2}{4N_m(N_m + 1)} \|r_m\|^2_{L^2(\Omega_m)} .$$

The expression on the right-hand side has the typical structure of the error indicators in the h-version of FEM (scaling with respect to the local mesh size, elemental residual measured in the L^2-norm); see, e. g., Verfürth (1996).

However, the use of this expression as an error indicator would lead to a non-optimal lower bound in lieu of (5.4.11), as far as the dependence upon N_m is concerned.

Estimates (5.4.10) and (5.4.11) will be proven in Sect. 5.4.3. The analysis can be extended to the case of variable coefficients α, β, γ (where also the error of approximating them by suitable polynomials has to be taken into account), as well as to the case of a SEM-NI discretization (where the quadrature errors also have to be taken into account).

Estimate (5.4.10) is the basis for any strategy of *adaptive* reduction of the discretization error, through the combination of mesh refinement and polynomial degree increment. A popular strategy consists of detecting and marking the elements which carry the bulk of the estimated error. Precisely, assuming for simplicity that the error coming from the polynomial approximation of the data f is already small, the marked elements are those for which

$$\eta_m^2 > \sigma \bar{\eta}^2 \, ,$$

where $\bar{\eta}^2 = \frac{1}{M} \sum_{j=1}^{M} \eta_j^2$ is the mean value of the error indicators and σ is a fixed parameter in $(0, 1)$. Next, for each marked element, one can either split the element into two elements of size $h_m/2$ while keeping the polynomial degree unchanged in each of them, or increase the polynomial degree by a few units while keeping the geometric element unchanged. Figure 5.15 provides an example of a (two-dimensional) adaptive discretization obtained by a similar strategy.

5.4.2 Multidimensional Analysis

We now move to the SEM problem (5.3.11). We assume that the functions α, β and γ are bounded in Ω (precisely, $\alpha, \gamma \in L^\infty(\Omega)$, $\beta \in (L^\infty(\Omega))^d$); furthermore, we assume that $\nabla \cdot \beta$ exists and satisfies

$$\tfrac{1}{2}\nabla\cdot\beta + \gamma \geq 0 \qquad \text{in } \Omega \, . \tag{5.4.13}$$

Under these assumptions, the bilinear form $a(u, v) = \sum_m a_{\Omega_m}(u, v)$ is continuous and coercive in $H_0^1(\Omega)$, i.e., it satisfies

$$|a(u, v)| \leq A_0 \|u\|_{H^1(\Omega)} \|v\|_{H^1(\Omega)} \qquad \text{for all } u, v \in H_0^1(\Omega)$$

and

$$a(v, v) \geq \alpha_* \|v\|_{H^1(\Omega)}^2 \qquad \text{for all } v \in H_0^1(\Omega) \, ,$$

for suitable constants $A_0 \geq \alpha_* > 0$, as in the one-dimensional case. Furthermore, we assume that $f \in L^2(\Omega)$ and $\psi \in L^2(\Gamma_N)$. Then, the SEM problem (5.3.11) has a unique solution, which satisfies the uniform stability estimate

$$\|u_\delta\|_{H^1(\Omega)} \leq \frac{1}{\alpha_*} \left(\|f\|_{L^2(\Omega)} + C\|\psi\|_{L^2(\Gamma_N)} \right) \, . \tag{5.4.14}$$

A similar result holds for the solution of the SEM-NI problem (5.3.13) under more restrictive assumptions on the coefficients. They have to be continuous in each element, and (5.4.13) has to be replaced by

$$\tfrac{1}{2}\nabla\cdot(I_{N_m}\beta) + \gamma \geq 0 \quad \text{in each element } \Omega_m .$$

Alternatively, condition (5.4.13) is sufficient provided the advection term is treated in the *skew-symmetric form*, i. e., the term $-\sum_m(\beta u_\delta, \nabla v)_{N_m,\Omega_m}$, which is part of the left-hand side, is replaced by

$$-\tfrac{1}{2}\sum_m(\beta u_\delta, \nabla v)_{N_m,\Omega_m} + \tfrac{1}{2}\sum_m(\beta\cdot\nabla u_\delta, v)_{N_m,\Omega_m}$$

$$+ \tfrac{1}{2}\sum_m((\nabla\cdot\beta)u_\delta, v)_{N_m,\Omega_m} \tag{5.4.15}$$

(see also CHQZ2, Sect. 6.4.3). Then, the stability estimate (5.4.14) takes the form

$$\|u_\delta\|_{H^1(\Omega)} \leq \frac{C}{\alpha_*}\sum_m\left(\|f\|_{C^0(\overline{\Omega}_m)} + \|\psi\|_{C^0(\Gamma_N\cap\partial\Omega_m)}\right) .$$

A Priori Error Analysis. Under suitable assumptions on the partition $\mathcal{T} = \{\Omega_m\}$ and the polynomial degree N_m in each element, an a priori error bound of the form

$$\|u-u_\delta\|_{H^1(\Omega)} \leq \frac{A_0}{\alpha_*}\left\{\sum_m C(s_m)h_m^{2(\min(N_m+1,s_m)-1)}N_m^{2(1-s_m)}\|u\|_{H^{s_m}(\Omega_m)}^2\right\}^{1/2} \tag{5.4.16}$$

holds. In particular, if $N_m = N$ is constant in the partition, and the diameters h_m of the elements satisfy $ch \leq h_m \leq c'h$ for all m, then setting $s = \min_m s_m$, one has

$$\|u - u_\delta\|_{H^1(\Omega)} \leq C_1(s)h^{\min(N+1,s)-1}N^{1-s}\|u\|_{H^s(\Omega)} . \tag{5.4.17}$$

Note that on the right-hand sides there appear norms rather than seminorms, unlike in the one-dimensional analog (5.4.4) or in the approximation estimates on the reference element $\hat{\Omega}$ given in CHQZ2, Sect. 5.8. Indeed, the mappings F_m between $\hat{\Omega}$ and the elements Ω_m are responsible for this change due to the chain rule of differentiation. However, if all F_m are affine, the seminorms are conserved on the subdomains.

Estimate (5.4.16) stems from Cea's Lemma (formula (5.4.2), now with $H^1(a,b)$ replaced by $H^1(\Omega)$) and the existence of an element $v_\delta = R_\delta u \in V_\delta$ such that

$$\|u - R_\delta u\|_{H^1(\Omega)} \leq \left\{\sum_m C(s_m)h_m^{2(\min(N_m+1,s_m)-1)}N_m^{2(1-s_m)}\|u\|_{H^{s_m}(\Omega_m)}^2\right\}^{1/2} \tag{5.4.18}$$

(see also CHQZ2, Sect. 6.4.1). The specific construction of the function $R_\delta u$ depends on geometric and functional properties of the discretization, as well as on the smoothness of u within the elements. Examples of such construction will be given in Sect. 5.4.3.

An error estimate like (5.4.16) may not be optimal in domains with corners due to the singularities induced on the exact solution by the geometry. Optimal error estimates (from above but also from below) for solutions in polygonal domains having corner singularities of the type $r^\lambda (\log r)^\mu$ (where r is the distance from a corner) have been established by Babuška and Guo (2001, 2002). They are based on approximation results in Jacobi-weighted Sobolev spaces, rather than in the usual Sobolev spaces. For the reference domain $\hat{\Omega}_C^2$, the Jacobi-weighted Sobolev space $H^{k,\beta}(\hat{\Omega}_C^2)$, with $k \geq 0$ and $\beta = (\beta_1, \beta_2) \in \mathbb{N} \times \mathbb{N}$, is the space of all L^2-functions v such that the norm

$$\|v\|_{H^{k,\beta}(\hat{\Omega}_C^2)} = \left(\sum_{|\alpha|=0}^{k} \int_{\hat{\Omega}_C^2} |D^\alpha v|^2 w_{\alpha,\beta}(\mathbf{x}) \, d\mathbf{x} \right)^{1/2}$$

is finite, where $w_{\alpha,\beta}(\mathbf{x}) = (1 - x_1^2)^{\alpha_1+\beta_1}(1 - x_2^2)^{\alpha_2+\beta_2}$.

A Posteriori Error Analysis. We now extend the a posteriori estimates (5.4.10) and (5.4.11) to the multidimensional case. We follow Melenk and Wohlmuth (2001), who in turn generalize results by Bernardi (1996). For simplicity, we assume that the SEM discretization (5.3.11) is applied to the Poisson problem with homogeneous Dirichlet conditions:

$$-\Delta u = f \quad \text{in } \Omega, \qquad u = 0 \quad \text{on } \partial\Omega, \qquad (5.4.19)$$

where $\Omega \subset \mathbb{R}^2$ is a polygonal domain. Let $\mathcal{T} = \{\Omega_m\}$ be a partition composed of parallelograms and triangles; let the partition and the polynomial degree distribution be shape-regular, in the sense of (5.4.29).

Let us first introduce the local error indicators. For any m, let $f_\delta^{(m)}$ denote the L^2-orthogonal projection of $f_{|\Omega_m}$ upon the local polynomial space $V_{N_m-1}(\Omega_m)$. The equation residual associated with the element Ω_m is defined as

$$r_m = f_\delta^{(m)} + \Delta u_\delta^{(m)},$$

where $u_\delta^{(m)}$ is the restriction of u_δ to Ω_m. The function r_m will be termed an *element residual*. In two or more dimensions, another residual comes into play which measures the jump of the normal derivative across the interface between two elements. Precisely, let $\{\Gamma_j\}$ be the collection of all internal edges of the partition, say, $\Gamma_j = \partial\Omega_m \cap \partial\Omega_{m'} \neq \emptyset$ for suitable m and m'. Let $\mathbf{n}_j$ denote a unit vector normal to Γ_j (orientation does not matter). Then, we define

$$r_j = \left[\frac{\partial u_\delta}{\partial n_j} \right]_{\Gamma_j},$$

which we term an *interface residual*. Element and interface residuals have to be weighted. Let $\hat{w}$ be the function on the reference domain $\hat{\Omega}$ (the unit square or the standard simplex) that satisfies $\hat{w}(\hat{\mathbf{x}}) = \mathrm{dist}(\hat{\mathbf{x}}, \partial\hat{\Omega})$. In the element Ω_m, we define the weight function w_m by $w_m = c_m \hat{w} \circ F_m^{-1}$, where the constant c_m is chosen so that

$$\int_{\Omega_m} w_m \, d\mathbf{x} = \int_{\Omega_m} 1 \, d\mathbf{x} \, .$$

Let us denote by $\|g\|_{L^2_\lambda(\Omega_m)}$ the weighted L^2-norm

$$\|g\|_{L^2_\lambda(\Omega_m)} = \left(\int_{\Omega_m} g^2(\mathbf{x}) \, w_m^\lambda(\mathbf{x}) \, d\mathbf{x} \right)^{1/2} \, ,$$

where $0 \leq \lambda \leq 1$ is a parameter. Similarly, on each interface Γ_j, let w_j denote the quadratic function that vanishes at the endpoints of Γ_j and satisfies

$$\int_{\Gamma_j} w_j \, ds = \int_{\Gamma_j} 1 \, ds \, .$$

Let us denote by $\|g\|_{L^2_\lambda(\Gamma_j)}$ the weighted L^2-norm

$$\|g\|_{L^2_\lambda(\Gamma_j)} = \left(\int_{\Gamma_j} g^2(s) \, w_j^\lambda(s) \, ds \right)^{1/2} \, .$$

We denote by h_j the length of Γ_j and we set $N_j = \max(N_m, N_{m'})$ if $\Gamma_j = \partial\Omega_m \cap \partial\Omega_{m'}$.

We are ready to introduce the local error indicators, which are associated with the elements Ω_m of the partition. Precisely, we set, for $0 \leq \lambda \leq 1$,

$$\eta_{m,\lambda}^2 = \frac{h_m^2}{N_m^2} \|r_m\|_{L^2_\lambda(\Omega_m)}^2 + \sum_{j \,:\, \Gamma_j \subset \partial\Omega_m} \frac{h_j}{N_j} \|r_j\|_{L^2_\lambda(\Gamma_j)}^2 \, . \tag{5.4.20}$$

Then, the following upper bounds for the SEM error hold: There exists a constant $C > 0$ independent of the discretization parameters such that, for $0 \leq \lambda \leq 1$,

$$\|u - u_\delta\|_{H^1(\Omega)}^2 \leq \frac{C}{\alpha_*} \sum_m \left(N_m^{2\lambda} \eta_{m,\lambda}^2 + \frac{h_m^2}{N_m^2} \|f - f_\delta^{(m)}\|_{L^2(\Omega_m)}^2 \right) \, . \tag{5.4.21}$$

On the other hand, the following local lower bounds for the error hold: For all $\varepsilon > 0$, there exists a constant $C(\varepsilon) > 0$ independent of the discretization parameters, such that, for $0 \leq \lambda \leq 1$,

$$\eta_{m,\lambda}^2 \leq C(\varepsilon) N_m^{\max(1-2\lambda+2\varepsilon,\,0)}$$

$$\times \left(N_m \|u - u_\delta\|_{H^1(\tilde{\Omega}_m)}^2 + \frac{h_m^2}{N_m^{2(1-\varepsilon)}} \|f - f_\delta\|_{L^2(\tilde{\Omega}_m)}^2 \right) \, , \tag{5.4.22}$$

where $\tilde{\Omega}_m = \bigcup \{\Omega_{m'} \,:\, \partial\Omega_m \cap \partial\Omega_{m'} \text{ is an edge}\}$.

We note that the error estimation resulting from these bounds is not as tight as in the one-dimensional case (compare with (5.4.10) and (5.4.11)). In particular, the upper bound is optimal only for $\lambda = 0$ (no power of N_m multiplies $\eta_{m,\lambda}^2$), whereas the lower bound is tighter for $\lambda > \frac{1}{2}$.

Guo (2005) provides nearly optimal error bounds at the expense of measuring the residuals in more sophisticated norms, such as those of the Jacobi-weighted Sobolev spaces already mentioned above.

The previous a posteriori analysis provides the theoretical foundation for *adaptive refinement* strategies. A popular criterion is the equidistribution of the error among the elements. Denoting by

$$\overline{\eta}_\lambda^2 = \frac{1}{M} \sum_m \eta_{m,\lambda}^2$$

(where M is the number of elements) the mean value of the error indicators, element Ω_m is marked for refinement if

$$\eta_{m,\lambda}^2 > \sigma \overline{\eta}_\lambda^2 ,$$

where $0 < \sigma < 1$ is a fixed threshold (a standard choice is $\sigma = 0.75$). An *h-refinement* (the element is divided into four sons) or an *N-refinement* (the degree N_m is increased by one) is then applied, depending on a predicted value of the local error at next step suggested by the a priori error estimates. We refer to Melenk and Wohlmuth (2001) for further details. Figure 5.15 shows the result of a few steps of successive refinements in a *L*-shaped domain with $f = 1$, starting from a coarse uniform mesh with constant polynomial degree equal to two. (In the actual implementation, the *h*-refinement allows

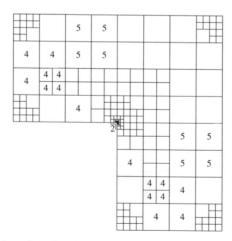

Fig. 5.15. Adapted mesh and polynomial degree distribution for the SEM approximation of a Dirichlet problem in an *L*-shaped domain. The polynomial degree is 3 wherever it is not explicitly indicated [Courtesy of J. Melenk and B. Wohlmuth]

for hanging nodes, not considered in the discussion above.) The adapted polynomial degree is shown in each element, except for those elements (the majority, indeed) in which it is equal to three.

5.4.3 Some Proofs

We conclude this section with the proofs of some results stated above.

Proof of the Interpolation Error Estimate (5.4.3). We proceed in three steps, by considering first the interpolation error on the reference interval $\hat{\Omega} = (-1, 1)$, then on an arbitrary element $\Omega_m = (\overline{x}_{m-1}, \overline{x}_m)$ and finally on the whole domain Ω.

(i) *Interpolation on the reference interval.* For every integer $N \geq 1$, let us denote by $\hat{I}_N : C^0(\overline{\hat{\Omega}}) \to \mathbb{P}_N$ the interpolation operator at the $N+1$ LGL nodes on $\hat{\Omega}$, say $\hat{x}_i^{(N)}$, $i = 0, \ldots, N$. Then, for every function $\hat{u} \in H^s(\hat{\Omega})$, $s \geq 1$,

$$\|\hat{u} - \hat{I}_N \hat{u}\|_{H^k(\hat{\Omega})} \leq \hat{C} N^{k-s} |\hat{u}|_{H^{s;N}(\hat{\Omega})}, \qquad k = 0, 1, \tag{5.4.23}$$

where $\hat{C}$ is a positive constant depending on k and s (but neither on N nor on $\hat{u}$). This result is given in CHQZ2, Sect. 5.4.3 (see formulas (5.4.33) and (5.4.35)). We recall, for the reader's convenience, the definition of the *s-to-N Sobolev seminorm* (see CHQZ2, formula (5.4.10)):

$$|v|_{H^{s;N}(\omega)} = \left(\sum_{j=\min(s,N+1)}^{s} \|v^{(j)}\|_{L^2(\omega)}^2 \right)^{1/2}, \tag{5.4.24}$$

where ω denotes any one-dimensional interval, v any function of $H^s(\omega)$, and $v^{(j)}$ the j-th derivative of v.

(ii) *Interpolation on an arbitrary element.* For every $m = 1, \ldots, M$, let $\Omega_m = (\overline{x}_{m-1}, \overline{x}_m)$ denote the current element of the partition of Ω. If N_m denotes the polynomial degree used in Ω_m, let $x_i^{(m)}$, $i = 0, \ldots, N_m$ denote the LGL nodes in $\overline{\Omega}_m$. If $F_m : \hat{\overline{\Omega}} \to \hat{\Omega}_m$ is the affine mapping (5.2.9), then $x_i^{(m)} = F_m(\hat{x}_i^{(N_m)})$, that is the $N_m + 1$ LGL nodes in $\overline{\Omega}_m$ are obtained as images of the $N_m + 1$ LGL nodes in $\hat{\Omega}$.

Now let $I_{N_m}^{(m)} : C^0(\overline{\Omega}_m) \to \mathbb{P}_{N_m}$ denote the interpolation operator in Ω_m. If for any function v in Ω_m we denote by $\hat{v}$ its image (according to F_m) in $\hat{\Omega}$, that is, $\hat{v} = v \circ F_m$, then

$$(u - I_{N_m}^{(m)} u)\hat{} = \hat{u} - \hat{I}_{N_m} \hat{u} . \tag{5.4.25}$$

It follows that

$$|u - I_{N_m}^{(m)} u|_{H^k(\Omega_m)}^2 = \left(\frac{2}{h_m} \right)^{2k-1} |\hat{u} - \hat{I}_{N_m} \hat{u}|_{H^k(\hat{\Omega})}^2, \qquad k = 0, 1 .$$

Using (5.4.23) with $s = s_m$ and $N = N_m$, we obtain

$$|u - I_{N_m}^{(m)} u|^2_{H^k(\Omega_m)} \leq \hat{C}^2 \left(\frac{2}{h_m}\right)^{2k-1} N_m^{2(k-s_m)} |\hat{u}|^2_{H^{s_m};N_m(\hat{\Omega})}, \quad k = 0, 1.$$

$$(5.4.26)$$

We distinguish two cases. If $N_m \geq s_m - 1$, then

$$|\hat{u}|^2_{H^{s_m};N_m(\hat{\Omega})} = |\hat{u}|^2_{H^{s_m}(\hat{\Omega})} = \left(\frac{h_m}{2}\right)^{2s_m-1} |u|^2_{H^{s_m}(\Omega_m)}, \quad (5.4.27)$$

whereas if $N_m < s_m - 1$, then

$$|\hat{u}|^2_{H^{s_m};N_m(\hat{\Omega})} = \sum_{j=N_m+1}^{s_m} \|\hat{u}^{(j)}\|^2_{L^2(\hat{\Omega})}$$

$$= \sum_{j=N_m+1}^{s_m} \left(\frac{h_m}{2}\right)^{2j-1} \|u^{(j)}\|^2_{L^2(\Omega_m)}$$

$$\leq \left(\frac{h_m}{2}\right)^{2N_m+1} |u|^2_{H^{s_m};N_m(\Omega_m)}.$$

Using these results with (5.4.26) it follows that

$$|u - I_{N_m}^{(m)} u|^2_{H^k(\Omega_m)}$$

$$\leq \hat{C} \left(\frac{2}{h_m}\right)^{2k-1} N_m^{2(k-s_m)} \left(\frac{h_m}{2}\right)^{2\min(N_m+1,s_m)-1} |u|^2_{H^{s_m};N_m(\Omega_m)}$$

$$= \hat{C} \left(\frac{h_m}{2}\right)^{2(\min(N_m+1,s_m)-k)} N_m^{2(k-s_m)} |u|^2_{H^{s_m};N_m(\Omega_m)}.$$

Since

$$|u - I_\delta u|^2_{H^k(\Omega)} = \sum_{m=1}^{M} |u - I_{N_m}^{(m)} u|^2_{H^k(\Omega_m)},$$

by summing up on $m = 1, \ldots, M$ and taking the square root on both sides we obtain the desired result (5.4.3).

Proof of the a Posteriori Error Bounds (5.4.10) and (5.4.11). We start by considering (5.4.10). By (5.4.1) and the coercivity of the bilinear form a defined in (5.2.4), we have for $v = u - u_\delta$

$$\alpha_* \|u - u_\delta\|^2_{H^1(a,b)} \leq a(u - u_\delta, v - v_\delta) \quad \text{for all } v_\delta \in V_\delta. \quad (5.4.28)$$

Assume that v_δ is chosen to satisfy $v_\delta(\bar{x}_m) = v(\bar{x}_m)$ for $m = 1, \ldots, M - 1$. Counter-integrating by parts in each element, we obtain

$$a(u - u_\delta, v - v_\delta) = \sum_{m=1}^{M} \int_{\Omega_m} \left(\mathcal{L}u^{(m)} - \mathcal{L}u_\delta^{(m)}\right) \left(v^{(m)} - v_\delta^{(m)}\right).$$

Since $\mathcal{L}u^{(m)} = f^{(m)} = f^{(m)}_{N_m} + (f^{(m)} - f^{(m)}_{N_m})$, we have

$$
a(u - u_\delta, v - v_\delta) = \sum_{m=1}^{M} \int_{\Omega_m} \left(r_m + (f^{(m)} - f^{(m)}_{N_m}) \right) \left(v^{(m)} - v^{(m)}_\delta \right)
$$

$$
= \sum_{m=1}^{M} \int_{\Omega_m} \left(r_m + (f^{(m)} - f^{(m)}_{N_m}) \right) \sqrt{w_m} \, \frac{v^{(m)} - v^{(m)}_\delta}{\sqrt{w_m}}
$$

$$
\leq \sum_{m=1}^{M} \left(\|r_m\|_{L^2_{w_m}(\Omega_m)} + \|f^{(m)} - f^{(m)}_{N_m}\|_{L^2_{w_m}(\Omega_m)} \right) \left\| \frac{v^{(m)} - v^{(m)}_\delta}{\sqrt{w_m}} \right\|_{L^2(\Omega_m)}.
$$

Let us now define $v^{(m)}_\delta$. Given a function $\hat{v} \in H^1(-1,1)$, define

$$
\hat{v}_N(\hat{x}) = \hat{v}(-1) + \int_{-1}^{\hat{x}} (P_{N-1}\hat{v}')(s) \, ds \,,
$$

where P_{N-1} denotes truncation of the Legendre series, i.e., the $L^2(-1,1)$-orthogonal projection upon $\mathbb{P}_{N-1}$. One can prove that $\hat{v}_N(\pm 1) = \hat{v}(\pm 1)$ and

$$
\int_{-1}^{1} \frac{(\hat{v} - \hat{v}_N)^2(\hat{x})}{1 - \hat{x}^2} \, d\hat{x} \leq \frac{1}{N(N+1)} \int_{-1}^{1} (\hat{v}')^2(\hat{x}) \, d\hat{x}
$$

(see, e.g., Schwab (1998), Theorem 3.14). Set $\hat{v}(\hat{x}) = v^{(m)}(F_m(\hat{x}))$, where F_m is the affine mapping (5.2.9), and define $v^{(m)}_\delta(x) = \hat{v}_{N_m}(F_m^{-1}(x))$. Then, the previous inequality yields

$$
\left\| \frac{v^{(m)} - v^{(m)}_\delta}{\sqrt{w_m}} \right\|_{L^2(\Omega_m)} \leq \frac{1}{\sqrt{N_m(N_m+1)}} \|v^{(m)}\|_{H^1(\Omega_m)} \,.
$$

Thus, by the Cauchy–Schwarz inequality we get

$$
a(u - u_\delta, v - v_\delta) \leq \left(\sum_{m=1}^{M} \frac{1}{N_m(N_m+1)} \|r_m\|^2_{L^2_{w_m}(\Omega_m)} \right)^{1/2} \|u - u_\delta\|_{H^1(a,b)}
$$

$$
+ \left(\sum_{m=1}^{M} \frac{1}{N_m(N_m+1)} \|f^{(m)} - f^{(m)}_{N_m}\|^2_{L^2_{w_m}(\Omega_m)} \right)^{1/2} \|u - u_\delta\|_{H^1(a,b)} \,,
$$

which, together with (5.4.28), proves (5.4.10).

Next, we consider (5.4.11). One has

$$
\|r_m\|^2_{L^2_{w_m}(\Omega_m)} = \int_{\Omega_m} r_m^2 w_m \, dx = \int_{\Omega_m} r_m v^{(m)} \, dx \,,
$$

with $v^{(m)} = r_m w_m \in H^1_0(\Omega_m)$. Writing $r_m = \mathcal{L}(u^{(m)} - u^{(m)}_\delta) - (f^{(m)} - f^{(m)}_{N_m})$

and integrating by parts, we have

$$\|r_m\|^2_{L^2_{w_m}(\Omega_m)} = a_{\Omega_m}(u^{(m)} - u^{(m)}_\delta, v^{(m)}) - \int_{\Omega_m} (f^{(m)} - f^{(m)}_{N_m}) r_m w_m \, dx$$

$$\leq A_0 \|u^{(m)} - u^{(m)}_\delta\|_{H^1(\Omega_m)} \|v^{(m)}\|_{H^1(\Omega_m)}$$

$$+ \|f^{(m)} - f^{(m)}_{N_m}\|_{L^2_{w_m}(\Omega_m)} \|r_m\|_{L^2_{w_m}(\Omega_m)} \,.$$

The proof will be concluded if we establish that $\|v^{(m)}\|_{H^1(\Omega_m)} \leq c\|r_m\|_{L^2_{w_m}(\Omega_m)}$ for a suitable constant $c > 0$ independent of δ. We observe that

$$\|v^{(m)}\|^2_{H^1(\Omega_m)} = \int_{\Omega_m} r^2_m w^2_m \, dx + 2 \int_{\Omega_m} (r'_m)^2 w^2_m \, dx + 2 \int_{\Omega_m} r^2_m (w'_m)^2 \, dx$$

$$\leq \tfrac{1}{4} h^2_m \int_{\Omega_m} r^2_m w_m \, dx + 2 \int_{\Omega_m} (r'_m)^2 w^2_m \, dx + 2h^2_m \int_{\Omega_m} r^2_m \, dx \,.$$

We now invoke the inverse inequalities on the reference interval:

$$\int_{-1}^{1} (r')^2 (1 - \hat{x}^2)^2 \, d\hat{x} \leq c' N^2 \int_{-1}^{1} r^2 (1 - \hat{x}^2) \, d\hat{x}$$

and

$$\int_{-1}^{1} r^2 \, d\hat{x} \leq c'' N^2 \int_{-1}^{1} r^2 (1 - \hat{x}^2) \, d\hat{x} \,,$$

which hold for all polynomials $r \in \mathbb{P}_N$ (see, e.g., Schwab (1998), Theorems 3.95 and 3.96). Transporting these inequalities on the element Ω_m via the affine mapping F_m, we get

$$\int_{\Omega_m} (r'_m)^2 w^2_m \, dx \leq c' N^2_m \int_{\Omega_m} r^2_m w_m \, dx \quad \text{and} \quad \int_{\Omega_m} r^2_m \, dx \leq c'' N^2_m \int_{\Omega_m} r^2_m w_m \, dx \,;$$

whence,

$$\|v^{(m)}\|^2_{H^1(\Omega_m)} \leq \left(\tfrac{1}{4}(b-a)^2 + 2c' + 2c'' \right) \|r_m\|_{L^2_{w_m}(\Omega_m)} \,.$$

Proofs of the Approximation Bound (5.4.18). The most classical situation in multidomain spectral methods is when all elements Ω_m are images of the same tensorial reference domain $\hat{\Omega}$ (the square $\hat{\Omega}^2_C$ in 2D, the cube $\hat{\Omega}^3_C$ in 3D) under smooth mappings F_m, and the polynomial degree $N_m = N$ is constant throughout the partition. Then, if $s_m > d/2$ so that $u_{|\Omega_m}$ is continuous (precisely, $H^{s_m}(\Omega_m) \subset C^0(\overline{\Omega}_m)$), we can define $R_\delta u$ by the conditions

$$(R_\delta u)_{|\Omega_m} = I^{(m)}_N (u_{|\Omega_m}) \qquad \text{for all } m \,,$$

where $I^{(m)}_N v \in V_N(\Omega_m)$ is the (mapped) interpolant of v at the (mapped)

LGL points of order N in Ω_m. Precisely,

$$I_N^{(m)} v = (\hat{I}_N \hat{v}) \circ F_m^{-1} \qquad \text{with } \hat{v} = v \circ F_m \,,$$

where $\hat{I}_N \hat{v} \in \mathbb{Q}_N(\hat{\Omega})$ is the interpolant of $\hat{v}$ at the tensor-product LGL-points of order N in $\hat{\Omega}$. Note that $R_\delta u$ is a continuous function across the element interfaces, since the (mapped) LGL points along each interface are shared by the contiguous subdomains. Now, if all mappings F_m are *uniformly regular*, i.e., if the ratio between the diameter of the largest ball contained in $\overline{\Omega}_m$ and the diameter h_m of Ω_m is bounded from below uniformly with respect to the partition, one has, for $s_m \geq (d+1)/2$,

$$\|u_{|\Omega_m} - I_N^{(m)}(u_{|\Omega_m})\|_{H^1(\Omega_m)} \leq C(s_m) h^{\min(N+1,s_m)-1} N^{1-s_m} \|u\|_{H^{s_m}(\Omega_m)}$$

(see CHQZ2, formula (5.8.27)), from which (5.4.18) immediately follows.

Another situation that leads to (5.4.18) is when each element is the image of the reference element under a multilinear mapping (in the tensorial case) or a linear mapping (in the simplicial case). Let us briefly sketch the proof in dimension 2 (full details can be found in Babuška and Guo (2002)). We assume that the partition $\mathcal{T}$ and the polynomial degree distribution are *shape-regular*, i.e., there exist constants $c, c' > 0$ independent of the discretization such that

$$c h_m \leq h_{m'} \leq c' h_m \quad \text{and} \quad c(N_m + 1) \leq (N_{m'} + 1) \leq c'(N_m + 1) \quad (5.4.29)$$

for all m, m' such that $\overline{\Omega}_m \cap \overline{\Omega}_{m'} \neq \emptyset$. We also assume that the same condition is satisfied by the smoothness indices $s_m > 0$, which appear in (5.4.18).

For each vertex $\mathbf{v}_i$ of the partition, let $\mathcal{M}_i$ denote the set of indices m of the elements that have $\mathbf{v}_i$ as a vertex. Let us set $\Omega^{(i)} = \bigcup_{m \in \mathcal{M}_i} \overline{\Omega}_m$; let us also set $h = \max_{m \in \mathcal{M}_i} h_m$, $N = \min_{m \in \mathcal{M}_i} N_m$ and $s = \min_{m \in \mathcal{M}_i} s_m$. There exists an affine mapping G_i that maps $\Omega^{(i)}$ into a patch $\hat{\Omega}^{(i)}$ of quadrilaterals and triangles, such that $\hat{\Omega}^{(i)} \subseteq \hat{\Omega}_C^2$ and area $\hat{\Omega}^{(i)} \sim 1$. Assuming that $u_{|\Omega^{(i)}} \in H^s(\Omega^{(i)})$, the transformed function $\hat{u} = u \circ G_i^{-1}$ satisfies $\hat{u} \in H^s(\hat{\Omega}^{(i)})$. By standard Sobolev space results, $\hat{u}$ can be extended to a function $\tilde{u} \in H^s(\hat{\Omega}_C^2)$ in a continuous way. Let $\nu \geq 1$ denote an integer to be fixed later as a function of N. By the single-domain approximation results (see CHQZ2, Sect. 5.8.2), there exists a polynomial $\tilde{u}_\nu \in \mathbb{Q}_\nu(\hat{\Omega}_C^2)$ such that

$$\|\tilde{u} - \tilde{u}_\nu\|_{H^1(\hat{\Omega}_C^2)} \leq C \nu^{1-s} \|\tilde{u}\|_{H^s(\hat{\Omega}_C^2)} \,.$$

We set $\hat{u}_\nu = \tilde{u}_{\nu|\hat{\Omega}^{(i)}}$ if $\mathbf{v}_i$ is an internal node, or a non-Dirichlet boundary node. Otherwise $\hat{u}_\nu$ is a suitable modification of $\tilde{u}_{\nu|\hat{\Omega}^{(i)}}$ to match the boundary condition if $\mathbf{v}_i$ is a Dirichlet boundary node; such a modification is made through suitable stable polynomial liftings (see, e.g., Bernardi and Maday (1997), Muñoz-Sola (1997)). In all cases, we set $u_\nu^{(i)} = \hat{u}_\nu \circ G_i \in \mathbb{Q}_\nu(\Omega^{(i)})$, which satisfies

$$\|u - u_\nu^{(i)}\|_{H^1(\Omega^{(i)})} \leq C h^{\nu-1} \nu^{1-s} \|u\|_{H^s(\Omega^{(i)})}$$

(for simplicity, we have assumed $\nu < s$). Finally, we blend all these local approximations. Let φ_i be the discrete characteristic function associated with the vertex $\mathbf{v}_i$ of the partition, i.e., the function that satisfies $\varphi_i(\mathbf{v}_i) = 1$, $\varphi_j(\mathbf{v}_j) = 1$, and on each element Ω_m is either linear (if Ω_m is a triangle) or bilinear (if Ω_m is a quadrilateral). The functions φ_i form a *partition of unity* in Ω, i.e., they satisfy $\sum_i \varphi_i \equiv 1$; this yields the identity $u = \sum_i \varphi_i u_{|\Omega^{(i)}}$. Thus, we are led to define $R_\delta u = \sum_i \varphi_i u_\nu^{(i)}$. We finally observe that $v_{i,m} = \varphi_i u_\nu^{(i)}|_{\Omega_m} \in \mathbb{Q}_{\nu+1}(\Omega_m)$ for all m. It is easily seen that if F_m maps the reference domain $\hat{\Omega}$ into Ω_m, then $\hat{v}_{i,m} = v_{i,m} \circ F_m \in \mathbb{Q}_{2(\nu+1)}(\hat{\Omega})$ if $\hat{\Omega}$ is the square $\hat{\Omega}_C^2$, or $\hat{v}_{i,m} \in \mathbb{Q}_{\nu+1}(\hat{\Omega}) \subset \mathbb{P}_{2(\nu+1)}(\hat{\Omega})$ if $\hat{\Omega}$ is the triangle $\hat{\Omega}_S^2$. In both cases, if ν is chosen as the largest integer for which $2(\nu+1) \leq N$, we have $\hat{v}_{i,m} \in \hat{P}_N$. Thus, (5.4.18) is proven.

5.5 Some Numerical Results for the SEM-NI Approximations

In this section we show several numerical results that substantiate the theoretical error estimates proven in the previous section. Moreover, we plot and compare the eigenfunctions of a single-domain G-NI and a four-domain SEM-NI approximation of the Laplace equation.

5.5.1 Error Decay vs. N and h

We consider the Dirichlet advection-diffusion-reaction problem

$$-\nabla \cdot (\alpha \nabla u - \boldsymbol{\beta} u) + \gamma u = f \quad \text{in } \Omega = (0,1)^2 \,,$$
$$u = g \quad \text{on } \partial \Omega \,. \tag{5.5.1}$$

The coefficients are $\alpha = 1$, $\boldsymbol{\beta} = (1,0)^T$, $\gamma = 1$. The right-hand side f and the boundary data g are chosen in such a way to generate exact solutions having different orders of Sobolev regularity. The corresponding SEM-NI approximations will consequently exhibit different rates of convergence with respect to the discretization parameters, according to the error estimate (5.4.17).

In Fig. 5.16 we report the error curves in the H^1-norm for different values of the (constant) polynomial degree N and length h of the element edges. The chosen exact solutions are $u(x,y) = x \sin(\pi y) + \chi_{[1/2,1]}(x - 1/2)^{3/2}$ (which is in $H^{2-\varepsilon}(\Omega)$ $\forall \varepsilon > 0$) in the top plots, $u(x,y) = x \sin(\pi y) + \chi_{[1/2,1]}(x - 1/2)^{7/2}$ (which is in $H^{4-\varepsilon}(\Omega)$ $\forall \varepsilon > 0$) in the middle plots, and $u(x,y) = \sin(\pi x) \sin(\pi y)$ (which is infinitely smooth) in the bottom plots. The function

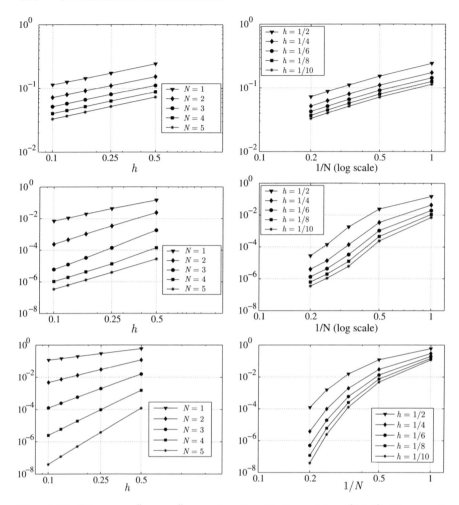

Fig. 5.16. The error $\|u - u_\delta\|_{H^1(\Omega)}$ for the elliptic problem (5.5.1). The exact solution is $u(x,y) = x\sin(\pi y) + \chi_{[1/2,1]}(x - 1/2)^{3/2}$ (*top*), $u(x,y) = x\sin(\pi y) + \chi_{[1/2,1]}(x - 1/2)^{7/2}$ (*middle*), $u(x,y) = \sin(\pi x)\sin(\pi y)$ (*bottom*)

$\chi_{[1/2,1]}$ is the characteristic function of the half domain $\Omega_R = \{1/2 \leq x \leq 1, 0 \leq y \leq 1\}$, i.e., it is equal to 1 in Ω_R and vanishes elsewhere. (In many of the error decay plots that we exhibit in the remainder of this book we change the plotting convention for the abscissas, so that the errors are decreasing to the left rather than to the right, as in all the single-domain plots that we exhibited heretofore.) The objective of these plots is to provide empirical verification of the theoretical estimate (5.4.17) of the preceding section. The plots exhibited in the top and middle rows of Fig. 5.16 display the expected algebraic convergence in h and N (nearly linear for the top and nearly

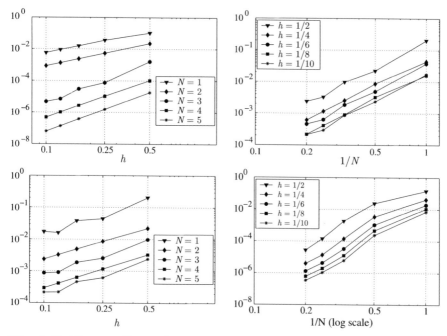

Fig. 5.17. The error $\|u - u_\delta\|_{L^2(\Omega)}$, for the advection problem (5.5.2). The exact solution is $u(x,y) = x\sin(\pi y) + \chi_{[1/2,1]}(x - 1/2)^{3/2}$ (*top*), $u(x,y) = x\sin(\pi y) + \chi_{[1/2,1]}(x - 1/2)^{7/2}$ (*bottom*)

quadratic for the middle), whereas the bottom row converges algebraically in h (with an exponent increasing with N) and exponentially in N.

We also illustrate the SEM-NI convergence for a pure advection problem, namely

$$\nabla \cdot (\boldsymbol{\beta} u) + \gamma u = f \quad \text{in } \Omega = (0,1)^2 \,,$$
$$u = g \quad \text{on } \partial\Omega_{\text{in}} = \{0\} \times (0,1) \,. \tag{5.5.2}$$

As before, the coefficients are $\boldsymbol{\beta} = (1,0)^T$ and $\gamma = 1$. Such a case was not explicitly covered in the previous section; however, an estimate similar to (5.4.17) holds, but in the L^2-norm rather than H^1-norm. In Fig. 5.17 we plot the L^2-error curves corresponding to the exact solution $u(x,y) = x\sin(\pi y) + \chi_{[1/2,1]}(x-1/2)^{3/2}$ (top plots) and $u(x,y) = x\sin(\pi y) + \chi_{[1/2,1]}(x-1/2)^{7/2}$ (bottom plots). The error behavior is again consistent with the theoretical prediction.

5.5.2 Eigenfunction Approximation

The next set of numerical results aims at comparing the single-domain and the multidomain approaches in the approximation of smooth solutions that oscillate to varying degrees. We consider the eigenvalue problem for the Laplace

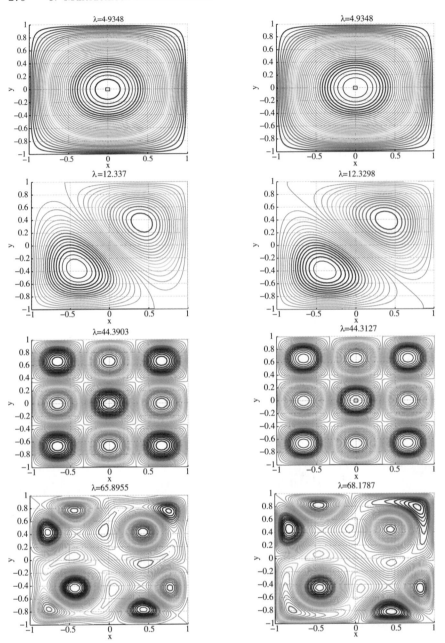

Fig. 5.18. Eigenfunctions of $M^{-1}A$, corresponding to the SEM-NI discretization of the Laplacian in $\Omega = (-1, 1)^2$ under Dirichlet boundary conditions. Single-domain with $N = 8$ (*left*), multidomain with $N = 4$ and 2×2 elements (*right*)

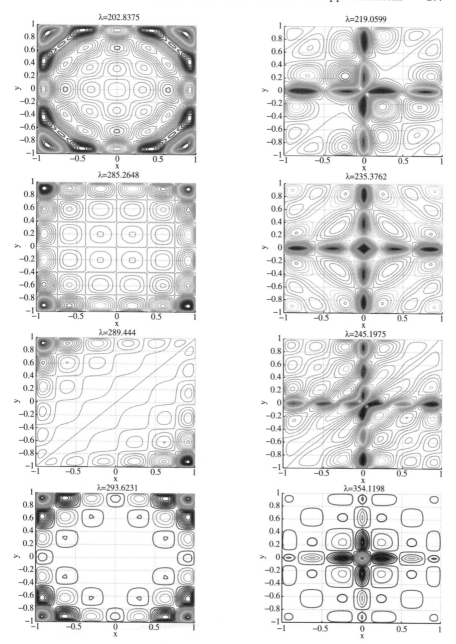

Fig. 5.19. Eigenfunctions of $M^{-1}A$, corresponding to the SEM-NI discretization of the Laplacian in $\Omega = (-1, 1)^2$ under Dirichlet boundary conditions. Single-domain with $N = 8$ (*left*), multidomain with $N = 4$ and 2×2 elements (*right*)

operator under Dirichlet boundary conditions in the domain $\Omega = (-1,1)^2$, i.e.,

$$-\Delta w = \lambda w \quad \text{in } \Omega \,,$$
$$w = 0 \quad \text{on } \partial\Omega \,. \tag{5.5.3}$$

The eigenfunctions of this problem are $w_k(x,y) = \sin k_1 \frac{\pi}{2}(x+1) \sin k_2 \frac{\pi}{2}(y+1)$ with corresponding eigenvalues $\lambda_k = \frac{\pi^2}{4}(k_1^2 + k_2^2)$, for $k = (k_1, k_2)$ and arbitrary $k_1, k_2 \geq 1$. The SEM-NI method is applied, leading to the generalized algebraic eigenvalue problem

$$A\mathbf{w} = \lambda M\mathbf{w} \,,$$

where A is the SEM-NI stiffness matrix considered in Sect. 5.3.4, and M is the SEM-NI (thus diagonal) mass matrix. We use either a single element and the polynomial degree 8 in each direction, or 2×2 equal elements and the polynomial degree 4 in each direction within each element. Thus, the total number of degrees of freedom, 49, is the same in both cases.

The level curves of some of the discrete eigenfunctions (normalized with respect to the discrete L^2-norm) are shown in Figs. 5.18 and 5.19, according to the increasing order of the corresponding eigenvalues. The single-domain results appear in the left-hand side columns, whereas the multidomain results appear in the right-hand side columns. The computed discrete eigenvalues are also reported on top of each plot. The top plots of Fig. 5.18 show the lowest frequency eigenfunctions, corresponding to $k = (1,1)$ (the exact eigenvalue being $\lambda_k = 4.9348\ldots$); conversely, the bottom plots of Fig. 5.19 show the highest frequency eigenfunctions, corresponding to $k = (7,7)$ (the exact eigenvalue being $\lambda_k = 241.81\ldots$). In the presence of eigenvalues with multiplicity greater than 1, the closest single-domain and multidomain discrete eigenfunctions are selected, with respect to the discrete L^2-norm.

The figures clearly document that low-frequency to mid-frequency eigenfunctions are correctly reproduced by both the single-domain and the multidomain discretizations; the lower frequency eigenfunctions are indeed virtually indistinguishable from each other. However, as the wavenumber increases, the multidomain computations becomes less and less accurate compared to the single-domain ones. For the higher portion of the discrete spectrum, the geometric structure of the domain decomposition rather than the operator dictates the eigenfunction patterns.

The observed behavior is consistent with the theoretical predictions discussed in Sect. 5.4.1 (albeit for the one-dimensional case). In particular, the single-domain approach is superior to the multidomain approach in the approximation of infinitely smooth, structured solutions, provided that there is enough resolution to correctly represent the features of the solution.

5.6 SEM for Stokes and Navier–Stokes Equations

The SEM for the approximation of the Stokes (or Navier–Stokes) equations is based again on the Galerkin principle. We confine ourselves to the simple

Stokes problem with homogeneous Dirichlet conditions on the velocity field (the generalization to other types of boundary conditions and the Navier–Stokes equations does not present any specific difficulty for the SEM formulation):

$$-\nu \triangle \mathbf{u} + \nabla p = \mathbf{g} \quad \text{in } \Omega \subset \mathbb{R}^d \quad (d = 2, 3),$$
$$\nabla \cdot \mathbf{u} = 0 \quad \text{in } \Omega, \tag{5.6.1}$$
$$\mathbf{u} = 0 \quad \text{on } \partial \Omega.$$

Setting $V = (H_0^1(\Omega))^d$ and $Q = \{q \in L^2(\Omega) : \int_\Omega q \, d\mathbf{x} = 0\}$, the weak formulation of this problem consists of seeking $\mathbf{u} \in V$ and $p \in Q$ such that

$$a(\mathbf{u}, \mathbf{v}) + b(\mathbf{v}, p) = (\mathbf{g}, \mathbf{v}) \quad \text{for all } \mathbf{v} \in V,$$
$$b(\mathbf{u}, q) = 0 \quad \text{for all } q \in Q, \tag{5.6.2}$$

where $a(\mathbf{u}, \mathbf{v}) = \nu(\nabla \mathbf{u}, \nabla \mathbf{v})$, $b(\mathbf{v}, q) = -(\nabla \cdot \mathbf{v}, q)$, $\mathbf{g} \in (L^2(\Omega))^d$ and the L^2-inner products are used here. This is a particular case of a saddle-point formulation (see Sect. 3.7).

5.6.1 SEM and SEM-NI Formulations

As done in Sect. 5.3, we introduce piecewise polynomial spaces on the partition $\mathcal{T} = \{\Omega_m\}$ of the computational domain Ω. Hybrid partitions composed of both quadrilaterals and triangles (parallelepipeds and tetrahedra in 3D) are admissible. An example of a hybrid partition is provided by Fig. 5.20. We will need two distinct spaces—one for velocity, the other for pressure. They are built starting from admissible pairs of velocity–pressure spaces on one or more reference domains. To be precise, for any m let us choose a pair of spaces $V_{N_m}(\hat{\Omega})$, $Q_{N_m}(\hat{\Omega})$ among those considered in Sect. 3.5.2. Here, $\hat{\Omega}$ is the (tensorial or simplicial) reference domain mapped onto Ω_m by the mapping F_m, whereas N_m is the parameter by which the polynomial velocities and pressures belonging to $V_{N_m}(\hat{\Omega})$ and $Q_{N_m}(\hat{\Omega})$, respectively, are selected. Let N_m^v be the largest integer such that

$$(\mathbb{Q}_{N_m^v})^d \subseteq V_{N_m}(\hat{\Omega}) \quad \text{or} \quad (\mathbb{P}_{N_m^v})^d \subseteq V_{N_m}(\hat{\Omega}), \tag{5.6.3}$$

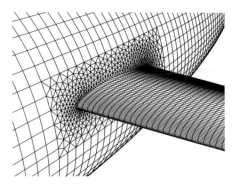

Fig. 5.20. Example of a hybrid spectral element partition

depending upon whether $\hat{\Omega}$ is a tensorial or simplicial domain. Similarly, let N_m^p be the largest integer such that, independently of the type of domain,

$$\mathbb{P}_{N_m^p} \subseteq Q_{N_m}(\hat{\Omega}) . \tag{5.6.4}$$

The maximal polynomial degrees N_m^v and N_m^p are expressed in terms of N_m and are usually different from each other. For instance, if the $\mathbb{Q}_{N_m} - \mathbb{Q}_{N_m-2}$ method is selected on the reference domain, we have $N_m^v = N_m$ and $N_m^p = N_m - 2$.

Next, these spaces generate local spaces $V_{N_m}(\Omega_m)$ and $Q_{N_m}(\Omega_m)$, via the mapping F_m. The conventional way to define them, which is used, e. g., when F_m is affine, is through a componentwise transformation; precisely, one sets

$$V_{N_m}(\Omega_m) = \{\mathbf{v} = (v_i) : v_i = \hat{v}_i \circ F_m^{-1}, \ 1 \leq i \leq d, \\ \text{for some } \hat{\mathbf{v}} = (\hat{v}_i) \in V_{N_m}(\hat{\Omega})\} \tag{5.6.5}$$

and

$$Q_{N_m}(\Omega_m) = \{q : q = \hat{q} \circ F_m^{-1} \text{ for some } \hat{q} \in Q_{N_m}(\hat{\Omega})\} . \tag{5.6.6}$$

However, if F_m is not affine, as in the case of curvilinear elements, the inf-sup condition (see Sect. 3.7.2) for the pair $V_{N_m}(\Omega_m)$, $Q_{N_m}(\Omega_m)$ thus defined may not be guaranteed to follow from the analogous property for the pair of spaces on the reference domain. In this case, a more elaborate definition of $V_{N_m}(\Omega_m)$ is needed. Following Chilton and Suri (2000), let us split $V_{N_m}(\hat{\Omega})$ as

$$V_{N_m}(\hat{\Omega}) = V_{N_m}^0(\hat{\Omega}) \oplus V_{N_m}^B(\hat{\Omega}) ,$$

where $V_{N_m}^0(\hat{\Omega}) = V_{N_m}(\hat{\Omega}) \cap (H_0^1(\hat{\Omega}))^d$ is the subspace of the *internal* velocities, which vanish on $\partial\hat{\Omega}$, whereas $V_{N_m}^B(\hat{\Omega})$ is a supplementary space made of *boundary* velocities. Then, the latter space generates a space $V_{N_m}^B(\Omega_m)$ of boundary velocities in Ω_m via the standard transform as in (5.6.5); this definition allows for an easy glueing of local velocities across element interfaces. On the other hand, the space $V_{N_m}^0(\hat{\Omega})$ generates the space $V_{N_m}^0(\Omega_m)$ of the internal velocities in Ω_m via the Piola transform (see Sect. E.3), i. e., one sets

$$V_{N_m}^0(\Omega_m) = \{\mathbf{v} : \mathbf{v} = J_m^{-1} DF_m \hat{\mathbf{v}} \text{ for some } \hat{\mathbf{v}} \in V_{N_m}^0(\hat{\Omega})\} , \tag{5.6.7}$$

where DF_m is the Jacobian matrix of F_m and $J_m = \det DF_m$. Finally, one sets

$$V_{N_m}(\Omega_m) = V_{N_m}^0(\Omega_m) \oplus V_{N_m}^B(\Omega_m) , \tag{5.6.8}$$

while $Q_{N_m}(\Omega_m)$ is again given by (5.6.6).

Once the local spaces are defined, one introduces the global spaces in a standard way, i. e., by setting

$$V_\delta = \{\mathbf{v} \in V : \mathbf{v}_{|\Omega_m} \in V_{N_m}(\Omega_m) \ \forall \Omega_m \in \mathcal{T}\} \tag{5.6.9}$$

and

$$Q_\delta = \{q \in Q : q_{|\Omega_m} \in Q_{N_m}(\Omega_m) \;\; \forall \Omega_m \in \mathcal{T}\}. \tag{5.6.10}$$

Note that the pressures of Q_δ are not necessarily continuous across the interelement boundaries, whereas the velocities of V_δ are continuous. Nodal or modal bases in V_δ and Q_δ are obtained from bases in each $V_{N_m}(\hat{\Omega})$ and $Q_{N_m}(\hat{\Omega})$; velocity basis functions are glued as in Sect. 5.3.2 to guarantee interelement continuity.

The SEM approximation to (5.6.1) consists of finding $\mathbf{u}_\delta \in V_\delta$, $p_\delta \in Q_\delta$ such that

$$\begin{aligned}
\sum_m \{a_{\Omega_m}(\mathbf{u}_\delta, \mathbf{v}) + b_{\Omega_m}(\mathbf{v}, p_\delta)\} &= \sum_m (\mathbf{g}, \mathbf{v})_{\Omega_m} && \forall \mathbf{v} \in V_\delta, \\
\sum_m b_{\Omega_m}(\mathbf{u}_\delta, q) &= 0 && \forall q \in Q_\delta,
\end{aligned} \tag{5.6.11}$$

where $a_{\Omega_m}(\mathbf{w}, \mathbf{v}) = \nu(\nabla \mathbf{w} \cdot \nabla \mathbf{v})_{\Omega_m}$, $b_{\Omega_m}(\mathbf{v}, q) = -(\nabla \cdot \mathbf{v}, q)_{\Omega_m}$ and $(\mathbf{g}, \mathbf{v})_{\Omega_m} = \int_{\Omega_m} \mathbf{g} \cdot \mathbf{v}$.

As discussed in Sects. 3.5 and 3.7 for the single-domain case, the well-posedness of this problem is essentially related to the absence of spurious pressure modes in Q_δ. We will prove in the next subsection that this occurs if no spurious pressure mode exists on the reference domain for any velocity–pressure pair $V_{N_m}(\hat{\Omega})$, $Q_{N_m}(\hat{\Omega})$ (provided each $V_{N_m}(\hat{\Omega})$ contains all polynomials of degree ≤ 2 and each $Q_{N_m}(\hat{\Omega})$ contains the constants).

Obviously, spurious pressure modes do exist if equal-order velocities and pressures are used, i.e., if the $\mathbb{Q}_N - \mathbb{Q}_N$ method (in tensorial elements) or the $\mathbb{P}_N - \mathbb{P}_N$ method (in simplicial elements) is chosen in all or some of the elements. In such a situation, some device to remove or prevent the onset of spurious pressure modes should complement the discretization. Filtering, regularization or stabilization techniques, mentioned in Sect. 3.5.1 for a single-domain discretization, can be easily adapted to the spectral element framework. In particular, spurious pressure modes can be avoided by properly stabilizing equations (5.6.11): this can be accomplished by using, for instance, *SUPG* or *GLS* (Galerkin least squares) strategies (see, e.g., Quarteroni and Valli (1994) and Gervasio and Saleri (1998)), or by enriching the local velocity spaces by suitable special bubbles (see Canuto, Russo and Van Kemenade (1998)), or, alternatively, by the *interior penalty* method (see Burman, Fernández and Hansbo (2006), Braack and Burman (2006), Braack et al. (2007), Schötzau, Schwab and Toselli (2003)).

The SEM-NI variant is obtained from (5.6.11) by replacing exact integrations by numerical Gaussian quadratures. Thus, the SEM-NI approximation to the Stokes problem (5.6.1) consists of seeking $\mathbf{u}_\delta \in V_\delta$, $p_\delta \in Q_\delta$ satisfying

$$\begin{aligned}
\sum_m \{a_{N_m,\Omega_m}(\mathbf{u}_\delta, \mathbf{v}) + b_{N_m,\Omega_m}(\mathbf{v}, p_\delta)\} &= \sum_m (\mathbf{g}, \mathbf{v})_{N_m,\Omega_m} && \forall \mathbf{v} \in V_\delta, \\
\sum_m b_{N_m,\Omega_m}(\mathbf{u}_\delta, q) &= 0 && \forall q \in Q_\delta,
\end{aligned}$$

$$\tag{5.6.12}$$

where the suffix N_m appended to the bilinear and linear forms denotes the use of numerical integration. In the relevant case in which the $\mathbb{Q}_{N_m} - \mathbb{Q}_{N_m-2}$ method is chosen in each (quadrilateral) element Ω_m, the velocity field can be represented by its values at the $(N_m + 1)^d$ LGL nodes, whereas the pressure can be represented by its values at $(N_m - 1)^d$ internal nodes, which can be either the internal LGL nodes or the LG (Legendre Gauss) nodes. In the latter case the values of the velocity need to be interpolated on the LG grid in the calculation of the form $b_{N_m,\Omega_m}(\mathbf{v}_\delta, q_\delta)$. The reader is referred to Sect. 3.5.1 for more details.

The discrete problems (5.6.11) or (5.6.12) give rise to algebraic systems with the typical saddle-point blockwise form (3.7.29). This time, however, the blocks A, B and B^T are the sum of elemental matrices. Solution strategies (and preconditioners) for this kind of systems will be addressed in Sect. 7.3.

The discretization, by SEM or SEM-NI, of the full Navier–Stokes equations is carried out along the same lines. Strategies for time-discretization are illustrated in Sect. 7.2. As the Reynolds number gets larger and larger, some form of stabilization of the convective term may be required. A simple choice consists of filtering the higher frequency modes of the velocity in each element. Fischer and Mullen (2001) advocate the use of a mild filter of the form $\theta \Pi_{N_m-1} + (1 - \theta)I$, $0 < \theta < 1$, where I is the identity and Π_{N_m-1} is a pseudo-projector, which first interpolates onto an $(N_m - 1)$-order LGL grid and then reinterpolates onto the usual N_m-order grid. We refer to Deville, Fischer and Mund (2002), Sect. 6.6, for further details. If a pressure stabilization device (such as those mentioned above) is incorporated into an equal-order velocity–pressure discretization, it usually provides a cure for the convection instability on the velocity as well. In addition, as documented by Canuto and van Kemenade (1996) for the bubble stabilization, it may significantly enhance the robustness of a continuation method used to handle the nonlinearity.

The pressure spaces considered so far consist of discontinuous functions across the element interfaces. The alternative choice of using continuous pressures is popular in the h-type finite-element community, particularly with the Taylor–Hood element $\mathbb{Q}_2 - \mathbb{Q}_1$ (on quadrilateral meshes). The high-order version is a SEM discretization based on generalized Taylor–Hood spaces $V_{N_m}(\hat{\Omega}) = \mathbb{Q}_{N_m}$ and $Q_{N_m}(\hat{\Omega}) = \mathbb{Q}_{N_m-1}$, with the constraint that pressures (as well as velocities) are continuous throughout the domain. Such a method is free of spurious modes; the inf-sup constant can be shown to be independent of the size of the elements (see, e. g., Brezzi and Fortin (1992)), but there is numerical evidence that it degrades when the polynomial degree in the elements increases (Ainsworth and Coggins (2002)). The latter authors propose a family of modified pairs of velocity and pressure spaces $V_{N_m}(\hat{\Omega})$, $Q_{N_m}(\hat{\Omega})$ with optimal approximation properties and the inf-sup condition bounded from below independently of the mesh size and the polynomial degree.

Finally, we mention that SEM and SEM-NI discretizations of the Navier–Stokes equations in vorticity–streamfunction formulation are studied in Bernardi, Girault and Maday (1992).

5.6.2 Stability and Convergence Analysis

The theoretical analysis of SEM and SEM-NI discretizations of the Stokes equations is based on general results for the approximation of saddle-point problems, presented in Sect. 3.7.2. We refer to (3.7.20) and to the following formulas, where the suffix N now has to be replaced by δ.

For simplicity, we confine ourselves to the SEM analysis, since SEM-NI poses only technical, but not substantial, overheads. Thus, the SEM approximation (5.6.11) fits into the abstract framework (3.7.20), where the (continuous) bilinear forms $a_\delta(\mathbf{u}_\delta, \mathbf{v}_\delta) = a(\mathbf{u}_\delta, \mathbf{v}_\delta)$ and $b_\delta(\mathbf{v}_\delta, q) = b(\mathbf{v}_\delta, q)$, as well as the (continuous) linear form $((\mathbf{g}, \mathbf{v}_\delta)) = (\mathbf{g}, \mathbf{v}_\delta)$, coincide with those appearing in the exact formulation of the Stokes problem (5.6.2).

Since the bilinear form $a(\mathbf{u}, \mathbf{v})$ is coercive on the whole of V, i.e.,

$$a(\mathbf{v}, \mathbf{v}) \geq \nu \|\mathbf{v}\|_V^2 \qquad \text{for all } \mathbf{v} \in V, \tag{5.6.13}$$

the stability of the SEM approximation crucially relies upon the fulfillment of the inf-sup condition (3.7.28).

Assume that each space $V_{N_m}(\hat{\Omega})$ contain $\mathbb{Q}_2$ (if $\hat{\Omega}$ is a tensorial domain) or $\mathbb{P}_2$ (if $\hat{\Omega}$ is a simplicial domain), that each $Q_{N_m}(\hat{\Omega})$ contain the constant functions, and that each pair $V_{N_m}^0(\hat{\Omega})$, $Q_{N_m}(\hat{\Omega})$ satisfy an inf-sup condition of the form

$$\sup_{\hat{\mathbf{v}} \in V_{N_m}^0(\hat{\Omega})} \frac{\int_{\hat{\Omega}} \hat{\nabla} \cdot \hat{\mathbf{v}} \, \hat{q} \, d\hat{\mathbf{x}}}{\|\hat{\mathbf{v}}\|_{(H^1(\hat{\Omega}))^d}} \geq \hat{\beta}_m \|\hat{q}\|_{L^2(\hat{\Omega})} \qquad \text{for all } \hat{q} \in Q_{N_m}(\hat{\Omega}) \cap L_0^2(\hat{\Omega})$$

$$\tag{5.6.14}$$

(with $L_0^2(\hat{\Omega}) = \{\hat{q} \in L^2(\hat{\Omega}) : \int_{\hat{\Omega}} \hat{q} \, d\hat{\mathbf{x}} = 0\}$), for some $\hat{\beta}_m > 0$ possibly depending on N_m. Under mild assumptions on the mappings F_m, one can prove that the pair V_δ, Q_δ defined by (5.6.9)–(5.6.10) satisfies (3.7.28) with

$$\beta_\delta = C \min_m \hat{\beta}_m > 0, \tag{5.6.15}$$

where $C > 0$ is a constant independent of δ. The main steps of the proof are illustrated below.

In the case of a partition composed of tensorial elements, the results of Sect. 3.7.3 imply that $\beta_\delta \sim cN^{(1-d)/2}$ if one of the pairs (3.5.10), (3.5.11), (3.5.14) or (3.5.15) are used in each element with constant $N_m = N$, whereas $\beta_\delta \geq \beta > 0$ independent of δ if one of the pairs (3.5.12) or (3.5.13) are used instead, even with different indices N_m.

The bounds (3.7.30)–(3.7.31) yield the stability estimate

$$\|\mathbf{u}_\delta\|_V + \beta_\delta \|p_\delta\|_Q \leq C \|\mathbf{g}\|_{(L^2(\Omega))^d}, \tag{5.6.16}$$

where $C > 0$ is a constant independent of δ.

As far as the convergence of the SEM approximation is concerned, the error bounds (3.7.32), (3.7.33) and (3.7.37) yield

$$\|\mathbf{u} - \mathbf{u}_\delta\|_V + \beta_\delta \|p - p_\delta\|_Q \leq C_1 \beta_\delta^{-1} \inf_{\mathbf{v}_\delta \in V_\delta} \|\mathbf{u} - \mathbf{u}_\delta\|_V + C_2 \inf_{q_\delta \in Q_\delta} \|p - q_\delta\|_Q .$$
(5.6.17)

Now, let us assume that for each m, $\mathbf{u}_{|\Omega_m} \in (H^{s_m}(\Omega_m))^d$ and $p_{|\Omega_m} \in H^{s_m-1}(\Omega_m)$ for some $s_m > 1$. Furthermore, recalling the definitions (5.6.3)–(5.6.4) of the maximal polynomial degrees N_m^v and N_m^p, we assume that the discretization is such that $\mathbf{u}$ and p can be approximated by elements $R_\delta^v \mathbf{u} \in V_\delta$ and $R_\delta^p p \in Q_\delta$ with approximation errors of the type (5.4.18) (with N_m replaced by N_m^v for the velocity, and N_m replaced by N_m^p, and s_m by $s_m - 1$, for the pressure). Then, (5.6.17) yields the error estimate

$$\|\mathbf{u} - \mathbf{u}_\delta\|_V + \beta_\delta \|p - p_\delta\|_Q$$

$$\leq C_3 \beta_\delta^{-1} \left\{ \sum_m h_m^{2\min(N_m^v, s_m-1)} (N_m^v)^{2(1-s_m)} \|\mathbf{u}\|^2_{(H^{s_m}(\Omega_m))^d} \right\}^{1/2}$$

$$+ C_4 \left\{ \sum_m h_m^{2\min(N_m^p+1, s_m-1)} (N_m^p)^{2(1-s_m)} \|p\|^2_{H^{s_m-1}(\Omega_m)} \right\}^{1/2} .$$
(5.6.18)

The negative presence of β_δ^{-1} on the right-hand side can be alleviated by interpolating between (5.6.17) and the uniform bound

$$\|\mathbf{u} - \mathbf{u}_\delta\|_V + \beta_\delta \|p - p_\delta\|_Q \leq C (\|\mathbf{u}\|_V + \|p\|_Q) ,$$
(5.6.19)

which follows from (3.7.32) and (3.7.33) by choosing $\mathbf{v}_\delta = \mathbf{0}$ and $q_\delta = 0$. See, e. g., Chilton and Suri (2001) for details. As an example, in the case in which $s_m \sim s$, $h_m \sim h$, $N_m \sim N$ for all m, and the $\mathbb{Q}_N - \mathbb{Q}_{N-2}$ method is used in each element (supposed of tensorial nature), one can prove that for all $\varepsilon > 0$ there exist constants $C_1(\varepsilon)$ and C_2 independent of N and h such that

$$\|\mathbf{u} - \mathbf{u}_\delta\|_V + N^{(1-d)/2} \|p - p_\delta\|_Q \leq C_1(\varepsilon) N^\varepsilon h^{\min(N,s-1)} N^{1-s} \|\mathbf{u}\|_{(H^s(\Omega))^d}$$

$$+ C_2 h^{\min(N-1,s-1)} N^{1-s} \|p\|_{H^{s-1}(\Omega)} .$$
(5.6.20)

Note that, in addition to the non-optimal dependence of the error upon N due to the non-uniform inf-sup condition, the dependence upon h is also non-optimal. Indeed, the gap of 2 between the polynomial degrees of velocities and pressures is larger than the gap of 1 between the orders of Sobolev regularity of these variables.

On the other hand, if the $\mathbb{Q}_N - \mathbb{P}_{N-1}$ method is used instead, one obtains the fully optimal error estimate

$$\|\mathbf{u} - \mathbf{u}_\delta\|_V + \|p - p_\delta\|_Q \leq C h^{\min(N,s-1)} N^{1-s} \left(\|\mathbf{u}\|_{(H^s(\Omega))^d} + \|p\|_{H^{s-1}(\Omega)} \right) ,$$
(5.6.21)

with C independent of N and h.

Finally, we mention that the theoretical analysis of SEM or SEM-NI discretizations of the full Navier–Stokes equations is based on the previous analysis for the Stokes problem and on general results for nonlinear problems, as detailed in Sect. 3.7 for the single-domain case.

Proof of the Global Inf-sup Condition. We provide the essential steps for deriving the inf-sup condition (3.7.28), with constant β_δ satisfying (5.6.15), from the inf-sup condition(s) (5.6.14) on the reference element(s) $\hat{\Omega}$.

We assume that the partitions $\mathcal{T}$ satisfy the usual hypotheses of shape regularity as made in Sect. 5.4.3. This implies that $\|q\|_{L^2(\Omega_m)} \sim ch_m^{d/2}\|\hat{q}\|_{L^2(\hat{\Omega})}$ if $q = \hat{q} \circ F_m^{-1}$. Furthermore, we assume that the mappings F_m are such that $\|\mathbf{v}\|_{(H^1(\Omega_m))^d} \sim ch_m^{-d/2}\|\hat{\mathbf{v}}\|_{(H^1(\hat{\Omega}))^d}$ if the vector fields $\mathbf{v}$ and $\hat{\mathbf{v}}$ are related by the Piola transform $\mathbf{v} = J_m^{-1}DF_m\hat{\mathbf{v}}$ (see (E.3.1)). This is true if the mapping F_m is affine or quadratic, or if it is a perturbation of such a map for $h_m \to 0$. Then, recalling that

$$\int_{\Omega_m} \nabla \cdot \mathbf{v}\, q \, d\mathbf{x} = \int_{\hat{\Omega}} \hat{\nabla} \cdot \hat{\mathbf{v}}\, \hat{q} \, d\hat{\mathbf{x}}$$

by (E.3.4), and setting $Q_{N_m}^0(\Omega_m) = \{q = \hat{q} \circ F_m^{-1} : \hat{q} \in Q_{N_m}(\hat{\Omega}) \cap L_0^2(\hat{\Omega})\}$, we easily deduce from (5.6.14) the local inf-sup conditions

$$\sup_{\mathbf{v} \in V_{N_m}^0(\Omega_m)} \frac{\int_{\Omega_m} \nabla \cdot \mathbf{v}\, q \, d\mathbf{x}}{\|\mathbf{v}\|_{(H^1(\Omega_m))^d}} \geq c\hat{\beta}_m\|q\|_{L^2(\Omega_m)} \qquad \text{for all } q \in Q_{N_m}^0(\Omega_m),$$

(5.6.22)

for some constant c independent of h_m and N_m.

Next, we apply a local-to-global argument, often referred to as the Boland–Nicolaides (1983) argument. Given any $q \in Q_\delta$, let us split it as $q = \bar{q} + q^0$, where $\bar{q}$ is the piecewise constant function, which in each element Ω_m takes the constant value $\bar{q}_m$ satisfying

$$\int_{\Omega_m} \bar{q}_m J_m^{-1} \, d\mathbf{x} = \int_{\Omega_m} q J_m^{-1} \, d\mathbf{x}.$$

It is easily seen that $\|q\|_{L^2(\Omega)}^2 \sim c\left(\|\bar{q}\|_{L^2(\Omega)}^2 + \sum_m \|q^0\|_{L^2(\Omega_m)}^2\right)$ and that $q_{|\Omega_m}^0 \in Q_{N_m}^0(\Omega_m)$ for all m. Since the pair formed by the space of continuous, piecewise (images of) quadratic velocities and the space of piecewise constant pressures is uniformly inf-sup stable (see, e. g., Brezzi and Fortin (1992)), there exists $\bar{\mathbf{v}} \in V_\delta$ such that

$$\int_\Omega \nabla \cdot \bar{\mathbf{v}}\, \bar{q} \, d\mathbf{x} \geq \bar{\beta}\|\bar{q}\|_Q^2 \qquad \text{with} \quad \|\bar{\mathbf{v}}\|_V \leq \bar{C}\|\bar{q}\|_Q,$$

for suitable constants $\bar{\beta}$ and $\bar{C}$ independent of δ. Furthermore, for any m, by (5.6.22) there exists $\mathbf{v}_m^0 \in V_{N_m}^0(\Omega_m)$ such that

$$\int_{\Omega_m} \nabla \cdot \mathbf{v}_m^0\, q^0 \, d\mathbf{x} \geq c\hat{\beta}_m\|q^0\|_{L^2(\Omega_m)}^2 \qquad \text{with} \quad \|\mathbf{v}_m^0\|_{(H^1(\Omega_m))^d} \leq C^0\|q^0\|_{L^2(\Omega_m)},$$

with C^0 independent of m. Let $\mathbf{v}^0 \in V_\delta$ be defined by $\mathbf{v}^0_{|\Omega_m} = \mathbf{v}^0_m$ for all m. Then, we set $\mathbf{v} = \lambda \bar{\mathbf{v}} + \mathbf{v}^0$ with $0 < \lambda < 1$. It is not difficult to check that $\|\mathbf{v}\|_V \le C'\|q\|_Q$ and there exist values of λ for which

$$\int_\Omega \nabla \cdot \mathbf{v} \, q \, d\mathbf{x} \ge C'' \min_m \hat{\beta}_m \|q\|_Q^2$$

for suitable constants C' and C'' independent of δ. This completes our proof.

5.6.3 Numerical Results

We end this section by showing a few numerical results obtained by solving the Stokes or the Navier–Stokes equations by the SEM-NI method.

Figure 5.21 reports the error curves obtained for the Kovasznay solution of the steady Stokes equations (see (3.7.44)–(3.7.45)). The computational domain is $\Omega = (-1,3) \times (0.5, 2.5)$. On the boundary of Ω we impose a Dirichlet condition. The numerical solution is obtained by the SEM-NI method with polynomials $\mathbb{Q}_N - \mathbb{Q}_{N-2}$ on staggered grids, using 8 or 32 elements of equal size. The numerical solution exhibits spectral accuracy for both problems.

The next test case concerns the motion of a regularized lid-driven cavity flow inside a plane square domain $\Omega = (0,1)^2$ with nonzero tangential velocity prescribed on the top boundary: $\mathbf{u}_\infty = (16x^2(1-x)^2, 0)^T$. We define the Reynolds number as $\text{Re} = D\|\mathbf{u}_\infty\|_\infty/\nu$, where $D = 1$ is the measure of the side of Ω, and $\|\mathbf{u}_\infty\|_\infty = 1$. On the vertical sides and on the bottom horizontal side a no-slip boundary condition is imposed. The time-dependent Navier–Stokes equations have been solved until the steady state was reached using the stabilized GLS method with equal-order polynomials for velocity and pressure (Gervasio and Saleri (1998), see above) combined with the Euler semi-implicit scheme for time advancement (3.2.24). The discretization parameters are $N_x = N_y = 8$, 6×6 squared spectral elements and a time-step

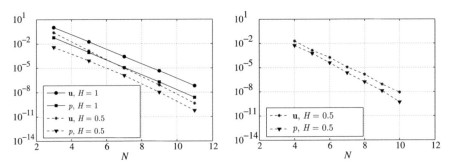

Fig. 5.21. Absolute errors in the H^1-norm for the velocity and in the L^2-norm for the pressure on the SEM-NI approximation to the Stokes (*left*) and Navier–Stokes (*right*) problems with the Kovasznay solution. $H = 1$ means 4×2 square elements, $H = 0.5$ means 8×4 square elements

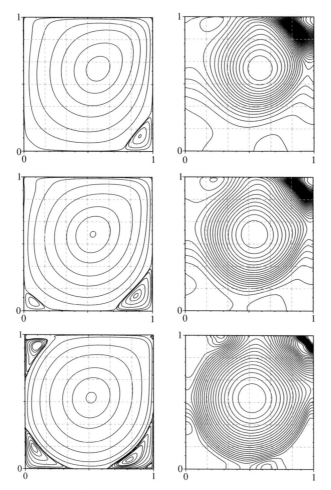

Fig. 5.22. Streamlines (*left*) and pressure contours (*right*) for the driven cavity flow with Re = 400 (*top*), Re = 1000 (*middle*), Re = 10000 (*bottom*)

$\Delta t = 10^{-1}$. Fig. 5.22 shows the streamlines and pressure contours for different Reynolds numbers. The excellent quality of the discrete solution is clearly documented. Contour lines pass smoothly across the domain decomposition (shown in the background). Furthermore, the built-in stabilization prevents the onset of spurious pressure modes.

Finally, we consider a laminar flow around a cylinder. The computational domain is the rectangle $\Omega = (0, 2.2) \times (0, 0.41)$ with a circular obstacle centered on $(0.2, 0.2)$ with diameter $D = 0.1$. Homogeneous Dirichlet boundary conditions are enforced on the horizontal sides and on the obstacle boundary, while an inflow Dirichlet condition is prescribed on the left vertical side: $\mathbf{u}(x, y) = [6y(0.41 - y)/0.1681, 0]^T$. The viscosity is $\nu = 10^{-3}$, so that the

Reynolds number is Re $= 100$. The outflow condition on the right vertical side is a free-stress condition $(\mathbf{n} \cdot \nabla)\mathbf{u} - p\mathbf{n} = \mathbf{0}$. The time interval for the simulation is $[0, 6]$. This is a nontrivial test case, considered in Schäfer and Turek (1996).

Let S denote the boundary of the circular obstacle, $\mathbf{n} = (n_x, n_y)$ the outward unit normal to S, u_τ the tangential component of the velocity field $\mathbf{u}$ on S, ρ the density of the fluid. The drag and lift forces are defined as

$$F_D(t) = \int_S \left[\rho\nu \frac{\partial u_\tau(t)}{\partial n} n_y - p(t)n_x \right] \mathrm{d}S \,, \quad F_L(t) = -\int_S \left[\rho\nu \frac{\partial u_\tau(t)}{\partial n} n_x + p(t)n_y \right] \mathrm{d}S \,,$$

whereas the drag and lift coefficients are

$$c_D(t) = \frac{2F_D(t)}{\rho\overline{U}^2 D} \,, \qquad c_L(t) = \frac{2F_L(t)}{\rho\overline{U}^2 D} \,,$$

where $\overline{U} = 1$ is the mean value of the velocity. After a transient, the periodic regime is reached and an interval of periodicity $[t_0, t_0 + 1/f]$ is defined based on the periodicity of the lift coefficient. Here t_0 is the time corresponding to the flow state with maximum lift coefficient and $f = f(c_L)$ is the frequency of the lift coefficient. We set $c_{L\,\max} = c_L(t_0) = \max_{[t_0, t_0+1/f]} c_L(t)$ and $c_{D\,\max} = \max_{[t_0, t_0+1/f]} c_D(t)$.

A SEM discretization based on stabilized $\mathbb{Q}_N - \mathbb{Q}_N$ elements is used with 116 quadrilateral elements. The degree N is varied between $N = 4$ and $N = 10$. In all cases, time-discretization is carried out using the BDF3 scheme with $\Delta t = 2.5 \cdot 10^{-3}$, plus an extrapolation formula of order three to linearize the convective term (see Chap. 7 and Table 7.1).

The computed values of the drag and lift coefficients, as well as the Strouhal number St $= fD/\overline{U}$ (the nondimensional frequency) and the difference Δp between the pressure ahead of and behind the cylinder (measured on the horizontal diameter) at half the period, are reported in Table 5.3. For

Table 5.3. Laminar flow around a cylinder. Comparison among different discretizations for different values of the polynomial degree N. The number of spectral elements is 116 for any discretization

N	d.o.f.	$c_{D\,\max}$	$c_{L\,\max}$	St	$\Delta p(t_0 + 1/(2f))$
4	5880	3.3598	1.2163	0.2899	2.5978
5	9090	3.1989	0.9813	0.2985	2.5161
6	12996	3.2436	1.0208	0.2985	2.5122
7	17598	3.2657	1.0213	0.2985	2.4979
8	22896	3.2557	1.0021	0.2985	2.4895
9	28890	3.2428	0.9916	0.2985	2.4837
10	35580	3.2335	0.9869	0.3030	2.4859
lower bound		3.2200	0.9900	0.2950	2.4600
upper bound		3.2400	1.0100	0.3050	2.5000

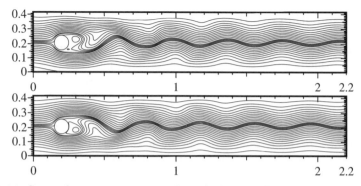

Fig. 5.23. Streamfunctions past a circular cylinder at different times: *top t* = 5.63, *bottom t* = 5.795

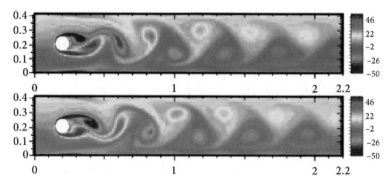

Fig. 5.24. Vorticity contours at different times: *top t* = 5.63, *bottom t* = 5.795

a comparison, shown are also the tight upper and lower bounds for the corresponding exact quantities, as carefully estimated in Schäfer and Turek (1996). The good quality of the SEM approximation is apparent, also considering the small number of degrees of freedom (d.o.f.) involved in the computations.

Figures 5.23 and 5.24 show the streamlines and vorticity contour lines obtained with $N = 8$.

5.7 The Mortar Element Method (MEM)

Bernardi, Maday and Patera (1994) introduced the MEM to generalize the SEM to geometrically nonconforming partitions, to subdomains with different resolutions (polynomial degrees) on subdomain interfaces, and also to allow the coupling of spectral element methods with other methods, such as, e. g., the *h*-version of the finite-element method. Its generality, however, goes beyond these two specific examples.

5.7.1 Formulation of MEM

As in Sect. 5.3.1, the computational domain Ω is a bounded open region of $\mathbb{R}^d$ ($d = 2, 3$), which can be decomposed into a partition $\mathcal{T} = \{\Omega_m\}$ of M open elements, so that $\overline{\Omega} = \bigcup_m \overline{\Omega}_m$, with $\Omega_k \cap \Omega_m = \emptyset$. Again, each Ω_k is the image of a reference element (a square or triangle in 2D, a cube, tetrahedron, prism or pyramid in 3D) through a mapping like those considered in Sect. 5.3.1. Moreover, if Ω_k is a boundary element, it can contribute to $\partial\Omega$ either by a whole face, or by a whole edge or by a vertex. However, the situation here can be more general than the one considered in Sect. 5.3.1 for SEM. In fact, two neighboring elements, $\overline{\Omega}_k$ and $\overline{\Omega}_m$, can share a vertex, a whole face or part of it, a whole edge or part of it, thus admitting geometrically nonconforming partitions, as in Fig. 5.25 or in Fig. 5.26.

Let us consider, for the sake of simplicity, the Poisson problem with homogeneous Dirichlet conditions (5.4.19), although no essential difference would arise if the more general problem (5.3.9) were considered. The idea is to approximate its weak form,

$$u \in H_0^1(\Omega): \quad \int_\Omega \nabla u \cdot \nabla v = \int_\Omega fv \quad \forall v \in H_0^1(\Omega), \quad (5.7.1)$$

by the discrete problem,

$$u_\delta \in V_\delta : \quad \sum_{m=1}^M (\nabla u_\delta, \nabla v_\delta)_{\Omega_m} = \sum_{m=1}^M (f, v_\delta)_{\Omega_m} \quad \forall \, v_\delta \in V_\delta, \quad (5.7.2)$$

where the space V_δ is a finite-dimensional space that approximates $H_0^1(\Omega)$ without necessarily being contained in $H^1(\Omega)$ or in $C^0(\overline{\Omega})$. More precisely,

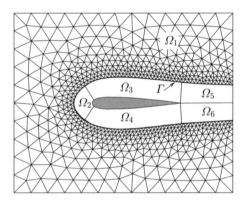

Fig. 5.25. Computational domain around a two-dimensional airfoil, partitioned into subregions that are either spectral elements (like Ω_m for $m = 2, \ldots, 6$) or subdomains triangulated with finite-elements (like Ω_1)

V_δ is a subspace of the space

$$Y_\delta = \{v_\delta \in L^2(\Omega) \,|\, v_{\delta|\Omega_m} \in Y_{m,\delta}, \ \forall \Omega_m \in \mathcal{T}\}, \tag{5.7.3}$$

where, for each m, $Y_{m,\delta}$ is a finite-dimensional subspace of $H^1(\Omega_m)$ that can be chosen according to several criteria. If Ω_m is a spectral element, then $Y_{m,\delta} = V_{N_m}(\Omega_m)$; see (5.3.3). In other instances a coupled finite-element/spectral element approach can be conveniently pursued. Then, one (or several) Ω_m could be further decomposed into elements forming a regular partition $\mathcal{T}_m = \{K_{m,i}\}$; in this case, $Y_{m,\delta}$ would be a finite-element space, e. g.,

$$Y_{m,\delta} = \{v_\delta \in C^0(\overline{\Omega}_m) \,|\, v_{\delta|K_{m,i}} \in \mathbb{P}_r(K_{m,i}) \,\forall i\},$$

where each $K_{m,i}$ is a triangle or a tetrahedron, and $\mathbb{P}_r(K_{m,i})$ is the space of all polynomials of global degree $\leq r$ (here $r \geq 1$ is the finite-element degree, and it is generally small). This is the case illustrated in Fig. 5.25, where we use several spectral elements, $\Omega_2, \ldots, \Omega_6$, in the near field around a two-dimensional airfoil (to capture better the boundary layer and the wake) and a single domain, Ω_1, made of finite-elements for the exterior field. The interface Γ is the one where the discontinuous solution undergoes a mortar condition.

While no requirement of compatibility is made for the restrictions of the functions of Y_δ on the element interfaces, the space V_δ will be made up of functions belonging to Y_δ that satisfy some kind of matching across the interfaces. Note that one or more subdomains could be, in turn, composed of patches of spectral elements across whose boundaries the functions are indeed continuous.

Let us detail the construction of V_δ in the pure spectral element case (see, e. g., Bernardi, Maday and Rapetti (2005) for the coupling spectral elements/finite-elements). Again, the index δ stands for the collection of the diameters h_m of the elements and of the degrees N_m of the polynomials therein.

To start with, assume for simplicity that Ω is a rectangle partitioned into two rectangles Ω_1 and Ω_2 sharing a common edge, Γ. This is a simple, geometrically conforming partition. If $v_\delta \in V_\delta$, let $v_\delta^{(1)} \in Y_{1,\delta} = V_{N_1}(\Omega_1)$ and $v_\delta^{(2)} \in Y_{2,\delta} = V_{N_2}(\Omega_2)$, denote its restrictions to Ω_1 and Ω_2, respectively. For one fixed index $m = 1$ or $m = 2$ (chosen independently of v_δ), the following integral matching conditions should be satisfied:

$$\int_\Gamma \left(v_\delta^{(1)} - v_\delta^{(2)}\right) \mu_\delta^{(m)} = 0 \qquad \forall\, \mu_\delta^{(m)} \in \tilde{\Lambda}_\delta^{(m)}, \tag{5.7.4}$$

where $\tilde{\Lambda}_\delta^{(m)}$ denotes a space of interface functions whose dimension is the same as that of the space of the restriction to Γ of the functions of $Y_{m,\delta}$.

One possibility (which is suitable only in the current case in which Γ is perpendicular to the boundary) is to use as test functions in (5.7.4) the traces on Γ of the functions of $V_{N_m}(\Omega_m)$. Thus, $\mu_\delta^{(m)}$ has the same polynomial degree as $v_\delta^{(m)}$ and vanishes at the endpoints of Γ (which lie on $\partial\Omega$). A more general choice consists of taking as $\mu_\delta^{(m)}$ polynomials without constraint on Γ but with two degrees less than that of $v_\delta^{(m)}$; thus, we set $\tilde{\Lambda}_\delta^{(m)} = \mathbb{P}_{N_m-2}(\Gamma)$.

Taking $m = 2$ in (5.7.4) amounts to letting Ω_1 play the role of *master* and Ω_2 that of *slave*. Then, (5.7.4) has to be intended as the way of generating the value of $v_{\delta|\Gamma}^{(2)}$ once $v_{\delta|\Gamma}^{(1)}$ is available. The alternative way, i. e., taking $m = 1$ in (5.7.4), is also admissible. In general, if $N_1 \neq N_2$, the method will produce different solutions, depending upon the choice of index m made in (5.7.4) (whereas, if $N_1 = N_2$, condition (5.7.4) for whatever m is equivalent to the continuity of v_δ across Γ, yielding $V_\delta \subset H_0^1(\Omega)$, i. e., the SEM).

The mathematical rationale behind the choice of the matching condition (5.7.4) (rather than a more "natural" condition of pointwise continuity at a suitable set of grid nodes on Γ) becomes clear from the convergence analysis for problem (5.7.2); see Sect. 5.7.3.

The construction of the mortar approximation in the case of a general partition of the domain Ω into an arbitrary number of elements (or subdomains), which may not be geometrically conforming, is more involved and requires special attention. Let us denote by $\mathcal{S}$ the so-called *skeleton* of the domain partition, that is, the union of all subdomain interfaces. We can represent $\mathcal{S}$ as the union of elementary components that are called *mortars*. Precisely,

$$\mathcal{S} = \bigcup_{j=1}^{J_0} \overline{\gamma}_j, \quad \text{with} \quad \gamma_j \cap \gamma_i = \emptyset \quad \text{if } j \neq i,$$

where each *mortar* γ_j is a whole edge in the case $d = 2$, or face in the case $d = 3$, of a specific element, say $\Omega_{m(j)}$. Thus, denoting by $\Gamma_{m,i}$ the edges or faces of each subdomain Ω_m, the mortar γ_j can be identified with one specific edge or side $\Gamma_{m(j),i(j)}$. For instance, in the case of Fig. 5.26, $j = 1, \ldots, 4$, and we can set $\gamma_1 = \Gamma_{1,4}$, $\gamma_2 = \Gamma_{4,2}$, $\gamma_3 = \Gamma_{3,1}$, $\gamma_4 = \Gamma_{5,1}$. Thus $\gamma_1 \subset \overline{\Omega}_1$, $\gamma_2 \subset \overline{\Omega}_4$, $\gamma_3 \subset \overline{\Omega}_2$ and $\gamma_4 \subset \overline{\Omega}_5$. Alternative choices could have been made as well, for instance, $\gamma_1 = \Gamma_{2,2}$, $\gamma_2 = \Gamma_{4,2}$, $\gamma_3 = \Gamma_{2,3}$ and $\gamma_4 = \Gamma_{3,3}$, or $\gamma_1 = \Gamma_{1,4}$, $\gamma_2 = \Gamma_{2,4}$, $\gamma_3 = \Gamma_{2,3}$, $\gamma_4 = \Gamma_{3,4}$ and $\gamma_5 = \Gamma_{5,1}$.

On each subdomain Ω_m we look for a discrete solution that is (the image of) a polynomial of degree $N_m \geq 2$, that is, we set $Y_{m,\delta} = V_{N_m}(\Omega_m)$. We indicate by $\Lambda_\delta^{(m,i)} = V_{N_m}(\Gamma_{m,i})$ the space of the traces of functions of $Y_{m,\delta}$ on $\Gamma_{m,i}$; in the case of rectangular elements, these are functions of $\mathbb{P}_{N_m}(\Gamma_{m,i})$. At last, we introduce the space $\tilde{\Lambda}_\delta^{(m,i)} = V_{N_m-2}(\Gamma_{m,i})$.

We can now define the discrete space V_δ as the space of functions v_δ such that

i) their restrictions $v_\delta^{(m)}$ to each Ω_m belong to $Y_{m,\delta}$ for $m = 1, \ldots, M$;

ii) they vanish on $\partial\Omega$;

iii) they satisfy the following matching conditions (also called *mortar* conditions): Let φ be the *mortar function* associated with v_δ, i. e., the function that on each $\gamma_j = \Gamma_{m(j),i(j)}$ coincides with the restriction of $v_\delta^{(m)}$

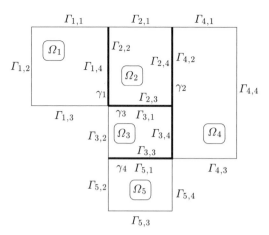

Fig. 5.26. Partition into five subdomains ($M = 5$): the skeleton is made of four mortars ($J_0 = 4$)

to γ_j; then, for all indices (m, i) such that $\Gamma_{m,i}$ is contained in $\mathcal{S}$ but $(m, i) \neq (m(j), i(j))$ for any j, we require that

$$\int_{\Gamma_{m,i}} (v_\delta^{(m)} - \varphi)(s)\psi(s)\, ds = 0 \quad \forall \psi \in \tilde{\Lambda}_\delta^{(m,i)} . \tag{5.7.5}$$

The latter conditions are of integral type. Note that V_δ is a subspace of the space $H_0^1(\Omega)$ (to which the exact solution belongs) if and only if all the polynomial degrees N_m are equal and if all the endpoints (in 2D) or edges (in 3D) of the $\Gamma_{m,i}$ are sitting on $\partial\Omega$.

Once the discrete space V_δ has been defined, the differential problem is approximated by (5.7.2). This is the MEM. More efficiently, the discrete problem can be written in a MEM-NI form, where exact integrals are replaced by their LGL approximations using $(N_m + 1)^d$ points in each subdomain Ω_m (assumed of tensorial type). This yields the problem

$$u_\delta \in V_\delta : \quad \sum_m (\nabla u_\delta, \nabla v_\delta)_{N_m, \Omega_m} = \sum_m (f, v_\delta)_{N_m, \Omega_m} , \tag{5.7.6}$$

where, as usual, $(\cdot, \cdot)_{N_m, \Omega_m}$ represents the analog in Ω_m of the Legendre Gauss–Lobatto inner product on the parent element $\hat{\Omega} = (-1, 1)^d$ using $N_m + 1$ Gauss–Lobatto nodes in each direction. One can also use a quadrature formula to compute the integrals in (5.7.5). However, the formula based on the standard LGL nodes would lead to matching conditions of pointwise (interpolation) type that are not optimal. Alternative quadratures should be used, see Maday, Rapetti and Wohlmuth (2002).

5.7.2 Algebraic Aspects of MEM

There are two techniques to convert the mortar scheme (5.7.6) into an algebraic system: The construction of a basis in the constrained space V_δ and the introduction of a Lagrange multiplier to enforce the matching conditions. Let us discuss both approaches.

For simplicity of exposition, let us first consider again the case of two subdomains. To generate a basis for the finite-dimensional space V_δ, we can proceed as follows. For $m = 1, 2$, let us denote by $\mathcal{N}_m$ the set of LGL nodes in the interior of Ω_m, and by $\mathcal{N}_\Gamma^{(m)}$ the set of nodes on Γ not sitting on $\partial\Omega$, whose cardinality will be indicated by K_m and $K_\Gamma^{(m)}$, respectively. Note that, in general, $\mathcal{N}_\Gamma^{(1)}$ and $\mathcal{N}_\Gamma^{(2)}$ can be totally unrelated.

Now, denote by $\varphi_k^{(1)}$, $k = 1, \dots, K_1$, the characteristic Lagrange polynomials in Ω_1 associated with the nodes of $\mathcal{N}_1$ (see Sect. 8.7.3); since they vanish on Γ, they can be extended by 0 in $\overline{\Omega}_2$. These extended functions are denoted by $\widetilde{\varphi}_k^{(1)}$, and can be taken as a first set of basis functions for V_δ.

Symmetrically, we can generate as many basis functions for V_δ as the number of nodes of $\mathcal{N}_2$ by extending by 0 in $\overline{\Omega}_1$ the characteristic Lagrange polynomials associated with these nodes. These new functions are denoted by $\widetilde{\varphi}_k^{(2)}$, $k = 1, \dots, K_2$.

Finally, always supposing that Ω_1 is the master domain and Ω_2 its slave, for every characteristic Lagrange polynomial $\varphi_{k,\Gamma}^{(1)}$ in $\overline{\Omega}_1$, $k = 1, \dots, K_\Gamma^{(1)}$, associated with a node in $\mathcal{N}_\Gamma^{(1)}$, we obtain a basis function $\widetilde{\varphi}_{k,\Gamma}$ as follows:

$$\widetilde{\varphi}_{k,\Gamma} = \begin{cases} \varphi_{k,\Gamma}^{(1)} & \text{in } \overline{\Omega}_1, \\ \widetilde{\varphi}_{k,\Gamma}^{(2)} & \text{in } \overline{\Omega}_2, \end{cases}$$

where

$$\widetilde{\varphi}_{k,\Gamma}^{(2)} = \sum_{j=1}^{K_\Gamma^{(2)}} \xi_j^{(k)} \varphi_{j,\Gamma}^{(2)};$$

$\varphi_{j,\Gamma}^{(2)}$ are the characteristic Lagrange polynomials in $\overline{\Omega}_2$ associated with the nodes of $\mathcal{N}_\Gamma^{(2)}$, and $\xi_j^{(k)}$ are unknown coefficients that are determined through the fulfillment of the matching equations (5.7.4). Precisely, if $\tilde{\Lambda}_\delta^{(2)}$ is the space of the traces of functions in $V_{N_2}(\Omega_2)$, they must satisfy

$$\int_\Gamma \left(\sum_{j=1}^{K_\Gamma^{(2)}} \xi_j^{(k)} \varphi_{j,\Gamma}^{(2)} - \varphi_{k,\Gamma}^{(1)} \right) \varphi_{l,\Gamma}^{(2)} = 0 \qquad \forall\, l = 1, \dots, K_\Gamma^{(2)}. \tag{5.7.7}$$

The system is uniquely solvable as its matrix is the mass matrix whose elements are $m_{lj} = (\varphi_{l,\Gamma}^{(2)}, \varphi_{j,\Gamma}^{(2)})$. If the projection is rather taken on the space of lower degree $\mathbb{P}_{N-2}$, then an inf-sup condition holds for the bilinear form of the scalar product between functions of $\mathbb{P}_N^0$ and $\mathbb{P}_{N-2}$. This guarantees the solvability of system (5.7.7). A basis for V_δ is therefore provided by the set of all functions $\tilde{\varphi}_k^{(1)}$, $k = 1, \ldots, K_1$, $\tilde{\varphi}_k^{(2)}$, $k = 1, \ldots, K_2$, and $\tilde{\varphi}_{k,\Gamma}$, $k = 1, \ldots, K_\Gamma^{(1)}$.

In the mortar method the interface matching is achieved through an L^2-interface projection, or, equivalently, by equating moments up to a fixed order, thus involving computation of interface integrals. In particular, from (5.7.4) we have two different kinds of integrals to evaluate (take, for instance, $m = 2$):

$$I_{12} := \int_\Gamma v_\delta^{(1)} \mu_\delta^{(2)}, \quad \text{and} \quad I_{22} := \int_\Gamma v_\delta^{(2)} \mu_\delta^{(2)}.$$

The computation of I_{22} raises no special difficulties because both functions $v_\delta^{(2)}$ and $\mu_\delta^{(2)}$ live on the same mesh, the one inherited from Ω_2. On the contrary, $v_\delta^{(1)}$ and $\mu_\delta^{(2)}$ are functions defined on different domains, and the computation of integrals like I_{12} requires proper quadrature rules. This process needs to be done with special care, especially for three-dimensional problems, for which subdomain interfaces are made up of faces, edges and vertices; otherwise the overall accuracy of the mortar approximation may be compromised.

The discrete problem (5.7.2) can also be reformulated as a saddle-point problem of the following form (see, e.g., Ben Belgacem (1999)). One looks for $u_\delta \in Y_\delta$ and $\lambda_\delta \in \tilde{\Lambda}_\delta^{(2)}$ satisfying

$$a(u_\delta, v_\delta) + b(v_\delta, \lambda_\delta) = \sum_{m=1}^2 \left(f, v_\delta^{(m)} \right)_{\Omega_m} \qquad \forall\, v_\delta \in Y_\delta\,,$$

$$b(u_\delta, \mu_\delta) = 0 \qquad \forall\, \mu_\delta \in \tilde{\Lambda}_\delta^{(2)}\,,$$

where

$$a(w_\delta, v_\delta) = \sum_{m=1}^2 \int_{\Omega_m} \nabla w_\delta^{(m)} \cdot \nabla v_\delta^{(m)}, \qquad b(v_\delta, \mu_\delta) = \int_\Gamma (v_\delta^{(1)} - v_\delta^{(2)}) \mu_\delta\,.$$

In this system, λ_δ plays the role of the Lagrange multiplier associated with the "constraint" (5.7.4).

Denoting by φ_j, $j = 1, \ldots, K$, where $K = (K_1 + K_\Gamma^{(1)}) + (K_2 + K_\Gamma^{(2)})$, a basis of Y_δ and by ψ_l, $l = 1, \ldots, K_\Gamma^{(2)}$, a basis of $\tilde{\Lambda}_\delta^{(2)}$, we introduce the matrices A, B whose entries are

$$A_{sj} = a(\varphi_j, \varphi_s)\,, \quad B_{ls} = b(\varphi_s, \psi_l)\,.$$

Defining by $\mathbf{u}$ and $\boldsymbol{\lambda}$ the vectors of the nodal values of u_δ and λ_δ, respectively, and by $\mathbf{f}$ the vector whose components are given by $\sum_{m=1}^2 (f, \varphi_s^{(m)})_{\Omega_m}$, $s =$

$1, \ldots, K$, we have the linear system

$$\begin{pmatrix} A & B^T \\ B & 0 \end{pmatrix} \begin{pmatrix} \mathbf{u} \\ \boldsymbol{\lambda} \end{pmatrix} = \begin{pmatrix} \mathbf{f} \\ \mathbf{0} \end{pmatrix} .$$

The matrix A is block-diagonal (with one block per subdomain Ω_m), each block corresponding to a problem for the Laplace operator with a Dirichlet boundary condition on $\partial\Omega_m \cap \partial\Omega$ and a Neumann boundary condition on $\partial\Omega_m \backslash \partial\Omega$.

After elimination of the degrees of freedom internal to the subdomains, the method leads to the reduced linear system (still of a saddle-point type)

$$\begin{pmatrix} S & C^T \\ C & 0 \end{pmatrix} \begin{pmatrix} \mathbf{u}_\Gamma \\ \boldsymbol{\lambda} \end{pmatrix} = \begin{pmatrix} \mathbf{g}_\Gamma \\ \mathbf{0} \end{pmatrix} ,$$

where the matrix S is block-diagonal, C is a so-called jump operator, $\mathbf{u}_\Gamma$ is the set of all nodal values at subdomain interfaces, and $\mathbf{g}_\Gamma$ is a suitable right-hand side.

An inf-sup condition that guarantees the nonsingularity of this block matrix is automatically satisfied owing to the special choice of the function spaces.

This system can be regarded as a Schur complement system to a nonconforming approximation. (Schur complement systems for domain decomposition methods will be introduced later in Sect. 6.4.3. In particular, see (6.4.16) for the case of two subdomains.) However, in the case at hand $\boldsymbol{\lambda}$ plays the role of a Lagrange multiplier (a flux) associated with the constraint of matching at subdomain interfaces. In fact, the m-th block of S is the analog of Σ_m (see (6.4.22)), and corresponds to a discretized Steklov–Poincaré operator on the subdomain Ω_m.

This strategy of eliminating internal unknowns in favor of interface unknowns will be systematically pursued in Sect. 6.4.

For a discussion of the implementation of spectral mortar element methods, see, e. g., Deville, Fischer and Mund (2002), Sect. 7.3.

An analysis of preconditioned conjugate gradient iterations for the solution of the systems above is carried out in Achdou, Maday and Widlund (1999), where it is shown that the condition number of the preconditioned system can be as small as $C \max_m (1 + \log N_m)^2$. An extensive analysis of the mortar finite-element method and its solution by parallel preconditioners is given in Wohlmuth (2001). See also Stefanica and Klawonn (1999) and Stefanica (2002) for FETI type preconditioners (see Sect. 6.4.4) of the saddle-point mortar system.

5.7.3 Analysis of MEM

The analysis of mortar methods is rather involved. For the sake of simplicity we will confine ourselves to the previously addressed Galerkin case (5.7.2) with only two subdomains (i. e., $M = 2$).

With this aim we introduce the quantity

$$||v||_* = (||v||^2_{0,\Omega} + ||\nabla v_{|\Omega_1}||^2_{0,\Omega_1} + ||\nabla v_{|\Omega_2}||^2_{0,\Omega_2})^{1/2}, \tag{5.7.8}$$

which is a norm, called the "broken norm", on the Hilbert space

$$H_* = \{v \in L^2(\Omega) \,|\, v_{|\Omega_1} \in H^1(\Omega_1), \; v_{|\Omega_2} \in H^1(\Omega_2)\}. \tag{5.7.9}$$

Using a generalized Poincaré inequality (see (A.10)), Bernardi, Maday and Patera (1994) prove that, for a suitable $\alpha^* > 0$ independent of δ,

$$\sum_{m=1}^{2} \int_{\Omega_m} |\nabla v_\delta|^2 \geq \alpha_* ||v_\delta||^2_* \quad \forall\, v_\delta \in V_\delta, \tag{5.7.10}$$

whence, the discrete problem (5.7.2) admits a unique solution by a straight-forward application of the Lax–Milgram Theorem (A.3). For any $v_\delta \in V_\delta$ we now have

$$\begin{aligned}
\alpha_* ||u_\delta - v_\delta||^2_* &\leq \sum_{m=1}^{2} \int_{\Omega_m} |\nabla(u_\delta - v_\delta)|^2 \\
&\leq \sum_{m=1}^{2} \int_{\Omega_m} \nabla u_\delta \cdot \nabla(u_\delta - v_\delta) - \sum_{m=1}^{2} \int_{\Omega_m} \nabla v_\delta \cdot \nabla(u_\delta - v_\delta) \\
&= \sum_{m=1}^{2} \int_{\Omega_m} f(u_\delta - v_\delta) - \sum_{m=1}^{2} \int_{\Omega_m} \nabla v_\delta \cdot \nabla(u_\delta - v_\delta).
\end{aligned} \tag{5.7.11}$$

Replacing f by $-\Delta u$ and integrating by parts on each Ω_m, we obtain

$$\sum_{m=1}^{2} \int_{\Omega_m} f(u_\delta - v_\delta) = \sum_{m=1}^{2} \int_{\Omega_m} \nabla u \cdot \nabla(u_\delta - v_\delta) - \int_\Gamma \frac{\partial u}{\partial n} [u_\delta - v_\delta]_\Gamma, \tag{5.7.12}$$

where $\frac{\partial}{\partial n}$ is the normal derivative on Γ pointing into Ω_2,

$$[w_\delta]_\Gamma = w^{(1)}_{\delta|\Gamma} - w^{(2)}_{\delta|\Gamma},$$

denotes the jump across Γ of a function $w_\delta \in V_\delta$, and the last integral should be intended in the sense of a duality pairing between a space of trace functions on Γ (namely, $H^{1/2}_{00}(\Gamma)$) and its dual, whenever u is not smooth enough. From (5.7.11) and (5.7.12) we have that

$$\alpha_* ||u_\delta - v_\delta||^2_* \leq ||u - v_\delta||_* ||u_\delta - v_\delta||_* + \left| \int_\Gamma \frac{\partial u}{\partial n} [u_\delta - v_\delta]_\Gamma \right|;$$

whence,

$$||u_\delta - v_\delta||_* \leq \frac{1}{\alpha_*} \left(||u - v_\delta||_* + \sup_{w_\delta \in V_\delta} \frac{|\int_\Gamma \frac{\partial u}{\partial n} [w_\delta]_\Gamma|}{||w_\delta||_*} \right).$$

Using the triangle inequality

$$||u - u_\delta||_* \leq ||u - v_\delta||_* + ||u_\delta - v_\delta||_* \,,$$

we arrive at the following inequality for the error $u - u_\delta$:

$$||u - u_\delta||_* \leq \left(1 + \frac{1}{\alpha_*}\right) \inf_{v_\delta \in V_\delta} ||u - v_\delta||_* + \frac{1}{\alpha_*} \sup_{w_\delta \in V_\delta} \frac{\left|\int_\Gamma \frac{\partial u}{\partial n}[w_\delta]_\Gamma\right|}{||w_\delta||_*} \,. \quad (5.7.13)$$

The approximation error of (5.7.2) is therefore bounded (up to a multiplicative constant) by the best approximation error of u (that is, the distance between the exact solution u and the finite-dimensional space V_δ) plus an extra error involving interface jumps. The latter would not appear in the framework of classical Galerkin approximations (like SEM) and is the price to pay for the violation (the so-called "variational crime" in the finite-element literature) of the conforming property; that is, for the fact that $V_\delta \not\subset H_0^1(\Omega)$.

The error estimate (5.7.13) is *optimal* if each one of the two terms on the right-hand side can be bounded by the norm of *local* errors arising from the approximations of u in Ω_1 and Ω_2, without the presence of terms that combine them in a multiplicative fashion. In this way, we can take advantage of the local regularity of the exact solution as well as the approximation properties enjoyed by the local subspaces $Y_{m,\delta}$ of $H^1(\Omega_m)$.

The second term is optimal owing to the special choice of the mortar. In fact, due to the orthogonality property (5.7.4), we can subtract from $\frac{\partial u}{\partial n}$ an arbitrary function ψ_δ in $\tilde{\Lambda}_\delta^{(m)}$, where m is the index chosen in (5.7.4). It is therefore important that such functions allow for an optimal approximation of the normal derivative of u at the interface. If they were chosen as traces on Γ of test functions in V_δ, they would vanish at the endpoints of Γ. Thus, they would provide an optimal approximation of $\frac{\partial u}{\partial n}$ only if this function would vanish therein—a too restrictive condition. This argument explains why in a more general setting the space of multipliers (i. e., test functions in (5.7.5)) is chosen in such a way that the values of its functions at the endpoints of $\Gamma_{m,i}$ are not fixed.

Taking as ψ_δ the $L^2(\Gamma)$-projection upon $\tilde{\Lambda}_\delta^{(m)} = \mathbb{P}_{N_m-2}(\Gamma)$ of $\frac{\partial u}{\partial n}$ (which we assume in $L^2(\Gamma)$), we get

$$\int_\Gamma \frac{\partial u}{\partial n}[w_\delta]_\Gamma = \int_\Gamma \left(\frac{\partial u}{\partial n} - \psi_\delta\right)[w_\delta]_\Gamma = \int_\Gamma \left(\frac{\partial u}{\partial n} - \psi_\delta\right)([w_\delta]_\Gamma - \eta_\delta) \quad (5.7.14)$$

for all $\eta_\delta \in \mathbb{P}_{N_m-2}(\Gamma)$. The latter equality follows from the property of orthogonal projection for ψ_δ. Using now the approximation results for the Legendre orthogonal projection in an interval (see CHQZ2, Sect. 5.4.2), we get the bound (for all $t \geq 0$)

$$\left|\int_\Gamma \frac{\partial u}{\partial n}[w_\delta]_\Gamma\right| \leq CN^{-(t+1/2)} \left\|\frac{\partial u}{\partial n}\right\|_{H^t(\Gamma)} ||w_\delta||_* \,. \quad (5.7.15)$$

If $u_{|\Omega_m} \in H^{s_m}(\Omega_m)$, then by setting $t = s_m - 3/2$ and using standard arguments, one easily obtains from (5.7.15) the following optimal error estimate:

$$\|u - u_\delta\|_* \leq C \sum_{m=1}^{2} N_m^{1-s_m} |u|_{H^{s_m}(\Omega_m)}, \quad s_m \geq 1. \tag{5.7.16}$$

The analysis for the general case (5.7.6) where M is arbitrary and LGL integration is used on each Ω_k is more complex. Here we only recall the main results.

The bilinear form $a_\delta(u_\delta, v_\delta)$ given by the left-hand side of (5.7.6) is coercive on H_* (whose definition extends (5.7.9) in an obvious manner) if each $\tilde{\Lambda}_\delta^{(m,i)}$ contains (the image of) $\mathbb{P}_{N_{m,i}^*}(\Gamma_{m,i})$ for some integer $N_{m,i}^*$ depending on the decomposition. In 2D, $N_{m,i}^*$ is the number of vertices of all subdomains inside $\Gamma_{m,i}$. We refer to Bernardi, Maday and Rapetti (2005) for more details.

Concerning the error estimate, if we assume that $u_{|\Omega_m} \in H^{s_m}(\Omega_m)$ for all m, a representative result for the MEM discretization has the form

$$\|u - u_\delta\|_* \leq C(1 + r_\delta)^{1/2} \sum_{m=1}^{M} N_m^{1-s_m} (\log N_m)^{p_m} \|u\|_{H^{s_m}(\Omega_m)}, \tag{5.7.17}$$

where r_δ and p_m are quantities depending on the decomposition and the discretization. They are both equal to 0 if the decomposition is geometrically conforming; usually r_δ is bounded independently of δ in situations of practical interest. Again we refer to Bernardi, Maday and Rapetti (2005) for details.

The convergence analysis for the MEM-NI leads to a comparable error bound. On the contrary, if the standard LGL quadrature formula is used to approximate the integrals in (5.7.5) (which is equivalent to enforcing the matching conditions pointwise at the internal LGL points), then the power of N_m in the previous estimate is usually reduced, i.e., it is non-optimal. (The case $m = 2$ considered above is indeed a special case for which the error estimate is optimal.) Indeed, in general (5.7.14) no longer holds, and a suboptimal estimate of the interface error replaces (5.7.15). As has already been mentioned, more accurate quadratures should be used.

Obviously, the dependence on the element diameter h_m can be made to appear explicitly in (5.7.17), getting an error estimate similar to the one for SEM, see (5.4.16). This is particularly relevant if each Ω_m is not just a single element, but rather a patch of elements that are conformally matched in the SEM manner.

5.7.4 Other Applications

The spectral mortar element method can be employed to solve other boundary-value problems such as, e.g., the incompressible Navier–Stokes equations. In this case the idea is to start from an admissible pair of velocity–pressure spaces on each subdomain Ω_m (see Sects. 3.5.2 and 5.6), then to guarantee continuity of the velocity components across the interface by mortar conditions similar to those reported in (5.7.5) for the scalar case. No continuity is required for the pressure solutions across subdomain interfaces.

As usual, the MEM-NI variant can be developed if the exact integrals are replaced by the LGL quadratures (see, e.g., Deville, Fischer, Mund (2002), Sect. 7.3.3).

The analysis of this method on the Stokes problem when a $\mathbb{Q}_N - \mathbb{Q}_{N-2}$ approximation is used at the subdomain level can be found in Ben Belgacem et al. (2000). In particular, it is proven by a Boland-Nicolaides argument (see Sect. 5.6.2; see also Ben Belgacem (2000)) that the inf-sup constant of the whole approximation is bounded from below by $\gamma_D \min_m \beta_m$, where β_m is the inf-sup constant on the subdomain Ω_m, while γ_D is a constant depending on the geometrical parameters of the subdomain partitions.

For applications of mortar methods to fluid dynamics we refer to, e.g., Mavriplis (1989), Anagnostou (1991), Kruse (1997) and Feng (2003).

The mortar approach can be applied within a spectral discontinuous Galerkin method if the partition is geometrically nonconforming. We refer to Sect. 5.9.2 for the discussion, with applications to hyperbolic problems.

Several MEM discretizations of fourth-order boundary-value problems, which may be part of the solution of the Navier–Stokes equations in streamfunction formulation, are investigated by Belhachmi (1997).

5.8 The Spectral Discontinuous Galerkin Method (SDGM) in 1D

In the *spectral discontinuous Galerkin* method (SDGM), the approximate solution is a piecewise-polynomial function (as in the SEM and the MEM case), which, however, is allowed to be discontinuous at the element interfaces. The jump of the solution across the subdomain interface is accounted for (and consequently controlled) in the integral formulation of the discrete problem. This section is devoted to presenting the general principles of SDGM on 1D problems including a linear scalar advection equation, linear hyperbolic systems, time-dependent linear problems and nonlinear conservation laws. Subsequently, we illustrate the extension to multidimensional problems and to elliptic problems. Then we survey the analysis of SDGM and finally discuss applications to Euler and Navier–Stokes problems. General background on the discontinuous Galerkin (DG) method for various types of discretizations is provided by the many articles in Cockburn, Karniadakis and Shu (2000).

5.8.1 Linear Advection Problems in 1D

In order to introduce the SDGM we consider the simple one-dimensional linear advection problem

$$\mathcal{L}u = (\beta u)' + \gamma u = f, \qquad a < x < b,$$
$$u(a) = g,$$

(5.8.1)

where $\beta(x) > 0$ and $\gamma(x) \geq 0$ are the two given functions, and g is a real number (the inflow data).

A spectral Galerkin approximation of (5.8.1) can be set up as follows. We look for a polynomial u_N (that approximates u) belonging to $\mathbb{P}_N$ that satisfies

$$-\int_a^b \beta u_N v_N' + \int_a^b \gamma u_N v_N + \beta u_N v_N|_{x=b}$$
$$= \int_a^b f v_N + \beta g v_N|_{x=a} \qquad \forall v_N \in \mathbb{P}_N.$$

(5.8.2)

Its G-NI version is given by

$$-(\beta u_N, v_N')_N + (\gamma u_N, v_N)_N + \beta u_N v_N|_{x=b}$$
$$= (f, v_N)_N + \beta g v_N|_{x=a} \qquad \forall v_N \in \mathbb{P}_N,$$

(5.8.3)

where now $(\cdot, \cdot)_N$ indicates the discrete inner product of $L^2(a, b)$ based on a Gaussian integration formula (Gauss or Gauss–Radau or Gauss–Lobatto). (However, when β is not a constant, the skew-symmetric decomposition of $\mathcal{L}$ should be used in (5.8.3) in order to guarantee numerical stability, see Sect. 5.4.15).

Both (5.8.2) and (5.8.3) enforce the inflow condition in a weak sense. The SDGM extends the weak enforcement from just the inflow boundary condition to the enforcement of the continuity of the solution at the interfaces of the subintervals that comprise the partition of the computational interval (a, b).

Retaining the notation of Sect. 5.2 for the partition of $\Omega = (a, b)$, let us now consider the space of piecewise polynomials:

$$W_\delta = \{v \in L^2(a, b) : v|_{\Omega_m} \in \mathbb{P}_{N_m}, \quad \forall m = 1, \ldots, M\},$$

(5.8.4)

where $\Omega_m = (\overline{x}_{m-1}, \overline{x}_m)$ is the m−th element. Note that no continuity is required at the endpoints $\overline{x}_m$.

Then we look for $u_\delta \in W_\delta$, which satisfies

$$\sum_{m=1}^M \left\{ -(\beta u_\delta, v_\delta')_m + (\gamma u_\delta, v_\delta)_m + \beta^- u_\delta^- v_\delta^-|_{\overline{x}_m} - \beta^+ u_\delta^- v_\delta^+|_{\overline{x}_{m-1}} \right\}$$
$$= \sum_{m=1}^M (f, v_\delta)_m, \qquad \forall v_\delta \in W_\delta,$$

(5.8.5)

where, if v is a discontinuous function at $\bar{x}_m$, $v^-(\bar{x}_m)$ and $v^+(\bar{x}_m)$ represent its left-hand side and right-hand side values, respectively. In the previous formula, for $m = 1$ the term $u_{\delta|\bar{x}_0}$ stands for the inflow data g. Note that we have enforced weakly the continuity condition $u_\delta^+|_{\bar{x}_{m-1}} = u_\delta^-|_{\bar{x}_{m-1}}$ for all $2 \leq m \leq M$.

We have indicated by $(u,v)_m$ either the exact integral $(u,v)_{\Omega_m} = \int_{\bar{x}_{m-1}}^{\bar{x}_m} uv$ or its numerical approximation $(u,v)_{N_m,\Omega_m}$ by the LGL integration formula on the interval Ω_m (see (5.2.16)). In the former case, (5.8.5) is the SDGM approximation of (5.8.1); in the latter it represents the SDGM-NI (i. e., the spectral discontinuous Galerkin method with numerical integration) for approximations of (5.8.1).

By reassembling the nodal terms we also have

$$\sum_{m=1}^{M} \left\{ -(\beta u_\delta, v_\delta')_m + (\gamma u_\delta, v_\delta)_m \right\} - \sum_{m=1}^{M-1} u_\delta^- [\beta v_\delta]_{\bar{x}_m} + \beta^- u_\delta^- v_\delta^- |_{x=b}$$

$$= \sum_{m=1}^{M} (f, v_\delta)_m + \beta^+ g v_\delta^+ |_{x=a} \, ,$$

where

$$[w]_{\bar{x}_m} = (w^+ - w^-)(\bar{x}_m) \tag{5.8.6}$$

denotes the jump of a discontinuous function w at the node $\bar{x}_m$ (another notation used in the sequel will be $[w]|_{\bar{x}_m}$).

Since v_δ is also discontinuous, we can reduce (5.8.5) to a set of M relations that hold on the M elements. For $m = 1, \ldots, M$ (setting $u_\delta^{(m)} = u_{\delta|\Omega_m}$ whenever the potential for confusion exists), we obtain

$$- (\beta u_\delta, v_{N_m}')_m + (\gamma u_\delta, v_{N_m})_m + \beta^- u_\delta^{(m)} v_{N_m}|_{\bar{x}_m} - \beta^+ u_\delta^{(m-1)} v_{N_m}|_{\bar{x}_{m-1}}$$
$$= (f, v_{N_m})_m \qquad \forall v_{N_m} \in \mathbb{P}_{N_m} \, . \tag{5.8.7}$$

The analogy between (5.8.7) and (5.8.2) (or (5.8.3)) is that on the element Ω_m the quantity $u_\delta^{(m-1)}(\bar{x}_{m-1})$ plays the same role of "inflow" boundary value that is played by the inflow boundary value g at $x = a$ in (5.8.2) or (5.8.3).

The SDGM—(5.8.7) with exact integration—can be reformulated in a different (albeit mathematically equivalent) form. By counter-integrating by parts the first integral we obtain

$$(Lu_\delta - f, v_{N_m})_m + \beta^+ [u_\delta] v_{N_m}|_{\bar{x}_{m-1}} = 0 \, . \tag{5.8.8}$$

This can be regarded as a way to satisfy the differential equation independently on each Ω_m and to enforce interelement continuity by a penalty term.

When the numerical integration is based on LGL (Legendre Gauss–Lobatto) quadrature, the SDGM-NI (5.8.7) can be given a nodal interpretation as a *generalized collocation* method as follows. Let $x_j^{(m)}$, $j = 0, \ldots, N_m$, be the $N_m + 1$ LGL nodes on the m-th element (see (5.2.11)), and denote by $w_j^{(m)}$ the corresponding LGL weights. Now, let v_{N_m} be the characteristic Lagrange polynomial $\psi_j^{(m)}$ associated with $x_j^{(m)}$, extended by zero outside Ω_m. Next, as introduced in Sect. 5.2.3, let $I_{N_m}^{(m)}$ denote the interpolation operator of degree N_m using the LGL nodes. Then integrating by parts the first integral in (5.8.7), we obtain, for all $j = 0, \ldots, N_m$,

$$\left((I_{N_m}^{(m)}(\beta u_\delta))' + \gamma u_\delta - f, \psi_j^{(m)} \right)_{N_m, \Omega_m} + \beta^+ [u_\delta] \psi_j^{(m)} |_{\bar{x}_{m-1}} = 0 .$$

This shares the same structure as (5.8.8); however, now the differential operator $\mathcal{L}$ has been replaced in Ω_m by its interpolation approximation $\mathcal{L}_\delta^{(m)} w = (I_{N_m}^{(m)}(\beta w))' + \gamma w = \mathcal{D}_{N_m}(\beta w) + \gamma w$.

Since $\psi_j^{(m)}(\bar{x}_{m-1}) = 0$ for all $j = 1, \ldots, N_m$, we obtain

$$\mathcal{L}_\delta^{(m)} u_\delta = f \quad \text{at} \quad x_j^{(m)}, \quad j = 1, \ldots, N_m . \tag{5.8.9}$$

For $j = 0$ we obtain the boundary equation

$$\mathcal{L}_\delta^{(m)} u_\delta = f - \frac{1}{w_0^{(m)}} \beta^+ [u_\delta] \quad \text{at} \quad \bar{x}_{m-1} = x_0^{(m)} , \tag{5.8.10}$$

which can be regarded as a weak enforcement, or penalization, of the interelement condition $u_\delta^{(m)} = u_\delta^{(m-1)}$ at the inflow point $\bar{x}_{m-1}$.

This is a type of penalty method as discussed for single-domain methods in CHQZ2, Sect. 6.6. The use of penalty methods for spectral domain decomposition purposes is discussed in Hesthaven (2000) and Gottlieb and Hesthaven (2001).

Although the discussion here was couched in terms of the LGL integration formula, one could as well have used the LG (Legendre Gauss) formula. Indeed, for certain multidimensional problems the LG formula is distinctly preferred, for the practical implementation reasons explained in Sect. 5.9.1. Nevertheless, for simplicity in presentation, we will continue to present the method in terms of the LGL formula.

5.8.2 Linear Hyperbolic Systems in 1D

Next, let us consider a 1D time-independent, linear hyperbolic system of the form

$$\mathcal{L}\mathbf{u} = A\frac{\partial \mathbf{u}}{\partial x} + D\mathbf{u} = \mathbf{f}, \quad a < x < b, \tag{5.8.11}$$

$$B_L \mathbf{u}(a) = \mathbf{g}_L, \quad B_R \mathbf{u}(b) = \mathbf{g}_R .$$

We assume that for $q \geq 2$ the following conditions, which define a *Friedrichs system*, are satisfied:

$A = A(x)$ is a bounded differentiable function in (a, b), taking values in the symmetric matrices; its first derivative $\partial_x A$ is also a bounded function in (a, b). In symbols, $A \in [L^\infty(a, b)]^{q \times q}$, with $\partial_x A \in [L^\infty(a, b)]^{q \times q}$.

$D = D(x)$ is a bounded function in (a, b), i.e., $D \in [L^\infty(a, b)]^{q \times q}$, taking values in the positive semi-definite matrices.

Moreover, we assume that $D + D^T - \partial_x A \geq \sigma_0 I$ for a suitable positive constant $\sigma_0 > 0$. B_L and B_R are two rectangular matrices; the number of rows of each matrix coincides with the number of positive eigenvalues of the matrix A at the corresponding boundary point. The following discussion relies upon the concepts and notation of the characteristic compatibility method (CCM) described in Sect. 4.2.1.

For any x, $A = A(x)$ is symmetric; thus, there exist $W, \Lambda \in \mathbb{R}^{q \times q}$ such that $A = W \Lambda W^T$, where Λ is the diagonal matrix consisting of the eigenvalues of A. To impose the boundary conditions, we switch to the characteristic variables $\mathbf{z} = W^T \mathbf{u}$.

We can apply a Galerkin formulation as in (4.2.33) on each element Ω_m, $m = 1, \ldots, M$, enforcing (weakly) the continuity of the characteristic variables at every end-point $\bar{x}_m$, $m = 2, \ldots, M - 1$. Then, summing up over $m = 1, \ldots, M$, the SDGM for linear hyperbolic systems on (a, b) can be formulated as follows. We look for $\mathbf{u}_\delta \in (W_\delta)^q$ such that, for all $\mathbf{v}_\delta \in (W_\delta)^q$,

$$
\sum_{m=1}^{M} \left(A \frac{\partial \mathbf{u}_\delta}{\partial x} + D \mathbf{u}_\delta, \mathbf{v}_\delta \right)_m
$$

$$
+ \sum_{m=2}^{M} \left\{ (\mathbf{y}_{\delta,p}^{(m)})^T \Lambda^+ [\mathbf{z}_{\delta,p}]_{\bar{x}_{m-1}} + (\mathbf{y}_{\delta,n}^{(m-1)})^T \Lambda^- [\mathbf{z}_{\delta,n}]_{\bar{x}_{m-1}} \right\}
$$

$$
+ (\mathbf{y}_{\delta,p}^{(1)})^T \Lambda^+ (\mathbf{z}_{\delta,p}^{(1)} - S_L \mathbf{z}_{\delta,n}^{(1)})|_{\bar{x}_0} - (\mathbf{y}_{\delta,n}^{(M)})^T \Lambda^- (\mathbf{z}_{\delta,n}^{(M)} - S_R \mathbf{z}_{\delta,p}^{(M)})|_{\bar{x}_M}
$$

$$
= \sum_{m=1}^{M} (\mathbf{f}, \mathbf{v}_\delta)_m + (\mathbf{y}_{\delta,p}^{(1)})^T \Lambda^+ \mathbf{z}_L|_{\bar{x}_0} - (\mathbf{y}_{\delta,n}^{(M)})^T \Lambda^- \mathbf{z}_R|_{\bar{x}_M},
$$

where $\mathbf{z}_\delta = W^T \mathbf{u}_\delta$ and $\mathbf{y}_\delta = W^T \mathbf{v}_\delta$ are the local (since all matrices depend on x) *characteristic variables* associated with $\mathbf{u}_\delta$ and $\mathbf{v}_\delta$, the matrices Λ^+ and Λ^- are the diagonal matrices of the positive (or negative) eigenvalues, the symbol $\mathbf{z}_{\delta,p}$ (or $\mathbf{z}_{\delta,n}$) is used to identify the vector of components of $\mathbf{z}_\delta$ associated with positive (or negative) eigenvalues, and the jumps of vectors are defined in a manner similar to (5.8.6). Whenever confusion might arise, we use the superscript (m) to denote restriction to Ω_m.

By introducing the matrices

$$
A_p = W \begin{pmatrix} \Lambda^+ & 0 \\ 0 & 0 \end{pmatrix} W^T \quad \text{and} \quad A_n = W \begin{pmatrix} 0 & 0 \\ 0 & \Lambda^- \end{pmatrix} W^T, \tag{5.8.12}
$$

called, respectively, the positive and negative parts of A, we can write the above formulation solely in terms of the physical unknowns. Indeed, let us note that

$$(\mathbf{y}_{\delta,p}^{(m)})^T \Lambda^+ [\mathbf{z}_{\delta,p}]_{\bar{x}_{m-1}} + (\mathbf{y}_{\delta,n}^{(m-1)})^T \Lambda^- [\mathbf{z}_{\delta,n}]_{\bar{x}_{m-1}}$$
$$= \left((\mathbf{v}_\delta^{(m)})^T A_p + (\mathbf{v}_\delta^{(m-1)})^T A_n\right) [\mathbf{u}_\delta]_{\bar{x}_{m-1}} .$$

Then, the SDG solution $\mathbf{u}_\delta \in (W_\delta)^q$ is such that for all $\mathbf{v}_\delta \in (W_\delta)^q$,

$$\sum_{m=1}^{M} (\mathcal{L}\mathbf{u}_\delta, \mathbf{v}_\delta)_m + \sum_{m=2}^{M} \left((\mathbf{v}_\delta^{(m)})^T A_p + (\mathbf{v}_\delta^{(m-1)})^T A_n\right) [\mathbf{u}_\delta]_{\bar{x}_{m-1}}$$
$$+ (\mathbf{v}_\delta^{(1)})^T A_p (\mathbf{u}_\delta^{(1)} - W\tilde{S}_L W^T \mathbf{u}_\delta^{(1)})|_{\bar{x}_0}$$
$$- (\mathbf{v}_\delta^{(M)})^T A_n (\mathbf{u}_\delta^{(M)} - W\tilde{S}_R W^T \mathbf{u}_\delta^{(M)})|_{\bar{x}_M} \qquad (5.8.13)$$
$$= \sum_{m=1}^{M} (\mathbf{f}, \mathbf{v}_\delta)_m + (\mathbf{v}_\delta^{(1)})^T A_p W\tilde{\mathbf{z}}_L|_{\bar{x}_0} - (\mathbf{v}_\delta^{(M)})^T A_n W\tilde{\mathbf{z}}_R|_{\bar{x}_M} ,$$

where we have set

$$\tilde{S}_R = \begin{pmatrix} 0 & 0 \\ S_R & 0 \end{pmatrix}, \quad \tilde{S}_L = \begin{pmatrix} 0 & S_L \\ 0 & 0 \end{pmatrix}, \quad \text{and} \quad \tilde{\mathbf{z}}_L = \begin{pmatrix} \mathbf{z}_L \\ 0 \end{pmatrix}, \quad \tilde{\mathbf{z}}_R = \begin{pmatrix} 0 \\ \mathbf{z}_R \end{pmatrix}.$$

If natural boundary conditions are imposed on the incoming characteristic variables, that is, if

$$\mathbf{z}^+(a) = \boldsymbol{\varphi}_L \quad \text{at } x = a, \qquad \mathbf{z}^-(b) = \boldsymbol{\varphi}_R \quad \text{at } x = b,$$

then the reflection matrices vanish: $S_L = 0$ and $S_R = 0$. Additionally, $C_L = I$ and $C_R = I$. Then, $\mathbf{z}_L = \boldsymbol{\varphi}_L$ and $\mathbf{z}_R = \boldsymbol{\varphi}_R$, as in Sect. 4.2.1.

Note that integrating the first term of (5.8.13) by parts yields:

$$\sum_{m=1}^{M} \left\{ (D\mathbf{u}_\delta, \mathbf{v}_\delta)_m - \left(\mathbf{u}_\delta, \frac{\partial}{\partial x}(A\mathbf{v}_\delta)\right)_m \right\}$$
$$+ \sum_{m=2}^{M} \left\{ -(\mathbf{v}_\delta^{(m)})^T A \, \mathbf{u}_\delta^{(m)}|_{\bar{x}_{m-1}} + (\mathbf{v}_\delta^{(m-1)})^T A \, \mathbf{u}_\delta^{(m-1)}|_{\bar{x}_{m-1}} \right.$$
$$\left. + \left((\mathbf{v}_\delta^{(m)})^T A_p + (\mathbf{v}_\delta^{(m-1)})^T A_n\right) [\mathbf{u}_\delta]_{\bar{x}_{m-1}} \right\}$$
$$= \sum_{m=1}^{M} (f, \mathbf{v}_\delta)_m + \text{ boundary terms.}$$

Since

$$\left((\mathbf{v}_\delta^{(m)})^T A_p + (\mathbf{v}_\delta^{(m-1)})^T A_n\right) [\mathbf{u}_\delta]_{\bar{x}_{m-1}}$$
$$= (\mathbf{v}_\delta^{(m)})^T A [\mathbf{u}_\delta]_{\bar{x}_{m-1}} - [\mathbf{v}_\delta]_{\bar{x}_{m-1}} A_n [\mathbf{u}_\delta]_{\bar{x}_{m-1}} ,$$

then

$$\sum_{m=1}^{M} \left\{ (D\mathbf{u}_\delta, \mathbf{v}_\delta)_m - \left(\mathbf{u}_\delta, \frac{\partial}{\partial x}(A\mathbf{v}_\delta) \right)_m \right\}$$

$$- \sum_{m=2}^{M} \left\{ [\mathbf{v}_\delta]^T_{\bar{x}_{m-1}} A \, \mathbf{u}_\delta^{(m-1)}|_{\bar{x}_{m-1}} + [\mathbf{v}_\delta]^T_{\bar{x}_{m-1}} A_n [\mathbf{u}_\delta]_{\bar{x}_{m-1}} \right\} \qquad (5.8.14)$$

$$= \sum_{m=1}^{M} (f, \mathbf{v}_\delta)_m + \text{ boundary terms}.$$

In short-hand notation, (5.8.14) can be rewritten as

$$\sum_{m=1}^{M} (\mathbf{u}_\delta, \mathcal{L}^* \mathbf{v}_\delta)_m - \sum_{m=1}^{M-1} \langle H_m(\mathbf{u}_\delta), \mathbf{v}_\delta \rangle_{\bar{x}_m} = \sum_{m=1}^{M} (f, \mathbf{v}_\delta)_m + \text{ boundary terms},$$

$$(5.8.15)$$

where $\mathcal{L}^*$ is the adjoint operator of $\mathcal{L}$, that is,

$$\mathcal{L}^* \mathbf{v}_\delta = -\frac{\partial}{\partial x}(A\mathbf{v}_\delta) + D^T \mathbf{v}_\delta \, ,$$

and

$$\langle H_m(\mathbf{u}_\delta), \mathbf{v}_\delta \rangle_{\bar{x}_m} = [\mathbf{v}_\delta]^T_{\bar{x}_m} A \, \mathbf{u}_\delta^{(m)}|_{\bar{x}_m} + [\mathbf{v}_\delta]^T_{\bar{x}_m} A_n [\mathbf{u}_\delta]_{\bar{x}_m} \qquad (5.8.16)$$

denotes the action of a "numerical flux" $H_m(\mathbf{u}_\delta)$ on $\mathbf{v}_\delta$ at $\bar{x}_m$. Specifically, this expression is the sum of the physical flux $A \, \mathbf{u}_\delta^{(m)}|_{\bar{x}_m}$ applied to the jump of $\mathbf{v}_\delta$ at $\bar{x}_m$ plus a correction term depending on the negative part of A and the jumps of both $\mathbf{v}_\delta$ and $\mathbf{u}_\delta$ at $\bar{x}_m$.

The SDGM-NI solution satisfies a problem like (5.8.13) (and (5.8.14)) with the exact integrals $(\cdot, \cdot)_m$ on each Ω_m replaced by LGL or LG integration formulas. Note that in the latter case (Legendre Gauss integration) all integrals are computed exactly if the matrices A and D are constant with respect to x.

Example 5.8.1. We consider the linear 2×2-system

$$u_t + c v_x = 0 \, ,$$
$$v_t + c u_x = 0 \, ,$$

for $x \in \Omega = (0, 1)$ and $0 < t < T$, with boundary conditions $u(0, t) = v(0, t) = e^{-i\omega t}$, with $\omega > 0$, and where c is a discontinuous coefficient

$$c(x) = \begin{cases} \frac{\omega}{k_1} & \text{in } \Omega_1 = (0, 0.5) \, , \\ \frac{\omega}{k_2} & \text{in } \Omega_2 = (0.5, 1) \, , \end{cases}$$

for two positive integers k_1 and k_2. The exact solution of (5.8.17) is

$$u(x, t) = v(x, t) = \begin{cases} e^{i(k_1 x - \omega t)} & \text{in } \Omega_1 \, , \\ e^{i(k_2 x - \omega t)} & \text{in } \Omega_2 \, . \end{cases}$$

In each element c is constant, so the first component u of the solution of (5.8.17) satisfies the wave equation:

$$\partial_{tt}u - c^2\partial_{xx}u = 0 . \tag{5.8.17}$$

Let us assume that u and v are of the following form

$$u(x,t) = e^{-i\omega t}u^*(x) \text{ and } v(x,t) = e^{-i\omega t}v^*(x), \text{ with } u^*(0) = 1 \text{ and } v^*(0) = 1 .$$

Then, the problem is equivalent to finding $(u^*, v^*) : \Omega \to \mathbb{C}^2$ such that

$$-i\omega u^* + cv_x^* = 0 ,$$
$$-i\omega v^* + cu_x^* = 0 .$$

Next, we split the complex functions into their real and imaginary parts,

$$u^*(x) = u^r(x) + iu^i(x) \quad \text{and} \quad v^*(x) = v^r(x) + iv^i(x) ,$$

so that finally the problem becomes: $\bar{\mathbf{u}} = (u^r, u^i, v^r, v^i)^T : \Omega \to \mathbb{R}^4$ such that

$$A\frac{\partial\bar{\mathbf{u}}}{\partial x} + D\bar{\mathbf{u}} = 0 , \tag{5.8.18}$$

where

$$A = \begin{pmatrix} 0 & 0 & 1 & 0 \\ 0 & 0 & 0 & 1 \\ 1 & 0 & 0 & 0 \\ 0 & 1 & 0 & 0 \end{pmatrix} \quad \text{and} \quad D = \frac{\omega}{c}\begin{pmatrix} 0 & 1 & 0 & 0 \\ -1 & 0 & 0 & 0 \\ 0 & 0 & 0 & 1 \\ 0 & 0 & -1 & 0 \end{pmatrix} .$$

The solution of (5.8.18) reads

$$u^r(x) = v^r(x) = \cos\left(\frac{\omega}{c}x\right) \quad \text{and} \quad u^i(x) = v^i(x) = \sin\left(\frac{\omega}{c}x\right) ,$$

and we see that $\bar{\mathbf{u}} \in [H^s(\Omega_i)]^4$, $i = 1, 2$, for all $s \geq 0$. We approximate this system by the SDGM (5.8.13), using a uniform partition of Ω with mesh size $h = 1/(2k)$ for $k \in \mathbb{N}$; since $x = 0.5$ is at an interelement boundary, the coefficients are elementwise constant. Figure 5.27 shows the approximations of u^r, u^i, v^r and v^i obtained for uniform grid-size $h = 1/4$ and constant polynomial

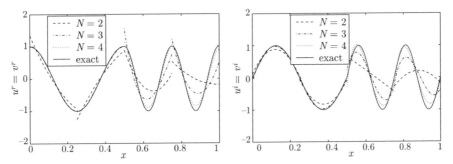

Fig. 5.27. The exact solution of problem (5.8.18) and its approximation by the SDG method for $h = 1/4$ and $N = 2, 3, 4$

degree $N = 2, 3, 4$. In fact, the numerical results exhibit exponential convergence in N and algebraic convergence in h; see Sect. 5.11 for the theoretical analysis and the numerical verification.

5.8.3 Time-Dependent Problems

We consider the time-dependent advection equation

$$\frac{\partial u}{\partial t} + \frac{\partial}{\partial x}(\beta u) + \gamma u = f , \qquad a < x < b , \quad t > 0 ,$$

$$u(a, t) = g(t), \qquad t > 0 , \qquad (5.8.19)$$

$$u(x, 0) = u_0(x) , \qquad a < x < b ,$$

under the same assumptions made on β and γ in (5.8.1). This time, $f = f(x, t)$, $u_0 = u_0(x)$ and $g = g(t)$ are given functions whose regularity is enough to guarantee that all operations carried out in the rest of this subsection are meaningful.

The SDG approximation to (5.8.19) consists of finding $u_\delta(t) \in W_\delta$ such that, for all $v_\delta \in W_\delta$ and for all $t > 0$,

$$\sum_{m=1}^{M} \left(\frac{\partial u_\delta}{\partial t}, v_\delta \right)_m$$

$$+ \sum_{m=1}^{M} \left\{ -\left(\beta u_\delta, \frac{\partial v_\delta}{\partial x} \right)_m + (\gamma u_\delta, v_\delta)_m + \beta^- u_\delta^- v_\delta^- |_{\bar{x}_m} - \beta^+ u_\delta^+ v_\delta^+ |_{\bar{x}_{m-1}} \right\}$$

$$= \sum_{m=1}^{M} (f, v_\delta)_m , \qquad (5.8.20)$$

where the same notation as in (5.8.5) has been adopted, still using the convention that $u_\delta^{(0)} = g$. The dependence of u_δ and f on t has been suppressed.
Equivalently, for all $m = 1, \ldots, M$, and for all $v_{N_m} \in \mathbb{P}_{N_m}$ one has

$$\left(\frac{\partial u_\delta}{\partial t}, v_{N_m} \right)_m - \left(\beta u_\delta, \frac{\partial v_{N_m}}{\partial x} \right)_m + (\gamma u_\delta, v_{N_m})_m$$

$$+ \beta^- u_\delta^{(m)} v_{N_m} |_{\bar{x}_m} - \beta^+ u_\delta^{(m-1)} v_{N_m} |_{\bar{x}_{m-1}} = (f, v_{N_m})_m , \qquad (5.8.21)$$

where again we set $u_\delta^{(m)} = u_{\delta|\Omega_m}$ whenever confusion might arise.
The algebraic form of (5.8.20) reads

$$M_\delta \frac{\mathrm{d}}{\mathrm{d}t} \mathbf{u}_\delta + L_\delta \mathbf{u}_\delta = \mathbf{f} , \qquad \mathbf{u}_\delta(0) = \mathbf{u}_\delta^0 .$$

The vector $\mathbf{u}_\delta$ collects the unknown coefficients of u_δ with respect to the chosen basis $\{\varphi_k\}$ in W_δ. The matrix M_δ is the mass matrix, i.e.,

$$(M_\delta)_{kj} = \sum_{m=1}^{M} (\varphi_k^{(m)}, \varphi_j^{(m)})_m ,$$

whereas the matrix L_δ accounts for the discretization of the spatial terms under the sum on the left-hand side of (5.8.20). Finally, $\boldsymbol{f}$ is the vector collecting the right-hand sides. The mass matrix M_δ can be made diagonal if we choose as basis functions in every element Ω_m the set of the first $N_m + 1$ Legendre orthogonal polynomials in Ω_m in the case of SDGM, or the characteristic Lagrange functions $\{\psi_k^{(m)}\}$ for SDGM-NI. Having a diagonal mass matrix is a desirable property if an explicit method is used to advance in time equation (5.8.21). We recall that no interelement continuity is required for the functions in W_δ; hence, the basis functions can be chosen with local support on each subinterval.

For time-discretization, following Cockburn and Shu (1991), we can apply the following second-order method (assume $\boldsymbol{f} = \boldsymbol{0}$ for simplicity):

$$M_\delta(\boldsymbol{u}_\delta^* - \boldsymbol{u}_\delta^n) = -\Delta t L_\delta \boldsymbol{u}_\delta^n \,,$$
$$M_\delta(\boldsymbol{u}_\delta^{**} - \boldsymbol{u}_\delta^*) = -\Delta t L_\delta \boldsymbol{u}_\delta^* \,,$$
$$\boldsymbol{u}_\delta^{n+1} = \frac{1}{2}(\boldsymbol{u}_\delta^n + \boldsymbol{u}_\delta^{**}) \,,$$

which is a convenient way of writing the second-order Runge–Kutta (RK) method (see (D.2.15)). If $N_m = 1$ for all m (although this would hardly be considered a multidomain spectral method), the scheme is stable in the $L^2(a,b)$ norm (for all $t > 0$), provided that the following restriction holds on the time step:

$$\Delta t \le \frac{1}{3}\frac{h}{\beta_{\max}} \,,$$

where $h = 1/M$ denotes the (constant) length of each subinterval and $\beta_{\max}$ is the maximum of $\beta(x)$. Numerical evidence suggests a bound of the type

$$\Delta t \le C\frac{h}{2N + 3}$$

if polynomials of degree N are used.

Another strategy is based on the use of the broad class of multistage and multistep schemes with the so called strong stability preserving (SSP) property, as proposed by Gottlieb, Shu and Tadmor (2001) (see also Hesthaven, Gottlieb and Gottlieb (2006)). Time discretizations typically used in conjunction with the SDGM for fluids applications are discussed in Sect. 5.12.

Now let us consider the case of time-dependent linear *hyperbolic systems* of the following form:

$$\frac{\partial \boldsymbol{u}}{\partial t} + \mathcal{L}\boldsymbol{u} = \frac{\partial \boldsymbol{u}}{\partial t} + A\frac{\partial \boldsymbol{u}}{\partial x} + D\boldsymbol{u} = \boldsymbol{f} \,, \qquad a < x < b \,, \quad t > 0 \,,$$
$$B_L \boldsymbol{u}(a,t) = \boldsymbol{g}_L(t), \; B_R \boldsymbol{u}(b,t) = \boldsymbol{g}_R(t) \,, \qquad t > 0 \,,$$
$$\boldsymbol{u}(x,0) = \boldsymbol{u}_0(x) \,, \qquad a < x < b \,,$$

$$(5.8.22)$$

where $A(x,t) \in \mathbb{R}^{q\times q}$ and $D(x,t) \in \mathbb{R}^{q\times q}$ are chosen as in problem (5.8.11).

For all $t > 0$, the SDGM for (5.8.22) consists in finding $\mathbf{u}_\delta(t) \in (W_\delta)^q$ such that for all $\mathbf{v}_\delta \in (W_\delta)^q$ one has (see (5.8.13) and recall the notation for W, A_p, A_n, etc.)

$$\sum_{m=1}^{M} \left(\frac{\partial \mathbf{u}_\delta}{\partial t} + \mathcal{L}\mathbf{u}_\delta, \mathbf{v}_\delta \right)_m + \sum_{m=2}^{M} \left((\mathbf{v}_\delta^{(m)})^T A_p + (\mathbf{v}_\delta^{(m-1)})^T A_n \right) [\mathbf{u}_\delta]_{\bar{x}_{m-1}}$$
$$+ (\mathbf{v}_\delta^{(1)})^T A_p (\mathbf{u}_\delta^{(1)} - W\tilde{S}_L W^T \mathbf{u}_\delta^{(1)})|_{\bar{x}_0} - (\mathbf{v}_\delta^{(M)})^T A_n (\mathbf{u}_\delta^{(M)} - W\tilde{S}_R W^T \mathbf{u}_\delta^{(M)})|_{\bar{x}_M}$$
$$= \sum_{m=1}^{M} (\mathbf{v}_\delta, \mathbf{f})_m + (\mathbf{v}_\delta^{(1)})^T A_p W \tilde{\mathbf{z}}_L|_{\bar{x}_0} - (\mathbf{v}_\delta^{(M)})^T A_n W \tilde{\mathbf{z}}_R|_{\bar{x}_M}.$$

The terms evaluated at the boundary points $a = \bar{x}_0$ and $b = \bar{x}_M$ appear as in the case for the steady-state SDG formulation.

Equivalently, we can write the SDG scheme in a manner similar to (5.8.15), i.e.,

$$\sum_{m=1}^{M} \left(\frac{\partial \mathbf{u}_\delta}{\partial t}, \mathbf{v}_\delta \right)_m + \sum_{m=1}^{M} (\mathbf{u}_\delta, \mathcal{L}^* \mathbf{v}_\delta)_m - \sum_{m=1}^{M-1} \langle H_m(\mathbf{u}_\delta), \mathbf{v}_\delta \rangle_{\bar{x}_m}$$
$$= \sum_{m=1}^{M} (\mathbf{f}, \mathbf{v}_\delta)_m + \text{boundary terms}, \tag{5.8.23}$$

In Fig. 5.28 we report results obtained for the system

$$\frac{\partial \mathbf{u}}{\partial t} + A \frac{\partial \mathbf{u}}{\partial x} = 0, \qquad 0 < x < 1, \quad t > 0, \tag{5.8.24}$$

with

$$A = \begin{pmatrix} 0 & 1 \\ 1 & 0 \end{pmatrix}.$$

The boundary conditions are imposed on the incoming characteristics and the initial conditions are drawn in Fig. 5.28 (see Sect. 4.2.1 where a similar example is discussed). The initial data are continuous, piecewise C^1-functions, such that the endpoints of the elements coincide with the jumps of the first derivatives. In Fig. 5.29 we show the dependence on N (the polynomial degree on each element Ω_m) of the computed solutions at the time-level $t = 0.6$, while in Fig. 5.30 we plot the solutions computed using a constant number of degrees of freedom $2(N + 1)/h = 24$ for different values of h and N. The results of Fig. 5.29 show convergence with respect to N, those of Fig. 5.30 indicate that the best strategy is N-refinement rather than h-refinement.

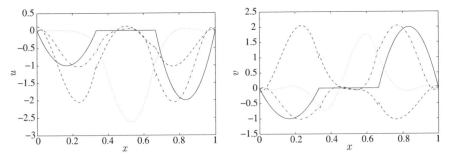

Fig. 5.28. The two components u and v of the solution of problem (5.8.24) at different time-steps computed by the SDG method with polynomial degree $N = 4$ and $h = 1/3$. The solid line is for $t = 0$, the dash-dotted line for $t = 0.06$, the dotted line for $t = 0.27$, and the dashed line for $t = 0.6$

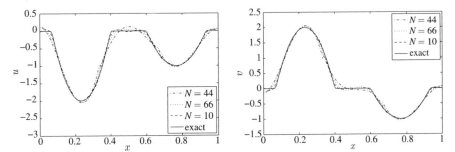

Fig. 5.29. Solutions u (*left*) and v (*right*) of problem (5.8.24) at $t = 0.6$ for different values of N (the polynomial degree used in each element) and fixed $h = 1/3$

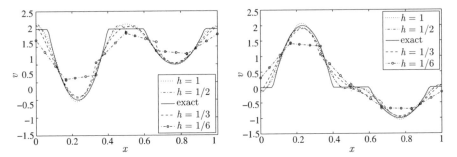

Fig. 5.30. Solutions u (*above*) and v (*below*) of problem (5.8.24) at $t = 0.6$ computed using a constant number of degrees of freedom $2(N + 1)/h = 24$

A key aspect of many numerical methods for hyperbolic problems with discontinuous solutions is the *Riemann problem*, which is a one-dimensional hyperbolic problem with initial data that is piecewise constant and has a single jump discontinuity, at, say, $x = x_d$. For a scalar problem such as (5.8.19), we would have

$$u_0(x) = \begin{cases} u_L & \text{if } a < x < x_d\,, \\ u_R & \text{if } x_d < x < b\,, \end{cases} \tag{5.8.25}$$

where u_L and u_R are constants. When β is a constant, $\gamma = 0$ and $f = 0$, the Riemann problem can be solved analytically. As the Riemann problem is very prominent in SDG methods, we write here the general Riemann problem for a constant-coefficient, linear, hyperbolic system:

$$\begin{aligned} &\frac{\partial \mathbf{u}}{\partial t} + A\mathbf{u} = \mathbf{0}\,, & a < x < b\,, \quad t > 0\,, \\ &B_L \mathbf{u}(a,t) = \mathbf{g}_L(t)\,, \ B_R \mathbf{u}(b,t) = \mathbf{g}_R(t)\,, & t > 0\,, \\ &\mathbf{u}(x,0) = \begin{cases} \mathbf{u}_L & a < x < x_d\,, \\ \mathbf{u}_R & x < x_d < b\,, \end{cases} \end{aligned} \tag{5.8.26}$$

where $\mathbf{u}_L$ and $\mathbf{u}_R$ are constants. This, too, can be solved analytically by diagonalizing the system, solving for each characteristic variable, and transforming back. The text by LeVeque (2002) contains a thorough discussion of exact and approximate solutions to Riemann problems.

Hu, Hussaini and Rasetarinera (1999) analyzed the dispersion and dissipation properties of the spectral discontinuous Galerkin method in the semi-discrete context of the one-dimensional scalar advection equation and the two-dimensional wave equation with periodic boundary conditions, discretized on triangular or quadrilateral elements. Their analyses were verified by results from full numerical solutions of the simple scalar advection equation and the Euler equations governing acoustic wave propagation. The study showed that the dispersion relation and the dissipation rate depend on the interfacial flux formula. Specifically, in the case of a scalar advection equation, the dissipation error is dominant relative to the dispersion error if an upwind flux is used. The choice of the order of the method is then determined by the accuracy limit imposed on the dissipation error; the dispersion relation is almost exactly satisfied for nondimensional wavenumbers up to a value equal to the order of the method. For the centered flux, the dissipation rate is exactly zero, but the range of wavenumbers for which the discrete dispersion relation accurately approximates the exact one is relatively small. The orientation of elements in a mesh introduces anisotropy in the phase speed as well as the damping rate. The anisotropy in the dissipation rate is more pronounced than that in the dispersion relation. See Hu, Hussaini and Rasetarinera (1999) for more details on the effects of the mesh upon numerical anisotropy. Stanescu, Kopriva and Hussaini (2000) relaxed the assumption of periodicity, thereby providing the spatial distribution of both amplitude and phase errors within the elements.

5.8.4 Nonlinear Conservation Laws in 1D

We consider now the case of a nonlinear *scalar* conservation law

$$\frac{\partial u}{\partial t} + \frac{\partial \mathcal{F}}{\partial x}(u) = 0, \qquad a < x < b, \quad t > 0, \tag{5.8.27}$$

where $\mathcal{F}(u)$ is a nonlinear function of u, with the initial condition $u_{|t=0} = u_0$ and boundary condition $u_{|x=a} = g$ (we are assuming that $x = a$ is the inflow point for all $t > 0$). We discretize this equation by the SDGM as follows, using the same notation as in the previous subsection. For every $t > 0$, we look for $u_\delta(t) \in W_\delta$ such that, for all $m = 1, \ldots, M$,

$$\left(\frac{\partial u_\delta}{\partial t}, v_\delta\right)_m - \left(\mathcal{F}(u_\delta), \frac{\partial v_\delta}{\partial x}\right)_m + H_m(u_\delta)v_\delta(\overline{x}_m)$$
$$- H_{m-1}(u_\delta)v_\delta(\overline{x}_{m-1}) = 0 \quad \forall v_\delta \in \mathbb{P}_{N_m}. \tag{5.8.28}$$

The initial data u_δ^0 is given by the set of equations

$$(u_\delta^0, v_\delta)_m = (u_0, v_\delta)_m \qquad \forall v_\delta \in \mathbb{P}_{N_m}, \quad \forall m = 1, \ldots, M.$$

The function H_m denotes the *numerical flux* at the node $\overline{x}_m$. Ideally this would be the exact solution to the Riemann problem (see Sect. 5.8.3) defined by the conservation law with discontinuous data at the node $\overline{x}_m$, i.e., $u_L = u_\delta^-(\overline{x}_m, t)$ and $u_R = u_\delta^+(\overline{x}_m, t)$. For nonlinear conservation laws, however, the Riemann problem is usually only solved approximately (even when an exact solution is obtainable), and we denote the resulting *numerical flux* by

$$H_m(u_\delta(t)) = H\left(u_\delta^-(\overline{x}_m, t), u_\delta^+(\overline{x}_m, t)\right). \tag{5.8.29}$$

Furthermore, we set $u_\delta^-(\overline{x}_0, t) = g(t)$, which is the inflow datum; on the other hand, the boundary flux $H_M(u_\delta)$ has to be properly designed to avoid spurious reflections at $x = b$.

Using an approximate solution to the Riemann problem introduces an error into the numerical solution that is in addition to the error due to the underlying discretization (SDGM in the present case). As we shall see in the subsequent numerical examples, this error appears to be negligible for all practical purposes—it does not interfere with the achievement of infinite-order convergence as N is increased.

Of all the possible types of numerical fluxes, the preference is for one that yields a *monotone* scheme when $N_m = 0$ (i.e., when piecewise constant polynomials are used), namely, a scheme for which the constant value $u_\delta^{(m)}$ at the new time-level t^{n+1} depends on $u_\delta^{(m-1)}$, $u_\delta^{(m)}$ and $u_\delta^{(m+1)}$ at the previous time-level t^n through an increasing monotone function. This preference rests on the fundamental result, due to N.N. Kuznetsov (1976), that monotone schemes are stable and converge to the solution with nondecreasing entropy.

The latter is the one among all weak solutions that is physically relevant (the one that, e. g., is the unique limit for $\nu \to 0$ of the solutions of the regularized equations obtained by adding a viscous term $-\nu \partial^2 u/\partial x^2$ to (5.8.27); see, e. g., LeVeque (2002), Chap. 11).

When $N_m = 0$, the scheme (5.8.28) becomes, for all $m = 1, \ldots, M$,

$$h_m \frac{d}{dt} u_\delta(t) + H_m(u_\delta(t)) - H_{m-1}(u_\delta(t)) = 0 .$$

The initial datum is $u_\delta^{0,(m)} = h_m^{-1} \int_{\bar{x}_{m-1}}^{\bar{x}_m} u_0$ on Ω_m.

In order for the scheme to be monotone, the numerical flux itself should be monotone. This amounts to requiring that the function $H = H(v, w)$ must be Lipschitz continuous with respect to both its arguments, nondecreasing with respect to v and not increasing with respect to w, and consistent with the physical flux $\mathcal{F}$, i. e., $H(\bar{u}, \bar{u}) = \mathcal{F}(\bar{u})$, for every constant $\bar{u}$. Some celebrated examples of monotone fluxes are (in increasing order of complexity) the *Godunov* flux

$$H(v, w) = \begin{cases} \min_{v \le u \le w} \mathcal{F}(u) & \text{if } v \le w , \\ \max_{w \le u \le v} \mathcal{F}(u) & \text{if } v > w , \end{cases} \qquad (5.8.30)$$

the *Lax–Friedrichs* flux

$$H(v, w) = \frac{1}{2}[\mathcal{F}(v) + \mathcal{F}(w) - \delta(w - v)] , \qquad (5.8.31)$$

where

$$\delta = \max_{\inf_x u_0(x) \le y \le \sup_x u_0(x)} |\mathcal{F}'(y)|$$

(note that if

$$\delta = \delta(v, w) = \max_{\min(v,w) \le y \le \max(v,w)} |\mathcal{F}'(y)| ,$$

then $H(v, w)$ is called a *Rusanov* flux), the *Roe* flux

$$H(v, w) = \frac{1}{2}[\mathcal{F}(v) + \mathcal{F}(w) - |A_{vw}|(w - v)] , \qquad (5.8.32)$$

where $A_{vw} = A(u_{vw})$, $|A_{vw}| = (A_{vw})_p - (A_{vw})_n$ denotes its positive part, and u_{vw} is the *Roe average* of v and w, i. e., a solution to

$$\mathcal{F}(w) - \mathcal{F}(v) = A_{vw}(w - v) , \qquad (5.8.33)$$

and the *Engquist–Osher* flux

$$H(v, w) = \int_0^v \max(\mathcal{F}'(u), 0) \, du + \int_0^w \min(\mathcal{F}'(u), 0) \, du + \mathcal{F}(0) . \quad (5.8.34)$$

The Lax–Friedrichs flux and the Roe flux are the ones most widely used in applications of SDGM.

In the scalar, linear case, in which $\mathcal{F}(u) = \beta u$, where β is continuous, all cases reduce to the *upwind* flux

$$H(v, w) = \beta \frac{v + w}{2} - \frac{|\beta|}{2}(w - v).$$

In that case it is easy to see that the formulations (5.8.28) and (5.8.21) do coincide. We point out that the SDG formulation (5.8.7) may be regarded as derived from the weak form of the exact problem (5.8.1) in which we replace the exact flux by the upwind numerical flux. Also, note that counter-integrating by parts the second term in (5.8.28) yields, for $m = 0, \ldots, M - 1$,

$$\left(\frac{\partial u_\delta}{\partial t} + \frac{\partial \mathcal{F}}{\partial x}(u_\delta), v_\delta \right)_m + (H_m(u_\delta) - \mathcal{F}(u_\delta^{(m)}))v_\delta|_{\bar{x}_m}$$

$$+ (\mathcal{F}(u_\delta^{(m)}) - H_{m-1}(u_\delta))v_\delta|_{\bar{x}_{m-1}} = 0. \qquad (5.8.35)$$

If H is an upwind flux, then the term at $x = \bar{x}_m$ disappears, and (5.8.35) can be regarded as the counterpart of the formula (5.8.8) that we have derived for the linear advection problem.

Consider now the case of a *system* of nonlinear conservation laws

$$\frac{\partial \mathbf{q}}{\partial t} + \frac{\partial}{\partial x} \mathcal{F}(\mathbf{q}) = \mathbf{0}, \qquad a < x < b, \quad t > 0, \qquad (5.8.36)$$

where $\mathbf{q}(x, t) \in \mathbb{R}^q$, $q \geq 2$, and $\mathcal{F}(\mathbf{q})$ is a nonlinear vector function called the flux. (Here we use the notation $\mathbf{q}$, rather than $\mathbf{u}$ as in the linear case, for the vector of unknowns in order to be consistent with the notation used in Chap. 4 for the dependent variables in the Euler equations.)

This system must be supplemented by a suitable set of boundary conditions at the endpoints $x = a$ and $x = b$ and by an initial condition $\mathbf{q}(x, 0) = \mathbf{q}_0(x)$, $a < x < b$.

For all $t > 0$, the SDGM for system (5.8.36) consists of looking for a polynomial vector $\mathbf{q}_\delta(t) \in (W_\delta)^q$ such that, for every $m = 1, \ldots, M$,

$$\left(\frac{\partial \mathbf{q}_\delta}{\partial t}, \mathbf{v}_\delta \right)_m - \left(\mathcal{F}(\mathbf{q}_\delta), \frac{\partial \mathbf{v}_\delta}{\partial x} \right)_m$$

$$+ (\mathbf{v}_\delta(\bar{x}_m))^T \mathbf{H}_m(\mathbf{q}_\delta) - (\mathbf{v}_\delta(\bar{x}_{m-1}))^T \mathbf{H}_{m-1}(\mathbf{q}_\delta) = 0 \qquad \forall \mathbf{v}_\delta \in (W_\delta)^q. \qquad (5.8.37)$$

For every m, $\mathbf{H}_m(\mathbf{q}_\delta)$ is a suitable (vector) numerical flux depending on $\mathbf{q}_\delta^-(\bar{x}_m, t)$ and $\mathbf{q}_\delta^+(\bar{x}_m, t)$.

For the extreme boundary elements Ω_1 and Ω_M, the boundary terms that arise from (5.8.37) are dealt with using the characteristic compatibility method as done in (5.8.14). The matrix A is represented by the local Jacobian matrix of the flux $\mathcal{F}$, say $A = \mathcal{F}' = \left(\frac{\partial \mathcal{F}_i}{\partial x_j} \right)$, that is, $A = \mathcal{F}'(\mathbf{q}_\delta(a,t))$ at $x = a$, while $A = \mathcal{F}'(\mathbf{q}_\delta(b,t))$ at $x = b$. The matrices A_n and A_p are those associated with A.

As noted earlier the numerical flux is usually obtained through the approximate solution of a Riemann problem at every $\overline{x}_{m-1}$, that is, a problem like (5.8.36) whose initial values are two constant states, $\mathbf{q}_\delta^L$ for $x < \overline{x}_{m-1}$ and $\mathbf{q}_\delta^R$ for $x > \overline{x}_{m-1}$. (Since finding the exact solution of the Riemann problem can be computationally expensive, a suitable alternative is to compute an approximate one, that is, to resort to an *approximate Riemann solver*, as previously discussed in connection with (5.8.28).) See LeVeque (2002) for a comprehensive description of the various numerical fluxes and thorough references. Here we just describe two among the most popular—the Lax–Friedrichs and the Roe fluxes.

In the former case, the generalization of the scalar Lax–Friedrichs flux (5.8.31) is

$$ \mathbf{H}_m(\mathbf{q}_\delta) = \frac{1}{2} \left(\mathcal{F}(\mathbf{q}_\delta^{(m)}) + \mathcal{F}(\mathbf{q}_\delta^{(m+1)}) \right)_{|\overline{x}_m} - \frac{1}{2} |\lambda_{\max}(A)| [\mathbf{q}_\delta]_{\overline{x}_m}, \qquad (5.8.38) $$

where $|\lambda_{\max}(A)|$ is the maximum (in absolute value) eigenvalue of A, while $[\mathbf{q}_\delta]_{\overline{x}_m}$ is the jump $\mathbf{q}_\delta^R - \mathbf{q}_\delta^L = \mathbf{q}_\delta^{(m+1)} - \mathbf{q}_\delta^{(m)}$ of the vector $\mathbf{q}_\delta$ at $\overline{x}_m$.

The Roe flux is generalized from (5.8.32) to the case of systems as follows. For $m = 2, \ldots, M-1$,

$$ \mathbf{H}_m(\mathbf{q}_\delta) = \frac{1}{2} \left(\mathcal{F}(\mathbf{q}_\delta^{(m)}) + \mathcal{F}(\mathbf{q}_\delta^{(m+1)}) \right)_{|\overline{x}_m} - \frac{1}{2} |A_{LR}| [\mathbf{q}_\delta]_{\overline{x}_m}. $$

The matrix A_{LR} is a local "linearization" of the flux $\mathcal{F}$ at $\overline{x}_m$, and $|A_{LR}|$ denotes its positive part, that is, $(A_{LR})_p - (A_{LR})_n$. A_{LR} is chosen in a problem-dependent way to satisfy the "linearized Rankine–Hugoniot condition"

$$ \mathcal{F}(\mathbf{q}_\delta^R) - \mathcal{F}(\mathbf{q}_\delta^L) = A_{LR}(\mathbf{q}_\delta^R - \mathbf{q}_\delta^L), \qquad (5.8.39) $$

as illustrated for the Euler equations in Sect. 5.12.

5.9 SDGM for Multidimensional Problems

The SDGM extends to multidimensional problems in fairly straightforward fashion, except of course, for the complexities of the interfaces in multiple dimensions. In this section we first present the basic formulations of SDGM for linear, hyperbolic problems and for nonlinear conservation laws for problems with geometrically conforming subdomains. Then, we outline the interface treatment in the more general, nonconforming case.

5.9.1 Multidimensional Formulation

Linear Problems. Consider first the multidimensional version of the advection equation (5.8.19):

$$\frac{\partial u}{\partial t} + \mathcal{L}u = f, \qquad \mathbf{x} \in \Omega \subset \mathbb{R}^d, \ t > 0,$$
$$u(x,t) = g(x,t), \qquad \mathbf{x} \in \partial\Omega^{\text{in}}, \ t > 0, \qquad (5.9.1)$$
$$u(x,0) = u_0(x), \qquad \mathbf{x} \in \Omega,$$

where $\mathcal{L}u = \nabla \cdot (\boldsymbol{\beta} u) + \gamma u$ with $\boldsymbol{\beta}$, γ, f, u_0 and g are given functions, and

$$\partial\Omega^{\text{in}} = \{\mathbf{x} \in \partial\Omega \ : \ \boldsymbol{\beta}(\mathbf{x}) \cdot \mathbf{n}(\mathbf{x}) < 0\}.$$

Here $\mathbf{n}$ denotes the outward unit normal to $\partial\Omega$. Consider a partition of Ω into M geometrically conforming disjoint subdomains Ω_m, i.e., for a partition in which all subdomains share either a complete edge (in 2D and 3D) or a complete face (in 3D), as in Figs. 5.5 and 5.6.

The SDGM for the approximation of (5.9.1) is defined as follows. For every $t > 0$, we look for $u_\delta(t) \in W_\delta$ such that $u_\delta(0) = u_{0,\delta}$ and, for all $v_\delta \in W_\delta$,

$$\left(\frac{\partial u_\delta}{\partial t}, v_\delta\right)_\Omega + \sum_{m=1}^{M} \left\{ a_{\Omega_m}(u_\delta, v_\delta) - \int_{\partial\Omega_m^{\text{in}}} \boldsymbol{\beta} \cdot \mathbf{n}_m \, [u_\delta] v_\delta^+ \, d\gamma \right\} = (f, v_\delta)_\Omega,$$
$$(5.9.2)$$

where

$$W_\delta = \{v \in L^2(\Omega) \ : \ v|_{\Omega_m} \in V_{N_m}(\Omega_m), \ m = 1, \ldots, M\}, \qquad (5.9.3)$$

$V_{N_m}(\Omega_m)$ being the space introduced in (5.3.3); $\mathbf{n}_m$ denotes the outward unit normal to $\partial\Omega_m$ and

$$\partial\Omega_m^{\text{in}} = \{\mathbf{x} \in \partial\Omega_m \ : \ \boldsymbol{\beta}(\mathbf{x}) \cdot \mathbf{n}_m(\mathbf{x}) < 0\}.$$

The bilinear form a_{Ω_m} is defined as

$$a_{\Omega_m}(u, v) = \int_{\Omega_m} (\nabla \cdot (\boldsymbol{\beta} u) + \gamma u)v,$$

while the jump function is

$$[u(\mathbf{x})] = \begin{cases} u^+(\mathbf{x}) - u^-(\mathbf{x}), & \mathbf{x} \notin \partial\Omega^{\text{in}}, \\ u^+(\mathbf{x}) - g(\mathbf{x}), & \mathbf{x} \in \partial\Omega^{\text{in}}, \end{cases}$$

where

$$u^\pm(\mathbf{x}) = \lim_{s \to 0^\pm} u(\mathbf{x} + s\boldsymbol{\beta}), \qquad \mathbf{x} \in \partial\Omega_m. \qquad (5.9.4)$$

Since the test functions are discontinuous, by taking for every $m = 1, \ldots, M$ a test function that vanishes identically outside Ω_m, we recover from (5.9.2) a set of uncoupled equations involving only integrals over Ω_m. The glueing term is the one involving $\partial \Omega_m^{\text{in}}$, which should now be regarded as a means to prescribe inflow boundary data for $u_\delta^{(m)} = u_\delta|_{\Omega_m}$. Specifically,

$$\left(\frac{\partial u_\delta}{\partial t}, v_{N_m}\right)_{\Omega_m} + a_{\Omega_m}(u_\delta, v_{N_m}) + \int_{\partial \Omega_m^{\text{in}}} |\boldsymbol{\beta} \cdot \mathbf{n}_m|[u_\delta]v_{N_m}^+ \, \mathrm{d}\gamma$$
$$= (f, v_{N_m})_{\Omega_m} \qquad \forall v_{N_m} \in V_{N_m}(\Omega_m).$$

This can be regarded as the counterpart of (5.8.21) (up to an integration by parts) for the time-dependent multidimensional case. Indeed, if, for instance, Ω_m is an internal element (no boundary edges), we obtain, for all $v_{N_m} \in V_{N_m}(\Omega_m)$, and with the obvious meaning of $\partial \Omega^{\text{out}}$

$$\left(\frac{\partial u_\delta}{\partial t}, v_{N_m}\right)_{\Omega_m} + (u_\delta, \mathcal{L}^* v_{N_m})_{\Omega_m}$$
$$+ \int_{\partial \Omega_m^{\text{out}}} (\boldsymbol{\beta} \cdot \mathbf{n}_m) u_\delta^+ v_{N_m} \, \mathrm{d}\gamma + \int_{\partial \Omega_m^{\text{in}}} (\boldsymbol{\beta} \cdot \mathbf{n}_m) u_\delta^- v_{N_m} \, \mathrm{d}\gamma = (f, v_{N_m})_{\Omega_m},$$

where $\mathcal{L}^* v = -\boldsymbol{\beta} \cdot \nabla v + \gamma v$ is the adjoint of the spatial operator $\mathcal{L}v$.

For the case in which the partition is geometrically conforming and the polynomial order along all common edges and faces is the same in the intersecting subdomains, the SDGM-NI version can be easily obtained by replacing exact integrals in Ω_m by their Gaussian quadrature approximations. However, in this case if $\boldsymbol{\beta}$ is not constant, the skew-symmetric decomposition of the bilinear form is the only one that rigorously guarantees stability, as seen in (5.4.15) and CHQZ2, Sects. 6.4.3 and 6.5.2.

When the numerical integration is based on the LGL quadrature, the SDGM-NI formulation can still be shown to be equivalent to a generalized collocation problem in which the inflow boundary conditions on $\partial \Omega_m^{\text{in}}$ are recovered in a weak form on every subdomain Ω_m.

However, for SDGM-NI on multidimensional problems, LG (Legendre Gauss) quadratures are typically used for a variety of reasons, which are similar to the reasons that the staggered grid method is preferred for multidimensional patching methods (see Sects. 4.6.1 and 5.13.3). Considering just the two-dimensional situation, some of the considerations are the following. (i) "Cross-points", which are corners shared by four domains in two-dimensions (see Fig. 5.51), are much easier to treat. (ii) When the LGL distribution is used, there is ambiguity about the normal direction to be used in the solution of the Riemann problem (see below). (iii) A computer code must account for the various cases of different numbers of domains sharing a corner, greatly increasing the bookkeeping burden. (iv) Inter-subdomain communication increases from just two subdomains to all the subdomains sharing the corner (impeding efficient parallel execution). (v) Accuracy is limited when the subdomain metrics are not continuous across the subdomain boundary. Finally, (vi) the mass matrices are not diagonal because the

LGL quadrature is not exact. Many of these reasons are even more compelling in three dimensions.

In the vector case, let us consider the hyperbolic system (4.2.27) with appropriate initial and boundary conditions. Again assuming a geometrically conforming partition Ω_m, $m = 1, \ldots, M$, of Ω, the SDGM consists of finding, for every $t > 0$, a vector $\mathbf{u}_\delta(t) \in (W_\delta)^q$ such that for all $\mathbf{v}_\delta \in (W_\delta)^q$ one has (in analogy with the 1D-case (5.8.15))

$$\sum_{m=1}^{M} \left(\frac{\partial \mathbf{u}_\delta}{\partial t}, \mathbf{v}_\delta \right)_{\Omega_m} + (\mathbf{u}_\delta, \mathcal{L}^* \mathbf{v}_\delta)_{\Omega_m}$$

$$- \int_{\partial \Omega_m \backslash \partial \Omega} \langle \mathbf{H}_m(\mathbf{u}_\delta), \mathbf{v}_\delta \rangle \, d\gamma = \sum_{m=1}^{M} (\mathbf{g}, \mathbf{v}_\delta)_{\Omega_m} + \text{boundary terms} \,,$$

where $\mathcal{L}^*$ is the adjoint operator of $\mathcal{L}\mathbf{u} = \sum_{j=1}^{d} A_j \frac{\partial \mathbf{u}}{\partial x_j} + C\mathbf{u}$, that is $\mathcal{L}^* \mathbf{v} = -\sum_{j=1}^{d} \frac{\partial}{\partial x_j}(A_j \mathbf{v}) + C^T \mathbf{v}$, while $\mathbf{H}_m$ is a (multidimensional) numerical flux, formally defined as is its 1D counterpart in (5.8.16). Precisely, for each point $\mathbf{x} \in \partial \Omega_m$, if $\mathbf{n}_m(\mathbf{x})$ denotes the outward normal unit vector to $\partial \Omega_m$ at the point $\mathbf{x}$, and ν_j are its components, we set $A(\mathbf{n}_m) = \sum_{j=1}^{d} \nu_j A_j(\mathbf{x})$ and call $A_p(\mathbf{n}_m)$ and $A_n(\mathbf{n}_m)$ its positive and negative parts, respectively. Then, the numerical flux is defined as

$$\langle \mathbf{H}_m(\mathbf{u}_\delta), \mathbf{v}_\delta \rangle_{|\mathbf{x}} = [\mathbf{v}_\delta]^T A(\mathbf{n}_m) \mathbf{u}_\delta^{(m)}|_{\mathbf{x}} + [\mathbf{v}_\delta]^T A_n(\mathbf{n}_m)[\mathbf{u}_\delta]|_{\mathbf{x}} \,.$$

Boundary terms are treated via the usual locally one-dimensional CCM approach described in Sect. 4.2.1.

Nonlinear Conservation Laws. Next, consider the case of nonlinear hyperbolic systems of conservation laws

$$\frac{\partial \mathbf{q}}{\partial t} + \nabla \cdot \mathcal{F}(\mathbf{q}) = \mathbf{0} \,, \qquad \mathbf{x} \in \Omega \,, \quad t > 0 \,, \tag{5.9.5}$$

where Ω is a domain of $\mathbb{R}^d$, $d \geq 2$, and $\mathbf{q} = \mathbf{q}(\mathbf{x}, t) \in \mathbb{R}^q$, for $q \geq 2$, is the unknown vector field, supplemented with suitable initial and boundary conditions. The flux $\mathcal{F}$ is a second-order tensor that depends upon $\mathbf{q}$.

By defining the matrices $A_i = A_i(\mathbf{q})$, $i = 1, \ldots, d$, with

$$(A_i(\mathbf{q}))_{rs} = \frac{\partial \mathcal{F}_{ri}}{\partial q_s}(\mathbf{q}), \qquad r, s = 1, \ldots, q \,, \tag{5.9.6}$$

we can rewrite (5.9.5) in the nonconservative form

$$\frac{\partial \mathbf{q}}{\partial t} + \sum_{i=1}^{d} A_i(\mathbf{q}) \frac{\partial \mathbf{q}}{\partial x_i} = \mathbf{0} \,. \tag{5.9.7}$$

This is a quasi-linear system, which reduces to (4.2.27) if the flux is linear. Note that the multidimensional Euler system is a special case of (5.9.5). An

SDG approximation of (5.9.5) can be set up by generalizing the approach followed for 1D systems of conservation laws.

We look for $\mathbf{q}_\delta \in (W_\delta)^q$, where W_δ is the space of (images of) polynomials introduced in (5.9.3) that satisfies, in every element Ω_m of the decomposition,

$$\left(\frac{\partial \mathbf{q}_\delta}{\partial t}, \mathbf{v}_\delta\right)_{\Omega_m} - (\mathcal{F}(\mathbf{q}_\delta), \nabla \mathbf{v}_\delta)_{\Omega_m}$$

$$+ \int_{\partial\Omega_m} (\mathbf{v}_\delta^{(m)})^T \mathbf{H}_m(\mathbf{q}_\delta; \mathbf{n}_m) = 0 \qquad \forall \mathbf{v}_\delta \in (W_\delta)^q . \qquad (5.9.8)$$

The vector function $\mathbf{H}_m$ is a suitable numerical normal flux. Its definition is more involved than in the scalar case. However, a quite common approach consists of constructing $\mathbf{H}_m$ through the solution of local 1D Riemann problems, which were addressed in Sect. 5.8.4.

For the assumed (for now) geometrically conforming partition, let Γ_{km} denote the common edge in 2D (or face in 3D) between Ω_k and Ω_m. Let $\mathbf{u}^{(k)} = \mathbf{q}_{\delta|\Gamma_{km}}^{(k)}$ and $\mathbf{u}^{(m)} = \mathbf{q}_{\delta|\Gamma_{km}}^{(m)}$ be the two restrictions to Γ_{km} of $\mathbf{q}_\delta^{(k)}$ and $\mathbf{q}_\delta^{(m)}$, respectively. Then, on Γ_{km} $\mathbf{u}^{(k)}$ and $\mathbf{u}^{(m)}$ are two (images of) polynomials of (possibly) different degree, N_k and N_m, respectively.

Assume, for instance, that $N_m \geq N_k$. In the 2D case, Γ_{km} is an edge (either straight or curved). We denote by $\mathbf{x}_j$, $j = 0, \ldots, N_m$, the $N_m + 1$ interpolation points on Γ_{km} that are used for the representation of $\mathbf{u}^{(m)}$. They can be the nodes on Γ_{km} obtained as images of either the LGL or the LG nodes on the reference interval $[-1, 1]$. Then, we generate the values of $\mathbf{u}^{(k)}$ at the nodes $\mathbf{x}_j$ either by interpolation or by L^2-projection on the space of (images of) polynomials of degree N_m (the two strategies yield the same result since $N_k \leq N_m$).

At every node $\mathbf{x}_j$ we now have two "constant" states, $\mathbf{u}^{(m)}(\mathbf{x}_j)$ and $\mathbf{u}^{(k)}(\mathbf{x}_j)$, which can be used as "left" and "right" states to solve an approximate 1D Riemann problem along the direction $\mathbf{n}^{(j)}$ normal to Γ_{km}. The components of the flux in the tangential direction are simply treated in an upwind fashion.

Let $\mathbf{H}_{mk}^{(l)}$ denote the corresponding numerical flux, and let $\mathbf{H}_{mk}$ be the interpolant of degree N_m that is obtained from the gridvalues $\{\mathbf{H}_{mk}^{(l)}, l = 0, \ldots, N_m\}$.

Now $\mathbf{H}_{mk}$, which is (the image of) a polynomial of degree N_m on Γ_{km}, is projected back by an L^2-projection to generate two numerical fluxes on Γ_{km}, one for Ω_m, say $\mathbf{H}_{mk}^{(m)}$, the other for Ω_k, say $\mathbf{H}_{mk}^{(k)}$.

In practice, all these interface treatments are more easily accomplished on reference elements. This requires an a priori transformation of the original equation (5.9.8) on $\hat{\Omega}_C^d$ or $\hat{\Omega}_S^d$ (see Sect. 5.3.1). One possible map is a Piola transform. This is discussed in Sect. E.3, where it is shown that the Piola transform guarantees conservation for exact integration. There are additional considerations when quadrature is employed; for the SGDM-NI method, Kopriva (2006) describes precisely how to implement mappings so as to achieve exact conservation for the discretized problem.

5.9.2 The Mortar Technique for Geometrical Nonconformities

A nonconforming partition of the domain is often a necessity in practical applications of SDGM. The left frame of Fig. 5.31 illustrates a simple partition of a square. The interface between Ω_1 and Ω_2 is geometrically conforming, whereas their interfaces with Ω_3 are nonconforming.

For the SEM, geometrically nonconforming interfaces are treated with the use of mortars, and in that context its developers dignified this approach by naming it a separate method—the mortar element method (see Sect. 5.7). The mortar technique has been adapted to SDGM by Kopriva, Woodruff and Hussaini (2002), following the earlier adaptation to multidomain collocation methods by Kopriva (1996a). In more recent SDGM, mortars are regarded as an intrinsic component of the method rather than a distinction requiring a different name for the method. At the end of the previous subsection, we described how to treat the interface between Ω_1 and Ω_2 when the order of approximation in the x-coordinate differed between the two subdomains. This can be viewed as a special type of mortar. Here we focus on the use of mortars for treating geometric nonconformities.

The right frame of Fig. 5.31 illustrates the mortars used for generating fluxes on both sides of the interface after solving locally one-dimensional Riemann problems. For geometrically nonconforming interfaces, the mortar approach to interface treatment is the same as that for order nonconformities; for every pair of adjoining edges (or faces), the mortar edge is chosen as the shortest edge and the higher polynomial degree of the two edges is used. (This is different from the case of elliptic problems, where the choice of the mortar can be made indifferently). Then, the two vector functions $\mathbf{u}_\delta^{(m)}$ and $\mathbf{u}_\delta^{(k)}$ are first L^2-projected onto the mortar. Next, local Riemann problems are solved at each grid point. Then a unique mortar flux is generated by interpolation,

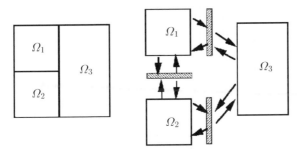

Fig. 5.31. Diagram of the mortar communication between three subdomains that subdivide a square. [From D.A. Kopriva (1996a)]

and this is projected back to the adjoining elements. See Kopriva, Woodruff and Hussaini (2002) for more details.

We now present a numerical example, taken from the work of Kopriva, Woodruff and Hussaini (2002), of an SDGM computation for a two-dimensional linear hyperbolic problem. These electromagnetics computations use *Maxwell's equations*, which can be written in the form

$$\frac{\partial \mathbf{q}}{\partial t} + \nabla \cdot \mathcal{F}(\mathbf{q}) = \mathbf{S}, \qquad \mathbf{x} \in \Omega, \quad t > 0,$$

with

$$\mathbf{q} = \begin{pmatrix} \mathbf{B} \\ \mathbf{D} \end{pmatrix}, \qquad \mathcal{F}_i = \begin{pmatrix} \mathbf{e}_i \times \mathbf{E} \\ -\mathbf{e}_i \times \mathbf{H} \end{pmatrix}, \qquad \mathbf{D} = \epsilon \mathbf{E}, \qquad \mathbf{B} = \mu \mathbf{H},$$

where $\mathbf{e}_i$ is a unit vector in the i-th coordinate direction, μ and ϵ are the permittivity and permeability, respectively, which are taken to be constant in this example, and $\mathbf{S}$ is a problem-dependent source term. The field variables are the electric field $\mathbf{E}$, the dielectric displacement $\mathbf{D}$, the magnetic field $\mathbf{B}$, and the magnetic intensity $\mathbf{H}$. In this particular problem, an incident sinusoidal plane electromagnetic wave is scattered by double slits (symmetrically placed about the origin) in an infinitesimally thin, perfectly conducting plate. See the left frame of Fig. 5.32 for the geometry and the computational domain. To the left of the plate, two waves are specified as the external boundary state. The first is the incident wave as defined, and the second is the wave that would be reflected from the conducting plate if the slits were not present.

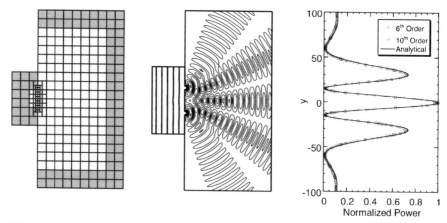

Fig. 5.32. Computational mesh (*left*), the diffraction contours (*middle*), and comparison of sixth and tenth-order SDGM results of the power spectrum (*right*) of SDGM electromagnetics computations using the two-dimensional Maxwell equations [From D.A. Kopriva, S.L. Woodruff, M.Y. Hussaini (2002)]

The left frame of Fig. 5.32 depicts the subdomains used in the computation. The subdomains are the size of one wavelength, except in the vicinity of the slits where they are much smaller in order to resolve the details of the geometry. To the right of the slits, a damping layer two wavelengths thick was used to damp the outgoing waves. The grid consisted of 262 elements wherein the solution was represented by Legendre polynomials of various degrees. This SDGM computation utilized LG quadrature, a mortar treatment to handle both the geometrically nonconforming interfaces and the different resolutions in some adjoining subdomains, and the relatively simple exact solution for the relevant Riemann problem (Mohammadian, Shankar and Hall (1991)). The physical domain of interest is surrounded by damping layers of two-wavelength thickness, which are shown shaded. (The damping layer serves a similar function in the electromagnetics problem as the buffer domain, discussed in Sect. 3.6.2, serves for incompressible flow, i.e., to permit waves to propagate through an artificial computational boundary without generating spurious reflected waves. See Stanescu et al. (1999) for details of the damping layer.)

The middle frame of Fig. 5.32 shows the instantaneous contours of the electric field phasor in the physical domain. The exact solution at large distances from the slit is approximated well by the analytical solution corresponding to treating the slits as electric dipoles. Figure 5.32 (right) provides a comparison of the computed radiated power (the modulus of the electric field phasor squared) with the asymptotic solution along a vertical line roughly in the middle of the rectangular region to the right of the slits. The SDGM solutions were computed with sixth-degree and tenth-degree Legendre polynomials (i.e., six and ten points per wavelength). In this case we see agreement to within the accuracy of the graph between the computed and analytic solutions, except for an error of about 8% near the top and bottom damping layers.

5.10 SDGM for Diffusion Equations

Thus far, we have discussed only the use of SDGM for first-order partial differential equations. The generalization of SDGM to second-order partial differential equations requires some extra care. For ease of exposition, we address only the case of the Poisson problem with homogeneous Dirichlet boundary conditions (5.4.19) in a domain Ω partitioned into the union of M disjoint subdomains Ω_m, $m = 1, \ldots, M$.

Heuristically, the SDGM can be derived as follows. Let v_δ be a test function belonging to

$$W_\delta^0 = \{v \in W_\delta : v|_{\partial\Omega} = 0\},$$

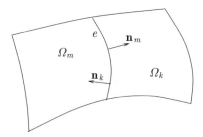

Fig. 5.33. An "edge" separating two adjoining subdomains (or elements)

where W_δ is defined in (5.9.3). Then

$$\sum_{m=1}^{M}(-\triangle u, v_\delta)_{\Omega_m} = \sum_{m=1}^{M}\left((\nabla u, \nabla v_\delta)_{\Omega_m} - \int_{\partial\Omega_m} v_\delta \nabla u \cdot \mathbf{n}_m\right), \quad (5.10.1)$$

where $\mathbf{n}_m$ still denotes the outward unit normal to $\partial\Omega_m$. Denoting by $\mathcal{E}_\delta$ the union of all internal edges, i. e., the interelement boundaries (boundary edges can be disregarded since v_δ vanishes there), we can rearrange terms to get

$$-\sum_{m=1}^{M}\int_{\partial\Omega_m} v_\delta \nabla u \cdot \mathbf{n}_m = -\sum_{e\in\mathcal{E}_\delta}\int_e (v_\delta^+\nabla u^+ \cdot \mathbf{n}^+ + v_\delta^-\nabla u^- \cdot \mathbf{n}^-)|_e, \quad (5.10.2)$$

where the signs "+" and "−" denote information coming from the different sides of the edge under examination. For instance, in the case of Fig. 5.33, we can take $\mathbf{n}^+ = \mathbf{n}_m$ and $\mathbf{n}^- = \mathbf{n}_k$. This corresponds to a given choice of the orientation of e.

We make use of the following notation concerning averages and jumps on edges:

$$\{v\} = \frac{v^+ + v^-}{2}, \qquad\qquad [v] = v^+\mathbf{n}^+ + v^-\mathbf{n}^-,$$

$$\{\!\!\{\nabla w\}\!\!\} = \frac{(\nabla w)^+ + (\nabla w)^-}{2}, \quad [\!\![\nabla w]\!\!] = (\nabla w)^+ \cdot \mathbf{n}^+ + (\nabla w)^- \cdot \mathbf{n}^-.$$

Then with the help of a little algebra, we obtain

$$
\begin{aligned}
v_\delta^+\nabla u^+ \cdot \mathbf{n}^+ + v_\delta^-\nabla u^- \cdot \mathbf{n}^- &= 2[v_\delta]\{\!\!\{\nabla u\}\!\!\} - (v_\delta^+\nabla u^- \cdot \mathbf{n}^+ + v_\delta^-\nabla u^+ \cdot \mathbf{n}^-) \\
&= 2[v_\delta]\{\!\!\{\nabla u\}\!\!\} + 2[\!\![\nabla u]\!\!]\{v_\delta\} \\
&\quad - (v_\delta^+\nabla u^+ \cdot \mathbf{n}^+ + v_\delta^-\nabla u^- \cdot \mathbf{n}^-) ;
\end{aligned}
$$

whence,

$$v_\delta^+\nabla u^+ \cdot \mathbf{n}^+ + v_\delta^-\nabla u^- \cdot \mathbf{n}^- = [v_\delta]\{\!\!\{\nabla u\}\!\!\} + [\!\![\nabla u]\!\!]\{v_\delta\}. \quad (5.10.3)$$

Using (5.10.2) and (5.10.3), we obtain from (5.10.1)

$$\sum_{m=1}^{M} (\nabla u, \nabla v_\delta)_{\Omega_m} - \sum_{e \in \mathcal{E}_\delta} \int_e ([v_\delta]\{\!\!\{\nabla u\}\!\!\} + [\![\nabla u]\!]\{v_\delta\}) = \sum_{m=1}^{M} (f, v_\delta)_{\Omega_m} \, .$$

The previous expression, combined with the observation that the term $[\![\nabla u]\!]\{v_\delta\}$ can be dropped without compromising consistency, motivates the following SDGM for approximating problem (5.4.19). We look for $u_\delta \in W_\delta^0$, which satisfies

$$\sum_{m=1}^{M} (\nabla u_\delta, \nabla v_\delta)_{\Omega_m} - \sum_{e} \int_e ([v_\delta]\{\!\!\{\nabla u_\delta\}\!\!\} + \sigma[u_\delta]\{\!\!\{\nabla v_\delta\}\!\!\})$$

$$= \sum_{m=1}^{M} (f, v_\delta)_{\Omega_m} \, , \qquad \forall v_\delta \in W_\delta^0 \, . \tag{5.10.4}$$

The newly added term $\sigma[u_\delta]\{\!\!\{\nabla v_\delta\}\!\!\}$, in which σ is a suitable constant, does not affect consistency and is added with the purpose of providing more generality and better stability properties. In particular, the special case of $\sigma = 0$ corresponds to the so-called *interior penalty* method, which unfortunately suffers from potential instability (see Babuška et al. (1999), Oden et al. (1998)). Note that for the sake of keeping the notation simple, we enforce the Dirichlet data "strongly" instead of weakly as would normally be done in this approach.

Several variants have been proposed in order to better stabilize the method in the context of finite-element approximations. Here we only give a short description of the most classical ones, referring the interested reader to the paper by Arnold, Brezzi, Cockburn and Marini (2002) for both a general overview and also the detailed stability and convergence analyses.

One variant consists of adding on the left-hand side of (5.10.4) (with $\sigma = 1$) a positive term that penalizes the jumps of u_δ, namely,

$$\sum_{e} \gamma |e|^{-1} \int_e [u_\delta][v_\delta] \, , \tag{5.10.5}$$

where γ is a suitable positive constant, and $|e|$ denotes the length of e.

Another popular variant, consists of using the Bassi and Rebay stabilization term

$$\sum_{e} \gamma \int_e r_e([u_\delta]) \, r_e([v_\delta]) \tag{5.10.6}$$

instead of (5.10.5), where r_e is a suitable operator, which takes the jump of a function on e and produces a new continuous function defined on the elements that share the edge e. See Bassi et al. (1997), as well as Arnold et al. (2002), for further details; see also Bassi and Rebay (1997).

Yet another variant consists of adding (5.10.5) plus replacing the averages $\{\!\{\nabla w\}\!\}$ in (5.10.4) by the *relaxed* averages

$$\{\!\{\nabla w\}\!\}_\theta = \theta \nabla w^+ + (1-\theta)\nabla w^-, \qquad 0 < \theta < 1.$$

An SDGM-NI version can be easily achieved by replacing volume integrals $(\cdot,\cdot)_{\Omega_m}$ by local LGL quadratures and proceeding similarly with the edge intervals.

The previous description generalizes to a problem in which the spatial operator is written as the divergence of a flux, where the flux depends upon ∇u. It is enough to replace ∇u in all previous formulas with the flux. Problems with both first-derivative and second-derivative terms, such as advection-diffusion problems, can be discretized by the SDGM by applying standard DG techniques to the first-derivative terms and the techniques described in this subsection to the second-derivative terms.

The analysis of DG methods for "high-order" finite-element approximation of second-order elliptic problems is given, e. g., in the following references: Houston, Schwab and Süli (2002), Rivière, Wheeler and Girault (2001), Oden, Babuška and Baumann (1998), Wihler, Frauenfelder and Schwab (2003).

The spectral element approximation of the Stokes equations (see Sect. 5.6.1) can be generalized to account for discontinuous velocities within an SDG framework. In particular, if the partition into spectral elements is geometrically conforming, or made by tensor-product elements possibly with hanging nodes, Schötzau, Schwab and Toselli (2002) show that the choice $V_{N_m}(\hat{\Omega}) = \mathbb{Q}_{N_m}$, $Q_{N_m}(\hat{\Omega}) = \mathbb{Q}_{N_m-1}$ as local spaces for velocities and pressures yields a stable method (i. e., free of spurious pressure modes). This result is somehow surprising, as the same choice of local spaces within the classical SEM framework (continuous velocities, discontinuous pressures) would generate spurious modes (see Sect. 3.5.1). The inf-sup constant for the SDG method is shown to be bounded from below by a quantity proportional to $(\max_m N_m)^{-1}$, although there is numerical evidence in 2D (see Toselli (2002)) that it is actually uniformly bounded from below. If the partition is geometrically nonconforming, some results can be found in Filippini and Toselli (2002), where the spaces $\mathbb{Q}_{N_m}$, $\mathbb{Q}_{N_M-2}$ have been used.

5.11 Analysis of SDGM

In this section we recall the main results concerning stability and convergence of spectral discontinuous Galerkin methods. We consider first the 1D linear advection equation (5.8.1) and its (continuous) Galerkin approximations (5.8.2) and (5.8.3). By assuming that the source term $f \in L^2(0,1)$ and that $\frac{1}{2}\beta' + \gamma \geq \alpha > 0$ for a suitable $\alpha > 0$, the solution of (5.8.1) satisfies the inequality

$$\alpha\|u\|^2_{L^2(a,b)} + \beta(b)u(b)^2 \leq \frac{1}{\alpha}\|f\|^2_{L^2(a,b)} + \beta(a)g^2, \tag{5.11.1}$$

which follows upon multiplying by u the first equation in (5.8.1) and integrating by parts.

Solutions of both (5.8.2) and (5.8.3) are stable and satisfy

$$\alpha\|u_N\|^2 + \beta(b)u_N(b)^2 \leq \frac{1}{\alpha}\|f\|^2 + \beta(a)g^2 , \qquad (5.11.2)$$

where $\|\cdot\|$ now represents the $L^2(a,b)$-norm in the case of problem (5.8.2) and the discrete norm $\|\cdot\|_N = \sqrt{(\cdot,\cdot)_N}$ in the case of problem (5.8.3).

To prove the stability of SDGM, we can proceed as follows. On each element, taking $v_{N_m} = u_\delta^{(m)}$ in (5.8.7), using integration-by-parts and the Young inequality yields

$$\alpha\|u_\delta^{(m)}\|^2_{L^2(\Omega_m)} + \frac{1}{2}\beta^-(u_\delta^{(m)})^2|_{\bar{x}_m} \leq \frac{1}{2}\beta^+(u_\delta^{(m-1)})^2|_{\bar{x}_{m-1}} + \left(f, u_\delta^{(m)}\right)_{\Omega_m} ,$$

which guarantees that the local L^2-norm and the outflow flux are safely controlled from the inflow flux and the elemental source term.

Summing up on all elements and using the Cauchy–Schwarz inequality yields

$$\alpha\|u_\delta\|^2_{L^2(a,b)} + \sum_{m=1}^{M-1}\frac{1}{2}(\beta^- - \beta^+)(u_\delta^{(m)})^2|_{\bar{x}_m} + \frac{1}{2}\beta^-\left(u_\delta^{(M)}\right)^2|_{x=b}$$

$$\leq \frac{1}{2}\beta^+(u_\delta^{(0)})^2|_{x=a} + \sum_{m=1}^{M}\left(f, u_\delta^{(m)}\right)_{\Omega_m} . \qquad (5.11.3)$$

If β is continuous, this is a stability estimate. A stability inequality can also be obtained for functions β that are nonincreasing at the interface points $\bar{x}_m$, $m = 1, \ldots, M - 1$. Similar results can be found for the SDGM-NI, provided that $(f, u_\delta^{(m)})_{\Omega_m}$ now reads $(f, u_\delta^{(m)})_{N_m,\Omega_m}$, and $\|u_\delta\|^2_{L^2(a,b)}$ is replaced by its discrete approximation

$$\|u_\delta\|^2_\delta = \sum_{m=1}^{M}(u_\delta^{(m)}, u_\delta^{(m)})_{N_m,\Omega_m} . \qquad (5.11.4)$$

Similar results hold for the SDG solution of the time-dependent problem (5.8.19). Precisely, for all $t > 0$ we obtain

$$\frac{1}{2}\|u_\delta(t)\|^2_{L^2(a,b)} + \int_0^t LH(\tau)\,d\tau \leq \frac{1}{2}\|u_\delta^0\|^2_{L^2(a,b)} + \int_0^t RH(\tau)\,d\tau ,$$

where we have denoted by $LH(\tau)$ and $RH(\tau)$, respectively, the left-hand-side and the right-hand-side of the inequality (5.11.3) at time τ.

These results are true also for the SDGM-NI. However, in this case the norm $\|\cdot\|_{L^2(a,b)}$ should be replaced by the discrete $L^2(a,b)$ norm $\|\cdot\|_\delta$.

We turn now to the multidimensional problem (5.9.1). We assume that $\beta(x)$ is smooth, and that there exists a constant $\alpha \geq 0$ such that

$$c(x) = \gamma(x) + \frac{1}{2}\nabla \cdot \beta(x) \geq \alpha \,.$$

Then the following stability inequality can be proven for the solution of the SDG approximation (5.9.2):

$$\|u_\delta(t)\|_{L^2(\Omega)}^2 + \int_0^t \|\|u_\delta(\tau)\|\|_{DG}^2 \, d\tau$$

$$\leq C\left[\|u_\delta^0\|_{L^2(\Omega)}^2 + \int_0^t \left(\|f(\tau)\|_{L^2(\Omega)}^2 + \int_{\partial\Omega^{\mathrm{in}}} |\beta \cdot \mathbf{n}|g^2 \, d\gamma\right) d\tau\right],$$

for every $t > 0$, where u_δ^0 is the approximation of the initial data. We have set

$$\|\|v\|\|_{DG}^2 = \sum_{m=1}^M \left(\|\sqrt{c}v\|_{L^2(\Omega_m)}^2 + \int_{\partial\Omega_m^{\mathrm{in}}} |\beta \cdot \mathbf{n}^m|[v]^2\right).$$

In the steady case, the following convergence results can be proven in the case of constant β and for partitions of Ω in parallelepipeds (see Houston, Schwab and Süli (2000)). If for all m, $u \in H^k(\Omega_m)$ for some $k \geq 1$, setting $h = \max_m h_m$, $N = \min_m N_m$, $s = \min(k, N+1)$ and $|u|_{s,\delta} = \left(\sum_{m=1}^M |u|_{H^s(\Omega_m)}^2\right)^{1/2}$, then

$$\|\|u - u_\delta\|\|_{DG} \leq C\left(\frac{h}{N+1}\right)^{s-1/2} |u|_{s,\delta} \,. \tag{5.11.5}$$

In Figure 5.34 we report the error behavior with respect to h or $1/N$ when f is chosen in such a way that the exact solution is $u(x,y) = x\sin(\pi y) + \chi_{[1/2,1]}(x - 1/2)^{3/2}$. Note that in this case $u \notin H^2(\Omega)$ yet $u \in H^{2-\varepsilon}(\Omega)$ for all $\varepsilon > 0$ (hence, the choice $k = 2$ in the previous estimate is indicative of the error behavior). Likewise, the results in Fig. 5.35 are for a case with the exact solution $u(x,y) = x\sin(\pi y) + \chi_{[1/2,1]}(x - 1/2)^{7/2}$. This time $u \notin H^4(\Omega)$ yet $u \in H^{4-\varepsilon}(\Omega)$ for all $\varepsilon > 0$ (hence, the choice $k = 4$ is indicative).

Similarly, for the SDGM (5.10.4) for the diffusion equation (5.4.19), the following error estimate holds:

$$\|\|u - u_\delta\|\|_{DG} \leq C\frac{h^{s-1}}{N^{s-3/2}}\left(\sum_{m=1}^M \|u\|_{H^s(\Omega_m)}^2\right)^{1/2}. \tag{5.11.6}$$

An estimate of the type (5.11.5) holds for the SGDM approximation to hyperbolic systems as well. In Fig. 5.36 we report the error behavior with respect to h and $1/N$ for the solution of problem (5.8.11) with

$$A = \begin{pmatrix} 0 & 1 \\ 1 & 0 \end{pmatrix}, \quad D = I, \quad f = \begin{pmatrix} 2\sinh(x) + x^2 - 1 \\ 2\cosh(x) + x \end{pmatrix},$$

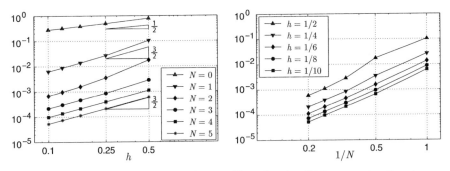

Fig. 5.34. Behavior of the error $\||u - u_\delta\||_{DG}$ for the SDG approximation to an $H^{2-\varepsilon}$ solution to (5.9.1) with respect to h (*left*) and N (*right*)

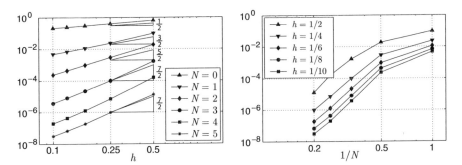

Fig. 5.35. Behavior of the error $\||u - u_\delta\||_{DG}$ for the SDG approximation to an $H^{4-\varepsilon}$ solution to (5.9.1) with respect to h (*left*) and N (*right*)

so that the exact solution is $\mathbf{u} = (e^x + x^2, e^{-x} - x)^T$, and $\mathbf{u} \in (C^\infty(0,1))^2$. Owing to the regularity of the solution, the error estimate is given by

$$\||\mathbf{u} - \mathbf{u}_\delta\||_{DG} \leq C \left(\frac{h}{N+1}\right)^{N+1/2} |\mathbf{u}|_{N+1,\delta} ,$$

and the numerical results in Fig. 5.36 conform to this estimate.

On a time-dependent advection problem like (5.9.1) the SDG method has been shown to have less dissipation and less dispersion error than the corresponding SEM solution (which is continuous at the interface).

The dispersion analysis is carried out in Ainsworth (2004). In order to give an idea of the result, consider the function

$$u(x,t) = e^{i(kx - \omega t)} , \qquad (5.11.7)$$

which satisfies the advection equation

$$\frac{\partial u}{\partial t} + \frac{k}{\omega}\frac{\partial u}{\partial x} = 0, \qquad x \in \mathbb{R}, \; t > 0, \; k, \omega > 0 , \qquad (5.11.8)$$

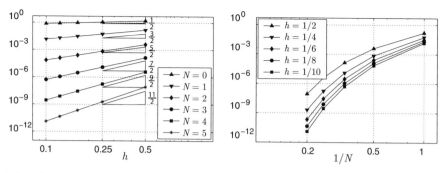

Fig. 5.36. Behavior of the error $\||\mathbf{u} - \mathbf{u}_\delta\||_{DG}$ for the SDG approximation to a $C^\infty(0,1)$ solution to (5.8.11) with respect to h (*left*) and N (*right*)

with $u(x,0) = \tilde{u}(x) = e^{ikx}$. The function (5.11.7) is called a Bloch-wave solution of (5.11.8), and it satisfies $u(x+h, t+\tau) = e^{i(hk-\omega\tau)}u(x,t)$. Note that $u(x,t) = e^{-i\omega t}\tilde{u}(x)$, and that the real and imaginary parts of $\tilde{u}$ are such that the vector function $\mathbf{u} = (\tilde{u}^r \ \tilde{u}^i)^T : (0,1) \to \mathbb{R}^2$ satisfies

$$A\,\frac{\partial \mathbf{u}}{\partial x} + D\mathbf{u} = \mathbf{0}\,,\tag{5.11.9}$$

with

$$D = \begin{pmatrix} 0 & \omega \\ -\omega & 0 \end{pmatrix} \quad \text{and} \quad A = \begin{pmatrix} \frac{\omega}{k} & 0 \\ 0 & \frac{\omega}{k} \end{pmatrix}\,.$$

The constant k is a measure of the characteristic frequency of $\tilde{u}$. The discrete approximation of (5.11.9) obtained by the SDG method exhibits a similar Bloch-wave structure. However, now the (discrete) characteristic frequency is $k_\delta \neq k$. (See Ainsworth (2004) for the precise form.) In Ainsworth (2004) it is shown that, for fixed h, the behavior of the phase error with respect to N exhibits three different regimes: oscillatory when $2N+1 < hk - C(hk)^{1/3}$, algebraic decay when $hk - C(hk)^{1/3} < 2N+1 < hk + o(hk)^{1/3}$, super-exponential decay when $2N+1 \gg hk$.

To illustrate the case of a small wavenumber, we take $k = 2\pi$ and $\omega = 2\pi$. Figure 5.37 shows, for different values of N and h, the dissipation and dispersion errors defined, respectively, as $|Re(\rho_N)|$, $|Im(\rho_N)|$, where ρ_N is the relative error

$$\rho_N = \frac{e^{ihk} - e^{ihk_\delta}}{e^{ihk}}\,.$$

To examine the large wavenumber case we take $k = \omega = 200$. Figure 5.38 shows the real and imaginary parts of ρ_N for $h = 1/8$ and $h = 1/4$ depending on the polynomial order N.

A few comments are in order about how discontinuous Galerkin methods (DG, in short) compare to classical, continuous Galerkin methods (CG, in short). For pure advection problems, or advection–reaction problems with dominant advection, DG methods have quasi-optimal error estimates with

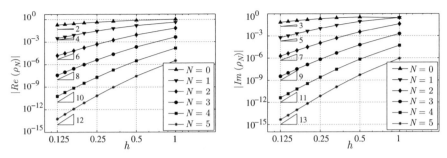

Fig. 5.37. Dissipation (*left*) and dispersion (*right*) errors for different values of N with respect to h, for a small wavenumber, with $k = 2\pi$, $\omega = 2\pi$

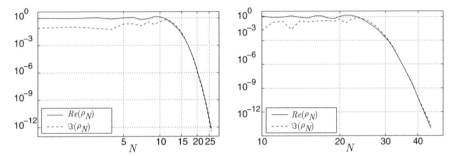

Fig. 5.38. Dissipation ($|Re(\rho_N)|$) and dispersion ($|Im(\rho_N)|$) errors for $h = 1/8$ (*left*) and $h = 1/4$ (*right*) with $\omega = k = 200$

respect to N and h; CG methods with SUPG (or GLS) stabilization would not. The result for DG is proven in Houston, Schwab and Süli (2002) in the case of a constant advection field. For discontinuous solutions, numerical evidence shows that DG solutions oscillate only on a scale of diameter h around discontinuities. Instead, CG methods without stabilization yield, on a single domain, oscillatory solutions throughout the domain (see CHQZ2, Sect. 7.2).

As an alternative to DG methods, one could use CG methods stabilized by integrals on element interfaces containing gradient jumps (such as $\gamma(h) \int_e [\nabla u_\delta][\nabla v_\delta]$). These are called CGIP methods (IP stands for "interior penalty"). For smooth solutions, CGIP and DG methods have similar precision with respect to N and h (see Burman and Ern (2007)). In general, CGIP methods are more efficient, yielding smaller errors for the same number of degrees of freedom. On non-smooth solutions, the two approaches have similar efficiency.

The SDGM for nonlinear hyperbolic systems with smooth solutions has been successfully used, e.g., by Eskilsson and Sherwin (2004), Sherwin and Peiró (2003), Sherwin et al. (2003). See also the fluid dynamics examples in the following section.

The SDGM with nonconforming partitions into subdomains can be regarded as a particular case of the hp-method studied in Toselli (2003).

5.12 SDGM for Euler and Navier–Stokes Equations

We close this discussion of spectral discontinuous Galerkin methods with the additional details needed for application to compressible Euler and Navier–Stokes problems, along with some illustrative examples for fluids applications.

Approximate Numerical Fluxes. Consider first the one-dimensional Euler equations in conservation form. In terms of (5.8.36) we have that $\mathbf{q}$ and $\mathcal{F}$ are

$$\mathbf{q} = \begin{pmatrix} \rho \\ \rho u \\ \rho E \end{pmatrix} \quad \text{and} \quad \mathcal{F} = \begin{pmatrix} \rho u \\ \rho u^2 \\ (\rho E + p)u \end{pmatrix}, \tag{5.12.1}$$

where ρ is the density, u is the fluid velocity, and E is the specific total energy. We also use $H = E + p/\rho$ for the specific total enthalpy. The local Jacobian matrix is

$$A = \begin{pmatrix} 0 & 1 & 0 \\ \dfrac{\gamma - 3}{2}u^2 & (3 - \gamma)u & (\gamma - 1) \\ -u\left(E + \dfrac{p}{\rho}\right) + \dfrac{(\gamma - 1)}{2}u^3 & \left(E + \dfrac{p}{\rho}\right) - (\gamma - 1)u^2 & \gamma u \end{pmatrix}. \tag{5.12.2}$$

Its eigenvalues are the same, of course, as those of the equivalent system (4.3.2), which are u, $u + c$ and $u - c$, where c is the sound speed given by (1.2.21). Hence, we have from (5.8.38) the simple result that the Lax–Friedrichs flux for the Euler equations is given by (5.8.38) with

$$|\lambda_{\max}(A)| = \max(|u_\delta|, |u_\delta + c_\delta|, |u_\delta - c_\delta|). \tag{5.12.3}$$

In the case of the Roe flux for the Euler equations, the approximate Jacobian matrix in (5.8.39) is

$$A_{LR} = \begin{pmatrix} 0 & 1 & 0 \\ \dfrac{\gamma - 3}{2}(u_\delta^{LR})^2 & (3 - \gamma)u_\delta^{LR} & (\gamma - 1) \\ -u_\delta^{LR}H_\delta^{LR} + \dfrac{(\gamma - 1)}{2}(u_\delta^{LR})^3 & H_\delta^{LR} - (\gamma - 1)(u_\delta^{LR})^2 & \gamma u_\delta^{LR} \end{pmatrix}, \tag{5.12.4}$$

where u_δ^{LR} and $H_\delta^{LR} = (E_\delta + p_\delta/\rho_\delta)^{LR}$ are to be determined. Substituting (5.12.1)–(5.12.4) into (5.8.39), one obtains three equations, the first one of which is a trivial identity. The second and third equations are solved for the unknowns u_δ^{LR} and H_δ^{LR} to yield (after cumbersome algebra)

$$u_\delta^{LR} = \frac{\sqrt{\rho_\delta^R} u_\delta^R + \sqrt{\rho_\delta^L} u_\delta^L}{\sqrt{\rho_\delta^R} + \sqrt{\rho_\delta^L}} \quad \text{and} \quad H_\delta^{LR} = \frac{\sqrt{\rho_\delta^R} H_\delta^R + \sqrt{\rho_\delta^L} H_\delta^L}{\sqrt{\rho_\delta^R} + \sqrt{\rho_\delta^L}}.$$

These are sometimes called the Roe-averaged velocity and total enthalpy, respectively. Obviously, u_δ^{LR} lies between u_δ^L and u_δ^R, and H_δ^{LR} between H_δ^L and H_δ^R. See, e.g., LeVeque (2002), Chap. 15, for additional examples of the determination of $\mathbf{q}_\delta^{LR}$ for other approximate solutions of the Riemann problem for the Euler equations. In the case of multidimensional problems, the momentum components tangential to the interface are simply upwinded.

As the SDGM using Legendre Gauss quadrature has no complications at corner or edge points, only locally one-dimensional Riemann problems involving the normal flux need to be solved. In the multidimensional case, the above formulas are applicable where u is now the component of the velocity in the normal direction at the interface.

Time-Discretizations. For genuine SDG methods (say, with $N_m > 3$) which are not TVD, a variety of conventional higher-order Runge–Kutta methods have been advocated in fluids applications, such as the low-storage methods described in Appendix C.3. See also Stanescu and Habashi (1998), Kopriva, Woodruff and Hussaini (2002), Chen and Hussaini (2007) for some alternatives that have been successfully utilized in SDGM fluids applications.

Numerous comments were made in Chap. 4 about how the lack of efficient implicit time-discretizations has hampered the use of single-domain spectral methods for compressible flows. The use of multidomain spectral methods eases the time-step restriction to a considerable extent. Nevertheless, an efficient implicit or semi-implicit scheme would yield further improvements. Some progress in this direction has been made by Rasetarinera and Hussaini (2001), who developed an efficient implicit SDG method to solve steady flows or unsteady aerodynamic problems where the temporal accuracy requirement is far less stringent than the stability condition. Their method combines the unconditionally stable backward Euler scheme for temporal discretization with a Legendre SDG discretization in each subdomain that employs the Osher (1984) flux at subdomain interfaces.

In order to solve the resulting implicit systems, Rasetarinera and Hussaini (2001) use the matrix-free Newton–Krylov–Schwarz algorithm that has emerged as a promising technique for the implicit solution of large-scale aerodynamics problems on parallel computers (Keyes (1995)). It combines the Newton–Krylov method as the nonlinear solver with the Krylov–Schwarz iterative method for the solution of the linear system arising from the Newton linearization. The Krylov–Schwarz method is especially well-suited for parallel implementation of the SDGM since each subdomain can be treated separately. In the matrix-free approach, the Jacobian vector prod-

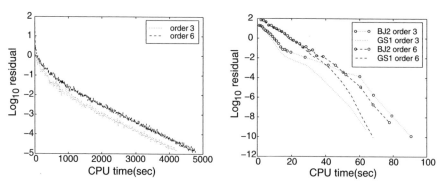

Fig. 5.39. Comparison of CPU time for explicit time-discretization (*left*) and various implicit time-discretizations for SDGM solutions (with $N = 3$ using 284 subdomains and $N = 6$ using 70 subdomains) of Mach 0.2 flow over a circular bump in a channel [Courtesy of P. Rasetarinera, M.Y. Hussaini (2001)

uct within the Krylov algorithm is approximated by the Fréchet derivative (Brown and Saad (1990)). This permits considerable savings in storage.

This method has been demonstrated on three prototypical two-dimensional aerodynamic problems—subsonic flow in a channel with a circular bump, transonic flow in a nozzle, and subsonic flow over a three-element airfoil. Both block Jacobi as well as one-level and two-level LU-symmetric Gauss–Seidel preconditioners were evaluated. Figure 5.39 displays a comparison between an explicit scheme (second-order Runge–Kutta) employing local time-stepping and the implicit scheme with two preconditioners for Mach 0.2 flow past a circular bump in a channel. The preconditioners were two-level block Jacobi (BJ2) and one-level LU-symmetric Gauss–Seidel (GS1). The latter preconditioner provided the best performance on this case, reaching a steady-state solution 50 times faster than the explicit method. The two-level procedure proved essential for achieving fast convergence for relatively large numbers of subdomains.

Numerical Examples. We now provide an example of a multidimensional computation using the Euler equations for a simple case with an analytical solution, in particular, the axisymmetric, calorically perfect, subsonic flow due to a line source. The analytic solution is obtained by the hodograph method, and can be found, e.g., in Courant and Friedrichs (1976) and Howarth (1953). Under a suitable scaling, the nondimensional density ρ and radial velocity u_r at a nondimensional radius r are found from the equations

$$ r = \frac{1}{2\pi\rho} \frac{1}{\sqrt{1 - \rho^{(\gamma-1)}}}, \qquad u_r = \sqrt{1 - \rho^{(\gamma-1)}}, \qquad (5.12.5) $$

and the Mach number is given by

$$\frac{1}{\rho} = \left[1 + \frac{\gamma - 1}{2}M^2\right]^{\frac{1}{\gamma - 1}}.$$

The computational domain for this SDGM numerical example is illustrated in the left half of Fig. 5.40; it is $(0.73, 1.46) \times (0, 0.67)$. The Mach number at the lower left corner, $(x, y) = (0.73, 0)$, is 0.6. This flow has been used as a test case for multidomain spectral methods since Kopriva (1991). The present example uses the exact solution for the requisite conditions on the domain boundary and applies the Riemann solver otherwise. The heavy lines in the figure indicate the subdomain boundaries. The subdomains have been deliberately skewed to verify the mapping procedures. The contours are those of the Mach number. The point source is located out of the field of view of the figure, down and to the left. The flow is subsonic in the computational domain. No distortion in the contours is evident at the subdomain boundaries. The right half of the figure demonstrates that exponential convergence is achieved as the degree of the polynomial in each subdomain is increased.

For application of SDGM to viscous, compressible flow, we rewrite the unforced, compressible Navier–Stokes equations in conservation form (1.3.11)–(1.3.13) as

$$\frac{\partial \mathbf{q}}{\partial t} + \nabla \cdot \boldsymbol{\mathcal{F}}^{\text{inv}}(\mathbf{q}) - \nabla \cdot \boldsymbol{\mathcal{F}}^{\text{vis}}(\mathbf{q}) = \mathbf{0} \quad \text{in } \Omega, \tag{5.12.6}$$

$$\mathbf{q}(\mathbf{x}, 0) = \mathbf{q}_0(\mathbf{x}) \quad \text{at } t = 0, \tag{5.12.7}$$

where $\mathbf{q} = (\rho, \rho u_1, \rho u_2, \rho u_3, \rho E)^T$ is the vector of conserved variables. The inviscid flux $\boldsymbol{\mathcal{F}}^{\text{inv}} = (\boldsymbol{\mathcal{F}}_1^{\text{inv}}, \boldsymbol{\mathcal{F}}_2^{\text{inv}}, \boldsymbol{\mathcal{F}}_3^{\text{inv}})^T$, and the viscous flux $\boldsymbol{\mathcal{F}}^{\text{vis}} = (\boldsymbol{\mathcal{F}}_1^{\text{vis}}, \boldsymbol{\mathcal{F}}_2^{\text{vis}}, \boldsymbol{\mathcal{F}}_3^{\text{vis}})^T$ are defined, respectively, as

$$\boldsymbol{\mathcal{F}}_i^{\text{inv}}(\mathbf{q}) = \begin{pmatrix} \rho u_i \\ \rho u_1 u_i + p\delta_{1i} \\ \rho u_2 u_i + p\delta_{2i} \\ \rho u_3 u_i + p\delta_{3i} \\ (\rho E + p)u_i \end{pmatrix},$$

and

$$\boldsymbol{\mathcal{F}}_i^{\text{vis}}(\mathbf{q}) = \begin{pmatrix} 0 \\ \tau_{1i} \\ \tau_{2i} \\ \tau_{3i} \\ \tau_{1i}u_1 + \tau_{2i}u_2 + \tau_{3i}u_3 + \dfrac{\mu}{\text{Pr}}e_{,x_i} \end{pmatrix},$$

where τ_{ij} are the components of the viscous stress tensor (1.3.2). Here Pr is the reference Prandtl number. Black (2000) and Collis (2002) appear to have

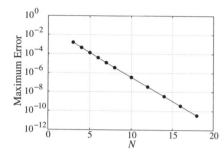

Fig. 5.40. Computational domain and the pressure field (*left*) for the point source flow computed with the SDGM method, and the convergence of the maximum error in the pressure as a function of the degree N of the polynomials in each subdomain [Courtesy of D.A. Kopriva]

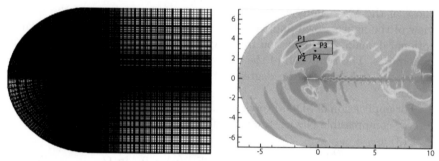

Fig. 5.41. Computational mesh (*left*) and the flow field for a noise minimization computation using SDGM with the Navier–Stokes equations [Courtesy of G. Chen, M.Y. Hussaini (2006)]

produced the first viscous flow computations using SDGM; they combined a standard hyperbolic approach to the inviscid flux $\mathcal{F}^{inv}(\mathbf{q})$ with the Bassi–Rebay approach to the diffusive flux $\mathcal{F}^{vis}(\mathbf{q})$ (see (5.10.6)).

Our numerical example is taken from Chen and Hussaini (2006), who used the Lax–Friedrichs approximate Riemann solver (5.8.38) with (5.12.3) together with the Bassi–Rebay approximation to the diffusive flux. The problem consists in the optimal control of unsteady blowing and suction at the trailing edge of a Blake (1975) airfoil to minimize noise in a far-field observation region. The objective function to be minimized is defined as the deviation of pressure from its target value, integrated over the observation region and over a certain time window. The two-dimensional compressible Navier–Stokes equations and their adjoint equations are solved by SDGM, using a fourth-order Runge–Kutta scheme for temporal integration due to Hagar (2000).

Figure 5.41 (left) shows the computational domain around the airfoil; the grid consists of 5450 non-overlapping body-conforming quadrilateral elements, in each of which the solution is represented by a Legendre polyno-

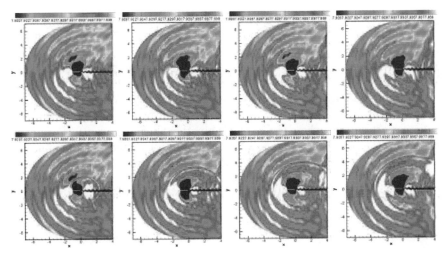

Fig. 5.42. Uncontrolled (*top*) and controlled (*bottom*) pressure field for trailing edge noise minimization using SDGM with the Navier–Stokes equations [Courtesy of G. Chen, M.Y. Hussaini (2006)]

mial of degree 4. The observation regions of interest (to identify the induced vortex-shedding noise along with four measurement stations) are shown in Fig. 5.41 (right) against the background of acoustic pressure contours at an arbitrary instant. The flow conditions correspond to a Mach number of 0.3 and a Reynolds number of 5000 based on the chord length of the Blake airfoil. Figure 5.42 shows the contours of acoustic pressure at different time-instants with and without control. With control (bottom row), the contours of acoustic pressure inside the observation region are noticeably reduced.

The SDGM of Chen and Hussaini (2006) actually used a modal basis rather than a nodal basis, as has been assumed in the description of the SDGM-NI method. The reason is that this control problem required the computation not only of the solution to the Navier–Stokes problem but also to the adjoint problem. This turned out to be much simpler to formulate with a modal basis. See Chen and Hussaini (2006) for details. Furthermore, the restriction to two dimensions and to relatively low-order was driven by the very large storage demands for the adjoint problem.

Shock Tracking. In Sects. 4.7 and 4.8 we discussed the applications of single-domain spectral methods to shock-fitting and shock capturing, respectively. We discuss spectral multidomain shock-fitting algorithms in Sect. 5.13.3. Here we take note of yet another approach to dealing with shocks with spectral methods, namely, shock tracking.

The general shock-tracking approach is based on the level-set methodology of Osher and Sethian (1988); see also Glimm et al. (2001). The spectral shock-tracking approach is based on the reasoning that adaptation is better than filtering, hyperviscosity or limiters (as are widely used in low-order shock-capturing methods). High-order elements with limiters devolve very quickly into large, first-order subdomains when the limiter is activated; the high-order degrees of freedom are then wasted, and the accuracy degraded.

Touil, Hussaini and Sussman (2007) developed a spectral shock-tracking algorithm that coupled SDGM with a level-set procedure. The computational domain is divided into two types of subdomains. Where the solution is smooth, it is represented by polynomials of degree N. In the vicinity of a discontinuity (i. e., where the level-set function Ψ changes sign), the subdomains are the so-called *Godunov subdomains*. In a Godunov subdomain, the solution $\mathbf{q}$ to the conservation law (5.9.5) is split into two smooth components $\mathbf{q}_1$ and $\mathbf{q}_2$ that solve formally identical problems for $\Psi < 0$ and for $\Psi > 0$; these components are coupled with the evolution equation for the level set:

$$\frac{\partial \Psi}{\partial t} + S(\mathbf{q}_1, \mathbf{q}_2, \mathbf{n})|\nabla \Psi| = \mathbf{0} \,, \tag{5.12.8}$$

where $\Psi = 0$ represents a shock wave, whose unit normal $\mathbf{n}$ is given by

$$\mathbf{n} = \frac{\nabla \Psi}{|\nabla \Psi|} \,, \tag{5.12.9}$$

and whose speed S is obtained from solving the appropriate Riemann problem with the two states. (This level-set procedure is the spectral version of the Aslam (2003) approach.) Each Godunov subdomain is then subdivided into 2^N subdomains (in the one-dimensional case), and each subdomain is treated with a first-order accurate method so that the virtual order of accuracy in a Godunov subdomain is first order with respect to $h = 2^{-N}$, which implies spectral accuracy with respect to the total number of degrees of freedom. One must note the attractive simplicity of the adaptation in this method, which is totally local to an element (unlike usual adaptive mesh refinement strategies), and is in large part due to the flexibility of the SDGM. The procedure is elegant in that the error is kept low by adapting the h–N parameters to suit the nature of the solution.

One of the Touil, Hussaini and Sussman (2007) demonstration cases was for a shock interacting with a density wave using the 1D Euler equations— the commonly used test problem from Shu and Osher (1989). Figure 5.43 demonstrates that both the location of the shock and the solution itself can be computed with an error that decays exponentially with the total number of degrees of freedom in the calculation. The single-domain and multidomain shock-fitting procedures discussed in Sects. 4.7 and 5.13.3, respectively, both determine the shock location and the solution behind the shock with spectral accuracy, but they have limited applicability. Shock-capturing methods have yet to demonstrate spectral accuracy for either the shock location or the solution. The SDGM level-set procedure offers the prospect of finding the shock location with spectral accuracy in general situations.

5.13 The Patching Method

As noted in the introduction to this chapter, the earliest spectral multidomain methods—the patching methods—are based on the strong form of the differential equation, and they invariably use collocation approximations in the subdomains. Although patching methods continued to be pursued in some quarters through at least the mid-1990s, spectral multidomain methods based on the weak form of the differential equations, such as the SEM, MEM and SDGM methods discussed previously, had become the predominant form used for applications of multidomain spectral methods by the close of the 1980s, as well has the focus of most of the theoretical developments of the subject. Accordingly, this text emphasizes the weak formulation of multidomain spectral methods. Nevertheless, in this section we provide the theoretical formulation of the patching methods for model elliptic and hyperbolic problems, compare patching methods with SEM-NI for a model elliptic problem, and highlight the major developments in patching techniques for hyperbolic systems, especially the Euler equations.

5.13.1 Formulation of Patching Methods

We outline the formulation of spectral patching methods in a computational domain that is partitioned into subdomains of quadrilateral type. The governing differential equation is first reformulated as a coupled system involving the same kind of differential equation in each subdomain with suitable "patching" conditions on the different solutions at subdomain interfaces. This coupled, strong form formulation underlies spectral collocation methods built on decomposed domains.

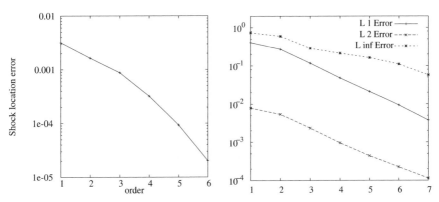

Fig. 5.43. Error as a function of polynomial degree N from SDGM shock-tracking solution to shock-entropy wave interaction problem. Shock location error (*left*) and global density error (*right*) [From H. Touil, M.Y. Hussaini, M. Sussman (2007)]

The first illustration is for the Poisson problem (5.4.19), where Ω is a d-dimensional domain ($d = 2, 3$) with a Lipschitz boundary $\partial\Omega$, whose outward unit normal is denoted by $\mathbf{n}$ and f is a given function of $L^2(\Omega)$. The same general principles apply to any symmetric, linear, elliptic boundary-value problem.

As the very simplest case, we assume that Ω is partitioned into two non-overlapping subdomains Ω_1 and Ω_2, and denote by Γ the subdomain interface $\overline{\Omega}_1 \cap \overline{\Omega}_2$ (see Fig. 5.44). We also assume that Γ is a Lipschitz $(d-1)$-dimensional manifold. We denote by u_i the restriction to Ω_i, $i = 1, 2$, of the solution u to (5.4.19), and by $\mathbf{n}^i$ the normal direction on $\partial\Omega_i \cap \Gamma$, oriented outward. For simplicity of notation we also set $\mathbf{n}^\Gamma = \mathbf{n}^1$.

It is easily seen that the Poisson problem (5.4.19) can be reformulated in the equivalent *multidomain form*:

$$-\triangle u_1 = f \qquad\qquad \text{in } \Omega_1 , \qquad\qquad (5.13.1\text{a})$$

$$u_1 = 0 \qquad\qquad \text{on } \partial\Omega_1 \cap \partial\Omega , \qquad\qquad (5.13.1\text{b})$$

$$u_1 = u_2 \qquad\qquad \text{on } \Gamma , \qquad\qquad (5.13.1\text{c})$$

$$\frac{\partial u_2}{\partial n^\Gamma} = \frac{\partial u_1}{\partial n^\Gamma} \qquad\qquad \text{on } \partial\Gamma , \qquad\qquad (5.13.1\text{d})$$

$$u_2 = 0 \qquad\qquad \text{on } \partial\Omega_2 \cap \partial\Omega , \qquad\qquad (5.13.1\text{e})$$

$$-\triangle u_2 = f \qquad\qquad \text{in } \Omega_2 . \qquad\qquad (5.13.1\text{f})$$

This is the formulation associated with the *patching* method (see box 1.2 of Fig. 6.3). Equations (5.13.1c) and (5.13.1d) are the *transmission conditions* for u_1 and u_2 on Γ.

The physical meaning of this split formulation is clear provided that the solution of problem (5.4.19) is smooth enough (say, $u \in C^1(\overline{\Omega})$). In a more general framework, the equivalence between (5.4.19) and (5.13.1) can be shown by resorting to the weak formulation of both problems (see Quarteroni and Valli (1994)).

The domain decomposition formulation (5.13.1) lends itself to a straightforward Legendre collocation approximation as follows. Assume for ease of notation that Ω is a two-dimensional rectangle split into two rectangles separated by a vertical interface Γ as shown in Fig. 5.44. Then, let $V_{N_m}(\Omega_m)$, $m = 1, 2$, be the space introduced in (5.3.3). We assume that

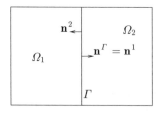

Fig. 5.44. A rectangle partitioned into two subdomains Ω_1 and Ω_2

$N_1 = (N_{1,x}, N_{1,y})$, $N_2 = (N_{2,x}, N_{2,y})$ with the restriction that $N_{1,y} = N_{2,y}$ in order to allow for functional continuity across the interface Γ. Then we look for two polynomials $u_N^{(m)} \in V_{N_m}(\Omega_m)$, $m = 1, 2$, that satisfy the following set of equations:

a) the differential equations

$$-\triangle u_N^{(m)} = f \qquad (5.13.2)$$

at the $(N_{m,x} - 1) \times (N_{m,y} - 1)$ internal LGL collocation nodes of Ω_m, $m = 1, 2$;

b) the boundary conditions

$$u_N^{(m)} = 0 \qquad (5.13.3)$$

at all the LGL nodes lying on $\partial\Omega_m \cap \partial\Omega$;

c) the continuity conditions

$$u_N^{(1)} = u_N^{(2)} \qquad (5.13.4)$$

and

$$\frac{\partial u_N^{(1)}}{\partial x} = \frac{\partial u_N^{(2)}}{\partial x} \qquad (5.13.5)$$

at the $N_{m,y} - 1$ LGL nodes lying on the interface Γ (excluding the two endpoints of Γ).

The set of equations (5.13.2)–(5.13.5) represents the *patching collocation method* (see box 2.2.1 of Fig. 6.3).

The situation is less simple in the case of "cross-points" or "T-points" (see e. g. Fig. 5.51). There the flux continuity should entail a linear combination between normal fluxes through the different edges merging at the point.

A similar kind of approach applies at every time-step of a time-dependent diffusion-advection-reaction problem advanced in time by an implicit scheme. The transmission conditions at the new time-level still enforce the continuity of the solution, as in (5.13.4), and that of the normal flux (instead of (5.13.5)). In the case of a spatial operator like in (5.5.1), the latter takes the form

$$\alpha \frac{\partial u_N}{\partial n^\Gamma}^{(1)} - \boldsymbol{\beta} \cdot \mathbf{n}^\Gamma u_N^{(1)} = \alpha \frac{\partial u_N}{\partial n^\Gamma}^{(2)} - \boldsymbol{\beta} \cdot \mathbf{n}^\Gamma u_N^{(2)} \qquad \text{on } \Gamma, \qquad (5.13.6)$$

where $\mathbf{n}^\Gamma$ is the unit normal vector on Γ directed from Ω_1 to Ω_2.

The second model problem used for these illustrations is the first-order problem (5.9.1) with an implicit time-discretization. This yields the first-order linear advection equation

$$L_0 u = \nabla \cdot (\boldsymbol{\beta} u) + \tilde{\gamma} u = \tilde{f} \qquad \text{in } \Omega, \qquad (5.13.7)$$

where $\tilde{\gamma}$ and $\tilde{f}$ depend on the reciprocal of the time-step. The appropriate interface condition is that the normal flux $(\boldsymbol{\beta} \cdot \mathbf{n}_\Gamma) u$ is continuous across Γ:

$$(\boldsymbol{\beta}_{|\Omega_1} \cdot \mathbf{n}^\Gamma) u^{(1)} = (\boldsymbol{\beta}_{|\Omega_2} \cdot \mathbf{n}^\Gamma) u^{(2)} \qquad \text{on } \Gamma, \qquad (5.13.8)$$

where, as usual, $u^{(m)}$ denotes the restriction of u in Ω_m, $m = 1, 2$. The remaining conditions on the interface boundary Γ as well as those on the domain boundary $\partial\Omega$ must respect the direction of information propagation. Set

$$\Gamma^{\mathrm{in}} = \{\mathbf{x} \in \Gamma : \boldsymbol{\beta}(\mathbf{x}) \cdot \mathbf{n}^\Gamma(\mathbf{x}) < 0\},$$
$$\Gamma^{\mathrm{out}} = \{\mathbf{x} \in \Gamma : \boldsymbol{\beta}(\mathbf{x}) \cdot \mathbf{n}^\Gamma(\mathbf{x}) > 0\}.$$

Then, provided that $(\boldsymbol{\beta} \cdot \mathbf{n}_\Gamma)$ is continuous across Γ, we impose

$$u^{(1)} = u^{(2)} \qquad \text{on } \Gamma^{\mathrm{in}} \cup \Gamma^{\mathrm{out}}. \tag{5.13.9}$$

We also define

$$\partial\Omega^{\mathrm{out}} = \{\mathbf{x} \in \partial\Omega : \boldsymbol{\beta}(\mathbf{x}) \cdot \mathbf{n}(\mathbf{x}) > 0\},$$

and, for $m = 1, 2$, we put

$$\partial\Omega_m^{\mathrm{in}} = \partial\Omega^{\mathrm{in}} \cap \partial\Omega_m, \quad \partial\Omega_m^{\mathrm{out}} = \partial\Omega^{\mathrm{out}} \cap \partial\Omega_m.$$

A nonstaggered-grid collocation scheme consists of (1) updating all collocation points of Ω_1, including those lying on Γ^{out}, via the advection equation, (2) enforcing the continuity condition $(\boldsymbol{\beta}_{|\Omega_i} \cdot \mathbf{n}^\Gamma) u_N^{(1)} = (\boldsymbol{\beta}_{|\Omega_2} \cdot \mathbf{n}^\Gamma) u_N^{(2)}$ at the collocation nodes of Γ^{in}, and then (3) doing the specular approach in Ω_2.

The extension of this approach to the case of hyperbolic systems is (almost) a straightforward application of the collocation method in the interior combined with appropriate use of the physical boundary conditions on Ω, the flux continuity conditions on Γ, and appropriate use of the CCM discussed in Sect. 4.2.1. Similar extensions can be made to the incompressible Stokes or Navier–Stokes equations, for which at every time-step we have to enforce on the interface Γ the continuity of the velocity field as well as that of the normal Cauchy stress. An extensive analysis is carried out in Quarteroni and Valli (1994). In the same reference the patching collocation method is generalized to the case in which a domain Ω is partitioned into several non-overlapping subdomains.

As discussed in Sect. 4.6.1, bi-characteristic boundary conditions are often needed at the corners of the subdomains. Multidomain patching methods using a staggered grid are discussed in Sect. 5.13.3.

One fluid dynamics application domain where patching methods are still utilized is linear and nonlinear stability analysis. As noted in Sect. 2.4, both the Macaraeg, Streett and Hussaini (1988) staggered-grid method and the Malik (1990) nonstaggered-grid method for compressible linear stability have multidomain patching implementations. (Actually, in the Macareag and Streett (1986) multidomain patching method for elliptic problems, it is not the solution and the first derivatives that are matched at the interface (as suggested by (5.13.1), but rather the solution and the fluxes.) So, too, does the

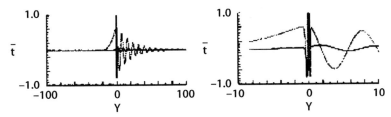

Fig. 5.45. Temperature perturbation for a supersonic mode in a Mach 4 compressible free shear layer, on the full domain (*left*) and on a close-up of the inner structure (*right*) [From Macaraeg, Streett, Hussaini (1988)]

Bertolotti and Herbert (1991) compressible parabolized stability equations (PSE) code. Multidomain methods are especially attractive for compressible stability computations because of the thin internal layers (see Fig. 2.24) and supersonic modes (see Fig. 2.17) that cannot be resolved at all efficiently with the grids associated with single-domain spectral methods and require a very large number of grid points with low-order methods. All of the above codes used the patching approach.

The detailed structure of the temperature perturbation for an unstable mode in a Mach 4 compressible free shear layer is shown in Fig. 5.45. (This corresponds to the pressure perturbation shown in the right frame of Fig. 2.17). This mode was computed by Macaraeg, Streett and Hussaini (1988) using a Chebyshev multidomain patching method using three domains tailored to the eigenfunction structure: $[-100, -1]$ with $N_1 = 24$, $[-1, -1]$ with $N_2 = 40$, and $[1, 100]$ with $N_3 = 104$. Such a computation with a single-domain method is not remotely competitive due to the huge number of grid points that would be required.

5.13.2 Comparison of Patching and SEM-NI

One property that the patching and the SEM-NI methods share is that the original differential equation (or its interpolation approximation) is satisfied at all collocation nodes that lie in the interior of the subdomains. However, while the patching method enforces the flux continuity (e. g., (5.13.5), (5.13.6) or (5.13.8)) pointwise at every collocation node lying on Γ, in the SEM-NI approach this continuity is only enforced in a weak sense by involving a linear combination with the residual of the differential equation (as done in (5.2.19) in the one-dimensional case). (The SDGM-NI method is similar to SEM-NI in this regard.)

The accumulated empirical evidence suggests that for elliptic problems, at least, the SEM-NI approach enjoys better accuracy and stability than the patching method. We provide a simple illustration here on the solution of the Poisson problem (5.4.19). The computational domain is $\Omega = (-1, 1)^2$ and the right-hand side is chosen so that the exact solution $u(x, y) =$

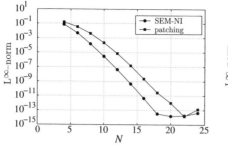 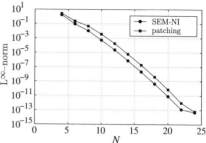

Fig. 5.46. The relative errors in the L^∞-norm versus the polynomial degree $N = N_x = N_y$ between the exact solution $u(x,y) = \sin(2\pi x)\sin(2\pi y)$ and the numerical solution u_N of (5.4.19) in $\Omega = (-1,1)^2$ obtained with the SEM-NI and patching approaches. The *left figure* refers to a decomposition into 2×2 subdomains, the *right one* into two aligned subdomains

$\sin(2\pi x)\sin(2\pi y)$. In Fig. 5.46 the relative errors $\|u - u_N\|_{L^\infty(\Omega)}/\|u\|_{L^\infty(\Omega)}$ are shown versus the polynomial degree N for two different decompositions: one into 2×2 elements with the cross-point located at $(0.15, 0.15)$, and another into two subdomains with a vertical interface located at $x = 0.6$. In the former case, at the cross-point the following flux characteristic is enforced (instead of (5.13.5))

$$\sum_{m=1}^{4} \left(\frac{\partial u_N^{(m)}}{\partial n_{m,x}} + \frac{\partial u_N^{(m)}}{\partial n_{m,y}} \right) = 0 \,,$$

where $n_{m,x}$ is the normal vector to the vertical side of Ω_m merging at the point, $n_{m,y}$ that of the horizontal side. The errors from the patching method are considerably larger than those obtained with the SEM-NI method.

The numerical stability of the methods is reflected in the eigenvalues of the matrix governing the algebraic problem produced by the discretization. In the case of the above numerical example, the matrix A is associated with the discretization of the Laplacian operator with homogeneous Dirichlet boundary conditions; the rows and columns associated with Dirichlet boundary conditions have been removed. The extreme eigenvalues (and their ratio) of the respective matrices A for the SEM-NI and patching approaches for the above numerical example are displayed in Fig. 5.47 as a function of H for fixed $N = 4$, whereas they are given in Fig. 5.48 as a function of N for fixed $H = 1$. From the left frame of Fig. 5.47 we observe that $\lambda_{\max} \simeq O(1)$ for SEM-NI, $\lambda_{\max} \simeq 87.9 \cdot H^{-1}$ for patching, $\lambda_{\min} \simeq 0.31 \cdot H^2$ for SEM-NI and $\lambda_{\min} \simeq 0.40 \cdot H^2$ for patching. Its right frame suggests that $\lambda_{\max}/\lambda_{\min} \simeq 48.18 H^{-2}$ for SEM-NI and $\lambda_{\max}/\lambda_{\min} \simeq 2.19 \cdot H^{-3}$ for patching. The corresponding scalings as a function of N extracted from the left frame of Fig. 5.47 are $\lambda_{\max} \simeq 1.90 \cdot N$ for SEM-NI, $\lambda_{\max} \simeq 4.10 \cdot N^2$ for patching, $\lambda_{\min} \simeq 4.58 \cdot N^{-2}$ for SEM-NI and $\lambda_{\min} \simeq 4.72 \cdot N^{-2}$ for patch-

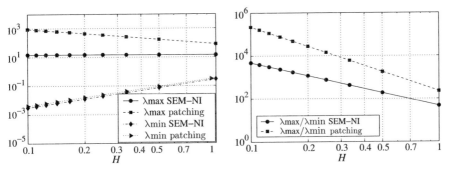

Fig. 5.47. Extreme eigenvalues for both the SEM-NI and the patching approaches with a polynomial degree of $N = 4$ for the Laplace operator in 2D. The modulus of the eigenvalues of largest and smallest magnitude is on the *left*, and their ratio is on the *right*

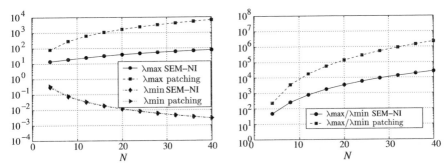

Fig. 5.48. Extreme eigenvalues for both the SEM-NI and the patching approaches with a domain size of $H = 1$ for the Laplace operator in 2D. The modulus of the eigenvalues of largest and smallest magnitude is on the *left*, and their ratio is on the *right*

ing, while its right frame yields $\lambda_{\max}/\lambda_{\min} \simeq 0.41 \cdot N^3$ for SEM-NI and $\lambda_{\max}/\lambda_{\min} \simeq 0.87 \cdot N^4$ for patching.

Finally, the full spectrum of the matrix A produced by the discretization is shown for both the SEM-NI and the patching approaches in Fig. 5.49 for several values of H and N. The domain Ω is partitioned into $M \times M$ square subdomains with sides of length $H = 2/M$.

Since the linear stability, secondary instability and PSE problems mentioned at the end of the previous subsection involve the solution of second-order differential equations, the evidence here suggests that SEM-NI might yield accurate results for this class of problems.

5.13.3 Collocation Methods for the Euler Equations

We now summarize the state-of-the-art in collocation methods for the compressible Euler equations, focusing on three topics: nonstaggered grids, stag-

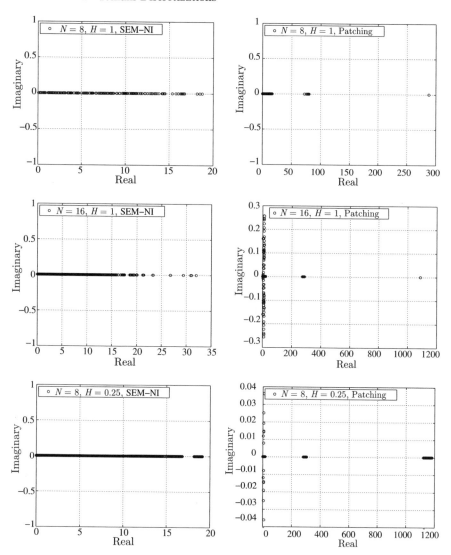

Fig. 5.49. Spectrum of A (discretization of the Laplace operator in 2D), 2×2 subdomains of equal size

gered grids, and shock-fitting. In the latter case, there have been a few developments in shock-fitting using the Navier–Stokes equations, which we also briefly mention.

Collocation Using a Nonstaggered Grid. The essential elements of collocation for the Euler equations on nonstaggered grids are extensively covered by Kopriva (1991). If it were not for the treatment of corner points (in 2D and 3D) and edge points (in 3D), these patching methods would just be a straightforward application of the principles of patching for the first-order problems discussed in Sect. 5.13.1, i. e., apply a standard collocation method in the interior of the subdomains, use the locally one-dimensional CCM to enforce flux continuity at subdomain interfaces, and apply a combination of the physical boundary conditions and the locally one-dimensional CCM at the domain boundaries. However, as noted already in Sect. 4.6.1, the corner points at the domain boundaries, and now also at the subdomain boundaries, usually need special treatment to avoid spurious oscillations and numerical instabilities.

For the Euler source flow problem discussed in Sect. 5.12, the many different types of multidomain patching grids used by Kopriva (1991) are illustrated in Fig. 5.50. The different types of interface and boundary points that can arise for this and other problems are illustrated in Fig. 5.51. In general, the locally one-dimensional CCM method has proven inadequate for the "cross", "T" and "subdomain-wall" points, and a bi-characteristic approach is needed (Kopriva (1991)). For the types of grids and interface points shown

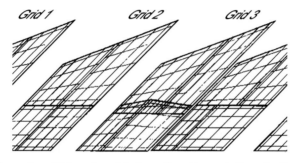

Fig. 5.50. Several multidomain grids for collocation solutions to the Euler source flow problem [From D.A. Kopriva (1991)]

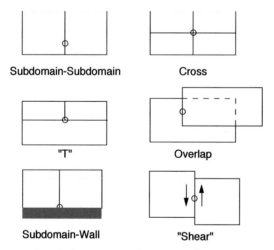

Fig. 5.51. Types of multidomain interface points [From D.A. Kopriva (1991)]

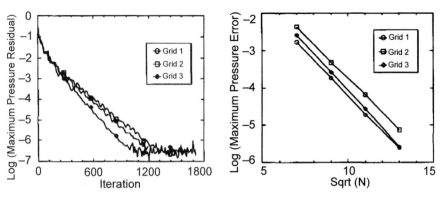

Fig. 5.52. Iterative convergence (*left*) and convergence (*right*) for collocation solution to Euler source problem on each of the grids shown in Fig. 5.50. The abscissa in the right frame is the square root of the total number of grid points [From D.A. Kopriva (1991)]

in Figs. 5.50 and 5.51, patching methods for the Euler source flow problem using these more complex boundary conditions at corner points have been shown, as illustrated in Fig. 5.52, to yield full convergence of both the residuals and the errors without the need for any type of filtering. (These were 32-bit computations.)

Nevertheless, the implementation of patching methods on nonstaggered grids requires the coding of a host of special cases, not just for the corner points, but also for all types of subdomain interface points (see Fig. 5.51). This becomes an overwhelming task in 3D when edge points must also be considered.

Collocation Using a Staggered Grid. Little wonder, then, that the use of staggered grids has proven to be far more attractive for collocation solutions of the Euler equations (Kopriva and Kolias 1996), Kopriva (1996a)). The basic principles of staggered grids in two dimensions are discussed in Sect. 4.6.1; see Fig. 4.9 for an illustration of a two-dimensional staggered grid for Euler computations. Obviously, for the nonconforming interfaces shown in Fig. 5.51, a mortar approach is needed. See the discussion in Sect. 5.9.2 for the general principles, and see Kopriva (1996a) for more details in the present context. The staggered-grid collocation method requires only the use of locally one-dimensional boundary and interface treatments using Riemann solvers), and it admits a very simple coding implementation uncluttered with the drudgery of accounting for all sorts of special cases. Strictly speaking, the staggered-grid method is a penalty method rather than a patching method, although the penalty term follows automatically from the discretization. The extension to three-dimensional problems and to viscous problems is rather straightforward; see, e. g., Jacobs, Kopriva and Mashayek (2005). There are neither corner nor edge points, and therefore there is no need for bi/tri-characteristic boundary/interface conditions.

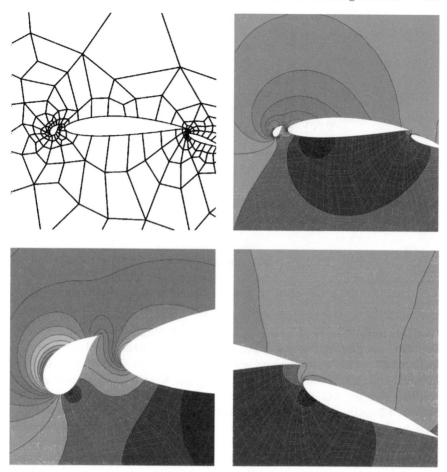

Fig. 5.53. Computational subdomains (*top left*) and computed pressure field in full flowfield (*top right*), near slat (*bottom left*) and near flap (*bottom right*) for staggered grid results for Mach 0.1 inviscid flow past a three-element airfoil [From D.A. Kopriva (1998)]

Figure 5.53 from Kopriva (1998) displays the results of a complex geometry computation using the staggered-grid collocation method. The computation is for Mach 0.1 flow past a three-element airfoil. There are a total of 275 subdomains in the mesh (see the top left frame of Fig. 5.53) and the order of approximation within each subdomain is $N = 8$. The computed flow is smooth everywhere, and the computation converges to full machine precision, again without recourse to any type of filtering. We should note that Kopriva (1998) describes a local time-stepping procedure that reduces the computation time by almost an order of magnitude compared with using the same time-step throughout the flow. (This use of local time-stepping is analogous to that long in use for finite-difference and finite-volume methods

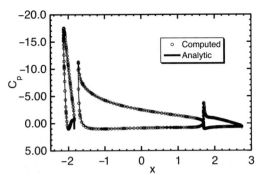

Fig. 5.54. Comparison of pressure coefficients for a three-element airfoil. The computed results are from a staggered-grid solution for Mach 0.1 flow, and the analytic results are for incompressible flow [From D.A. Kopriva (1998)]

for steady flow problems, but infrequently exploited for spectral solutions of the Euler equations.) As verification of the spectral solution, Fig. 5.54 shows a comparison of the spectral collocation result with an analytical solution (Suddhoo and Hall (1985)) for the corresponding incompressible flow case.

Multidomain Shock Fitting. The final patching methods that we discuss here are the extensions of the spectral shock-fitting method discussed in Sect. 4.7 to multiple domains and to viscous flows. The basic principles of multidomain spectral shock-fitting using patching for inviscid flow are described by Kopriva (1992). The major complication regarding the descriptions in Sect. 4.7 and earlier in this section is the treatment of those interface points that are located on the shock boundary. At such a point the two subdomains each have their own value for the normal to the shock, and these do not necessarily agree. The conflict is resolved in the usual way by using the shock normal and shock acceleration that are computed on the upstream side of the interface to update the position and flow variables at the corner point on the shock boundary.

Figure 5.55 illustrates the multidomain grid and the convergence of several known quantities for Mach 6 flow past a 10° cone. The exact solution to this class of problems was given by Taylor and Maccoll (1933). The right frame of Fig. 5.55 demonstrates that spectral accuracy is achieved for the shock slope and the flow variables. (These were 32-bit computations.) Even for this simple problem the multidomain approach is advantageous because the minimum grid spacing in the direction along the cone is significantly larger than it would be with a single domain; hence, a much larger time-step can be employed. Examples of multidomain shock-fitted solutions to more complicated inviscid flows can be found in Kopriva (1992, 1999).

The spectral shock-fitting method has been extended to viscous flow for both single-domain (Kopriva (1993)) and multidomain (Kopriva (1996b)) methods. Of course, the "shocks" in viscous flows are not infinitesimally

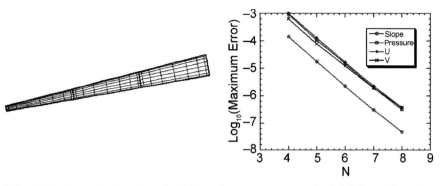

Fig. 5.55. Computational mesh (*left*) and convergence results (*right*) as a function of the number of grid points N in the wall-normal direction for spectral shock-fitting solutions to inviscid flow past a sharp cone [From D.A. Kopriva (1992)]

thin as they are for inviscid flows. This method is only applicable in the range of Mach and Reynolds numbers where the shock thickness can be neglected. (Moretti and Salas (1969) provide a detailed discussion of where shock-fitting is appropriate for viscous flows, and they describe the shock-fitting approach for finite-difference discretizations.) In shock-fitting for viscous flows the shock is still treated as a discontinuity, and the viscous terms in the locally one-dimensional CCM treatment at the shock are ignored. The wall boundary conditions are now no-slip for the velocity, either isothermal or adiabatic conditions for temperature, and the pressure is determined from the CCM. The outflow boundary now necessarily contains a subsonic region near the wall. In this portion of the boundary, the pressure is specified (from the inviscid solution to the problem), velocities are computed from the CCM, and the streamwise gradients in the viscous fluxes are ignored. The subdo-

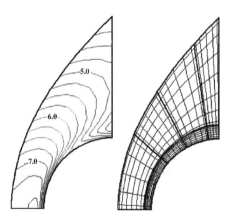

Fig. 5.56. Computational mesh (*right*) and computed temperature (*left*) for shock-fitted solution to viscous flow past a cylinder [From D.A. Kopriva (1996b)]

main interface conditions for the viscous problem are an extension of the inviscid treatment with a penalty term added for the viscous fluxes (Kopriva (1996)).

Figure 5.56 shows the results of computations for Mach 5.73 flow past a cylinder with an adiabatic wall. There are sharp gradients in the near-wall region, and the flexibility afforded by the multidomain approach allows finer resolution in that part of the domain. The computed solutions are smooth at all interfaces, and no filtering is needed to obtain an efficient, converged solution.

Despite these successes of the spectral patching method for inviscid flow, we still hazard the guess that the SDGM will eventually emerge as the high-order method of choice for such problems.

5.14 3D Applications in Complex Geometries

We close this description of multidomain spectral discretization methods with illustrations of some large-scale applications that have been made with these methods in complex geometries, starting with some fluid dynamics examples and then continuing with examples in other application domains to emphasize the broad applicability of modern spectral methods in computational engineering and science.

5.14.1 The Spectral Element Method: Application to Incompressible Flow

The spectral element method (SEM) has been applied to a wide variety of challenging fluid dynamics applications, almost invariably using numerical integration (SEM-NI). One can find many examples in the texts by Deville, Fischer and Mund (2002) and Karniadakis and Sherwin (2005). We provide just one example here in Fig. 5.57, taken from Fischer et al. (2007), of a direct numerical simulation of the transitional flow of blood near an arteriovenous graft. These grafts connect a high-pressure vessel (the artery) to a low-pressure one (the vein). The resulting high flow rates normally produce a weakly turbulent state, which can lead to eventual failure of the graft. The figure illustrates the coherent structures in the transitional and weakly turbulent flow downstream of the graft. This particular SEM applied to the Navier–Stokes equations used the $\mathbb{Q}_N - \mathbb{Q}_{N-2}$ method, semi-implicit time-discretization, and an algebraic splitting method (such as is discussed in Sect. 7.4). The computation shown in the figure used 2640 elements with $N = 12$ in each dimension, corresponding to 4.5×10^6 grid points. These computations take about a day of wall-clock time per cardiac cycle using 256 nodes of a TSC1 supercomputer; multiple cardiac cycles must be run to obtain the requisite statistical data. These SEM calculations are used to

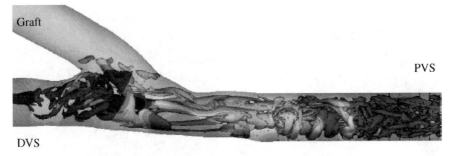

Graft

PVS

DVS

Fig. 5.57. SEM-NI solution of blood flow near an arteriovenous graft that joins an artery to a vein. The simulation includes the flow coming through both the graft and the distal venous segment (DVS) and their joint flow into the proximal venous segment (PVS). The figure illustrates the coherent structures in this transitional flow [From P.F. Fischer, F. Loth, S.E. Lee, S.-W. Lee, D.S. Smith, H.S.Bassiouny (2007)]

examine how changes to the graft geometry can reduce the adverse impact of weakly turbulent flows on the integrity of the graft.

5.14.2 The Spectral Discontinuous Galerkin Method: Application to Compressible Flow

An example of a nonlinear, three-dimensional computation performed with the SDGM is provided in Fig. 5.58. This is from the work of Stanescu, Hussaini and Farrasat (2002). The problem is to predict noise propagation and scattering off the surface of an aircraft, given the engine tone noise source as a combination of spinning modes on a circular disc surface appropriately situated inside the nacelle. To that end, the three-dimensional compressible Euler equations are solved by SDGM on a computational domain consisting of a truncated half-space (assuming symmetry of the problem about a vertical plane that bisects the aircraft along the fuselage). The computational domain consists of a physical domain abutted by damping layers, which ensure that the physical domain of interest remains uncontaminated by reflections. The computational domain is covered by an unstructured grid of $103,105$ non-overlapping hexahedral elements with a sixth-order Legendre polynomial representation of the solution in each element. Thus the total number of Legendre Gauss points was 22×10^6. The temporal discretization employed a low-storage, third-order Runge–Kutta method. The computations occupied one node (32 processors running at 1.1 GHz clock speed) of an IBM Regatta-type SP4 machine. Each run lasted about 10 days.

Figure 5.58 (left) shows the hexahedral representation of the underlying aircraft surface geometry with a Legendre Gauss point distribution with sixth-order elements. Figure 5.58 (right) shows a snapshot of the sound pressure level contours for the spinning mode $(18, 0)$ on the surface of the aircraft

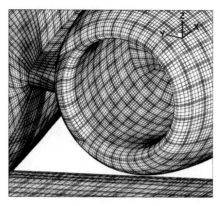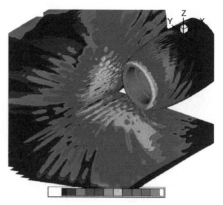

Fig. 5.58. Close-up of the computational mesh (*left*) and the sound pressure field (*right*) in the vicinity of an aircraft engine for an SDGM computation using the three-dimensional Euler equations [Courtesy of D. Stanescu, J. Xu, M.Y. Hussaini, F. Farassat (2002)]

at nondimensional time $t = 44$. The sound pressure level at a point is measured in decibels (dB) and is equal to 20 times the logarithm to the base 10 of the ratio of root-mean-square sound pressure to the reference sound pressure $(2 \times 10^{-5}\,\mathrm{Pa})$. This computation demonstrates that three-dimensional SDG computations for physically significant aerodynamic problems can be computed with relatively modest computational resources.

This and the previous applications of the SDGM indicate that it has clear advantages compared with low-order methods on the one hand, and the patching multidomain spectral methods discussed in Sect. 5.13 on the other hand. These advantages lie in the fact that it has i) higher accuracy per node than low-order methods, ii) compactness and robustness, ii) easy accommodation of complex geometry, iii) easy implementation of boundary conditions, iv) solution adaptivity (in terms of grid refinement and/or local enhancement of formal order of accuracy, and v) easy parallelizability. The disadvantages are i) highly restrictive explicit time-stepping and ii) higher computational time per node. The former disadvantage can be alleviated by implicit time-stepping (Rasetarinera and Hussaini (2001)), such as the one described in Sect. 5.9.2, and the latter can be advantageous on computers with limited inter-node communication speeds. Thus, these methods are bound to become increasingly popular in complex flow physics problems.

5.14.3 The Spectral Element Method: Application to Thermoelasticity

We now shift to demonstrations of multidomain spectral methods on significant engineering applications outside the domain of fluid dynamics. The first of these two non-fluids examples is taken from a practical application in

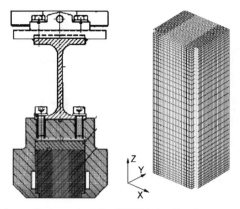

Fig. 5.59. *Left*: Schematic layout of the LHC injection beam stopper (TDI). Cyan and light blue areas identify the position of the absorbing block. *Right*: Spectral element partition of an absorber block (spectral nodes inside the elements not shown) [Courtesy of L. Bruno, G. Fotia, Y. Kadi, F. Maggio, L. Massidda, F. Mura]

the analysis and design optimization of a component for a new particle accelerator, the Large Hadron Collider (LHC) at CERN in Geneva. The LHC target dump injection (TDI) beam stopper is a mobile structure composed of 20 suitably supported absorbing blocks used to adjust the trajectory of the proton beam and to protect other LHC elements in case of failure.

The absorption of high-energy proton beams subjects the TDI beam stopper to rapid transient thermomechanical loading. Transient thermomechanical analyses are required for various loading conditions on each block (see Fig. 5.59 for a two-dimensional layout of the TDI and for a three-dimensional mesh of a single block). The numerical computations account for the coupled thermal and structural responses to the loads. Typical time-dependent computations for the analysis and design optimization of the TDI were performed with a SEM using an unstructured hexahedral mesh with 26,928 elements along with an explicit second-order (leap frog) time-discretization. The polynomial degree in each element was 4, corresponding to hexahedral elements with 125 nodes. The total number of nodes was 1,767,405, and the total number of degrees of freedom was larger than 7,000,000. Free-stress boundary conditions at all external surfaces of the absorber block were enforced. Figure 5.60 illustrates the transient dynamical response, in terms of the distribution of kinetic energy at the times $t = 12\mu s$ and $16\,\mu s$, of a boron nitride block for a near surface incident proton beam.

The results of numerous simulations of the baseline design using the multidomain spectral method showed exceedingly high stress levels in some of the absorber blocks. The design was revised accordingly—a new material was selected for the blocks that would guarantee their structural integrity, and modifications of the geometry were adopted to optimize the performance of

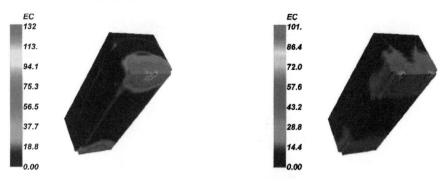

Fig. 5.60. Computed distribution of kinetic energy at times $t = 12\,\mu s$ (*left*) and $t = 16\,\mu s$ (*right*) [Courtesy of L. Bruno, G. Fotia, Y. Kadi, F. Maggio, L. Massidda, F. Mura]

the system. The assembly and installation of the optimized system at CERN began in 2005.

5.14.4 The Spectral Element Method: Structural Dynamics Analysis of the Roman Colosseum

Our final example is the use of the SEM to investigate the structural dynamic response of a large, complex historical structure—the Colosseum in Rome—to impact forces representative of the vibrations induced by road and railway traffic. The mathematical problem that has been solved is given by the equilibrium equations for an elastic bounded medium in $\mathbb{R}^3$:

$$\rho \frac{\partial^2 \mathbf{u}}{\partial t^2} = \text{div } \boldsymbol{\sigma}(\mathbf{u}),$$

where $\mathbf{u}$ is the displacement of the medium, t is the time, ρ is the density and the divergence term on the right hand side denotes the contribution of the internal forces resulting from the stress $\boldsymbol{\sigma}$.

The materials comprising the Colosseum (see Fig. 5.61) are travertine (outer wall), tufa, brick and concrete (interior), with Poisson ratios ranging from 0.1 to 0.2 and densities ranging from 1800 to 2400 kg/m^3. The relationship between $\boldsymbol{\sigma}$ and the strain tensor $\boldsymbol{\epsilon}$ is assumed to be governed by Hooke's law (corresponding to small, linear displacements of the structure). The computational domain (the Colosseum building) is approximately 48 meters high, 189 meters long, and 156 meters wide.

The use of an unstructured hexahedral mesh for the SEM permits the accurate representation of important architectural details and material properties distributions. The spatial discretization parameters were as follows: 107,487 spectral elements were used, the polynomial degree in each element was $N = 3$, corresponding to 64-node hexahedral elements. This yields

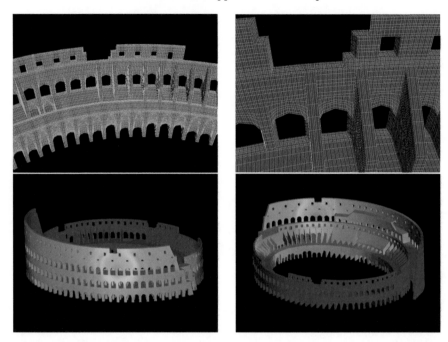

Fig. 5.61. Close-up of the computational mesh of the outer wall of the Colosseum (*top, left* and *right*) and computed unitary dilatation at time $t = 0.5$ s, outer view (*bottom, left*) and inner view (*bottom, right*) [Courtesy of F. Bettio, G. Fotia, F. Maggio, L. Massidda, G. Siddi]

3,413,502 LGL nodes and more than 10,000,000 degrees of freedom. The time-discretization was the explicit, second-order, leap frog method. The number of time-steps was 100,000, with $\Delta t = 5 \cdot 10^{-6}$ s. A free surface condition was set on the whole boundary. The initial condition was $\mathbf{u} = \mathbf{0}$. The response was calculated for an external horizontal force of Ricker type, that is, a traction $\mathbf{T}(\mathbf{x}, t) = \mathbf{g}(\mathbf{x})h(t)$ applied on a small area of the internal part of the building (corresponding to the red area in Fig. 5.61, bottom), where $\mathbf{g}(\mathbf{x})$ is a Gaussian and $h(t) = [1 - 2\beta(t - t_0)^2]e^{-\beta(t - t_0)^2}$. The cutoff frequency $\sqrt{\beta}$ was equal to 7.1 Hz. This translates into a nonhomogeneous Neumann condition on a portion of the domain boundary. Figure 5.61 shows the computed dilatation (unitary change in volume) at time $t = 0.5$ s. This computation required about 42 hours of computing time on 8 four-processor nodes (32 total CPU's) of a Linux Cluster (ADM Opteron Dualcore Processor 256, 1800 MHz, 1 MB cache, 4 GigaByte memory, DDR 400 MHz).

The computation illustrated here is used in a combined experimental and computational dynamical characterization that allows the health monitoring and damage risk assessments of historical monuments.

6. Solution Strategies for Spectral Methods in Complex Domains

6.1 Introduction

In this chapter we address the issue of how to solve efficiently the algebraic systems arising from the spectral approximations in complex geometries that we have introduced earlier in Chap. 5. The discussion and the attendant details will mostly be given just for SEM discretizations to the Poisson problem (5.4.19) considered in Sect. 5.2; see the SEM problems (5.2.6), (5.3.11) or their SEM-NI variants (5.2.15), (5.3.13). However, most of the approaches apply equally well to other discretizations such as patching collocation methods, and to more general elliptic problems, in particular to the Stokes equations for incompressible viscous flows. Methods specifically suited to collocation methods are addressed in Sect. 6.5, while those for Stokes and Navier–Stokes equations will be addressed in Sect. 7.3 and 7.4 of the next chapter.

6.2 On Domain Decomposition Preconditioners

In order to unify our notation, in this chapter we will denote by

$$A\mathbf{u} = \mathbf{f} \qquad\qquad (6.2.1)$$

the linear system arising from the SEM discretization (or its SEM-NI variant) that we have mentioned above. The system matrix is indicated with the letter A to conform with the conventional notation in numerical linear algebra. Let us recall that the computational domain Ω is partitioned into non-overlapping elements (or subdomains) Ω_m, as indicated in Sect. 5.3. Unless otherwise specified, in our presentation Ω_m will coincide with a single spectral element. However, more generally, Ω_m could be a subdomain made of a patch of spectral elements, say $\Omega_m = \bigcup_k e_k^{(m)}$; see Fig. 6.1 for two instances. Indeed, the latter is a more convenient situation in view of parallel implementations.

The diameter of Ω_m will be indicated by H_m. (In the case of a single spectral element, $H_m = h_m$.) If, for the sake of exposition, we assume that all the Ω_m are quadrilaterals (in 2D) or parallelepipeds (in 3D), then A is the stiffness matrix associated with the sum of the bilinear forms, $\sum_m a_{\Omega_m}(\cdot,\cdot)$

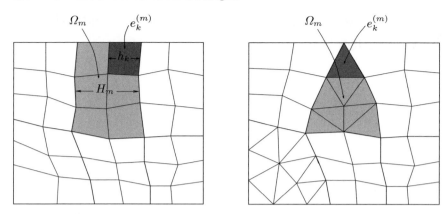

Fig. 6.1. A partition of Ω into subdomains Ω_m that are clusters of spectral elements $\{e_k^{(m)}\}$: with all quadrilaterals to the *left*, with hybrid quadrilateral and triangular elements to the *right*

or $\sum_m a_{N_m, \Omega_m}(\cdot, \cdot)$, M is the number of elements (or subdomains), $\mathbf{f}$ is the right-hand side, and $\mathbf{u}$ is the vector of the nodal values of the solution u_δ at the n LGL nodes that do not lie on the boundary of the domain. Then, n is the sum of the $n_I = \sum_m (N_m - 1)^d$ nodes internal to the elements plus the n_Γ nodes belonging to the interfaces between elements. We are not counting the nodes on $\partial\Omega$ as u_δ is specified there in the case of Dirichlet boundary conditions. Otherwise the sum should include those boundary nodes where Neumann boundary conditions hold.

Similar considerations hold for partitions like those of Fig. 6.1.

For the efficient solution of (6.2.1) several approaches can be pursued. When A is positive definite, iterative methods based on Krylov iterations can be set up. If A is also symmetric (this is the case of problem (5.4.19) or of problem (5.3.9) if $\beta = 0$), the conjugate gradient method is the method of choice. However, A is typically ill conditioned. In fact, as already pointed out in Sect. 5.3.4, for a uniform polynomial degree N and a uniform element diameter H, for a second-order elliptic problem, its iterative condition number, which indeed coincides with the spectral condition number (see (C.1.10)–(C.1.13)), behaves as

$$\mathcal{K}(A) \simeq N^3 H^{-2}$$

(see Figs. 5.9, 5.10 and Fig. 6.2). Here we are considering the case where the sizes of subdomains are comparable to those of the spectral elements, that is there are two constants $c', c'' > 0$ such that $c' h_{\min} \leq H_m \leq c'' h_{\max}$, for all m. Then, the set up of efficient preconditioners is in order. Unfortunately, diagonal preconditionings are of little help in alleviating the severe dependence of the convergence rate on N and H (see, e. g., Deville et al. (2002), Sect. 4.5.5). As pointed out in Sect. 5.3.5, several kinds of finite-element-based preconditioners can be built up using shape functions centered at the Gaussian nodes

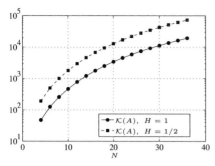 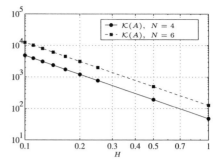

Fig. 6.2. (*Left*) The iterative/spectral condition number of the SEM-NI discretization matrix of the Laplacian in 2D, versus the (constant) polynomial degree N. Here H denotes the (uniform) length of each spectral element edge. When $H = 1$, $\mathcal{K}(A) \simeq 0.41 \cdot N^3$, and when $H = 1/2$, $\mathcal{K}(A) \simeq 1.61 \cdot N^3$. (*Right*) The same quantity versus the element side $H = 2/M$. When $N = 4$, $\mathcal{K}(A) \simeq 4.81 \cdot H^{-2}$, and when $N = 6$, $\mathcal{K}(A) \simeq 124.31 \cdot H^{-2}$

used in each spectral element. Either linear elements on simplices (triangles in 2D, tetrahedra in 3D) or bilinear (resp. trilinear) in 2D (resp. 3D) on rectangles (resp. hexahedra) are appropriate. See Figs. 5.11 and 5.12. Indeed, for both kinds of families, the condition number of the preconditioned matrix can be bounded uniformly with respect to N and H (see the several tables provided in Sect. 5.3.5).

This property is known as FEM-SEM spectral equivalence.

However, due to the splitting of the computational domain Ω into elements (or subdomains) Ω_m, domain decomposition (DD) preconditioners are in general more efficient.

An *efficient DD preconditioner* is, on the one hand, *optimal* with respect to the dependence on the polynomial degree N, i.e., the spectral condition number of the preconditioned system is uniformly bounded with respect to N, and, on the other hand, amenable to an efficient parallel implementation of the resulting algorithm. Ideally, besides being optimal it should also be *scalable*, i.e., it should give rise to a convergence rate that is independent of the number of subdomains (that is, on H when considering uniform partitions of a given domain Ω).

There are two basic approaches to set up DD preconditioners, yielding the so-called Schwarz or Schur (families of) preconditioners, respectively. Typically, the former involves using overlapping subdomains (which can be obtained, e. g., by extending slightly every original element Ω_m) then solving at each step a family of independent Dirichlet problems on each subdomain—see Sect. 6.3. The latter are based on the Schur complement, which, in domain decomposition applications, is derived from the global system (6.2.1) by eliminating the n_I interior unknowns and reducing therefore (6.2.1) to a smaller system of order n_Γ involving solely the unknowns on the interface. We will address this type of preconditioner in Sect. 6.4. As we will see, both strategies require enrichment by means of a coarse-level preconditioner in order to achieve scalability.

Both the Schwarz and Schur preconditioning approaches often exploit the fact that a good preconditioner of a good preconditioner remains a good preconditioner. According to this observation, a possible strategy is that at a first stage the system (6.2.1) is preconditioned by the finite-element matrix A_{FE} using low-order finite elements on the Ω_m. Then, A_{FE} is in its turn replaced by one of the effective DD preconditioners (of either Schwarz or Schur type) which are available for finite-element approximations (see Smith, Bjørstad and Gropp (1996), Quarteroni and Valli (1999), Toselli and Widlund (2005), Wohlmuth (2001)).

From a pure mathematical point of view, all the previous methods can be reformulated under a general framework involving space decomposition and projection operators (see Sect. 6.3.5) and are called "Schwarz methods" (see e. g., the monograph by Toselli and Widlund (2005), Chap. 2). However, we prefer keeping two distinct names (Schwarz and Schur) to identify methods based on partitions *with* or *without* overlap, respectively.

Finally, we warn the reader that although the search for efficient DD preconditioners of the algebraic system has nowadays become the standard route to DD methods, there are nonetheless other approaches that arise from diferent perspectives and are still successfully used in several contexts (e. g., in multiphysics problems).

Fig. 6.3 has the purpose of providing an indicative (albeit neither exhaustive nor rigorous) overview.

On the lower part we show the way to introduce DD preconditioners in the context of iterative methods that are set up for the solution of the algebraic system: either the one (6.2.1) governed by the global stiffness matrix A, or the above mentioned Schur complement system that is obtained from (6.2.1) by formally eliminating all unknowns associated with nodal points that do not sit on subdomain interfaces.

On the upper part we show DD methods that are derived "directly" from the physical problem (prior to its spectral discretization). Precisely, the idea is (i) to begin with the differential problem or with its equivalent formulations as either its split (patching) form that makes use of transmission conditions at interfaces, or else its Steklov–Poincaré (reformulation) involving solely the unknown value of the exact solution at interfaces, next (ii) to develop iterative methods involving subproblems in the subdomains (these are often called *iterative substructuring* methods in the DD literature), and then (iii) to approximate every subproblem by any of the (standard) single-domain spectral solvers discussed in the previous Chaps. 1–4 and in CHQZ2.

In Fig. 6.3 the solid line $\boxed{A} \longrightarrow \boxed{B}$ means: "block A is used to solve block B", while the bidirectional line $\longleftrightarrow$ stands for equivalence.

Obviously, the situation is not as crystallized as one might conclude just from this short introduction. In fact, several analogies, and sometimes even

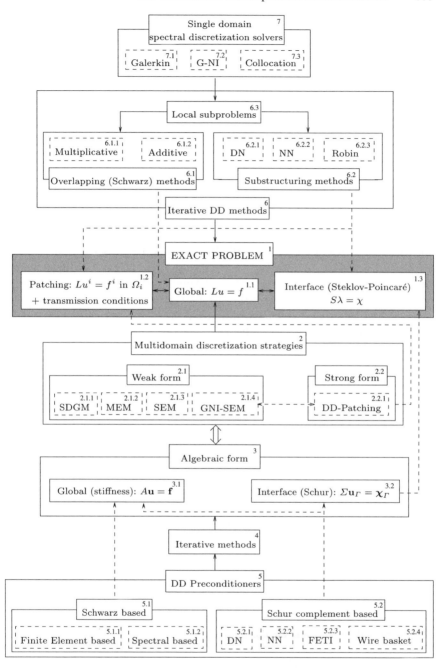

Fig. 6.3. The roadmap of spectral domain decomposition methods

precise relationships, do exist between several blocks of this picture (these analogies/relationships are indicated in the figure by the presence of dotted lines).

Due to these numerous interrelations, there are several different routes for telling the whole story. This is the reason why, in the discussion that follows, we prefer not simply to proceed "block-wise" monotonically from block 1 downward till block 5, and then from block 1 upward to 6 and 7.

We will start from the method that is the noble ancestor of modern DD methods, and proceed following a path that optimizes intellectual "ergonomicity".

6.3 (Overlapping) Schwarz Alternating Methods

The Schwarz method has been originally proposed by H. Schwarz in 1869 as an iterative method for proving the existence of solutions to elliptic equations in domains that are the union of simpler ones where existence can be easily proved (Schwarz (1869)). Although it can still be used as a solution method in general domains, nowadays it is more successfully used as a domain-decomposition preconditioner for conjugate gradient (or Krylov) iterations of algebraic systems arising from the numerical discretization of boundary-value problems.

As already mentioned, a distinguishing feature of the Schwarz method is that it works on a subdivision of the computational domain into subdomains that *overlap*.

In the spectral element context, a typical way to obtain an overlapping partition consists of starting from a subdivision $\{\Omega_m\}$ in spectral elements and extend these by an amount δ^* in each direction to create overlapping subdomains $\tilde{\Omega}_m$ (see Fig. 6.4).

More precisely, δ^* is the minimum distance between the portions of the boundaries of $\tilde{\Omega}_m$ and Ω_m not included in $\partial\Omega$, for every m. If δ^* is proportional to H, the overlap is said to be *generous*, while it is said to be *small* when δ^* has the order of the grid-size (the finite-element size h or the distance between two LGL nodes in SEM discretizations). More particularly, the overlap is said to be *minimum* if it involves just a single layer of elements outside Ω_m.

To simplify notation, in this section, when no ambiguity occurs, we will still indicate by Ω_m the "extended" subdomains. Moreover, for the sake of exposition, at this stage we confine ourselves to the simple Poisson problem (5.4.19).

To begin with, assume that Ω is a 2D domain, and let Ω_1 and Ω_2 be two overlapping subdomains whose union gives Ω. Then, we define $\Gamma_1 = \partial\Omega_1 \setminus \partial\Omega$ and $\Gamma_2 = \partial\Omega_2 \setminus \partial\Omega$ (see e.g., Fig. 6.5).

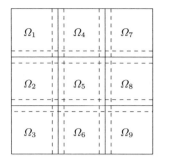 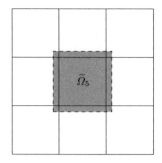

Fig. 6.4. Subdivision of a 2D rectangular region Ω into nine non-overlapping sub-domains Ω_m (*left*), and the extended overlapping subdomain $\widetilde{\Omega}_5$ (*right*)

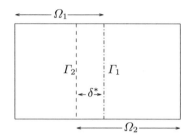

Fig. 6.5. An overlapping decomposition of a rectangle using two rectangles

The classical Schwarz alternating method was designed to generate a sequence that converges to the solution of (5.4.19) (see box 6.1 of Fig. 6.3) and is defined as follows.

Given $\hat{u}_2^{(0)}$, for $k \geq 0$ solve

$$
\begin{aligned}
-\triangle \hat{u}_1^{(k+1)} &= f && \text{in } \Omega_1, \\
\hat{u}_1^{(k+1)} &= \hat{u}_2^{(k)} && \text{on } \Gamma_1, \\
\hat{u}_1^{(k+1)} &= 0 && \text{on } \partial\Omega_1 \setminus \Gamma_1,
\end{aligned}
\tag{6.3.1}
$$

$$
\begin{aligned}
-\triangle \hat{u}_2^{(k+1)} &= f && \text{in } \Omega_2, \\
\hat{u}_2^{(k+1)} &= \hat{u}_1^{(k+1)} && \text{on } \Gamma_2, \\
\hat{u}_2^{(k+1)} &= 0 && \text{on } \partial\Omega_2 \setminus \Gamma_2 .
\end{aligned}
\tag{6.3.2}
$$

The functions $\hat{u}_m$ are defined solely on Ω_m.

This is a sequential version that is called the *multiplicative Schwarz alternating* method. A "parallel" version is easy to obtain by using in (6.3.2) $\hat{u}_1^{(k)}$ instead of $\hat{u}_1^{(k+1)}$ on Γ_2. The corresponding scheme is termed the *additive Schwarz alternating* method. The reason behind the use of this terminology (multiplicative versus additive) is explained below, after Theorem 6.3.1.

The subproblems (6.3.1) and (6.3.2) can be approximated by using suitable single-domain spectral solvers, such as Galerkin, G-NI or collocation solvers. In this way the Schwarz method could be regarded as an iterative method to solve the differential problem (see box 6.1 in Fig. 6.3).

In spite of its simple formulation and ease of implementation, the Schwarz alternating method converges slowly in the case of a small overlap. Indeed, by setting $e_m^{(k)} = u_{|\Omega_m} - \hat{u}_m^{(k)}$, for $m = 1, 2$, for both multiplicative and additive cases, it can be proven with the help of the maximum principle that

$$\|e^{(k)}\|_{L^\infty(\Omega_m)} \leq [\rho(\delta^*)]^{2k} \|e^{(0)}\|_{L^\infty(\Omega_m)}, \quad m = 1, 2 ,$$

for a suitable constant $0 < \rho(\delta^*) < 1$ that depends on δ^* and decreases monotonically when the size of the overlap increases. (As usual, $\|v\|_{L^\infty(\Omega_m)}$ stands for the maximum norm of the function v in Ω_m).

For instance, in the multiplicative case, if Ω is is the one-dimensional interval $(-1, 1)$, we obtain

$$\rho(\delta^*) = \frac{1 - \delta^*/2}{1 + \delta^*/2} ,$$

which shows that increasing the overlap is helpful from the point of view of convergence, although the computational efficiency decreases.

The method (6.3.1)–(6.3.2) can be given a Richardson interpretation by means of suitable projection operators. Besides being essential for carrying out a convergence analysis, this interpretation is crucial as it suggests the way to generalize this method to more interesting forms. Indeed, later in this section we will show how to re-shape the Schwarz method as an efficient preconditioner for the algebraic system (6.2.1) (see box 5.1 in Fig. 6.3).

Let us define in a formal manner the following operators (for their precise definition the reader is referred to the proof of Theorem 6.3.1). For every function $g \in L^2(\Omega)$, let Gg be the solution of the Poisson problem whose right-hand side is g:

$$\begin{aligned} -\triangle(Gg) &= g &&\text{in } \Omega, \\ Gg &= 0 &&\text{on } \partial\Omega. \end{aligned} \tag{6.3.3}$$

Moreover, for $m = 1, 2$, and for every function $w \in H_0^1(\Omega)$, let $P_m w \in H_0^1(\Omega)$ be the solution of the following problem:

$$\begin{aligned} -\triangle(P_m w) &= -\triangle w &&\text{in } \Omega_m, \\ P_m w &= 0 &&\text{on } \overline{\Omega} \setminus \Omega_m . \end{aligned} \tag{6.3.4}$$

Finally, we set

$$Q_{\text{mul}} = P_1 + P_2 - P_2 P_1 \quad \text{and} \quad Q_{\text{add}} = P_1 + P_2 . \tag{6.3.5}$$

The following result holds. Its proof is deferred until Sect. 6.3.5; however, we assure the reader that understanding the proof is not essential for understanding the forthcoming algorithmic developments.

Theorem 6.3.1 *An iteration of the multiplicative Schwarz alternating method can be regarded as a Richardson iteration for the problem*

$$Q_{\mathrm{mul}} u = g_{\mathrm{mul}} , \qquad (6.3.6)$$

where $g_{\mathrm{mul}} = Q_{\mathrm{mul}} G f$, that is,

$$u^{(k+1)} = u^{(k)} + g_{\mathrm{mul}} - Q_{\mathrm{mul}} u^{(k)} . \qquad (6.3.7)$$

Similarly, an iteration of the additive Schwarz alternating method can be regarded as a Richardson iteration for the problem

$$Q_{\mathrm{add}} u = g_{\mathrm{add}} , \qquad (6.3.8)$$

where $g_{\mathrm{add}} = Q_{\mathrm{add}} G f$, that is,

$$u^{(k+1)} = u^{(k)} + g_{\mathrm{add}} - Q_{\mathrm{add}} u^{(k)} . \qquad (6.3.9)$$

Then, the error at the k-th iteration, $e^{(k)} = u - u^{(k)}$, satisfies the recursive relation

$$e^{(k+1)} = (I - P_2)(I - P_1) e^{(k)}, \quad k \geq 0$$

in the multiplicative case, and

$$e^{(k+1)} = (I - P_1 - P_2) e^{(k)}, \quad k \geq 0$$

in the additive case.

The adjectives multiplicative and additive arise from the multiplicative structure of Q_{mul} and the additive structure of Q_{add}, respectively. Hybrid (additive-multiplicative) versions exist as well, at least in the finite-element context (see, e.g., Toselli and Widlund (2005)).

6.3.1 Algebraic Form of Schwarz Methods for Finite-Element Discretization

Schwarz iterative methods can be adapted for the solution of algebraic systems arising from the numerical approximation of boundary-value problems. To start with, let us consider a system like (6.2.1) which we now suppose to be associated with a finite-element approximation of the Poisson problem. The matrix A should now be regarded as a finite-element stiffness matrix, say A^{fe}, associated with continuous piecewise linear (or bilinear in 2D or trilinear in 3D) finite elements based at the LGL nodes. The Schwarz preconditioner that we will set up for this matrix will then be applied directly to precondition the original spectral matrix (6.2.1) (see box 5.1.1 of Fig. 6.3).

We still assume that Ω is decomposed in two overlapping subdomains Ω_1 and Ω_2 as indicated in Fig. 6.5. We denote by n the total number of interior

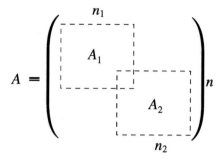

Fig. 6.6. The submatrices A_1 and A_2 of the stiffness matrix A

nodes of Ω and by n_1 and n_2 the interior nodes of Ω_1 and Ω_2. Note that $n \le n_1 + n_2$ and that the equality holds only if the overlap reduces to a single layer (of elements) among two columns of nodes. Let $I = \{1, \ldots, n\}$ be the set of indices of nodes of Ω, while I_1 and I_2 are those related to Ω_1 and Ω_2, respectively. Then $I = I_1 \cup I_2$, while $I_1 \cap I_2 = \emptyset$ only in the case of a single layer of overlap.

We reorder the nodes in three blocks in such a way that the first block corresponds to the nodes of $\Omega_1 \setminus \Omega_2$, the second to $\Omega_1 \cap \Omega_2$, and the third to $\Omega_2 \setminus \Omega_1$.

The stiffness matrix A contains two submatrices, say A_1 and A_2, that correspond to the local stiffness matrices associated with the Dirichlet problems in Ω_1 and Ω_2, respectively (see Fig. 6.6).

We warn the reader that the analysis that we are about to carry out applies only to the case of Galerkin methods using *local* basis functions. As such, we cannot extend it to either the SEM (or SEM-NI) or the patching collocation discretization of the Poisson problem.

The two submatrices can be obtained as follows:

$$A_1 = R_1 A \, R_1^T \in \mathbb{R}^{n_1 \times n_1} \qquad \text{and} \qquad A_2 = R_2 A \, R_2^T \in \mathbb{R}^{n_2 \times n_2} ,$$

where R_m and R_m^T $(m = 1, 2)$ are restriction and prolongation operators whose matrix representations is is as follows

$$R_1^T = \begin{pmatrix} 1 \ldots 0 \\ \vdots \ddots \vdots \\ 0 \ldots 1 \\ \mathbf{0} \end{pmatrix} \in \mathbb{R}^{n \times n_1} \qquad \text{and} \qquad R_2^T = \begin{pmatrix} \mathbf{0} \\ 1 \ldots 0 \\ \vdots \ddots \vdots \\ 0 \ldots 1 \end{pmatrix} \in \mathbb{R}^{n \times n_2} .$$

If $\mathbf{v}$ is a vector of $\mathbb{R}^n$, $R_1 \mathbf{v}$ is a vector of $\mathbb{R}^{n_1}$ whose components coincide with the first n_1 components of $\mathbf{v}$. If $\mathbf{v}$ is a vector of $\mathbb{R}^{n_1}$, $R_1^T \mathbf{v}$ is a vector of dimension n whose last $n - n_1$ components are null.

With the help of these matrices, one iteration of the multiplicative Schwarz method of the system (6.2.1) can be represented as follows:

$$\mathbf{u}^{k+1/2} = \mathbf{u}^k + R_1^T A_1^{-1} R_1 (\mathbf{f} - A\mathbf{u}^k) , \tag{6.3.10}$$

$$\mathbf{u}^{k+1} = \mathbf{u}^{k+1/2} + R_2^T A_2^{-1} R_2 (\mathbf{f} - A\,\mathbf{u}^{k+1/2}) . \tag{6.3.11}$$

The equation (6.3.10) can be regarded as the discretization of problem (6.3.1) with the solution extended by zero outside $\overline{\Omega}_1$, while (6.3.11) is the discrete counterpart of (6.3.2) after extending by zero the solution outside $\overline{\Omega}_2$.

Setting

$$Q_m = R_m^T A_m^{-1} R_m , \quad P_m = R_m^T A_m^{-1} R_m A = Q_m A , \quad m = 1, 2 , \tag{6.3.12}$$

we obtain $\mathbf{u}^{k+1/2} = (I - P_1)\mathbf{u}^k + P_1\mathbf{u}$, and therefore

$$\mathbf{u}^{k+1} = (I - P_2)\mathbf{u}^{k+1/2} + P_2\mathbf{u} = (I - P_2)(I - P_1)\mathbf{u}^k + (P_1 + P_2 - P_2 P_1)\mathbf{u} .$$

Similarly, one iteration of the additive Schwarz method becomes

$$\mathbf{u}^{k+1} = \mathbf{u}^k + (R_1^T A_1^{-1} R_1 + R_2^T A_2^{-1} R_2)(\mathbf{f} - A\mathbf{u}^k) , \tag{6.3.13}$$

and therefore

$$\mathbf{u}^{k+1} = (I - P_1 - P_2)\mathbf{u}^k + (P_1 + P_2)\mathbf{u} . \tag{6.3.14}$$

Thus, for the multiplicative Schwarz method, from (6.3.10) and (6.3.11) we obtain

$$\begin{aligned} \mathbf{u}^{k+1} &= \mathbf{u}^k + Q_1(\mathbf{f} - A\,\mathbf{u}^k) + Q_2[\mathbf{f} - A(\mathbf{u}^k + Q_1(\mathbf{f} - A\mathbf{u}^k))] \\ &= \mathbf{u}^k + (Q_1 + Q_2 - Q_2 Q_1)(\mathbf{f} - A\mathbf{u}^k) \end{aligned}$$

in analogy with (6.3.7). Similarly, from (6.3.13) it follows that one iteration of the additive Schwarz method becomes

$$\mathbf{u}^{k+1} = \mathbf{u}^k + (Q_1 + Q_2)(\mathbf{f} - A\mathbf{u}^k) , \tag{6.3.15}$$

in analogy with (6.3.9). The last formula can be easily extended to the case of a decomposition of Ω into $M \geq 2$ overlapping subdomains $\{\Omega_m\}$ (see, e. g., Fig. 6.4), yielding

$$\mathbf{u}^{k+1} = \mathbf{u}^k + \left(\sum_{m=1}^{M} Q_m \right) (\mathbf{f} - A\mathbf{u}^k) . \tag{6.3.16}$$

Although multiplicative Schwarz methods generally converge in fewer iterations, additive Schwarz is more popular (especially in the case of many subdomains) because it is fully parallel. Moreover, it does not require reevaluation of any part of the residual, an operation that is costly in high-order methods where the formation of local stiffness matrices is impractical.

6.3.2 The Schwarz Method as an Algebraic Preconditioner

By defining

$$P_{\text{as}} = \left(\sum_{m=1}^{M} Q_m \right)^{-1}, \tag{6.3.17}$$

it follows from (6.3.16) that *one iteration of the additive Schwarz method corresponds to a preconditioned Richardson iteration for system (6.2.1) with preconditioner P_{as}*. For this reason P_{as} is called *additive Schwarz preconditioner*.

Equivalently, if we define the preconditioned matrix

$$Q_{\text{add}} = P_{\text{as}}^{-1} A = \sum_{m=1}^{M} P_m, \tag{6.3.18}$$

then *one iteration of the additive Schwarz method corresponds to a Richardson iteration for the system $Q_{\text{add}} \mathbf{u} = \mathbf{g}_{\text{add}}$, where $\mathbf{g}_{\text{add}} = P_{\text{as}}^{-1} \mathbf{f}$*.

The latter result can be regarded as the algebraic counterpart of (6.3.9).

For $m = 1, 2$, P_m is a symmetric, nonnegative matrix with respect to the A-scalar product

$$(\mathbf{w}, \mathbf{v})_A = (A\mathbf{w}, \mathbf{v}) \qquad \forall \mathbf{w}, \mathbf{v} \in \mathbb{R}^n .$$

Indeed, for $m = 1, 2$,

$$\begin{aligned}
(P_m \mathbf{w}, \mathbf{v})_A &= (A P_m \mathbf{w}, \mathbf{v}) = (R_m^T A_m^{-1} R_m A \, \mathbf{w}, A \, \mathbf{v}) \\
&= (A\mathbf{w}, R_m^T A_m^{-1} R_m A \, \mathbf{v}) \\
&= (\mathbf{w}, P_m \mathbf{v})_A \qquad \forall \mathbf{v}, \mathbf{w} \in \mathbb{R}^n .
\end{aligned}$$

Moreover,

$$\begin{aligned}
(P_m \mathbf{v}, \mathbf{v})_A &= (A P_m \mathbf{v}, \mathbf{v}) = (R_m^T A_m^{-1} R_m A \, \mathbf{v}, A \, \mathbf{v}) \\
&= (A_m^{-1} R_m A \, \mathbf{v}, R_m A \, \mathbf{v}) \geq 0 \qquad \forall \mathbf{v} \in \mathbb{R}^n .
\end{aligned}$$

We can also prove that the preconditioned matrix Q_{add} of the additive Schwarz method is symmetric and positive definite with respect to the A-scalar product.

We prove first the symmetry: for all $\mathbf{u}, \mathbf{v} \in \mathbb{R}^n$, owing to the symmetry of A and P_m we obtain

$$\begin{aligned}
(Q_{\text{add}} \mathbf{u}, \mathbf{v})_A &= (A Q_{\text{add}} \mathbf{u}, \mathbf{v}) = (Q_{\text{add}} \mathbf{u}, A \, \mathbf{v}) = \sum_m (P_m \mathbf{u}, A \, \mathbf{v}) \\
&= \sum_m (P_m \mathbf{u}, \mathbf{v})_A = \sum_m (\mathbf{u}, P_m \mathbf{v})_A = (\mathbf{u}, Q_{\text{add}} \mathbf{v})_A .
\end{aligned}$$

Concerning the positivity, taking $\mathbf{u} = \mathbf{v}$ in the previous identities we obtain

$$(Q_{\text{add}}\mathbf{v}, \mathbf{v})_A = \sum_m (P_m\mathbf{v}, \mathbf{v})_A = \sum_m (R_m^T A_m^{-1} R_m A \,\mathbf{v}, A\mathbf{v})$$

$$= \sum_m (A_m^{-1}\mathbf{q}_m, \mathbf{q}_m) \geq 0, \tag{6.3.19}$$

where $\mathbf{q}_m = R_m A\,\mathbf{v}$. It follows that $(Q_{\text{add}}\mathbf{v}, \mathbf{v})_A = 0$ if and only if $\mathbf{q}_m = 0 \;\forall m$. However, since the subdomains overlap, the latter conditions imply that all the components of the vector $A\mathbf{v}$ must be zero. Since A is positive definite, this occurs if and only if $\mathbf{v} = 0$.

A more efficient iterative method for (6.2.1) is obtained if we use preconditioned *conjugate gradient* (instead of Richardson) iterations with the same additive Schwarz preconditioner P_{as}.

Unfortunately, P_{as} is not scalable since the condition number of the preconditioned matrix Q_{add} increases as the size of the subdomains reduces (Quarteroni and Valli (1999), Toselli and Widlund (2005)).

This is not surprising since the exchange of information occurs only between neighboring subdomains; as a matter of fact, only local solves are involved by the application of $(P_{\text{as}})^{-1}$. This weakness can be overcome at the expense of introducing a *coarse* global problem set over the whole domain, which guarantees a mechanism of global communication among all subdomains. This correction can be achieved by resorting to a general multilevel strategy, by analogy with the classical multigrid method approach (with the difference that now the role of the coarse grid is played by that of a coarse subdomain partition).

In our case, a coarse term $Q_H = R_H^T A_H^{-1} R_H$ can be added to Q_{add}. A_H is the stiffness matrix that refers to the SEM approximation of the original problem (with non-overlapping subdomains) using (mapped) piecewise linear (or bilinear or trilinear) elements—thus with a subspace X_δ given by (5.3.4) with $N_m = 1$ for every m. See Fig. 6.7 for an example in a simple case of the Dirichlet problem on a square domain partitioned into 9 uniform spectral elements. Finally, R_H^T is the prolongation (by zero) matrix, and R_H is the algebraic transpose of R_H^T (a restriction matrix). More precisely, P_{as} is replaced by

$$P_{\text{as},H} = \left(Q_H + \sum_{m=1}^{M} Q_m \right)^{-1}, \tag{6.3.20}$$

that is,

$$P_{\text{as},H} = \left(R_H^T A_H^{-1} R_H + \sum_{m=1}^{M} R_m^T A_m^{-1} R_m \right)^{-1}. \tag{6.3.21}$$

The new preconditioned matrix

$$Q_{\text{add},H} = (P_{\text{as},H})^{-1} A \tag{6.3.22}$$

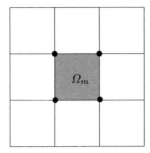 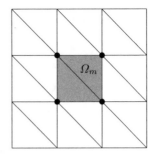

Fig. 6.7. Degrees of freedom for the coarse stiffness matrix A_H used for Schwarz preconditioners in the bilinear (*left*) and linear (*right*) case

has now a condition number that in two dimensions is bounded by

$$\mathcal{K}(Q_{\text{add},H}) \leq C \left(1 + \max_m \frac{H_m}{\delta_m} \right) , \qquad (6.3.23)$$

where H_m is the local coarse-mesh size, and δ_m is the subdomain overlap (see Dryja and Widlund (1995)).

Note that the condition number is uniformly bounded with respect to N and H in the case of generous overlap. Moreover, when applied to an elliptic operator with coefficients that change on the different subdomains, the constant C in general does depend on the jump of the coefficients.

Solving the original system (6.2.1) by an iterative method using $P_{\text{as},H}$ as preconditioner is equivalent to solving by the same iterative method (without preconditioning) the system

$$Q_{\text{add},H}\mathbf{u} = \mathbf{g}_{\text{add},H} , \qquad (6.3.24)$$

where $\mathbf{g}_{\text{add},H} = P_{\text{as},H}^{-1}\mathbf{f}$. We note that

$$\mathbf{g}_{\text{add},H} = \mathbf{g}_H + \sum_{m=1}^{M} \mathbf{g}_m , \qquad (6.3.25)$$

where

$$\mathbf{g}_H = R_H^T A_H^{-1} R_H \mathbf{f}$$

and

$$\mathbf{g}_m = R_m^T A_m^{-1} R_m \mathbf{f} \quad \text{for } m = 1, \dots, M .$$

Therefore, when using the preconditioned conjugate gradient method with the additive Schwarz preconditioner $P_{\text{as},H}$ for problem (6.2.1), at each iteration the computation of the preconditioned residual requires the solution of M uncoupled local finite-element Dirichlet problems in the subdomains $\Omega_1, \ldots, \Omega_M$ plus that of a global coarse problem (with mapped bilinear or trilinear elements) on the original non-overlapping partition of the domain Ω.

We conclude this section by recalling the preconditioned conjugate gradient (PCG) method (see (C.2.20)) and adapting it to the case of the additive Schwarz preconditioner.

Additive Schwarz PCG Method

Choose an initial guess $\mathbf{u}^0$, and compute $\mathbf{r}^0 = \mathbf{f}$, $\mathbf{z}^0 = \mathbf{r}^0$, $\mathbf{p}^0 = \mathbf{r}^0$. Then, for $k = 0, \ldots$, until convergence, do

$$\alpha_k = \frac{(\mathbf{r}^k, \mathbf{z}^k)}{(A\,\mathbf{p}^k, \mathbf{p}^k)} \,,$$

$$\mathbf{u}^{k+1} = \mathbf{u}^k + \alpha_k \mathbf{p}^k \,,$$

$$\mathbf{r}^{k+1} = \mathbf{r}^k - \alpha_k A\,\mathbf{p}^k \,,$$

$$P_{\mathrm{as},H} \mathbf{z}^{k+1} = \mathbf{r}^{k+1} \,,$$

$$\beta_{k+1} = \frac{(\mathbf{r}^{k+1}, \mathbf{z}^{k+1})}{(\mathbf{r}^k, \mathbf{z}^k)} \,,$$

$$\mathbf{p}^{k+1} = \mathbf{z}^k + \beta_{k+1} \mathbf{p}^k \,.$$

More precisely, in view of (6.3.20), for $\mathbf{r}^{k+1}$ given, $\mathbf{z}^{k+1}$ can be computed by:

> *For $m = 1, \ldots, M$ do in parallel*
>   *restrict the residual over Ω_m:* $\mathbf{r}_m^{k+1} = R_m \mathbf{r}^{k+1}$;
>
>   *solve the local Dirichlet problem:*
> $A_m \mathbf{z}_m^{k+1} = \mathbf{r}_m^{k+1}$;
>
> *Then:*
>
>   *solve* $A_H \mathbf{z}_H^{k+1} = R_H \mathbf{r}^{k+1}$;
>   *compute the global residual:*
> $\mathbf{z}^{k+1} = R_H^T \mathbf{z}_H^{k+1} + \sum_{m=1}^M R_m^T \mathbf{z}_m^{k+1}$.

6.3.3 Additive Schwarz Preconditioners for High-Order Methods

Let us return to system (6.2.1) which, as previously pointed out, can be considered as being associated with either the SEM (or SEM-NI) or the patching collocation discretization of the Poisson problem (or another self-adjoint elliptic problem).

If the stiffness matrix A is symmetric, it can be successfully preconditioned by one of its finite-element counterparts, say A_{fem}, that we have illustrated in Sect. 5.3.5, yielding the preconditioned matrices that we have called P_2 and P_3 or $P_{2,S}$ and $P_{3,S}$.

Now, a good preconditioner of a good preconditioner is still a good preconditioner. In fact, we note that if between A_{fem} and A and between $P_{\text{as},H}$ and A_{fem} we have the following set of inequalities:

$$\exists C_1, C_2 > 0 \text{ s.t. } C_1(A_{\text{fem}}\mathbf{v}, \mathbf{v}) \leq (A\,\mathbf{v}, \mathbf{v}) \leq C_2(A_{\text{fem}}\mathbf{v}, \mathbf{v}),$$

$$\exists C_3, C_4 > 0 \text{ s.t. } C_3(P_{\text{as},H}\mathbf{v}, \mathbf{v}) \leq (A_{\text{fem}}\mathbf{v}, \mathbf{v}) \leq C_4(P_{\text{as},H}\mathbf{v}, \mathbf{v})$$

for all $\mathbf{v} \in \mathbb{R}^n$, then obviously

$$C_1 C_3(P_{\text{as},H}\mathbf{v}, \mathbf{v}) \leq (A\,\mathbf{v}, \mathbf{v}) \leq C_2 C_4(P_{\text{as},H}\mathbf{v}, \mathbf{v}) \qquad \forall \mathbf{v} \in \mathbb{R}^n .$$

Owing to the fact that both matrices A and $P_{\text{as},H}$ are symmetric and positive definite, from these inequalities it can be deduced that (see, e. g., Quarteroni and Valli (1994), Theorem 2.5.1)

$$\mathcal{K}(P_{\text{as},H}^{-1}A) \leq \frac{C_2 C_4}{C_1 C_3} ,$$

where $\mathcal{K}(B)$ is the iterative condition number (see (C.1.10)) of the matrix B.

In view of that, we can use CG iterations on the spectral system (6.2.1) using instead of A_{fem} a suitable finite-element additive Schwarz preconditioner. The latter has a form like (6.3.20) and can be built up by considering overlapping subdomains $\tilde{\Omega}_m$ that are obtained by extending the original non-overlapping partition Ω_m, $m = 1, \ldots, M$ by an amount δ^* in each direction as indicated, e. g., in Fig. 6.4. An instance is provided in Fig. 6.8, where each $\tilde{\Omega}_m$ is obtained from Ω_m by adding one layer of Legendre Gauss-Lobatto (LGL) points from the neighboring elements. This is a case with minimum overlap. (More generous overlap can be called into play, as we will see later.)

In each $\tilde{\Omega}_m$ the piecewise bilinear (in 2D) or trilinear (3D) finite-element stiffness matrix A_m is constructed with respect to the internal nodes. Alternatively, linear (instead of bilinear or trilinear) elements can be used after dividing each LGL rectangle (or parallelepiped) into two triangles (or 6 tetrahedra, respectively). These alternatives were discussed in Sect. 5.3.5.

The coarse stiffness matrix A_H is obtained by using mapped piecewise bilinear (or trilinear) finite elements on the coarse triangulation represented by the original non-overlapping "macro-elements" Ω_m, $m = 1, \ldots, M$ (see Fig. 6.7, left). Alternatively, one can use linear elements on the triangulation obtained by subdividing each element into 2 triangles (2D) or 6 tetrahedra (3D) (see Fig. 6.7, right).

Casarin (1997) has proven that this preconditioned system achieves the bound (6.3.23) with a constant independent of the spectral polynomial degree N. In particular, when we overlap with a single layer of nodes, $\delta^* \sim H/N^2$, then the condition number behaves as $O(N^2)$.

To illustrate this behavior, we consider the SEM-NI discretization of the Poisson equation in $\Omega = (-1, 1)^2$ with Dirichlet boundary conditions. The

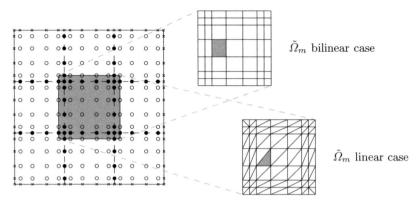

Fig. 6.8. Overlapping subdomains (with minimum overlap) in two dimensions. Indicated are the LGL nodes on each Ω_m (*left*) and the finite-element grid constructed on the nodes belonging to the extended domain $\tilde{\Omega}_m$ (*right*)

domain Ω is partitioned into $M^{1/2} \times M^{1/2}$ spectral elements (here coinciding with the subdomains) that are squares with edges of length $H = 2/M^{1/2}$. On each element, we use polynomials of degree N in each variable.

The additive Schwarz preconditioner (6.3.21) is used; the coarse matrix is $A_H = K_{FE,app}$ (CHQZ2, p. 235), i.e., the matrix of bilinear finite elements with numerical integration, associated to the coarse grid.

R_H^T is the matrix that interpolates a composite bilinear function, defined on the coarse grid, onto the SEM fine mesh, while $R_H = (R_H^T)^T$. Moreover, R_m are the restriction matrices R_m^T that are defined in Sect. 6.3.2, and A_m are the local stiffness matrices, $A_m = K_{FE,app}$ (CHQZ2, p. 235) built on bilinear finite elements with numerical integration, associated with the mesh of the enlarged subdomain element $\tilde{\Omega}_m$. The latter is obtained by extending Ω_m so as to include one or two layers of LGL nodes from neighboring spectral elements. In the former case (minimum overlap) the preconditioner is denoted by $P_{as,H}^m$, in the latter (small overlap) by $P_{as,H}^s$. Homogeneous Dirichlet conditions are imposed on $\partial\tilde{\Omega}_m \setminus \partial\Omega$, for $m = 1, \ldots, M$.

In Fig. 6.9, we report the PCG iterations required to solve our problem for an exact solution $u(x, y) = \cos(\pi x) \cos(\pi y)$, with a tolerance $tol = 10^{-12}$ and initial guess $\mathbf{u}_0 = \mathbf{0}$, for both $P = I$ and $P = P_{as,H}^m$ or $P_{as,H}^s$. On the left are the iterations versus the elements size H, and on the right are the iterations versus the polynomial degree N, for $H = 0.25$ (8×8 squared subdomains).

In Fig. 6.10 the condition number $\mathcal{K}(P^{-1}A)$ is shown versus both polynomial degree N and elements size H, for $P = I$ and $P = P_{as,H}^m$ or $P_{as,H}^s$.

We note that although the asymptotic behavior of the curves does not change, using small rather than minimum overlap nonetheless abates the condition number of the preconditioned system by a reasonable percentage and lowers the number of CG-iterations significantly.

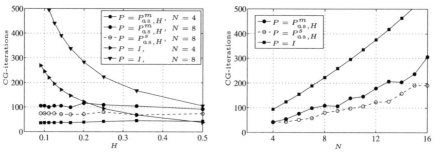

Fig. 6.9. CG iterations required to solve the Poisson problem, with a tolerance $tol = 10^{-12}$ are shown for $P = I$, $P = P^m_{as,H}$ and $P = P^s_{as,H}$. On the *left* the iterations versus the element size H, and on the *right* the iterations versus the polynomial degree N for $H = 0.25$ (8×8 spectral elements). The computational domain is, $\Omega = (-1,1)^2$, the exact solution is $u(x,y) = \cos(\pi x)\cos(\pi y)$, and the initial guess for PCG is $\mathbf{u}_0 = \mathbf{0}$. The *curves with empty circles* refer to the case of small overlap, that is, when the overlapping subdomains are obtained by extending every spectral element to include two layers of LGL nodes

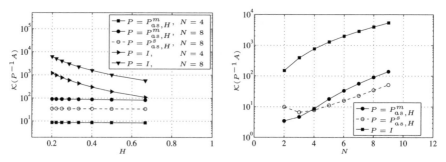

Fig. 6.10. The condition number $\mathcal{K}(P^{-1}A) = \lambda_{\max}(P^{-1}A)/\lambda_{\min}(P^{-1}A)$ for $P = I$, $P = P^m_{as,H}$ and $P = P^s_{as,H}$ versus the element size H (*left*), and versus the polynomial degree N (*right*), for $H = 0.25$ (8×8 spectral elements). The *curves with empty circles* refer to the case of small overlap, that is, when the subdomains are obtained by extending every spectral element to include two layers of LGL nodes

A similar approach was employed by Fischer to precondition the approximate pressure Schur complement (see (7.4.4), (7.4.7)) that arises from the $\mathbb{Q}_N - \mathbb{Q}_{N-2}$ SEM approximation of the Navier–Stokes equations (see Sect. 7.4.3).

Alternative approaches are based on the use of additive Schwarz preconditioners built on local *spectral* matrices constructed on the overlapping subdomains (as in box 5.1.2 of Fig. 6.3). This formulation can be set up by exploiting the abstract form of the Schwarz preconditioner method that we have derived at the end of the previous section. An example with minimum (a single LGL layer) overlap is provided in Casarin (1996), Sect. 3.5. The coarse-grid space is constructed as before, while the "local" subspaces V_m are subspaces of the SE-space V_δ (see (5.3.10)) vanishing at all the LGL nodes belonging to $\Omega \setminus \tilde{\Omega}_m$, on and outside $\partial\tilde{\Omega}_m$.

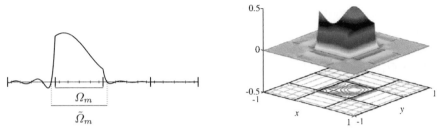

Fig. 6.11. A local function according to the approach of Casarin (1996), in 1D (*left*) and 2D (*right*)

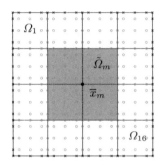

Fig. 6.12. The overlapping subdomain $\tilde{\Omega}_m$ whose center is the vertex $\overline{\mathbf{x}}_m$

The restriction and extension matrices R_m and R_m^T remain the same, while the local matrices A_m now become $A_m = R_m A_{SEM} R_m^T$, where A_{SEM} is the global SEM-NI stiffness matrix.

This time the supports of the local functions are not contained in $\tilde{\Omega}_m$ (see Fig. 6.11 for an example), while the same condition number bound (6.3.23) can still be proven.

We now describe some examples of Schwarz methods with generous overlap.

For simplicity we will confine ourselves to a partition of Ω into subdomains whose edges are aligned with the Cartesian axes, as in Fig. 6.12, and where the functions of V_δ on each element Ω_m are polynomials of $\mathbb{Q}_N(\Omega_m)$ (with the same degree on each Ω_m).

The first example is taken from Pavarino (1994a). We decompose the spectral space V_δ as $V_\delta = \sum_{i=0}^{M_0} V_i$, where M_0 is the number of internal vertices ($M_0 = 9$ in Fig. 6.12). Here, V_0 is the space $V_{1,M}$ of continuous piecewise bilinear functions on the grid defined by the spectral elements Ω_m, $m = 1, \ldots, M$; on the other hand, for every $m = 1, \ldots, M_0$ corresponding to an internal vertex $\overline{\mathbf{x}}_m$ of the original spectral elements, we set $V_m = V_N(\tilde{\Omega}_m) = \mathbb{Q}_N(\tilde{\Omega}_m)$ (see (5.3.3)), where now $\tilde{\Omega}_m$ denotes the extended subdomain obtained by the union of the four spectral elements merging at the vertex $\overline{\mathbf{x}}_m$ (see Fig. 6.12).

Then the algorithm consists of solving by the CG iterative method the equation

$$Q_{\text{add}} u_\delta = g_{\text{add}} , \qquad (6.3.26)$$

where $Q_{\text{add}} = \sum_{m=0}^{M_0} P_m$, and $P_m : V_\delta \to V_m$ is the projection operator defined by

$$a_\delta(P_m w, v) = a_\delta(w, v) \quad \forall v \in V_m .$$

Moreover, $g_{\text{add}} = \sum_{m=0}^{M_0} P_m u_\delta$. It is proven in Pavarino (1994a) that the spectral condition number $\mathcal{K}(Q_{\text{add}})$ is bounded uniformly with respect to N and H. This result is consistent with the estimate (6.3.23) as in this case $\delta_m \simeq H_m \; \forall m = 1, \dots, M$.

Other optimal alternating Schwarz preconditioners with generous overlap for the SEM on triangles (in 2D) and tetrahedra (in 3D) have been proposed and analyzed in Schöberl et al. (2007).

Overlapping Schwarz preconditioners for spectral elements with Fekete nodal points are investigated in Pasquetti et al. (2007b).

Schwarz methods with generous overlap are easy to implement on unstructured spectral element partitions (less obvious is the use of minimum overlap in these cases). An instance is provided in Pavarino and Warburton (2000), where hybrid unstructured spectral element partitions using triangular and quadrilateral elements in 2D, and tetrahedra and prismatic elements in 3D, have been used.

In this case the subdomains Ω_m are obtained by clustering together several spectral elements. Then, the overlapping subdomains $\tilde{\Omega}_m$ are obtained by extending Ω_m with *one layer of spectral elements* as shown in Fig. 6.13 for a 2D case and in Fig. 6.14 for a 3D case.

The coarse-grid matrix is built on the linear finite-element shape functions associated with the vertices of the spectral elements (see, e. g., Fig. 6.7).

The local matrices are instead obtained from the SEM stiffness matrices built on the spectral elements that form every extended subdomain $\tilde{\Omega}_m$.

Since this is a generous overlap, the condition number of the preconditioned matrix (6.3.22) is bounded uniformly with respect to H and N. See Tables 6.1 and 6.2.

Table 6.1. Ω is a rectangular domain in the plane shown in Fig. 6.13, $\mathcal{T}$ is an unstructured partition with 780 triangular and quadrilateral spectral elements, $M = 4, 8, 16$ subdomains and N varies from 2 to 12

	$M = 4$		$M = 8$		$M = 16$	
N	iter.	cond. no.	iter.	cond. no.	iter.	cond. no.
2	16	3.02	21	4.49	23	5.05
3	16	3.00	22	4.51	23	5.04
6	16	3.01	22	4.50	24	4.98
9	16	3.01	22	4.50	23	4.59
12	16	3.01	22	4.50	23	4.59

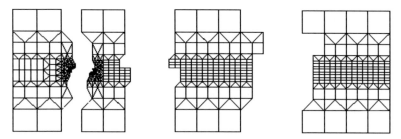

Fig. 6.13. Domain partition into 4, 8, 16 subdomains, for the simulation of Table 6.1 [courtesy of L.F. Pavarino and T. Warburton]

Table 6.2. $\Omega = (-1,1) \times (-1,1) \times (0,2)$ is a cubic domain, $\mathcal{T}$ is a structured mesh with $3072 = 8^3 \cdot 6$ tetrahedral spectral elements, $M = 8$ subdomains and N varies from 3 to 8; see Figure 6.14

N	iter.	cond. no.	N	iter.	cond. no.
3	29	5.75	6	29	5.75
4	29	5.75	7	29	5.75
5	29	5.75	8	29	5.76

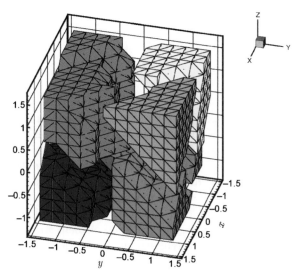

Fig. 6.14. Domain partition into eight subdomains, for the 3D simulation of Table 6.2 [courtesy of L.F. Pavarino and T. Warburton]

Efficient ways to construct coarse and local operators in the SEM context are described in Fischer et al. (2000) and in Deville et al. (2002), pp. 345–348.

Further bibliographic references on Schwarz preconditioners are given at the end of Sect. 7.4.2.

6.3.4 FEM-SEM Spectral Equivalence

This section is devoted to a rigorous justification of the *FEM-SEM equivalence* mentioned in Sect. 5.3.5.

We begin with an abstract result. For $m = 1, \ldots, M$, assume that $A_m \in \mathbb{R}^{n_m \times n_m}$ is a symmetric, positive-definite matrix, and that $B_m \in \mathbb{R}^{n_m \times n_m}$ be a suitable symmetric, positive-definite preconditioner of A_m. This means that there exist two positive constants c_m and C_M such that

$$c_m(B_m\mathbf{v}, \mathbf{v}) \leq (A_m\mathbf{v}, \mathbf{v}) \leq C_m(B_m\mathbf{v}, \mathbf{v}) \qquad \forall \mathbf{v} \in \mathbb{R}^{n_m} . \qquad (6.3.27)$$

Assume that

$$A = \sum_{m=1}^{M} R_m^T A_m R_m \qquad \text{and} \qquad B = \sum_{m=1}^{M} R_m^T B_m R_m .$$

Then,

$$c_{\min}(B\mathbf{v}, \mathbf{v}) \leq (A\,\mathbf{v}, \mathbf{v}) \leq C_{\max}(B\mathbf{v}, \mathbf{v}) \qquad \forall \mathbf{v} \in \mathbb{R}^n , \qquad (6.3.28)$$

with $c_{\min} = \min_m c_m$ and $C_{\max} = \max_m C_m$. These inequalities easily follow from the identity

$$(A\,\mathbf{v}, \mathbf{v}) = \sum_{m=1}^{M} (A_m\mathbf{v}_m, \mathbf{v}_m) \qquad \text{with } \mathbf{v}_m = R_m\mathbf{v} ,$$

from the companion identity involving B, and from (6.3.27). Thus, optimal local preconditioners induce optimal global preconditioners.

We apply this result to the case in which A is a SEM matrix (either the stiffness SEM-NI matrix A, or its Schur complement Σ introduced in Sect. 6.4.3) and B is obtained by a finite-element approximation using piecewise linear (or bilinear, or trilinear) finite-element functions at the local LGL nodes. Property (6.3.27) follows from a similar property that holds in the reference domain $\hat{\Omega} = (-1, 1)^d$, $d = 1, 2, 3$. Indeed, let $\hat{A}$ denote the G-NI stiffness matrix on $\hat{\Omega}$, and let $\hat{A}_{\text{fem}}$ be the finite-element stiffness matrix based on the piecewise linear/bilinear/trilinear shape functions associated with the tensor-product mesh whose vertices are the LGL nodes. Then, there exist constants $\hat{c}$ and $\hat{C}$ such that

$$\hat{c}(\hat{A}_{\text{fem}}\hat{\mathbf{v}}, \hat{\mathbf{v}}) \leq (\hat{A}\hat{\mathbf{v}}, \hat{\mathbf{v}}) \leq \hat{C}(\hat{A}_{\text{fem}}\hat{\mathbf{v}}, \hat{\mathbf{v}}) , \qquad \forall \hat{\mathbf{v}} \in \mathbb{R}^{\hat{n}} \qquad (6.3.29)$$

($\hat{n}$ being the order of the matrices). This is called the FEM–G-NI equivalence; it is documented in CHQZ2, Table 4.8 and proven in CHQZ2, Sect. 7.4. Note that

$$(\hat{A}_{\text{fem}}\hat{\mathbf{v}}, \hat{\mathbf{v}}) = \hat{a}(\hat{v}_h, \hat{v}_h) \qquad \text{and} \qquad (\hat{A}\hat{\mathbf{v}}, \hat{\mathbf{v}}) = \hat{a}_N(\hat{v}_N, \hat{v}_N) ,$$

where $\hat{v}_h$ ($\hat{v}_N$, resp.) is the finite-element function (the global polynomial function, resp.) which interpolates the values $\{\hat{v}_j\}$ at the LGL nodes. Also note that, e. g., in one dimension for the operator $\mathcal{L}u = -u_{xx}$, one has

$$\hat{a}(\hat{v}_h, \hat{v}_h) = \frac{h_m}{2} a_m(v_h, v_h) \quad \text{and} \quad \hat{a}_N(\hat{v}_N, \hat{v}_N) = \frac{h_m}{2} a_N(v_N, v_N) \ ,$$

where v_h and v_N are the images of $\hat{v}_h$ and $\hat{v}_N$, respectively, in Ω_m. In the multidimensional case, the bilinear forms scale similarly. We conclude that property (6.3.27) follows from (6.3.29) with constants $c_m \sim \hat{c}$ and $C_m \sim \hat{C}$.

The result (6.3.28), in which $c_{\min}$ and $C_{\max}$ are independent of both N and H, is precisely what is referred to as the FEM–SEM-NI equivalence.

6.3.5 Analysis of Schwarz Methods

We start this theoretical section by proving Theorem 6.3.1 on the convergence of the classical multiplicative Schwarz alternating method (6.3.1)–(6.3.2) and of its additive variant applied to the Poisson problem (5.4.19).

Proof of Theorem 6.3.1. We start by considering the *multiplicative* case. Let us set

$$V = H_0^1(\Omega), \ V_m^0 = H_0^1(\Omega_m) \text{ and } V_m^* = \{v \in V \,|\, v = 0 \text{ in } \Omega \setminus \overline{\Omega}_m\}, \ m = 1, 2 \ .$$

Then define

$$u^{(k+1/2)} = \begin{cases} \hat{u}_1^{(k+1)} & \text{in } \Omega_1 \ , \\ \hat{u}_2^{(k)}|_{\Omega_2 \setminus \Omega_1} & \text{in } \Omega \setminus \Omega_1 \ , \end{cases}$$

$$u^{(k+1)} = \begin{cases} \hat{u}_2^{(k+1)} & \text{in } \Omega_2 \ , \\ \hat{u}_1^{(k+1)}|_{\Omega_1 \setminus \Omega_2} & \text{in } \Omega \setminus \Omega_2 \ . \end{cases}$$

It can be verified that $u^{(k+1/2)} = u^{(k)} + \tilde{w}_1^{(k)}$, where

$$\tilde{w}_1^{(k)} = \begin{cases} w_1^{(k)} & \text{in } \Omega_1 \ , \\ 0 & \text{in } \Omega \setminus \Omega_1 \ , \end{cases}$$

and the function $w_1^{(k)} \in V_1^0$ satisfies the following Dirichlet problem in Ω_1:

$$a_{\Omega_1}\left(w_1^{(k)}, v_1\right) = (f, v_1)_{\Omega_1} - a_{\Omega_1}(u^{(k)}, v_1) \qquad \forall v_1 \in V_1^0 \ . \tag{6.3.30}$$

Similarly, $u^{(k+1)} = u^{(k+1/2)} + \tilde{w}_2^{(k)}$ where

$$\tilde{w}_2^{(k)} = \begin{cases} w_2^{(k)} & \text{in } \Omega_2 \ , \\ 0 & \text{in } \Omega \setminus \Omega_2 \ , \end{cases}$$

and $w_2^{(k)} \in V_2^0$ is the solution of the following Dirichlet problem in Ω_2:

$$a_{\Omega_2}(w_2^{(k)}, v_2) = (f, v_2)_{\Omega_2} - a_{\Omega_2}(u^{(k+1/2)}, v_2) \qquad \forall v_2 \in V_2^0 \ . \tag{6.3.31}$$

With obvious choice of notation, we rewrite the weak form of the Poisson problem (5.4.19) as

$$u \in V: \quad a(u,v) = (f,v)_\Omega \qquad \forall v \in V .$$

Then, from (6.3.30), for all $v \in V_1^*$ we obtain

$$a(u^{(k+1/2)} - u^{(k)}, v) = a(\tilde{w}_1^{(k)}, v) = a_{\Omega_1}(w_1^{(k)}, v_{|\Omega_1})$$
$$= (f, v_{|\Omega_1})_{\Omega_1} - a_{\Omega_1}(u^{(k)}, v_{|\Omega_1})$$
$$= (f, v)_\Omega - a(u^{(k)}, v) = a(u - u^{(k)}, v) .$$

Similarly, starting from (6.3.31) we obtain

$$a(u^{(k+1)} - u^{(k+1/2)}, v) = a(u - u^{(k+1/2)}, v) \quad \text{for all } v \in V_2^* .$$

For $m = 1, 2$, let $P_m^* : V \to V_m^*$ be the orthogonal projection operator with respect to the scalar product $a(.,.)$, i.e.,

$$\forall w \in V, \quad P_m^* w \in V_m^* : \quad a(P_m^* w - w, v) = 0 \quad \forall v \in V_m^* . \tag{6.3.32}$$

Then,

$$u^{(k+1/2)} - u^{(k)} = P_1^*(u - u^{(k)}) \tag{6.3.33}$$

and

$$u^{(k+1)} - u^{(k+1/2)} = P_2^*(u - u^{(k+1/2)}) . \tag{6.3.34}$$

Now let J_m be the injection operator $J_m : V_m^* \to V$, and define $P_m = J_m P_m^* : V \to V$. Then, $P_m w$ satisfies (6.3.4) in the sense of $\mathcal{D}'(\Omega_m)$. Moreover, let $G : H^{-1}(\Omega) \to V$ be the operator defined in (6.3.3). Then $u = Gf$.

From (6.3.33) and (6.3.34) we obtain, upon setting $L = -\triangle$,

$$u^{(k+1/2)} = u^{(k)} + P_1^*(u - u^{(k)}) = (I - P_1)u^{(k)} + P_1 Gf$$
$$= u^{(k)} + P_1 G(f - Lu^{(k)}) , \tag{6.3.35}$$

$$u^{(k+1)} = (I - P_2)u^{(k+1/2)} + P_2 Gf = u^{(k+1/2)} + P_2 G(f - Lu^{(k+1/2)}) . \tag{6.3.36}$$

Therefore,

$$u^{(k+1)} = (I - P_2)[(I - P_1)u^{(k)} + P_1 Gf] + P_2 Gf$$
$$= (I - P_2)(I - P_1)u^{(k)} + (I - P_2)P_1 Gf + P_2 Gf , \tag{6.3.37}$$

that is,

$$u^{(k+1)} = u^{(k)} + Q_{\text{mul}}(Gf - u^{(k)}) .$$

By defining $g_{\text{mul}} = Q_{\text{mul}} Gf$ we obtain (6.3.7). Note that the right-hand side is the residual of the problem $Q_{\text{mul}} u = g_{\text{mul}}$. This proves (6.3.6).

Now it is not difficult to see that if instead one considers the *additive* Schwarz alternating method, then at the k-th iteration the solution satisfies the recurrence relation (6.3.9). This time, $g_{\text{add}} = \sum_{m=1}^{2} g_{\text{add}}^{(m)}$, with $g_{\text{add}}^{(m)} = P_m G f$ for $m = 1, 2$. Owing to the definition of P_m we can obtain $g_{\text{add}}^{(m)}$ by solving the following Dirichlet problem:

$$a\left(g_{\text{add}}^{(m)}, v_m\right) = a(Gf, v_m) = a(u, v_m) = (f, v_m)_{\Omega_m} \qquad \forall v_m \in V_m^* .$$

Thus, the computation of g_{add} requires the solution of two independent Dirichlet problems in Ω_1 and Ω_2. This formulation is the one that underlies the successful additive Schwarz domain-decomposition preconditioner that was presented in Sect. 6.3.1.

6.3.6 A General Theoretical Framework for the Analysis of DD Iterations

A very general framework can be set up in order to analyze the Schwarz method as well as the domain decomposition methods based on the precon-ditioners of the Schur complement, which we describe in Sect. 6.4.

Let V_δ be a finite-dimensional Hilbert space with inner product $a(\cdot, \cdot)$, and let F be a linear functional on V_δ. We assume that the discrete problem to be solved has the abstract form: find $u_\delta \in V_\delta$ such that

$$a(u_\delta, v_\delta) = F(v_\delta) \qquad \forall v_\delta \in V_\delta ,$$

or, in operator form,

$$A_\delta \, u_\delta = F_\delta , \tag{6.3.38}$$

where

$$A_\delta : V_\delta \to V_\delta' \qquad \text{satisfies} \qquad \langle A_\delta \, u_\delta, v_\delta \rangle = a(u_\delta, v_\delta) \quad \forall v_\delta \in V_\delta .$$

Let V_δ be decomposed into $M + 1$ subspaces

$$V_\delta = V_\delta^0 + V_\delta^1 + \ldots + V_\delta^M .$$

This means that for every function $v_\delta \in V_\delta$ there exist $M + 1$ functions $v_\delta^{(m)} \in V_\delta^m$ such that $v_\delta = v_\delta^{(0)} + \ldots + v_\delta^{(M)}$.

For each subspace, assume that there is a symmetric, coercive bilinear form $a_m(\cdot, \cdot) : V_\delta^m \times V_\delta^m \to \mathbb{R}$, and define the operator $P_m : V_\delta \to V_\delta^m$ by

$$a_m(P_m v, w) = a(v, w) \qquad \forall w \in V_\delta^m .$$

If $a_m(v, w) = a(v, w)$, then P_m coincides with the orthogonal projection on V_δ^m with respect to the inner product $a(\cdot, \cdot)$. The operator

$$P = P_0 + P_1 + \ldots + P_M$$

is sometimes referred to as the *Schwarz operator*, in view of the formal analogy with the stiffness matrix preconditioned by the additive Schwarz preconditioner.

Now replace the original problem (6.3.38) by the following (preconditioned) problem:

$$Pu_\delta = g_\delta, \qquad \text{where} \quad g_\delta = \sum_{m=0}^{M} g_m, \qquad (6.3.39)$$

and g_m is the solution of the local problem

$$a_m(g_m, v) = a\,(u_\delta, v) = F(v) \qquad \forall v \in V_\delta^m\,.$$

The following theorem is a basic result due to P.L. Lions, J. Bramble, M. Dryja and O. Widlund. For a proof see, e. g., Dryja and Widlund (1995).

Theorem 6.3.2 *Let there exist*

i)   *a constant C_0 such that for all $v \in V_\delta$ there is a decomposition $v = \sum_{m=0}^{M} v_m$, with $v_m \in V_\delta^m$, such that the following property of stable decomposition holds*

$$\sum_{m=0}^{M} a_m(v_m, v_m) \le C_0^2 a(v, v)\,;$$

ii)   *a constant ω such that for $m = 0, \ldots, M$ the following property of local stability holds*

$$a(v, v) \le \omega a_m(v, v) \qquad \forall v \in V_m\,;$$

iii)   *constants ε_{ij} for $i, j = 1, \ldots, M$ such that the following strengthened Cauchy-Schwarz inequality holds*

$$a(v_i, v_j) \le \varepsilon_{ij}\sqrt{a(v_i, v_i)}\sqrt{a(v_j, v_j)} \qquad \forall v_i \in V_i, v_j \in V_j\,.$$

Then

$$C_0^{-2} a(v, v) \le a(Pv, v) \le \omega(\rho(E) + 1)a(v, v) \qquad \forall v \in V, \qquad (6.3.40)$$

where $\rho(E)$ is the spectral radius of the $M \times M$ matrix E whose entries are the ε_{ij}.

From (6.3.40) we obtain an upper bound for the condition number of P:

$$\mathcal{K}(P) \le C_0^2 \omega(\rho(E) + 1)\,. \qquad (6.3.41)$$

In particular, when solving problem (6.3.39) by conjugate gradient iterations, $C_0\sqrt{\omega}(\rho(E) + 1)^{1/2}$ provides an upper bound for the number of iterations.

In the case of the additive version of the Schwarz method for finite elements described in Sects. 6.3.1 and 6.3.2, the following identifications apply. Problem (6.3.39) is given by (6.3.24), where $Q_{\mathrm{add},H}$ is defined in (6.3.22) and the right-hand side in (6.3.25). We have, therefore,

$$P_m = R_m^T A_m^{-1} R_m A \quad \text{for } m = 1, \dots, M$$

and $P_0 = P_H = Q_H A$. Moreover, $a_m(\cdot, \cdot) = a_{\Omega_m}(\cdot, \cdot)$, V_δ^m is the subspace of V_m^* made of piecewise linear finite elements, while $V_\delta^0 = V_{1,M}(\Omega)$.

A similar interpretation applies for the spectral Schwarz preconditioner described in Section 6.3.3. In that case (6.3.39) is given by (6.3.26), and the bilinear form $a(\cdot, \cdot)$ is given by $a_\delta(\cdot, \cdot)$.

6.4 Schur Complement Iterative Methods

Domain decomposition methods based on *non-overlapping* subdomains are generally amenable to iterative procedures for an *interface* equation that is associated with the given differential problem. This interface problem can be defined in terms of the Schur complement system (see box 3.2 of Fig. 6.3) of the stiffness matrix taken with respect to the nodal values sitting at domain interfaces. In turn, the Schur complement matrix Σ can be regarded as a direct approximation of a (pseudo-)differential operator, the so-called Steklov–Poincaré operator S, which can be associated to the given differential problem (see box 1.3 of Fig. 6.3).

The properties of Σ can be deduced from those of S. Thus, we start by introducing the Steklov–Poincaré operator and its properties in Sects. 6.4.1 and 6.4.2. We assure the reader that reading these sections is not essential for the understanding of the subsequent discussion on the algorithms. Then, we introduce the Schur complement matrix and discuss how to set up efficient preconditioners for Σ.

6.4.1 The Steklov–Poincaré Interface Problem

Let us refer to the model problem (5.4.19). Its multidomain formulation was given in (5.13.1). The same arguments, however, apply to more general second-order boundary-value problems.

Let λ denote the unknown value of u on Γ. We consider the two Dirichlet problems

$$\begin{aligned} -\triangle w_m &= f && \text{in } \Omega_m , \\ w_m &= 0 && \text{on } \partial\Omega_m \cap \partial\Omega , \\ w_m &= \lambda && \text{on } \Gamma , \end{aligned} \qquad (6.4.1)$$

for $m = 1, 2$. Since our problem is linear, its solution can be split into two terms,

$$w_m = u_m^0 + u_m^*, \tag{6.4.2}$$

where u_m^0 and u_m^* are the solutions of the following Dirichlet problems:

$$
\begin{aligned}
-\Delta u_m^0 &= 0 && \text{in } \Omega_m, \\
u_m^0 &= 0 && \text{on } \partial\Omega_m \cap \partial\Omega, \\
u_m^0 &= \lambda && \text{on } \Gamma,
\end{aligned}
\tag{6.4.3}
$$

and

$$
\begin{aligned}
-\Delta u_m^* &= f && \text{in } \Omega_m, \\
u_m^* &= 0 && \text{on } \partial\Omega_m.
\end{aligned}
\tag{6.4.4}
$$

For each $m = 1, 2$, u_m^0 is called the *harmonic extension* of λ into Ω_m, and it will be denoted $\mathcal{H}_m\lambda$. On the other hand, since u_m^* depends solely on f, we can write $\mathcal{G}_m f$ instead of u_m^*. Thus,

$$w_m = \mathcal{H}_m\lambda + \mathcal{G}_m f. \tag{6.4.5}$$

By comparison of (5.13.1) with (6.4.1), it follows that

$$w_m = u_m \text{ for } m = 1, 2 \text{ if and only if } \frac{\partial w_1}{\partial n^\Gamma} = \frac{\partial w_2}{\partial n^\Gamma} \text{ on } \Gamma. \tag{6.4.6}$$

For every η defined on Γ, define

$$S_m\eta = \frac{\partial}{\partial n^m}\mathcal{H}_m\eta, \quad m = 1, 2, \tag{6.4.7}$$

where $\frac{\partial}{\partial n^m}$ denotes the normal derivative to Ω_m on Γ. S_m is called the *local Steklov–Poincaré operator* and also the *Dirichlet-to-Neumann map* in Ω_m. In fact, when applied to a Dirichlet data η on Γ, it returns the normal derivative on Γ (a Neumann data) of the harmonic extension of η in Ω_m.

Now we define $S = S_1 + S_2$. Then,

$$S\eta = \frac{\partial}{\partial n^\Gamma}\mathcal{H}_1\eta - \frac{\partial}{\partial n^\Gamma}\mathcal{H}_2\eta = \sum_{m=1}^{2} \frac{\partial}{\partial n^m}\mathcal{H}_m\eta. \tag{6.4.8}$$

Using the decomposition (6.4.5), we can easily see that (6.4.6) holds if and only if

$$S\lambda = \chi \quad \text{on} \quad \Gamma, \tag{6.4.9}$$

where

$$\chi = \frac{\partial}{\partial n^\Gamma}\mathcal{G}_2 f - \frac{\partial}{\partial n^\Gamma}\mathcal{G}_1 f = -\sum_{m=1}^{2} \frac{\partial}{\partial n^m}\mathcal{G}_m f. \tag{6.4.10}$$

S is the *Steklov–Poincaré operator*, while (6.4.9) is the interface Steklov–Poincaré equation associated with problem (5.4.19) (this is box 1.3 in Fig. 6.3).

6.4.2 Properties of the Steklov–Poincaré Operator

The operator S_m operates from the trace space $\Lambda = H_{00}^{1/2}(\Gamma)$ (which collects the traces on Γ of the functions in $H_0^1(\Omega)$) and its dual Λ'; see Appendix A.

The properties of S_m and S can be better analyzed after applying Green's formula to obtain

$$\langle S_m \eta, \mu \rangle = a_{\Omega_m}(\mathcal{H}_m \eta, \mathcal{H}_m \mu) \quad \forall \eta, \mu \in \Lambda .$$

Here $\langle \cdot, \cdot \rangle$ denotes the pairing between Λ' and Λ. Then,

$$\langle S\eta, \mu \rangle = \sum_{m=1}^{2} a_{\Omega_m}(\mathcal{H}_m \eta, \mathcal{H}_m \mu) . \tag{6.4.11}$$

From this representation it follows easily that S is a positive operator since $a(\cdot, \cdot)$ is coercive. Moreover, it is symmetric if $a(\cdot, \cdot)$ is also symmetric.

A crucial property for the analysis of iterative substructuring methods is that the harmonic extension of the same trace λ in two adjoining subdomains, say Ω_m and Ω_n, have equivalent energies, that is, their norms are equivalent: $\|\mathcal{H}_m \lambda\|_{H^1(\Omega_m)} \simeq \|\mathcal{H}_m \lambda\|_{H^1(\Omega_n)}$. This property is also true for the finite-element harmonic extension of a piecewise polynomial finite-element function λ_h on the interface Γ with an equivalence constant independent of h (see, e. g., Quarteroni and Valli (1999), Theorem 4.1.3). The property remains valid in the spectral context, owing to the possibility of constructing a stable polynomial extension of a polynomial λ_N on Γ (see Maday (1989), Bernardi and Maday (1990), Ben Belgacem (1994) and Muñoz Sola (1997)).

6.4.3 The Schur Complement Matrix

As already anticipated, the Schur complement matrix Σ of the stiffness matrix A in (6.2.1) can be regarded as the finite-dimensional approximation of the Steklov–Poincaré operator. In this subsection, we describe the Schur complement matrix for the SEM or the SEM-NI in some detail, but we stress that for the patching collocation method it can be formulated similarly.

The Case of Two Subdomains

We consider first the case of two subdomains Ω_1 and Ω_2 divided by the interface Γ. Each subdomain could either coincide with a spectral element or represent a cluster of spectral elements (as in Fig. 6.1).

It is convenient to distinguish between the nodes belonging to Γ and those lying on the interior of Ω_1 and Ω_2. We denote the corresponding vectors of the physical unknowns by $\mathbf{u}_\Gamma$, $\mathbf{u}_1$ and $\mathbf{u}_2$, respectively, and their lengths by n_Γ, n_1 and n_2 (see Fig. 6.15 for two illustrative examples, both with two

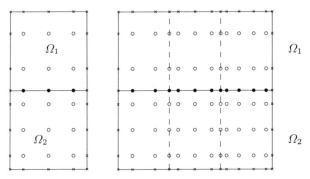

Fig. 6.15. (*Left*) A subdomain partition with two subdomains in which $n_\Gamma = 3$, $n_1 = 6$ and $n_2 = 9$. Each subdomain is made of a single spectral element. (*Right*) A subdomain partition with two subdomains (each one being a cluster of three spectral elements) in which $n_\Gamma = 11$, $n_1 = 22$, $n_2 = 33$. For both figures, LGL nodes are used within each spectral element

subdomains and with bold nodes indicating the nodes belonging to Γ). The vector corresponding to the given function f will be denoted by $\mathbf{f}$.

As we have already seen, the SEM problem (5.2.5) can be written in the algebraic form (6.2.1), where A is the $n \times n$ stiffness matrix, with $n = n_\Gamma + n_1 + n_2$.

Our system can be written in block form as

$$\begin{pmatrix} A_{11} & 0 & A_{1\Gamma} \\ 0 & A_{22} & A_{2\Gamma} \\ A_{\Gamma 1} & A_{\Gamma 2} & A_{\Gamma\Gamma} \end{pmatrix} \begin{pmatrix} \mathbf{u}_1 \\ \mathbf{u}_2 \\ \mathbf{u}_\Gamma \end{pmatrix} = \begin{pmatrix} \mathbf{f}_1 \\ \mathbf{f}_2 \\ \mathbf{f}_\Gamma \end{pmatrix}, \qquad (6.4.12)$$

where we have used the following notation: for $m = 1, 2$

$$(A_{mm})_{jl} = a_{\Omega_m}(\varphi_l^{(m)}, \varphi_j^{(m)}), \quad j, l = 1, \ldots, n_m, \qquad (6.4.13)$$

$a_{\Omega_m}(\cdot, \cdot)$ is the restriction of the bilinear form $a(\cdot, \cdot)$ to Ω_m (as in (5.2.7), or as in (5.3.12) for the case of advection-diffusion-reaction problems in 1D and 2D, respectively), and $\varphi_j^{(m)}$ are the basis functions associated with the nodes lying in Ω_m. Moreover,

$$(A_{\Gamma\Gamma})_{rs} = a_{\Omega_1}(\varphi_s^{(\Gamma)}, \varphi_r^{(\Gamma)}) + a_{\Omega_2}(\varphi_s^{(\Gamma)}, \varphi_r^{(\Gamma)}), \quad r, s = 1, \ldots, n_\Gamma, \quad (6.4.14)$$

where $\varphi_r^{(\Gamma)}$ are the Lagrangian basis functions associated with the nodes lying on Γ, and

$$(A_{m\Gamma})_{lr} = a_{\Omega_m}(\varphi_r^{(\Gamma)}, \varphi_l^{(m)}), \quad l = 1, \ldots, n_m, \quad r = 1, \ldots, n_\Gamma, \qquad (6.4.15)$$

while $A_{\Gamma m}$ denotes the transpose of $A_{m\Gamma}$, $m = 1, 2$.

After eliminating $\mathbf{u}_1$ and $\mathbf{u}_2$, we obtain the reduced system

$$\Sigma \mathbf{u}_\Gamma = \chi_\Gamma , \tag{6.4.16}$$

with

$$\chi_\Gamma = \mathbf{f}_\Gamma - A_{\Gamma 1} A_{11}^{-1} \mathbf{f}_1 - A_{\Gamma 2} A_{22}^{-1} \mathbf{f}_2 \tag{6.4.17}$$

and

$$\Sigma = A_{\Gamma\Gamma} - A_{\Gamma 1} A_{11}^{-1} A_{1\Gamma} - A_{\Gamma 2} A_{22}^{-1} A_{2\Gamma} \tag{6.4.18}$$

(see box 3.2 of Fig. 6.3). Should the solution $\mathbf{u}_\Gamma$ of (6.4.16) be available, the subdomain components $\mathbf{u}_1$ and $\mathbf{u}_2$ can be immediately recovered from (6.4.12) at the expense of two uncoupled local solves. Indeed,

$$A_{mm} \mathbf{u}_m = \mathbf{f}_m - A_{m\Gamma} \mathbf{u}_\Gamma , \qquad m = 1, 2 . \tag{6.4.19}$$

We can split the matrix $A_{\Gamma\Gamma}$ as follows:

$$A_{\Gamma\Gamma} = A_{\Gamma\Gamma}^{(1)} + A_{\Gamma\Gamma}^{(2)} , \tag{6.4.20}$$

where $A_{\Gamma\Gamma}^{(m)}$ denotes the contribution from the subdomain Ω_m, $m = 1, 2$. Then we can write

$$\Sigma = \Sigma_1 + \Sigma_2 , \tag{6.4.21}$$

with

$$\Sigma_m = A_{\Gamma\Gamma}^{(m)} - A_{\Gamma m} A_{mm}^{-1} A_{m\Gamma} , \quad m = 1, 2 . \tag{6.4.22}$$

The matrix Σ is called the *Schur complement matrix* and can be regarded as a discretization of the Steklov–Poincaré operator S introduced in (6.4.8) (and Σ_m approximates the operator S_m introduced in (6.4.7)). In turn, (6.4.16) is called the *Schur complement system* and provides a finite-dimensional approximation of the interface problem (6.4.9). Σ is symmetric and positive definite, as results from (6.4.18).

For each $m = 1, 2$, the matrix Σ_m is full: to compute its entries is expensive because we would need as many local solves A_{mm}^{-1} as the number of nodes on Γ. Indeed, for every node k on Γ, let $\mathbf{v}^{(k)}$ be the vector of length n_Γ whose components are $v_j^{(k)} = \delta_{jk}$. Then, denoting by $\mathbf{w}^{(k;m)}$ the solution of length n_m of the system $A_{mm} \mathbf{w}^{(k;m)} = -A_{m\Gamma} \mathbf{v}^{(k)}$, $m = 1, 2$, the entries of the matrices Σ_m are given by

$$(\Sigma_m)_{kj} = w_j^{(k;m)} , \qquad k, j = 1, \ldots, n_\Gamma . \tag{6.4.23}$$

Methods based on the explicit computation of the entries of Σ through (6.4.23) are direct methods called *substructuring*. They are common practice in the engineering community and correspond to having all interiors nodes eliminated by *static condensation* (see, for example, Przemieniecki (1985)). (This is what is sometimes called the *influence-matrix* approach in the spectral community, see Peyret (2002)).

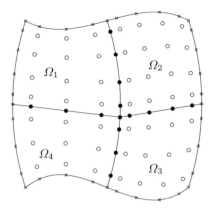

Fig. 6.16. A subdomain partition with four subdomains (each one being made of a single spectral element) in which $n_\Gamma = 15$, $n_1 = 12$, $n_2 = 16$, $n_3 = 12$, $n_4 = 9$, $J_1 = 8$, $J_2 = 9$, $J_3 = 8$, $J_4 = 7$

Most often, however, especially when the number of nodes lying on the interface Γ is large, since Σ is full and ill-conditioned, Krylov methods (e. g., the conjugate gradient method, or the GMRES method) are used for an iterative solution of the Schur complement system (6.4.16). For the computation of the residual, each matrix-vector multiplication with Σ involves two uncoupled subdomain solves, as in (6.4.19), which can be performed in parallel.

The Case of Many Subdomains

The case of partitions with many subdomains can be dealt with similarly. Let Ω be partitioned into M non-overlapping subdomains Ω_m of diameter H_m with interface Γ separating them: $\Gamma = \cup_{m=1}^{M} \Gamma_m$, $\Gamma_m = \partial\Omega_m \setminus \partial\Omega$. Let $I = \cup_{m=1}^{M} n_i$ denote the indices corresponding to the internal nodes (see Fig. 6.16).

Then, with obvious notation, we can write the algebraic problem (6.2.1) blockwise as follows:

$$
\begin{pmatrix} A_{II} & A_{I\Gamma} \\ A_{\Gamma I} & A_{\Gamma\Gamma} \end{pmatrix} \begin{pmatrix} \mathbf{u}_I \\ \mathbf{u}_\Gamma \end{pmatrix} = \begin{pmatrix} \mathbf{f}_I \\ \mathbf{f}_\Gamma \end{pmatrix} ,
\tag{6.4.24}
$$

where $A_{\Gamma I} = A_{I\Gamma}^T$. Since the interior nodes in each subdomain will be decoupled from the interior nodes in other subdomains, we will have

$$
A_{II} = \text{blockdiag}\,(A_{mm}) = \begin{pmatrix} A_{11} & \cdots & 0 \\ \cdot & \cdots & \cdot \\ 0 & \cdots & A_{MM} \end{pmatrix} .
\tag{6.4.25}
$$

The blocks A_{mm} are the principial submatrices of the local stiffness matrices that are associated with either Dirichlet or Neumann problems in the subdomains Ω_m. More precisely, let us define

$$(A_{mm})_{jl} = a_{\Omega_m}(\varphi_l, \varphi_j), \qquad 1 \le j, l \le n_m,$$
$$(A_{m\Gamma})_{lr} = a_{\Omega_m}(\psi_r, \varphi_l), \qquad 1 \le r \le J_m, 1 \le l \le n_m, \qquad (6.4.26)$$
$$(A_{\Gamma\Gamma}^{(m)})_{rs} = a_{\Omega_m}(\psi_s, \psi_r), \qquad 1 \le r, s \le J_m,$$

where φ_j, $j = 1, \ldots, N_m$, are the characteristic Lagrange basis functions associated with the internal nodes $\mathbf{x}_j$ in Ω_m, while ψ_r, $r = 1, \ldots, J_m$, are those associated with the nodes $\mathbf{y}_r$ belonging to Γ_m. Then the matrix

$$A_m = \begin{pmatrix} A_{mm} & A_{m\Gamma} \\ A_{\Gamma m} & A_{\Gamma\Gamma}^{(m)} \end{pmatrix} \qquad (6.4.27)$$

is the matrix associated with the spectral (Galerkin or G-NI) discretization of the Laplace problem in Ω_m with Neumann condition on the physical boundary Γ_m (and the homogeneous Dirichlet condition on the physical boundary $\partial\Omega \cap \partial\Omega_m$).

Similarly,

$$\mathcal{D}_m = \begin{pmatrix} A_{mm} & A_{m\Gamma} \\ 0 & I_\Gamma^{(m)} \end{pmatrix}, \qquad (6.4.28)$$

where $I_\Gamma^{(m)}$ is the identity matrix of order J_m, is the matrix associated with the spectral discretization of the Poisson problem in Ω_m with the Dirichlet condition prescribed on Γ_m (and the homogeneous Dirichlet condition on $\partial\Omega \cap \partial\Omega_m$).

The Schur complement matrix Σ associated with the n_Γ interface variables $\mathbf{u}_\Gamma$ of (6.4.24) is

$$\Sigma = A_{\Gamma\Gamma} - A_{\Gamma I} A_{II}^{-1} A_{I\Gamma}, \qquad (6.4.29)$$

and the associated interface problem reads (as in box 3.2 of Fig. 6.3)

$$\Sigma \mathbf{u}_\Gamma = \boldsymbol{\chi}_\Gamma, \qquad (6.4.30)$$

where the right-hand side is the vector of $\mathbb{R}^{n_\Gamma}$

$$\boldsymbol{\chi}_\Gamma = \mathbf{f}_\Gamma - A_{\Gamma I} A_{II}^{-1} \mathbf{f}_I. \qquad (6.4.31)$$

With the aim of representing Σ as a sum of local contributions, we start by splitting the stiffness matrix A into the contributions given by each subdomain Ω_m, $m = 1, \ldots, M$. Since

$$a(\varphi_l, \varphi_j) = \sum_{m=1}^{M} a_{\Omega_m}(\varphi_l, \varphi_j),$$

it is easily checked that A can be split as

$$A = \sum_{m=1}^{M} R_m^T A_m R_m ,\qquad (6.4.32)$$

where A_m is the local stiffness matrix introduced in (6.4.27), R_m is the restriction matrix from the full vector in Ω to local vectors in $\Omega_m \cup \Gamma_m$; and consequently, the prolongation matrix R_m^T implies prolongation by 0 on the nodes external to $\Omega_m \cup \Gamma_m$.

In a similar way,

$$\Sigma = \sum_{m=1}^{M} R_{\Gamma_m}^T \Sigma_m R_{\Gamma_m} .\qquad (6.4.33)$$

Here, $\Sigma_m = A_{\Gamma\Gamma}^{(m)} - A_{\Gamma m} A_{mm}^{-1} A_{m\Gamma}$, $m = 1, \ldots, M$, and R_{Γ_m} is the restriction matrix from the vector of coefficient unknowns related to the nodes on Γ to only those associated with Γ_m, and $R_{\Gamma_m}^T$ is the matrix that extends by 0 a nodal vector from Γ_m to Γ.

Application of the Schur Complement Matrix Σ *For a given $\mathbf{x}_\Gamma$, the vector $\mathbf{y}_\Gamma = \Sigma \mathbf{x}_\Gamma$ can be computed as follows.*

For $m = 1, \ldots, M$ do in parallel:

$$\mathbf{x}_m = R_{\Gamma_m} \mathbf{x}_\Gamma ,$$
$$\mathbf{z}_m = A_{m\Gamma} \mathbf{x}_m ,$$
$$\text{solve: } A_{mm} \mathbf{w}_m = \mathbf{z}_m ,$$
$$\tilde{\mathbf{y}}_m = A_{\Gamma\Gamma}^{(m)} \mathbf{x}_m - A_{\Gamma m} \mathbf{w}_m ,$$
$$\mathbf{y}_m = R_{\Gamma_m}^T \tilde{\mathbf{y}}_m ,$$

then sum the local contributions: $\mathbf{y}_\Gamma = \sum_{m=1}^{M} \mathbf{y}_m$.

Σ is block-wise full, and so the direct computation of its entries using a formula that generalizes (6.4.23) to the case of many subdomains would be prohibitive. Direct methods are therefore set aside in favor of iterative methods. However, as already anticipated, the matrix Σ is ill-conditioned. Actually, if the Legendre Gauss–Lobatto integration is used to compute the integrals in (6.4.22), the spectral condition number of Σ is bounded by

$$\mathcal{K}(\Sigma) \le \widehat{C}_0 N / H_{min}^2 \qquad (6.4.34)$$

for a suitable constant $\widehat{C}_0$ independent of N and H_{min}, the latter being the linear size of the smallest subdomain (see Melenk (2002), Theorem 2.2). We report in Fig. 6.17 (left) the spectral condition number of Σ for several values of N (and four subdomains), while on the right we consider the case of a fixed $N(= 4)$ and a variable number of subdomains. The challenging

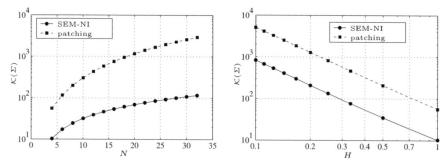

Fig. 6.17. (*Left*) The spectral condition number $\mathcal{K}(\Sigma)$ of the Schur complement matrix versus the polynomial degree N, for both the SEM-NI approach and the patching approach. $H = 1$ is fixed. We have $\mathcal{K}(\Sigma) \simeq 3.62 \cdot N$ for SEM-NI and $\mathcal{K}(\Sigma) \simeq 2.87 \cdot N^2$ for patching. (*Right*) The spectral condition number of the Schur complement matrix versus the elements side H, for both the SEM-NI approach and the patching approach. $N = 4$ is fixed. We have $\mathcal{K}(\Sigma) \simeq 8.30 \cdot H^{-2}$ for SEM-NI and $\mathcal{K}(\Sigma) \simeq 51.21 \cdot H^{-2}$ for the patching approach

point is therefore the construction of a suitable preconditioner for the Schur complement Σ such that the convergence rate of the preconditioned iterative method is (ideally) independent of both the polynomial degree N_m and the subdomain size H_m for every m. This issue is addressed in Sect. 6.4.4.

When an iterative method is applied to (6.4.16) (such as, e.g., the Richardson or the conjugate gradient method), the action of Σ on a given vector has to be computed. In view of (6.4.33), this amounts to inverting M independent matrices A_{ii} (or, more exactly, to solve M independent Dirichlet problems), one on each subdomain.

6.4.4 DD Preconditioners for the Schur Complement Matrix

(This corresponds to box 5.2 on Fig. 6.3) To start with, consider again the case of a decomposition with only two subdomains. Then, the local Schur complements Σ_m, $m = 1, 2$, introduced in (6.4.21) satisfy

$$\mathcal{K}\left(\Sigma_m^{-1}\Sigma\right) \leq \hat{C}, \quad m = 1, 2. \tag{6.4.35}$$

This result follows from Sect. 6.4.2: more precisely, it can be obtained from the property that the discrete harmonic extensions in Ω_1 and Ω_2 of a polynomial function on Γ have an equivalent energy norm. Both Σ_1 and Σ_2 are therefore *optimal preconditioners* of Σ.

In the (very) special case of a scalar second-order elliptic boundary-value problem with constant coefficients, using the same polynomial degree in Ω_1 and Ω_2 and two subdomains of equal size, it follows that $\Sigma_1 = \Sigma_2 = \frac{1}{2}\Sigma$. Thus, either Σ_1 or Σ_2 are ideal preconditioners for Σ, and they actually guarantee that the associated preconditioned iterative methods converge in just two iterations.

More generally, for decompositions with many subdomains, we will denote by P, with a suitable upper index, a preconditioner of the Schur complement matrix Σ. The Schur complement Σ_h obtained from the finite-element stiffness matrix built on the LGL nodes, say A_{fem}, is an optimal preconditioner for Σ also in the case of many subdomains (see Casarin (1996), p. 29,30, and Toselli and Widlund (2005), Lemma 7.10). This *FEM-SEM equivalence* property of the finite-element and spectral Schur complements is inherited from the FEM-SEM equivalence of the FEM and SEM-NI stiffness matrices, A_{fem} and A, respectively, that was discussed in Sect. 5.3.4. However, the action of the inverse of Σ_h is too expensive to produce a preconditioner for large problems. We can, however, use a good preconditioner, say $P = \hat{\Sigma}_h$, for Σ_h, since by a Rayleigh quotient argument we can then show that $\hat{\Sigma}_h$ remains a good preconditioner for Σ. Since the triangulation $\mathcal{T}_h$ made of the local LGL nodes of the SEM decomposition is not shape-regular, the classical FEM results on Schwarz and Schur preconditioners established in the literature do not apply in a straightforward manner. Estimates for nonregular triangulations are established in Casarin (1996), Sect. 3.3.

Alternatively, one can produce preconditioners P moving directly from Σ. To start with, let us consider the preconditioner

$$P^{DN} = \Sigma_2 , \qquad (6.4.36)$$

which is called the *Dirichlet–Neumann* preconditioner. Indeed, at each step of a preconditioned iterative procedure, the computation of the preconditioned residual $\mathbf{z}^k = (P^{DN})^{-1}\mathbf{r}^k$ involves the solution of a Dirichlet problem in Ω_1 and a Neumann problem in Ω_2. Actually, if $\mathbf{r}^k = \boldsymbol{\chi}_\Gamma - \Sigma\mathbf{u}_\Gamma^k$, then

$$\mathbf{z}^k = \Sigma_2^{-1}\boldsymbol{\chi}_\Gamma - \left(\Sigma_2^{-1}\Sigma_1 + I\right)\mathbf{u}_\Gamma^k .$$

(Note that the computation of $\Sigma_2^{-1}\boldsymbol{\chi}_\Gamma$ can be made off-line at the beginning of the iterative process).

We recall that, for $m = 1, 2$, Σ_m is associated with the so-called Dirichlet-to-Neumann map, since for any given interface nodal vector $\mathbf{v}_\Gamma$, $\Sigma_m\mathbf{v}_\Gamma$ returns the nodal values of the normal derivative on Γ of the discrete harmonic extension of $\mathbf{v}_\Gamma$. Accordingly, Σ_m^{-1} is associated with a Neumann-to-Dirichlet map and involves the solution of a discrete Neumann problem in Ω_m. Algebraically, the action of Σ_m^{-1} on a given vector $\mathbf{q}$ can be calculated without explicitly forming Σ_m because

$$\Sigma_m^{-1}\mathbf{q} = \left(0 \; I_\Gamma^{(m)}\right) A_m^{-1} \begin{pmatrix} 0 \\ I_\Gamma^{(m)} \end{pmatrix} \mathbf{q} , \qquad (6.4.37)$$

where $I_\Gamma^{(m)}$ is the identity matrix, and A_m is the stiffness matrix introduced in (6.4.28), which corresponds to solving a discrete (spectral) Laplace problem in Ω_m with a Neumann boundary condition on the interior interface Γ_m and a homogeneous boundary condition on $\partial\Omega_m \cap \partial\Omega$. Σ_m can be singular if $\partial\Omega_m$

does not share any component with that part of $\partial\Omega$ where Dirichlet data are prescribed. In that case instead of (6.4.37) we could use

$$\Sigma_m^\dagger \mathbf{q} = \left(0 \ I_\Gamma^{(m)} \right) A_m^\dagger \begin{pmatrix} 0 \\ I_\Gamma^{(m)} \end{pmatrix} \mathbf{q} , \tag{6.4.38}$$

where $A_m^\dagger$ denotes either the pseudo-inverse of A_m or the exact inverse of the regularized matrix $\tilde{A}_m = A_m + \frac{1}{H_m^2} M_m$, where H_m is the diameter of Ω_m, and M_m is the local mass matrix in Ω_m.

Dirichlet–Neumann Preconditioner *Given a residual* $\mathbf{r}_\Gamma$, *the vector* $\mathbf{z}_\Gamma = (P^{DN})^{-1}\mathbf{r}_\Gamma$ *can be computed as follows.*

$$\text{Compute: } \mathbf{r} = (0 \ I)^T \mathbf{r}_\Gamma ,$$
$$\text{solve: } A_2 \mathbf{w}_2 = \mathbf{r} \quad (\text{or else } \tilde{A}_2 \mathbf{w}_2 = \mathbf{r}) ,$$
$$\mathbf{z}_\Gamma = (0 \ I)\mathbf{w}_2 .$$

Another preconditioner for Σ reads

$$P^{NN} = \left(\sigma_1 \Sigma_1^{-1}\sigma_1 + \sigma_2 \Sigma_2^{-1}\sigma_2 \right)^{-1} , \tag{6.4.39}$$

where σ_1 and σ_2 are positive weights; it is called the *Neumann–Neumann* preconditioner. This time the computation of a preconditioned residual requires the solution of one Dirichlet problem and one Neumann problem in each subdomain.

Both preconditioners, P^{DN} and P^{NN}, can be generalized to the case of partitions using many subdomains. For example, P^{NN} can be generalized as follows:

$$\left(P^{NN} \right)^{-1} = \sum_{m=1}^{M} R_{\Gamma_m}^T D_m \Sigma_m^\dagger D_m R_{\Gamma_m} , \tag{6.4.40}$$

where D_m is a diagonal weighting matrix with positive entries, while R_{Γ_m} and $R_{\Gamma_m}^T$ are the matrices introduced in (6.4.33).

For instance, in Pavarino and Widlund (1996) and Pavarino (1997), D_m is the the diagonal matrix with nonzero elements only for the components on $\partial\Omega_m$ (given by $1/8,1/4,1/2$ for the vertex, edge and face components, respectively). Note that in SEM, 8, 4 and 2 are the number of elements sharing a vertex, an edge node or a face node (they would reduce to 4 and 2 in two dimensions).

The condition number of $\left(P^{NN} \right)^{-1} \Sigma$ deteriorates when the number M of subdomains becomes large. However, it grows sublinearly with respect to N (see Fig. 6.18, left). As was recommended for the Schwarz preconditioner, a global coarse correction needs to be introduced.

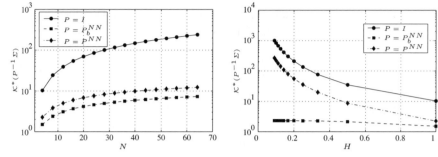

Fig. 6.18. *(Left)* The condition number $\mathcal{K}^*(P^{-1}\Sigma) = |\lambda_{\max}(P^{-1}\Sigma)|/|\lambda_{\min}(P^{-1}\Sigma)|$ versus the polynomial degree N. The element size $H = 1$ is fixed. When $P = I$ the condition number $\mathcal{K}^*(\Sigma)$ grows like CN with $C \simeq 3.77$. For $P = P^{NN}$ and $P = P_b^{NN}$, $\mathcal{K}^*(P^{-1}\Sigma)$ grows like $C(1 + \log N)^2$ with $C \simeq 0.46$ for P^{NN} and $C \simeq 0.28$ for $P = P_b^{NN}$. *(Right)* The condition number $\mathcal{K}^*(P^{-1}\Sigma)$ versus the spectral element side H (for a fixed polynomial degree $N = 4$). We have $\mathcal{K}^*(\Sigma) \simeq 8.30 \cdot H^{-2}$ (for $P = I$), $\mathcal{K}^*(P^{-1}\Sigma) \simeq 2.20 \cdot H^{-2}$ for $P = P^{NN}$ and $\mathcal{K}^*(P^{-1}\Sigma) \leq 2.40$ for $P = P_b^{NN}$

Neumann–Neumann Preconditioner *Given a residual $\mathbf{r}_\Gamma$, the vector $\mathbf{z}_\Gamma = (P^{NN})^{-1}\mathbf{r}_\Gamma$ can be computed as follows.*

For $m = 1, \ldots, M$, do in parallel

$$\textit{restrict the residual over } \Omega_m : \mathbf{r}_m = R_{\Gamma_m}\mathbf{r}_\Gamma \,,$$
$$\textit{define } \mathbf{r}_m^* = D_m\mathbf{r}_m \,,$$
$$\textit{compute } \tilde{\mathbf{r}}_m = (0\ I)^T\mathbf{r}_m^* \,,$$
$$\textit{solve } \tilde{\mathbf{w}}_m = A_m^\dagger\tilde{\mathbf{r}}_m \,,$$
$$\textit{define } \mathbf{w}_m^* = D_m\tilde{\mathbf{w}}_m \,,$$
$$\tilde{\mathbf{z}}_m = (0\ I)\mathbf{w}_m^* \,,$$
$$\mathbf{z}_m = R_{\Gamma_m}^T\tilde{\mathbf{z}}_m \;;$$

then compute the global vector $\mathbf{z}_\Gamma = \sum_{m=1}^{M}\mathbf{z}_m$.

The *balancing Neumann–Neumann* preconditioner consists of adding to P^{NN} a coarse-grid correction constructed by using the piecewise-constant coarse-grid space. The corresponding operator A_H involves one unknown (a constant function) per subdomain. With obvious notation the inverse of the new preconditioner reads

$$\left(P_b^{NN}\right)^{-1} = (I - \Sigma_H^\dagger\Sigma)(P^{NN})^{-1}(I - \Sigma\Sigma_H^\dagger) + \Sigma_H^\dagger \,, \tag{6.4.41}$$

where $\Sigma_H^\dagger = R_\Gamma^T A_H^\dagger R_\Gamma$.

The matrix R_Γ is applied to the vector of *all* the interface values, say $\mathbf{u}_\Gamma$, and provides a vector whose dimension is equal to the number of subdomains.

For each $m = 1, \ldots, M$, $(R_\Gamma \mathbf{u}_\Gamma)_m$ is the weighted sum of all the components of $\mathbf{u}_\Gamma$ lying on $\partial \Omega_m$, the weight being given by the inverse of the number of subdomains sharing that interface node. Note that $A_H^\dagger$ denotes the pseudo-inverse of the matrix A_H (the latter could in fact be singular).

Balancing Neumann–Neumann Preconditioner *Given a residual* $\mathbf{r}^k = \chi_\Gamma - \Sigma \mathbf{u}_\Gamma^k$, *the vector* $\mathbf{z}^k = (P_b^{NN})^{-1} \mathbf{r}^k$ *can be computed as:*

$$\mathbf{z}^k = \mathbf{z}^{k,1/4} + (P^{NN})^{-1} \mathbf{z}^{k,1/2} + \Sigma_H^\dagger \mathbf{z}^{k,3/4} \ ,$$

where $(P^{NN})^{-1}$ *is computed as in Algorithm 6.4.4,* $\Sigma_H^\dagger = R_\Gamma^T A_H^\dagger R_\Gamma$, *and*

$$\mathbf{z}^{k,1/4} = \Sigma_H^\dagger \mathbf{r}^k \ ,$$
$$\mathbf{z}^{k,1/2} = \mathbf{r}^k - \Sigma \mathbf{z}^{k,1/4} \ ,$$
$$\mathbf{z}^{k,3/4} = -\Sigma \mathbf{z}^{k,1/2} \ .$$

This coarse-grid correction has the effect of drastically reducing the condition number of the preconditioned matrix. In fact,

$$\mathcal{K}^* \left((P_b^{NN})^{-1} \Sigma \right) \simeq C \left(1 + \log^2 \frac{NH}{h} \right) \tag{6.4.42}$$

(see Mandel (1993) for h-type finite-element approximations, Pavarino and Widlund (1996) and Pavarino (1997) for SEM/hp approximations; see also Toselli and Widlund (2005), Sects. 7.4.2, 7.5.1 and 7.5.2). The constant C is independent of both the polynomial degree N and the subdomain diameter H (see Fig. 6.18). Note that $(H/h)^d$ provides a measure of the number of elements of a subdomain, while N^d is a measure of the degrees of freedom associated with a single element. Thus, $(NH/h)^d$ measures the number of degrees of freedom associated with a single subdomain.

In Fig. 6.19 we report the number of PCG (Preconditioned Conjugate Gradient) iterations for solving the Schur complement system (6.4.30) associated with the SEM-NI discretization of the Poisson problem (5.4.19). The square domain $\Omega = (-1,1)^2$ is partitioned into $M \times M$ square subdomains of side $H = 2/M$. The stopping criterion is that the relative norm of the residual is less than 10^{-12}.

An alternative approach to those described above is the so-called *FETI method*, (see, e.g., C. Farhat (1992)) for problems in structural analysis. It can be derived by formulating the Schur complement system with respect to the flux φ (the normal derivative of $\frac{\partial u}{\partial n}$ on Γ) instead of λ (the value of the solution u on Γ). In the case of two subdomains, this means that instead of (6.4.9), we consider the "dual" interface problem

$$S^* \varphi = \psi \ ,$$

where $S^* = S_1^{-1} + S_2^{-1}$, and ψ is a suitable right-hand side depending on f. The dual Schur complement system is defined accordingly, and so are the preconditioners. For instance, the inverse of a FETI preconditioner can be obtained from the right-hand side of (6.4.40) by replacing $\tilde{\Sigma}_m^{-1}$ with $\tilde{\Sigma}_m$ (see Toselli and Widlund (2005), Chap. 7 (see also Chap. 1), Toselli and Vasseur (2002), Toselli and Vasseur (2003)). Several variants exist; one, called FETI-DP (Dual-Primal) is particularly efficient (see, e. g., Toselli and Widlund (2005), Sect. 6.4). The condition number of the preconditioned problem still scales as in (6.4.42).

Extensive numerical studies are reported in Klawonn et al. (2006) and in Pavarino (2007). It is shown that for the SEM on the Poisson problem in a square uniformly subdivided into square spectral elements of edge h, gathered together to obtain square subdomains of edge H, the condition number is uniformly bounded with respect to the number of subdomains (for fixed values of N and H/h), grows as $\log^2(H/h)$ (for fixed values of N), and grows as $\log^2(N)$ for a fixed number of subdomains.

Similar results hold for the *wire-basket* preconditioner, which can be viewed as a block-diagonal preconditioner after transforming the spectral element Schur complement Σ into a convenient basis (see Casarin (1996), Sect. 3.4).

Further comments and bibliographical references on Schur preconditioners are provided at the end of Sect. 7.4.2.

6.4.5 Preconditioners for the Stiffness Matrix Derived from Preconditioners for the Schur Complement Matrix

In this Section, we show how to obtain a preconditioner for the stiffness matrix A once a preconditioner for the Schur complement Σ is available (this motivates the presence of a dashed connection between boxes 5.2 and 3.1 in Fig. 6.3).

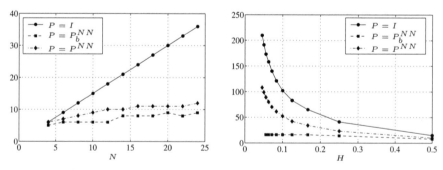

Fig. 6.19. (*Left*) Number of PCG-iterations versus the polynomial degree N; the subdomain size is $H = 1$ (i. e., four subdomains are used). (*Right*) Number of PCG-iterations versus the spectral element side H; the polynomial degree is $N = 4$ in every subdomain

After (6.4.12) and (6.4.20), the stiffness matrix in the case of a partition with two subdomains reads

$$A = \begin{pmatrix} A_{11} & 0 & A_{1\Gamma} \\ 0 & A_{22} & A_{2\Gamma} \\ A_{\Gamma 1} & A_{\Gamma 2} & A_{\Gamma\Gamma}^{(1)} + A_{\Gamma\Gamma}^{(2)} \end{pmatrix} . \tag{6.4.43}$$

It can be expressed as $A = LDL^T$, where, denoting by I_1, I_2 and I_Γ the identity matrices of dimension n_1, n_2 and n_Γ, respectively, the factors L and D have the form

$$L = \begin{pmatrix} I_1 & 0 & 0 \\ 0 & I_2 & 0 \\ A_{\Gamma 1} A_{11}^{-1} & A_{\Gamma 2} A_{22}^{-1} & I_\Gamma \end{pmatrix} , \tag{6.4.44}$$

$$D = \begin{pmatrix} A_{11} & 0 & 0 \\ 0 & A_{22} & 0 \\ 0 & 0 & \Sigma_1 + \Sigma_2 \end{pmatrix} .$$

Equivalently, we have the following block LU decomposition:

$$A = LU ,$$

where we have set

$$U = DL^T = \begin{pmatrix} A_{11} & 0 & A_{1\Gamma} \\ 0 & A_{22} & A_{2\Gamma} \\ 0 & 0 & \Sigma_1 + \Sigma_2 \end{pmatrix} . \tag{6.4.45}$$

Assume that a convenient preconditioner P is available for the Schur complement matrix Σ. Then we can define

$$Q = L\tilde{U} , \tag{6.4.46}$$

where L is given in (6.4.44), and $\tilde{U}$ is obtained from U in (6.4.45) by approximating Σ with P, that is,

$$\tilde{U} = \begin{pmatrix} A_{11} & 0 & A_{1\Gamma} \\ 0 & A_{22} & A_{2\Gamma} \\ 0 & 0 & P \end{pmatrix} . \tag{6.4.47}$$

Note that the blocks of Q coincide with those of A, except for the block $(3,3)$, which equals

$$(Q)_{33} = A_{\Gamma 1} A_{11}^{-1} A_{1\Gamma} + A_{\Gamma 2} A_{22}^{-1} A_{2\Gamma} + P .$$

(In the special case in which P is the Dirichlet–Neumann preconditioner Σ_2—see (6.4.36)—we obtain $(Q)_{33} = A_{\Gamma 1} A_{11}^{-1} A_{1\Gamma} + A_{\Gamma\Gamma}^{(2)}$.)

We claim that Q is a good preconditioner of A.

Let λ be an eigenvalue of $Q^{-1}A$, and $\mathbf{w} \in \mathbb{R}^n$ a corresponding eigenvector. Write $\mathbf{w} = (\mathbf{w}_1, \mathbf{w}_2, \mathbf{w}_\Gamma)^T$, with obvious notation. From

$$A\,\mathbf{w} = \lambda Q\mathbf{w}\,,$$

we have

$$(1 - \lambda)(A_{11}\mathbf{w}_1 + A_{1\Gamma}\mathbf{w}_\Gamma) = \mathbf{0}\,,$$
$$(1 - \lambda)(A_{22}\mathbf{w}_2 + A_{2\Gamma}\mathbf{w}_\Gamma) = \mathbf{0}\,,$$
$$(1 - \lambda)(A_{\Gamma 1}\mathbf{w}_1 + A_{\Gamma 2}\mathbf{w}_2) + A_{\Gamma\Gamma}\mathbf{w}_\Gamma$$
$$= \lambda(P\mathbf{w}_\Gamma + A_{\Gamma 1}A_{11}^{-1}A_{1\Gamma}\mathbf{w}_\Gamma + A_{\Gamma 2}A_{22}^{-1}A_{2\Gamma}\mathbf{w}_\Gamma)\,.$$

From (6.4.18) the last equation can be rewritten as

$$(1 - \lambda)(A_{\Gamma 1}\mathbf{w}_1 + A_{\Gamma 2}\mathbf{w}_2 + A_{\Gamma 1}A_{11}^{-1}A_{1\Gamma}\mathbf{w}_\Gamma + A_{\Gamma 2}A_{22}^{-1}A_{2\Gamma}\mathbf{w}_\Gamma) + \Sigma\mathbf{w}_\Gamma$$
$$= \lambda P\mathbf{w}_\Gamma.$$

Therefore, if $\lambda \neq 1$ we have

$$A_{11}\mathbf{w}_1 + A_{1\Gamma}\mathbf{w}_\Gamma = \mathbf{0}\,,$$
$$A_{22}\mathbf{w}_2 + A_{2\Gamma}\mathbf{w}_\Gamma = \mathbf{0}\,;$$

hence, $\mathbf{w}_\Gamma \neq \mathbf{0}$ and

$$\Sigma\mathbf{w}_\Gamma = \lambda P\mathbf{w}_\Gamma\,.$$

We conclude that the matrix $Q^{-1}A$ has the same eigenvalues as $P^{-1}\Sigma$ plus the eigenvalue 1 (which is also an eigenvalue of $P^{-1}\Sigma$ provided that the corresponding eigenvector $\mathbf{w}$ satisfies $\mathbf{w}_\Gamma \neq \mathbf{0}$).

If we assume that P is spectrally equivalent to Σ, that is, there exist two positive constants K_1 and K_2, independent of N, such that

$$K_1[P\boldsymbol{\eta}, \boldsymbol{\eta}] \leq [\Sigma\boldsymbol{\eta}, \boldsymbol{\eta}] \leq K_2[P\boldsymbol{\eta}, \boldsymbol{\eta}] \qquad \forall\, \boldsymbol{\eta} \in \mathbb{R}^{N_\Gamma}\,, \tag{6.4.48}$$

then we derive from the characterization of the eigenvalues of $Q^{-1}A$ that

$$\mathcal{K}(Q^{-1}A) \leq \frac{\widetilde{K}_2}{\widetilde{K}_1}\,, \tag{6.4.49}$$

where

$$\widetilde{K}_1 = \min(1, K_1), \qquad \widetilde{K}_2 = \max(1, K_2)\,, \tag{6.4.50}$$

and therefore we conclude that Q is spectrally equivalent to A.

From the computational viewpoint, the solution of the preconditioned system

$$Q\widetilde{\mathbf{x}} = \widetilde{\mathbf{b}}$$

entails the solution of the two-block triangular systems

$$L\widetilde{\mathbf{y}} = \widetilde{\mathbf{b}}, \quad \widetilde{U}\widetilde{\mathbf{x}} = \widetilde{\mathbf{y}}\,,$$

that is,

$$\tilde{\mathbf{y}}_1 = \tilde{\mathbf{b}}_1, \quad \tilde{\mathbf{y}}_2 = \tilde{\mathbf{b}}_2, \quad \tilde{\mathbf{y}}_\Gamma = \tilde{\mathbf{b}}_\Gamma - A_{\Gamma 1} A_{11}^{-1} \tilde{\mathbf{b}}_1 - A_{\Gamma 2} A_{22}^{-1} \tilde{\mathbf{b}}_2 \qquad (6.4.51)$$

and

$$P\tilde{\mathbf{x}}_\Gamma = \tilde{\mathbf{y}}_\Gamma, \quad A_{11}\tilde{\mathbf{x}}_1 = \tilde{\mathbf{b}}_1 - A_{1\Gamma}\tilde{\mathbf{x}}_\Gamma, \quad A_{22}\tilde{\mathbf{x}}_2 = \tilde{\mathbf{b}}_2 - A_{2\Gamma}\tilde{\mathbf{x}}_\Gamma . \qquad (6.4.52)$$

Both steps (6.4.51) and (6.4.52) require the solution of two independent Dirichlet problems, one for each subdomain. Furthermore, step (6.4.52) requires the solution of a problem associated with P. The whole process is therefore amenable to independent solves in Ω_1 and Ω_2, provided P splits into independent solves as well.

In conclusion, the preconditioner Q inherits all the good properties enjoyed by P in terms of both parallelism and spectral equivalence.

Note that in the case where we take the Dirichlet–Neumann preconditioner $P = \Sigma_2$, step (6.4.52) implies the solution of

$$\Sigma_2 \tilde{\mathbf{x}}_\Gamma = \tilde{\mathbf{y}}_\Gamma ,$$

which corresponds to a Neumann problem in Ω_2.

Similar conclusions hold in the case of partitions with many subdomains. Indeed, generalizing what was done for a two-domain subdivision, the stiffness matrix A in (6.4.24) can be factored as $A = LDL^T$, this time with

$$L = \begin{pmatrix} I_{\Omega \backslash \Gamma} & 0 \\ A_{\Gamma I} A_{II}^{-1} & I_\Gamma \end{pmatrix}, \quad D = \begin{pmatrix} A_{II} & 0 \\ 0 & \Sigma \end{pmatrix} . \qquad (6.4.53)$$

Equivalently, we have $A = LU$, with

$$U = \begin{pmatrix} A_{II} & A_{I\Gamma} \\ 0 & \Sigma \end{pmatrix} .$$

As previously noted, should P be a preconditioner of the Schur complement Σ, then the matrix (6.4.46) with

$$\tilde{U} = \begin{pmatrix} A_{II} & A_{I\Gamma} \\ 0 & P \end{pmatrix} , \qquad (6.4.54)$$

is a preconditioner of the stiffness matrix A. The eigenvalues of $Q^{-1}A$ are the same as those of $P^{-1}\Sigma$ plus the eigenvalue 1.

When an iterative method is used to solve the linear system associated with $Q^{-1}A$, at each iteration a system with the matrix Q has to be solved. Using the factorization (6.4.46), this requires first solving the L system by forward substitution, which involves the inversion of A_{mm} (the diagonal blocks in (6.4.25)) for each $m = 1, \ldots, M$, i.e., the solution of M independent Dirichlet problems.

As for the $\tilde{U}$ system, this is solved by backward substitution, starting from the block P, followed by the block A_{II}, which again yields the solution of M independent Dirichlet problems in the subdomains.

We can also use a preconditioner for the internal blocks by replacing A_{II} with $\tilde{A}_{II}$ in D and similarly in L:

$$\hat{P}^{-1} = (\tilde{L}\tilde{U})^{-1}(\tilde{U}\tilde{L})^{-1} = \begin{pmatrix} \tilde{A}_{II} & -A_{I\Gamma} \\ 0 & P^{-1} \end{pmatrix} \begin{pmatrix} I & 0 \\ -A_{\Gamma I}\tilde{A}_{II}^{-1} & I \end{pmatrix}.$$

6.5 Solution Algorithms for Patching Collocation Methods

We face now the task of solving the problems obtained using the patching collocation method (see Sects. 5.9 and 5.11) by iterative DD procedures. To keep our presentation simple we will refer to the Poisson problem (5.4.19), reformulated in a DD framework (5.13.1) and to its collocation approximation (5.13.2)–(5.13.5).

The traditional iterative methods for this problem are referred to as *iterative substructuring* methods. Typically, they introduce a sequence of subproblems in Ω_1 and Ω_2 for which the transmission conditions (5.13.4) and (5.13.5) provide, respectively, Dirichlet or Neumann data at the internal boundary Γ.

This can be accomplished in several ways, some of which are presented below.

To start with, we will illustrate our idea on the differential problem (we are therefore referring to box 6.2 of Fig. 6.3 and, more specifically, to the dashed line connecting box 6.2.1 to box 1.2).

In general, two sequences of functions $\{u_1^k\}$, $\{u_2^k\}$ are generated starting from an initial guess u_1^0, u_2^0, and they will converge to u_1 and u_2, respectively.

1. The Dirichlet–Neumann method

Given λ^0, solve for each $k \geq 0$:

$$\begin{aligned} -\triangle u_1^{k+1} &= f & &\text{in } \Omega_1, \\ u_1^{k+1} &= 0 & &\text{on } \partial\Omega_1 \cap \partial\Omega, \\ u_1^{k+1} &= \lambda^k & &\text{on } \Gamma, \end{aligned} \tag{6.5.1}$$

then

$$\begin{aligned} -\triangle u_2^{k+1} &= f & &\text{in } \Omega_2, \\ u_2^{k+1} &= 0 & &\text{on } \partial\Omega_2 \cap \partial\Omega, \\ \frac{\partial u_2^{k+1}}{\partial n^\Gamma} &= \frac{\partial u_1^{k+1}}{\partial n^\Gamma} & &\text{on } \Gamma, \end{aligned} \tag{6.5.2}$$

and update

$$\lambda^{k+1} = \theta u_{2|\Gamma}^{k+1} + (1 - \theta)\lambda^k , \qquad (6.5.3)$$

θ being a positive acceleration parameter.

For practical purposes, the differential problems on each subdomain must be approximated (see box 7 of Fig. 6.3). If we use, for instance, the collocation method (box 7.3), then the Dirichlet–Neumann iterations (6.5.1)–(6.5.3) give rise to the following discrete iterative substructuring method:

for a given λ^0 on Γ, solve for each $k \geq 0$

$$-\triangle u_{1,N}^{k+1} = f \quad \text{at all internal collocation nodes of } \Omega_1,$$

$$u_{1,N}^{k+1} = 0 \quad \text{at the collocation nodes lying on } \partial\Omega_1 \cap \partial\Omega ,$$

$$u_{1,N}^{k+1} = \lambda^k \quad \text{at the collocation nodes on } \Gamma$$

$$\text{(except those on } \partial\Omega) ; \qquad (6.5.4)$$

$$-\triangle u_{2,N}^{k+1} = f \qquad \text{at all internal collocation nodes of } \Omega_2,$$

$$u_{2,N}^{k+1} = 0 \qquad \text{at the collocation nodes lying on } \partial\Omega_2 \cap \partial\Omega ,$$

$$\frac{\partial u_N^{(2)}}{\partial n^\Gamma}^{k+1} = \frac{\partial u_N^{(1)}}{\partial n^\Gamma}^{k+1} \qquad \text{at the collocation nodes on } \Gamma$$

$$\text{(except those on } \partial\Omega) ; \qquad (6.5.5)$$

$$\lambda^{k+1} = \theta u_{2,N}^{k+1} + (1 - \theta)\lambda^k \qquad \text{at the collocation nodes on } \Gamma$$

$$\text{(except those on } \partial\Omega) . \qquad (6.5.6)$$

This can be regarded as an iterative substructuring method for solving the DD patching problems (5.13.2)–(5.13.5) (with an ideal dashed line connecting boxes 7.3 and 6.2.1 to box 2.2.1, which we forego drawing to avoid increasing the complexity of Fig. 6.3).

It is worthwhile pointing out that a similar iterative scheme could be set up to solve the SEM-NI problem. On the same Poisson problem we end up with a scheme that coincides with (6.5.4)–(6.5.6) except for the Neumann continuity (6.5.5), which would be satisfied in a weak manner (by penalty, e. g., as in (5.2.19)).

2. The Neumann–Neumann method

This method was considered by Bourgat et al. (1989); a previous version had been investigated already in Agoshkov and Lebedev (1985).

In this case, for each $k \geq 0$ we have to solve, for $m = 1, 2,$

$$-\triangle u_m^{k+1} = f \qquad \text{in } \Omega_m,$$

$$u_m^{k+1} = 0 \qquad \text{on } \partial\Omega_m \cap \partial\Omega, \qquad (6.5.7)$$

$$u_m^{k+1} = \lambda^k \qquad \text{on } \Gamma ,$$

and then

$$-\triangle\psi_m^{k+1} = 0 \qquad \text{in } \Omega_m,$$
$$\psi_m^{k+1} = 0 \qquad \text{on } \partial\Omega_m \cap \partial\Omega,$$
$$\frac{\partial\psi_m^{k+1}}{\partial n^\Gamma} = \frac{\partial u_1^{k+1}}{\partial n^\Gamma} - \frac{\partial u_2^{k+1}}{\partial n^\Gamma} \qquad \text{on } \Gamma, \tag{6.5.8}$$

and update

$$\lambda^{k+1} = \lambda^k - \theta(\sigma_1\psi_{1|\Gamma}^{k+1} - \sigma_2\psi_{2|\Gamma}^{k+1}). \tag{6.5.9}$$

As before, $\theta > 0$ is an acceleration parameter, σ_1 and σ_2 are two positive averaging coefficients, while λ^0 is a given datum.

3. The Robin method

This time, for each $k \geq 0$, we solve (see Agoshkov (1988) and P.-L. Lions (1990) for the presentation and the analysis of the algorithm)

$$-\triangle u_1^{k+1} = f \qquad \text{in } \Omega_1,$$
$$u_1^{k+1} = 0 \qquad \text{on } \partial\Omega_1 \cap \partial\Omega,$$
$$\frac{\partial u_1^{k+1}}{\partial n^\Gamma} + \gamma_1 u_1^{k+1} = \frac{\partial u_2^k}{\partial n^\Gamma} + \gamma_1 u_2^k \qquad \text{on } \Gamma, \tag{6.5.10}$$

and then

$$-\triangle u_2^{k+1} = f \qquad \text{in } \Omega_2,$$
$$u_2^{k+1} = 0 \qquad \text{on } \partial\Omega_2 \cap \partial\Omega,$$
$$\frac{\partial u_2^{k+1}}{\partial n^\Gamma} - \gamma_2 u_2^{k+1} = \frac{\partial u_1^{k+1}}{\partial n^\Gamma} - \gamma_2 u_1^{k+1} \qquad \text{on } \Gamma, \tag{6.5.11}$$

where u_2^0 is given, and γ_1 and γ_2 are nonnegative acceleration parameters satisfying $\gamma_1 + \gamma_2 > 0$. For the sake of parallelization, in (6.5.11) we could also consider u_1^k instead of u_1^{k+1}, assigning in that case also u_1^0.

Let us show that *the previous iterative schemes can be interpreted as a preconditioned Richardson procedure for the interface Steklov–Poincaré problem* (this motivates the presence of a dashed line connecting box 6.2 with box 1.3 in Fig. 6.3). For the sake of exposition, we refer to the differential formulation (6.5.1)–(6.5.3); however, the same kind of result can be proved for their approximation. We will make use of the notation of Sect. 6.4.1. Let us start with a step of the Dirichlet–Neumann method. At every iteration $k + 1$ we set $w_2^{k+1} = u_2^{k+1} - \mathcal{G}_2 f$. Choosing $\mathbf{n}^\Gamma = \mathbf{n}^1 = -\mathbf{n}^2$, we have

$$-\triangle w_2^{k+1} = 0 \qquad \text{in } \Omega_2,$$
$$w_2^{k+1} = 0 \qquad \text{on } \partial\Omega_2 \cap \partial\Omega,$$
$$\frac{\partial w_2^{k+1}}{\partial n^2} = \frac{\partial u_1^{k+1}}{\partial n^\Gamma} + \frac{\partial \mathcal{G}_2 f}{\partial n^\Gamma} \qquad \text{on } \Gamma.$$

Noting that $u_1^{k+1} = \mathcal{H}_1 \lambda^k + \mathcal{G}_1 f$, it follows that

$$
\begin{aligned}
u_{2|\Gamma}^{k+1} = w_{2|\Gamma}^{k+1} &= S_2^{-1} \left(-\frac{\partial \mathcal{H}_1 \lambda^k}{\partial n^\Gamma} - \frac{\partial \mathcal{G}_1 f}{\partial n^\Gamma} + \frac{\partial \mathcal{G}_2 f}{\partial n^\Gamma} \right) \\
&= S_2^{-1}(-S_1 \lambda^k + \chi) \, .
\end{aligned}
$$

Therefore,

$$
\begin{aligned}
\lambda^{k+1} &= \lambda^k + \theta[S_2^{-1}(-S_1 \lambda^k + \chi) - \lambda^k] \\
&= \lambda^k + \theta S_2^{-1}(-S \lambda^k + \chi) \, .
\end{aligned}
$$

(The definitions of the *extension operators* $\mathcal{H}_m$ and $\mathcal{G}_m$, $m = 1, 2$, were given in Sect. 6.4.1.) This is a Richardson iteration on the Steklov–Poincaré equation (6.4.9) using the operator S_2 as a preconditioner. This interpretation plays a central role in the convergence analysis of the Dirichlet–Neumann method.

We note here the analogy with the role of the Dirichlet–Neumann preconditioner P^{DN} that we have introduced in (6.4.36). Precisely, one step of the Dirichlet–Neumann iterative scheme at the discrete level would be equivalent to one step of the Richardson method applied to the Schur complement system (6.4.30) using $P^{DN} = \Sigma_2$ as preconditioner.

Also the Neumann–Neumann scheme can be interpreted as a preconditioned Richardson scheme. In fact, we have already noted that $u_m^{k+1} = \mathcal{H}_m \lambda^k + \mathcal{G}_m f$; therefore,

$$
\begin{aligned}
\psi_{1|\Gamma}^{k+1} &= S_1^{-1} \left(\frac{\partial \mathcal{H}_1 \lambda^k}{\partial n^\Gamma} + \frac{\partial \mathcal{G}_1 f}{\partial n^\Gamma} - \frac{\partial \mathcal{H}_2 \lambda^k}{\partial n^\Gamma} - \frac{\partial \mathcal{G}_2 f}{\partial n^\Gamma} \right) \\
&= -S_1^{-1}(-S \lambda^k + \chi) \, .
\end{aligned}
$$

Similarly, recalling that $\mathbf{n}^\Gamma = -\mathbf{n}^2$,

$$
\psi_{2|\Gamma}^{k+1} = S_2^{-1}(-S \lambda^k + \chi) \, .
$$

Therefore,

$$
\lambda^{k+1} = \lambda^k + \theta(\sigma_1 S_1^{-1} + \sigma_2 S_2^{-1})(\chi - S \lambda^k) \, ,
$$

which is a Richardson iteration for the Steklov–Poincaré equation (6.4.9) with the operator $(\sigma_1 S_1^{-1} + \sigma_2 S_2^{-1})^{-1}$ as a preconditioner.

At the discrete level this becomes the Neumann–Neumann preconditioner (6.4.39). (Here σ_i corresponds to σ_i^2 in (6.4.39).)

An iterative Dirichlet–Neumann (or Neumann–Neumann, or Robin) method can be still set up for other boundary-value problems as well. For instance, when addressing the approximation of the Stokes equations by spectral collocation methods, at the $(k + 1)$-th step the Stokes equations are collocated at the internal collocation points of Ω_1, and the velocity-continuity condition $\mathbf{u}_{1,N}^{k+1} = \theta\mathbf{u}_{2,N}^k + (1 - \theta)\mathbf{u}_{1,N}^k$ is satisfied at the collocation nodes lying on Γ. The Stokes equations are then collocated at the internal collocation nodes of Ω_2, and the continuity of normal stresses $\mathbf{T}(\mathbf{u}_{2,N}^{k+1}, p_{2,N}^{k+1}) \cdot \mathbf{n}^\Gamma = \mathbf{T}(\mathbf{u}_{1,N}^{k+1}, p_{1,N}^{k+1}) \cdot \mathbf{n}^\Gamma$ is satisfied at the collocation nodes of Γ.

Special care must be devoted to the pressure treatment in Ω_1. Actually, in the case of a pure Dirichlet problem (i.e., when $\Gamma_N = \emptyset$), then while solving the collocation problem in Ω_1, we must impose the integral condition $\int_{\Omega_1} p_{1,N}^{k+1} = -\int_{\Omega_2} p_{2,N}^k$ to make sure that the limit solution satisfies $\int_{\Omega_1} p_{1,N} + \int_{\Omega_2} p_{2,N} = 0$ (to get rid of the physical spurious mode $p_N^* = const$).

The adaptation of this iterative Dirichlet–Neumann technique to the complete Navier–Stokes equations is straightforward, as the role played by the interface conditions is the same. See Quarteroni and Valli (1999).

7. General Algorithms for Incompressible Navier–Stokes Equations

7.1 Introduction

This chapter is again devoted to the time-dependent, incompressible Navier–Stokes equations, whose numerical discretization has already been considered in Chap. 3 in the case of single-domain spectral methods and in Sect. 5.6 for multidomain spectral methods. Whereas both the formulations and solution algorithms for single-domain spectral methods as well as the formulations for multidomain spectral methods are by now well established and we can provide the reader with firm guidance on these subjects, such is not the case for the challenging task of selecting efficient solution algorithms for multidomain spectral methods. Hence, in this chapter we provide primarily a survey of the various alternatives that have been proposed along with some general considerations for selecting amongst the alternatives.

Here, we assume the equations to be set in a general domain and submitted to general boundary conditions, and we focus our attention on two fundamental aspects of their numerical treatment: the design of accurate time-discretization schemes and the design of efficient algebraic solvers. Together with the spatial discretization strategies already discussed, they are the fundamental ingredients that define a complete discretization process for the Navier–Stokes equations. A simplified roadmap of this chapter is illustrated in Fig. 7.1.

Throughout this chapter, we will use the nonconservative form (3.1.1)–(3.1.2) of the equations, although most of the considerations apply (with obvious modifications and no extra difficulty) to the other forms presented in Sect. 3.1. The equations are supplemented by the initial conditions (3.1.3) and by boundary conditions that, wherever periodicity is not imposed, prescribe either the velocity field, say $\mathbf{u} = \mathbf{g}$, or the normal component of the stress tensor, say $\nu \frac{\partial \mathbf{u}}{\partial n} - p\mathbf{n} = \mathbf{s}$.

An overview of the time-discretization techniques most commonly used with spectral methods has been given in Sect. 3.2.2. We have noticed therein that at each time-step these schemes either produce a coupled (generalized) Stokes problem, or a cascade of fractional steps in which elliptic problems involving as unknowns separately the velocity (Helmholtz or advection-diffusion equations) or the pressure (Poisson equation) have to be solved. Sect. 7.3 will be devoted to an extensive discussion of efficient algebraic methods for solving the generalized Stokes problem, both in a single domain and in multiple

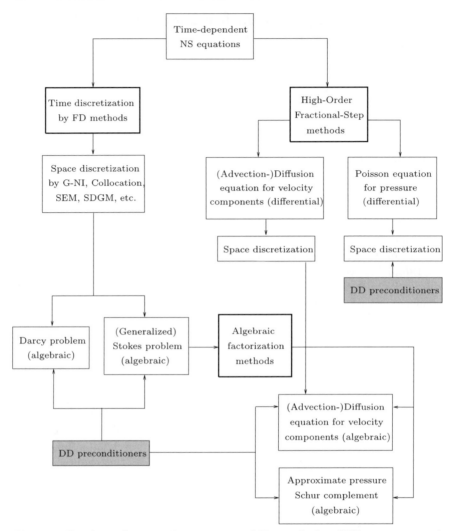

Fig. 7.1. Roadmap for time-discretization of Navier–Stokes (NS) equations and associated algebraic solvers

domains. On the other hand, Sect. 7.2 will be devoted to present several high-order fractional-step methods. Indeed, as seen in Sect. 3.2.4, the low-order Chorin–Temam scheme suffers from a severe splitting error, which can be alleviated by more sophisticated splitting techniques. In Sect. 7.4 we will address the case of algebraic fractional-step methods, which can be regarded as alternatives to the differential splitting schemes and are more flexible for the treatment of boundary conditions.

Notation. Throughout this chapter, the suffix N will indicate the set of all spatial discretization parameters, i. e., the polynomial degree in a single-domain spectral method, and the set of polynomial degrees N_m and element sizes h_m in a multidomain spectral method. Thus, N will denote here what is indicated by δ in Chaps. 5 and 6.

7.2 High-Order Fractional-Step Methods

Fractional-step methods, also called splitting methods, have received broad attention due to their distinguishing feature of allowing independent computation of the velocity and pressure fields. The boundary conditions that we consider in discussing this class of methods are no-slip (i. e., $\mathbf{g} = \mathbf{0}$) on a portion $\partial\Omega_0$ of the boundary (possibly coinciding with the whole boundary) and periodicity on the remaining part. Comments on more general boundary conditions are postponed to the end of the section.

As mentioned in Sect. 3.2.4, the progenitor of splitting methods is the Chorin–Temam method, which we recall here for the reader's convenience. When going from t^n to t^{n+1}, the following two steps are accomplished:

i) *first step: preliminary velocity*

$$\frac{1}{\Delta t}(\hat{\mathbf{u}}^{n+1} - \mathbf{u}^n) - \nu\triangle\hat{\mathbf{u}}^{n+1} = \mathbf{f}^{n+1} - C(\mathbf{u}^n) \qquad \text{in } \Omega ,$$

$$\hat{\mathbf{u}}^{n+1} = \mathbf{0} \qquad \text{on } \partial\Omega_0 , \tag{7.2.1}$$

where $C(\mathbf{u}^n)$ accounts for the convection term, which is often treated explicitly, e. g., $C(\mathbf{u}^n) = (\mathbf{u}^n \cdot \nabla)\mathbf{u}^n$;

ii) *second step: projection*

$$\frac{1}{\Delta t}(\mathbf{u}^{n+1} - \hat{\mathbf{u}}^{n+1}) + \nabla p^{n+1} = 0 \qquad \text{in } \Omega ,$$

$$\nabla \cdot \mathbf{u}^{n+1} = 0 \qquad \text{in } \Omega , \tag{7.2.2}$$

$$\mathbf{u}^{n+1} \cdot \mathbf{n} = 0 \qquad \text{on } \partial\Omega_0 .$$

This step, which yields a Darcy problem for the unknowns $\mathbf{u}^{n+1}$ and p^{n+1}, is in fact reducible to a Poisson problem for the pressure:

$$\triangle p^{n+1} = \frac{1}{\Delta t}\nabla \cdot \hat{\mathbf{u}}^{n+1} \qquad \text{in } \Omega ,$$

$$\frac{\partial p^{n+1}}{\partial n} = 0 \qquad \text{on } \partial\Omega_0 , \tag{7.2.3}$$

followed by an update of the velocity field:

$$\mathbf{u}^{n+1} = \hat{\mathbf{u}}^{n+1} - \frac{1}{\Delta t}\nabla p^{n+1} . \tag{7.2.4}$$

The latter equality, which follows from the Helmholtz decomposition principle, can be regarded as a correction (by a pressure gradient) of $\hat{\mathbf{u}}^{n+1}$ in order to ensure that the new velocity $\mathbf{u}^{n+1}$, also called end-of-step velocity, is divergence-free. The term projection is used because one has that

$$\mathbf{u}^{n+1} = P_H\hat{\mathbf{u}}^{n+1} , \tag{7.2.5}$$

where P_H is the orthogonal projection operator upon the space

$$H = \{\mathbf{v} \in (L^2(\Omega))^d |\ \nabla \cdot \mathbf{v} = 0 \text{ in } \Omega,\ \mathbf{v} \cdot \mathbf{n} = 0 \text{ on } \partial\Omega_0\}$$

with respect to the $L^2(\Omega)$ norm, that is,

$$(\mathbf{u}^{n+1}, \mathbf{v}) = (\hat{\mathbf{u}}^{n+1}, \mathbf{v}) \qquad \forall \mathbf{v} \in H ,$$

where $(\cdot, \cdot)$ indicates the L^2-inner product in Ω.

In spite of its simplicity, this method suffers from low accuracy in time. More precisely, the following error bound holds (provided $\mathbf{u}$ and p are smooth enough):

$$\max_n \|\mathbf{u}(t^n) - \mathbf{u}^n\| + \max_n \|\mathbf{u}(t^n) - \hat{\mathbf{u}}^n\| \leq C\Delta t , \tag{7.2.6}$$

$$\max_n \|p(t^n) - p^n\| + \max_n \|\nabla(\mathbf{u}(t^n) - \hat{\mathbf{u}}^n)\| \leq C\Delta t^{1/2} , \tag{7.2.7}$$

where $\| \cdot \|$ is the norm of $L^2(\Omega)$. The splitting error $O(\Delta t)$ on the velocity field is irreducible, as it cannot be improved by using a more accurate time-discretization scheme for the first step (7.2.1). In particular, the no-slip boundary condition is not satisfied exactly at the new time-level due to the presence of an $O(\Delta t)$ splitting error in the tangential component; this generates an unphysical pressure boundary layer of thickness $O(\nu\sqrt{\Delta t})$ (see Rannacher (1992)).

As anticipated in Sect. 3.2.4, several improvements of the Chorin–Temam method, still based on the differential form, have been proposed. Among them, we mention the *pressure-correction* method, consisting of adding an explicit incremental term for the pressure in the momentum equation.

One possible realization of an incremental pressure-correction scheme, due to van Kan (1986), is as follows:

$$\frac{1}{2\Delta t}(3\hat{\mathbf{u}}^{n+1} - 4\mathbf{u}^n + \mathbf{u}^{n-1}) - \nu\triangle\hat{\mathbf{u}}^{n+1} + \nabla p^n$$
$$= \mathbf{f}^{n+1} - C(\mathbf{u}^{n-1}, \mathbf{u}^n) \qquad\qquad \text{in } \Omega , \tag{7.2.8}$$
$$\hat{\mathbf{u}}^{n+1} = \mathbf{0} \qquad\qquad\qquad\qquad \text{on } \partial\Omega_0 .$$

The time derivative has been discretized by the second-order BDF2 formula, while $C(\mathbf{u}^{n-1}, \mathbf{u}^n)$ indicates a second-order discretization for the convection term, e. g., the fully explicit one based on the second-order Adams–Bashforth extrapolation (see Appendix D)

$$C(\mathbf{u}^{n-1}, \mathbf{u}^n) = \frac{3}{2}\mathbf{u}^n \cdot \nabla\mathbf{u}^n - \frac{1}{2}\mathbf{u}^{n-1} \cdot \nabla\mathbf{u}^{n-1} .$$

The projection step this time becomes

$$\frac{1}{2\Delta t}(3\mathbf{u}^{n+1} - 3\hat{\mathbf{u}}^{n+1}) + \nabla(p^{n+1} - p^n) = \mathbf{0} \qquad \text{in } \Omega ,$$
$$\nabla \cdot \mathbf{u}^{n+1} = 0 \qquad \text{in } \Omega , \tag{7.2.9}$$
$$\hat{\mathbf{u}}^{n+1} \cdot \mathbf{n} = 0 \qquad \text{on } \partial\Omega_0 .$$

Property (7.2.5) still holds.

If properly initialized, this method is more accurate than (7.2.1)–(7.2.2). Indeed, the error estimates (7.2.6)–(7.2.7) improve to

$$\left(\sum_n \Delta t \|\mathbf{u}(t^n) - \mathbf{u}^n\|^2\right)^{1/2} + \left(\sum_n \Delta t \|\mathbf{u}(t^n) - \mathbf{u}^n\|^2\right)^{1/2} \leq C\Delta t^2, \quad (7.2.10)$$

$$\max_n \|p(t^n) - p^n\| + \max_n \|\nabla(\mathbf{u}(t^n) - \hat{\mathbf{u}}^n)\| \leq C\Delta t. \quad (7.2.11)$$

In spite of the fact that a second-order discretization scheme was used in (7.2.8), the error for the pressure (and that for the velocity gradient) in (7.2.11) is only first order. This is because the solution is still affected by a pressure boundary layer, which is due to the unphysical boundary condition $\frac{\partial p}{\partial n}^{n+1} - \frac{\partial p}{\partial n}^n = 0$ on $\partial\Omega$ (and therefore $\frac{\partial p}{\partial n}^k = 0$ $\forall k \geq 1$ by recursion, if p^1 is initialized by (7.2.3)) associated with the projection step (7.2.9).

A modification of the projection step (7.2.9) due to Timmermans, Minev and van de Vosse (1996) avoids this unpleasant effect. The first step (7.2.8) remains unchanged, whereas (7.2.9) is modified as follows:

$$\frac{1}{2\Delta t}(3\mathbf{u}^{n+1} - 3\hat{\mathbf{u}}^{n+1}) + \nabla(p^{n+1} - p^n) = -\nu\nabla(\nabla \cdot \hat{\mathbf{u}}^{n+1}) \quad \text{in } \Omega,$$

$$\nabla \cdot \mathbf{u}^{n+1} = 0 \quad \text{in } \Omega,$$

$$\hat{\mathbf{u}}^{n+1} \cdot \mathbf{n} = 0 \quad \text{on } \partial\Omega_0.$$
$$(7.2.12)$$

By summing up (7.2.8) and (7.2.12) one obtains

$$\frac{1}{2\Delta t}(3\mathbf{u}^{n+1} - 4\mathbf{u}^n + \hat{\mathbf{u}}^{n-1}) + \nu\nabla \times \boldsymbol{\omega}^{n+1} + \nabla p^{n+1}$$
$$= \mathbf{f}^{n+1} - C(\mathbf{u}^{n-1}, \mathbf{u}^n) \quad \text{in } \Omega,$$

$$\nabla \cdot \mathbf{u}^{n+1} = 0 \quad \text{in } \Omega,$$

$$\hat{\mathbf{u}}^{n+1} = \mathbf{0} \quad \text{on } \partial\Omega_0,$$
$$(7.2.13)$$

with $\boldsymbol{\omega}^{n+1} = \nabla \times \mathbf{u}^{n+1}$, where we have used the identity

$$\triangle\mathbf{v} = \nabla(\nabla \cdot \mathbf{v}) - \nabla \times (\nabla \times \mathbf{v}) \quad (7.2.14)$$

and the fact that

$$\nabla \times (\nabla \times \hat{\mathbf{u}}^{n+1}) = \nabla \times \boldsymbol{\omega}^{n+1},$$

which follows from (7.2.12). From (7.2.13) we deduce that

$$\frac{\partial p}{\partial n}^{n+1} = \left(\mathbf{f}^{n+1} - C(\mathbf{u}^{n-1}, \mathbf{u}^n) - \nu\nabla \times \boldsymbol{\omega}^{n+1}\right) \cdot \mathbf{n} \quad \text{on } \partial\Omega_0. \quad (7.2.15)$$

Note that C does not give any contribution if we have homogeneous Dirichlet boundary conditions on $\partial\Omega_0$.

The scheme (7.2.8) and (7.2.12) (or, equivalently, (7.2.13)) is given the name of an *incremental pressure-correction* method *in rotational form*. This time the only splitting error is due to the inexact fulfillment of the tangential component of the velocity field on $\partial\Omega_0$. This scheme still satisfies an error estimate like (7.2.10). Moreover, instead of (7.2.11) we have

$$
\left(\sum_n \Delta t \|\nabla(\mathbf{u}(t^n) - \mathbf{u}^n)\|^2\right)^{1/2} + \left(\sum_n \Delta t \|\nabla(\mathbf{u}(t^n) - \hat{\mathbf{u}}^n)\|^2\right)^{1/2}
$$

$$
+ \left(\sum_n \Delta t \|p(t^n) - p^n\|^2\right)^{1/2} \leq C\Delta t^{3/2}, \quad (7.2.16)
$$

with $C\Delta t^2$ on the right-hand side in the case of a fully periodic flow.

This *rotational form* of the boundary condition on the pressure is a consistent one and was first reported by Orszag, Israeli and Deville (1986).

A rotational form of the boundary condition (without the incremental form of the pressure, though) can be found in Karniadakis and Sherwin (1995), pp. 253–254. If the forcing term $\mathbf{f}$ is zero, at the first stage one solves

$$
\frac{1}{\Delta t}(\hat{\mathbf{u}}^{n+1} - \mathbf{u}^n) - \frac{1}{2}\nu\triangle(\hat{\mathbf{u}}^{n+1} + \mathbf{u}^n) = -C(\mathbf{u}^{n-1}, \mathbf{u}^n) \qquad \text{in } \Omega,
$$

$$
\hat{\mathbf{u}}^{n+1} = \mathbf{0} \qquad \text{on } \partial\Omega_0,
$$

while the projection step becomes

$$
\frac{1}{\Delta t}(\mathbf{u}^{n+1} - \hat{\mathbf{u}}^{n+1}) = -\nabla p^{n+1} \qquad \text{in } \Omega,
$$

$$
\nabla \cdot \mathbf{u}^{n+1} = 0 \qquad \text{in } \Omega,
$$

$$
\frac{\partial p^{n+1}}{\partial n} = -\nu\mathbf{n} \cdot (\nabla \times \boldsymbol{\omega})^{n+1} \qquad \text{on } \partial\Omega_0.
$$

A potentially more accurate splitting scheme is one based on a higher-order discretization of the time derivative. Several possibilities exist. We report below a three-stage scheme based on a BDF treatment of the linear terms combined with an explicit extrapolation of the nonlinear convection terms (see, e. g., Karniadakis and Sherwin (2005), p. 428).

Let $J \geq 1$ be the number of steps, and $\gamma_0, \alpha_0, \alpha_1, \ldots, \alpha_{J-1}$ the coefficients, of the BDF formula

$$
\left.\frac{dy}{dt}\right|_{t=t^{n+1}} \simeq \frac{1}{\Delta t}\left(\gamma_0 y^{n+1} - \sum_{q=0}^{J-1} \alpha_q y^{n-q}\right) \qquad (7.2.17)
$$

for the discretization of the time derivative $\frac{dy}{dt}$ at time t^{n+1}.

Table 7.1. Coefficients for the BDF time-discretization formulas and the extrapolation of the nonlinear term

Coefficient	1st order	2nd order	3rd order
γ_0	1	3/2	11/6
α_0	1		3
α_1	0	-1/2	-3/2
α_2	0		1/3
β_0	1		3
β_1	0	-1	-3
β_2	0	0	1

Moreover, let $\beta_0, \ldots, \beta_{J-1}$ be the coefficients of the extrapolation formula

$$z^{n+1} \simeq z_{\text{ext}}^{n+1} = \sum_{q=0}^{J-1} \beta_q z^{n-q} . \tag{7.2.18}$$

For several values of J, the coefficients are provided in Table 7.1.

The first stage consists of solving the explicit problem

$$\frac{1}{\Delta t}\left(\hat{\mathbf{u}}^{n+1} - \sum_{q=0}^{J-1} \alpha_q \mathbf{u}^{n-q}\right) = \mathbf{f}^{n+1} - \sum_{q=0}^{J-1} \beta_q[(\mathbf{u}\cdot\nabla)\mathbf{u}]^{n-q} \quad \text{in } \Omega . \tag{7.2.19}$$

The second stage is the projection step

$$\begin{aligned}
\frac{1}{\Delta t}\left(\hat{\hat{\mathbf{u}}}^{n+1} - \hat{\mathbf{u}}^{n+1}\right) + \nabla p^{n+1} &= 0 & \text{in } \Omega, \\
\nabla \cdot \hat{\hat{\mathbf{u}}}^{n+1} &= 0 & \text{in } \Omega, \\
\hat{\hat{\mathbf{u}}}^{n+1} \cdot \mathbf{n} &= 0 & \text{on } \partial\Omega_0 .
\end{aligned} \tag{7.2.20}$$

The third stage is the diffusion step

$$\begin{aligned}
\frac{1}{\Delta t}(\gamma_0 \mathbf{u}^{n+1} - \hat{\hat{\mathbf{u}}}^{n+1}) - \nu\triangle\mathbf{u}^{n+1} &= 0 & \text{in } \Omega, \\
\mathbf{u}^{n+1} &= 0 & \text{on } \partial\Omega_0 .
\end{aligned} \tag{7.2.21}$$

Since $\hat{\hat{\mathbf{u}}}^{n+1}$ is divergence-free (which amounts to splitting $\hat{\mathbf{u}}^{n+1}$ in (7.2.20) according to the Helmholtz decomposition principle), equation (7.2.20) yields

$$\triangle p^{n+1} = \nabla \cdot \left(\frac{\hat{\mathbf{u}}^{n+1}}{\Delta t}\right) \quad \text{in } \Omega . \tag{7.2.22}$$

Taking the normal component on $\partial\Omega_0$ of the vector equation (7.2.20) would yield the Neumann boundary condition on p^{n+1}

$$\frac{\partial p}{\partial n}^{n+1} = \mathbf{f}^{n+1} \cdot \mathbf{n} - [\mathbf{u}\cdot\nabla\mathbf{u}]_{\text{ext}}^{n+1} \cdot \mathbf{n} .$$

However, another strategy would be to use a so-called *consistent* boundary condition: one takes the normal component of the exact momentum equation (3.1.1), then uses the identity (7.2.14) and obtains (see Karniadakis and Sherwin (2005), p. 428)

$$\frac{\partial p^{n+1}}{\partial n} = \mathbf{f}^{n+1} \cdot \mathbf{n} - \left\{ \frac{\partial \mathbf{u}^{n+1}}{\partial t} + [(\mathbf{u} \cdot \nabla)\mathbf{u}]_{\text{ext}}^{n+1} + \nu \sum_{q=0}^{J-1} \beta_q (\nabla \times \boldsymbol{\omega})^{n-q} \right\} \cdot \mathbf{n} ,$$

$$(7.2.23)$$

where the formula (7.2.18) has been used to approximate the diffusion and convection terms. In the case of homogeneous Dirichlet conditions, the term $\frac{\partial \mathbf{u}}{\partial t}^{n+1}$ is null; otherwise, it is evaluated analytically by taking the exact derivative of the Dirichlet data. This still yields a pressure-correction scheme in rotational form which in some literature is named the *KIO scheme* (from Karniadakis, Israeli and Orszag (1991)). Once p^{n+1} is computed, $\hat{\mathbf{u}}^{n+1}$ can be retrieved from (7.2.20). Equation (7.2.21) is an elliptic equation for $\mathbf{u}^{n+1}$. Note that with this scheme the end-of-step velocity $\mathbf{u}^{n+1}$ satisfies the complete boundary conditions but is not divergence-free (unlike $\hat{\mathbf{u}}^{n+1}$).

A two-stage variant is obtained by merging (7.2.21) and (7.2.19), giving rise to the following scheme:

$$\frac{1}{\Delta t} \left(\gamma_0 \hat{\mathbf{u}}^{n+1} - \sum_{q=0}^{J-1} \alpha_q \mathbf{u}^{n-q} \right) - \nu \triangle \hat{\mathbf{u}}^{n+1}$$

$$= \mathbf{f}^{n+1} - \sum_{q=0}^{J-1} \beta_q [(\mathbf{u} \cdot \nabla)\mathbf{u}]^{n-q} \quad \text{in } \Omega ,$$

$$(7.2.24)$$

$$\hat{\mathbf{u}}^{n+1} = \mathbf{0} \quad \text{on } \partial\Omega_0 ,$$

$$\frac{1}{\Delta t} \left(\mathbf{u}^{n+1} - \hat{\mathbf{u}}^{n+1} \right) + \nabla p^{n+1} = \mathbf{0} \quad \text{in } \Omega ,$$

$$\nabla \cdot \mathbf{u}^{n+1} = 0 \quad \text{in } \Omega , \qquad (7.2.25)$$

$$\mathbf{u}^{n+1} \cdot \mathbf{n} = 0 \quad \text{on } \partial\Omega_0 .$$

The Darcy problem (7.2.25) is still reducible to the pressure Poisson problem (7.2.22) with the Neumann condition (7.2.23).

An incremental pressure-correction version is one that introduces a pressure term in (7.2.24) and a pressure increment in (7.2.25), e. g.,

$$\frac{1}{\Delta t} \left(\gamma_0 \hat{\mathbf{u}}^{n+1} - \sum_{q=0}^{J-1} \alpha_q \mathbf{u}^{n-q} \right) - \nu \triangle \hat{\mathbf{u}}^{n+1} + (\nabla p)_{\text{ext}}^{n+1}$$

$$= \mathbf{f}^{n+1} - \sum_{q=0}^{J-1} \beta_q [(\mathbf{u} \cdot \nabla)\mathbf{u}]^{n-q} \quad \text{in } \Omega ,$$

$$\hat{\mathbf{u}}^{n+1} = \mathbf{0} \quad \text{on } \partial\Omega_0 ,$$

$$\frac{1}{\Delta t}\left(\mathbf{u}^{n+1} - \hat{\mathbf{u}}^{n+1}\right) + \nabla(p^{n+1} - p^n) = -\nu\nabla(\nabla \cdot \hat{\mathbf{u}}^{n+1}) \qquad \text{in } \Omega\,,$$

$$\nabla \cdot \mathbf{u}^{n+1} = 0 \qquad \text{in } \Omega\,,$$

$$\mathbf{u}^{n+1} \cdot \mathbf{n} = 0 \qquad \text{on } \partial\Omega_0\,.$$

The end-of-step velocity $\mathbf{u}^{n+1}$ is now divergence-free; however, the tangential component of its trace on $\partial\Omega_0$ is not zero. On the other hand, the intermediate velocity field $\hat{\mathbf{u}}^{n+1}$ is not divergence-free, but it satisfies the full Dirichlet conditions. In fact, the two sequences $\{\mathbf{u}^n\}$ and $\{\hat{\mathbf{u}}^n\}$ satisfy the same kind of error estimates; hence, either one can be retained as a suitable candidate to approximate the exact solution $\{\mathbf{u}(t^n)\}$.

As a general matter, the issue of treating boundary conditions of Neumann or mixed Dirichlet–Neumann type in the framework of fractional-step methods is non-trivial.

An alternative approach to operator splitting relies on the algebraic manipulation of the linear system arising from the spatial and temporal discretizations of the Navier–Stokes equations. This yields the so-called *algebraic factorization* methods. They allow a more flexible handling of boundary conditions since the latter are directly accounted for in the original unsplit spatial discretization. We postpone their presentation until Sect. 7.4, after the linear systems on which they operate have been introduced in the forthcoming section.

7.3 Solution of the Algebraic System Associated with the Generalized Stokes Problem

In Sect. 3.7.2 we have analyzed several spectral discretizations of the Stokes equations in single domains with several kinds of boundary conditions (periodic, Dirichlet, Neumann, mixed). Galerkin or G-NI (Galerkin with numerical integration) methods yield discrete problems that can be written in the general form (3.7.20), whose algebraic counterpart is (3.7.29). Indeed, (3.7.29) is a particular case of an algebraic system that can be written as

$$\mathcal{A}\begin{pmatrix}\mathbf{u}\\\mathbf{p}\end{pmatrix} = \begin{pmatrix}\mathbf{f}\\\mathbf{g}\end{pmatrix}, \quad \text{with} \quad \mathcal{A} = \begin{pmatrix}H_N & G_N\\D_N & \mathbf{0}\end{pmatrix}. \tag{7.3.1}$$

This is a general form of a system associated with the finite-dimensional discretization (by finite elements, finite volumes, spectral methods, ...) of saddle-point problems. Here the vectors $\mathbf{f}$ and $\mathbf{g}$ collect the contributions of forcing terms and boundary conditions. The block G_N is associated with the discretization of the (pressure) gradient, D_N with the (velocity) divergence. The matrix H_N depends on the acceleration, diffusion and (possibly) linearized convection terms. For instance, when we use the time-discretization methods (3.2.24) or (3.2.25) (both with an explicit treatment of the convective nonlinear term), the matrix H_N represents the spatial discretization of the Helmholtz operator and reads

$$H_N = \frac{\delta}{\Delta t} M_N + L_N, \qquad (7.3.2)$$

with $\delta = 1$ for (3.2.24) and $\delta = 11/6$ for (3.2.25).

In particular, (7.3.1) encompasses all spectral discretizations, including collocation, described in Sect. 3.5.1 whose general form (prior to time-discretization) is given in (3.5.4)–(3.5.5). The presence of one (or several) directions of periodicity does not create any additional trouble.

The form (7.3.1) also includes all the multidomain spectral discretization methods (like the SEM, MEM, SDGM, and their -NI variants with numerical integration) introduced and investigated in Chap. 5.

For the solution of (7.3.1), iterative methods are generally preferred to monolithic direct methods. In particular, if H_N is symmetric, the conjugate gradient method is the method of choice, except for systems of very moderate size. Alternatives include the GMRES method and the Bi-CGSTAB method (see Appendix C). In such circumstances, the challenging point is finding a suitable preconditioner.

In the remaining of this section we discuss two different approaches to solving (7.3.1). The first one, detailed in the subsection below, is based on using preconditioned iterative techniques that act on this coupled system. The second one, presented in Sect. 7.3.2, works on the Schur complement system for the pressure, which is obtained by formally eliminating the velocity unknowns in (7.3.1). Both approaches may take advantage of the domain decomposition philosophy by employing preconditioners built on a partition of the computational domain into subdomains. This topic is addressed in Sect. 7.3.3. Finally, there is a third approach to the solution of (7.3.1) that consists of replacing the block matrix of the system by an inexact LU decomposition, and then solving the L and U steps. Such an approach shares many similarities with the operator-splitting approach described earlier in this chapter, and, as already anticipated, leads to the algebraic factorization methods. Sect. 7.4 will be devoted to this topic.

7.3.1 Preconditioners for the Generalized Stokes Matrix $\mathcal{A}$

We consider the system (7.3.1) under the assumption $D_N = G_N^T$, that is, with the block matrix

$$\mathcal{A} = \begin{pmatrix} H_N & G_N \\ G_N^T & 0 \end{pmatrix}. \qquad (7.3.3)$$

Direct methods are in general not advisable due to stability issues (arising from the indefiniteness of A) and to the large amount of fill-in when forming the lower and upper (L and U) factors of $\mathcal{A}$.

Iterative methods are more popular, but they call for effective preconditioners. Options include multigrid preconditioners, or more frequently, block preconditioners based on the so-called *pressure Schur complement matrix*

$$S_N = G_N^T H_N^{-1} G_N \,. \tag{7.3.4}$$

In some circumstances, typically when stabilization techniques are used for the continuity equation, the zero block of $\mathcal{A}$ is indeed replaced by a nonnull matrix, say C_N. Then, the pressure Schur complement matrix S_N would become

$$S_N = G_N^T H_N^{-1} G_N - C_N \,. \tag{7.3.5}$$

Otherwise, what follows remains unchanged.

Quite often, block preconditioners are derived by noticing that $\mathcal{A}$ admits the following LDU factorization:

$$\mathcal{A} = \begin{pmatrix} I & 0 \\ G_N^T H_N^{-1} & I \end{pmatrix} \begin{pmatrix} H_N & 0 \\ 0 & -S_N \end{pmatrix} \begin{pmatrix} I & H_N^{-1} G_N \\ 0 & I \end{pmatrix} \,.$$

The preconditioners that we are going to derive depend on how the LDU decomposition is grouped and how approximations to the Schur complement are created. One preconditioner family is based on the grouping $L(DU)$, the other on the grouping $(LD)U$ (and is called of pressure-correction type). We start with the first family and consider therefore the factorization

$$\mathcal{A} = \begin{pmatrix} I & 0 \\ G_N^T H_N^{-1} & I \end{pmatrix} \begin{pmatrix} H_N & G_N \\ 0 & -S_N \end{pmatrix} = L(DU) \,.$$

The two matrices

$$P_D = \begin{pmatrix} H_N & 0 \\ 0 & -S_N \end{pmatrix} (= D) \quad \text{and} \quad P_T = \begin{pmatrix} H_N & G_N \\ 0 & -S_N \end{pmatrix} (= DU)$$

represent two optimal preconditioners of $\mathcal{A}$, one block diagonal, the other block triangular. However, both are too expensive due to the presence of the Schur complement S_N. Instead, one could use

$$\hat{P}_D = \begin{pmatrix} \hat{H}_N & 0 \\ 0 & -\hat{S}_N \end{pmatrix} \quad \text{or} \quad \hat{P}_T = \begin{pmatrix} \hat{H}_N & G_N \\ 0 & -\hat{S}_N \end{pmatrix}$$

where $\hat{H}_N$ and $\hat{S}_N$ represent suitable (inexpensive) spectrally equivalent preconditioners of H_N and S_N, respectively.

In the time-independent case, $\hat{H}_N$ can be one of the optimal preconditioners of L_N that we have introduced in CHQZ2, Chap. 5 in the single-domain case (e. g., the finite-element preconditioners), and in Chaps. 5 and 6 in the domain decomposition case; see Sects. 5.3.5 and 6.2. As for $\hat{S}_N$, several preconditioners of S_N that are effective also in the time-dependent case, are illustrated in Sect. 7.3.2.

The second family is based on the factorization

$$A = \begin{pmatrix} H_N & 0 \\ G_N^T & -S_N \end{pmatrix} \begin{pmatrix} I & H_N^{-1}G_N \\ 0 & I \end{pmatrix} = (LD)U . \tag{7.3.6}$$

This factorization will be extensively addressed in Sect. 7.3.2. It inspires a classical preconditioner, the so-called *SIMPLE (left) preconditioner* (Patankar (1980))

$$P_{\text{SIMPLE}} = \begin{pmatrix} H_N & 0 \\ G_N^T & -\hat{S}_N \end{pmatrix} \begin{pmatrix} I & \hat{D}^{-1}G_N \\ 0 & I \end{pmatrix} = (LD)\hat{U} \tag{7.3.7}$$

which is based on replacing H_N with its diagonal $\hat{D}$ and S_N with $\hat{S}_N = G_N^T \hat{D}^{-1} G_N$. Two variants, SIMPLEC and SIMPLER, are available; see, e. g., Wesseling (2001). P_{SIMPLEC} is obtained from (7.3.7) by replacing $\hat{D}$ with the absolute value of the row sums of H_N, whereas

$$P_{\text{SIMPLER}} = \begin{pmatrix} I & 0 \\ G_N^T \hat{D}^{-1} & I \end{pmatrix} \begin{pmatrix} H_N & G_N \\ 0 & -\hat{S}_N \end{pmatrix} .$$

In particular, when H_N is symmetric,

$$P_{\text{SIMPLER}} = P_{\text{SIMPLE}}^T \hat{U}^T E \hat{U} P_{\text{SIMPLE}} ,$$

where E is the diagonal of $(LD) + (LD)^T - A$. Note that the latter preconditioner can be regarded as an approximation of the $L(DU)$ grouping (7.3.5).

Another preconditioner reads

$$P_{ac} = \begin{pmatrix} I & -\omega G_N \\ 0 & I \end{pmatrix} \begin{pmatrix} H_N + \omega G_N G_N^T & 0 \\ 0 & -I/\omega \end{pmatrix} \begin{pmatrix} I & 0 \\ -\omega G_N^T & I \end{pmatrix} ,$$

where $\omega > 0$ is a small parameter. P_{ac} can be regarded as an inexact approximation to the stabilized matrix

$$A_\omega = \begin{pmatrix} H_N & G_N \\ G_N^T & -I/\omega \end{pmatrix} .$$

Note that this is known as an *artificial compressibility* approximation due to the perturbation $-(1/\omega)p_N$ that appears in the continuity equation. Note, moreover, that the term $G_N G_N^T$ represents the discrete counterpart of the differential operator *grad · div*.

In the case of nonsymmetric H_N, Bai et al. (2003) and Benzi, Golub and Liesen (2005) have proposed a preconditioner based on rewriting the generalized Stokes system after changing the sign in the continuity equation, yielding

$$A = \begin{pmatrix} H_N & G_N \\ -G_N^T & 0 \end{pmatrix} .$$

This matrix features eigenvalues with positive real sign.

Let us denote by $H_N^s = (H_N + H_N^T)/2$ the *symmetric part* of H_N, and by $H_N^{ss} = (H_N - H_N^T)/2$ the *skew-symmetric part* of H_N. Then, one sets

$$\mathcal{H} = \begin{pmatrix} H_N^s & 0 \\ 0 & 0 \end{pmatrix}, \qquad \mathcal{S} = \begin{pmatrix} H_N^{ss} & G_N \\ -G_N^T & 0 \end{pmatrix},$$

and introduces the so-called HSS (*Hermitian and Skew-Hermitian Splitting*) preconditioner

$$P_\alpha = \frac{1}{2\alpha}(\mathcal{H} + \alpha I)(\mathcal{S} + \alpha I)$$

where $\alpha > 0$ is a given parameter. We refer to Simoncini and Benzi (2004/05) for the analysis.

7.3.2 Conditioning and Preconditioning for the Pressure Schur Complement Matrix

Several solution approaches lead to the so-called pressure Schur complement system

$$D_N H_N^{-1} G_N \mathbf{p} = \boldsymbol{\varphi} \ (= D_N H_N^{-1} \mathbf{f} - \mathbf{g}) , \qquad (7.3.8)$$

which is obtained from (7.3.1) upon eliminating $\mathbf{u}$. For time-dependent problems, the right-hand side $\boldsymbol{\varphi}$ depends on the time-level, and (7.3.8) has to be faced at every time-level.

For Galerkin and G-NI Legendre approximations, $D_N = G_N^T$, and (7.3.8) takes the simpler form

$$S_N \mathbf{p} = \boldsymbol{\varphi}, \quad \text{with} \quad S_N = G_N^T H_N^{-1} G_N . \qquad (7.3.9)$$

The pressure Schur complement matrix S_N (introduced in (7.3.4)), sometimes called the Uzawa matrix, is symmetric provided H_N is symmetric (which is the case of, e. g., (7.2.8)). Moreover, since H_N is positive semi-definite, S_N is at least positive semi-definite. Indeed, for any $\mathbf{q}$,

$$\mathbf{q}^T S_N \mathbf{q} = (G_N \mathbf{q})^T H_N^{-1} G_N \mathbf{q} \geq 0 .$$

Thus, S_N is positive definite (up to a constant vector) if and only if there are no spurious pressure modes. This conclusion is drawn in Sect. 3.7 in the framework of the analysis of the so-called inf-sup condition.

Under the previous assumptions, it is advisable to solve (7.3.9) by an iterative method (especially if the dimension of S_N is not too small), and in particular by the conjugate gradient method owing to the symmetry. In that case, the use of a convenient preconditioner is mandatory in view of the poor conditioning of S_N, as illustrated in the upper curve of Fig. 7.2 for a Legendre G-NI method. In this respect, it is useful to address the extreme situations in which $1/\Delta t$ is either very small (or even zero, in the time-independent case) or very large.

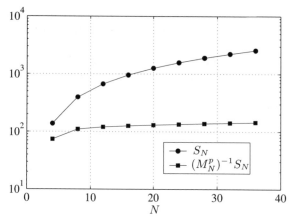

Fig. 7.2. The spectral condition number for the pressure Schur complement $S_N = G_N^T L_N^{-1} G_N$ (upper curve) and the preconditioned pressure Schur complement $(M_N^p)^{-1} S_N$, versus the polynomial degree N

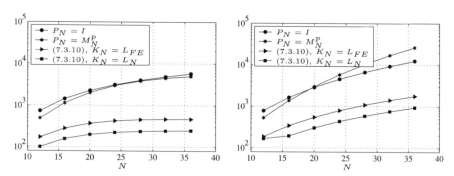

Fig. 7.3. The spectral condition number for the preconditioned pressure Schur complement $P^{-1} S_N$ for different preconditioners and $\delta/\Delta t = 10$ (*left*) and $\delta/\Delta t = 1000$ (*right*)

In the former case, since $H_N = L_N$, the analysis of the spectral properties of S_N can be deduced from the discussion in Sect. 3.5.1. The conclusion is that the pressure mass matrix M_N^p is a convenient preconditioner provided the inf-sup constant β_N is independent of N (or moderately decaying with N). Indeed, its spectral condition number

$$\kappa \left((M_N^p)^{-1} G_N^T L_N^{-1} G_N \right)$$

is proportional to β_N^{-2}, as shown in Sect. 3.7.4 (see Fig. 3.29, left, lower curve).

When $1/\Delta t$ is large (which is the typical case for time-discretization methods) this interesting property no longer holds. This is due to the fact that in this case the matrix S_N approaches $\Delta t/\delta$ times the *pseudo-Laplacian* $\triangle_N = G_N^T M_N^{-1} G_N$. The condition number of the latter blows up as N^4 since this operator behaves like a second-order operator (see Chap. 5). Thus,

M_N^p cannot act as a suitable preconditioner since it approaches the identity operator. Rather, Δ_N can be better preconditioned by a matrix K_N that represents in turn an efficient preconditioner of L_N. For an efficient realization of this preconditioner we refer to Fischer and Rønquist (1994). At this stage, any intermediate case between the two extrema just discussed can be handled by using a "trade-off" preconditioner P_N such that

$$P_N^{-1} = (M_N^p)^{-1} + \frac{\delta}{\Delta t} K_N^{-1} , \qquad (7.3.10)$$

as advocated by Cahouet and Chabard (1988). A preconditioner of this form enjoys good preconditioning properties in all regimes. See also Bramble and Pasciak (1997). In Fig. 7.2 we report the spectral condition number of the preconditioned matrix $P_N^{-1} S_N$; S_N is the pressure Schur complement (7.3.4) with H_N given in (7.3.2), for different kinds of preconditioners. The left-hand figure refers to the case $\delta/\Delta t = 10$, the right-hand figure to $\delta/\Delta t = 1000$.

A different preconditioner for S_N is proposed in Elman et al. (2002):

$$P_N = L_N^p (H_N^p)^{-1} M_N^p ,$$

where L_N^p and H_N^p represent discrete approximations to scaled Laplacian and advection-diffusion operators defined on the pressure space. This, however, requires the set up of suitable "pressure boundary conditions" that have to be devised according to the physical boundary conditions on the (Navier-) Stokes problem at hand. For discussion and analysis we refer the reader to Elman et al. (2002).

Additional preconditioners can be constructed for S_N based on domain decomposition techniques. This is the subject of the next subsection.

7.3.3 Domain Decomposition Preconditioners for the Stokes and Navier–Stokes Equations

We consider the approximation of the Stokes equations (5.6.1) provided by either the SEM problem (5.6.11) or its SEM-NI variant (5.6.12). The associated linear system takes the blockwise form

$$\begin{pmatrix} H_N & G_N \\ G_N^T & 0 \end{pmatrix} \begin{pmatrix} \mathbf{u} \\ \mathbf{p} \end{pmatrix} = \begin{pmatrix} \mathbf{f} \\ \mathbf{0} \end{pmatrix} . \qquad (7.3.11)$$

We recall that the inf-sup condition in matrix form is

$$\mathbf{q}^T G_N^T H_N^{-1} G_N \mathbf{q} \geq \beta_N^2 \mathbf{q}^T M_N^p \mathbf{q} \qquad \forall \mathbf{q} ,$$

where M_N^p is the pressure mass matrix. This matrix was introduced in Sect. 7.3.2 for the single-domain case. As in the single-domain situation, β_N^2 scales as $\lambda_{\min} ((M_N^p)^{-1} G_N^T H_N^{-1} G_N)$.

One way of solving (7.3.11) consists of eliminating $\mathbf{u}$ and getting the *pressure Schur complement system* $G_N^T H_N^{-1} G_N \mathbf{p} = G_N^T H_N^{-1} \mathbf{f}$ that we have already introduced in Sect. 7.3.2; see (7.3.9). This system can be solved by applying an iterative method (e. g.,the conjugate gradient method) and using a domain decomposition algorithm in the stage of computing H_N^{-1}.

Another approach consists of building up a Schur complement problem for all the degrees of freedom sitting at the interface nodes. For that, we reorder the vector of unknowns in the following sequence: $\mathbf{u}_I$ interior velocities, $\mathbf{p}_I$ interior pressures with zero averages, $\mathbf{u}_\Gamma$ interface velocities, $\mathbf{p}_0$ constant pressures in each Ω_m, $m = 1, \ldots, M$, M being the number of subdomains that define a disjoint partition of Ω. By reordering accordingly the matrix in (7.3.11) as

$$
\begin{pmatrix} H_N & G_N \\ G_N^T & 0 \end{pmatrix} = \begin{pmatrix} (H_N)_{II} & (G_N)_{II} & \vdots & (H_N)_{\Gamma I}^T & 0 \\ (G_N)_{II}^T & 0 & \vdots & (G_N)_{I\Gamma}^T & 0 \\ \cdots & \cdots & \cdots & \cdots \\ (H_N)_{\Gamma I} & (G_N)_{I\Gamma} & \vdots & (H_N)_{\Gamma\Gamma} & (G_N)_0 \\ 0 & 0 & \vdots & (G_N)_0^T & 0 \end{pmatrix} = \begin{pmatrix} K_{II} & K_{\Gamma I}^T \\ K_{\Gamma I} & K_{\Gamma\Gamma} \end{pmatrix}
$$

and eliminating the interior unknowns, we obtain the global (velocity and pressure) Schur complement system (in saddle-point form)

$$
\Sigma \begin{pmatrix} \mathbf{u}_\Gamma \\ \mathbf{p}_0 \end{pmatrix} = \begin{pmatrix} \tilde{\mathbf{f}} \\ 0 \end{pmatrix} , \tag{7.3.12}
$$

where

$$
\Sigma = K_{\Gamma\Gamma} - K_{\Gamma I} K_{II}^{-1} K_{\Gamma I}^T = \begin{pmatrix} \Sigma_\Gamma & (G_N)_0 \\ (G_N)_0^T & 0 \end{pmatrix}
$$

and

$$
\begin{pmatrix} \tilde{\mathbf{f}} \\ 0 \end{pmatrix} = \begin{pmatrix} \mathbf{f}_\Gamma \\ 0 \end{pmatrix} - K_{\Gamma I} K_{II}^{-1} \begin{pmatrix} \mathbf{f}_I \\ 0 \end{pmatrix} .
$$

The global Schur complement system can be solved by a (Krylov) iterative method with a suitable domain decomposition preconditioner (see below).

After computing $\mathbf{u}_\Gamma$ and $\mathbf{p}_0$, the internal variables $\mathbf{u}_I$ and $\mathbf{p}_I$ can be obtained by solving the system

$$
K_{II} \begin{pmatrix} \mathbf{u}_I \\ \mathbf{p}_I \end{pmatrix} = \begin{pmatrix} \mathbf{b}_I \\ 0 \end{pmatrix} - K_{\Gamma I}^T \begin{pmatrix} \mathbf{u}_\Gamma \\ \mathbf{p}_0 \end{pmatrix} = \begin{pmatrix} \mathbf{b}_I - (H_N)_{\Gamma I}^T \mathbf{u}_\Gamma \\ -(G_N)_{I\Gamma}^T \mathbf{u}_\Gamma \end{pmatrix} .
$$

Note that $\mathbf{u}_I$ and $\mathbf{p}_I$ do not depend on $\mathbf{p}_0$.

Using a further permutation that reorders, subdomain per subdomain, the interior velocities and pressures we notice that K_{II}^{-1} represents the solution of

M independent Stokes problems, one for each subdomain Ω_m with Dirichlet data on $\partial\Omega_m$. Thus,

$$K_{II}^{-1} = \begin{pmatrix} \left(K_{II}^{(1)}\right)^{-1} & & 0 \\ & \ddots & \\ 0 & & \left(K_{II}^{(M)}\right)^{-1} \end{pmatrix} .$$

Consequently, computing the action of Σ on a given vector $\mathbf{v}$ (to get the residual in an iterative procedure) involves the solution of M decoupled Stokes problems. Equivalently, $\Sigma\mathbf{v}$ is computed by subassembling the actions of the subdomain Schur complements Σ_m defined for Ω_m by

$$\Sigma_m = K_{\Gamma\Gamma}^{(m)} - K_{\Gamma I}^{(m)} \left(K_{II}^{(m)}\right)^{-1} K_{\Gamma I}^{(m)} = \begin{pmatrix} \Sigma_{\Gamma}^{(m)} & (G_N)_0^{(m)^T} \\ (G_N)_0^{(m)^T} & 0 \end{pmatrix} ,$$

with obvious choice of notation.

The following block-diagonal and lower block-triangular preconditioners for Σ are proposed by Pavarino and Widlund (1999):

$$P_D = \begin{pmatrix} \hat{\Sigma}_{\Gamma} & 0 \\ 0 & (\hat{G}_N)_0 \end{pmatrix} , \qquad P_T = \begin{pmatrix} \hat{\Sigma}_{\Gamma} & 0 \\ (G_N)_0^T & -\hat{M}_0 \end{pmatrix} , \qquad (7.3.13)$$

where $\hat{\Sigma}_{\Gamma}$ and $\hat{M}_0$ denote good preconditioners for Σ_{Γ} and the coarse pressure mass matrix M_0, respectively. One possible choice of $\hat{\Sigma}_{\Gamma}$ is provided by the matrix

$$\hat{\Sigma}_{\Gamma} = \text{blockdiag}\left(\hat{\Sigma}_{WB}, \hat{\Sigma}_{WB}, \hat{\Sigma}_{WB}\right) ,$$

where $\hat{\Sigma}_{WB}$ is a scalar *wire-basket preconditioner* for the Schur complement matrix associated with the Laplace equation (see Pavarino and Widlund (1996)).

Another possibility consists of using

$$\hat{\Sigma}_{\Gamma} = \text{blockdiag}\left(\hat{\Sigma}_{NN}, \hat{\Sigma}_{NN}, \hat{\Sigma}_{NN}\right) ,$$

where $\hat{\Sigma}_{NN} = P_{*\text{coarse}}^{NN}$ is the scalar Neumann–Neumann preconditioner for the Schur complement matrix associated with the Laplace equation (see (6.4.40)).

In both cases, the preconditioned matrices $(P^D)^{-1}\Sigma$ have a condition number that is bounded by $C(1 + \log N)^2 \beta_N^{-1}$, provided $\hat{M}_0$ is equal or spectrally equivalent to M_0 for a suitable constant C independent of N and M. For the analysis (and more details), see Pavarino and Widlund (1999) and Klawonn (1998a, 1998b).

Another example of a Neumann–Neumann preconditioner that is built up directly on the saddle-point problem (7.3.12) is proposed in Pavarino and Widlund (2002). The preconditioned Schur complement matrix reads $T = T_0 + (I - T_0) \sum_{m=1}^{M} T_m (I - T_0)$, where T_0 involves the solution of a coarse Stokes problem in Ω while T_m that of a local Stokes problem in Ω_m. The condition number of T is bounded by $C(1 + \beta_0^{-1})\beta_N^{-2}(1 + \log N)^2$, where β_0 is the inf-sup constant of the coarse Stokes problem.

An alternative approach consists of constructing a preconditioner of the global Stokes matrix in (7.3.11). Following Klawonn and Pavarino (2000), the simplest way is to use a block-diagonal preconditioner P_D such that

$$P_D^{-1} = \begin{pmatrix} \hat{H}_N^{-1} & 0 \\ 0 & \hat{M}_N^{-1} \end{pmatrix} , \tag{7.3.14}$$

where $\hat{H}_N$ and $\hat{M}_N$ are preconditioners of H_N and M_N^p, respectively.

The diagonal blocks can be, for instance, defined as overlapping additive Schwarz preconditioners

$$\hat{H}_N^{-1} = R_{H,u}^T A_H^{-1} R_H + \sum_{m=1}^{M} R_{m,u}^T A_m^{-1} R_{m,H} ,$$

$$\hat{M}_N^{-1} = R_{H,p}^T M_H^{-1} R_{H,p} + \sum_{m=1}^{M} R_{m,p}^T M_m^{-1} R_{m,p} ,$$

where

$$A_m = R_{m,u} H_N R_{m,u}^T , \qquad M_m = R_{m,p} M_N^p R_{m,p}^T ,$$

and $R_{m,u}$ and $R_{m,p}$ denote restriction operators in Ω_m acting, respectively, on the degrees of freedom of velocity and pressure; A_H and M_H denote the velocity stiffness matrix and the pressure mass matrix acting on the coarse decomposition.

A second preconditioner has the block-triangular form

$$P_T^{-1} = \begin{pmatrix} \hat{H}_N & 0 \\ B & -\hat{M}_N \end{pmatrix}^{-1} = \begin{pmatrix} \hat{H}_N^{-1} & 0 \\ \hat{M}_N^{-1} G_N \hat{H}_N^{-1} & -\hat{M}_N^{-1} \end{pmatrix} .$$

If $\hat{H}_N$ and $\hat{M}_N$ are chosen as in (7.3.14), the action of P_T^{-1} on a vector is only marginally more expensive than that of P_D^{-1}; however, P_T is no longer symmetric.

Following Klawonn and Pavarino (1998), another possibility consists of applying overlapping Schwarz preconditioners to the whole saddle-point matrix in (7.3.11). Unlike the block preconditioners P_D and P_T that eliminate all the coupling or half the coupling, respectively, global preconditioners maintain all the coupling between velocities and pressures in the local and coarse problems. Global Schwarz preconditioners have the form

$$P^{-1} = R_H^T K_H^{-1} R_H + \sum_{m=1}^{M} R_m^T K_m^{-1} R_m , \tag{7.3.15}$$

where the matrices $R_m^T K_m^{-1} R_m$ represent the local Stokes problem, while $R_H^T K_H^{-1} R_H$ (the coarse term) represents a global Stokes problem.

Finally, we mention the approach by Fischer (1997) and Rønquist (1999), which consists of applying additive Schwarz preconditioners as in Sect. 6.3.2 directly to the pressure Schur complement system. See also Sect. 7.4.3.

7.4 Algebraic Factorization Methods

Algebraic factorization methods, also called algebraic fractional-step methods or algebraic splitting methods, originate from replacing the matrix $\mathcal{A}$ in (7.3.1), or, more precisely, the one in (7.3.3), by an inexact factorization, then re-interpreting the associated problem as a fractional-step method carried out on the given algebraic system. To start with, let us reconsider the exact block-LU factorization of the matrix $\mathcal{A}$ given in (7.3.6). Exploiting this factorization for solving system (7.3.1) through a separate computation of velocity and pressure corresponds actually to the Uzawa methods described in Sect. 7.3. A reduction in computational cost can be achieved by using an approximation of (7.3.3). A quite general framework is given by the following *inexact block factorization* (see Quarteroni, Saleri and Veneziani (2000)):

$$\widehat{\mathcal{A}} = \begin{pmatrix} H_N & 0 \\ G_N^T & -G_N^T \mathcal{L}_N G_N \end{pmatrix} \begin{pmatrix} I & \mathcal{U}_N G_N \\ 0 & I \end{pmatrix}. \qquad (7.4.1)$$

where $\mathcal{L}_N$ and $\mathcal{U}_N$ represent two (possibly different) approximations of H_N^{-1}. $\widehat{\mathcal{A}}$ can be regarded as an approximation of the exact $(LD)U$ factorization of $\mathcal{A}$; see (7.3.4). For the algorithm to be efficient, it is essential that $\mathcal{L}_N$ be easier to invert than H_N. Stemming from this inexact factorization, the solution of problem (7.3.1) with $\widehat{\mathcal{A}}$ in place of $\mathcal{A}$ yields the solution of the two steps

L-step: $\begin{cases} H_N \widetilde{\mathbf{u}} = \mathbf{f}, & \text{(intermediate velocity)} \\ -G_N^T \mathcal{L}_N G_N \widetilde{\mathbf{p}} = \mathbf{g} - G_N^T \widetilde{\mathbf{u}}, & \text{(pressure)} \end{cases}$

U-step: $\mathbf{u} + \mathcal{U}_N G_N \mathbf{p} = \widetilde{\mathbf{u}}$ (end-of-step velocity).

7.4.1 Chorin–Temam and Yosida Algebraic Factorization Methods

In this framework, two main possibilities have been explored:

1. $\mathcal{L}_N = \mathcal{U}_N = (\frac{\delta}{\Delta t} M_N)^{-1}$,   (Chorin–Temam)
2. $\mathcal{L}_N = (\frac{\delta}{\Delta t} M_N)^{-1}$ and $\mathcal{U}_N = H_N^{-1}$ (Yosida).

The former choice (which corresponds to disregarding L_N in the computation of the inverse of H_N—see (7.3.2)) yields a scheme that can be regarded as the algebraic counterpart of the Chorin–Temam scheme because of the formal analogy with the original differential splitting method (see Perot (1993), Quarteroni et al. (2000)).

The latter case corresponds to the so-called *Yosida* method (see Quarteroni, Saleri and Veneziani (1999)). The main difference between these two possibilities is that in the former (algebraic Chorin–Temam) scheme the inexactness of the splitting affects only the (discrete) momentum equation; thus the solution satisfies discrete mass conservation exactly, whereas in the Yosida method the discrete momentum equation is exactly satisfied but the discrete continuity equation is perturbed.

In view of the Neumann expansion

$$
H_N^{-1} = \frac{\Delta t}{\delta}\left(I + \frac{\Delta t}{\delta}M_N^{-1}L_N\right)^{-1}M_N^{-1} = \sum_{j=0}^{\infty}\left(-\frac{\Delta t}{\delta}M_N^{-1}L_N\right)^{j}\left(\frac{\delta}{\Delta t}M_N\right)^{-1},
$$

we see that $(\delta/\Delta t M_N)^{-1}$ can be regarded as a first-order (in time) approximation of H_N^{-1}. The splitting error (that is the error between the exact solution of system (7.3.1) and the solution obtained using the inexact factorization) of these schemes is therefore limited to second order for the velocity, provided a minimally second order temporal discretization is used. A potential route to reducing the splitting error is to take more terms of the Neumann expansion. This strategy carries, however, an increase in the computational costs, and may lead to an unstable method (see Couzy (1995) and Veneziani (2003)).

In the case of the Yosida method, improvement in the accuracy can be obtained using the different strategy of considering the following generalization of the factorization (7.4.1) (see Saleri and Veneziani (2005), Gervasio, Saleri and Veneziani (2006)):

$$
\widehat{\mathcal{A}} = \begin{pmatrix} H_N & 0 \\ G_N^T & -\frac{\Delta t}{\delta}G_N^T M_N^{-1}G_N \end{pmatrix} \begin{pmatrix} I & H_N^{-1}G_N \\ 0 & Q_N \end{pmatrix}
$$

$$
= \begin{pmatrix} H_N & G_N \\ G_N^T & G_N^T(H_N^{-1}G_N - \frac{\Delta t}{\delta}M_N^{-1}G_N Q_N) \end{pmatrix},
$$

where Q_N is a square matrix (whose dimension is that of the discrete pressure variable) that is selected to increase the time accuracy of the corresponding splitting scheme. Note that the momentum equation has not changed (hence the linear momentum is still conserved). This factorization leads to algebraic splitting methods given by the following steps:

L-step $\begin{cases} H_N\widetilde{\mathbf{u}} = \mathbf{f}, & \text{(intermediate velocity)} \\ -\frac{\Delta t}{\delta}G_N^T M_N^{-1}G_N\widetilde{\mathbf{p}} = \mathbf{g} - G_N^T\widetilde{\mathbf{u}}, & \text{(intermediate pressure)} \end{cases}$

U-step $\begin{cases} Q_N\mathbf{p} = \widetilde{\mathbf{p}}, & \text{(end-of-step pressure)} \\ \mathbf{u} + H_N^{-1}G_N\mathbf{p} = \widetilde{\mathbf{u}} & \text{(end-of-step velocity)}. \end{cases}$

Observe that now also the pressure is first predicted and then corrected. For this reason the methods obtained exploiting this factorization are referred to as *pressure-correction* methods, and they can be considered as the algebraic counterparts of the differential method proposed by Timmermans for spectral element approximations (see Timmermans, Minev and Van De Vosse (1996) for the presentation of the method and Guermond and Shen (2003) for its stability analysis).

Q_N can be selected in order to minimize the splitting error associated with $\widehat{A}$. In Saleri and Veneziani (2005) the following expression for Q_N is proposed:

$$Q_N = B_N^{-1} S_N^{(\text{app})}, \tag{7.4.2}$$

where

$$B_N = G_N^T \frac{\Delta t}{\delta} M_N^{-1} H_N \frac{\Delta t}{\delta} M_N^{-1} G_N, \tag{7.4.3}$$

and

$$S_N^{(\text{app})} = G_N^T \frac{\Delta t}{\delta} M_N^{-1} G_N \tag{7.4.4}$$

is the approximate pressure Schur complement matrix. The Yosida method corresponding to (7.4.2) is accurate in time up to order three for the velocity provided a third-order BDF scheme is used to advance the momentum equation in time. In Gervasio et al. (2006), an improved version is proposed by replacing B_N in (7.4.2) by

$$\widehat{B}_N = B_N (S_N^{(\text{app})})^{-1} B_N + B_N - G_N^T \left(\frac{\Delta t}{\delta} M_N^{-1} H_N \right)^2 \frac{\Delta t}{\delta} M_N^{-1} G_N. \tag{7.4.5}$$

This choice gives rise to a fourth-order method provided it is associated with a fourth-order BDF scheme.

In these methods, the pressure-correction step requires the solution of systems whose matrix is the approximate Schur complement $S_N^{(\text{app})}$. This is evident when (7.4.2) is considered. The computation of $\mathbf{p}$ through (7.4.5) can be done with the following algorithm:

A step of the Yosida pressure-correction algorithm

For two given vectors $\mathbf{f}$ *and* $\mathbf{g}$

$$
i) \quad
\begin{aligned}
&solve && H_N \tilde{\mathbf{u}} = \mathbf{f} \\
&solve && -G_N^T \frac{\Delta t}{\delta} M_N^{-1} G_N \tilde{\mathbf{p}} = \mathbf{g} - G_N^T \tilde{\mathbf{u}} \\
&compute && \mathbf{z} = B_N \tilde{\mathbf{p}} \\
&solve && S_N^{(\text{app})} \mathbf{p} = \mathbf{z}
\end{aligned}
$$

$$
ii) \quad
\begin{aligned}
&compute && \mathbf{p}_B = B_N \mathbf{p} + \mathbf{z} - G_N^T \left(\frac{\Delta t}{\delta} M_N^{-1} H_N \right)^2 \frac{\Delta t}{\delta} M_N^{-1} G_N \tilde{\mathbf{p}} \\
&solve && S_N^{(\text{app})} \mathbf{p} = \mathbf{p}_B
\end{aligned}
$$

$$
iii) \quad solve \quad H_N (\mathbf{u} - \tilde{\mathbf{u}}) = -G_N \mathbf{p}.
$$

The entire step yields the fourth-order Yosida method, whereas dropping stage ii) produces the third-order Yosida method.

7.4.2 Numerical Results for Yosida Schemes

We will now present some numerical results for the second-order Yosida scheme as well as its variants (7.4.2), (7.4.3) and (7.4.4) of third order and (7.4.2), (7.4.4) and (7.4.5) of fourth order.

We consider the square domain $\Omega = (-1,1)^2$ and the Navier–Stokes problem with Dirichlet boundary conditions whose exact solution is

$$\mathbf{u}(x,y,t) = (\sin(x)\sin(y+t), \cos(x)\cos(y+t))^T ,$$
$$p(x,y,t) = \cos(x)\sin(y+t) . \tag{7.4.6}$$

We use a (single-domain) G-NI approximation for the space discretization based on the $\mathbb{Q}_N - \mathbb{Q}_{N-2}$ approach, together with Yosida schemes of order

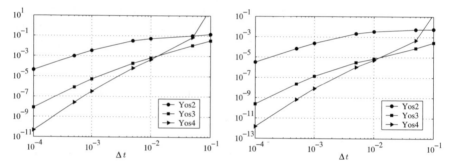

Fig. 7.4. The errors (versus Δt) $err(\mathbf{u})$ for the velocity (*left*) and $err(p)$ for the pressure (*right*), for $N = 16$, $\nu = 10^{-1}$. 'Yos2', 'Yos3' and 'Yos4' stand for the Yosida schemes of order 2, 3 and 4, respectively

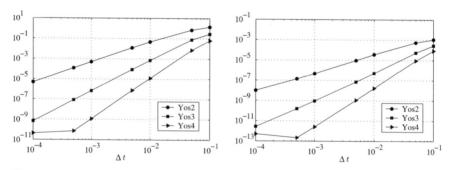

Fig. 7.5. The errors (versus Δt) $err(\mathbf{u})$ for the velocity (*left*) and $err(p)$ for the pressure (*right*), for $N = 16$, $\nu = 10^{-5}$. 'Yos2', 'Yos3' and 'Yos4' stand for the Yosida schemes of order 2, 3 and 4, respectively

two, three and four (Yos2, Yos3, and Yos4 in short). A BDF scheme of order two, three and four (respectively) is used for the time-discretization, together with an extrapolation scheme of order two, three and four (respectively) for the computation of the convective velocity. In Figs. 7.4 and 7.5 the errors in the $L^2_d(H^1)$-norm for the velocity (left) and the $L^2_d(L^2)$-norm for the pressure (right) are shown versus the time step Δt for the three different versions of the Yosida scheme. The polynomial degree is $N = 16$. Two different values of the viscosity have been considered: $\nu = 10^{-1}$ and $\nu = 10^{-5}$. Here L^2_d means the discrete L^2-norm on the time interval $(0, T)$ based on the composite mid point rule. Precisely,

$$err(\mathbf{u}) = \left(\sum_{n=0}^{N_{\Delta t}-1} \Delta t \|\mathbf{u}(t^n) - \mathbf{u}_N(t^n)\|^2_{H^1(\Omega)} \right)^{1/2} ,$$

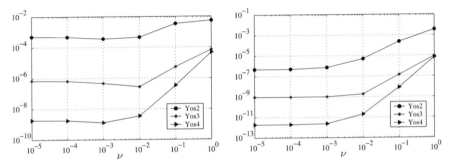

Fig. 7.6. The errors (versus ν) $err(\mathbf{u})$ for the velocity (*left*) and $err(p)$ for the pressure (*right*), for $N = 16$, $\Delta t = 10^{-3}$

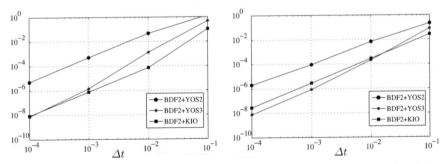

Fig. 7.7. Comparison between the fractional-step method (7.2.17)–(7.2.19), (7.2.23) and the Yosida algebraic factorization method, both associated with the BDF2 discretization method in time. Errors $err(\mathbf{u})$ (*left*) and $err(p)$ (*right*) at the time-level $T = 1$ are plotted versus the time step Δt. All curves on $err(\mathbf{u})$ decrease as Δt^2, those on $err(p)$ as $\Delta t^{3/2}$ for BDF2+YOS2, as Δt^2 for BDF2+YOS3 and BDF+KIO

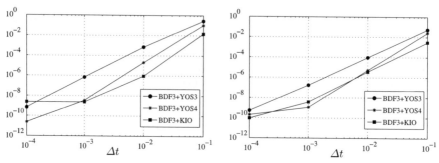

Fig. 7.8. Comparison between the fractional-step method (7.2.17)–(7.2.19), (7.2.23) and the Yosida algebraic factorization method, both associated with the BDF3 discretization method in time. Errors $err(\mathbf{u})$ (*left*) and $err(p)$ (*right*) at the time-level $T = 1$ are plotted versus the time step Δt. All curves on $err(\mathbf{u})$ decrease as Δt^3, those on $err(p)$ as $\Delta t^{5/2}$ for BDF3+YOS3, as Δt^3 for BDF3+KIO and BDF3+YOS4

while

$$err(p) = \left(\sum_{n=0}^{N_{\Delta t}-1} \Delta t \|p(t^n) - p_N(t^n)\|_{L^2(\Omega)}^2 \right)^{1/2} \, ,$$

where $N_{\Delta t}$ indicates the number of time-steps.

The same errors are plotted in Fig. 7.6 versus ν for $N = 16$ and $\Delta t = 10^{-3}$.

In Figs. 7.7 and 7.8 a comparison is made between the results obtained using the Yosida method of order 2 (BDF2+YOS2) and the fractional-step method (7.2.17)–(7.2.19), (7.2.23) (KIO scheme). The number of steps for the extrapolation is $J = 2$ for BDF2 (Fig. 7.7) and $J = 3$ for BDF3 (Fig. 7.8). The computational domain is $\Omega = (-1, 1)^2$, the spatial discretization is the G-NI method (a single element) using polynomials of degree $N = 16$ for the velocity field and $N = 2$ for the pressure. The exact solution is still (7.4.6), and the viscosity is $\nu = 10^{-5}$.

7.4.3 Preconditioners for the Approximate Pressure Schur Complement

The computational efficiency of the algebraic factorization methods depend primarily on the use of efficient solvers for systems like

$$S_N^{(\mathrm{app})} \mathbf{p} = \mathbf{p}_B \, . \tag{7.4.7}$$

The matrix $S_N^{(p)}$ is (a multiple of) the approximate pressure Schur complement $G_N^T M_N^{-1} G_N$, also called *consistent Poisson operator*. Efficient solvers for (7.4.7) use either Cholesky factorizations of $S_N^{(p)}$ computed in a pre-processing stage or preconditioned iterative methods.

Fischer and coworkers have investigated different approaches based on i) finite-element preconditioning, ii) two-level (i. e., with coarse grid) Schwarz preconditioning, and iii) the spectral element multigrid method (with several kinds of smoothers).

i) The former preconditioner is the usual finite-element stiffness matrix, say A_{fe}, using bilinear (in 2D) or trilinear (in 3D) piecewise polynomials; the grid points are those used for the spectral pressure field. This yields a preconditioned matrix like that denoted by P_3 in Sect. 5.3.5. In turn, at each conjugate gradient (CG) iteration the linear system associated with A_{fe} can be solved by either incomplete Cholesky factorization, the multifrontal method, or the overlapping Schwarz method described in Sect. 6.3.2.

ii) The two-level additive Schwarz preconditioner (see Fischer (1997)) is built up directly on the original spectral elements extended by adding a single layer of nodal points. As seen in Sect. 6.3.2 and Sect. 7.3.3 (see (7.3.15)), at each conjugate gradient (CG) iteration we have to compute first the spectral residual of the consistent Poisson problem, say $\mathbf{r}^{(k)} = \mathbf{p}_B - S_N^{(\mathrm{app})} \mathbf{p}^{(k)}$, and then compute the preconditioned residual

$$\mathbf{z}^{(k)} = P_{as,H}^{-1} \mathbf{r}^{(k)} = \sum_{m=1}^{M} R_m^T A_{m,fe}^{-1} R_m \mathbf{r}^{(k)} + R_H^T A_{H,fe}^{-1} R_H \mathbf{r}^{(k)} . \quad (7.4.8)$$

In the case for which the 4×4 LG nodes are used in each element for the pressure, the extended, overlapping subdomains are indicated in Fig. 7.9, left.

The local matrices $A_{m,fe}$ are the stiffness matrices in the extended elements $\tilde{\Omega}_m$ built on the internal grid points (with homogeneous Dirichlet

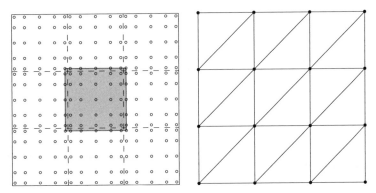

Fig. 7.9. Overlapping subdomains obtained by extension of the spectral elements. Indicated with an *empty circle* are the GL nodes used for the spectral pressure field. On the *right*, the *triangles* (and *grid points*) used for the coarse term of the Schwarz preconditioner are indicated

conditions at the boundary points). The global matrix $A_{H,fe}$ is the stiffness matrix built on the coarse grid (indicated on the right-hand side of
Fig. 7.9), with homogeneous Dirichlet boundary conditions on $\partial\Omega$.
The fine-scale errors are accounted for by the local solves, whereas the remaining errors, due to incorrect boundary conditions on the local solves,
are corrected by the coarse-grid solve.
Efficient construction of the local matrices $A_{m,fe}$ from the tensor product
of 1D finite-element functions based on the GL nodes for pressure is
addressed in Fischer, Miller and Tufo (2000). See also Deville, Fischer
and Mund (2002), Sect. 7.2.3.

iii) A further possibility is offered by the spectral element multigrid solver,
based on intra-element nested grids. Precisely, the number M of elements
is kept the same, whereas the polynomial degree in each element is reduced by two (say) when going from one level to the next coarser level,
$N \to N/2 \to \ldots$. Finite-element smoothers are not appropriate because
the aspect ratios of the finite elements get large near the boundary of the
spectral elements (due to the concentration of the GL, or LGL, nodes
near the element boundaries). A better smoother is represented by the
(one-level) Schwarz preconditioner (that in (7.4.8) without the coarse-
grid term). However, the local stiffness matrices on the extended (overlapping) elements is a spectral matrix, like in Casarin (1997b) (see also
Sect. 6.3.3) with a neighboring matrix (see Fischer and Lottes (2005)).

Bibliographical Notes and Concluding Remarks

Spectral Schwarz preconditioners for SEM with local grid refinement are analyzed in Pavarino (1994b); see also Korneev and Jensen (1997, 1999). Overlapping Schwarz methods for unstructured and hybrid spectral elements are
introduced and analyzed in Pavarino and Warburton (2000). The spectral
elements are affine images of a reference triangle or square in two dimensions, and of a reference tetrahedron, pyramid, prism or cube in three dimensions. Efficient strategies to solve by parallel direct methods different kinds
of coarse problems are proposed and analyzed in Tufo and Fischer (2001).
Additive Schwarz algorithms for hp-SEM on triangular meshes are discussed
in Korneev et al. (2002).

In Klawonn and Pavarino (2000), a comparison of Schwarz preconditioners
and block preconditioners for saddle-point problems is carried out. See also
Casarin (2001).

In Fischer (1997) several Schwarz methods are considered for the solution
of the unsteady Navier–Stokes equations by SEM or SEM-NI.

Additive Schwarz preconditioners based on finite-element discretization
of the Laplacian can be used for the pressure solver. The same approach has
also been proposed in Casarin (1995), Casarin (1997b), Rønquist (1996).

Moreover, in Fischer (1997) a preconditioner for the pressure matrix is
derived from local finite-element discretization of the Laplacian based on

triangulation of the spectral element vertices (see also Sect. 7.4.3). It is shown that the convergence of the preconditioned Schwarz iterations deteriorates in the presence of high aspect ratios for the elements, and that a remedy consists of allowing more generous overlap. One sweep of the Schwarz method is also used as smoother for multigrid solution to saddle-point problems (see Casarin (1997b), Lottes and Fischer (2005)). Multigrid (2-grid) methods with a Jacobi smoother were previously used by Rønquist and Patera (1987b) and Maday and Muñoz (1989).

Applications to transitional flows using nonconforming (mortar) elements are presented in Fischer et al. (2002).

Schwarz finite-element preconditioners for the pressure matrix obtained from the SEM-NI approximation of the Navier–Stokes equations on unstructured elements based on Fekete points are investigated in Warburton et al. (2000) and Warburton (2000). Schwarz preconditioners for SEM approximations of the Poisson equation using Fekete points are proposed by Pasquetti et al. (2007a).

For the solution of nonlinear problems such as the steady Navier–Stokes equations or the fully implicit time-discretization of the unsteady Navier–Stokes or Euler equations, the Newton–Krylov method can be used in combination with a domain decomposition preconditioner of either Schwarz or Schur type. A Newton–Krylov–Schwarz approach has been used by Rasetarinera et al. (2001) for the semi-implicit solution of the Euler equations by spectral discontinuous Galerkin methods.

As for efficiency, the ultimate word is still premature. For scalar computation, multigrid, or, more generally, multiplicative multilevel algorithms can be very efficient.

On the other hand, on parallel platforms, multilevel domain decomposition preconditioners, like additive Schwarz preconditioners, or else preconditioners for the Schur matrix, such as balanced Neumann–Neumann, FETI, or wire basket preconditioners, are in general more efficient. In particular, Schur preconditioners seem to be more popular in solid mechanics (especially when severe material inhomogeneities are present), whereas multilevel additive Schwarz methods feature very good performance on fluid dynamics problems.

This, however, depends on so many factors (data structure, programming paradigms, parallel platforms, etc.) that an ultimate conclusion is hard to draw.

8. Spectral Methods Primer

The aim of this chapter is to provide the reader with the essential concepts and formulas that are needed for understanding the discretization formulations and implementing the algorithms presented in this book. Nearly all of this material is drawn directly from CHQZ2, to which the reader is referred for thorough derivations, further details and extensive references. The reader interested in the theoretical foundations of spectral methods will need to consult CHQZ2 itself, especially Chaps. 5–7.

We cover here in Sects. 8.1–8.7 the fundamentals of spectral approximations in the reference domain of the relevant polynomial family, then discuss mappings in Sect. 8.8, which are employed extensively in many of the applications surveyed in this text, and finally in Sect. 8.9 we overview the essentials of constructing specific discretizations of differential equations.

8.1 The Fourier System

The set of functions

$$\phi_k(x) = e^{ikx} \tag{8.1.1}$$

is an orthogonal system (termed the *Fourier system*) over the interval $(0, 2\pi)$:

$$\int_0^{2\pi} \phi_k(x)\overline{\phi_l(x)}\, \mathrm{d}x = 2\pi\delta_{kl} = \begin{cases} 0 & \text{if } k \neq l, \\ 2\pi & \text{if } k = l, \end{cases} \tag{8.1.2}$$

(the overline on $\phi_l(x)$ denotes its complex conjugate, and δ_{kl} is the Kronecker delta-function).

The *Fourier coefficients* of an absolutely integrable function u are

$$\hat{u}_k = \frac{1}{2\pi} \int_0^{2\pi} u(x) e^{-ikx}\, \mathrm{d}x, \qquad k = 0, \pm 1, \pm 2, \dots . \tag{8.1.3}$$

These relations associate with u a sequence of complex numbers called the *Fourier transform* of u. The *Fourier cosine transform* and the *Fourier sine transform* of u are defined as

$$a_k = \frac{1}{2\pi} \int_0^{2\pi} u(x) \cos kx\, \mathrm{d}x, \qquad k = 0, \pm 1, \pm 2, \dots , \tag{8.1.4}$$

and

$$b_k = \frac{1}{2\pi} \int_0^{2\pi} u(x) \sin kx \, dx \,, \qquad k = 0, \pm 1, \pm 2, \dots \,. \qquad (8.1.5)$$

The three Fourier transforms of u are related by the formula $\hat{u}_k = a_k - ib_k$ for $k = 0, \pm 1, \pm 2, \dots$. Moreover, if u is a real-valued function, a_k and b_k are real numbers, and $\hat{u}_{-k} = \overline{\hat{u}_k}$.

The Fourier Series. The Fourier series of the function u is formally defined as

$$Su = \sum_{k=-\infty}^{\infty} \hat{u}_k \phi_k \,. \qquad (8.1.6)$$

The convergence of the series occurs naturally in the space $L^2(0, 2\pi)$ of the complex-valued, square-integrable functions in $[0, 2\pi]$, i.e., functions such that the norm

$$\|u\| = \left(\int_0^{2\pi} |u(x)|^2 \, dx \right)^{1/2} \qquad (8.1.7)$$

is finite. In such a space, an inner product is defined as

$$(u, v) = \int_0^{2\pi} u(x)\overline{v(x)} \, dx \,. \qquad (8.1.8)$$

The Fourier system $\{\phi_k\}$ is an orthogonal *basis* in $L^2(0, 2\pi)$. Indeed, for any u in this space, we have

$$u = \sum_{k=-\infty}^{\infty} \hat{u}_k \phi_k \,, \qquad \hat{u}_k = (2\pi)^{-1}(u, \phi_k) \,, \qquad (8.1.9)$$

and the series converges in $L^2(0, 2\pi)$, i.e., in the mean-square sense. This means that

$$\left\| u - \sum_{k=-N}^{N} \hat{u}_k \phi_k \right\| \to 0 \qquad \text{as } N \to \infty \,. \qquad (8.1.10)$$

Classical convergence (e.g., pointwise convergence) of the series occurs if u has additional smoothness properties.

The *Parseval identity* holds:

$$\|u\|^2 = 2\pi \sum_{k=-\infty}^{\infty} |\hat{u}_k|^2 \,. \qquad (8.1.11)$$

Truncation and Projection. For any even integer N, the trigonometric polynomial

$$P_N u(x) = \sum_{k=-N/2}^{N/2-1} \hat{u}_k e^{ikx} \qquad (8.1.12)$$

is the N-th order *truncated Fourier series* of u. (This form of truncation, as opposed to the one in (8.1.10), is the one most commonly applied in spectral methods that exploit the Fast Fourier Transform)

Let S_N be the space of the trigonometric polynomials of degree up to $N/2$, defined as

$$S_N = \text{span}\{e^{ikx} \mid -N/2 \leq k \leq N/2 - 1\} . \tag{8.1.13}$$

By the orthogonality relation (8.1.2) one has

$$(P_N u, v) = (u, v) \quad \text{for all } v \in S_N . \tag{8.1.14}$$

This shows that $P_N u$ is the orthogonal projection of u upon the space of the trigonometric polynomials of degree up to $N/2$. Equivalently, $P_N u$ is the closest element to u in S_N with respect to the norm (8.1.7).

An integral representation of $P_N u$ is given by

$$P_N u(x) = \frac{1}{2\pi} \int_0^{2\pi} D_N(x-y) u(y) \, dy , \tag{8.1.15}$$

where (assuming here a symmetric truncation $-N/2 \leq k \leq N/2$)

$$D_N(\xi) = 1 + 2\sum_{k=1}^{N/2} \cos k\xi = \begin{cases} \dfrac{\sin((N+1)\xi/2)}{\sin(\xi/2)} , & \xi \neq 2j\pi , \\ N+1 , & \xi = 2j\pi , \end{cases} \quad (j \in \mathbb{Z})$$

is the *Dirichlet kernel*. It can be considered an approximation of the (2π-periodic) delta-function in the space S_N.

Decay of the Fourier Coefficients. If u is m-times continuously differentiable in $[0, 2\pi]$ ($m \geq 1$), and if its j-th derivative $u^{(j)}$ is periodic for all $j \leq m - 2$, then

$$\hat{u}_k = O(k^{-m}) , \qquad k = \pm 1, \pm 2, \dots . \tag{8.1.16}$$

Consequently, the k-th Fourier coefficient of a function that is infinitely differentiable and periodic with all its derivatives on $[0, 2\pi]$ decays faster than any negative power of k. In this case the error between a function u and its approximation by its N-th order truncated Fourier series decays faster than algebraically in $1/N$. This property is commonly called *spectral accuracy*, or *infinite-order accuracy*, and we say that the approximation exhibits *infinite-order convergence*.

If in addition the function is analytic in a strip of the complex plane around $[0, 2\pi]$, then $\hat{u}_k$ decays exponentially in k, and the error in the approximation decays exponentially fast; we then say that the approximation exhibits *exponential convergence*. (In the past the term exponential convergence had also been used to characterize spectral methods. However, this term is no longer in common use as a descriptor of spectral accuracy for infinitely differentiable functions. In this text, we only use the term exponential convergence in the context of particular functions for which the convergence is actually exponentially fast.)

For instance, the function

$$u(x) = \sin(x/2) \tag{8.1.17}$$

is infinitely differentiable in $[0, 2\pi]$, but $u'(0^+) \neq u'(2\pi^-)$. Its Fourier coefficients are

$$\hat{u}_k = \frac{2}{\pi} \frac{1}{1 - 4k^2} \ .$$

The function

$$u(x) = \frac{3}{5 - 4\cos x} \tag{8.1.18}$$

is analytic and periodic with all its derivatives around $[0, 2\pi]$. Its Fourier coefficients are

$$\hat{u}_k = 2^{-|k|} \ , \qquad k = 0, \pm 1, \ldots \ .$$

Discrete Fourier Expansion and Interpolation. The equally spaced nodes in $[0, 2\pi]$:

$$x_j = \frac{2\pi j}{N} \ , \qquad j = 0, \ldots, N - 1 \ , \tag{8.1.19}$$

are *Gaussian quadrature nodes* for trigonometric polynomials. Indeed,

$$\frac{2\pi}{N} \sum_{j=0}^{N-1} v(x_j) = \int_0^{2\pi} v(x)\,\mathrm{d}x \qquad \text{for all } v \in S_{2N} \ . \tag{8.1.20}$$

The *discrete Fourier coefficients* of a complex-valued continuous function u in $[0, 2\pi]$ with respect to these points are

$$\tilde{u}_k = \frac{1}{N} \sum_{j=0}^{N-1} u(x_j) e^{-ikx_j} \ , \qquad k = -N/2, \ldots, N/2 - 1 \ . \tag{8.1.21}$$

The orthogonality relation

$$\frac{1}{N} \sum_{j=0}^{N-1} e^{-ipx_j} = \begin{cases} 1 & \text{if } p = Nm, \ m = 0, \pm 1, \pm 2, \ldots \ , \\ 0 & \text{otherwise} \ , \end{cases} \tag{8.1.22}$$

yields the inversion formula

$$u(x_j) = \sum_{k=-N/2}^{N/2-1} \tilde{u}_k e^{ikx_j} \ , \qquad j = 0, \ldots, N - 1 \ . \tag{8.1.23}$$

Consequently, the polynomial

$$I_N u(x) = \sum_{k=-N/2}^{N/2-1} \tilde{u}_k e^{ikx} \tag{8.1.24}$$

is the $N/2$-degree trigonometric interpolant of u at the nodes (8.1.19), i.e., $I_N u(x_j) = u(x_j)$, $j = 0, \ldots, N-1$. This polynomial is also called the *discrete Fourier series* of u.

The $\tilde{u}_k$'s depend only on the N values of u at the nodes (8.1.19). The *discrete Fourier transform* (DFT) is the mapping (8.1.21) between the N complex numbers $u(x_j)$, $j = 0, \ldots, N-1$, and the N complex numbers $\tilde{u}_k$, $k = -N/2, \ldots, N/2 - 1$. The inverse DFT is given by (8.1.23). From a computational point of view, they can be accomplished by the Fast Fourier Transform algorithm (see Appendix A.10).

Another form of the interpolant $I_N u$ is given by

$$I_N u(x) = \sum_{j=0}^{N-1} u(x_j)\psi_j(x) , \tag{8.1.25}$$

with

$$\psi_j(x) = \frac{1}{N} \sum_{k=-N/2}^{N/2-1} e^{ik(x-x_j)} . \tag{8.1.26}$$

The functions ψ_j are the trigonometric polynomials in S_N that satisfy

$$\psi_j(x_l) = \delta_{jl} , \qquad j, l = 0, \ldots, N-1 . \tag{8.1.27}$$

They are the *discrete delta-functions* at the nodes (8.1.19), also termed the *characteristic Lagrange trigonometric polynomials* at these nodes. See Fig. 8.1 for an illustration of some of the discrete delta-functions for $N = 8$.

The bilinear form

$$(u, v)_N = \frac{2\pi}{N} \sum_{j=0}^{N-1} u(x_j)\overline{v(x_j)} \tag{8.1.28}$$

coincides with the inner product (8.1.8) if u and v are polynomials of degree $N/2$, due to (8.1.22):

$$(u, v)_N = (u, v) \quad \text{for all } u, v \in S_N . \tag{8.1.29}$$

As a consequence, (8.1.28) is an inner product on S_N, and

$$\|u\|_N = \sqrt{(u, u)_N} = \sqrt{(u, u)} = \|u\| \tag{8.1.30}$$

is the associated norm.

The interpolant $I_N u$ of a continuous function u satisfies the identity

$$(I_N u, v)_N = (u, v)_N \quad \text{for all } v \in S_N , \tag{8.1.31}$$

which shows that $I_N u$ is the orthogonal projection upon S_N with respect to the discrete inner product.

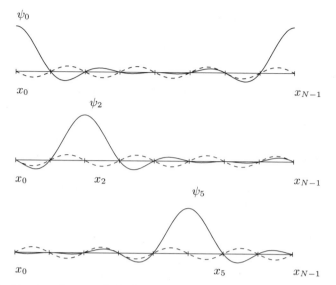

Fig. 8.1. Fourier characteristic Lagrange polynomials (Fourier discrete delta-functions) for $N = 8$. $Re(\psi_j)$ is drawn with a *continuous line*, and $Im(\psi_j)$ is drawn with a *dashed line*

Aliasing. The discrete Fourier coefficients can be expressed also in terms of the exact Fourier coefficients of u. Indeed, using (8.1.21) one gets

$$\tilde{u}_k = \hat{u}_k + \sum_{\substack{m=-\infty \\ m\neq 0}}^{+\infty} \hat{u}_{k+Nm} \, , \qquad k = -N/2, \ldots, N/2 - 1 \, . \qquad (8.1.32)$$

This shows that the k-th mode of the trigonometric interpolant of u depends not only on the k-th mode of u, but also on all the modes of u that *alias* the k-th mode on the discrete grid. The $(k+Nm)$-th wavenumber aliases the k-th wavenumber on the grid; they are indistinguishable at the nodes since $\phi_{k+Nm}(x_j) = \phi_k(x_j)$. The phenomenon is illustrated in Fig. 8.2. Shown there are three sine waves with frequencies $k = 6$, -2, and -10. Superimposed upon each wave are the eight grid-point values of the function. In each case these grid-point values coincide with the $k = -2$ wave.

An equivalent formulation of (8.1.32) is

$$I_N u = P_N u + R_N u \, , \qquad (8.1.33)$$

with

$$R_N u = \sum_{k=-N/2}^{N/2-1} \left(\sum_{\substack{m=-\infty \\ m\neq 0}}^{\infty} \hat{u}_{k+Nm} \right) \phi_k \, . \qquad (8.1.34)$$

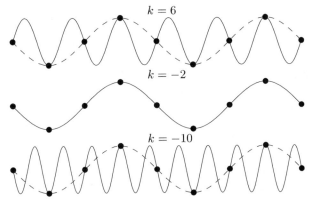

Fig. 8.2. Three sine waves that have the same $k = -2$ interpretation on an eight-point grid. The nodal values are denoted by the *filled circles*. The actual sine waves are denoted by the *solid curves*. Both the $k = 6$ and the $k = -10$ waves are misinterpreted as a $k = -2$ wave (*dashed curves*) on the coarse grid

The error $R_N u$ between the interpolating polynomial and the truncated Fourier series is called the *aliasing error*. It is orthogonal to the *truncation error* $u - P_N u$, so that

$$\|u - I_N u\|^2 = \|u - P_N u\|^2 + \|R_N u\|^2 . \tag{8.1.35}$$

Hence, the error due to the interpolation is always larger than the error due to the truncation of the Fourier series. However, the truncation and interpolation errors decay at the same rate if the function u is sufficiently smooth. In other words, aliasing is not a problem for well-resolved approximations.

Differentiation. Differentiation may be accomplished either in transform space or in physical space.

Differentiation in transform space consists of simply multiplying each Fourier coefficient by the imaginary unit times the corresponding wavenumber. If u is represented as in (8.1.9), then

$$u' = \sum_{k=-\infty}^{\infty} ik\hat{u}_k \phi_k , \tag{8.1.36}$$

provided u' is in $L^2(0, 2\pi)$. The function

$$(P_N u)' = P_N u' , \tag{8.1.37}$$

is called the *Fourier projection derivative* of u; note that truncation and differentiation commute.

Differentiation in physical space is based upon the values of the function u at the Fourier nodes (8.1.19). The function

$$\mathcal{D}_N u = (I_N u)' \tag{8.1.38}$$

is called the *Fourier interpolation derivative* of u. In general, we have

$$\mathcal{D}_N u \neq I_N(u') \qquad \text{as well as} \qquad \mathcal{D}_N u \neq P_N u' , \qquad (8.1.39)$$

unless $u \in S_N$.

There are two equivalent ways to compute the values $(\mathcal{D}_N u)_j$ at the grid points x_j of the interpolation derivative in terms of the values u_l of the function at the same points. The first method uses the direct and inverse DFT and performs the differentiation in transform space:

$$(\mathcal{D}_N u)_j = \sum_{k=-N/2}^{N/2-1} \tilde{u}_k^{(1)} e^{2ikj\pi/N} , \qquad j = 0, 1, \ldots, N-1 , \qquad (8.1.40)$$

where

$$\tilde{u}_k^{(1)} = ik\tilde{u}_k = \frac{ik}{N} \sum_{l=0}^{N-1} u(x_l) e^{-2ikl\pi/N} , \qquad k = -N/2, \ldots, N/2-1 . \quad (8.1.41)$$

The second method utilizes the matrix-vector multiply in physical space:

$$(\mathcal{D}_N u)_j = \sum_{l=0}^{N-1} (D_N)_{jl} u_l , \qquad (8.1.42)$$

where the entries of the matrix D_N, termed the *Fourier interpolation derivative matrix*, are given by

$$(D_N)_{jl} = \psi_l'(x_j) = \frac{1}{N} \sum_{k=-N/2}^{N/2-1} ik e^{2ik(j-l)\pi/N} \qquad (8.1.43)$$

(i. e., the entries of the interpolation derivative matrix are the values of the derivative of the characteristic Lagrange polynomials (8.1.26) at the grid points).

Neglecting the $k = -N/2$ term in this sum (as it makes a purely imaginary contribution if u is a real function), we get the closed form expression

$$(D_N)_{jl} = \begin{cases} \frac{1}{2}(-1)^{j+l} \cot\left[\frac{(j-l)\pi}{N}\right] , & j \neq l , \\ 0 , & j = l . \end{cases} \qquad (8.1.44)$$

The skew symmetry of this real matrix is evident. Its eigenvalues are ik, $k = -N/2+1, \ldots, N/2-1$. The eigenvalue 0 has double multiplicity; its two eigenvectors consist of the grid values of the functions 1 and $\cos(Nx/2)$.

Similarly, an explicit expression for the second-derivative matrix, again neglecting the $k = -N/2$ term, is

$$\left(D_N^{(2)}\right)_{jl} = \begin{cases} \frac{1}{4}(-1)^{j+l}N + \dfrac{(-1)^{j+l+1}}{2\sin^2\left[\frac{(j-l)\pi}{N}\right]} , & j \neq l , \\ -\dfrac{(N-1)(N-2)}{12} , & j = l . \end{cases} \tag{8.1.45}$$

Analogous formulas can be derived if one starts from the quadrature points

$$x_j = \frac{2j}{N+1}\pi , \qquad j = 0,\ldots,N . \tag{8.1.46}$$

Gibbs Phenomenon and Filtering. The Gibbs phenomenon describes the characteristic oscillatory behavior of the truncated Fourier series or the discrete Fourier series of a piecewise smooth function in the neighborhood of a point of discontinuity.

Consider, for instance, the periodic function (square wave)

$$\phi(x) = \begin{cases} 1 , & 0 \leq x < \pi , \\ -1 , & \pi \leq x < 2\pi . \end{cases} \tag{8.1.47}$$

From the integral representation (8.1.15) it can be proven that, in a neighborhood of the origin,

$$P_N\phi(x) \simeq \frac{1}{\pi}\int_0^x D_N(y)dy \quad \text{as } N \to \infty .$$

The oscillatory nature of the Dirichlet kernel D_N induces overshoots and undershoots in $P_N\phi$ around the value $\phi = 1$ on the right-hand side of the origin, and around the value $\phi = -1$ on the left-hand side of the origin. As $N \to \infty$, they concentrate near $x = 0$ but do not damp. For instance, at the points $z_N = \frac{2\pi}{N+1}$ one has $(P_N\phi)(z_N) \sim 1.17898\ldots$ for N large enough. In addition, away from discontinuities, the convergence of $P_N\phi$ to ϕ is very slow. One has, e.g., $(\phi - P_N\phi)(\pi/2) \sim 4/N$ as $N \to \infty$. See Fig. 8.3, top left, for an illustration of the Gibbs phenomenon in the approximation of a square wave.

Since the Gibbs phenomenon is related to the slow decay of the Fourier coefficients of a discontinuous function, it is natural to use smoothing procedures that attenuate the higher order coefficients. The truncated Fourier series $P_N u$ is replaced by the smoothed series

$$S_N u = \sum_{k=-N/2}^{N/2} \sigma_k \hat{u}_k e^{ikx} . \tag{8.1.48}$$

The smoothing factors σ_k are usually defined as

$$\sigma_k = \sigma(2k\pi/N) , \qquad k = -N/2,\ldots,N/2 , \tag{8.1.49}$$

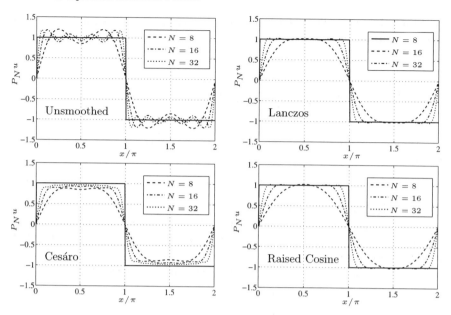

Fig. 8.3. Several smoothings for the square wave

where $\sigma = \sigma(\theta)$ is a *filtering function*, or simply a *filter*, of order p, i. e., a real, even function that satisfies the following three conditions: (i) σ is $(p-1)$-times continuously differentiable in $\mathbb{R}$, for some $p \geq 1$; (ii) $\sigma(\theta) = 0$ if $|\theta| \geq \pi$; (iii) $\sigma(0) = 1$, $\sigma^{(j)}(0) = 0$ for $1 \leq j \leq p - 1$.

Classical, low-order smoothing procedures are defined by the following filters:
the *Lanczos* filter (first-order)

$$\sigma(0) = 1 , \qquad \sigma(\theta) = \frac{\sin \theta}{\theta} , \qquad \theta \in (0, \pi] ; \qquad (8.1.50)$$

the *Cesáro* filter (first-order)

$$\sigma(\theta) = 1 - \frac{\theta}{\pi} , \qquad \theta \in [0, \pi] ; \qquad (8.1.51)$$

and the *raised cosine* filter (second-order)

$$\sigma(\theta) = \frac{1 + \cos \theta}{2} , \qquad \theta \in [0, \pi] . \qquad (8.1.52)$$

See again Fig. 8.3 for the effects of some of these filters on a square wave. Commonly used higher order filters include the *sharpened raised cosine* filter (eighth-order)

$$\sigma(\theta) = \sigma_0(\theta)^4 [35 - 84\sigma_0(\theta) + 70\sigma_0(\theta)^2 - 20\sigma_0(\theta)^3] \qquad (8.1.53)$$

(where σ_0 denotes the raised cosine filter);

the *exponential* filter of order p, for p even,

$$\sigma(\theta) = e^{-\alpha\theta^p} , \qquad \alpha > 0 , \qquad (8.1.54)$$

and the *Vandeven* filter of order p

$$\sigma(\theta) = 1 - \frac{(2p-1)!}{(p-1)!} \int_0^{\theta/\pi} [t(1-t)]^{p-1} \mathrm{d}t . \qquad (8.1.55)$$

See Fig. 4.3 for illustrations of these last two filters.

8.2 General Jacobi Polynomials in the Interval $(-1,1)$

Let α, $\beta > -1$ be real parameters. The Jacobi polynomials $\phi_k(x) = P_k^{(\alpha,\beta)}(x)$ of indices α, β and degree k are the eigenfunctions of the singular Sturm–Liouville problem

$$\frac{\mathrm{d}}{\mathrm{d}x}\left(r\,\frac{\mathrm{d}u}{\mathrm{d}x}\right) + \lambda w\, u = 0 \qquad \text{in the interval } (-1,1) , \qquad (8.2.1)$$

with $r(x) = (1-x)^{1+\alpha}(1+x)^{1+\beta}$, and $w(x) = (1-x)^{\alpha}(1+x)^{\beta}$ (the *Jacobi weight*). The corresponding eigenvalues are $\lambda_k = k(k+\alpha+\beta+1)$. The Jacobi polynomials are mutually orthogonal over the interval $(-1,1)$ with respect to the weight function w:

$$\int_{-1}^{1} \phi_k(x)\phi_m(x)w(x)\,\mathrm{d}x = 0 , \qquad m \neq k . \qquad (8.2.2)$$

Under the normalization $P_k^{(\alpha,\beta)}(1) = \binom{k+\alpha}{k}$, one has the expression

$$P_k^{(\alpha,\beta)}(x) = \frac{1}{2^k} \sum_{l=0}^{k} \binom{k+\alpha}{l}\binom{k+\beta}{k-l}(x-1)^l(x+1)^{k-l} . \qquad (8.2.3)$$

The *Rodriguez formula* provides an alternative representation, namely,

$$P_k^{(\alpha,\beta)}(x) = \frac{(-1)^k}{2^k k!}(1-x)^{-\alpha}(1+x)^{-\beta}\frac{\mathrm{d}^k}{\mathrm{d}x^k}\left((1-x)^{\alpha+k}(1+x)^{\beta+k}\right) . \quad (8.2.4)$$

Jacobi polynomials satisfy the recursion relation

$$P_0^{(\alpha,\beta)}(x) = 1 , \quad P_1^{(\alpha,\beta)}(x) = \frac{1}{2}[(\alpha-\beta) + (\alpha+\beta+2)x] ,$$

$$a_{1,k}P_{k+1}^{(\alpha,\beta)}(x) = a_{2,k}P_k^{(\alpha,\beta)}(x) - a_{3,k}P_{k-1}^{(\alpha,\beta)}(x) , \qquad (8.2.5)$$

where

$$a_{1,k} = 2(k+1)(k+\alpha+\beta+1)(2k+\alpha+\beta) \,,$$
$$a_{2,k} = (2k+\alpha+\beta+1)(\alpha^2-\beta^2) + x\Gamma(2k+\alpha+\beta+3)/\Gamma(2k+\alpha+\beta) \,,$$
$$a_{3,k} = 2(k+\alpha)(k+\beta)(2k+\alpha+\beta+2) \,.$$

(Here and below, Γ denotes the classical Gamma function; see, e. g., Abramowitz and Stegun (1972)).

The derivative of a Jacobi polynomial can be expressed as

$$b_{1,k}(x)\frac{\mathrm{d}}{\mathrm{d}x}P_k^{(\alpha,\beta)}(x) = b_{2,k}(x)P_k^{(\alpha,\beta)}(x) + b_{3,k}(x)P_{k-1}^{(\alpha,\beta)}(x) \,, \qquad (8.2.6)$$

where

$$b_{1,k}(x) = (2k+\alpha+\beta)(1-x^2) \,, \qquad b_{2,k}(x) = k(\alpha-\beta-(2k+\alpha+\beta)x) \,,$$
$$b_{3,k}(x) = 2(k+\alpha)(k+\beta) \,.$$

Another useful formula that relates Jacobi polynomials and their derivatives is

$$\frac{\mathrm{d}^m}{\mathrm{d}x^m}P_k^{(\alpha,\beta)}(x) = 2^{-m}\frac{\Gamma(k+m+\alpha+\beta+1)}{\Gamma(k+\alpha+\beta+1)}P_{k-m}^{(\alpha+m,\beta+m)}(x) \,; \qquad (8.2.7)$$

in particular, one has

$$\frac{\mathrm{d}}{\mathrm{d}x}P_k^{(\alpha,\beta)}(x) = \tfrac{1}{2}(k+1+\alpha+\beta)P_{k-1}^{(\alpha+1,\beta+1)}(x) \,. \qquad (8.2.8)$$

Jacobi polynomials for which $\alpha = \beta$ are called *ultraspherical polynomials* and are denoted simply by $P_k^{(\alpha)}(x)$. The choice $\alpha = 0$ yields the Legendre polynomials considered in Sect. 8.4:

$$L_k(x) = P_k^{(0)}(x) \,, \qquad (8.2.9)$$

whereas the ultraspherical polynomials with $\alpha = -1/2$ are proportional to the Chebyshev polynomials considered in Sect. 8.3:

$$T_k(x) = \frac{2^{2k}(k!)^2}{(2k)!}P_k^{(-1/2)}(x) \,. \qquad (8.2.10)$$

A different normalization of the ultraspherical polynomials leads to the *Gegenbauer polynomials* C_k^ν, which are defined as

$$C_k^\nu(x) = \frac{\Gamma(\nu+\tfrac{1}{2})\Gamma(2\nu+k)}{\Gamma(\nu+k+\tfrac{1}{2})\Gamma(2\nu)}P_k^{(\nu-1/2)}(x) \,. \qquad (8.2.11)$$

Some representative Jacobi polynomials are illustrated in Fig. 8.4.

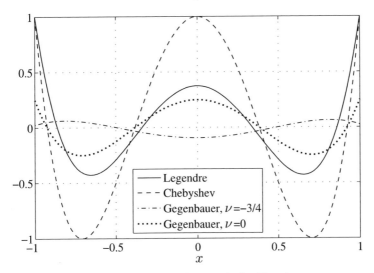

Fig. 8.4. Jacobi polynomials for $N = 4$

The Jacobi Series. Truncation and Projection. Each Jacobi system $\{\phi_k\}_{k=0,1,\ldots}$ is an orthogonal basis in the space $L_w^2(-1, 1)$ of the functions v for which the norm

$$\|v\|_w = \left(\int_{-1}^{1} |v(x)|^2 w(x) \, dx \right)^{1/2} \tag{8.2.12}$$

is finite. The associated inner product is

$$(u, v)_w = \int_{-1}^{1} u(x) v(x) w(x) \, dx \ . \tag{8.2.13}$$

When $w \equiv 1$ (Legendre weight), we will often use the simpler and more classical notation $L^2(-1, 1)$ instead of $L_w^2(-1, 1)$. The formal series of a function $u \in L_w^2(-1, 1)$ in terms of the system $\{\phi_k\}$ is

$$Su = \sum_{k=0}^{\infty} \hat{u}_k \phi_k \ ,$$

where the expansion coefficients $\hat{u}_k$, termed the *Jacobi coefficients* of u, are defined as

$$\hat{u}_k = \frac{1}{\|\phi_k\|_w^2} (u, \phi_k)_w \ , \tag{8.2.14}$$

with

$$\|\phi_k\|_w^2 = \frac{2^{\alpha+\beta+1}}{2k+\alpha+\beta+1} \frac{\Gamma(k+\alpha+1)\Gamma(k+\beta+1)}{k!\Gamma(k+\alpha+\beta+1)} \ . \tag{8.2.15}$$

Equation (8.2.14) represents the *polynomial transform* of u. For an integer $N > 0$, the truncated series of u of order N is the polynomial

$$P_N u = \sum_{k=0}^{N} \hat{u}_k \phi_k . \tag{8.2.16}$$

Due to (8.2.2), $P_N u$ is the orthogonal projection of u upon $\mathbb{P}_N(-1, 1)$ (the space of all algebraic polynomials of degree $\leq N$ restricted to the interval $(-1, 1)$) in the inner product (8.2.13), i. e.,

$$(P_N u, v)_w = (u, v)_w \quad \text{for all } v \in \mathbb{P}_N(-1, 1) . \tag{8.2.17}$$

The completeness of the system $\{\phi_k\}$ is equivalent to the property that, for all $u \in L^2_w(-1, 1)$,

$$\|u - P_N u\|_w \to 0 \quad \text{as } N \to \infty . \tag{8.2.18}$$

Gauss-Type Quadrature Formulas and Discrete Inner Products.
Let w be a Jacobi weight and let $\{\phi_k\}$ be the corresponding system of orthogonal polynomials. Several Gauss-type quadrature formulas of the form

$$\sum_{j=0}^{N} f(x_j) w_j \sim \int_{-1}^{1} f(x) w(x) \, dx \tag{8.2.19}$$

can be defined; these involve $N + 1$ distinct nodes $x_0 < x_1 < \cdots < x_N$ in the interval $[-1, 1]$; the corresponding weights w_j are always strictly positive. Their degree of precision (or exactness), i. e., the highest degree of polynomials integrated exactly, depends on the number of nodes placed at the endpoints of the interval.

The *Jacobi Gauss* quadrature formula uses as nodes the roots of the $(N + 1)$-th orthogonal polynomial ϕ_{N+1}; these all lie in $(-1, 1)$. Its degree of precision is $2N + 1$.

A *Jacobi Gauss–Radau* formula uses as nodes the roots of the polynomial of the form $\phi_{N+1} + a\phi_N$ that vanishes at $x = -1$; thus, $x_0 = -1$. Its degree of precision is $2N$. A specularly symmetric formula has $x_N = 1$, rather than x_0, as the boundary node.

Finally, the *Jacobi Gauss–Lobatto* formula uses as nodes the roots of the polynomial $(1 - x^2)\phi'_N$; consequently, $x_0 = -1$ and $x_N = 1$. Its degree of precision is $2N - 1$.

For any of the above formulas and for any u, v continuous on $[-1, 1]$, we set

$$(u, v)_N = \sum_{j=0}^{N} u(x_j) v(x_j) w_j . \tag{8.2.20}$$

The degree of precision of the formula implies that

$$(u, v)_N = (u, v)_w \quad \text{if } uv \in \mathbb{P}_{2N+\delta}(-1, 1) , \tag{8.2.21}$$

where $\delta = 1, 0, -1$ for Gauss, Gauss–Radau or Gauss–Lobatto integration, respectively. In particular, $(u, v)_N$ is invariably an inner product on $\mathbb{P}_N(-1, 1)$, often referred to as a *discrete inner product*. The corresponding norm is

$$\|u\|_N = \sqrt{(u, u)_N} \ . \tag{8.2.22}$$

Due to (8.2.21), the polynomials ϕ_k for $0 \le k \le N$ are orthogonal also with respect to the discrete inner product; precisely, for $k, m = 0, \ldots, N$ one has

$$(\phi_m, \phi_k)_N = \gamma_k \delta_{km} \ , \qquad \gamma_k = \|\phi_k\|_N^2 \ . \tag{8.2.23}$$

Explicit formulas for γ_k for the more common orthogonal polynomials are supplied in Sects. 8.3 and 8.4.

Discrete Polynomial Transform and Interpolation. The quadrature formula (8.2.19) allows for the approximate computation of the Jacobi coefficients of a continuous function u defined in the interval $[-1, 1]$. The coefficients

$$\tilde{u}_k = \frac{1}{\gamma_k}(u, \phi_k)_N \ , \qquad k = 0, \ldots, N \ , \tag{8.2.24}$$

are called the *discrete Jacobi coefficients* of u (compare with (8.2.14)). Since

$$u(x_j) = \sum_{k=0}^{N} \tilde{u}_k \phi_k(x_j) \ , \qquad j = 0, \ldots, N \ , \tag{8.2.25}$$

equations (8.2.24) and (8.2.25) enable one to transform between physical space $\{u(x_j)\}$ and transform space $\{\tilde{u}_k\}$. Such a transformation for orthogonal polynomials is the analog of the transformation (8.1.21) and (8.1.23) for trigonometric polynomials. We shall call it the *discrete polynomial transform* associated with the weight w and the nodes $x_0, \ldots, x_N$.

By (8.2.25), the polynomial

$$I_N u = \sum_{k=0}^{N} \tilde{u}_k \phi_k \tag{8.2.26}$$

satisfies

$$I_N u(x_j) = u(x_j) \ , \qquad 0 \le j \le N \ , \tag{8.2.27}$$

i. e., it is the unique polynomial in $\mathbb{P}_N(-1, 1)$ interpolating u at the quadrature nodes. For any continuous v, (8.2.27) gives

$$(I_N u, v)_N = (u, v)_N \ . \tag{8.2.28}$$

This shows that the interpolant $I_N u$ is the orthogonal projection of u upon $\mathbb{P}_N(-1, 1)$ with respect to the discrete inner product (8.2.20).

Another expression for $I_N u$ is

$$I_N u = \sum_{l=0}^{N} u(x_l)\psi_l \,, \tag{8.2.29}$$

where ψ_l denotes the l-th *characteristic Lagrange polynomial* relative to the given set of nodes, i.e., the unique polynomial that satisfies

$$\psi_l \in \mathbb{P}_N(-1,1) \,, \qquad \psi_l(x_j) = \delta_{jl} \quad \text{for } j = 0,\dots,N \,. \tag{8.2.30}$$

The characteristic Lagrange polynomials for the Legendre Gauss–Lobatto nodes are illustrated in Fig. 8.5. The general expression for such polynomials is

$$\psi_l(x) = \prod_{\substack{j \neq l \\ 0 \leq j, l \leq N}} \frac{(x - x_j)}{(x_l - x_j)} \,. \tag{8.2.31}$$

For numerical stability reasons, often the Lagrangian polynomials are reformulated in *barycentric form* as

$$\psi_l(x) = \frac{\dfrac{\lambda_l}{x - x_l}}{\displaystyle\sum_{k=0}^{N} \dfrac{\lambda_k}{x - x_k}} \,, \qquad \lambda_l = \frac{1}{\displaystyle\prod_{k \neq l}(x_l - x_k)} \,. \tag{8.2.32}$$

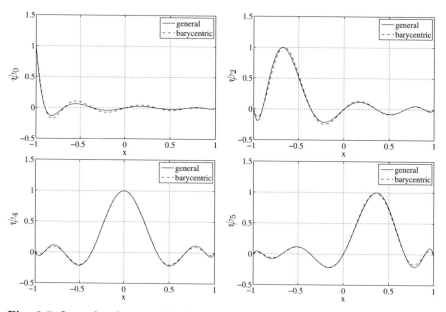

Fig. 8.5. Legendre characteristic Lagrange polynomials (discrete delta-functions) relative to the Gauss–Lobatto nodes for $N = 8$

Differentiation. Differentiation can be accomplished in transform space or in physical space, according to the representation of the function.

Differentiation in transform space consists of computing the Jacobi expansion of the derivative of a function in terms of the Jacobi expansion of the function itself. If $u = \sum_{k=0}^{\infty} \hat{u}_k \phi_k$, u' can be (formally) represented as

$$u' = \sum_{k=0}^{\infty} \hat{u}_k^{(1)} \phi_k \,, \tag{8.2.33}$$

where the coefficients $\hat{u}_k^{(1)}$ can be expressed in terms of the coefficients $\hat{u}_l$ using (8.2.6). Explicit formulas for the more common orthogonal polynomials are supplied in the forthcoming sections.

The approximation of u' given by $(P_N u)'$ is termed the *Jacobi projection derivative* of u'. Note that in general $(P_N u)' \neq P_{N-1} u'$, i.e., differentiation and Jacobi truncation do not commute.

Differentiation in physical space is accomplished by replacing truncation by interpolation. Given a set of $N+1$ Gaussian nodes in $[-1, 1]$, the polynomial

$$\mathcal{D}_N u = (I_N u)' \tag{8.2.34}$$

is called the *Jacobi interpolation derivative* of u relative to the chosen set of quadrature nodes. In general, $\mathcal{D}_N u \neq (P_N u)'$. The physical values of the interpolation derivative can be expressed as linear combinations of the physical values of the function, i.e.,

$$(\mathcal{D}_N u)(x_j) = \sum_{l=0}^{N} (D_N)_{jl} u(x_l) \,, \qquad j = 0, \ldots, N \,. \tag{8.2.35}$$

By (8.2.29), the coefficients are given by $(D_N)_{jl} = \psi_l'(x_j)$; they form the entries of the *first-derivative interpolation matrix* D_N.

8.3 Chebyshev Polynomials

The Chebyshev polynomials of the first kind, $T_k(x)$, $k = 0, 1, \ldots$, are the eigenfunctions of the singular Sturm–Liouville problem

$$\left(\sqrt{1 - x^2} T_k'(x) \right)' + \frac{k^2}{\sqrt{1 - x^2}} T_k(x) = 0 \,, \tag{8.3.1}$$

which is (8.2.1) for $\alpha = \beta = -1/2$. For any k, $T_k(x)$ is even if k is even, and odd if k is odd. If T_k is normalized so that $T_k(1) = 1$, then

$$T_k(x) = \cos k\theta \,, \qquad \theta = \arccos x \,. \tag{8.3.2}$$

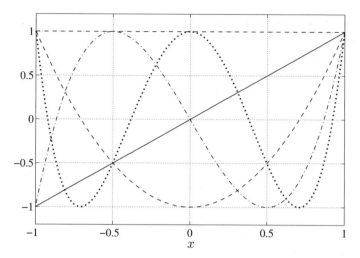

Fig. 8.6. Chebyshev polynomials for $N = 0, \ldots, 4$

Thus, the Chebyshev polynomials are nothing but cosine functions after a change of independent variable. See Fig. 8.6 for the lowest order Chebyshev polynomials.

The Chebyshev polynomials can be expanded in power series as

$$T_k(x) = \frac{k}{2} \sum_{l=0}^{[k/2]} (-1)^k \frac{(k-l-1)!}{l!(k-2l)!} (2x)^{k-2l} , \tag{8.3.3}$$

where $[k/2]$ denotes the integral part of $k/2$. The trigonometric formula $\cos(k+1)\theta + \cos(k-1)\theta = 2\cos\theta\cos k\theta$ gives the recursion relation

$$T_{k+1}(x) = 2xT_k(x) - T_{k-1}(x) , \tag{8.3.4}$$

with $T_0(x) = 1$ and $T_1(x) = x$, which is (8.2.5) taking into account (8.2.10). On the other hand, the trigonometric identity $2\sin\theta\cos k\theta = \sin(k+1)\theta - \sin(k-1)\theta$ yields the relation

$$2T_k(x) = \frac{1}{k+1}T'_{k+1}(x) - \frac{1}{k-1}T'_{k-1}(x) , \qquad k \geq 1 . \tag{8.3.5}$$

Some properties of the Chebyshev polynomials are

$$|T_k(x)| \leq 1 , \quad -1 \leq x \leq 1 , \tag{8.3.6}$$

$$T_k(\pm 1) = (\pm 1)^k , \tag{8.3.7}$$

$$|T'_k(x)| \leq k^2 , \quad -1 \leq x \leq 1 , \tag{8.3.8}$$

$$T'_k(\pm 1) = (\pm 1)^{k+1}k^2 , \tag{8.3.9}$$

$$\int_{-1}^{1} T_k^2(x) \frac{\mathrm{d}x}{\sqrt{1-x^2}} = c_k \frac{\pi}{2} , \tag{8.3.10}$$

where

$$c_k = \begin{cases} 2 \,, & k = 0 \,, \\ 1 \,, & k \geq 1 \,. \end{cases} \tag{8.3.11}$$

The Chebyshev expansion of a function $u \in L_w^2(-1, 1)$ is

$$u(x) = \sum_{k=0}^{\infty} \hat{u}_k T_k(x) \,, \quad \hat{u}_k = \frac{2}{\pi c_k} \int_{-1}^{1} u(x) T_k(x) w(x) \, dx \,. \tag{8.3.12}$$

If we define the even periodic function $\bar{u}$ by $\bar{u}(\theta) = u(\cos\theta)$, then

$$\bar{u}(\theta) = \sum_{k=0}^{\infty} \hat{u}_k \cos k\theta \,,$$

i. e., the Chebyshev series for u corresponds to a cosine series for $\bar{u}$.

Quadrature Formulas and Discrete Transforms. Explicit formulas for the Gaussian quadrature points and weights are as follows:

Chebyshev Gauss (CG).

$$x_j = \cos\frac{(2j+1)\pi}{2N+2} \,, \quad w_j = \frac{\pi}{N+1} \,, \quad j = 0, \ldots, N \,; \tag{8.3.13}$$

Chebyshev Gauss–Radau (CGR).

$$x_j = \cos\frac{2\pi j}{2N+1} \,, \quad w_j = \begin{cases} \dfrac{\pi}{2N+1} \,, & j = 0 \,, \\[2mm] \dfrac{2\pi}{2N+2} \,, & j = 1, \ldots, N \,; \end{cases} \tag{8.3.14}$$

Chebyshev Gauss–Lobatto (CGL).

$$x_j = \cos\frac{\pi j}{N} \,, \quad w_j = \begin{cases} \dfrac{\pi}{2N} \,, & j = 0, N \,, \\[2mm] \dfrac{\pi}{N} \,, & j = 1, \ldots, N-1 \,. \end{cases} \tag{8.3.15}$$

(Note that the Chebyshev quadrature points as just defined are ordered from right to left.)

The normalization factors γ_k introduced in (8.2.23) are given by

$$\gamma_k = \frac{\pi}{2} c_k \qquad \text{for } k < N \,,$$

$$\gamma_N = \begin{cases} \dfrac{\pi}{2} & \text{for Gauss and Gauss–Radau formulas} \,, \\[2mm] \pi & \text{for the Gauss–Lobatto formula} \,. \end{cases} \tag{8.3.16}$$

For the most commonly used Gauss–Lobatto points, the matrix representing the transformation from physical space to Chebyshev transform space (see (8.2.24)) is available in the simple form

$$C_{kj} = \frac{2}{N\bar{c}_j\bar{c}_k} \cos \frac{\pi jk}{N} , \qquad (8.3.17)$$

where

$$\bar{c}_j = \begin{cases} 2 , & j = 0, N , \\ 1 , & j = 1, \ldots, N - 1 . \end{cases} \qquad (8.3.18)$$

Likewise, the inverse transformation (see (8.2.25)) is represented by

$$(C^{-1})_{jk} = \cos \frac{\pi jk}{N} . \qquad (8.3.19)$$

Both transforms may be evaluated by the Fast Fourier Transform (Appendix B).

The structure of the aliasing error is similar to (8.1.32); precisely,

$$\tilde{u}_k = \hat{u}_k + \sum_{\substack{j=2mN\pm k \\ j>N}} \hat{u}_j . \qquad (8.3.20)$$

Differentiation. The derivative of a function u expanded in Chebyshev polynomials according to (8.3.12) can be represented formally as

$$u' = \sum_{k=0}^{\infty} \hat{u}_k^{(1)} T_k . \qquad (8.3.21)$$

From (8.3.5) one has

$$2k\hat{u}_k = c_{k-1}\hat{u}_{k-1}^{(1)} - \hat{u}_{k+1}^{(1)} , \qquad k \geq 1 , \qquad (8.3.22)$$

or, equivalently,

$$c_k\hat{u}_k^{(1)} = \hat{u}_{k+2}^{(1)} + 2(k+1)\hat{u}_{k+1} , \qquad k \geq 0 . \qquad (8.3.23)$$

This yields

$$\hat{u}_k^{(1)} = \frac{2}{c_k} \sum_{\substack{p=k+1 \\ p+k \text{ odd}}}^{\infty} p\hat{u}_p , \qquad k \geq 0 . \qquad (8.3.24)$$

The coefficients of the second derivative, $u'' = \sum_{k=0}^{\infty} \hat{u}_k^{(2)} T_k$, are

$$\hat{u}_k^{(2)} = \frac{1}{c_k} \sum_{\substack{p=k+2 \\ p+k \text{ even}}}^{\infty} p(p^2 - k^2)\hat{u}_p , \qquad k \geq 0 . \qquad (8.3.25)$$

Concerning differentiation in physical space, the characteristic Lagrange polynomials (8.2.31) at the Chebyshev Gauss–Lobatto points (8.3.15) can be expressed as

$$\psi_l(x) = \frac{(-1)^{l+1}(1-x^2)T_N'(x)}{\bar{c}_l N^2(x-x_l)} . \tag{8.3.26}$$

From this expression, one gets the following form of the Chebyshev first-derivative interpolation matrix (see (8.2.35))

$$(D_N)_{jl} = \begin{cases} \dfrac{\bar{c}_j}{\bar{c}_l} \dfrac{(-1)^{j+l}}{x_j - x_l} , & j \neq l , \\[2ex] -\dfrac{x_l}{2(1-x_l^2)} , & 1 \leq j = l \leq N-1 , \\[2ex] \dfrac{2N^2+1}{6} , & j = l = 0 , \\[2ex] -\dfrac{2N^2+1}{6} , & j = l = N . \end{cases} \tag{8.3.27}$$

A numerically more stable form replaces the first expression in (8.3.27) by

$$-\frac{\bar{c}_j}{2\bar{c}_l} \frac{(-1)^{j+l}}{\sin[(j+l)\pi/2N]\sin[(j-l)\pi/2N]}$$

and the second one by

$$-\frac{x_j}{2\sin^2(j\pi/N)} .$$

Alternatively, using the barycentric form (8.2.32) of the Lagrange polynomials, the resulting representation of D_N is

$$(D_N)_{jl} = \begin{cases} \dfrac{\delta_l}{\delta_j} \dfrac{(-1)^{j+l}}{x_j - x_l} , & j \neq l , \\[2ex] -\displaystyle\sum_{i=0,i\neq j}^{N} \dfrac{\delta_i}{\delta_j} \dfrac{(-1)^{i+j}}{x_j - x_i} , & j = l , \end{cases} \tag{8.3.28}$$

where $\delta_l = 1/2$ if $l = 0$ or N, $\delta_l = 1$ otherwise.

8.4 Legendre Polynomials

The Legendre polynomials, $L_k(x)$, $k = 0, 1, \ldots$, are the eigenfunctions of the singular Sturm–Liouville problem

$$((1-x^2)L_k'(x))' + k(k+1)L_k(x) = 0 , \tag{8.4.1}$$

which is (8.2.1) for $\alpha = \beta = 0$. $L_k(x)$ is even if k is even and odd if k is odd.

If $L_k(x)$ is normalized so that $L_k(1) = 1$, then, for any k,

$$L_k(x) = \frac{1}{2^k} \sum_{l=0}^{[k/2]} (-1)^l \binom{k}{l} \binom{2k - 2l}{k} x^{k-2l} , \tag{8.4.2}$$

where $[k/2]$ denotes the integral part of $k/2$. The Legendre polynomials satisfy the recursion relation (see (8.2.5))

$$L_{k+1}(x) = \frac{2k + 1}{k + 1} x L_k(x) - \frac{k}{k + 1} L_{k-1}(x) , \tag{8.4.3}$$

where $L_0(x) = 1$ and $L_1(x) = x$. Other useful formulas are

$$(1 - x^2) L_k'(x) = k L_{k-1}(x) - kx L_k(x) , \qquad k \geq 1 , \tag{8.4.4}$$

(see (8.2.6)) and

$$(2k + 1) L_k(x) = L_{k+1}'(x) - L_{k-1}'(x) , \qquad k \geq 0 . \tag{8.4.5}$$

See Fig. 8.7 for the lowest order Legendre polynomials.

Relevant properties of the Legendre polynomials are

$$|L_k(x)| \leq 1 , \quad -1 \leq x \leq 1 , \tag{8.4.6}$$

$$L_k(\pm 1) = (\pm 1)^k , \tag{8.4.7}$$

$$|L_k'(x)| \leq \tfrac{1}{2} k(k + 1) , \quad -1 \leq x \leq 1 , \tag{8.4.8}$$

$$L_k'(\pm 1) = (\pm 1)^{k+1} \tfrac{1}{2} k(k + 1) , \tag{8.4.9}$$

$$\int_{-1}^{1} L_k^2(x) \, dx = (k + \tfrac{1}{2})^{-1} . \tag{8.4.10}$$

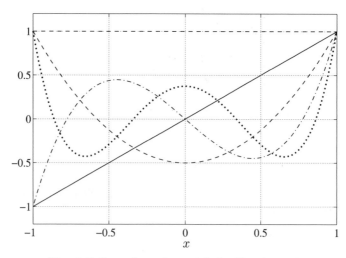

Fig. 8.7. Legendre polynomials for $N = 0, \ldots, 4$

The expansion of any $u \in L^2(-1,1)$ in terms of the Legendre polynomials is

$$u(x) = \sum_{k=0}^{\infty} \hat{u}_k L_k(x) , \quad \hat{u}_k = \left(k + \tfrac{1}{2}\right) \int_{-1}^{1} u(x) L_k(x) \, dx . \tag{8.4.11}$$

Quadrature Formulas and Discrete Norms. While explicit formulas for the quadrature nodes introduced in (8.2.19) are not known, the quadrature weights can be expressed as follows:

Legendre Gauss (LG).

$$w_j = \frac{2}{(1 - x_j^2)[L'_{N+1}(x_j)]^2} , \quad j = 0, \ldots, N ; \tag{8.4.12}$$

Legendre Gauss–Radau (LGR).

$$w_0 = \frac{2}{(N+1)^2} , \quad w_j = \frac{1}{(N+1)^2} \frac{1 - x_j}{[L_N(x_j)]^2} , \quad j = 1, \ldots, N ; \tag{8.4.13}$$

Legendre Gauss–Lobatto (LGL).

$$w_j = \frac{2}{N(N+1)} \frac{1}{[L_N(x_j)]^2} , \quad j = 0, \ldots, N . \tag{8.4.14}$$

The normalization factors γ_k introduced in (8.2.23) are given by

$$\gamma_k = \left(k + \tfrac{1}{2}\right)^{-1} \quad \text{for } k < N ,$$

$$\gamma_N = \begin{cases} \left(N + \tfrac{1}{2}\right)^{-1} & \text{for Gauss and Gauss–Radau formulas ,} \\ 2/N & \text{for the Gauss–Lobatto formula .} \end{cases} \tag{8.4.15}$$

Differentiation. If $u = \sum_{k=0}^{\infty} \hat{u}_k L_k$, then its first derivative can be represented as $u' = \sum_{k=0}^{\infty} \hat{u}_k^{(1)} L_k$. Formula (8.4.5) yields the recursion relation

$$\hat{u}_k = \frac{\hat{u}_{k-1}^{(1)}}{2k - 1} - \frac{\hat{u}_{k+1}^{(1)}}{2k + 3} , \quad k \geq 1 , \tag{8.4.16}$$

which implies

$$\hat{u}_k^{(1)} = (2k + 1) \sum_{\substack{p=k+1 \\ p+k \text{ odd}}}^{\infty} \hat{u}_p , \quad k \geq 0 . \tag{8.4.17}$$

For the second derivative, $u'' = \sum_{k=0}^{\infty} \hat{u}_k^{(2)} L_k$, we have

$$\hat{u}_k^{(2)} = \left(k + \tfrac{1}{2}\right) \sum_{\substack{p=k+2 \\ p+k \text{ even}}}^{\infty} [p(p+1) - k(k+1)] \hat{u}_p , \quad k \geq 0 . \tag{8.4.18}$$

Concerning differentiation in physical space, for the commonly used Gauss–Lobatto points a closed form of the Legendre first-derivative interpolation matrix (see (8.2.35)) is given by

$$
(D_N)_{jl} = \begin{cases}
\dfrac{L_N(x_j)}{L_N(x_l)} \dfrac{1}{x_j - x_l}\,, & j \neq l\,, \\[2ex]
-\dfrac{(N+1)N}{4}\,, & j = l = 0\,, \\[2ex]
\dfrac{(N+1)N}{4}\,, & j = l = N\,, \\[2ex]
0 & \text{otherwise}\,.
\end{cases}
\tag{8.4.19}
$$

Alternative expressions with more favorable round-off error properties are available.

8.5 Modal and Nodal Boundary-Adapted Bases on the Interval

A basis of the space $\mathbb{P}_N(-1,1)$ is termed *boundary-adapted* if it contains two functions that are nonzero at precisely one endpoint of the interval (these are called the *vertex basis functions*) and $N-1$ functions that vanish at both endpoints (these are called *bubble functions*, or *internal basis functions*). If we drop the vertex functions from a boundary-adapted basis, we get a basis for the subspace $\mathbb{P}_N^0(-1,1)$ of the polynomials vanishing at the endpoints.

An example of a boundary-adapted basis is provided by the characteristic Lagrange basis ψ_l, $l = 0, \ldots, N$, associated with a set of Jacobi Gauss–Lobatto points in $[-1,1]$ (see (8.2.30)). The vertex functions are ψ_0 and ψ_N, whereas the bubble functions are ψ_l for $l = 1, \ldots, N-1$. Such a basis is also an example of *nodal* basis, i.e., a basis in which each function is responsible for reproducing the value of the polynomial at one particular node in the interval.

Boundary-adapted bases of *modal* type, i.e., such that each basis function provides one particular pattern of oscillation (of higher and higher frequency), are also useful. A naive example is obtained by taking simple linear combinations of orthogonal polynomials. If $\phi_k(x)$ denotes either $T_k(x)$ or $L_k(x)$, then a simple boundary-adapted modal basis is as follows:

$$
\begin{aligned}
\eta_0(x) &= \frac{1}{2}\bigl(\phi_0(x) - \phi_1(x)\bigr) = \frac{1-x}{2}\,, \\[1ex]
\eta_1(x) &= \frac{1}{2}\bigl(\phi_0(x) + \phi_1(x)\bigr) = \frac{1+x}{2}\,, \\[1ex]
\eta_k(x) &= \begin{cases}
\phi_0(x) - \phi_k(x), & k \text{ even } \geq 2\,, \\
\phi_1(x) - \phi_k(x), & k \text{ odd } \geq 3\,.
\end{cases} \qquad 2 \leq k \leq N\,.
\end{aligned}
\tag{8.5.1}
$$

Here the vertex functions are η_0 and η_1, whereas the bubble functions are the remaining ones. The drawback in applications to differential problems is that neither these functions nor their derivatives are orthogonal in the (weighted) L^2-inner product. In other words, neither the *mass matrix* M, whose entries are $M_{hk} = (\eta_h, \eta_k)_w$, $0 \le h, k \le N$, nor the *stiffness matrix* K, whose entries are $K_{hk} = (\eta_h', \eta_k')_w$, $0 \le h, k \le N$, are diagonal.

A better suited boundary-adapted modal basis in the Legendre case is given by

$$\eta_0(x) = \frac{1}{2}\big(L_0(x) - L_1(x)\big) = \frac{1-x}{2} \ ,$$
$$\eta_1(x) = \frac{1}{2}\big(L_0(x) + L_1(x)\big) = \frac{1+x}{2} \ , \tag{8.5.2}$$
$$\eta_k(x) = \sqrt{\frac{2k-1}{2}} \int_x^1 L_{k-1}(s)\,ds, \qquad 2 \le k \le N \ .$$

(A similar basis involving the Chebyshev polynomials can be constructed as well.) Recalling (8.4.5), one easily gets

$$\eta_k(x) = \frac{1}{\sqrt{2(2k-1)}}\big(L_{k-2}(x) - L_k(x)\big), \qquad 2 \le k \le N \ ; \tag{8.5.3}$$

another useful expression for η_k is

$$\eta_k(x) = \frac{\sqrt{2(2k-1)}}{k}\left(\frac{1-x}{2}\right)\left(\frac{1+x}{2}\right)P_{k-2}^{(1,1)}(x), \qquad 2 \le k \le N \ , \tag{8.5.4}$$

where $P_k^{(1,1)}$ is the k-th Jacobi polynomial of indices $\alpha = \beta = 1$ (see Sect. 8.2). The basis is designed to yield a diagonal stiffness matrix (apart from the first row and column). Precisely, the nonzero entries of the matrix are

$$K_{10} = K_{01} = -\frac{1}{2} \ , \qquad K_{hh} = \begin{cases} \frac{1}{2}, & h = 0 \ \text{or} \ h = 1 \ , \\ 1, & 2 \le h \le N \ . \end{cases} \tag{8.5.5}$$

The mass matrix has a nearly pentadiagonal structure. Precisely, the nonzero entries of such a symmetric matrix, in its upper triangular part, are

$$M_{00} = M_{11} = \frac{2}{3} \ , \qquad M_{01} = \frac{1}{3} \ , \tag{8.5.6a}$$
$$M_{02} = M_{12} = \frac{1}{\sqrt{6}} \ , \qquad M_{03} = -M_{13} = -\frac{1}{3\sqrt{10}} \ , \tag{8.5.6b}$$

and, for $2 \le h \le k \le N$,

$$M_{hk} = \begin{cases} \dfrac{2}{(2h-3)(2h+1)} \ , & k = h \ , \\[4mm] -\dfrac{1}{(2h+1)\sqrt{(2h-1)(2h+3)}} \ , & k = h+2 \ . \end{cases} \tag{8.5.6c}$$

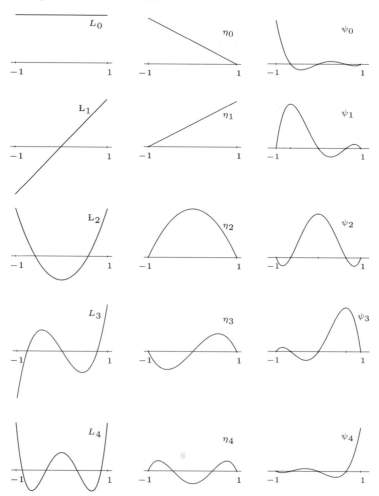

Fig. 8.8. Various Legendre basis functions on the interval $(-1,1)$, for $N = 4$: the modal orthogonal basis $\{L_k\}$ (*left*), the modal boundary-adapted basis $\{\eta_k\}$ given by (8.5.2) (*center*), the nodal basis at the Gauss–Lobatto points $\{\psi_k\}$ (*right*)

A comparison of the behavior of the members of the three bases mentioned above is given in Fig. 8.8 for $N = 4$.

8.6 Orthogonal Systems in Unbounded Domains

Laguerre Polynomials and Laguerre Functions. For any $\alpha > -1$, the *Laguerre polynomials* $l_k^{(\alpha)}(x)$, $k \geq 0$, are the eigenfunctions of the singular

Sturm–Liouville problem in $(0, +\infty)$:

$$\left(x^{\alpha+1} e^{-x} \left(l_k^{(\alpha)} \right)'(x) \right)' + kx^\alpha e^{-x} l_k^{(\alpha)}(x) = 0 \ . \tag{8.6.1}$$

They are orthogonal in $(0, +\infty)$ with respect to the weight $w(x) = x^\alpha e^{-x}$; precisely, using the normalization $l_k^{(\alpha)}(0) = \binom{k+\alpha}{k}$, one has

$$\int_0^{+\infty} l_k^{(\alpha)}(x) l_m^{(\alpha)}(x) x^\alpha e^{-x} \, dx = \Gamma(\alpha+1) \binom{k+\alpha}{k} \delta_{km} \ , \quad k, m \geq 0 \ . \tag{8.6.2}$$

In the particular case $\alpha = 0$, the polynomials $l_k(x) = l_k^{(0)}(x)$ satisfy $l_k(0) = 1$ and are orthonormal in $(0, +\infty)$.

The analog of the Rodriguez formula is

$$l_k^{(\alpha)}(x) = \frac{1}{k!} x^{-\alpha} e^x \frac{d^k}{dx^k} (x^{k+\alpha} e^{-x}) \ . \tag{8.6.3}$$

The Laguerre polynomials satisfy the recursion relation

$$l_{k+1}^{(\alpha)}(x) = (2k + \alpha + 1 - x) l_k^{(\alpha)}(x) - (k + \alpha) l_{k-1}^{(\alpha)}(x) \ , \tag{8.6.4}$$

where $l_0^{(\alpha)}(x) = 1$ and $l_1^{(\alpha)}(x) = \alpha + 1 - x$. The derivative of a Laguerre polynomial satisfies the relations

$$\frac{d}{dx} l_k^{(\alpha)}(x) = -l_{k-1}^{(\alpha+1)}(x) \tag{8.6.5}$$

and

$$x \frac{d}{dx} l_k^{(\alpha)}(x) = k l_k^{(\alpha)}(x) - l_{k-1}^{(\alpha)}(x) \ . \tag{8.6.6}$$

Any function $v \in L_w^2(0, +\infty)$ can be expanded in a Laguerre series as $v = \sum_k \hat{v}_k^{(\alpha)} l_k^{(\alpha)}$. The convergence of the series (in weighted square mean) is faster than algebraic, provided all the derivatives of the function belong to $L_w^2(0, +\infty)$.

For approximating functions that vanish at $+\infty$, it may be more appropriate to expand in the *Laguerre functions* defined as $\mathcal{L}_k(x) = e^{-x/2} l_k^{(0)}(x)$. Thanks to (8.6.2), they satisfy

$$\int_0^{+\infty} \mathcal{L}_k(x) \mathcal{L}_m(x) \, dx = \delta_{km} \ , \quad k, m \geq 0 \ , \tag{8.6.7}$$

and thus form an orthonormal basis in $L^2(0, +\infty)$.

Hermite Polynomials and Hermite Functions. The *Hermite polynomials* $H_k(x)$, $k \geq 0$, are the eigenfunctions of the singular Sturm–Liouville problem in $(-\infty, +\infty)$:

$$\left(e^{-x^2} H'_k(x)\right)' + 2ke^{-x^2} H_k(x) = 0 . \qquad (8.6.8)$$

They are orthogonal in $(-\infty, +\infty)$ with respect to the weight $w(x) = e^{-x^2}$; precisely, they satisfy

$$\int_{-\infty}^{+\infty} H_k(x) H_m(x) e^{-x^2} \, \mathrm{d}x = \sqrt{\pi}\, 2^k \, k! \, \delta_{km} , \quad k, m \geq 0 . \qquad (8.6.9)$$

The analog of the Rodriguez formula is

$$H_k(x) = (-1)^k e^{x^2} \frac{\mathrm{d}^k}{\mathrm{d}x^k} e^{-x^2} . \qquad (8.6.10)$$

The Hermite polynomials satisfy the recursion relation

$$H_{k+1}(x) = 2x H_k(x) - 2k H_{k-1}(x) , \quad k \geq 1 , \qquad (8.6.11)$$

where $H_0(x) = 1$ and $H_1(x) = 2x$. The derivative of an Hermite polynomial satisfies the relation

$$\frac{\mathrm{d}}{\mathrm{d}x} H_k(x) = 2k H_{k-1}(x) . \qquad (8.6.12)$$

A related family of Hermite polynomials is given by

$$He_k(x) = (1/\sqrt{2^k}) H_k(x/\sqrt{2}) , \quad k \geq 0 . \qquad (8.6.13)$$

Such polynomials are orthogonal with respect to the weight $\tilde{w}(x) = e^{-x^2/2}$.

The *Hermite functions* are defined as $\mathcal{H}_k(x) = e^{-x^2/2} H_k(x)$. Thanks to (8.6.9), they are orthogonal in $L^2(-\infty, +\infty)$:

$$\int_{-\infty}^{+\infty} \mathcal{H}_k(x) \mathcal{H}_m(x) \, \mathrm{d}x = \delta_{km} , \quad k, m \geq 0 , \qquad (8.6.14)$$

and form an orthonormal basis of this space.

8.7 Multidimensional Expansions

We deal with expansions in Cartesian-product domains, such as the square in 2D and the cube in 3D, as well as expansions in simplicial domains, such as the triangle in 2D and the tetrahedron in 3D. Expansions in such 3D domains as prisms or pyramids can then be obtained by a proper blending of the two previous expansions.

8.7.1 Tensor-Product Expansions

The most natural way to build a multidimensional expansion is to take tensor products of one-dimensional expansions; the resulting functions are defined on Cartesian products of intervals and inherit the one-dimensional features. Precisely, given d families $\{\phi_{k_l}^{(l)}\}_{k_l}$ of one-dimensional basis functions on intervals (a_l, b_l), the family $\{\phi_{\mathbf{k}}(\mathbf{x})\}_{\mathbf{k}}$ defined as

$$\phi_{\mathbf{k}}(\mathbf{x}) = \prod_{l=1}^{d} \phi_{k_l}^{(l)}(x_l), \quad \mathbf{k} = (k_1, \ldots, k_d), \ \mathbf{x} = (x_1, \ldots, x_d), \qquad (8.7.1)$$

is a multidimensional basis on the domain $\Omega = \prod_{l=1}^{d}(a_l, b_l)$. Examples are the multidimensional Fourier basis

$$\phi_{\mathbf{k}}(\mathbf{x}) = \prod_{l=1}^{d} e^{ik_l x_l} = e^{i\mathbf{k}\cdot\mathbf{x}}$$

defined on the periodic box $\Omega = (0, 2\pi)^d$, or the 3D Fourier-Chebyshev basis

$$\phi_{\mathbf{k}}(\mathbf{x}) = e^{i(k_1 x_1 + k_2 x_2)} T_{k_3}(x_3)$$

defined on $\Omega = (0, 2\pi)^2 \times (-1, 1)$.

Orthogonality of each one-dimensional family with respect to a weight $w_l(x_l)$ implies orthogonality of the tensor-product family with respect to the weight $w(\mathbf{x}) = \prod_{l=1}^{d} w_l(x_l)$. On the other hand, if each individual factor $\phi_{k_l}^{(l)}$ is a characteristic Lagrange polynomial relative to a family of quadrature points in $[a_l, b_l]$, then $\phi_{\mathbf{k}}$ is a characteristic Lagrange polynomial relative to the family of tensorized quadrature points in $\overline{\Omega}$.

The one-dimensional results on the precision of the quadrature rules and the decay rates of the coefficients extend to the tensor-product case as well.

First-order partial differentiation in wavenumber or in physical space can be accomplished by applying one-dimensional differentiation matrices to the coefficient vector in standard tensor-product fashion. For instance, if D_N denotes the one-dimensional Fourier interpolation derivative matrix (8.1.43), and I_N denotes the identity matrix of order N, then

$$D_N^{x_1} = D_N \otimes I_N \otimes I_N , \quad D_N^{x_2} = I_N \otimes D_N \otimes I_N , \quad D_N^{x_3} = I_N \otimes I_N \otimes D_N \quad (8.7.2)$$

are the 3D Fourier interpolation partial derivative matrices.

Boundary-adapted bases in each spatial direction tensorize to produce a boundary-adapted basis in Ω. A nodal basis in $(-1, 1)^d$ is given by

$$\psi_{\mathbf{j}}(\mathbf{x}) = \prod_{l=1}^{d} \psi_{j_l}(x_l) , \qquad (8.7.3)$$

where ψ_j is one of the N-degree characteristic Lagrange polynomials (8.2.30) relative to the Legendre Gauss–Lobatto points in $[-1, 1]$. On the other hand, a modal basis in $(-1, 1)^d$ is given by

$$\eta_{\mathbf{k}}(\mathbf{x}) = \prod_{l=1}^{d} \eta_{k_l}(x_l) , \qquad (8.7.4)$$

where η_k is one of the polynomials introduced in (8.5.2).

Any boundary-adapted basis is formed by bubble functions vanishing on the boundary $\partial\Omega$, by vertex functions not vanishing at precisely one vertex, by edge functions not vanishing at precisely one edge, by face functions not

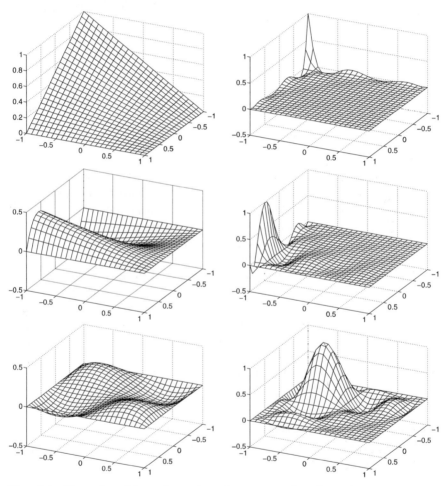

Fig. 8.9. Examples of boundary-adapted tensor-product basis functions on the square $(-1, 1)^2$ for $N = 4$: modal (*left*), nodal (*right*); vertex (*top*), edge (*center*), bubble (*bottom*). See also Fig. 8.8

vanishing at precisely one face, and so on. For instance, in 2D the bases (8.7.3) and (8.7.4) contain $(N-1)^2$ bubble functions, 4 vertex functions and $4(N-1)$ edge functions. Some of these functions are represented in Fig. 8.9. As is typical, nodal basis functions are more localized than modal basis functions, but are more oscillatory.

8.7.2 Expansions on Simplicial Domains

Spectral polynomial approximations on elementary non-tensor-product domains, such as triangles or tetrahedra, can be built according to two different strategies, which are described in the following two subsections.

Collapsed Coordinates and Warped Tensor-Product Expansions. We describe this approach, often associated with the name of Dubiner, in two dimensions. Let us introduce the reference triangle

$$\hat{\Omega}_S^2 = \{(x_1, x_2) \in \mathbb{R}^2 \ : \ -1 < x_1, x_2 \ ; \ x_1 + x_2 < 0\}$$

as well as the reference square

$$\hat{\Omega}_C^2 = \{(\xi_1, \xi_2) \in \mathbb{R}^2 \ : \ -1 < \xi_1, \xi_2 < 1\} \, .$$

The mapping

$$(x_1, x_2) \mapsto (\xi_1, \xi_2), \qquad \xi_1 = 2\frac{1+x_1}{1-x_2} - 1, \quad \xi_2 = x_2 \, , \qquad (8.7.5)$$

is a bijection between $\hat{\Omega}_S^2$ and $\hat{\Omega}_C^2$. Its inverse is given by

$$(\xi_1, \xi_2) \mapsto (x_1, x_2), \qquad x_1 = \frac{1}{2}(1+\xi_1)(1-\xi_2) - 1, \quad x_2 = \xi_2 \, . \quad (8.7.6)$$

Note that the mapping $(x_1, x_2) \mapsto (\xi_1, \xi_2)$ sends the ray in $\hat{\Omega}_S^2$ issuing from the upper vertex $(-1, 1)$ and passing through the point $(x_1, -1)$ into the vertical segment in $\hat{\Omega}_C^2$ of the equation $\xi_1 = x_1$ (see Fig. 8.10). Consequently, the transformation becomes singular at the upper vertex of the triangle, although it stays bounded as one approaches the vertex. The determinant of the Jacobian of the inverse transformation is given by

$$\left| \frac{\partial(x_1, x_2)}{\partial(\xi_1, \xi_2)} \right| = \frac{1-\xi_2}{2} \, . \qquad (8.7.7)$$

We term (ξ_1, ξ_2) the *collapsed Cartesian coordinates* of the point on the triangle whose regular Cartesian coordinates are (x_1, x_2).

For $\mathbf{k} = (k_1, k_2)$, define the *warped tensor-product basis* function on $\hat{\Omega}_C^2$:

$$\Phi_{\mathbf{k}}(\xi_1, \xi_2) = \Psi_{k_1}(\xi_1)\Psi_{k_1, k_2}(\xi_2), \qquad (8.7.8)$$

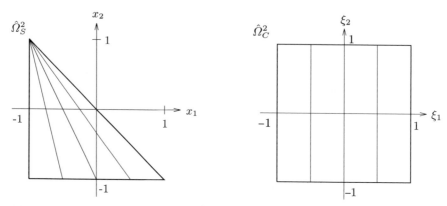

Fig. 8.10. The reference triangle $\hat{\Omega}_S^2$ is mapped onto the reference square $\hat{\Omega}_C^2$. Oblique segments are transformed into vertical segments

where

$$\Psi_{k_1}(\xi_1) = P_{k_1}^{(0,0)}(\xi_1), \quad \Psi_{k_1,k_2}(\xi_2) = (1-\xi_2)^{k_1} P_{k_2}^{(2k_1+1,0)}(\xi_2), \quad (8.7.9)$$

which is a polynomial of degree k_1 in ξ_1 and $k_1 + k_2$ in ξ_2 ($P_k^{(\alpha,\beta)}$ denotes as usual a Jacobi polynomial). By applying the mapping (8.7.5) one obtains the function defined on $\hat{\Omega}_S^2$:

$$\begin{aligned}
\varphi_\mathbf{k}(x_1, x_2) &= \Phi_\mathbf{k}(\xi_1, \xi_2) \\
&= P_{k_1}^{(0,0)}\left(2\frac{1+x_1}{1-x_2} - 1\right)(1-x_2)^{k_1} P_{k_2}^{(2k_1+1,0)}(x_2).
\end{aligned} \quad (8.7.10)$$

It is easily seen that $\varphi_\mathbf{k}$ is a polynomial of global degree k_1+k_2 in the variables x_1, x_2. Furthermore, thanks to the orthogonality of Jacobi polynomials, one has, for $\mathbf{k} \neq \mathbf{h}$,

$$\int_{\hat{\Omega}_S^2} \varphi_\mathbf{k}(x_1,x_2)\varphi_\mathbf{h}(x_1,x_2)\,dx_1 dx_2 = \frac{1}{2}\int_{-1}^{1} P_{k_1}^{(0,0)}(\xi_1)P_{h_1}^{(0,0)}(\xi_1)\,d\xi_1$$

$$\times \int_{-1}^{1} P_{k_2}^{(2k_1+1,0)}(\xi_2)P_{h_2}^{(2h_1+1,0)}(\xi_2)(1-\xi_2)^{k_1+h_1+1}\,d\xi_2 = 0\,.$$

We conclude that the set $\{\varphi_\mathbf{k} \; : \; 0 \leq k_1, k_2 \text{ and } k_1+k_2 \leq N\}$ is an orthogonal modal basis of the space

$$\mathbb{P}_N = \text{span}\left\{x_1^i x_2^j \; : \; 0 \leq i, j \text{ and } i+j \leq N\right\} \quad (8.7.11)$$

of the polynomials of global degree $\leq N$ in the variables x_1, x_2.

While orthogonality simplifies the structure of the mass matrix, it complicates the enforcement of boundary conditions, or of matching conditions between subdomains. This difficulty can be surmounted by building a new modal basis, say $\{\varphi_{\mathbf{k}}^{ba}\}$, where "ba" stands for boundary adapted; it consists of boundary functions (3 vertex functions plus $3(N-1)$ edge functions) and internal functions ($\frac{1}{2}(N-2)(N-1)$ bubbles). Each basis function retains the same "warped tensor-product" structure as above. Indeed, it is enough to replace the one-dimensional Jacobi basis $P_k^{(\alpha,0)}(\xi)$ (with $\alpha = 0$ or $2k+1$) by the boundary-adapted basis given by the two boundary functions $\dfrac{1+\xi}{2}$ and $\dfrac{1-\xi}{2}$ and by the $N-1$ bubbles $\left(\dfrac{1+\xi}{2}\right)\left(\dfrac{1-\xi}{2}\right)P_{k-2}^{(\alpha,\beta)}(\xi)$, $k = 2,\ldots,N$, (for suitable $\alpha,\beta \geq 1$ fixed). Note that the choice $\alpha = \beta = 1$ yields the boundary-adapted basis η_k, $k = 0,\ldots,N$, defined in (8.5.1)–(8.5.4) (up to a normalization factor).

These univariate functions are then combined as in (8.7.8) to form the 2D basis. To be precise, the vertex functions, expressed in the (ξ_1,ξ_2)-coordinates, are

$$\Phi^{V_1}(\xi_1,\xi_2) = \left(\frac{1-\xi_1}{2}\right)\left(\frac{1-\xi_2}{2}\right) \qquad (\text{vertex } V_1 = (-1,-1)) \,,$$

$$\Phi^{V_2}(\xi_1,\xi_2) = \left(\frac{1+\xi_1}{2}\right)\left(\frac{1-\xi_2}{2}\right) \qquad (\text{vertex } V_2 = (+1,-1)) \,,$$

$$\Phi^{V_3}(\xi_1,\xi_2) = \frac{1+\xi_2}{2} \qquad (\text{vertex } V_3 = (-1,+1)) \,;$$

the edge functions are defined as

$$\Phi_{k_1}^{V_1 V_2}(\xi_1,\xi_2) = \left(\frac{1-\xi_1}{2}\right)\left(\frac{1+\xi_1}{2}\right) P_{k_1-2}^{(\beta,\beta)}(\xi_1)\left(\frac{1-\xi_2}{2}\right)^{k_1} \,, \quad 2 \leq k_1 \leq N \,,$$

$$\Phi_{k_2}^{V_1 V_3}(\xi_1,\xi_2) = \left(\frac{1-\xi_1}{2}\right)\left(\frac{1-\xi_2}{2}\right)\left(\frac{1+\xi_2}{2}\right) P_{k_2-2}^{(\beta,\beta)}(\xi_2) \,, \quad 2 \leq k_2 \leq N \,,$$

$$\Phi_{k_2}^{V_2 V_3}(\xi_1,\xi_2) = \left(\frac{1+\xi_1}{2}\right)\left(\frac{1-\xi_2}{2}\right)\left(\frac{1+\xi_2}{2}\right) P_{k_2-2}^{(\beta,\beta)}(\xi_2) \,; \quad 2 \leq k_2 \leq N \,;$$

finally, the bubble functions are defined for $k_1, k_2 \geq 2$ and $k_1 + k_2 \leq N$, as

$$\Phi_{k_1,k_2}^{B}(\xi_1,\xi_2) = \left(\frac{1-\xi_1}{2}\right)\left(\frac{1+\xi_1}{2}\right) P_{k_1-2}^{(\beta,\beta)}(\xi_1)$$

$$\times \left(\frac{1-\xi_2}{2}\right)^{k_1}\left(\frac{1+\xi_2}{2}\right) P_{k_2-2}^{(2k_1-1+\delta,\beta)}(\xi_2) \,.$$

The choice $\beta = \delta = 2$ yields orthogonality among the bubble functions (and certain boundary functions). However, usually the choice $\beta = 1$, $\delta = 0$

is preferred. Indeed, thanks to property (8.2.8), it guarantees a good compromise in the sparsity pattern of both mass and stiffness matrices; furthermore, it leads to a more favorable conditioning of the stiffness matrix associated with a second-order operator.

Expansions in 3D simplicial domains are illustrated in CHQZ2, Sect. 2.9 and are thoroughly covered in Karniadakis and Sherwin (2005).

Non-Tensor-Product Expansions. Boundary-adapted nodal bases $\{\psi_j\}$ in the reference triangle $\hat{\Omega}_S^2$ are uniquely determined by suitable distributions of interpolation nodes $\{x_j\}$, with the condition that the nodes on each side of $\partial\hat{\Omega}_S^2$ are the (mapped) one-dimensional Gauss–Lobatto nodes. For a basis in $\mathbb{P}_N$, whose dimension is $\frac{1}{2}(N+1)(N+2)$, the $3N$ boundary nodes are thus fixed, whereas the remaining internal nodes are usually chosen according to some optimization criterion. Among the existing constructions, we mention the Stieltjes–Hesthaven electrostatic points, the Chen-Babuška points, and the Fekete points. We refer to CHQZ2, Sect. 2.9.2 for further details. None of these families are known analytically, and their numerical computation becomes more and more delicate as N increases.

8.8 Mappings

While the families of orthogonal polynomials are generally defined on $[-1, 1]$, one-dimensional physical domains can be on an arbitrary domain $[a, b]$ that can be finite, semi-infinite, or the entire real line. In general, some type of mapping is required between the computational, or reference, domain $[-1, 1]$ and the physical domain $[a, b]$. In this section we first survey the types of mappings that have been employed in classical spectral methods, and then comment on approaches to mappings in more than one dimension. We restrict ourselves here to mappings involving scalar variables. Mappings for vector variables are treated in Sect. E.3. Furthermore, we survey here just the basic, general principles of the use of mappings with spectral methods, while some special cases are discussed in Chaps. 2–4. See Boyd (2001), Chaps. 16 and 17, for extensive coverage of this subject, especially on how the choice of mapping affects the convergence rate in spectral methods. We concentrate on mappings used in conjunction with Chebyshev expansions, although we make a few comments on mappings useful for Fourier expansions.

For the one-dimensional case, we denote the original, physical variable by $x \in [a, b]$ and the computational variable by $\xi \in [-1, 1]$ (the reference domain for Chebyshev polynomials), and write the mapping as

$$x = F(\xi), \quad \xi = F^{-1}(x) = G(x), \quad \xi \in [-1, 1], \quad x \in [a, b]. \quad (8.8.1)$$

Any function $u(x)$ is then expanded as

$$u(x) = \sum_{k=0}^{N} \hat{u}_k T_k(\xi(x)) . \tag{8.8.2}$$

For collocation methods, the impact of the mapping upon the derivative of a function $u(x)$ appears in the computational domain as

$$\frac{du}{dx} = \frac{d\xi}{dx} \frac{du}{d\xi} , \tag{8.8.3}$$

where $d\xi/dx$ is the Jacobian of the mapping. Although this Jacobian can sometimes be computed analytically, one can always utilize the discretization to approximate the Jacobian as

$$\frac{d\xi}{dx} \approx \frac{1}{\mathcal{D}_N^\xi[x(\xi)]} , \tag{8.8.4}$$

where the superscript ξ on $\mathcal{D}_N$ denotes that this is the interpolation differentiation operator (see (8.2.34)) with respect to ξ. Indeed, many practitioners prefer to compute the Jacobian numerically in all cases. An additional consideration is that the accuracy of the results are dependent upon how well the mapping is resolved by the discretization. Hence, the more severe the mapping, the more grid points are needed.

8.8.1 Finite Intervals

If $[a, b] \neq [-1, 1]$, then at a minimum the following affine mapping is required:

Affine Mapping.

$$x = a + \frac{b-a}{2}(\xi + 1) , \quad \xi = \frac{2}{b-a}(x - a) - 1 , \quad \xi \in [-1, 1] , \quad x \in [a, b] . \tag{8.8.5}$$

But for any finite interval, a nonlinear mapping may be desired to achieve a local clustering of grid points or to relax the time-step restriction. The mapping (8.8.1) places more points in x (relative to their space in the reference variable ξ) where $dx/d\xi$ is small and fewer where $dx/d\xi$ is large.

For orthogonal polynomial approximations on the reference interval $[-1, 1]$, the grid points are naturally clustered near the endpoints of the interval. If the function $u(x)$ has significant variation in the vicinity of an interior point x_c of $(-1, 1)$, then a mapping such as

Polynomial Clustering Mapping.

$$\frac{\xi - \xi_c}{\Delta_\xi} = x + \epsilon \tanh\left(\frac{x - x_c}{\Delta_x}\right) , \qquad \xi, x \in [-1, 1] , \qquad (8.8.6)$$

where

$$\xi_c = \epsilon \frac{\tanh\left(\frac{1+x_c}{\Delta_x}\right) - \tanh\left(\frac{1-x_c}{\Delta_x}\right)}{2 + \epsilon \tanh\left(\frac{1+x_c}{\Delta_x}\right) + \tanh\left(\frac{1-x_c}{\Delta_x}\right)} ,$$

$$\Delta_\xi = \frac{2}{2 + \epsilon \tanh\left(\frac{1+x_c}{\Delta_x}\right) + \tanh\left(\frac{1-x_c}{\Delta_x}\right)} ,$$

(8.8.7)

will cluster points near x_c. (This mapping must be inverted numerically.) The other free parameters Δ_x and ϵ control the width of the region and the strength of the clustering, respectively.

Another motivation for a nonlinear mapping is to relax the time-step restriction. The mapping

Kosloff–Tal-Ezer Mapping.

$$x = \frac{\sin^{-1}[(1 - \beta)\xi]}{\sin^{-1}(1 - \beta)} , \qquad \xi = \frac{\sin\left[x \sin^{-1}(1 - \beta)\right]}{1 - \beta} , \qquad \xi, x \in [-1, 1] ,$$

(8.8.8)

for $0 < \beta < 1$, was proposed by Kosloff and Tal-Ezer (1993) to relax the natural clustering near the endpoints in Chebyshev collocation methods. In the limit $\beta \to 1$, this produces the usual Chebyshev distribution, whereas in the limit $\beta \to 0$, the result is a uniform distribution in x of the grid points. In the latter limit, or even in cases for which $\beta \approx 1/N^2$, the time-step limitation is indeed much relaxed from that associated with the usual Chebyshev distribution, but spectral accuracy is then lost. Figure 8.11 (left) illustrates the effect of (8.8.8) for three values of β. The grid spacings near the endpoints are 0.0048, 0.0082 and 0.0466 for $\beta = 1$, $1 - \cos(1/2)$ and $1 - \cos(1/N)$, respectively. See Boyd (2001) for further discussion of the trade-off between the time-step restriction and accuracy.

Fourier approximations use a uniform grid on the reference interval $(0, 2\pi)$. For a case in which the solution has strong variation in the vicinity of, say, $x = 0$ or $x = \pi$, Boyd (2001) recommends the mapping

Fourier Clustering Mapping.

$$x = \pi + 2 \tan^{-1}\left[L \tan\left(\frac{\xi - \pi}{2}\right)\right] ,$$

$$\xi = \pi + 2 \tan^{-1}\left[\frac{1}{L} \tan\left(\frac{x - \pi}{2}\right)\right] , \qquad \xi, x \in [0, 2\pi] ,$$

(8.8.9)

to provide clustering in x near 0. Here L is a length scale that controls the clustering: near 0 if $L < 1$, near π if $L > 1$, and the identify map if $L = 1$. See Fig. 8.11 (right) for an illustration.

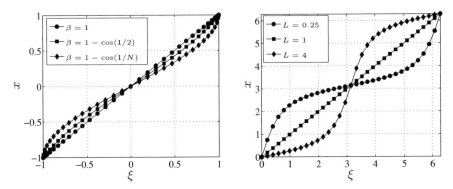

Fig. 8.11. (*Left*) Mapping of the computational coordinate $\xi \in [-1,1]$ into the physical coordinate $x \in [-1,1]$ to relax the normal Chebyshev clustering near the endpoints in the physical domain. The symbols denote the images of the standard Chebyshev collocation points for $N = 32$. (*Right*) Mapping of the computational coordinate $\xi \in [0, 2\pi]$ into the physical coordinate $x \in [0, 2\pi]$ to cluster points in the physical domain. The symbols denote the images of the standard Fourier collocation points for $N = 32$

8.8.2 Semi-Infinite Intervals

There are three general approaches to dealing with semi-infinite intervals such as $[0, \infty)$: semi-infinite stretching, domain truncation and truncated stretching.

Semi-infinite stretching retains the semi-infinite interval as the approximation domain, but uses a nonlinear mapping between $[-1, 1)$ and $[0, +\infty)$. There are two general categories of semi-infinite stretchings, given by the following formulas:

(Semi-Infinite) Algebraic Mapping.

$$x = L\frac{1+\xi}{1-\xi}\,, \quad \xi = \frac{x-L}{x+L}\,, \qquad \xi \in [-1, 1)\,, \quad x \in [0, +\infty)\,; \qquad (8.8.10)$$

(Semi-Infinite) Exponential Mapping.

$$x = -L \log\left(\frac{1-\xi}{2}\right)\,, \quad \xi = 1 - 2e^{-x/L}\,, \qquad \xi \in [-1, 1)\,, \quad x \in [0, +\infty)\,. \qquad (8.8.11)$$

The functions $T_k(x(\xi))$ resulting from the algebraic mapping are rational functions of ξ, referred to as rational Chebyshev functions. Unlike expansions on a finite domain, spectral approximations on a semi-infinite domain have two discretization parameters—the length scale L in addition to the usual series truncation parameter N. As a general rule, the length scale L needs to be increased with N in order to obtain spectral accuracy (see Boyd (2001)). Numerous authors (e.g., Grosch and Orszag (1977), Boyd (1982), Herbert

(1984)) have reported that, in practice, algebraic mappings are more accurate and more robust (less sensitive to the scale factor L) than exponential ones.

Simple domain truncation consists of approximating the function on the interval $[0, x_{max}]$ instead of $[0, +\infty)$ and using the affine mapping (8.8.5):

(Semi-Infinite) Domain Truncation.

$$x = \frac{x_{max}}{2}(\xi+1) , \quad \xi = \frac{2x}{x_{max}} - 1 , \qquad \xi \in [-1,1] , \quad x \in [0, x_{max}] , \quad (8.8.12)$$

Obviously, domain truncation can be combined with nonlinear mapping to produce truncated stretchings, e.g.,

(Semi-Infinite) Truncated Algebraic Mapping.

$$x = L\frac{1+\xi}{\xi_{max} - \xi} , \qquad \xi = \frac{x\xi_{max} - L}{x + L} ,$$

$$\xi_{max} = \frac{2L + x_{max}}{x_{max}} , \qquad \xi \in [-1,1] , \quad x \in [0, x_{max}] ; \qquad (8.8.13)$$

(Semi-Infinite) Truncated Exponential Mapping.

$$x = -L \log\left(\frac{(\xi_{max} - 2)\xi + \xi_{max}}{2}\right) , \qquad \xi = \xi_{max} - 2e^{-x/L} ,$$

$$\xi_{max} = 1 + e^{-x_{max}/L} , \qquad \xi \in [-1,1] , \quad x \in [0, x_{max}] . \qquad (8.8.14)$$

The truncated mappings introduce a third parameter, x_{max}, that needs to increase with N in order to obtain spectral accuracy.

Both the full semi-infinite mapping and the transformations truncated at $x_{max} = 15$ are illustrated in Fig. 8.12. The length scales used for this

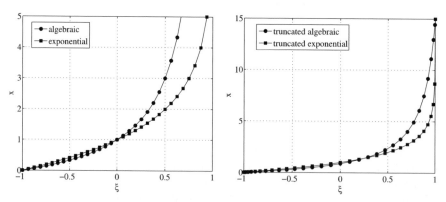

Fig. 8.12. Mapping of the computational coordinate ξ into the physical coordinate $x \in [0, +\infty)$ (*left*) and into $x \in [0, 15]$ (*right*) for algebraic and exponential transformations. The symbols denote the images of the Chebyshev Gauss–Lobatto points for $N = 32$

figure—$L = 1$ for the algebraic map and $L = 1/\log(2)$ for the exponential map—were chosen so that the full semi-infinite mappings have half of their collocation points in the interval $[0, 1]$. Clearly, the algebraic mapping places more collocation points at larger values of x, while the exponential mapping has better resolution near $x = 0$.

This terminology for mappings used with spectral methods dates from the work of Grosch and Orszag (1977) and refers to the rate at which ξ approaches 1 as $x \to +\infty$. The constant L sets the length scale of the mappings. Boyd (1982) provides an extensive discussion and guidelines for the choice of length scale. In general, the algebraic mapping is best suited to approximation of functions that decay relatively slowly, e. g., algebraically in $1/x$ as $x \to +\infty$, whereas the exponential mapping is more appropriate for more rapidly decaying functions, e. g., decaying exponentially in x.

A wide variety of nonlinear mappings have been used in combination with domain truncation. One specific example that arises in Sect. 2.2 is the

(Semi-Infinite) Truncated Hyperbolic Tangent Mapping.

$$x = \frac{x_{\max}}{2} \frac{[1 - \tanh(\sigma)](1 + \xi)}{1 - \tanh[\sigma(1 + \xi)/2]} , \qquad \xi \in [-1, 1] , \quad x \in [0, x_{\max}] . \quad (8.8.15)$$

This mapping does not have a closed form inverse.

Spalart (1984) observed that the use of the exponential mapping (8.8.11) for a function that decays faster than exponentially (as a Gaussian, for example) results in an inefficient distribution of grid points. Because of the clustering of nodes at $\xi = -1$ and $\xi = 1$, there will be more nodes for large x than are required to resolve the function. Spalart proposed replacing (8.8.11) with

Half Exponential Mapping.

$$x = -L \log \xi , \quad \xi = e^{-x/L} , \qquad \xi \in (0, 1] , \quad x \in [0, +\infty) . \quad (8.8.16)$$

Then the function $u(x(\xi))$ and all of its derivatives are zero at $\xi = 0$. Hence, $u(x(\xi))$ may be extended smoothly to a function on $[-1, 0)$. (In some cases, just the odd or just the even Chebyshev polynomials are appropriate expansion functions.) The grid points are clustered near $\xi = 1$ ($x = 0$) and are coarsely distributed near $\xi = 0$ ($x = +\infty$). Likewise, for an exponentially decaying function, Chebyshev expansions may be combined with the map

Half Algebraic Mapping.

$$x = L\frac{\xi}{1 - \xi} , \quad \xi = \frac{x}{x + L} , \qquad \xi \in [0, 1) , \quad x \in [0, +\infty) . \quad (8.8.17)$$

8.8.3 The Real Line

Similar considerations apply to expansions on the real line $(-\infty, +\infty)$ as on semi-infinite intervals. We'll focus here just on the mappings for the full real line. For Chebyshev expansions the counterparts of (8.8.10)–(8.8.12) are

(Real Line) Algebraic Mapping.

$$x = L\frac{\xi}{\sqrt{1-\xi^2}}, \quad \xi = \frac{x}{\sqrt{x^2 + L^2}}, \quad \xi \in (-1,1), \quad x \in (-\infty, +\infty);$$

$$(8.8.18)$$

(Real Line) Exponential Mapping.

$$x = L\tanh^{-1}\xi, \quad \xi = \tanh\left(\frac{x}{L}\right), \quad \xi \in (-1,1), \quad x \in (-\infty, +\infty);$$

$$(8.8.19)$$

(Real Line) Domain Truncation.

$$x = x_{\max}\,\xi, \quad \xi = \frac{x}{x_{\max}}, \quad \xi \in [-1,1], \quad x \in [-x_{\max}, x_{\max}]. \quad (8.8.20)$$

For (8.8.18) and (8.8.19) one expects infinite-order accuracy, even if $u(-\infty) \neq u(+\infty)$, provided that the derivatives of u decay sufficiently fast, i.e., algebraic decay with (8.8.18) and exponential decay with (8.8.19), and of, course, provided that $u(x)$ is analytic at $x = \pm\infty$. As for the semi-infinite interval, the functions $T_k(x(\xi))$ resulting from the algebraic mapping on the real line are rational Chebyshev functions, and the length scale L needs to be increased with N in order to have spectral accuracy (see Boyd (2001), Chap. 17). Domain truncation can be combined with nonlinear mappings, similar to (8.8.13)–(8.8.14), for the real line case as well.

For the real line, mappings combined with Fourier expansions have seen considerable use. Cain et al. (1984) suggested the use of two such mappings in conjunction with Fourier series:

(Real Line) Half Cotangent.

$$x = -L\cot(\xi/2), \quad \xi = -2\cot^{-1}(x/L), \quad \xi \in (0, 2\pi), \quad x \in (-\infty, +\infty);$$
$$(8.8.21)$$

(Real Line) Full Cotangent.

$$x = -L\cot\xi, \quad \xi = -\cot^{-1}(x) \quad \xi \in (0, \pi), \quad x \in (-\infty, +\infty). \quad (8.8.22)$$

In the former case, infinite-order accuracy is only achieved if the function $u(x)$ and all of its derivatives exist and match at $x = -\infty$ and $x = +\infty$. The reason is that the function $u(x(\xi))$ is implicitly extended periodically by the use of Fourier series, and continuity of $u(x(\xi))$ and all its derivatives is required for spectral accuracy.

The latter case is applicable for functions that approach different limits (but exponentially fast) at $x = \pm\infty$, such as $u(x) = \tanh x$. Here, $u(x(\xi))$ is extended to $\xi \in (\pi, 2\pi)$ by reflection. When coupled with Fourier series, this yields infinite-order accuracy.

Boyd (1987) has discussed the use of the mapping (8.8.22) on just $(0, \pi)$ in conjunction with a sine and cosine expansion (as opposed to the complex

Fourier series on $(0, 2\pi)$). He noted that if just the cosine expansion is used, then $u(x)$ must at least have exponential decay (or special symmetries). If the decay is only algebraic and no special symmetries are present, then only algebraic convergence is possible with the cosine expansion.

8.8.4 Multidimensional Mappings on Finite Domains

In two dimensions, a fairly simple procedure exists for mapping the reference domain, in this case the square $\hat{\Omega}_C^2$, into a general quadrilateral Ω with curved boundaries. The basic geometry is illustrated in Fig. 8.13. Let the four sides of the quadrilateral be denoted by Γ_i, for $i = 1, 2, 3, 4$ and those of the reference square by $\hat{\Gamma}_i$. One uses parametrizations π_i from the interval $[-1, 1]$ to Γ_i to construct the mapping F from $\hat{\Omega}_C^2$ to $\overline{\Omega}$, such that $F(\hat{\Gamma}_i) = \Gamma_i$ for all i. Gordon and Hall (1973a, 1973b) described a variety of mappings. The simplest is a linear blending mapping, for which F can be expressed in terms of the π_i as

$$
\begin{aligned}
F(\xi, \eta) = {} & \frac{1-\eta}{2}\pi_3(\xi) + \frac{1+\eta}{2}\pi_1(\xi) \\
& + \frac{1-\xi}{2}\left[\pi_2(\eta) - \frac{1+\eta}{2}\pi_2(1) - \frac{1-\eta}{2}\pi_2(-1)\right] \\
& + \frac{1+\xi}{2}\left[\pi_4(\eta) - \frac{1+\eta}{2}\pi_4(1) - \frac{1-\eta}{2}\pi_4(-1)\right] .
\end{aligned}
\tag{8.8.23}
$$

(We assume that the arcs Γ_1 and Γ_3 are oriented from left to right and the arcs Γ_2 and Γ_4 from bottom to top.)

The Gordon–Hall transformation can be easily extended to three dimensions. A straightforward implementation is as follows: Let $\hat{\Omega}_C^3 = (-1, 1)^3$ be the reference cube with coordinates (ξ, η, ζ), and let $\hat{a}_i$ ($i = 1, \ldots, 8$) and $\hat{\Sigma}_i$ ($i = 1, \ldots, 6$) denote its vertices and faces, respectively, numbered as shown in Fig. 8.14. Let $\Omega \subset \mathbb{R}^3$ be a hexahedron, with faces Σ_i ($i = 1, \ldots, 6$), that is the image of $\hat{\Omega}_C^3$ under a smooth transformation F. We assume that we

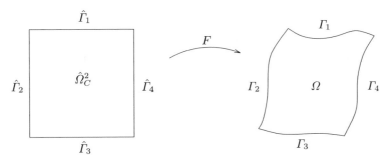

Fig. 8.13. Mapping of the reference square $\hat{\Omega}_C^2 = (-1, 1)^2$ into a quadrilateral Ω with curved boundaries

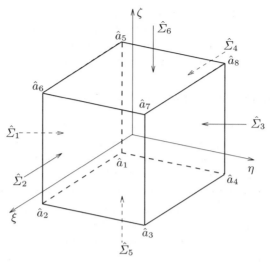

Fig. 8.14. Vertices and faces of the unit reference cube $\hat{\Omega}_C^3 = (-1, 1)^3$

know each mapping $\pi_i : [-1, 1]^2 \to \mathbb{R}^3$ from the reference square to the face Σ_i, that is, the image of the face $\hat{\Sigma}_i$ under the transformation; with obvious notation, we have $\pi_1 = \pi_1(\xi, \zeta)$, $\pi_2 = \pi_2(\eta, \zeta)$, $\pi_3 = \pi_3(\xi, \zeta)$, $\pi_4 = \pi_4(\eta, \zeta)$, $\pi_5 = \pi_5(\xi, \eta)$, $\pi_6 = \pi_6(\xi, \eta)$. The vertices of Ω can be obtained as $a_1 = \pi_1(-1, -1)$, $a_2 = \pi_1(1, -1)$, $a_3 = \pi_3(1, -1)$, $a_4 = \pi_3(-1, -1)$, $a_5 = \pi_1(-1, 1)$, $a_6 = \pi_1(1, 1)$, $a_7 = \pi_3(1, 1)$, and $a_8 = \pi_3(-1, 1)$. Then, we can define $F \colon \hat{\Omega}_C^3 \to \overline{\Omega}$ as follows:

$$
F(\xi, \eta, \zeta) = \frac{1-\xi}{2}\pi_4(\eta, \zeta) + \frac{1+\xi}{2}\pi_2(\eta, \zeta) + \frac{1-\eta}{2}\pi_1(\xi, \zeta)
$$

$$
+ \frac{1+\eta}{2}\pi_3(\xi, \zeta) + \frac{1-\zeta}{2}\pi_5(\xi, \eta) + \frac{1+\zeta}{2}\pi_6(\xi, \eta)
$$

$$
- \frac{1-\xi}{2}\frac{1-\eta}{2}\frac{1-\zeta}{2}\pi_1(-1, -1) - \frac{1+\xi}{2}\frac{1-\eta}{2}\frac{1-\zeta}{2}\pi_1(1, -1)
$$

$$
- \frac{1+\xi}{2}\frac{1+\eta}{2}\frac{1-\zeta}{2}\pi_3(1, -1) - \frac{1-\xi}{2}\frac{1+\eta}{2}\frac{1-\zeta}{2}\pi_3(-1, -1)
$$

$$
- \frac{1-\xi}{2}\frac{1-\eta}{2}\frac{1+\zeta}{2}\pi_1(-1, 1) - \frac{1+\xi}{2}\frac{1-\eta}{2}\frac{1+\zeta}{2}\pi_1(1, 1)
$$

$$
- \frac{1+\xi}{2}\frac{1+\eta}{2}\frac{1+\zeta}{2}\pi_3(1, 1) - \frac{1-\xi}{2}\frac{1+\eta}{2}\frac{1+\zeta}{2}\pi_3(-1, 1).
$$

$$(8.8.24)$$

More efficient implementations, in which the vertices, edges and faces are accounted for hierarchically, are available; see, e.g., Deville et al. (2002).

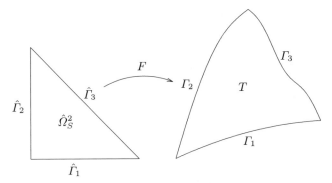

Fig. 8.15. Mapping of the reference triangle $\hat{\Omega}_S^2$ into a triangle T with curved boundaries

In the event that the domain Ω is actually a subdomain in a multidomain spectral method (see Chap. 5), the use of an isoparametric description of the curves Γ_i may be desirable. Here one chooses the curves Γ_i so that they are exactly parametrizable by polynomials of the same order as the discretization within Ω.

The idea underlying the Gordon–Hall transformations provides the guidance needed to define a transformation between a reference triangle and a triangular domain with possibly curved sides, or between a nontensorial reference domain such as a prism, a pyramid or a tetrahedron, and a similar domain with possibly curved faces and edges.

Let us consider the 2D situation. Let $\hat{\Omega}_S^2 = \{(\xi, \eta) \in \mathbb{R}^2 \; : \; -1 < \xi, \eta \; ; \; \xi + \eta < 0\}$ be the reference triangle, and let $\hat{\Gamma}_1$ ($\hat{\Gamma}_2$, $\hat{\Gamma}_3$, resp.) denote the side of $\hat{\Omega}_S^2$ whose equation is $\eta = -1$ ($\xi = -1$, $\xi + \eta = 0$, resp.) (see Fig. 8.15). Let T be a triangular domain in the plane, with possibly curved sides, Γ_1, Γ_2, Γ_3, such that parametrizations $\pi_i : [-1, 1] \to \Gamma_i$ ($i = 1, 2, 3$) of the sides are known. We assume that the three vertices of T are described by $\pi_1(-1) = \pi_2(-1)$, $\pi_1(1) = \pi_3(1)$ and $\pi_2(1) = \pi_3(-1)$.

A mapping $F : \hat{\Omega}_S^2 \to \overline{T}$ that extends smoothly the boundary mappings is given by

$$F(\xi, \eta) = \frac{1 - \eta}{2} \pi_1(\xi) - \frac{1 + \xi}{2} \pi_1(-\eta) + \frac{1 - \xi}{2} \pi_2(\eta) - \frac{1 + \eta}{2} \pi_2(-\xi)$$

$$+ \left(1 + \frac{\xi + \eta}{2} \right) \pi_3(\xi) - \frac{1 + \xi}{2} \pi_3(1 + \xi + \eta) \qquad (8.8.25)$$

$$+ \frac{1 + \xi}{2} \pi_1(1) + \frac{\xi + \eta}{2} \pi_1(-1) \, .$$

As for the tensorial case, a common practice for parametrizing the sides of T is to use isoparametric interpolation, i. e., polynomials of the same order as the basis chosen in $\hat{\Omega}_S^2$.

8.9 Basic Spectral Discretization Methods

This section illustrates the standard spectral discretization schemes in the context of a model one-dimensional boundary-value problem. Given that non-periodic boundary conditions are used, these examples all utilize expansions in algebraic (Jacobi) polynomials. The same principles apply to the use of expansions in trigonometric (Fourier) polynomials, except that the tau method is not applicable to that case. Complete coverage of spectral discretization methods in a single domain is provided in CHQZ2, to which we refer for all details.

We consider the model second-order differential equation

$$-\frac{d}{dx}\left(\alpha\frac{du}{dx}\right) + \beta\frac{du}{dx} + \gamma u = f \qquad (8.9.1)$$

in the interval $I = (-1,1)$, coupled with homogeneous Dirichlet boundary conditions

$$u(-1) = u(1) = 0 . \qquad (8.9.2)$$

(Other boundary conditions will be considered in Sect. 8.9.5.) We assume that the coefficients α, β, γ are given functions such that the boundary-value problem (8.9.1)–(8.9.2) has a unique solution for any sufficiently smooth data f. For instance, this situation occurs if the coefficients are bounded and smooth in I, α satisfies $\alpha \geq \alpha_0$ in I for some constant $\alpha_0 > 0$, and β and γ satisfy $-\frac{1}{2}\frac{d\beta}{dx} + \gamma \geq 0$ in I. Let us introduce the differential operator $\mathcal{L} = -\frac{d}{dx}\left(\alpha\frac{d}{dx}\right) + \beta\frac{d}{dx} + \gamma$, so that (8.9.1) can be written as $\mathcal{L}u = f$.

Equation (8.9.1) is in *strong form*, i. e., the PDE is required to be satisfied at each point in its domain. This is not the only way the PDE may be enforced. Other forms may be more suitable for the numerical discretization. A *weak form* of the PDE is obtained by requiring that the integral of both sides of the PDE against all functions in an appropriate space X of test functions be equal; precisely, we multiply both sides of (8.9.1) by each test function v and, possibly, by a fixed weight function w, and we integrate in space, to obtain

$$\int_{-1}^{1} \mathcal{L}u(x)v(x)w(x)\,dx = \int_{-1}^{1} f(x)v(x)w(x)\,dx \qquad \text{for all } v \in X . \quad (8.9.3)$$

This is often referred to as an *integral form* of the PDE. Equivalently, these relations amount to requiring that the residual, $r(u) = f - \mathcal{L}u$, be orthogonal to all test functions v in the weighted inner product $(u,v)_w = \int_{-1}^{1} u(x)v(x)w(x)\,dx$.

Another weak form of the PDE that is often used is obtained (under the assumption that the weight w is the constant 1) by performing an integration-by-parts on the second-order term in the left-hand side of (8.9.3), yielding

$$-\int_{-1}^{1} \frac{d}{dx}\left(\alpha\frac{du}{dx}\right) v\,dx = \int_{-1}^{1} \alpha\frac{du}{dx}\frac{dv}{dx}\,dx - \left[\alpha\frac{du}{dx}v\right]_{-1}^{1}.$$

If all test functions are now required to vanish at the endpoints of the interval, consistent with the boundary conditions (8.9.2), we obtain the new weak form (also called the *variational form*)

$$\int_{-1}^{1}\left(\alpha\frac{du}{dx}\frac{dv}{dx} + \beta\frac{du}{dx}v + \gamma u\,v\right)dx = \int_{-1}^{1} f(x)v(x)w(x)\,dx \quad \text{for all } v \in V,$$

(8.9.4)

where V denotes the space of all test functions satisfying $v(\pm 1) = 0$.

Equation (8.9.3) is meaningful if u is twice differentiable, whereas the test functions need not be differentiable. In contrast, (8.9.4) requires less regularity on the solution, at the expense of increasing the regularity requirement on the test functions. All three formulations are equivalent if the solution is smooth enough. The weak formulations, however, can accommodate less regular solutions. In addition, other kinds of boundary conditions, such as those involving the first derivative of u, can be accounted for in the integration-by-parts procedure, leading to a *weak enforcement* of these boundary conditions.

Let us now turn to the numerical discretization of problem (8.9.1)–(8.9.2). We aim at approximating u by an N-degree algebraic polynomial u^N. Let $\mathbb{P}_N(I)$ denote the linear space of all polynomials of degree up to N, restricted to the interval I; let X_N denote the subspace of $\mathbb{P}_N(I)$ of the polynomials vanishing at the endpoints of the interval. Then, taking into account (8.9.2), we will look for u^N in X_N. A spectral method for solving (8.9.1)–(8.9.2) is completely determined by specifying a way to represent u^N in terms of suitable basis functions, and a way to enforce (8.9.1) in an approximate manner. At the algebraic level, it produces a linear system of the form

$$L\mathbf{u} = \mathbf{b},$$

(8.9.5)

where $\mathbf{u}$ is the vector collecting the unknowns that represent u^N, $\mathbf{b}$ is a known vector depending on the data f, and L is the matrix corresponding to the equations defined by the method.

8.9.1 Tau Method

Historically, this was the first method of spectral type used for nonperiodic problems. It is now used relatively rarely, and then only for problems in which the coefficients are constant or polynomials of very low degree; in the constant-coefficient case, (8.9.1) becomes

$$-\alpha\frac{d^2u}{dx^2} + \beta\frac{du}{dx} + \gamma u = f.$$

(8.9.6)

Let us introduce a system $\{\phi_k,\ k = 0, 1, \dots\}$ of orthogonal polynomials with respect to a weight function w; they satisfy $\deg\phi_k = k$ and $(\phi_k, \phi_h)_w = c_k\delta_{kh}$

$(k, h = 0, 1, \dots)$ for suitable constants $c_k > 0$. Examples are the Chebyshev system $\{T_k, \ k = 0, 1, \dots\}$, for which $w(x) = (1 - x^2)^{-1/2}$ (see Sect. 8.3), the Legendre system $\{L_k, \ k = 0, 1, \dots\}$, for which $w(x) \equiv 1$ (see Sect. 8.4), and, more generally, any Jacobi system $\{P_k^{(\lambda,\mu)}, \ k = 0, 1, \dots\}$, for which $w(x) = (1 - x)^{\lambda}(1 + x)^{\mu}$, $\lambda, \mu > -1$ (see Sect. 8.2).

The polynomials ϕ_k, $0 \leq k \leq N$, form a modal basis of $\mathbb{P}_N(I)$. The discrete solution is therefore represented as

$$u^N(x) = \sum_{k=0}^{N} \hat{u}_k \phi_k(x) , \tag{8.9.7}$$

where the unknowns $\hat{u}_k = (u^N, \phi_k)_w / (\phi_k, \phi_k)_w$ are the expansion coefficients of u^N along the chosen basis. The boundary conditions (8.9.2) impose two linear combinations upon the coefficients of u^N, namely,

$$\sum_{k=0}^{N} \hat{u}_k \phi_k(\pm 1) = 0 . \tag{8.9.8}$$

The differential equation is enforced by resorting to the weak form (8.9.3), i.e., by requiring that the residual $r(u^N) = f - \mathcal{L}u^N$ be orthogonal to all polynomials of degree up to $N - 2$. This means that

$$(\mathcal{L}u^N, \phi_h)_w = (f, \phi_h)_w \qquad \text{for } 0 \leq h \leq N - 2 , \tag{8.9.9}$$

i.e.,

$$\alpha \hat{u}_h^{(2)} + \beta \hat{u}_h^{(1)} + \gamma \hat{u}_h = \hat{f}_h , \qquad 0 \leq h \leq N - 2 , \tag{8.9.10}$$

where $\hat{u}_h^{(2)}$, $\hat{u}_h^{(1)}$ and $\hat{f}_h$ are the expansion coefficients of $\dfrac{d^2 u^N}{dx^2}$, $\dfrac{du^N}{dx}$ and f, respectively. The coefficients $\hat{u}_h^{(2)}$ and $\hat{u}_h^{(1)}$ can be expressed as linear combinations of the coefficients $\hat{u}_k$ (see (8.3.24) and (8.3.25) for the Chebyshev system, and (8.4.17) and (8.4.18) for the Legendre system). Thus, (8.9.10) with (8.9.8) form a linear system of $N+1$ equations in the $N+1$ unknowns $\mathbf{u} = (\hat{u}_0, \dots, \hat{u}_N)^T$, which we represent as (8.9.5) with $\mathbf{b} = (\hat{f}_0, \dots, \hat{f}_{N-2}, 0, 0)^T$. The matrix L is nonsingular; hence, $\mathbf{u}$ and consequently u^N is uniquely determined. Sect. 4.1 in CHQZ2 discusses efficient methods for solving this linear system.

Finally, we note that setting $Y_N = \mathbb{P}_{N-2}(I)$, the tau discretization of problem (8.9.1)–(8.9.2) can be equivalently written as

$$u^N \in X_N , \qquad (\mathcal{L}u^N, v)_w = (f, v)_w \qquad \text{for all } v \in Y_N .$$

This shows that the tau method is a particular form of a *Petrov–Galerkin method*. This term refers to a method in which the solution is sought in a space X_N of trial functions while the equation is enforced by requiring the orthogonality of the residual to a space Y_N of test functions (having the same dimension of X_N but possibly different from it).

8.9.2 Collocation Method

If some of the coefficients of the equation are variable, the tau method is much less efficient because L is full (unless the coefficients are polynomials of very low degree). In this case, the collocation method is an efficient alternative. It relies on the strong form (8.9.1) of the differential equation. In order to define it, let us introduce the nodes x_j and weights w_j $(j = 0, \ldots, N)$ of the N-order Gauss–Lobatto quadrature formula for some Jacobi weight $w(x) = (1 - x)^\alpha (1 + x)^\beta$ (see Sect. 8.2). We recall that the nodes x_0 and x_N coincide with the endpoints of the interval $[-1, 1]$, and that the quadrature formula is exact for all polynomials of degree $\leq 2N - 1$, i.e.,

$$\sum_{j=0}^{N} v(x_j) w_j = \int_{-1}^{1} v(x)\, w(x)\, \mathrm{d}x \qquad \text{for all } v \in \mathbb{P}_{2N-1}(I) . \qquad (8.9.11)$$

Let ψ_l, $l = 0, \ldots, N$, denote the characteristic Lagrange polynomials at the Gauss–Lobatto nodes (see (8.2.30)); these form a nodal, boundary-adapted basis in $\mathbb{P}_N(I)$. For any continuous function v in $[-1, 1]$, let $I_N v$ be the unique polynomial in $\mathbb{P}_N(I)$ that interpolates v at the Gauss–Lobatto nodes (see (8.2.29)).

A collocation method for problem (8.9.1)–(8.9.2) is defined as follows. The discrete solution u^N is represented as

$$u^N(x) = \sum_{l=1}^{N-1} u_l \psi_l(x) , \qquad (8.9.12)$$

where the first and last basis functions have been dropped in order to match the boundary conditions. The unknown values $u_l = u^N(x_l)$ at the internal Gauss–Lobatto nodes are defined by collocating the differential equation at these points. A naive implementation would simply enforce $\mathcal{L}u^N = f$ therein. However, this would require the exact differentiation of the product $\alpha \dfrac{\mathrm{d}u^N}{\mathrm{d}x}$, which may be cumbersome if the coefficient α is arbitrary. On the other hand, differentiating the interpolant of this product is easy. Thus, the expression $\mathcal{L}u^N$ is replaced by

$$\mathcal{L}_N u^N = -\mathcal{D}_N \left(\alpha \frac{\mathrm{d}u^N}{\mathrm{d}x} \right) + \beta \frac{\mathrm{d}u^N}{\mathrm{d}x} + \gamma u^N , \qquad (8.9.13)$$

where $\mathcal{D}_N v = \dfrac{\mathrm{d}}{\mathrm{d}x} I_N v$ denotes the interpolation derivative of the function v (see (8.2.34)). The collocation equations are

$$\mathcal{L}_N u^N(x_j) = f(x_j) , \qquad 1 \leq j \leq N - 1 . \qquad (8.9.14)$$

In matrix terms, the operator $\mathcal{L}_N$ is expressed as $L_N = -D_N A D_N + B D_N + G$, where D_N indicates the interpolation derivative matrix at the

Gauss–Lobatto nodes (see (8.2.35)), $A = \mathrm{diag}\,\boldsymbol{\alpha}$ with $\boldsymbol{\alpha} = (\alpha(x_j))_{j=0,\dots,N}$, and the diagonal matrices B and G are defined similarly. Let L denote the matrix of order $N - 1$ obtained by dropping the first and last rows and columns of L_N (this reduction corresponds to enforcing the homogeneous Dirichlet conditions and to collocating the differential equation at the internal nodes only). Then, the collocation equations above are translated into the linear system (8.9.5), where $\mathbf{u} = (u_1, \dots, u_{N-1})^T$ is the vector collecting the values of u^N at the internal Gauss–Lobatto nodes, whereas $\mathbf{b} = (f(x_1), \dots, f(x_{N-1}))^T$.

8.9.3 Galerkin Method

Let us fix a boundary-adapted basis $\{\eta_k,\ k = 1, \dots, N - 1\}$ in X_N, such as the one obtained from the modal basis (8.5.2) or the nodal basis (8.2.30) after removing the vertex functions (ϕ_0 and ϕ_N, or ψ_0 and ψ_N) to match the boundary conditions.

The Galerkin approach is based on one of the weak forms (8.9.3) or (8.9.4) of the differential equation. If we start from the former one, the Galerkin method consists of enforcing the orthogonality of the residual $r(u^N) = f - \mathcal{L}u^N$ to all basis functions in X_N, i.e.,

$$(\mathcal{L}u^N, \eta_h)_w = (f, \eta_h)_w \qquad \text{for } 1 \le h \le N - 1 . \tag{8.9.15}$$

By taking any linear combination of such equations, the Galerkin method can be equivalently written as

$$(\mathcal{L}u^N, v)_w = (f, v)_w \qquad \text{for all } v \in X_N . \tag{8.9.16}$$

In other words, in a Galerkin method, the space Y_N of test functions coincides with the space X_N of trial functions. The set of equations (8.9.15) can be expressed as a linear system of order $N - 1$ of the form (8.9.5), if we represent u^N in terms of the chosen basis as $u^N(x) = \sum_{k=1}^{N-1} u_k \eta_k(x)$. Precisely, by linearity (8.9.15) becomes

$$\sum_{k=1}^{N-1} u_k (\mathcal{L}\eta_k, \eta_h)_w = (f, \eta_h)_w , \qquad 1 \le h \le N - 1 ,$$

which is (8.9.5) if we set $L = (L_{hk})$ with $L_{hk} = (\mathcal{L}\eta_k, \eta_h)_w$ for $1 \le h, k \le N - 1$, $\mathbf{u} = (u_1, \dots, u_{N-1})^T$ and $\mathbf{b} = (b_h)$ with $b_h = (f, \eta_h)_w$ for $1 \le h \le N - 1$.

It is customary to term L the *stiffness matrix* of the Galerkin method and to indicate it by K. In the case of constant coefficients α, β and γ, the use of the modal basis (8.5.2) yields a banded stiffness matrix, with bandwidth at most equal to 5 (see (8.5.5)–(8.5.6) or CHQZ2, Sect. 3.8). On the other hand, the nodal basis (8.2.30) leads to a full matrix.

If the Legendre weight $w(x) \equiv 1$ (and consequently a Legendre Galerkin method) is used, then the Galerkin method can be viewed as a discretization of the weak form (8.9.4) of the differential problem. The expression

$$a(u, v) = \int_{-1}^{1} \left(\alpha \frac{du}{dx} \frac{dv}{dx} + \beta \frac{du}{dx} v + \gamma u v \right) dx \qquad (8.9.17)$$

is a bilinear form (i.e., it is linear with respect to each argument u and v) defined on a space of sufficiently smooth functions (the space $H_0^1(I)$, see Sect. A.8), which obviously includes all spaces X_N. Then, (8.9.16) is equivalently written as

$$a(u^N, v) = (f, v) \qquad \text{for all } v \in X_N . \qquad (8.9.18)$$

The previous expression of $a(u, v)$ shows in particular that, whenever $\beta = 0$, the bilinear form a is symmetric, i.e., $a(u, v) = a(v, u)$ for all u and v; in this case, the stiffness matrix L turns out to be symmetric. If in addition, $\alpha \geq \alpha_0$ and $\gamma \geq 0$ in $(-1, 1)$, we have $a(v, v) \geq \alpha_0 \int_{-1}^{1} \left(\frac{dv}{dx} \right)^2 dx > 0$ for all v not identically vanishing in $(-1, 1)$. This implies that L is positive definite. This feature, which holds for the multidimensional version of our problem as well, can be exploited to design efficient iterative solvers for the corresponding algebraic linear system (see CHQZ2, Sect. 4.5). We will come back to the integration-by-parts argument while discussing the enforcement of other types of boundary conditions.

It is worth observing that another matrix may come into play in Galerkin methods: it is the *mass matrix* $M = (M_{hk})$ with $M_{hk} = (\eta_k, \eta_h)_w$ for $1 \leq h, k \leq N - 1$. It is always symmetric and positive definite. It occurs, e.g., in the Galerkin discretization (in space) of the time-dependent version of (8.9.1), i.e.,

$$\frac{du}{dt} + \mathcal{L}u = f$$

(again with (8.9.2) as boundary conditions), which reads as follows:

$$\left(\frac{du^N}{dt}, \eta_h \right)_w + (\mathcal{L}u^N, \eta_h)_w = (f, \eta_h)_w , \qquad 1 \leq h \leq N - 1 . \qquad (8.9.19)$$

In algebraic terms, this is expressed as

$$M \frac{d\mathbf{u}}{dt} + L\mathbf{u} = \mathbf{b} . \qquad (8.9.20)$$

Another situation of interest is the eigenvalue problem

$$\mathcal{L}u = \lambda u ,$$

which is discretized by the Galerkin method as

$$(\mathcal{L}u^N, \eta_h)_w = \lambda(u^N, \eta_h)_w , \qquad 1 \leq h \leq N - 1 . \qquad (8.9.21)$$

This leads to the generalized algebraic eigenvalue problem

$$L\mathbf{u} = \lambda M\mathbf{u} . \qquad (8.9.22)$$

8.9.4 Galerkin with Numerical Integration (G-NI) Method

The presence of variable coefficients and general right-hand sides suggests that we modify the Galerkin method by evaluating the integrals via a high-precision quadrature formula. This leads to a so-called G-NI method. Usually, the Gauss–Lobatto formula introduced in (8.9.11) is employed; correspondingly, the nodal basis $\{\psi_j, \, j = 1, \ldots, N - 1\}$ is chosen to represent functions in X_N.

The G-NI method is almost invariably applied in a Legendre setting. In this case, the L^2-inner product $(u, v) = \int_{-1}^{1} u(x) \, v(x) \, dx$ is replaced by the discrete inner product $(u, v)_N = \sum_{j=0}^{N} u(x_j)v(x_j)w_j$ (see (8.2.20)), which by (8.9.11) coincides with (u, v) if the function uv belongs to $\mathbb{P}_{2N-1}(I)$. Correspondingly, the bilinear form $a(u, v)$ is replaced by

$$a_N(u, v) = \left(\alpha \frac{du}{dx}, \frac{dv}{dx} \right)_N + \left(\beta \frac{du}{dx}, v \right)_N + (\gamma u, v)_N \, . \tag{8.9.23}$$

Then, the G-NI approach defines $u^N \in X_N$ as the solution of

$$a_N(u^N, v) = (f, v)_N \qquad \text{for all } v \in X_N$$

(compare to (8.9.18)) or, equivalently, of

$$a_N(u^N, \psi_j) = (f, \psi_j)_N \, , \qquad 1 \le j \le N - 1 \, . \tag{8.9.24}$$

Under the chosen Dirichlet boundary conditions, the method is equivalent to the collocation method. Indeed, recalling (8.9.11), we have, for all $v \in X_N$,

$$\left(\alpha \frac{du^N}{dx}, \frac{dv}{dx} \right)_N = \left(I_N \left(\alpha \frac{du^N}{dx} \right), \frac{dv}{dx} \right)_N = \int_{-1}^{1} I_N \left(\alpha \frac{du^N}{dx} \right) \frac{dv}{dx} \, dx$$

$$= - \int_{-1}^{1} \frac{d}{dx} I_N \left(\alpha \frac{du^N}{dx} \right) v \, dx = - \left(\frac{d}{dx} I_N \left(\alpha \frac{du^N}{dx} \right), v \right)_N \, .$$

Consequently,

$$a_N(u^N, v) = (\mathcal{L}_N u^N, v)_N \, ,$$

where $\mathcal{L}_N u^N$ is defined in (8.9.13). Recalling the definition of the nodal basis functions, the G-NI equations (8.9.24) can be equivalently written as

$$\mathcal{L}_N u^N(x_j)w_j = f(x_j)w_j \, , \qquad 1 \le j \le N - 1 \, ; \tag{8.9.25}$$

these relations are precisely (8.9.14) up to multiplication by the quadrature weights. In algebraic terms, the system (8.9.24) is written in the form (8.9.5), where $L = (L_{jl})$ with $L_{jl} = a_N(\psi_l, \psi_j)$ for $1 \le j, l \le N - 1$, $\mathbf{u} = (u_1, \ldots, u_{N-1})^T$ and $\mathbf{b} = (b_j)$ with $b_j = f(x_j)w_j$ for $1 \le j \le N - 1$. It is easily seen that the G-NI linear system is obtained from the collocation linear system by left multiplication by the matrix $M_N = \text{diag}\,(w_1, \ldots, w_{N-1})$. This is the *discrete mass matrix*, derived by applying the Gauss–Lobatto quadrature formula to the elements of the standard mass matrix M relative to the nodal basis $\{\psi_j \, , j = 1, \ldots, N - 1\}$.

8.9.5 Other Boundary Conditions

We now briefly illustrate how the previous schemes should be modified to handle boundary conditions different from (8.9.2).

Let us first consider inhomogeneous Dirichlet conditions,

$$u(-1) = g_- , \qquad u(1) = g_+ . \tag{8.9.26}$$

For the tau method, (8.9.8) becomes

$$\sum_{k=0}^{N} \hat{u}_k \phi_k(\pm 1) = g_\pm ; \tag{8.9.27}$$

hence, it is enough to define the right-hand side of the corresponding linear system (8.9.5) as $\mathbf{b} = (\hat{f}_0, \ldots, \hat{f}_{N-2}, g_-, g_+)^T$.

For all other methods, since the basis $\{\eta_k , k = 1, \ldots, N-1\}$ chosen in $\mathbb{P}_N(I)$ is boundary adapted (see Sect. 8.5), we enforce (8.9.27) by looking for a polynomial $u^N \in \mathbb{P}_N(I)$ of the form

$$u^N(x) = \sum_{k=1}^{N-1} u_k \eta_k(x) + g_- \eta_0(x) + g_+ \eta_N(x) = u_0^N + u_g^N ,$$

with the function $u_0^N = \sum_{k=1}^{N-1} u_k \eta_k$ in X_N, i.e., vanishing at the endpoints of the interval. Next, u^N is required to satisfy Eqs. (8.9.14) in the collocation case, or (8.9.15) in the Galerkin case, or (8.9.24) in the G-NI case. The unknown u_0^N is therefore determined by the equations

$$\mathcal{L}_N u_0^N(x_j) = f(x_j) - \mathcal{L}_N u_g^N(x_j) , \qquad 1 \le j \le N-1 , \tag{8.9.28}$$

in the collocation case, or

$$(\mathcal{L}u_0^N, \eta_h)_w = (f, \eta_h)_w - (\mathcal{L}u_g^N, \eta_h)_w , \qquad 1 \le h \le N-1 , \tag{8.9.29}$$

in the Galerkin case, or

$$a_N(u_0^N, \psi_j) = (f, \psi_j)_N - a_N(u_g^N, \psi_j) , \qquad 1 \le j \le N-1 , \tag{8.9.30}$$

in the G-NI case. For all schemes, the matrix of the linear system to be solved is the same as for the homogeneous boundary conditions; the boundary data influence the right-hand sides only.

Let us now assume that a homogeneous Dirichlet condition is prescribed at $x = -1$, whereas a nonhomogeneous Neumann or Robin condition is prescribed at $x = 1$; precisely, let us enforce

$$u(-1) = 0 , \qquad \left(\alpha \frac{du}{dx} + \sigma u \right)(1) = g , \tag{8.9.31}$$

with $\sigma \geq 0$ (the case $\sigma = 0$ corresponds to the Neumann condition). For such boundary conditions, the relevant subspace of $\mathbb{P}_N(I)$ is the N-dimensional space X_N of the polynomials that vanish at $x = -1$. The discrete solution u^N belongs to this space.

For the tau method, one still represents u^N as in (8.9.7). The boundary conditions are translated into the equations

$$\sum_{k=0}^{N} \hat{u}_k \phi_k(-1) = 0 \,, \qquad \sum_{k=0}^{N} \hat{u}_k \left(\alpha \frac{d\phi_k}{dx}(1) + \sigma \phi_k(1) \right) = g \,. \qquad (8.9.32)$$

For the other methods, the representation of u^N is

$$u^N(x) = \sum_{k=1}^{N} u_k \eta_k(x) \,,$$

which accounts for the Dirichlet condition at $x = -1$, leaving the function still undetermined at $x = 1$.

In the collocation case, in addition to the internal equations (8.9.14), one enforces the second condition (8.9.32).

For the Galerkin method (with the Legendre weight $w(x) \equiv 1$) this condition is accounted for in the definition of the bilinear form $a(u, v)$. Indeed, in the present situation one has

$$-\int_{-1}^{1} \frac{d}{dx} \left(\alpha \frac{du}{dx} \right) v \, dx = \int_{-1}^{1} \alpha \frac{du}{dx} \frac{dv}{dx} \, dx - \left(\alpha \frac{du}{dx} v \right)(1) \qquad \text{for } v \in X_N \,.$$

The boundary condition at $x = 1$ yields $\alpha \frac{du}{dx}(1) = g - \sigma u(1)$; whence

$$(\mathcal{L}u, v) = \int_{-1}^{1} \left(\alpha \frac{du}{dx} \frac{dv}{dx} + \beta \frac{du}{dx} v + \gamma u \, v \right) dx + \sigma u(1) v(1) - g v(1) \,.$$

Setting

$$a(u, v) = \int_{-1}^{1} \left(\alpha \frac{du}{dx} \frac{dv}{dx} + \beta \frac{du}{dx} v + \gamma u \, v \right) dx + \sigma u(1) v(1) \,,$$

the boundary-value problem (8.9.1) and (8.9.31) is translated into the weak equations

$$a(u, v) = (f, v) + g v(1) \,,$$

which hold for all sufficiently smooth test functions v vanishing at $x = -1$. One can prove that if the solution u of these weak equations is sufficiently smooth, then it is also the strong solution of (8.9.1) and satisfies the boundary conditions (8.9.31). Therefore, it is natural to define the Galerkin method by requiring that the polynomial solution u^N satisfy the equations

$$a(u^N, \eta_k) = (f, \eta_k) + g\eta_k(1) , \qquad 1 \le k \le N .\qquad (8.9.33)$$

In general, u^N will not fulfill exactly the boundary condition at $x = 1$, but only within an error that tends to 0 as N tends to ∞. Note that the form $a(u, v)$ is symmetric, implying the symmetry of the matrix L of the Galerkin method. This is not the case for the collocation matrix. The weak enforcement of Neumann or Robin boundary conditions leads to a symmetric linear system, for which more efficient solution techniques than for a nonsymmetric system are available.

Finally, the G-NI method is defined as

$$a_N(u^N, \psi_j) = (f, \psi_j)_N + g\psi_j(1) , \qquad 1 \le j \le N ,\qquad (8.9.34)$$

where, as in the case of pure Dirichlet conditions, integrals appearing in the Galerkin equations have been replaced by the corresponding Gauss–Lobatto quadrature formulas. In this case, too, the method can be given a pointwise (strong) interpretation. Choosing in (8.9.34) $j = 1, \ldots, N - 1$ (i.e., the test functions associated with the internal Gauss–Lobatto nodes) one gets again the equations (8.9.25), showing that the G-NI method collocates the differential equation at these nodes. On the other hand, choosing $j = N$ above and integrating by parts backwards in the expression $a_N(u^N, \psi_N)$, one readily gets the relation

$$(\mathcal{L}_N u^N(1) - f(1))w_N + \alpha \frac{\mathrm{d}u^N}{\mathrm{d}x}(1) + \sigma u^N(1) - g = 0 .\qquad (8.9.35)$$

Taking into account that $w_N \sim 2/N^2$ as $N \to \infty$, this shows that the weak enforcement of the boundary condition at $x = 1$ is equivalent to a particular *penalty* strategy for handling this condition.

Appendix A. Basic Mathematical Concepts

A.1 Hilbert and Banach Spaces

(a) Hilbert Spaces

Let X be a real vector space. An *inner product* on X is a function $X \times X \to \mathbb{R}$, denoted by (u, v), that satisfies the following properties:

(i) $(u, v) = (v, u)$ for all $u, v \in X$;
(ii) $(\alpha u + \beta v, w) = \alpha(u, w) + \beta(v, w)$ for all $\alpha, \beta \in \mathbb{R}$ and all $u, v, w \in X$;
(iii) $(u, u) \geq 0$ for all $u \in X$;
(iv) $(u, u) = 0$ implies $u = 0$.

Two elements $u, v \in X$ are said to be *orthogonal* in X if $(u, v) = 0$. The inner product (u, v) defines a *norm* on X by the relation

$$\|u\| = (u, u)^{1/2} \quad \text{for all } u \in X .$$

The *distance* between two elements $u, v \in X$ is the positive number $\|u - v\|$. A *Cauchy sequence* in X is a sequence $\{u_k \mid k = 0, 1, \ldots\}$ of elements of X that satisfies the following property:

for each positive number $\varepsilon > 0$, there exists an integer $N = N(\varepsilon) > 0$ such that the distance $\|u_k - u_m\|$ between any two elements of the sequence is smaller than ε provided both k and m are larger than $N(\varepsilon)$.

A sequence in X is said to *converge* to an element $u \in X$ if the distance $\|u_k - u\|$ tends to 0 as k tends to ∞.

A *Hilbert space* is a vector space equipped with an inner product for which all the Cauchy sequences are convergent.

Examples

(i) $\mathbb{R}^n$ endowed with the Euclidean product $(\mathbf{u}, \mathbf{v}) = \displaystyle\sum_{i=1}^{n} u_i v_i$ is a finite-dimensional Hilbert space.

(ii) If $[a, b] \subset \mathbb{R}$ is an interval, the space $L^2(a, b)$ (see (A.6)) is an infinite-dimensional Hilbert space for the inner product

$$(u, v) = \int_a^b u(x)v(x) \, \mathrm{d}x \ .$$

If X is a complex vector space, the inner product on X will be a complex-valued function. Then condition (i) has to be replaced by (i$'$)

$$(u, v) = \overline{(v, u)} \quad \text{for all } u, v \in X \ .$$

(b) Banach Spaces

The concept of Banach space extends that of Hilbert space. Given a vector space X, a *norm* on X is a function $X \to \mathbb{R}$, denoted by $\|u\|$, that satisfies the following properties:

$$\begin{aligned} \|u + v\| &\leq \|u\| + \|v\| && \text{for all } u, v \in X \ ; \\ \|\lambda u\| &= |\lambda| \|u\| && \text{for all } u \in X \ , \text{ and all } \lambda \in \mathbb{R} \ ; \\ \|u\| &\geq 0 && \text{for all } u \in X \ ; \\ \|u\| &= 0 && \text{if and only if } u = 0 \ . \end{aligned}$$

A *Banach space* is a vector space equipped with a norm for which all the Cauchy sequences are convergent.

Examples

(i) $\mathbb{R}^n$ endowed with the norm $\|\mathbf{u}\| = \left(\sum_{i=1}^n |u_i|^p \right)^{1/p}$ (with $1 \leq p < +\infty$) is a finite-dimensional Banach space.

(ii) If $[a, b] \subset \mathbb{R}$ is an interval and $1 \leq p < +\infty$, the space $L^p(a, b)$ (see again (A.6)) is an infinite-dimensional Banach space for the norm

$$\|u\| = \left(\int_a^b |u(x)|^p \mathrm{d}x \right)^{1/p} \ .$$

(c) Dual Spaces

Let X be a Hilbert or a Banach space. A linear form $F : X \to \mathbb{R}$ is said to be *continuous* if there exists a constant $C > 0$ such that

$$|F(u)| \leq C \|u\| \quad \text{for all } u \in X \ .$$

The set of all the linear continuous forms on X is a vector space. We can define a norm on this space by setting

$$\|F\|_{X'} = \sup_{\substack{u \in X \\ u \neq 0}} \frac{F(u)}{\|u\|} .$$

The vector space of all the linear continuous forms on X is called the *dual space* of X and is denoted by X'. Endowed with the previous norm, it is itself a Banach space.

The bilinear form from $X' \times X$ into $\mathbb{R}$ defined by

$$\langle F, u \rangle = F(u)$$

is called the *duality pairing* between X and X'.

(d) The Riesz Representation Theorem

If X is a Hilbert space, the dual space X' can be canonically identified with X (hence, it is a Hilbert space). In fact, the Riesz representation theorem states that for each linear continuous form F on X, there exists a unique element $u \in X$ such that

$$\langle F, v \rangle = (u, v) \quad \text{for all } v \in X .$$

Moreover, $\|F\|_{X'} = \|u\|_X$. The Lax-Milgram Theorem (A.3) extends this result to the case in which (u, v) is replaced by a non-symmetric bilinear form $a(u, v)$.

A.2 The Cauchy-Schwarz Inequality

Let X be a Hilbert space, endowed with the inner product (u, v) and the associated norm $\|u\|$ (see (A.1.a)). The Cauchy-Schwarz inequality states that

$$|(u, v)| \leq \|u\| \, \|v\| \quad \text{for all } u, v \in X .$$

Of particular importance in the analysis of numerical methods for partial differential equations is the Cauchy-Schwarz inequality in the Lebesgue space $L^2(\Omega)$, where Ω is a domain in $\mathbb{R}^n$ (see (A.9.h)). The previous inequality becomes:

$$\left| \int_\Omega u(\mathbf{x}) v(\mathbf{x}) \, d\mathbf{x} \right| \leq \left(\int_\Omega u^2(\mathbf{x}) \, d\mathbf{x} \right)^{1/2} \left(\int_\Omega v^2(\mathbf{x}) \, d\mathbf{x} \right)^{1/2}$$

for all functions $u, v \in L^2(\Omega)$.

A.3 The Lax-Milgram Theorem

Let V be a real Hilbert space (see (A.1.a)). Let $a : V \times V \to \mathbb{R}$ be a bilinear continuous form on V, i.e., a satisfies

(i) $a(\lambda u + \mu v, w) = \lambda a(u, w) + \mu a(v, w)$ and
    $a(u, \lambda v + \mu w) = \lambda a(u, v) + \mu a(u, w)$
    for all $u, v, w \in V$ and all $\lambda, \mu \in \mathbb{R}$;

(ii)   there exists a constant $\beta > 0$ such that
    $|a(u, v)| \leq \beta \|u\|_V \|v\|_V$ for all $u, v \in V$.

Assume that the form a is V-*coercive*, or V-*elliptic*, i.e.,

(iii) there exists a constant $\alpha > 0$ such that

$$a(u, u) \geq \alpha \|u\|_V^2 \quad \text{for all } u \in V .$$

Then for each form $F \in V'$ (the dual space of V, see (A.1.c)), there exists a unique solution $u \in V$ to the variational problem

$$a(u, v) = F(v) \quad \text{for all } v \in V .$$

Moreover, the following inequality holds:

$$\|u\|_V \leq \frac{1}{\alpha} \|F\|_{V'} .$$

A.4 Dense Subspace of a Normed Space

Let X be a Hilbert or a Banach space with norm $\|v\|$. Let $S \subset X$ be a subspace of X. S is said to be *dense* in X if for each element $v \in X$ there exists a sequence $\{v_n \mid n = 0, 1, \ldots\}$ of elements $v_n \in S$, such that

$$\|v - v_n\| \longrightarrow 0 \quad \text{as } n \longrightarrow \infty .$$

Thus, each element of X can be approximated arbitrarily well by elements of S, in the distance induced by the norm of X.

For example, the subspace $C^0([a, b])$ of the continuous functions on a bounded, closed interval $[a, b]$ of the real line, is dense in $L^2(a, b)$, the space of the measurable square-integrable functions on (a, b). Indeed, for each function $v \in L^2(a, b)$ and each $n > 0$, one can find a continuous function $v_n \in C^0([a, b])$ such that

$$\int_a^b |v(x) - v_n(x)|^2 \mathrm{d}x \leq \frac{1}{n^2} .$$

A.5 The Spaces $C^m(\overline{\Omega})$, $m \ge 0$

Let Ω be an open subset of $\mathbb{R}^d$, with sufficiently smooth boundary. Let us denote by $\overline{\Omega}$ the closure of Ω. For each multi-index $\alpha = (\alpha_1, \dots, \alpha_d)$ of nonnegative integers, set $|\alpha| = \alpha_1 + \cdots + \alpha_d$ and $D^\alpha v = \partial^{|\alpha|} v / \partial x_1^{\alpha_1} \dots \partial x_d^{\alpha_d}$.

We denote by $C^m(\overline{\Omega})$ the vector space of the functions $v : \overline{\Omega} \to \mathbb{R}$ such that for each multi-index α with $0 \le |\alpha| \le m$, $D^\alpha v$ exists and is continuous on $\overline{\Omega}$. Since a continuous function on a closed, bounded set is bounded there, one can set

$$\|v\|_{C^m(\overline{\Omega})} = \sup_{0 \le |\alpha| \le m} \sup_{\mathbf{x} \in \overline{\Omega}} |D^\alpha v(\mathbf{x})| .$$

This is a norm for which $C^m(\overline{\Omega})$ is a Banach space (see (A.1.b)).

The space $C^\infty(\overline{\Omega})$ is the space of the infinitely differentiable functions on $\overline{\Omega}$. Thus, a function v belongs to $C^\infty(\overline{\Omega})$ if and only if it belongs to $C^m(\overline{\Omega})$ for all $m > 0$.

A.6 The Spaces $L^p(\Omega)$, $1 \le p \le +\infty$

Let Ω denote a bounded, open domain in $\mathbb{R}^d$, for $d \ge 1$.

For $p < +\infty$, we denote by $L^p(\Omega)$ the space of the measurable functions $u : \Omega \to \mathbb{R}$ such that $\int_\Omega |u(\mathbf{x})|^p d\mathbf{x} < +\infty$. It is a Banach space for the norm

$$\|u\|_{L^p(\Omega)} = \left(\int_\Omega |u(\mathbf{x})|^p d\mathbf{x} \right)^{1/p} .$$

Let $L^\infty(\Omega)$ be the Banach space of the measurable functions $u : \Omega \to \mathbb{R}$ that are bounded outside a set of measure zero, equipped with the norm

$$\|u\|_{L^\infty(\Omega)} = \operatorname{ess\,sup}_{\mathbf{x} \in \Omega} |u(\mathbf{x})| .$$

The space $L^2(\Omega)$ is a Hilbert space for the inner product

$$(u, v) = \int_\Omega u(\mathbf{x}) v(\mathbf{x}) d\mathbf{x} ,$$

which induces the norm

$$\|u\|_{L^2(\Omega)} = \left(\int_\Omega |u(\mathbf{x})|^2 d\mathbf{x} \right)^{1/2} .$$

One can define spaces $L^p(\Omega)$ of complex functions in a straight-forward manner.

A.7 Infinitely Differentiable Functions and Distributions

Let Ω be a bounded, open domain in $\mathbb{R}^d$, for $d = 1, 2$ or 3. If $\alpha = (\alpha_1, \ldots, \alpha_d)$ is a multi-index of nonnegative integers, let us set

$$D^\alpha v = \frac{\partial^{\alpha_1 + \cdots + \alpha_d} v}{\partial x_1^{\alpha_1} \cdots \partial x_d^{\alpha_d}} .$$

We denote by $\mathscr{D}(\Omega)$ the vector space of all the infinitely differentiable functions $\phi : \Omega \to \mathbb{R}$, for which there exists a closed set $K \subset \Omega$ such that $\phi \equiv 0$ outside K.

We say that a sequence of functions $\phi_n \in \mathscr{D}(\Omega)$ *converges in* $\mathscr{D}(\Omega)$ to a function $\phi \in \mathscr{D}(\Omega)$ as $n \to \infty$, if there exists a common closed set $K \subset \Omega$ such that all the ϕ_n vanish outside K, and $D^\alpha \phi_n \to D^\alpha \phi$ uniformly on K as $n \to \infty$, for all multi-indices α.

(a) Distributions

Let T be a linear form on $\mathscr{D}(\Omega)$, i.e., a linear mapping $T : \mathscr{D}(\Omega) \to \mathbb{R}$. We shall denote the value of T on the element $\phi \in \mathscr{D}(\Omega)$ by $\langle T, \phi \rangle$. T is said to be *continuous* if for each sequence $\phi_n \in \mathscr{D}(\Omega)$ that converges in $\mathscr{D}(\Omega)$ to a function $\phi \in \mathscr{D}(\Omega)$ as $n \to \infty$, one has

$$\langle T, \phi_n \rangle \longrightarrow \langle T, \phi \rangle \quad \text{as } n \longrightarrow \infty .$$

A *distribution* is a linear continuous form on $\mathscr{D}(\Omega)$. The set of all the distributions on Ω is a vector space denoted by $\mathscr{D}'(\Omega)$.

Examples

(i) Each integrable function $f \in L^1(\Omega)$ (see (A.6)) can be identified with the distribution T_f defined by

$$\langle T_f, \phi \rangle = \int_\Omega f(\mathbf{x}) \phi(\mathbf{x}) d\mathbf{x} \quad \text{for all } \phi \in \mathscr{D}(\Omega) .$$

(ii) Let $\mathbf{x}_0 \in \Omega$. The linear form on $\mathscr{D}(\Omega)$,

$$\langle \delta_{\mathbf{x}_0}, \phi \rangle = \phi(\mathbf{x}_0) \quad \text{for all } \phi \in \mathscr{D}(\Omega) ,$$

is a distribution, which is commonly (but improperly) called the "Dirac function".

We notice that if T_1 and T_2 are two distributions, then they are "equal in the sense of distributions" if

$$\langle T_1, \phi \rangle = \langle T_2, \phi \rangle \quad \text{for all } \phi \in \mathscr{D}(\Omega) .$$

(b) Derivative of Distributions

Let α be a nonnegative multi-index and set $m = \alpha_1 + \cdots + \alpha_d$. For each distribution $T \in \mathscr{D}'(\Omega)$ let us consider the linear form on $\mathscr{D}(\Omega)$:

$$\langle D^\alpha T, \phi \rangle = (-1)^m \langle T, D^\alpha \phi \rangle \quad \text{for all } \phi \in \mathscr{D}(\Omega) .$$

This linear form is continuous on $\mathscr{D}(\Omega)$; hence, it is a distribution, which is called the α-*distributional derivative* of T.

It follows that each integrable function $u \in L^1(\Omega)$ is infinitely differentiable in the sense of distributions, and the following Green's formula holds:

$$\langle D^\alpha u, \phi \rangle = (-1)^m \int_\Omega u(\mathbf{x}) D^\alpha \phi(\mathbf{x}) \mathrm{d}\mathbf{x} \quad \text{for all } \phi \in \mathscr{D}(\Omega) .$$

If u is m-times continuously differentiable in Ω, then the α-distributional derivative of u coincides with the classical derivative of index α. In general, a distributional derivative of an integrable function can be an integrable function or merely a distribution. We say that the α-distributional derivative of an integrable function $u \in L^1(\Omega)$ is an *integrable function* if there exists $g \in L^1(\Omega)$ such that

$$\langle D^\alpha u, \phi \rangle = \int_\Omega g(\mathbf{x}) \phi(\mathbf{x}) \mathrm{d}\mathbf{x} \quad \text{for all } \phi \in \mathscr{D}(\Omega) .$$

Examples

(i) Consider the function $u(x) = \frac{1}{2}|x|$ in the interval $(-1, 1)$. Note that u is not classically differentiable at the origin. The first derivative of u in the distributional sense is represented by the step function

$$v(x) = \begin{cases} 1/2 & \text{if } x > 0 \\ -1/2 & \text{if } x < 0 . \end{cases}$$

(ii) Consider the function v now defined. Note that the classical derivative is zero at all the points $x \neq 0$. The first derivative of v in the sense of distributions is the "Dirac function" δ_0 at the origin. This distribution cannot be represented by an integrable function.

Functions having a certain number of distributional derivatives that can be represented by integrable functions play a fundamental role in the modern theory of partial differential equations. The spaces of these functions are named Sobolev spaces (see (A.8)).

(c) Periodic Distributions

Let $\Omega = (0, 2\pi)^d$, for $d = 1, 2$ or 3. We define the space $C_p^\infty(\overline{\Omega})$ as the vector space of the functions $u : \overline{\Omega} \to \mathbb{C}$ that have derivatives of any order continuous in the closure $\overline{\Omega}$ of Ω, and 2π-periodic in each space direction. A sequence $\phi_n \in C_p^\infty(\overline{\Omega})$ *converges in* $C_p^\infty(\overline{\Omega})$ to a function $\phi \in C_p^\infty(\overline{\Omega})$ if $D^\alpha\phi_n \to D^\alpha\phi$ uniformly on $\overline{\Omega}$, as $n \to \infty$ for all nonnegative multi-indices α.

A *periodic distribution* is a linear form $T : C_p^\infty(\overline{\Omega}) \to \mathbb{C}$ that is continuous, i.e., such that

$$\langle T, \phi_n \rangle \longrightarrow \langle T, \phi \rangle \quad \text{as } n \longrightarrow \infty ,$$

whenever $\phi_n \to \phi$ in $C_p^\infty(\overline{\Omega})$.

The *derivative* of index α of a periodic distribution T is the periodic distribution $D^\alpha T$ defined by

$$\langle D^\alpha T, \phi \rangle = (-1)^m \langle T, D^\alpha\phi \rangle \quad \text{for all } \phi \in C_p^\infty(\overline{\Omega})$$

(where $m = \alpha_1 + \cdots + \alpha_d$).

Note that each function in $\mathscr{D}(\Omega)$ also belongs to $C_p^\infty(\overline{\Omega})$. Thus, it is easily seen that each periodic distribution is indeed a distribution in classical sense.

A.8 Sobolev Spaces and Sobolev Norms

We introduce hereafter some relevant Hilbert spaces, which occur in the numerical analysis of boundary-value problems. They are spaces of square-integrable functions, which possess a certain number of derivatives (in the sense of distributions) representable as square-integrable functions.

(a) The Spaces $H^m(a, b)$ and $H^m(\Omega)$, $m \geq 0$

Let (a, b) be a bounded interval of the real line, and let $m \geq 0$ be an integer.

We define $H^m(a, b)$ to be the vector space of the functions $v \in L^2(a, b)$ such that all the distributional derivatives of v of order up to m can be represented by functions in $L^2(a, b)$. In short,

$$H^m(a, b) = \left\{ v \in L^2(a, b) : \text{ for } 0 \leq k \leq m, \frac{\mathrm{d}^k v}{\mathrm{d}x^k} \in L^2(a, b) \right\} .$$

$H^m(a, b)$ is endowed with the inner product

$$(u, v)_m = \sum_{k=0}^m \int_a^b \frac{\mathrm{d}^k u}{\mathrm{d}x^k}(x) \frac{\mathrm{d}^k v}{\mathrm{d}x^k}(x) \mathrm{d}x$$

for which $H^m(a, b)$ is a Hilbert space. The associated norm is

$$\|v\|_{H^m(a,b)} = \left(\sum_{k=0}^m \left\| \frac{\mathrm{d}^k v}{\mathrm{d}x^k} \right\|_{L^2(a,b)}^2 \right)^{1/2} .$$

The Sobolev spaces $H^m(a, b)$ form a hierarchy of Hilbert spaces, in the sense that $\cdots \subset H^{m+1}(a, b) \subset H^m(a, b) \subset \cdots \subset H^0(a, b) = L^2(a, b)$, each inclusion being continuous. Clearly, if a function u has m classical continuous derivatives in $[a, b]$, then u belongs to $H^m(a, b)$—in other words, $C^m([a, b]) \subset H^m(a, b)$ with continuous inclusion. Conversely, if u belongs to $H^m(a, b)$ for $m \geq 1$, then u has $m - 1$ classical continuous derivatives in $[a, b]$, i.e., $H^m(a, b) \subset C^{m-1}([a, b])$ with continuous inclusion. This is an example of the so-called "Sobolev imbedding theorems". As a matter of fact, $H^m(a, b)$ can be equivalently defined as

$$H^m(a, b) = \left\{ v \in C^{m-1}([a, b]) : \frac{\mathrm{d}}{\mathrm{d}x} v^{(m-1)} \in L^2(a, b) \right\} ,$$

where the last derivative is in the sense of distributions.

Functions in $H^m(a, b)$ can be approximated arbitrarily well by infinitely differentiable functions in $[a, b]$, in the distance induced by the norm of $H^m(a, b)$. In other words,

$$C^\infty([a, b]) \text{ is dense in } H^m(a, b)$$

(see (A.4) for the definition of density of a subspace).

Now, let $\Omega \subset \mathbb{R}^d$, for $d \geq 2$, be an open, bounded set with sufficiently smooth boundary. Given a multi-index $\alpha = (\alpha_1, \ldots, \alpha_d)$ of nonnegative integers, we set $|\alpha| = \alpha_1 + \cdots + \alpha_d$ and

$$D^\alpha v = \frac{\partial^{|\alpha|} v}{\partial x_1^{\alpha_1} \cdots \partial x_d^{\alpha_d}} .$$

The previous definition of Sobolev spaces can be extended to higher space dimensions as follows. We define

$$H^m(\Omega) = \{v \in L^2(\Omega) : \text{ for each nonnegative multi-index } \alpha \text{ with } |\alpha| \leq m,$$
$$\text{the distributional derivative } D^\alpha v \text{ belongs to } L^2(\Omega)\}.$$

This is a Hilbert space for the inner product

$$(u, v)_m = \sum_{|\alpha| \leq m} \int_\Omega D^\alpha u(\mathbf{x}) D^\alpha v(\mathbf{x}) \mathrm{d}\mathbf{x} ,$$

which induces the norm

$$\|v\|_{H^m(\Omega)} = \left(\sum_{|\alpha| \leq m} \|D^\alpha v\|_{L^2(\Omega)}^2 \right)^{1/2} .$$

Functions in $H^m(\Omega)$ for $m \geq 1$ need not have their derivatives of order $m - 1$ continuous in Ω. However, for all $m > d/2$, the weaker Sobolev inclusion

$H^m(\Omega) \subset C^k(\overline{\Omega})$ (where $k = [m - d/2]$ is the integer part of $m - d/2$) holds. On the other hand, as in the 1D case

$$C^\infty(\overline{\Omega}) \text{ is dense in } H^m(\Omega) .$$

Sobolev spaces $H^s(\Omega)$ of non-integer order s can be defined, e.g., by "interpolation" (see Bergh and Löfstrom (1976)) between the two Sobolev spaces $H^m(\Omega)$ and $H^{m+1}(\Omega)$, with $m < s < m + 1$.

Sobolev spaces $H^s(\Gamma)$, where Γ is a sufficiently smooth $(d-1)$-dimensional manifold in $\mathbb{R}^d$, can be defined as well.

(b) The Spaces $H_0^1(a, b)$ and $H_0^1(\Omega)$

Dirichlet conditions are among the simplest and most common boundary conditions to be associated with a differential operator. Therefore, the subspaces of the Sobolev spaces H^m spanned by the functions satisfying homogeneous Dirichlet boundary conditions play a fundamental role.

Since the functions of $H^1(a, b)$ are continuous up to the boundary by the Sobolev imbedding theorem, it is meaningful to introduce the following subspace of $H^1(a, b)$:

$$H_0^1(a, b) = \{v \in H^1(a, b) : v(a) = v(b) = 0\} .$$

This is a Hilbert space for the same inner product of $H^1(a, b)$. It is often preferable to endow $H^1(a, b)$ with a different, although equivalent, inner product. This is defined as

$$[u, v] = \int_a^b \frac{du}{dx}(x) \frac{dv}{dx}(x) dx .$$

By the Poincaré inequality, it is indeed an inner product on $H_0^1(a, b)$. The associated norm, denoted by

$$\|v\|_{H_0^1(a,b)} = \left(\int_a^b \left| \frac{dv}{dx} \right|^2 dx \right)^{1/2} ,$$

is equivalent to the $H^1(a, b)$-norm, in the sense that there exists a constant $C > 0$ such that, for all $v \in H_0^1(a, b)$,

$$C\|v\|_{H^1(a,b)} \leq \|v\|_{H_0^1(a,b)} \leq \|v\|_{H^1(a,b)} .$$

Again, this follows from the Poincaré inequality (A.10).

The functions of $H_0^1(a, b)$ can be approximated arbitrarily well in the norm of this space not only by infinitely differentiable functions on $[a, b]$, but also by infinitely differentiable functions that vanish identically in a neighborhood of $x = a$ and $x = b$. In other words

$$\mathcal{D}((a, b)) \text{ is dense in } H^1(a, b)$$

(see (A.4) and (A.7)).

We turn now to more space dimensions. If Ω is a bounded domain in $\mathbb{R}^d$ with sufficiently smooth boundary, the functions of $H^1(\Omega)$ need not be continuous on the closure of Ω. Thus, their pointwise values on the boundary $\partial\Omega$ of Ω need not be defined. However, it is possible to extend the trace operator $v \mapsto v|_{\partial\Omega}$ (classically defined for functions $v \in C^0(\overline{\Omega})$) so as to be a linear continuous mapping between $H^1(\Omega)$ and $L^2(\partial\Omega)$, the space of the square-integrable functions on $\partial\Omega$. (More precisely, the image of the trace operator is a proper subspace of $L^2(\partial\Omega)$, indicated by $H^{1/2}(\partial\Omega)$.) With this in mind, it is meaningful to define $H_0^1(\Omega)$ as the subspace of $H^1(\Omega)$ of the functions whose trace at the boundary is zero. Precisely we set

$$H_0^1(\Omega) = \{v \in H^1(\Omega) : v|_{\partial\Omega} = 0\} \ .$$

This is a Hilbert space for the inner product of $H^1(\Omega)$, or for the inner product

$$[u, v] = \int_\Omega \nabla u(\mathbf{x}) \cdot \nabla v(\mathbf{x}) \, d\mathbf{x} \ .$$

The associated norm is denoted by

$$\|v\|_{H_0^1(\Omega)} = \left(\int_\Omega |\nabla v|^2 d\mathbf{x} \right)^{1/2}$$

and is equivalent to the $H^1(\Omega)$-norm, by the Poincaré inequality (A.10).

Concerning the approximation of the functions of $H_0^1(\Omega)$ by infinitely smooth functions, the following result holds:

$$\mathcal{D}(\Omega) \text{ is dense in } H_0^1(\Omega) \ .$$

The dual spaces (see (A.1.c)) of the Hilbert spaces of type H_0^1 now defined are usually denoted by H^{-1}. Thus, $H^{-1}(a, b)$ is the dual space of $H_0^1(a, b)$, $H^{-1}(\Omega)$ is the dual space of $H_0^1(\Omega)$.

Finally let us mention that for $m \geq 2$, one can define the subspaces $H_0^m(a, b)$ of $H^m(a, b)$ (and similarly for $H_0^m(\Omega)$) of the functions of $H^m(a, b)$ whose derivatives of order up to $m - 1$ vanish on the boundary of the domain of definition. Again, these spaces are Hilbert spaces for the inner product of $H^m(a, b)$, or for an equivalent inner product that only involves the derivatives of order m.

(c) The Spaces $H_p^m(0, 2\pi)$ and $H_p^m(\Omega)$, $m \geq 0$

In the analysis of Fourier methods, the natural Sobolev spaces are those of periodic functions. In this framework, functions are complex valued, and their derivatives are taken in the sense of the periodic distributions (see (A.7)). We set

$$H_p^m(0, 2\pi) = \left\{ v \in L^2(0, 2\pi) : \text{ for } 0 \le k \le m, \text{ the derivative } \frac{d^k v}{dx^k} \text{ in the} \right.$$

$$\text{sense of periodic distribution}$$

$$\left. \text{belongs to } L^2(0, 2\pi) \right\} .$$

$H_p^m(0, 2\pi)$ is a Hilbert space for the inner product

$$(u, v)_m = \sum_{k=0}^{m} \int_0^{2\pi} \frac{d^k u}{dx^k}(x) \overline{\frac{d^k v}{dx^k}}(x) dx ,$$

whose associated norm is

$$\|v\|_{H_p^m(0,2\pi)} = \left(\sum_{k=0}^{m} \left\| \frac{d^k v}{dx^k} \right\|_{L^2(0,2\pi)}^2 \right)^{1/2} .$$

The space $H_p^m(0, 2\pi)$ coincides with the space of the functions $v : [0, 2\pi] \to \mathbb{C}$ that have $m-1$ continuously differentiable, 2π-periodic derivatives on $[0, 2\pi]$, and such that the periodic distributional derivative $(d/dx)v^{(m-1)}$ can be represented by a function of $L^2(0, 2\pi)$. The space $C_p^\infty([0, 2\pi])$ is dense in $H_p^m(0, 2\pi)$.

If $\Omega = (0, 2\pi)^d$ for $d = 2$ or 3, we set

$$H_p^m(\Omega) = \{ v \in L^2(\Omega) : \text{ for each multi-index } \alpha \text{ with } |\alpha| \le m,$$

$$\text{the derivative } D^\alpha v \text{ in the sense of periodic}$$

$$\text{distributions belongs to } L^2(\Omega)\}.$$

This is a Hilbert space for the inner product

$$(u, v)_m = \sum_{|\alpha| \le m} \int_\Omega D^\alpha u(\mathbf{x}) \overline{D^\alpha v(\mathbf{x})} d\mathbf{x} ,$$

with associated norm

$$\|v\|_{H_p^m(\Omega)} = \left(\sum_{|\alpha| \le m} \|D^\alpha v\|_{L^2(\Omega)}^2 \right)^{1/2} .$$

The space $C_p^\infty(\overline{\Omega})$ is dense in $H_p^m(\Omega)$. Note that since a periodic distribution is also a distribution, each space $H_p^m(0, 2\pi)$ (resp. $H_p^m(\Omega)$) is a subspace of the space $H^m(0, 2\pi)$ (resp. $H^m(\Omega)$).

A.9 The Sobolev Inequality

Let $(a, b) \subset \mathbb{R}$ be a bounded interval of the real line. For each function $u \in H^1(a, b)$ (see (A.8)) the following inequality holds:

$$\|u\|_{L^\infty(a,b)} \leq \left(\frac{1}{b-a} + 2\right)^{1/2} \|u\|_{L^2(a,b)}^{1/2} \|u\|_{H^1(a,b)}^{1/2} \, .$$

A.10 The Poincaré Inequality

Let v be a function of $H^1(a, b)$ (see (A.8)). We know that v is continuous on $[a, b]$. Assume that at a point $x_0 \in [a, b]$, $v_0(x_0) = 0$. The Poincaré inequality states that there exists a constant C (depending upon the interval length $b - a$) such that

$$\|v\|_{L^2(a,b)} \leq C\|v'\|_{L^2(a,b)} \, , \qquad (A.10.1)$$

i.e., the L^2-norm of the function is bounded by the L^2-norm of the derivative. The Poincaré inequality applies to functions belonging to $H_0^1(a, b)$, for which $x_0 = a$ or b, and also to functions of $H^1(a, b)$ that have zero average on (a, b), since necessarily such functions change sign in the domain.

In space dimension $d \geq 2$, the functions to which the Poincaré inequality applies must vanish on a manifold of dimension $d - 1$. Confining ourselves to the case of functions vanishing on the boundary $\partial\Omega$ of the domain of definition Ω, one has

$$\|v\|_{L^2(\Omega)} \leq C\|\nabla v\|_{(L^2(\Omega))^d} \quad \text{for all } v \in H_0^1(\Omega) \, . \qquad (A.10.2)$$

The same result holds if the domain Ω is simply connected and v only vanishes on a portion of $\partial\Omega$ of positive measure.

The smallest constant $C = C_P$ for which (A.10.1) or (A.10.2) holds is termed the Poincaré constant of the domain. Often, this term refers to any constant C appearing in (A.10.1) or in (A.10.2).

Appendix B. Fast Fourier Transforms

Basics

The Fast Fourier Transform (FFT) is a recursive algorithm for evaluating the discrete Fourier transform and its inverse. The FFT is conventionally written for the evaluation of

$$\tilde{u}_k = \sum_{j=0}^{N-1} u_j e^{+2\pi ijk/N} , \qquad k = 0, 1, \ldots, N-1 , \tag{B.1.a}$$

or

$$\tilde{u}_k = \sum_{j=0}^{N-1} u_j e^{-2\pi ijk/N} , \qquad k = 0, 1, \ldots, N-1 , \tag{B.1.b}$$

where u_j, $j = 0, 1, \ldots, N-1$ are a set of complex data. The FFT quickly became a widely used tool in signal processing after its description by Cooley and Tukey (1965). (As noted later by Cooley, Lewis and Welch (1969), most essential components of the FFT date back to the 1920s.) The Cooley–Tukey algorithm enables the sums in (B.1) to be evaluated in $5N \log_2 N$ real operations (when N is a power of 2), instead of the $8N^2$ real operations required by the straightforward sum. Moreover, calculation of (B.1) via the FFT incurs less error due to round-off than the direct summation method (Cooley et al. (1969)).

Many versions of the FFT are now in existence. The review by Temperton (1983) contains an especially clear description of a simple yet efficient one. It allows N to be of the form

$$N = 2^p 3^q 4^r 5^s 6^t \tag{B.2}$$

and has the operation count

$$N \left(5p + 9\tfrac{1}{3}q + 8\tfrac{1}{2}r + 13\tfrac{3}{5}s + 13\tfrac{1}{3}t - 6 \right) . \tag{B.3}$$

No additional flexibility is gained by the inclusion of the factors 4 and 6. The algorithm is, however, more efficient when these factors are included. Not only is the operation count lower—for example, by 15% when $N = 64$—but, due to the higher ratio of arithmetic operations to memory accesses, most Fortran compilers generate more efficient code for the larger factors. For the sake of simplicity, however, throughout this book we use $(5 \log_2 N - 6)N$ as the operation count for the complex FFT; moreover, the lower order term linear in N is usually omitted.

We should also mention the book by Brigham (1974), which is devoted entirely to the Fast Fourier Transform, and the FFTW package by Frigo and Johnson (2005), which received the 1999 Wilkinson Prize for Numerical Software. (The FFTW software is available at http://www.fftw.org/.)

Use in Spectral Methods

In applications of Fourier spectral methods, the sums that one must evaluate are

$$\tilde{u}_k = \frac{1}{N} \sum_{j=0}^{N-1} u_j e^{-2\pi ijk/N} , \qquad k = -\frac{N}{2}, -\frac{N}{2} + 1, \ldots, \frac{N}{2} - 1 , \qquad \text{(B.4)}$$

and

$$u_j = \sum_{k=-N/2}^{N/2-1} \tilde{u}_k e^{2\pi ijk/N} , \qquad j = 0, 1, \ldots, N - 1 \qquad \text{(B.5)}$$

(see (8.1.21) and (8.1.23)). From (B.4) it is apparent that, for integers p and k,

$$\tilde{u}_{k+pN} = \tilde{u}_k . \qquad \text{(B.6)}$$

When the array $(u_0, u_1, \ldots, u_{N-1})$ is fed into a standard FFT for evaluating (B.1.b) it returns, in effect, the array

$$(N\tilde{u}_0, N\tilde{u}_1, \ldots, N\tilde{u}_{N/2-1}, N\tilde{u}_{-N/2}, N\tilde{u}_{-N/2+1}, \ldots, N\tilde{u}_{-1}) .$$

Conversely, when this array (without the factor N) is fed into the standard FFT for evaluating (B.1.a) (with the plus sign), the array $(u_0, u_1, \ldots, u_{N-1})$ is returned.

In most applications of spectral methods, the direct use of the complex FFT (B.1) is needlessly expensive. This is true, for example if the function u_j is real or if a cosine transform (for a Chebyshev spectral method) is desired. These issues have been addressed by Orszag (1971a, Appendix II) and by Brachet et al. (1983, Appendix C). A summary of some of the relevant transformations follows.

Real Transforms

The simplest case occurs when many real transforms are desired at once, as arises for multidimensional problems. They can be computed pairwise.

Suppose that u_j^1 and u_j^2, $j = 0, 1, \ldots, N - 1$, are two sets of real data. Then one can define

$$v_j = u_j^1 + iu_j^2 \tag{B.7}$$

and compute $\tilde{v}_k$ according to (B.4) by the standard N-point complex FFT. Then the transforms $\tilde{u}_k^1$ and $\tilde{u}_k^2$ can be extracted according to

$$\tilde{u}_k^1 = \frac{1}{2}(\tilde{v}_k + \bar{\tilde{v}}_{-k})$$

$$\tilde{u}_k^2 = -\frac{i}{2}(\tilde{v}_k - \bar{\tilde{v}}_{-k}) \qquad , \qquad k = 0, 1, \ldots, \frac{N}{2} - 1 . \tag{B.8}$$

(The Fourier coefficients of real data for negative k are related to those for positive k by $\tilde{u}_{-k} = \bar{\tilde{u}}_k$.) This process is readily reversed. In fact, if one is performing a Fourier interpolation derivative, one need not even bother with the separation (B.8) in Fourier space, since

$$\frac{du^1}{dx}\bigg|_j + i\,\frac{du^2}{dx}\bigg|_j = \sum_{k=-N/2}^{N/2-1} ik\tilde{v}_k . \tag{B.9}$$

If only a single real transform is desired, then one may follow the prescription given by Orszag (1971a). Let $M = N/2$ and define

$$v_j = u_{2j} + iu_{2j+1} , \qquad j = 0, 1, \ldots, M - 1 . \tag{B.10}$$

Then take an M-point transform of v_j, set $\tilde{v}_M = \tilde{v}_0$, and extract the desired coefficients via

$$\tilde{u}_k = \frac{1}{2}(\tilde{v}_k + \bar{\tilde{v}}_{M-k}) - \frac{i}{2}e^{2\pi ik/N}(\tilde{v}_k - \bar{\tilde{v}}_{M-k}) , \qquad k = 0, 1, \ldots, M - 1 . \tag{B.11}$$

For both of these approaches the cost of a single, real-to-half-complex transform is essentially $(5/2)N \log_2 N$.

Chebyshev Transforms

The discrete Chebyshev transforms based on the Gauss–Lobatto points (8.3.15) are given by

$$\tilde{u}_k = \frac{2}{N\bar{c}_k} \sum_{j=0}^{N} \frac{1}{\bar{c}_j} u_j \cos\frac{\pi jk}{N} , \qquad k = 0, 1, \ldots, N , \tag{B.12}$$

(see (8.2.24) and (8.3.17)) and

$$u_j = \sum_{k=0}^{N} \tilde{u}_k \cos\frac{\pi jk}{N} , \qquad j = 0, 1, \ldots, N \tag{B.13}$$

(see (8.2.25) and (8.3.19)). Suppose that the transform (B.12) is desired for two real sets of data u_j^1 and u_j^2. Then define the complex data v_j by

$$v_j = \begin{cases} u_j^1 + iu_j^2 , & j = 0, 1, \ldots, N , \\ v_{2N-j} , & j = N+1, N+2, \ldots, 2N-1 , \end{cases} \tag{B.14}$$

and by periodicity (with period $2N$) for other integers j. Next, define $\tilde{v}_k$, $k = 0, 1, \ldots, N$, by (B.12) and define $\tilde{V}_k$, $k = 0, 1, \ldots, 2N-1$, by (B.1.a) with N replaced by $2N$. It is readily shown that

$$\tilde{V}_k = \frac{1}{N\bar{c}_k}\tilde{v}_k , \qquad k = 0, 1, \ldots, N , \tag{B.15}$$

and that

$$\tilde{V}_k = \sum_{l=0}^{N-1} v_{2l}e^{2\pi ikl/N} + e^{\pi ik/N}\sum_{l=0}^{N-1} v_{2l+1}e^{2\pi ikl/N} . \tag{B.16}$$

Now, define w_j by

$$w_j = v_{2j} + i(v_{2j+1} - v_{2j-1}) , \qquad j = 0, 1, \ldots, N-1 , \tag{B.17}$$

and compute $\tilde{w}_k$ according to the complex FFT (B.1.a). We have

$$\begin{aligned} \tilde{w}_k &= \sum_{l=0}^{N-1} v_{2l}e^{2\pi ikl/N} + i(1 - e^{2\pi ik/N})\sum_{l=0}^{N-1} v_{2l+1}e^{2\pi ikl/N} , \\ \tilde{w}_{N-k} &= \sum_{l=0}^{N-1} v_{2l}e^{2\pi ikl/N} - i(1 - e^{2\pi ik/N})\sum_{l=0}^{N-1} v_{2l+1}e^{2\pi ikl/N} . \end{aligned} \tag{B.18}$$

Consequently,

$$\begin{aligned} \tilde{v}_0 &= \frac{1}{N}\sum_{j=0}^{N}\frac{1}{\bar{c}_j}v_j , \\ \tilde{v}_k &= \frac{1}{N}\left[\left(\frac{1}{2} + \frac{1}{4\sin\frac{\pi k}{N}}\right)\tilde{w}_k + \left(\frac{1}{2} - \frac{1}{4\sin\frac{\pi k}{N}}\right)\tilde{w}_{N-k}\right] , \\ \tilde{v}_N &= \frac{1}{N}\sum_{j=0}^{N}(-1)^j\frac{1}{\bar{c}_j}v_j . \end{aligned} \tag{B.19}$$

The desired real coefficients $\tilde{u}_k^1$ and $\tilde{u}_k^2$ are the real and imaginary parts, respectively, of the $\tilde{v}_k$. Thus, the discrete Chebyshev transform (B.12) can be computed in $\frac{5}{2}N\log_2 N + 4N$ real operations per transform, assuming that a large number of such transforms are computed. The inverse discrete

Chebyshev transform (B.13) can be evaluated with only minor modifications to the algorithm given by (B.14), (B.17) and (B.19).

Discrete sine transforms can be handled in a similar manner: (B.14) (with v_{2N-j} replaced by $-v_{2N-j}$) and (B.17) are retained as is the central equation in (B.19) with the coefficient of $\tilde{w}_{N-k}$ having the opposite sign; the entire $\tilde{v}_k$ term is multiplied by i and one sets $\tilde{v}_0 = \tilde{v}_N = 0$. Swarztrauber (1986) described how real cosine and sine transforms can be computed without the pre- and post-processing costs incurred by (B.17) and (B.19).

Other Cosine Transforms

In some applications, such as the use of a staggered grid in Navier–Stokes calculations (see Sect. 3.4) and in simulations of flows with special symmetries (Brachet et al. (1983)), discrete Chebyshev transforms with respect to the Gauss points (see (8.3.13) but with $N-1$ in place of N) are required. Consider

$$\tilde{u}_k = \frac{2}{N} \sum_{j=0}^{N-1} u_j \cos \frac{(2j+1)\pi k}{2N} , \qquad k = 0, 1, \ldots, N-1 . \qquad \text{(B.20)}$$

Brachet et al. (1983) have provided prescriptions for computing efficiently this and related sums. Put

$$v_j = \begin{cases} u_{2j} , & j = 0, 1, \ldots, \dfrac{N}{2} - 1 , \\[2ex] u_{2N-2j-1} , & j = \dfrac{N}{2}, \dfrac{N}{2}+1, \ldots, N-1 , \end{cases} \qquad \text{(B.21)}$$

and compute $\tilde{v}_k$ according to (B.1.a). Then $\tilde{u}_k$ may be extracted via

$$\tilde{u}_k = \frac{1}{N}[e^{2\pi ik/2N}\tilde{v}_k + e^{-2\pi ik/2N}\tilde{v}_{N-k}] , \qquad k = 0, 1, \ldots, N-1 . \qquad \text{(B.22)}$$

The corresponding inverse Chebyshev transform

$$u_j = \sum_{k=0}^{N-1} \tilde{u}_k \cos \frac{(2j+1)\pi k}{2N} \qquad \text{(B.23)}$$

can be evaluated by reversing these steps.

For some problems, the Chebyshev expansion may be over the interval $[0, 1]$ instead of $[-1, 1]$. Moreover, it may also be useful to use only the odd (or even) polynomials (Spalart (1984); see also Sect. 8.8.2). Spalart (1986, pers. comm.) explained how to employ the FFT for an expansion over $[0, 1]$ in terms of just the odd Chebyshev polynomials. The collocation points are

$$x_j = \cos \frac{(2j+1)\pi}{2N} , \qquad j = 0, 1, \ldots, N-1 , \qquad \text{(B.24)}$$

the series expansion is

$$u^N(x) = \sum_{k=0}^{N-1} \tilde{u}_k T_{2k+1}(x) \ , \tag{B.25}$$

and the discrete transforms are

$$\tilde{u}_k = \frac{2}{N} \sum_{j=0}^{N-1} u_j \cos \frac{(2k+1)(2j+1)\pi}{4N} \ , \qquad k = 0, 1, \ldots, N-1 \ , \tag{B.26}$$

and

$$u_j = \sum_{k=0}^{N-1} \tilde{u}_k \cos \frac{(2k+1)(2j+1)\pi}{4N} \ , \qquad j = 0, 1, \ldots, N-1 \ . \tag{B.27}$$

(In order for a half-interval Chebyshev expansion to be spectrally accurate, one needs $u(x)$ and all of its derivatives to vanish at $x = 0$.) Spalart's trick for evaluating (B.27) is to define

$$\tilde{v}_k = \frac{\tilde{u}_k + \tilde{u}_{k-1}}{2 \cos \left(\dfrac{k\pi}{2N} \right)} \ , \qquad k = 0, 1, \ldots, N \ , \tag{B.28}$$

where $\tilde{u}_{-1} = \tilde{u}_N = 0$, to compute v_j according to (B.13), and then to extract u_j via

$$u_j = \frac{\tilde{v}_j + \tilde{v}_{j+1}}{2 \cos \dfrac{(2j+1)\pi}{4N}} \ , \qquad j = 0, 1, \ldots, N-1 \ . \tag{B.29}$$

(Note however that this transform is not suitable for use with the Gauss–Lobatto points.)

Appendix C. Iterative Methods for Linear Systems

In this appendix, we review some of the most important iterative methods for the solution of a linear system of the form

$$Lu = f . \tag{C.0.1}$$

The discussion will be at a tutorial level. For an extensive presentation and a thorough analysis the reader may refer to Golub and Van Loan (1996), Saad (1996), Greenbaum (1997), Van der Vorst (2003), and to the ample literature cited therein.

C.1 A Gentle Approach to Iterative Methods

A particularly simple iterative scheme is the Richardson (1910) method. Given an initial guess $\mathbf{v}^0$ to $\mathbf{u}$, subsequent approximations are obtained via

$$\mathbf{v}^{n+1} = \mathbf{v}^n + \omega \mathbf{r}^n , \tag{C.1.1}$$

where ω is a relaxation parameter and

$$\mathbf{r}^n = \mathbf{f} - L\mathbf{v}^n \tag{C.1.2}$$

is the residual associated with $\mathbf{v}^n$. The error obeys the relation

$$\left(\mathbf{v}^{n+1} - \mathbf{u}\right) = G\left(\mathbf{v}^n - \mathbf{u}\right) , \tag{C.1.3}$$

where the iteration matrix G of the Richardson scheme is given by

$$G = I - \omega L . \tag{C.1.4}$$

The iterative scheme is convergent if the spectral radius ρ of G is less than 1. In the case of the Richardson scheme this condition is equivalent to

$$|1 - \omega\lambda| < 1 , \tag{C.1.5}$$

for all the eigenvalues λ of L. The simultaneous fulfillment of these inequalities is possible only if all the eigenvalues of L have nonzero real parts of constant sign. A particularly relevant case is that of a matrix with all real and strictly positive eigenvalues; symmetric and positive-definite matrices enjoy this property, but these are not necessary conditions. For example, the matrices generated by Chebyshev or Legendre collocation discretizations of second-order problems have all real and strictly positive eigenvalues. In such a situation, we have $0 < \lambda_{\min} \leq \lambda_{\max}$, where $\lambda_{\min}$ and $\lambda_{\max}$ are the extreme eigenvalues of L. The convergence condition (C.1.5) is satisfied for $0 < \omega < \omega_{\max}$, where

$$\omega_{\max} = 2/\lambda_{\max} . \tag{C.1.6}$$

The best choice of ω is that which minimizes ρ. It is obtained from the relation

$$(1 - \omega\lambda_{\max}) = -(1 - \omega\lambda_{\min}) , \tag{C.1.7}$$

for then the largest values of $1 - \omega\lambda$ are equal in magnitude and have opposite sign (see, e.g., Quarteroni and Valli (1994)). The optimal relaxation parameter is thus

$$\omega_{\text{opt}} = \frac{2}{\lambda_{\max} + \lambda_{\min}} . \tag{C.1.8}$$

It produces the spectral radius

$$\rho = \frac{\lambda_{\max} - \lambda_{\min}}{\lambda_{\max} + \lambda_{\min}} . \tag{C.1.9}$$

Note that the dependence upon the extreme eigenvalues enters only in the combination

$$\mathcal{K} = \mathcal{K}(L) = \frac{\lambda_{\max}}{\lambda_{\min}} . \tag{C.1.10}$$

We shall call this ratio the *iterative condition number* of L to distinguish it from the generic definition of condition number,

$$\kappa_{\|\cdot\|}(L) = \|L\| \, \|L^{-1}\| , \tag{C.1.11}$$

and the *spectral condition number*

$$\kappa_2(L) = \left[\frac{\lambda_{\max}\left(L^T L\right)}{\lambda_{\min}\left(L^T L\right)} \right]^{1/2} . \tag{C.1.12}$$

In the case that L is symmetric and positive definite, this becomes

$$\kappa_2(L) = \frac{\lambda_{\max}\left(L\right)}{\lambda_{\min}\left(L\right)} = \mathcal{K} . \tag{C.1.13}$$

For some nonsymmetric discretization matrices that have real positive eigenvalues, such as those mentioned above, the spectral and the iterative condition numbers might differ. In terms of $\mathcal{K}$, (C.1.9) becomes

$$\rho = \frac{\mathcal{K} - 1}{\mathcal{K} + 1} . \tag{C.1.14}$$

Define the *rate of convergence* $\mathcal{R}$ to be

$$\mathcal{R} = -\log \rho , \tag{C.1.15}$$

and denote its reciprocal by $\mathcal{J}$. The latter quantity measures the number of iterations required to reduce the error by a factor of e. This immediately follows from the error bound

$$\|\mathbf{v}^n - \mathbf{u}\|_L \leq \rho^n \|\mathbf{v}^0 - \mathbf{u}\|_L ,$$

which holds with $\|\mathbf{v}\|_L = (\mathbf{v}^T L \mathbf{v})^{1/2}$. Clearly, the larger the convergence rate that a method has for a problem, the fewer iterations that are required to obtain a solution to a given accuracy. For the Richardson method described above, the number of iterations increases as

$$\mathcal{J} \cong \frac{1}{2}\mathcal{K} . \tag{C.1.16}$$

The basic Richardson method (C.1.1) can be improved and extended in several ways. The discussion thus far concerned only the *stationary* Richardson method. In a *non-stationary* Richardson method, the parameter ω in (C.1.1) is allowed to depend on n, i.e., to change in the course of iterations, in order to speed up the convergence.

For a *static* non-stationary Richardson (NSR) method one cycles through a fixed number k of parameters. Using the minimax property of Chebyshev polynomials, the following expressions for the optimal parameters can be derived:

$$\omega_j = \frac{2/\lambda_{\min}}{(\mathcal{K} - 1) \cos \dfrac{(2j - 1)\pi}{2k} + (\mathcal{K} + 1)} , \qquad j = 1, \ldots, k , \tag{C.1.17}$$

yielding the effective spectral radius

$$\rho = \frac{1}{T_k \left(\dfrac{\mathcal{K} + 1}{\mathcal{K} - 1} \right)^{1/k}} . \tag{C.1.18}$$

Both ω_j (for all j) and ρ depend on $\mathcal{K}$. However, this approach suffers from the same limitation as the basic Richardson method—information must be available on the eigenvalues of L in order to compute $\mathcal{K}$.

A broad family of *dynamic* non-stationary Richardson methods are based on an optimality strategy that does not require the knowledge of the extreme eigenvalues. We address dynamic non-stationary Richardson methods in Sect. C.2.

The primary cause of the inefficiency of the Richardson method is that the convergence rate decreases as the iterative condition number increases; in spectral methods, the condition number typically increases with the approximation parameter N. This can be alleviated by *preconditioning* the problem, in effect solving

$$H^{-1}Lu = H^{-1}f$$

rather than (C.0.1). (This is called *left preconditioning*. Other options are available as well, such as *right preconditioning* and *symmetric preconditioning*; see (C.2.15) and (C.2.18), respectively.)

A preconditioned version of (C.1.1) is

$$H\left(\mathbf{v}^{n+1} - \mathbf{v}^{n}\right) = \omega\mathbf{r}^{n} . \tag{C.1.19}$$

One obvious requirement for H is that this equation can be solved inexpensively, i.e., in fewer operations than are required to evaluate $L\mathbf{v}^n$. The effective iteration matrix is now

$$G = I - \omega H^{-1}L . \tag{C.1.20}$$

The second requirement on the preconditioning matrix is that H^{-1} be a good approximation to L^{-1}, i.e., that the new iterative condition number $\mathcal{K}(H^{-1}L)$ be much smaller than $\mathcal{K}(L)$. In such circumstances, the new spectral radius ρ is much smaller than that of the non-preconditioned Richardson method.

This property can be rigorously justified whenever L and H are both symmetric and positive definite. Indeed, denoting by $H^{1/2}$ the square root of H, (C.1.19) can be written equivalently as

$$\mathbf{w}^{n+1} = \mathbf{w}^{n} + \omega(H^{-1/2}\mathbf{f} - H^{-1/2}LH^{-1/2}\mathbf{w}^{n})$$

with $\mathbf{w}^n = H^{1/2}\mathbf{v}^n$, showing that (C.1.19) is nothing but a Richardson iteration applied to the symmetric and positive-definite matrix $H^{-1/2}LH^{-1/2}$. Since this matrix is similar to $H^{-1}L$, we have

$$\mathcal{K}(H^{-1/2}LH^{-1/2}) = \mathcal{K}(H^{-1}L) .$$

The discussion so far has presumed that the eigenvalues of $H^{-1}L$ are confined to the interval $[\lambda_{\min}, \lambda_{\max}]$ on the positive real-axis. However, the Richardson iteration schemes can work on problems for which the eigenvalues are complex but have positive real parts. If we still use a real ω, then it should obey the following restriction for convergence:

$$\omega < 2\frac{Re(\lambda_i)}{|\lambda_i|^2} ,$$

for all eigenvalues λ_i of $H^{-1}L$ (see, e.g., Quarteroni and Valli (1994), Sect. 2.4). One could also use a complex ω, in which case the iterations can be performed entirely in real arithmetic according to

$$\mathbf{v}^{n+1} = \mathbf{v}^n + 2Re(\omega)H^{-1}\mathbf{r}^n + |\omega|^2 H^{-1}LH^{-1}\mathbf{r}^n . \tag{C.1.21}$$

The value of the optimal parameter ω_{opt} is obtained by solving a minimax problem in complex arithmetic.

C.2 Descent Methods for Symmetric Problems

Unlike the stationary Richardson method discussed previously, descent methods have no parameters such as ω that require knowledge of the extreme eigenvalues $\lambda_{\min}$ and $\lambda_{\max}$ of the matrix L or of $H^{-1}L$, where H is a suitable preconditioner. The principle is to adjust the current guess $\mathbf{v}^n$ via

$$H(\mathbf{v}^{n+1} - \mathbf{v}^n) = \alpha_n \mathbf{r}^n , \tag{C.2.1}$$

where $\mathbf{r}^n = \mathbf{f} - L\mathbf{v}^n$ is the residual, and the scalar α_n—the dynamic relaxation parameter—is chosen according to some optimality criterium, as described below. In this section we will assume that both L and H are symmetric and positive definite (but the reader should be aware that these iterative methods may work even if this condition is not satisfied).

The most natural option for defining α_n is to minimize the Euclidean norm of the new residual $\mathbf{r}^{n+1}$; another option is to minimize the so-called H-norm of the new preconditioned residual $\mathbf{p}^{n+1} = H^{-1}\mathbf{r}^{n+1}$, i.e., the quantity $\|\mathbf{p}^{n+1}\|_H = (H\mathbf{p}^{n+1}, \mathbf{p}^{n+1})^{1/2} = \|\mathbf{r}^{n+1}\|_{H^{-1}}$. Both options are referred to as preconditioned minimum residual Richardson (PMRR) methods and will be denoted by PMRR$_2$ and PMRR$_H$, respectively. An additional option is to minimize the L-norm of the new error $\mathbf{e}^{n+1} = \mathbf{u} - \mathbf{v}^{n+1}$, i.e., the quantity $\|\mathbf{e}^{n+1}\|_L = (L\mathbf{e}^{n+1}, \mathbf{e}^{n+1})^{1/2}$. This is referred to as a preconditioned steepest descent Richardson (PSDR) method.

The corresponding algorithms can be written compactly as follows:

Preconditioned Richardson Methods

Initialize

$$\mathbf{v}^0, \quad \mathbf{r}^0 = \mathbf{f} - L\mathbf{v}^0 , \quad H\mathbf{p}^0 = \mathbf{r}^0 .$$

Iterate

α_n defined according to one of the rows of table C.1 ,

$$\mathbf{v}^{n+1} = \mathbf{v}^n + \alpha_n \mathbf{p}^n ,$$

$$\mathbf{r}^{n+1} = \mathbf{r}^n - \alpha_n L\mathbf{p}^n , \tag{C.2.2}$$

$$H\mathbf{p}^{n+1} = \mathbf{r}^{n+1} .$$

Table C.1. The three different strategies for Richardson iterations (PMRR)

Name of method	acceleration parameter	method minimizes
PMRR$_2$	$\alpha_n = \dfrac{(\mathbf{r}^n, L\mathbf{p}^n)}{(L\mathbf{p}^n, L\mathbf{p}^n)}$	$\|\mathbf{r}^{n+1}\|$
PMRR$_H$	$\alpha_n = \dfrac{(\mathbf{p}^n, L\mathbf{p}^n)}{(L\mathbf{p}^n, H^{-1}L\mathbf{p}^n)}$	$\|\mathbf{p}^{n+1}\|_H$
PSDR	$\alpha_n = \dfrac{(\mathbf{p}^n, \mathbf{r}^n)}{(\mathbf{p}^n, L\mathbf{p}^n)}$	$\|\mathbf{e}^{n+1}\|_L$

Note that for non-preconditioned iterations, then $H = I$ and $\mathbf{p}^n = \mathbf{r}^n$ in Table C.1. (In particular, PMRR$_2$ and PMRR$_H$ coincide if $P = I$.)

For PMRR$_H$ iterations the following estimate holds for the preconditioned residual:

$$\|\mathbf{p}^n\|_H \leq \left(\frac{\mathcal{K}-1}{\mathcal{K}+1}\right)^n \|\mathbf{p}^0\|_H \, , \tag{C.2.3}$$

where $\mathcal{K}$ still denotes the iterative condition number of $H^{-1}L$, while for PSDR iterations we have

$$\|\mathbf{e}^n\|_L \leq \left(\frac{\mathcal{K}-1}{\mathcal{K}+1}\right)^n \|\mathbf{e}^0\|_L \tag{C.2.4}$$

(see Quarteroni and Valli (1994), Sect. 2.4). Note that when $H = I$ (no preconditioning), the PSDR method reduces to the classical steepest descent (or gradient) algorithm. Also note that, in both cases, the number of iterations required for convergence is proportional to

$$\mathcal{J} \cong \frac{1}{2}\mathcal{K} \, . \tag{C.2.5}$$

When the eigenvalues of the preconditioned matrix $H^{-1}L$ are complex but with dominant real parts, a surrogate for $\mathcal{K}$ that is still representative of the convergence behavior of the Richardson iterations is

$$\mathcal{K}^* = \frac{\max_j |\lambda_j|}{\min_j |\lambda_j|} \, . \tag{C.2.6}$$

A substantial improvement in the convergence rate can be achieved by using *conjugate direction methods* in place of PMRR or PSDR. The two most common conjugate direction methods are known as the conjugate gradient method and the conjugate residual method. These methods were proposed by

Hestenes and Stiefel in 1952 as direct methods for solving symmetric, positive-definite linear systems. For such problems, the conjugate direction methods produce the exact answer (in the absence of round-off errors) in a finite number of steps. In the late 1960s and early 1970s these methods began to be considered seriously as iterative, rather than direct, solution schemes that can produce a very accurate result in a small number of iterations.

In a non-preconditioned conjugate direction method the update of the iterate is generalized from (C.2.1) to

$$\mathbf{v}^{n+1} = \mathbf{v}^n + \alpha_n \mathbf{p}^n \ . \tag{C.2.7}$$

In the conjugate gradient version, the directions satisfy the orthogonality property

$$\left(\mathbf{p}^{n+1}, L\mathbf{p}^n\right) = 0 \ . \tag{C.2.8}$$

The scheme is initialized with an initial guess $\mathbf{v}^0$. The initial direction vector is chosen to be $\mathbf{p}^0 = \mathbf{r}^0$, where $\mathbf{r}^0$ is the initial residual. Subsequent iterations are made according to the following formulas:

Conjugate Gradient (CG) *Method*

$$\alpha_n = \frac{(\mathbf{r}^n, \mathbf{r}^n)}{(\mathbf{p}^n, L\mathbf{p}^n)} \ ,$$

$$\mathbf{v}^{n+1} = \mathbf{v}^n + \alpha_n \mathbf{p}^n \ ,$$

$$\mathbf{r}^{n+1} = \mathbf{r}^n - \alpha_n L\mathbf{p}^n \ , \tag{C.2.9}$$

$$\beta_n = \frac{\left(\mathbf{r}^{n+1}, \mathbf{r}^{n+1}\right)}{(\mathbf{r}^n, \mathbf{r}^n)} \ ,$$

$$\mathbf{p}^{n+1} = \mathbf{r}^{n+1} + \beta_n \mathbf{p}^n \ .$$

In (C.2.9) the formula for the familiar scalar α_n results from the requirement that $\mathbf{v}^{n+1}$ minimize the energy norm of the error, and the formula for the additional scalar β_n is an equivalent, yet numerically preferred, form of the expression $\beta_n = -(\mathbf{r}^{n+1}, L\mathbf{p}^n)/(\mathbf{p}^n, L\mathbf{p}^n)$, which follows from the requirement (C.2.8).

The following orthogonality properties hold:

$$\left(\mathbf{r}^k, \mathbf{r}^l\right) = 0, \quad \left(\mathbf{p}^k, L\mathbf{p}^l\right) = 0 \quad \text{for } k \neq l \ . \tag{C.2.10}$$

The first of these implies that $\mathbf{r}^m = 0$ for some $m \leq nd$, where nd is the order of the matrix L. This explains the claim that the exact solution is obtained in a finite number of iterations. However, the presence of rounding errors leads to some contamination of the residual and direction vectors. The second orthogonality relation shows that the CG method does far more than the original requirement (C.2.8); indeed, we say that the directions $\{\mathbf{p}^k\}$ are *L-conjugated*.

The favorable convergence properties of this method are reflected by the estimate for the energy error (which improves the one in (C.2.4)):

$$\|\mathbf{e}^n\|_L \leq 2 \left(\frac{\sqrt{\mathcal{K}} - 1}{\sqrt{\mathcal{K}} + 1} \right)^n \|\mathbf{e}^0\|_L . \tag{C.2.11}$$

The number of iterations required for convergence is therefore proportional to

$$\mathcal{J} = \frac{1}{2}\sqrt{\mathcal{K}} . \tag{C.2.12}$$

This is a decided improvement over the result (C.2.5). Of course, the CG method is more costly per iteration, both in CPU time and storage.

The *conjugate residual method* is similar, but now the orthogonality property is

$$\left(L\mathbf{p}^{n+1}, L\mathbf{p}^n \right) = 0 , \tag{C.2.13}$$

and the requirement on $\mathbf{v}^{n+1}$ is that it minimize the Euclidean norm of the residual.

Let us now include a symmetric preconditioning, denoted as usual by H, in these descent methods. It is tempting to write (C.0.1) as either

$$\tilde{L}\mathbf{u} = \tilde{\mathbf{f}} \quad \text{with} \quad \tilde{L} = H^{-1}L \quad \text{and} \quad \tilde{\mathbf{f}} = H^{-1}\mathbf{f} \tag{C.2.14}$$

or

$$\tilde{L}\tilde{\mathbf{u}} = \mathbf{f}, \quad \text{where} \quad \tilde{L} = LH^{-1} \quad \text{and} \quad \tilde{\mathbf{u}} = H\mathbf{u}, \tag{C.2.15}$$

and then apply the preceding formulas to either (C.2.14) or (C.2.15). However, $\tilde{L}$ is not necessarily symmetric and positive definite (unless L and H^{-1} commute). We can, however, choose Q such that

$$H = QQ^T , \tag{C.2.16}$$

and use

$$\tilde{L}\tilde{\mathbf{u}} = \tilde{\mathbf{f}} , \tag{C.2.17}$$

with

$$\tilde{L} = Q^{-1}LQ^{-T}, \quad \tilde{\mathbf{f}} = Q^{-1}\mathbf{f}, \quad \tilde{\mathbf{u}} = Q^T\mathbf{u} . \tag{C.2.18}$$

We also use

$$\tilde{\mathbf{v}} = Q^T\mathbf{v}, \quad \tilde{\mathbf{p}} = Q^T\mathbf{p}, \quad \tilde{\mathbf{r}} = Q^{-1}\mathbf{r} . \tag{C.2.19}$$

This ensures that the matrix $\tilde{L}$ is symmetric and positive definite. After inserting (C.2.18) into the preceding schemes and then manipulating the expressions into computationally convenient forms, we arrive at the following:

Preconditioned Conjugate Gradient (PCG) *Method*

Initialize

$$\mathbf{v}^0, \quad \mathbf{r}^0 = \mathbf{f} - L\mathbf{v}^0, \quad H\mathbf{z}^0 = \mathbf{r}^0, \quad \mathbf{p}^0 = \mathbf{z}^0.$$

Iterate

$$\alpha_n = \frac{(\mathbf{r}^n, \mathbf{z}^n)}{(\mathbf{p}^n, L\mathbf{p}^n)},$$
$$\mathbf{v}^{n+1} = \mathbf{v}^n + \alpha_n \mathbf{p}^n,$$
$$\mathbf{r}^{n+1} = \mathbf{r}^n - \alpha_n L\mathbf{p}^n,$$
$$H\mathbf{z}^{n+1} = \mathbf{r}^{n+1}, \qquad\qquad (C.2.20)$$
$$\beta_n = \frac{(\mathbf{r}^{n+1}, \mathbf{z}^{n+1})}{(\mathbf{r}^n, \mathbf{z}^n)},$$
$$\mathbf{p}^{n+1} = \mathbf{z}^{n+1} + \beta_n \mathbf{p}^n.$$

Preconditioned Conjugate Residual (PCR) *Method*

Initialize

$$\mathbf{v}^0, \quad \mathbf{r}^0 = \mathbf{f} - L\mathbf{v}^0, \quad H\mathbf{z}^0 = \mathbf{r}^0, \quad \mathbf{p}^0 = \mathbf{z}^0.$$

Iterate

$$\alpha_n = \frac{(\mathbf{r}^n, L\mathbf{p}^n)}{(L\mathbf{p}^n, L\mathbf{p}^n)},$$
$$\mathbf{v}^{n+1} = \mathbf{v}^n + \alpha_n \mathbf{p}^n,$$
$$\mathbf{r}^{n+1} = \mathbf{r}^n - \alpha_n L\mathbf{p}^n,$$
$$H\mathbf{z}^{n+1} = \mathbf{r}^{n+1}, \qquad\qquad (C.2.21)$$
$$\beta_n = -\frac{(L\mathbf{z}^{n+1}, L\mathbf{p}^n)}{(L\mathbf{p}^n, L\mathbf{p}^n)},$$
$$\mathbf{p}^{n+1} = \mathbf{z}^{n+1} + \beta_n \mathbf{p}^n.$$
$$L\mathbf{p}^{n+1} = L\mathbf{z}^{n+1} + \beta_n L\mathbf{p}^n.$$

The preconditioned conjugate gradient method minimizes the L-norm of the error; thus, the associated error satisfies (C.2.11). However, now the relevant condition number is that of $Q^{-1}LQ^{-T}$ (which coincides with that of $H^{-1}L$) rather than that of L.

For the CG and CR methods, their orthogonality properties are lost when applied to nonsymmetric problems. In this case they are more properly called the *truncated conjugate gradient* (TCG) and *truncated conjugate residual* (TCR) methods. Their preconditioned versions are abbreviated as the PTCG and PTCR methods, and they are given by (C.2.20) and (C.2.21), respectively.

Although the descent methods described in this section may work for nonsymmetric problems, the methods in the following section are generally preferable for the general case.

C.3 Krylov Methods for Nonsymmetric Problems

The subject of iterative schemes for nonsymmetric problems has received much attention since the 1980s. The descent methods that we discuss in this subsection are but a small subset of the schemes that have been proposed.

Since the matrix L is not symmetric, we can use either one of the transformations (C.2.14)–(C.2.15) or (C.2.16)–(C.2.19). The preconditioned matrix $\tilde{L}$ determines the performance of Krylov methods.

When the Richardson method (C.1.1) is applied to the solution of the linear system (C.0.1), the residual, $\mathbf{r}^n = \mathbf{f} - L\mathbf{v}^n$, at the n-th iteration can be related to the initial residual as

$$\mathbf{r}^n = \prod_{j=0}^{n-1}(I - \omega_j L)\mathbf{r}^0 = p_n(L)\mathbf{r}^0 \ , \tag{C.3.1}$$

where ω_j is the relaxation parameter at the j-th step, while $p_n(L)$ indicates a polynomial in L of degree n.

Let us introduce the space

$$K_m(L; \mathbf{w}) = \text{span}\{\mathbf{w}, L\mathbf{w}, \dots, L^{m-1}\mathbf{w}\} \ , \qquad m \geq 1 \ , \tag{C.3.2}$$

called the *Krylov space of order* m associated with the matrix L and the vector $\mathbf{w}$. Then, $\mathbf{r}^n \in K_{n+1}(L; \mathbf{r}^0)$. From (C.1.1) we obtain

$$\mathbf{v}^n = \mathbf{v}^0 + \sum_{j=1}^{n-1}\omega_j \mathbf{r}^j \ ;$$

thus,

$$\mathbf{v}^n - \mathbf{v}^0 \in K_n(L; \mathbf{r}^0)$$

and

$$\mathbf{v}^n - \mathbf{v}^0 = p_{n-1}(L)\mathbf{r}^0.$$

More generally, methods can be devised in such a way that

$$\mathbf{v}^n - \mathbf{v}^0 = q_{n-1}(L)\mathbf{r}^0 \ , \tag{C.3.3}$$

where q_{n-1} is a polynomial chosen so that $\mathbf{v}^n$ represents the "best" approximation of the solution, $\mathbf{u}$, in $\tilde{K}_n = \mathbf{v}^0 + K_n(L; \mathbf{r}^0)$. Any such method is called a *Krylov method*.

For any fixed $m \geq 1$, an orthonormal basis $\{\mathbf{w}_i\}$ for $K_m(L; \mathbf{w})$ can be computed using the so-called *Arnoldi algorithm*. Setting $\mathbf{w}_1 = \mathbf{v}/\|\mathbf{w}\|$, we apply the Gram-Schmidt procedure: for $k \geq 1$,

$$g_{ik} = \mathbf{w}_i^T L\mathbf{w}_k \ , \qquad i = 1, \dots, k \ , \tag{C.3.4}$$

$$\mathbf{z}_k = L\mathbf{w}_k - \sum_{i=1}^{k}g_{ik}\mathbf{w}_i \ , \tag{C.3.5}$$

$$g_{k+1,k} = \|\mathbf{z}_k\| \ . \tag{C.3.6}$$

Should $\mathbf{z}_k = \mathbf{0}$ the process terminates, and we say that a *breakdown* of the algorithm has occurred. Otherwise, we set

$$\mathbf{w}_{k+1} = \frac{\mathbf{z}_k}{\|\mathbf{z}_k\|} , \qquad (C.3.7)$$

and the algorithm continues, incrementing k by 1.

If the algorithm terminates at the step m, then $\{\mathbf{w}_1, \ldots, \mathbf{w}_m\}$ forms a basis for $K_m(L; \mathbf{v})$. In such a case, denoting by $W_m \in \mathbb{R}^{n \times m}$ the matrix whose columns are the vectors $\mathbf{w}_i$, we obtain

$$W_m^T L W_m = G_m, \qquad W_{m+1}^T L W_m = \hat{G}_m , \qquad (C.3.8)$$

where $\hat{G}_m \in \mathbb{R}^{(m+1) \times m}$ is an upper-Hessenberg matrix whose entries are the g_{ij}, while $G_m \in \mathbb{R}^{m \times m}$ is the restriction of $\hat{G}_m$ to the first m rows and m columns. In our application, the Krylov space will be invariably constructed for $\mathbf{v} = \mathbf{r}^0$.

This algorithm for generating an orthonormal basis for a Krylov space of any order is the foundation for solving the linear system (C.0.1) by a Krylov method. The most natural approach would be to search for $\mathbf{v}^n$ as the vector that minimizes the error $\|\mathbf{v}^n - \mathbf{u}\|$ in $\tilde{K}_n$. However, since $\mathbf{u}$ is unknown, this method would not work in practice. Two alternative strategies that are workable are

1. Compute $\mathbf{v}^n$ by enforcing that the residual $\mathbf{r}^n$ be orthogonal to any vector in $K_n(L; \mathbf{r}^0)$, i.e.,

$$\mathbf{v}^T (\mathbf{f} - L\mathbf{v}^n) = 0 \qquad \forall \mathbf{v} \in K_n(L; \mathbf{r}^0) . \qquad (C.3.9)$$

This leads to the so-called *full orthogonalization method* (FOM).

2. Compute $\mathbf{v}^n \in \tilde{K}_n$ by minimizing the norm of the residual $\mathbf{r}^n$, i.e.,

$$\|\mathbf{f} - L\mathbf{v}^n\| = \min_{\mathbf{v} \in \tilde{K}_n} \|\mathbf{f} - L\mathbf{v}\| , \qquad (C.3.10)$$

which yields the *generalized minimum residual method* (GMRES).

Note that

$$\mathbf{v}^n = \mathbf{v}^0 + W_n \mathbf{q}^n, \qquad (C.3.11)$$

where $\mathbf{q}^n$ has to be chosen according to the selected optimality criterion ((C.3.9) or (C.3.10)).

Then,

$$\mathbf{r}^n = \mathbf{r}^0 - LW_n \mathbf{q}^n ,$$

since $\mathbf{r}^0 = \mathbf{w}_1 \|\mathbf{r}_0\|$. From (C.3.8) it follows that

$$\mathbf{r}^n = W_{n+1}(\|\mathbf{r}^0\|\mathbf{e}_1 - \hat{G}_n \mathbf{q}^n) , \qquad (C.3.12)$$

where $\mathbf{e}_1$ is the first unit vector of the canonical basis of $\mathbb{R}^{n+1}$. Thus, in the GMRES method, the solution at step n is computed through (C.3.11) where

$$\mathbf{q}^n \text{ minimizes } \| \,(\|\mathbf{r}^0\|\mathbf{e}_1 - \hat{G}_n \mathbf{q}) \,\| \text{ with respect to } \mathbf{q}. \qquad (C.3.13)$$

Note that the matrix W_{n+1} appearing in (C.3.12) does not change the value of $\|\mathbf{r}^0\|$ since it is an orthogonal matrix.

Clearly, the GMRES method will be the more effective the smaller the number of iterations, particularly since at each step one has to solve a least-squares problem (C.3.13). The GMRES method in exact arithmetic enjoys the so-called finite-termination property, i.e., it terminates after at most nd iterations, where again nd denotes the order of the matrix L. Premature stops are due to a breakdown in the Arnoldi orthonormalization algorithm. This breakdown occurs only if the computed solution $\mathbf{v}^n$ coincides with the exact solution $\mathbf{u}$ for some $n < nd$. However, unless acceptable convergence is reached after just a few iterations, the GMRES method requires prohibitive computational costs for the orthogonalization and excessive storage for the retention of the Krylov subspace bases.

A popular variant consists of restarting GMRES after each m iteration steps. This algorithm is referred to as GMRES(m); the nonrestarted version is sometimes called *full GMRES*. As pointed out in van der Vorst (2003), there is no simple rule to determine a suitable value of m; in fact, the speed of convergence of GMRES(m) may vary drastically for nearby values of m. In some cases, a superlinear convergence behavior of the full GMRES iterations is observed.

The convergence analysis of GMRES is not trivial, and we report just some of the more elementary results here. If L is positive definite, i.e., its symmetric part L_S has positive eigenvalues, then the n-th residual decreases according to the following bound:

$$\|\mathbf{r}^n\| \leq \sin^n(\beta)\|\mathbf{r}^0\| \,, \tag{C.3.14}$$

where $\cos(\beta) = \lambda_{\min}(L_S)/\|L\|$ with $\beta \in [0, \pi/2)$. As usual, $\|\cdot\|$ denotes the Euclidean vector or matrix norm. Moreover, GMRES(m) converges for all $m \geq 1$. In order to obtain a bound on the residual at a step $n \geq 1$, let us assume that the matrix L is diagonalizable:

$$L = T \Lambda T^{-1} \,,$$

where Λ is the diagonal matrix of eigenvalues, $\{\lambda_j\}_{j=1,\ldots,nd}$, and $T = (\boldsymbol{\omega}^1, \ldots, \boldsymbol{\omega}^{nd})$ is the matrix whose columns are the right eigenvectors of L. Under these assumptions, the residual norm after n steps of GMRES satisfies

$$\|\mathbf{r}^n\| \leq \kappa_2(T)\delta\|\mathbf{r}^0\| \,,$$

where $\kappa_2(T) = \|T\|_2\|T^{-1}\|_2$ is the condition number of T defined in (C.1.12), and

$$\delta = \min_{p \in \mathbb{P}_n, p(0)=1} \max_{1 \leq i \leq nd} |p(\lambda_i)| \,.$$

Moreover, suppose that the initial residual is dominated by m eigenvectors, i.e., $\mathbf{r}^0 = \sum_{j=1}^m \alpha_j \boldsymbol{\omega}^j + \mathbf{e}$, with $\|\mathbf{e}\|$ small in comparison to $\|\sum_{j=1}^m \alpha_j \boldsymbol{\omega}^j\|$, and assume that if some complex $\boldsymbol{\omega}^j$ appears in the previous sum, then its conjugate $\overline{\boldsymbol{\omega}}^j$ appears as well. Then

$$\|\mathbf{r}^n\| \leq \kappa_2(T)c_n\|\mathbf{e}\| ,$$

$$c_n = \max_{p>n} \prod_{j=1}^{n} \left| \frac{\lambda_p - \lambda_j}{\lambda_j} \right| .$$

Very often, c_n is of order one; hence, n steps of GMRES reduce the residual norm to the order of $\|\mathbf{e}\|$ provided that $\kappa_2(T)$ is not too large.

In general, as highlighted from the previous estimate, the eigenvalue information alone is not enough, and information on the eigensystem is also needed. If the eigensystem is orthogonal, as for normal matrices, then $\kappa_2(T) = 1$, and the eigenvalues are descriptive for convergence. Otherwise, upper bounds for $\|\mathbf{r}^n\|$ can be provided in terms of both spectral and pseudospectral information, as well as the so-called *field of values* of L:

$$\mathcal{F}(L) = \{\mathbf{v}^H L\mathbf{v} \mid \|\mathbf{v}\| = 1\} ,$$

where the superscript H denotes the Hermitian transpose. If $0 \notin \mathcal{F}(L)$, then the estimate (C.3.14) can be improved by replacing $\lambda_{\min}(L_S)$ with $\mathrm{dist}(0, \mathcal{F}(L))$.

An extensive discussion of convergence of GMRES and GMRES(m) can be found in Saad (1996) and van der Vorst (2003).

The GMRES method can of course be implemented for a preconditioned system. We provide here an implementation of the preconditioned GMRES method with a left preconditioner H.

Preconditioned GMRES (PGMRES) *Method*

Initialize
  $\mathbf{v}^0$, $H\mathbf{r}^0 = \mathbf{f} - L\mathbf{v}^0$, $\beta = \|\mathbf{r}^0\|$, $\mathbf{v}^1 = \mathbf{r}^0/\beta$.

Iterate

  For $j = 1, \ldots, n$ *Do*
    Compute $H\mathbf{w}^j = L\mathbf{v}^j$
    For $i = 1, \ldots, j$ *Do*
      $g_{ij} = (\mathbf{v}^i)^T \mathbf{w}^j$
      $\mathbf{w}^j = \mathbf{w}^j - g_{ij}\mathbf{v}_i$
    End Do
    $g_{j+1,j} = \|\mathbf{w}^j\|$ (C.3.15)
    (*if* $g_{j+1,j} = 0$ set $n = j$ and *Goto* (1))
    $\mathbf{v}^{j+1} = \mathbf{w}^j/g_{j+1,j}$
  End Do
  $W_n = [\mathbf{v}^1, \ldots, \mathbf{v}^n]$, $\hat{G}_n = \{g_{ij}\}$, $1 \leq j \leq n$, $1 \leq i \leq j+1$;
  (1) Compute $\mathbf{q}^n$, the minimizer of $\|\beta\mathbf{e}_1 - \hat{G}_n\mathbf{q}\|$
  Set $\mathbf{v}^n = \mathbf{v}^0 + W_n\mathbf{q}^n$

More generally, as proposed by Saad (1996), a variable preconditioner H_n can be used at the n-th iteration, yielding the so-called *flexible GMRES* method. The use of a variable preconditioner is especially interesting in those situations where the preconditioner is not explicitly given, but implicitly defined, for instance, as an approximate Jacobian in a Newton iteration or by a few steps of an inner iteration process. Another meaningful case is the one of domain-decomposition preconditioners (of either Schwarz or Schur type) where the preconditioning step involves one or several substeps of local solves in the subdomains (see Chap. 6).

Several considerations for the practical implementation of GMRES, its relation with FOM, how to restart GMRES, and the Householder version of GMRES can be found in Saad (1996).

A different approach to iterative methods for nonsymmetric matrices consists of generalizing the conjugate gradient method through a specific characterization of the properties satisfied by the residual.

The property that the residual vectors $\mathbf{r}^n$ generated by the CG method satisfy a three-term recurrence is lost when L is not symmetric. The *biconjugate gradient (Bi-CG)* method introduced by Fletcher in 1976 constructs a residual $\mathbf{r}^k$ orthogonal to another row of vectors $\tilde{\mathbf{r}}^0, \tilde{\mathbf{r}}^1, \ldots, \tilde{\mathbf{r}}^{n-1}$, and, vice versa, $\tilde{\mathbf{r}}^n$ is orthogonal with respect to $\mathbf{r}^0, \mathbf{r}^1, \ldots, \mathbf{r}^{n-1}$. This method enjoys the finite-termination property, but there is no minimization property as in CG or GMRES for the intermediate steps. When this method converges, both $\{\mathbf{r}^n\}$ and $\{\tilde{\mathbf{r}}^n\}$ converge towards zero but only the convergence of the $\{\tilde{\mathbf{r}}^n\}$ is exploited. Based on this observation, Sonneveld in 1989 proposed a modification called the *conjugate gradient-squared (CGS)* method that focuses more strongly on the $\{\mathbf{r}^n\}$ vectors. CGS generates residual vectors $\mathbf{r}^n$ given by

$$\mathbf{r}^n = p_n^2(L)\mathbf{r}^0 ,$$

where $p_n(L)$ is that n-th degree polynomial in L for which $p_n(L)\mathbf{r}^0$ is equal to the residual at the n-th step obtained by means of the Bi-CG method.

In the Bi-CGStab method, introduced by van der Vorst (1992), instead of simply squaring the Bi-CG polynomial, as in CGS, the more general form

$$\mathbf{r}^n = q_n(L)p_n(L)\mathbf{r}^0 , \tag{C.3.16}$$

is used, where now $q_n(x) = \prod_{i=1}^n (1 - \omega_i x)$, and ω_i are suitable constants chosen in such a way that $\|\mathbf{r}^n\|$ is minimized with respect to ω_i.

The preconditioned algorithm can be described as follows:

Preconditioned Bi-CGStab (PBi-CGStab) *Method*

Initialize

$$\mathbf{v}^0, \quad \mathbf{r}^0 = \mathbf{f} - L\mathbf{v}^0, \quad \text{choose } \tilde{\mathbf{r}}^0 \text{ s.t. } (\tilde{\mathbf{r}}^0, \mathbf{r}^0) \neq 0, \text{ (e.g., } \tilde{\mathbf{r}}^0 = \mathbf{r}^0)$$

Iterate

$$\rho_{n-1} = (\mathbf{r}^{n-1}, \tilde{\mathbf{r}}^0)$$
$$if\ \rho_{n-1} = 0$$
$$\quad then\ \text{the method fails}$$
$$end\ if$$
$$if\ n = 1$$
$$\quad then\ \mathbf{p}^n = \mathbf{r}^{n-1}$$
$$\quad else\ \beta_{n-1} = (\rho_{n-1}/\rho_{n-2})(\alpha_{n-1}/\omega_{n-1})$$
$$\qquad \mathbf{p}^n = \mathbf{r}^{n-1} + \beta_{n-1}(\mathbf{p}^{n-1} - \omega_{n-1}\mathbf{w}^{n-1})$$
$$end\ if$$
$$H\hat{\mathbf{p}} = \mathbf{p}^n$$
$$\mathbf{w}^n = L\hat{\mathbf{p}}$$
$$\alpha_n = \rho_{n-1}/(\mathbf{w}^n, \tilde{\mathbf{r}}^0)$$
$$\mathbf{s} = \mathbf{r}^{n-1} - \alpha_n\mathbf{w}^n \tag{C.3.17}$$
$$if\ \|\mathbf{s}\|\ \text{small enough}$$
$$\quad then\ \mathbf{v}^n = \mathbf{v}^{n-1} + \alpha_n\hat{\mathbf{p}};\ \text{quit}$$
$$end\ if$$
$$H\hat{\mathbf{s}} = \mathbf{s}$$
$$\mathbf{t} = L\hat{\mathbf{s}}$$
$$\omega_n = (\mathbf{t}, \mathbf{s})/(\mathbf{t}, \mathbf{t})$$
$$\mathbf{v}^n = \mathbf{v}^{n-1} + \alpha_n\hat{\mathbf{p}} + \omega_n\hat{\mathbf{s}}$$
$$if\ \mathbf{v}^n\ \text{is accurate enough}$$
$$\quad then\ \text{quit}$$
$$end\ if$$
$$\mathbf{r}^n = \mathbf{s} - \omega_n\mathbf{t}$$

For continuation it is necessary that $\omega_n \neq 0$.

For an unfavorable choice of $\tilde{\mathbf{r}}^0$, ρ_n or $(\mathbf{w}^n, \tilde{\mathbf{r}}^0)$ can be 0 or very small. In this case one has to restart, e.g., with $\tilde{\mathbf{r}}^0$ and $\mathbf{v}^0$ given by the last available values of $\mathbf{r}^n$ and $\mathbf{v}^n$. In exact arithmetic, Bi-CGStab is also a finite termination method (i.e., $\mathbf{v}^n = \mathbf{u}$ for some $n \leq nd$). Its theoretical convergence properties are similar to those of CGS; however, it converges more smoothly, i.e., the oscillations of the residuals (with n) of Bi-CGStab are in general less pronounced than those of CGS.

It is clear from the previous algorithm description that a weakness of Bi-CGStab is that a breakdown occurs if an ω_n is equal to zero (but also a very small ω_n may be troublesome).

Another non-ideal property is that the q_n polynomial in (C.3.16) has only real roots by construction, whereas optimal reduction polynomials for matrices with complex eigenvalues may also have complex roots. These considerations have led to the introduction of a variant, called Bi-CGStab(2), in which q_n is constructed as a product of quadratic factors. For its derivation and analysis, the reader is referred, e.g., to van der Vorst (2003).

Unfortunately, for a general nonsymmetric matrix, Krylov methods are not guaranteed to converge, but neither are any other known iterative methods. As noted earlier, GMRES(m) does have a convergence guarantee if L_S has positive eigenvalues.

Appendix D. Time Discretizations

In this appendix we will make some general comments about time-discretizations, survey classical methods for ODEs and their stability regions, and highlight some low-storage time-discretization formulas that have been widely used in conjunction with spectral methods. Some standard references from the extensive literature on numerical methods for ODEs are the books by Gear (1971), Lambert (1991), Shampine (1994), Hairer, Norsett and Wanner (1993), Hairer and Wanner (1996) and Butcher (2003).

D.1 Notation and Stability Definitions

The typical evolution equation can be written

$$\frac{\partial u}{\partial t} = f(u, t) , \quad t > 0 ,$$

$$u(0) = 0 ,$$

(D.1.1)

where the (generally) nonlinear operator f contains the spatial part of the PDE. The unknown u can be either a scalar function or a vector function (the latter case occurs, e.g., for the Euler or Navier–Stokes equations in fluid dynamics).

After space discretization (by one of the several methods of spectral type considered in this book), for all times $t > 0$ the exact solution $u(t)$ is replaced by a function that is a polynomial $u^N(t)$ for single-domain classical spectral methods or a piecewise (mapped) polynomial function $u_\delta(t)$ for multidomain spectral methods. In both cases, this function is represented according to a chosen basis; let $\mathbf{u}(t)$ denote the vector of the unknown coefficients in this representation. Then, the spatial discretization can be written in the algebraic form

$$\frac{d\mathbf{u}}{dt} = \mathbf{f}(\mathbf{u}, t) , \quad t > 0 ,$$

$$\mathbf{u}(0) = \mathbf{u}_0 ,$$

(D.1.2)

where $\mathbf{f}$ is the vector-valued function governing the semi-discrete problem. For Galerkin-type methods (such as, e.g., SEM, MEM, SDGM and their

-NI variants), $\mathbf{f}$ may incorporate the matrix M^{-1}, where M denotes a mass matrix.

For time-dependent, linear PDEs, (D.1.2) reduces to

$$\frac{d\mathbf{u}}{dt} = -L\mathbf{u} + \mathbf{b}, \qquad t > 0,$$
$$\mathbf{u}(0) = \mathbf{u}_0, \tag{D.1.3}$$

where L is the matrix representing the spatial discretization by the chosen spectral method. This is also called a method-of-lines approach or a continuous-in-time discretization.

A corresponding, representative scalar model problem is

$$\frac{du}{dt} = \lambda u, \tag{D.1.4}$$

where λ is a complex number, which for (D.1.2) is "representative" of the partial derivative of f with respect to u (in the scalar case) or of the eigenvalues of the Jacobian matrix $(\partial f_i/\partial u_j)$ in the vector case, and which for (D.1.3) is representative of the eigenvalues of $-L$.

In most applications of spectral methods to partial differential equations, the spatial discretization is spectral but the temporal discretization uses conventional finite differences. In describing the time-discretizations, we denote the time-step by Δt, the n-th time-level by $t_n = n\Delta t$, the approximate solution at time-step n by $\mathbf{u}^n$, and set $\mathbf{f}^n = \mathbf{f}(\mathbf{u}^n, t^n)$.

If the spatial discretization is presumed fixed, then we use the term stability in its ODE context. The time-discretization is said to be *stable* (sometimes called *zero-stable*) if there exist positive constants σ, ε and $C(T)$, independent of Δt, such that, for all $T > 0$ (perhaps limited by a maximal $T_{\max}$ depending on the problem) and for all $0 \leq \Delta t < \sigma$,

$$\|\mathbf{u}^n - \mathbf{v}^n\| \leq C(T)\|\mathbf{u}^0 - \mathbf{v}^0\| \qquad \text{for } 0 \leq t_n \leq T \tag{D.1.5}$$

provided that $\|\mathbf{u}^0 - \mathbf{v}^0\| < \varepsilon$, where $\|\mathbf{u}^n\|$ is some spatial norm of $\mathbf{u}^n$. The constant $C(T)$ is permitted to grow with T. Here, $\mathbf{v}^n$ is the solution obtained by the same numerical method corresponding to a (perturbed) initial data $\mathbf{v}^0$. On a linear problem (hence in particular, for the problems (D.1.3) or (D.1.4)), property (D.1.5) can be equivalently replaced by

$$\|\mathbf{u}^n\| \leq C(T)\|\mathbf{u}^0\| \qquad \text{for } 0 \leq t_n \leq T. \tag{D.1.6}$$

For many problems involving integration over long time intervals, a method which admits the temporal growth allowed by the estimate (D.1.5) is undesirable. As one example, take a problem of the form (D.1.2) for which $(\partial \mathbf{f}/\partial \mathbf{u})(\mathbf{w}, t)$ is negative for all $\mathbf{w}$ and t, or more generally, for which $\mathbf{f}$ satisfies the *right Lipschitz condition*: there exists $\mu < 0$ such that

$$\langle \mathbf{f}(\mathbf{u}, t) - \mathbf{f}(\mathbf{v}, t), \mathbf{u} - \mathbf{v} \rangle \leq \mu \|\mathbf{u} - \mathbf{v}\|^2 \qquad \text{for all } \mathbf{u}, \mathbf{v}, t,$$

where $\langle \cdot, \cdot \rangle$ is a suitable scalar product and $\| \cdot \|$ its associated norm. In these cases,

$$\|\mathbf{u}(t) - \mathbf{v}(t)\| \leq e^{\mu t} \|\mathbf{u}(0) - \mathbf{v}(0)\| .$$

(Such problems are referred to as dissipative Cauchy problems in the ODE literature.) The ODEs resulting from spectral spatial discretizations of the heat equation or the time-dependent Stokes equations for incompressible flows (with homogeneous boundary data and zero source term) fall into this category. In this case one desires that the time-discretization be *asymptotically stable*, i.e., that instead of (D.1.5) it satisfy the stronger requirement

$$\|\mathbf{u}^n - \mathbf{v}^n\| \to 0 \quad \text{as} \quad t_n \to +\infty , \tag{D.1.7}$$

As another example for which the above notion of stability is too weak, consider ODEs resulting from the spatial discretization of linear, spatially periodic, purely hyperbolic systems. For these problems, asymptotic stability for the time-discretization is undesirable since the exact solution is undamped in time. Instead, we rather desire a time-discretization which is *temporally stable*, for which we merely require that

$$\|\mathbf{u}^n\| \leq \|\mathbf{u}^0\| \qquad \text{for all } n \geq 1 . \tag{D.1.8}$$

The notion of *weak instability* is sometimes used in a loose sense for schemes which admit solutions to periodic hyperbolic problems which grow with time, but for which the growth rate decreases with Δt. For example, the constant $C(T)$ in (D.1.5) might have the form

$$C(T) = e^{\alpha (\Delta t)^p T} ,$$

where $\alpha > 0$ and p is a positive integer. For such weakly unstable schemes, the longer the time interval of interest, i.e., the larger is T, the smaller must Δt be chosen to keep the spurious growth of the solution within acceptable bounds.

Another notion that is relevant to periodic, hyperbolic problems is that of *reversible* (or *symmetric*) time-discretizations. These are schemes for which the solution may be marched forward from t^n to t^{n+1} and then backwards to t^n with the starting solution at t^n recovered exactly (except for round-off errors).

Two final definitions are in order for our subsequent discussion. The *absolute stability region* (often referred to just as the *stability region*), say $\mathcal{A}$, of a numerical method is customarily defined for the scalar model problem (D.1.4) to be the set of all complex numbers $\alpha = \lambda \Delta t$ such that any sequence $\{u^n\}$ generated by the method with such λ and Δt satisfies $\|u^n\| \leq C$ as $t_n \to \infty$, for a suitable constant C. Furthermore, a method is called *A-stable* if the region of absolute stability includes the region $Re(\lambda \Delta t) < 0$. We warn the reader that in some books the absolute stability region is defined as the set of all $\lambda \Delta t$ such that $\|u^n\| \to 0$ as $t_n \to \infty$. This new region, say

$\mathcal{A}^0$, would not necessarily coincide with $\mathcal{A}$. In general, if $\mathcal{A}^0$ is non-empty, $\mathcal{A}$ is its closure. However, there are cases for which $\mathcal{A}^0$ is empty (e.g., the midpoint or leap-frog method) and $\mathcal{A}$ is not ($\mathcal{A} = \{z = \alpha i, \ -1 \leq \alpha \leq 1\}$ for the midpoint method). Finally, we note that zero-stable methods are those for which $\mathcal{A}$ contains the origin $z = 0$ of the complex plane.

As noted by Reddy and Trefethen (1990, 1992), having the eigenvalues scaled by the time-step Δt falling within the absolute stability region of the ODE method is not always sufficient for stability of the computation. They present a stability criterion utilizing the so-called ϵ-pseudospectra. However, as discussed by Trefethen (2000, Chap. 10), in almost all cases the "rule-of-thumb" condition involving the standard eigenvalues is acceptable.

On the other hand, we may be interested in the behavior of the computed solution as both the spatial and temporal discretizations are refined. We now define stability by an estimate of the form (D.1.5) where C is independent of Δt, ε and the spatial discretization parameter N or δ, the norm is independent of N or δ, but σ will in general be a function of N or δ. The functional dependence of σ upon N or δ which is necessary to obtain an estimate of the form (D.1.5) is termed the *stability limit* of the numerical method. If σ is in fact independent of N or δ, then the method is called *unconditionally stable*. Clearly, a necessary condition for the fully discrete problem to be stable is that the semi-discrete problem be spatially stable. Likewise, a *temporal* stability limit for the fully discrete scheme for a hyperbolic system is the functional dependence of σ upon N or δ which is necessary to obtain an estimate of the form (D.1.8).

D.2 Standard ODE Methods

In this section we furnish as a convenience the basic formulas and diagrams for the absolute stability regions for those time-discretizations of (D.1.2) that are most commonly used in conjunction with spectral discretizations in space. Among the factors which influence the choice of a time-discretization are the accuracy, stability, storage requirements, and work demands of the methods. The storage and work requirements of a method can be deduced in a straightforward manner from the definition of the method and the nature of the PDE. The accuracy of a method follows from a truncation error analysis, and the stability for a given problem is intimately connected with the spectrum of the spatial discretization. In this section we will describe some of the standard methods for ODEs and relate their stability regions to the spectra of the advection and diffusion operators. Bear in mind that in many problems different time-discretizations are used for different spatial terms in the equation. The illustrations of the spectra of the spectral differentiation, mass and stiffness matrices furnished in CHQZ2 and throughout this book combined with the stability diagrams in this section suffice for general conclusions to be drawn on appropriate choices of time-discretization methods and time-step limits for temporal stability.

For the reader's convenience, Table D.1 provides the numerical values of the intersections of the absolute stability regions with the negative real axis and the positive imaginary axis for all methods discussed in this section.

D.2.1 Leap Frog Method

The *leap frog* (LF) method (also called *midpoint* method) is a second-order, two-step scheme given by

$$\mathbf{u}^{n+1} = \mathbf{u}^{n-1} + 2\Delta t \mathbf{f}^n \ . \tag{D.2.1}$$

This produces solutions of constant norm for the model problem provided that $\lambda \Delta t$ is on the imaginary axis and that $|\lambda \Delta t| \leq 1$ (see Table D.1). Thus, leap frog is a suitable explicit scheme for problems with purely imaginary eigenvalues. It also is a reversible, or symmetric, method. However, since it is only well-behaved on a segment in the complex $\lambda \Delta t$-plane for the model problem, extra care is needed in practical situations.

The most obvious application is to periodic advection problems, for the eigenvalues of the Fourier approximation to d/dx are imaginary. The difficulty with the leap frog method is that the solution is subject to a temporal oscillation with period $2\Delta t$. This arises from the extraneous (spurious) solution to the temporal difference equations. The oscillations can be controlled by every so often averaging the solution at two consecutive time-levels.

Leap frog is quite inappropriate for problems whose spatial eigenvalues have nonzero real parts. This certainly includes the approximation of diffusion operators. Leap frog is also not viable for advection operators with nonperiodic boundary conditions, since the discrete spectra of Chebyshev and Legendre approximations to the standard advection operator have appreciable real parts.

D.2.2 Adams–Bashforth Methods

This is a class of explicit multistep methods which includes the simple *forward Euler* (FE) method

$$\mathbf{u}^{n+1} = \mathbf{u}^n + \Delta t \mathbf{f}^n \ , \tag{D.2.2}$$

the popular *second-order Adams–Bashforth* (AB2) method

$$\mathbf{u}^{n+1} = \mathbf{u}^n + \tfrac{1}{2}\Delta t \left[3\mathbf{f}^n - \mathbf{f}^{n-1} \right] \ , \tag{D.2.3}$$

the still more accurate *third-order Adams–Bashforth* (AB3) method

$$\mathbf{u}^{n+1} = \mathbf{u}^n + \tfrac{1}{12}\Delta t \left[23\mathbf{f}^n - 16\mathbf{f}^{n-1} + 5\mathbf{f}^{n-2} \right] \ , \tag{D.2.4}$$

and the *fourth-order Adams–Bashforth* (AB4) method

$$\mathbf{u}^{n+1} = \mathbf{u}^n + \tfrac{1}{24}\Delta t \left[55\mathbf{f}^n - 59\mathbf{f}^{n-1} + 37\mathbf{f}^{n-2} - 9\mathbf{f}^{n-3} \right] . \qquad (D.2.5)$$

These methods are not reversible.

The stability regions $\mathcal{A}$ of these methods are shown in Fig. D.1 (left) and the stability boundaries along the axes are given in Table D.1. Note that the size of the stability region decreases as the order of the method increases. Note also that except for the origin, no portion of the imaginary axis is included in the stability regions of the first and second-order methods, whereas the third- and fourth-order versions do have some portion of the imaginary axis included in their stability regions. Nevertheless, the AB2 method is weakly unstable, i.e., for a periodic, hyperbolic problem the acceptable Δt decreases at T increases.

As is evident from Fig. D.1 (left), higher order AB methods are temporally stable for Fourier approximations to periodic advection problems. Let the upper limit of the absolute stability region along the imaginary axis be denoted by c. Then the temporal stability limit is

$$\frac{N}{2}\Delta t \le c, \qquad \text{or} \qquad \Delta t \le \frac{c}{\pi}\Delta x . \qquad (D.2.6)$$

The limit on Δt is smaller by a factor of π than the corresponding limit for a second-order finite-difference approximation in space. The Fourier spectral approximation is more accurate in space because it represents the high-frequency components much more accurately than the finite-difference method. The artificial damping of the high-frequency components which is produced by finite-difference methods enables the stability restriction on the time-step to be relaxed.

Chebyshev and Legendre approximations to advection problems appear to be temporally stable under all Adams–Bashforth methods for sufficiently small Δt; precisely, for $\Delta t \le CN^{-2}$ for a suitable constant C. (For simplicity, this and the subsequent stability limits refer to a single-domain discretization. For multidomain methods, the limits on Δt should also scale with the size of the subdomains, in a way that depends on the specific spatial discretization method that is being used). Since the spatial eigenvalues all have negative real parts, the failure of the AB2 method to include the imaginary axis in its absolute stability region does not preclude temporal stability.

The temporal stability limits for Adams–Bashforth methods for Fourier, Chebyshev and Legendre approximations to diffusion equations are easy to deduce since their spatial eigenvalues (i.e., the eigenvalues of the matrix $-L$) are real, negative and limited in modulus as indicated, e.g., in CHQZ2, Chap. 4. Combining this information with the stability bounds along the negative real axis as provided in Table D.1, one gets that Δt should be limited by a constant times N^{-2} for Fourier approximations, by a constant times N^{-4} for Chebyshev or Legendre collocation approximations, and by a constant times N^{-3} for Legendre G-NI approximations.

D.2.3 Adams–Moulton Methods

A related set of implicit multistep methods are the Adams–Moulton methods. They include the *backward Euler* (BE) method

$$\mathbf{u}^{n+1} = \mathbf{u}^n + \Delta t \mathbf{f}^{n+1} , \tag{D.2.7}$$

the *Crank–Nicolson* (CN) method

$$\mathbf{u}^{n+1} = \mathbf{u}^n + \tfrac{1}{2}\Delta t[\mathbf{f}^{n+1} + \mathbf{f}^n] , \tag{D.2.8}$$

the *third-order Adams–Moulton* (AM3) method

$$\mathbf{u}^{n+1} = \mathbf{u}^n + \tfrac{1}{12}\Delta t[5\mathbf{f}^{n+1} + 8\mathbf{f}^n - \mathbf{f}^{n-1}] , \tag{D.2.9}$$

and the *fourth-order Adams–Moulton* (AM4) method

$$\mathbf{u}^{n+1} = \mathbf{u}^n + \tfrac{1}{24}\Delta t[9\mathbf{f}^{n+1} + 19\mathbf{f}^n - 5\mathbf{f}^{n-1} + \mathbf{f}^{n-2}] . \tag{D.2.10}$$

Forward Euler (FE) (see D.2.2), backward Euler (BE) and Crank–Nicolson (CN) methods are special cases of θ-methods, defined as

$$\mathbf{u}^{n+1} = \mathbf{u}^n + \Delta t[\theta \mathbf{f}^{n+1} + (1 - \theta)\mathbf{f}^n] , \tag{D.2.11}$$

for $0 \leq \theta \leq 1$. Precisely, they correspond to the choice $\theta = 0$ (FE), $\theta = 1$ (BE) and $\theta = 1/2$ (CN). All θ-methods except for FE are implicit. All θ-methods are first-order accurate, except for CN, which is second-order. For each $\theta < \frac{1}{2}$, the absolute stability region is the circle in the left half-plane $Re(\lambda\Delta t) \leq 0$ with center $z = (2\theta - 1)^{-1}$ and radius $r = (1 - 2\theta)^{-1}$. The stability region of the CN method coincides with the half-plane $Re(\lambda\Delta t) \leq 0$. For each $\theta > \frac{1}{2}$, the absolute stability region is the exterior of the open circle in the right half-plane $Re(\alpha) > 0$ with center $z = (2\theta - 1)^{-1}$ and radius $r = (2\theta - 1)^{-1}$. Thus, all θ-methods for $\frac{1}{2} \leq \theta \leq 1$ are A-stable.

The absolute stability regions of the third- and fourth-order Adams–Moulton methods are displayed in Fig. D.1 (right) and the stability boundaries along the axes are given in Table D.1. In comparison with the explicit Adams–Bashforth method of the same order, an Adams–Moulton method has a smaller truncation error (by factors of five and nine for second and third-order versions), a larger stability region, and requires one fewer levels of storage. However, it does require the solution of an implicit set of equations. The CN method is reversible; the others are not.

The CN method is commonly used for diffusion problems. In Navier–Stokes calculations, it is frequently applied to the viscous and pressure gradient components. Although CN is absolutely stable for the former and temporally stable for the latter, it has the disadvantage that it damps high-frequency components very weakly, whereas in reality these components decay very rapidly.

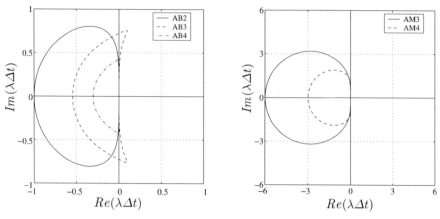

Fig. D.1. Absolute stability regions of Adams–Bashforth (*left*) and Adams–Moulton (*right*) methods

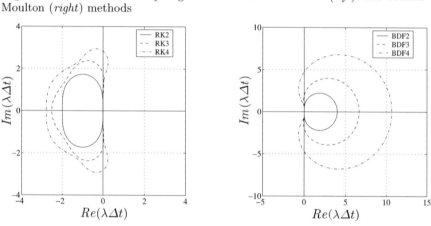

Fig. D.2. Absolute stability regions of backwards-difference formulas (*left*) and Runge–Kutta methods (*right*). The BDF methods are absolutely stable on the exteriors (and boundaries) of the regions enclosed by the curves, whereas the RK methods are absolutely stable on the interiors (and boundaries) of the regions enclosed by the *curves*

The Adams–Moulton methods of third and higher order are only conditionally stable for advection and diffusion problems. The stability limits implied by Fig. D.1 indicate that the stability limit of a high-order Adams–Moulton method is roughly ten times as large for a diffusion problem as the stability limit of the corresponding Adams–Bashforth method. In addition, AM3 and AM4 are weakly unstable for Fourier approximations to advection problems, since the origin is the only part of the imaginary axis which is included in their absolute stability regions.

Table D.1. Intersections of absolute stability regions with the negative real axis (*left*) and with the positive imaginary axis (*right*)

Method	$\mathcal{A} \cap \mathbb{R}_-$	$\mathcal{A} \cap i\mathbb{R}_+$
Leap frog (midpoint)	$\{0\}$	$[0, 1]$
Forward Euler	$[-2, 0]$	$\{0\}$
Crank–Nicolson	$(-\infty, 0]$	$[0, +\infty)$
Backward Euler	$(-\infty, 0]$	$[0, +\infty)$
θ-method, $\theta < 1/2$	$[2/(2\theta - 1), 0]$	$\{0\}$
θ-method, $\theta \geq 1/2$	$(-\infty, 0]$	$[0, +\infty)$
AB2	$(-1, 0]$	$\{0\}$
AB3	$[-6/11, 0]$	$[0, 0.723]$
AB4	$[-3/10, 0]$	$[0, 0.43]$
AM3	$[-6, 0]$	$\{0\}$
AM4	$[-3, 0]$	$\{0\}$
BDF2	$(-\infty, 0]$	$[0, +\infty)$
BDF3	$(-\infty, 0]$	$[0, 1.94)$
BDF4	$(-\infty, 0]$	$[0, 4.71)$
RK2	$[-2, 0]$	$\{0\}$
RK3	$[-2.51, 0]$	$[0, 1.73]$
RK4	$[-2.79, 0]$	$[0, 2.83]$

D.2.4 Backwards-Difference Formulas

Another class of implicit time discretizations is based upon backwards-difference formulas. These include the *first-order backwards-difference scheme* (BDF1), which is identical to backward Euler, the *second-order backwards-difference scheme* (BDF2)

$$\mathbf{u}^{n+1} = \tfrac{1}{3}[4\mathbf{u}^n - \mathbf{u}^{n-1}] + \tfrac{2}{3}\Delta t \mathbf{f}^{n+1} , \qquad (D.2.12)$$

the *third-order backwards-difference scheme* (BDF3)

$$\mathbf{u}^{n+1} = \tfrac{1}{11}[18\mathbf{u}^n - 9\mathbf{u}^{n-1} + 2\mathbf{u}^{n-2}] + \tfrac{6}{11}\Delta t \mathbf{f}^{n+1} , \qquad (D.2.13)$$

and the *fourth-order backwards-difference scheme* (BDF4)

$$\mathbf{u}^{n+1} = \tfrac{1}{25}[48\mathbf{u}^n - 36\mathbf{u}^{n-1} + 16\mathbf{u}^{n-2} - 3\mathbf{u}^{n-3}] + \tfrac{12}{25}\Delta t \mathbf{f}^{n+1} . \qquad (D.2.14)$$

The absolute stability regions of these methods are displayed in Fig. D.2 (left), and the stability boundaries along the axes are given in Table D.1. The stability regions are much larger than those of the corresponding AM methods.

D.2.5 Runge–Kutta Methods

Runge–Kutta methods are single-step, but multistage, time-discretizations. The modified Euler version of a second-order Runge–Kutta (RK2) method can be written

$$\mathbf{u}^{n+1} = \mathbf{u}^n + \frac{1}{2}\Delta t[\mathbf{f}(\mathbf{u}^n, t^n) + \mathbf{f}(\mathbf{u}^n + \Delta t\mathbf{f}(\mathbf{u}^n, t^n), t^n + \Delta t)] \ . \qquad \text{(D.2.15)}$$

A popular third-order Runge–Kutta (RK3) method is

$$\begin{aligned}
\mathbf{k}_1 &= \mathbf{f}\left(\mathbf{u}^n, t_n\right) \ , \\
\mathbf{k}_2 &= \mathbf{f}\left(\mathbf{u}^n + \tfrac{1}{2}\Delta t\, \mathbf{k}_1, t_n + \tfrac{1}{2}\Delta t\right) \ , \\
\mathbf{k}_3 &= \mathbf{f}\left(\mathbf{u}^n + \tfrac{3}{4}\Delta t\, \mathbf{k}_2, t_n + \tfrac{3}{4}\Delta t\right) \ , \\
\mathbf{u}^{n+1} &= \mathbf{u}^n + \tfrac{1}{9}\Delta t[2\mathbf{k}_1 + 3\mathbf{k}_2 + 4\mathbf{k}_3] \ .
\end{aligned} \qquad \text{(D.2.16)}$$

The classical fourth-order Runge–Kutta (RK4) method is

$$\begin{aligned}
\mathbf{k}_1 &= \mathbf{f}\left(\mathbf{u}^n, t_n\right) \ , \\
\mathbf{k}_2 &= \mathbf{f}\left(\mathbf{u}^n + \tfrac{1}{2}\Delta t\, \mathbf{k}_1, t_n + \tfrac{1}{2}\Delta t\right) \ , \\
\mathbf{k}_3 &= \mathbf{f}\left(\mathbf{u}^n + \tfrac{1}{2}\Delta t\, \mathbf{k}_2, t_n + \tfrac{1}{2}\Delta t\right) \ , \\
\mathbf{k}_4 &= \mathbf{f}\left(\mathbf{u}^n + \Delta t\, \mathbf{k}_3, t_n + \Delta t\right) \ , \\
\mathbf{u}^{n+1} &= \mathbf{u}^n + \tfrac{1}{6}\Delta t[\mathbf{k}_1 + 2\mathbf{k}_2 + 2\mathbf{k}_3 + \mathbf{k}_4] \ .
\end{aligned} \qquad \text{(D.2.17)}$$

All Runge–Kutta methods of a given order have the same stability properties. The absolute stability region are given in Fig. D.2 (right), and the stability boundaries along the axes are given in Table D.1. Note that the stability region expands as the order increases. Note also that RK2 methods are afflicted with the same weak instability as the AB2 scheme. When storage is not an issue, then the classical RK4 method is commonly used. Otherwise, the low-storage versions of third and fourth-order methods, such as those described in Sect. D.3, have been preferred.

In the event that $\mathbf{f}$ contains no explicit dependence upon t, the following formulation, due to Jameson et al. (1981), applies:

Set

$$\mathbf{u} = \mathbf{u}^n$$

For $k = s, \ 1, \ -1$

$$\mathbf{u} \leftarrow \mathbf{u}^n + \frac{1}{k}\Delta t\mathbf{f}(\mathbf{u}) \qquad \text{(D.2.18)}$$

End For

$$\mathbf{u}^{n+1} = \mathbf{u} \ .$$

It yields a Runge–Kutta method of order s (for linear problems) and requires at most three levels of storage.

D.3 Low-Storage Schemes

When high-order discretization schemes such as spectral methods are employed in space, the primary contributor to the error in the fully discrete approximation is usually the temporal discretization error unless the time-discretization itself is at least third order or the time-step is very small. When computations are constrained by memory limitations, a premium is placed on minimizing storage demands. This has made special low-storage Runge–Kutta methods very attractive for large-scale problems. Several popular low-storage Runge–Kutta methods are available that permit third-order or fourth-order temporal accuracy to be obtained with only two levels of storage. Such economies are not available for multistep methods.

We shall note here some of the low-storage Runge–Kutta methods that have been widely used for large-scale spectral computations. The description shall be given for the ODE

$$\frac{d\mathbf{u}}{dt} = \mathbf{g}(\mathbf{u}, t) + \mathbf{l}(\mathbf{u}, t) \, , \tag{D.3.1}$$

where $\mathbf{g}(\mathbf{u}, t)$ is treated with a low-storage Runge–Kutta method and $\mathbf{l}(\mathbf{u}, t)$ is treated implicitly with the Crank–Nicolson method. Such mixed explicit/ implicit time-discretizations are very common for incompressible Navier–Stokes computations, for which $\mathbf{g}(\mathbf{u}, t)$ represents (nonlinear) advection and $\mathbf{l}(\mathbf{u}, t)$ (linear) diffusion.

The general representation of a low-storage Runge–Kutta/Crank–Nicolson method requiring only 2 levels of storage (for $\mathbf{u}$ and $\mathbf{h}$) is

$$\mathbf{h} = \mathbf{0}$$

$$\mathbf{u} = \mathbf{u}^n$$

$For \ k = 1 \ to \ K$

$$t^k = t^n + \alpha_k \Delta t$$

$$t^{k+1} = t^n + \alpha_{k+1} \Delta t$$

$$\mathbf{h} \leftarrow \mathbf{g}(\mathbf{u}, t^k) + \beta_k \mathbf{h} \tag{D.3.2}$$

$$\mu = \frac{1}{2} \Delta t (\alpha_{k+1} - \alpha_k)$$

$$\mathbf{v} - \mu \mathbf{l}(\mathbf{v}, t^{k+1}) = \mathbf{u} + \gamma_k \Delta t \mathbf{h} + \mu \mathbf{l}(\mathbf{u}, t^k)$$

$$\mathbf{u} \leftarrow \mathbf{v}$$

$End \ For$

$$\mathbf{u}^{n+1} = \mathbf{u}$$

Table D.2. Coefficients of low-storage Runge–Kutta/Crank–Nicolson schemes

	Williamson 3rd-order	Carpenter-Kennedy 4th-order
α_1	0	0
α_2	1/3	0.1496590219993
α_3	3/4	0.3704009573644
α_4	1	0.6222557631345
α_5	–	0.9582821306748
α_6	–	1
β_1	0	0
β_2	−5/9	−0.4178904745
β_3	−153/128	−1.192151694643
β_4	–	−1.697784692471
β_5	–	−1.514183444257
γ_1	1/3	0.1496590219993
γ_2	15/16	0.3792103129999
γ_3	8/15	0.8229550293869
γ_4	–	0.6994504559488
γ_5	–	0.1530572479681

(note that the penultimate instruction in the loop indicates that $\mathbf{v}$ is the solution of the implicit equation on the left-hand side).

Table D.2 lists the values of these parameters for one third-order scheme, due to Williamson (1980), and one fourth-order scheme from Carpenter and Kennedy (1994). The stability limits (on the imaginary axis) for these schemes are 1.73 for the third-order scheme and 3.34 for the fourth-order scheme. Both of these have been widely used for the time-discretization in applications of spectral methods. Both references contain a family of low-storage methods. Another low-storage family popular in the spectral methods community originated with A. Wray (unpublished), and was extended by Spalart et al. (1993).

Appendix E. Supplementary Material

E.1 Numerical Solution of the Generalized Eigenvalue Problem

The generalized eigenvalue problems produced by linear stability methods (regardless of their spatial discretizations) typically have the form

$$A\mathbf{q} = \omega B\mathbf{q} \tag{E.1.1}$$

for temporal stability and

$$A_0\mathbf{q} + \alpha A_1\mathbf{q} + \alpha^2 A_2\mathbf{q} = \mathbf{0} \tag{E.1.2}$$

for spatial stability in formulations using second-order equations, such as (1.5.5)–(1.5.8) for incompressible flow and (1.5.19) for compressible flow. (When higher-order equations, such as the Orr–Sommerfeld equation (1.5.10) are used, then clearly (E.1.2) contains additional terms with higher powers of α.)

Consider first the temporal eigenvalue problem. In some applications, the complete spectrum of (E.1.1) is desired. The LAPACK package (Anderson et al. (1999)) contains numerous routines for finding all the eigenvalues and some or all of the eigenvectors of such generalized eigenvalue problems. One must be aware that the computed spectrum often contains spurious eigenvalues due entirely to the numerical discretization of the problem. (See Fig. 2.12 and the related discussion.) Frequently, these spurious eigenvalues have the largest absolute values. Care is needed in selecting the eigenvalue(s) of physical interest.

In other cases, one is interested in only a few, or perhaps just one, eigenvalue of (E.1.1). In many cases, a good guess for ω is available. This occurs, for example, when computing a neutral curve—the locus (α, Re) for which $Im\{\omega\} = 0$. A simple approach to computing just a single eigenvalue is inverse Rayleigh iteration. Suppose that ω_0 is an approximate value of ω, and that $\mathbf{x}^n$ and $\mathbf{y}^n$ are current approximations to the eigenvectors of (E.1.1) and to the adjoint problem, respectively. These approximations are updated via

$$(A - \omega_0 B)\mathbf{x}^{n+1} = B\mathbf{x}^n ,$$
$$(A - \omega_0 B)^* \mathbf{y}^{n+1} = B^* \mathbf{y}^n .$$

(E.1.3)

The eigenvalue ω is then approximated by

$$\omega \cong \omega^{n+1} = \frac{(\mathbf{y}^{n+1})^H A\mathbf{x}^{n+1}}{(\mathbf{y}^{n+1})^H B\mathbf{x}^{n+1}} ,$$

(E.1.4)

and, of course, $\mathbf{x}^{n+1}$ and $\mathbf{y}^{n+1}$ are improved approximations to the right and left eigenvectors corresponding to ω. (The superscript H denotes the Hermitian transpose.)

A more sophisticated approach, which is both more efficient and returns a set of eigenvalues, uses the Arnoldi algorithm (Saad (1980)), which is a Krylov subspace method. Equation (E.1.1) is rewritten as

$$C\mathbf{q} = \mu\mathbf{q} ,$$

(E.1.5)

with

$$C = (A - \omega_0 B)^{-1} B, \qquad \mu = \frac{1}{\omega - \omega_0} .$$

(E.1.6)

The Arnoldi iteration starts with an initial guess $\mathbf{q}_1$ of unit L^2-norm and then computes $m - 1$ additional vectors by

$$\hat{\mathbf{q}}_{j+1} = C\mathbf{q}_j - \sum_{i=1}^{j} h_{ij}\mathbf{q}_i ,$$
$$h_{j+1,j} = \|\hat{\mathbf{q}}_{j+1}\|_2 ,$$
$$\mathbf{q}_{j+1} = \hat{\mathbf{q}}_{j+1}/h_{j+1,j} ,$$

(E.1.7)

for $j = 1, \ldots, m - 1$, where $h_{ij} = (\mathbf{q}_i)^H C\mathbf{q}_j$ (and so $\mathbf{q}_{j+1}$ is C-orthogonal to all earlier $\mathbf{q}_j$ with respect to the symmetric part of C). Then the QR algorithm (Wilkinson (1965)) applied to the Hessenberg matrix with elements h_{ij} yields approximations to the eigenvalues of C with largest absolute values. Note that the LU decomposition of $(A - \omega_0 B)$ is only computed once. For Chebyshev collocation discretizations, transform methods may be used to reduce the cost of computing the $B\mathbf{q}_j$ contribution to the $C\mathbf{q}_j$ term in the iteration. See Nayar and Ortega (1993) for more details in the context of fluid dynamic stability problems. The ARPACK package (Lehoucq et al. (1998)) contains routines for various Arnoldi iterative methods that are applicable to finding some some of the eigenvalues and eigenvectors of such generalized eigenvalue problems.

For spatial eigenvalue problems, the companion matrix approach (Bridges and Morris (1984)) is widely used. For the particular case of (E.1.2) the companion matrix is

$$\tilde{A}\tilde{\mathbf{q}} = \alpha \tilde{B}\tilde{\mathbf{q}} , \tag{E.1.8}$$

where

$$\tilde{A} = \begin{pmatrix} -A_1 & -A_2 \\ I & 0 \end{pmatrix} , \qquad \tilde{B} = \begin{pmatrix} A_0 & 0 \\ 0 & I \end{pmatrix} \quad \text{and} \quad \tilde{\mathbf{q}} = \begin{pmatrix} \alpha \mathbf{q} \\ \mathbf{q} \end{pmatrix} . \tag{E.1.9}$$

The matrices for this eigenvalue problem are twice the size of those in the temporal problem. Similarly, for the Orr–Sommerfeld equation, (E.1.2) is a quartic in the eigenvalue α, and the companion matrix is four times the size. Standard algorithms may be applied to extract the full spectrum. Bridges and Morris (1984) discuss a factorization approach to produce a subset of the eigenvalues, and Bridges and Vaserstein (1986) describe several methods based on Newton iteration. The Arnoldi method may be applied to the companion matrix, but the Jacobi-Davidson algorithm (Sleijpen and van der Vorst (1996)) works with the original equation (E.1.2) without requiring the use of the larger companion matrix (Sleijpen et al. (1996)). See Heeg and Geurts (1998) for an application to fluid dynamic stability. For a more comprehensive overview of methods for solving the generalized algebraic eigenvalue problems that arise from discretizations of fluid dynamic linear stability equations, see Theofilis (2003).

E.2 Tau Correction for the Kleiser–Schumann Method

The tau approximation to (3.4.17)–(3.4.19) can be written

$$\nu \tilde{u}_m^{(2)} - \lambda \tilde{u}_m - ik\tilde{p}_m = -\tilde{r}_{x,m} - \tilde{\tau}_{x,m}, \quad m = 0, \ldots, N , \tag{E.2.1}$$

$$\hat{u}(\pm 1) = 0 , \tag{E.2.2}$$

$$\nu \tilde{v}_m^{(2)} - \lambda \tilde{v}_m - \tilde{p}_m^{(1)} = -\tilde{r}_{y,m} - \tilde{\tau}_{y,m}, \quad m = 0, \ldots, N , \tag{E.2.3}$$

$$\hat{v}(\pm 1) = 0 , \tag{E.2.4}$$

$$\tilde{d}_m \equiv ik\tilde{u}_m + \tilde{v}_m^{(1)} = 0, \quad m = 0, \ldots, N . \tag{E.2.5}$$

The tau terms $\tilde{\tau}_{x,m}$ and $\tilde{\tau}_{y,m}$ vanish for $0 \leqslant m \leqslant N - 2$. The application of the discrete divergence to (E.2.1) and (E.2.3) yields

$$\nu \tilde{d}_m^{(2)} - \lambda \tilde{d}_m - \tilde{p}_m^{(2)} + k^2 \tilde{p}_m = -\tilde{r}_m - (ik\tilde{\tau}_{x,m} + \tilde{\tau}_{y,m}^{(1)}), \quad m = 0, \ldots, N , \tag{E.2.6}$$

where

$$\tilde{r}_m = ik\tilde{r}_{x,m} + \tilde{r}_{y,m}^{(1)}, \quad m = 0, \ldots, N . \tag{E.2.7}$$

But, (E.2.5) is equivalent to

$$\tilde{d}_m = 0 \quad m = 0, \ldots, N - 2, \qquad \hat{d}(\pm 1) = 0 . \tag{E.2.8}$$

Hence, from (E.2.6) the discrete A-problem is

$$\tilde{p}_m^{(2)} - k^2\tilde{p}_m = \tilde{r}_m + \tilde{\sigma}_m^{(1)}, \quad m = 0, \ldots, N-2, \qquad \hat{v}'(\pm 1) = 0$$
$$\nu\tilde{v}_m^{(2)} - \lambda\tilde{v}_m - \tilde{p}_m^{(1)} = -\tilde{r}_{y,m} - \tilde{\sigma}_m, \quad m = 0, \ldots, N, \qquad \hat{v}(\pm 1) = 0 ,$$

$$\text{(E.2.9)}$$

and the discrete B-problem is

$$\tilde{p}_m^{(2)} - k^2\tilde{p}_m\tilde{r}_m + \tilde{\sigma}_m^{(1)} , \quad m = 0, \ldots, N-2, \qquad \hat{p}(\pm 1) = \hat{p}_\pm ,$$
$$\nu\tilde{v}_m^{(2)} - \lambda\tilde{v}_m - \tilde{p}_m^{(1)} = -\tilde{r}_{y,m} - \tilde{\sigma}_m, \quad m = 0, \ldots, N, \qquad \hat{v}(\pm 1) = 0 ,$$

$$\text{(E.2.10)}$$

where we use $\tilde{\sigma}_m = \tilde{\tau}_{y,m}$. If not for the "tau correction" embodied by the $\tilde{\sigma}_n$ and $\tilde{\sigma}_n^{(1)}$ terms, the influence-matrix solution procedure would be a straightforward application of the Helmholtz equation techniques discussed in CHQZ2, Sect. 4.1.2. Kleiser and Schumann (1980) describe how to solve the discrete B-problem. Define the $\overline{B}_1$-problem by

$$\tilde{p}_m^{(2)} - k^2\tilde{p}_m = \tilde{r}_m, \quad m = 0, \ldots, N-2 ,$$
$$\hat{p}(\pm 1) = \hat{p}_{b\pm} ,$$
$$\nu\tilde{v}_m^{(2)} - \lambda\tilde{v}_m - \tilde{p}_m^{(1)} = -\tilde{r}_{y,m}, \quad m = 0, \ldots, N-2 ,$$
$$\hat{v}(\pm 1) = 0 ,$$

$$\text{(E.2.11)}$$

and the $\overline{B}_0$-problem by

$$\tilde{p}_m^{(2)} - k^2\tilde{p}_m = \frac{2}{\bar{c}_m}m', \quad m = 0, \ldots, N-2 ,$$
$$\hat{p}(\pm 1) = 0 ,$$
$$\nu\tilde{v}_m^{(2)} - \lambda\tilde{v}_m = \tilde{p}_m^{(1)}, \quad m = 0, \ldots, N-2 ,$$
$$\hat{v}(\pm 1) = 0 ,$$

$$\text{(E.2.12)}$$

where

$$m' = \begin{cases} N-1, & m \text{ even,} \\ N, & m \text{ odd} \end{cases}$$

$$\text{(E.2.13)}$$

(assuming N is even). Furthermore, define $\tilde{\sigma}_{1_m}$ and $\tilde{\sigma}_{0_m}$ for $m = N-1, N$ as the tau terms that must be added to the v-momentum equations in (E.2.11) and (E.2.12), respectively, for them to hold for $m = N-1, N$. One can show that

$$\tilde{\sigma}_m = \tilde{\sigma}_{1_m}/(1 - \tilde{\sigma}_{0_m}), \quad m = N-1, N ,$$

$$\text{(E.2.14)}$$

and that

$$\tilde{p}_m = \tilde{p}_{1_m} + \tilde{\sigma}_{m'}\tilde{p}_{0_m} ,$$
$$\tilde{v}_m = \tilde{v}_{1_m} + \tilde{\sigma}_{(m+1)'}\tilde{v}_{0_m} , \quad m = 0, \ldots, N .$$

$$\text{(E.2.15)}$$

The solution to the original B-problem is achieved by

(1) Solving (E.2.12) for $\tilde{p}_0$, $\tilde{v}_0$, and evaluating $\tilde{\sigma}_{0_m}$ for $m = N - 1, N$ from the v-momentum equation of (E.2.12).
(2) Solving (E.2.11) for $\tilde{p}_1$, $\tilde{v}_1$, and evaluating $\tilde{\sigma}_{1_m}$ for $m = N - 1, N$ from the v-momentum equation of (E.2.11).
(3) Determining $\tilde{\sigma}_m$ from (E.2.14) and $\tilde{\sigma}_m^{(1)}$ from the standard recurrence relation (8.3.23).
(4) Determining $\tilde{p}$, $\tilde{v}$ from (E.2.15).

Step (1) is redundant for the second B-problem in the influence-matrix calculation. Hence, for each wavenumber k the tau solution to the A-problem can be found at the cost of five complex Helmholtz equation solutions. (This can be reduced to four if one wishes to store $\tilde{p}_0$ and $\tilde{v}_0$.) To this cost must be added the cost of solving for $\tilde{u}$ from either (E.2.1) and (E.2.2) or (E.2.5). The cost is negligible in the latter case.

If the tau correction is simply ignored, then the computed solution will not satisfy all of (E.2.1)–(E.2.5). If (E.2.1)–(E.2.2) are used to determine $\tilde{u}$, then the solution will have a nonzero divergence in the interior. If (E.2.5) is used instead, then the momentum equation will not be satisfied, and the numerical experience is that catastrophic numerical instability occurs (Kleiser (1986)).

E.3 The Piola Transform

Let $\hat{\Omega}$ and Ω be two domains in $\mathbb{R}^d$, $d \geq 2$, with coordinates $\hat{\mathbf{x}} \in \hat{\Omega}$ and $\mathbf{x} \in \Omega$, respectively. Let $F : \hat{\Omega} \to \Omega$ be a differentiable and invertible mapping with differentiable inverse $F^{-1} : \Omega \to \hat{\Omega}$. Let $DF(\hat{\mathbf{x}})$ denote the Jacobian matrix of F at the point $\mathbf{x} \in \hat{\Omega}$, and let $J(\hat{\mathbf{x}}) = |DF(\hat{\mathbf{x}})| \neq 0$ denote its determinant.

Given a vector field $\hat{\mathbf{v}}$ in $\hat{\Omega}$, we define the vector field $\mathbf{v}$ in Ω by

$$\mathbf{v} = J^{-1}DF\hat{\mathbf{v}} , \tag{E.3.1}$$

i.e.,

$$\mathbf{v}(\mathbf{x}) = J^{-1}(\hat{\mathbf{x}})DF(\hat{\mathbf{x}})\hat{\mathbf{v}}(\hat{\mathbf{x}}), \qquad \text{if } \hat{\mathbf{x}} = F^{-1}(\mathbf{x}) .$$

Equivalently, we can express $\hat{\mathbf{v}}$ in terms of $\mathbf{v}$ as

$$\hat{\mathbf{v}} = J\,DF^{-1}\mathbf{v} . \tag{E.3.2}$$

The mappings $\hat{\mathbf{v}} \to \mathbf{v}$ and $\mathbf{v} \to \hat{\mathbf{v}}$ are the direct and inverse *Piola transforms* associated with F. They have the remarkable property that the divergences of the (supposed smooth) vector fields $\hat{\mathbf{v}}$ and $\mathbf{v}$ are related by the formula

$$\nabla_{\mathbf{x}} \cdot \mathbf{v} = J^{-1}\nabla_{\hat{\mathbf{x}}} \cdot \hat{\mathbf{v}} , \tag{E.3.3}$$

which, in particular, proves that $\hat{\mathbf{v}}$ *is divergence-free if and only if* $\mathbf{v}$ *is.* An equivalent integral formulation of (E.3.3) is

$$\int_\Omega \nabla_{\mathbf{x}} \cdot \mathbf{v}\, \varphi \, d\mathbf{x} = \int_{\hat{\Omega}} \nabla_{\hat{\mathbf{x}}} \cdot \hat{\mathbf{v}}\, \hat{\varphi} \, d\hat{\mathbf{x}} \qquad \text{if } \hat{\varphi} = \varphi \circ F \,, \tag{E.3.4}$$

which holds for all integrable scalar fields φ in Ω. This implies the *conservation property*

$$\int_\omega \nabla_{\mathbf{x}} \cdot \mathbf{v}\, d\mathbf{x} = \int_{F^{-1}(\omega)} \nabla_{\hat{\mathbf{x}}} \cdot \hat{\mathbf{v}}\, d\hat{\mathbf{x}} \tag{E.3.5}$$

for all open domains ω contained in Ω, as can be seen by taking as φ in (E.3.4) the characteristic function of ω, i.e., the function which has the value 1 if $\mathbf{x} \in \omega$ and 0 otherwise.

Identities (E.3.3) and (E.3.4) can be proven as follows. Assume that φ is continuously differentiable in Ω and vanishing on $\partial\Omega$. Then, by the divergence theorem and the chain rule applied to $\nabla_{\mathbf{x}}\varphi$, one has

$$\int_\Omega \nabla_{\mathbf{x}} \cdot \mathbf{v}\, \varphi \, d\mathbf{x} = -\int_\Omega \mathbf{v} \cdot \nabla_{\mathbf{x}}\varphi \, d\mathbf{x} = -\int_{\hat{\Omega}} J^{-1}\hat{\mathbf{v}}^T (DF)^T (DF^{-1})^T \nabla_{\hat{\mathbf{x}}}\hat{\varphi} \, J \, d\hat{\mathbf{x}}$$

$$= -\int_{\hat{\Omega}} \hat{\mathbf{v}} \cdot \nabla_{\hat{\mathbf{x}}}\hat{\varphi} \, d\hat{\mathbf{x}} = \int_{\hat{\Omega}} \nabla_{\hat{\mathbf{x}}} \cdot \hat{\mathbf{v}}\, \hat{\varphi} \, d\hat{\mathbf{x}} = \int_\Omega (\nabla_{\hat{\mathbf{x}}} \cdot \hat{\mathbf{v}})\, \varphi \, J^{-1} \, d\mathbf{x} \,.$$

Formula (E.3.3) follows directly since φ can be chosen to be nonzero at any interior point. The validity of (E.3.4) for any integrable φ (not necessarily vanishing at the boundary) can be established by a density argument.

In order to implement the inverse Piola transform (E.3.2), one needs the inverse matrix DF^{-1}. Let us detail its computation in 3D. Denote by

$$\mathbf{a}_j = \frac{\partial \mathbf{x}}{\partial \hat{x}_j} \,, \qquad j = 1, 2, 3 \,, \tag{E.3.6}$$

the column vectors of the matrix DF. Then, it is easily seen that the rows $\mathbf{a}^i = \nabla_{\mathbf{x}}\hat{x}_i$, $i = 1, 2, 3$, of the matrix DF^{-1} satisfy the relations

$$\mathbf{a}^i = J^{-1}(\mathbf{a}_j \times \mathbf{a}_k)^T \,, \qquad (i, j, k) \text{ cyclic} \,, \tag{E.3.7}$$

where the Jacobian determinant itself can be computed in the form

$$J = \mathbf{a}_i \cdot (\mathbf{a}_j \times \mathbf{a}_k) \,, \qquad (i, j, k) \text{ cyclic} \,. \tag{E.3.8}$$

The entries of $\mathbf{a}^i$ can be equivalently written as

$$J(\mathbf{a}^i)_n = J(DF^{-1})_{i,n} = -\hat{\mathbf{e}}_i \cdot \nabla_{\hat{\mathbf{x}}} \times (x_l \nabla_{\hat{\mathbf{x}}} x_m) \,, \qquad (n, m, l) \text{ cyclic} \,, \tag{E.3.9}$$

where $\hat{\mathbf{e}}_i$ are the vectors of the canonical basis. Note that the inverse Jacobian only appears in (E.3.2) in the combination $J(DF^{-1})$.

In many situations, only the values of F at a set of interpolation nodes in $\hat{\Omega}$ (such as the LGL or the LG nodes) are used to implement the transformation numerically. Then, as shown in detail by Kopriva (2006), the discrete version of (E.3.9), rather than of (E.3.7), should be used in order to guarantee the conservation property at the discrete level. Precisely, denoting by

$$\mathcal{D}_{\hat{x}_j} = \frac{\partial}{\partial \hat{x}_j} \hat{I}_N$$

the interpolation partial derivative with respect to $\hat{x}_j$ (where $\hat{I}_N$ is the interpolation operator at the selected set of nodes), and by $\tilde{\nabla}_{\hat{x}} = \nabla_{\hat{x}} \hat{I}_N$ the interpolation gradient operator, one replaces (E.3.6) and (E.3.9) by

$$\tilde{\mathbf{a}}_j = \mathcal{D}_{\hat{x}_j} \mathbf{x} , \qquad j = 1, 2, 3 , \tag{E.3.10}$$

and

$$(\tilde{J}\mathbf{a}^i)_n = (\widetilde{JDF^{-1}})_{i,n} = -\hat{\mathbf{e}}_i \cdot \tilde{\nabla}_{\hat{x}} \times (x_l \tilde{\nabla}_{\hat{x}} x_m) , \qquad (n, m, l) \text{ cyclic} , \tag{E.3.11}$$

respectively. In this way, if

$$\eta_t + \nabla_{\mathbf{x}} \cdot \mathcal{F} = 0$$

is a conservation law in Ω, and if one sets $\tilde{\eta} = \tilde{J}\eta$ and $\tilde{\mathcal{F}} = \widetilde{JDF^{-1}}\mathcal{F}$, one gets the conservation law in $\hat{\Omega}$:

$$\tilde{\eta}_t + \nabla_{\hat{x}} \cdot \tilde{\mathcal{F}} = 0 .$$

Finally, in the context of enforcing the discrete conservation laws in integral form, as used for example in discontinuous Galerkin spectral methods, there are additional subtleties with exactly how the integral form of the flux term is written. There are some forms of the integral form of the flux term that are not integrated exactly with the Gauss–Lobatto formula, whereas they are integrated exactly with the Gauss formula. This makes the Gauss formula more generally recommendable for ensuring conservation in the discrete equations. See Kopriva (2006) for the details.

References

S. Abarbanel, P. Dutt, D. Gottlieb (1989): Splitting methods for low Mach number Euler and Navier Stokes equations. Comput. & Fluids **17**(1), 1–12

Y. Achdou, Y. Maday, O.B. Widlund (1999): Iterative substructuring preconditioners for mortar element methods in two dimensions. SIAM J. Numer. Anal. **36**(2), 551–580

N.A. Adams, L. Kleiser (1996): Subharmonic transition to turbulence in a flat-plate boundary layer at Mach number 4.5. J. Fluid Mech. **317**, 301–335

V.I. Agoshkov (1988): 'Poincaré-Steklov's operators and domain decomposition methods in finite dimensional spaces'. In: *First International Symposium on Domain Decomposition Methods for Partial Differential Equations*, ed. by R. Glowinski et al. (SIAM, Philadelphia) pp. 73–112

V.I. Agoshkov, V.I. Lebedev (1985): 'Poincaré-Steklov operators and the methods of partition of the domain in variational problems'. In: *Computational Processes and Systems, N. 2*, ed. by G.I. Marchuk (Nauka, Moscow) pp. 173–227 [in Russian]

M. Ainsworth (2004): Dispersive and dissipative behaviour of high order discontinuous Galerkin finite element methods. J. Comput. Phys. **198**(1), 106–130

M. Ainsworth, P. Coggins (2002): A uniformly stable family of mixed hp-finite elements with continuous pressure for incompressible flow. IMA J. Numer. Anal. **22**(2), 307–327

M. Ainsworth, T.J. Oden (1992): A procedure for a posteriori error estimation for hp finite element methods. Comput. Meth. Appl. Mech. Engrg. **101**(1–3), 73–96

M. Anagnostou (1991): Nonconforming Sliding Spectral Element Methods for Unsteady Incompressible Navier–Stokes Equations. PhD thesis, Massachusetts Institute of Technology, Cambridge, MA

P.W. Anderson (1963): *Concepts in Solids: Lectures on the Theory of Solids* (W.A. Benjamin, New York)

E. Anderson, Z. Bai, C. Bischof, S. Blackford, J. Demmel, J. Dongarra, J. Du Croz, A. Greenbaum, S. Hammarling, A. McKenney, D. Sorensen (1999): *LAPACK Users' Guide*, 3rd edition (SIAM, Philadelphia)

Anon. (1949): 'The NBS-NACA Tables of Thermal Properties of Gases'. (National Bureau of Standards)

Anon. (1953): 'Equations, Tables and Charts for Compressible Flow'. NACA Report – 1135 (NACA)

D.N. Arnold, F. Brezzi, B. Cockburn, D. Marini (2002): Unified analysis of discontinuous Galerkin methods for elliptic problems. SIAM J. Numer. Anal. **39**(5), 1749–1779

T.D. Aslam (2003): A level set algorithm for tracking discontinuities in hyperbolic conservation laws II: Systems of Equations. J. Sci. Comput. **19**(1–3), 37–62

I. Babuška, C.E. Baumann, J.T. Oden (1999): A discontinuous hp finite element method for diffusion problems: 1-D analysis. Comput. Math. Appl. **37**(9), 103–122

I. Babuška, B. Guo (2001): Direct and inverse approximation theorems for the p-version of the finite element method in the framework of weighted Sobolev spaces. Part I: Approximability of functions in the weighted Besov spaces. SIAM J. Numer. Anal. **39**(5), 1512–1538

I. Babuška, B. Guo (2002): Direct and inverse approximation theorems for the p-version of the finite element method in the framework of weighted Besov spaces. Part II: Optimal rate of convergence of the p-version finite element solutions. Math. Mod. Methods Appl. Sci. **12**, 689–719

I. Babuška, T. Strouboulis (2001): *The Finite Element Method and Its Reliability* (Clarendon Press, Oxford)

I. Babuška, M. Suri (1994): The p and $h - p$ versions of the finite element method, basic principles and properties. SIAM Review **36**(4), pp. 578–632

Z.-Z. Bai, G.H. Golub, M.K. Ng (2003): Hermitian and skew-Hermitian splitting methods for non-Hermitian positive definite linear systems. SIAM J. Matrix Anal. Appl. **24**(3), 603–626

C. Basdevant (1983): Technical improvements for direct numerical simulation of homogeneous three-dimensional turbulence. J. Comput. Phys. **50**(2), 209–214

F. Bassi, S. Rebay (1997): A high-order accurate discontinuous finite element method for the numerical simulation of the compressible Navier–Stokes equation. J. Comput. Phys. **131**(2), 267–279

F. Bassi, S. Rebay, G. Mariotti, S. Pedinotti, M. Savini (1997): A high-order accurate discontinuous finite element method for inviscid and viscous turbomachinery flows, in *Proceedings of the 2nd*

European Conference on Turbomachinery, Fluid Dynamics and Thermodynamics, ed. by R. Decuypere and G.Dibelius, Technologisch Institut, Antwerpen, Belgium, pp. 99–108

G.K. Batchelor (2000): *An Introduction to Fluid Dynamics* (Cambridge Univ. Press, Cambridge)

Z. Belhachmi (1997): Nonconforming mortar element methods for the spectral discretization of two-dimensional fourth-order problems. SIAM J. Numer. Anal. **34**(4), 1545–1573

Y.Y. Belov, N.N. Yanenko (1971): Influence of viscosity on the smoothness of solutions of incompletely parabolic systems. Math. Notes Acad. Sci. USSR **10**, 480–483

F. Ben Belgacem (1994): Polynomial extension of compatible polynomial traces in three dimensions. Comput. Meth. Appl. Mech. Engrg. **116**(1–4), 235–241.

F. Ben Belgacem (1999): The mortar finite element method with Lagrange multipliers. Numer. Math. **85**(2), 173–197

F. Ben Belgacem (2000): The mixed mortar finite element method for the incompressible Stokes problem: convergence analysis. SIAM J. Numer. Anal. **37**(4), 1085–1100

F. Ben Belgacem, C. Bernardi, N. Chorfi, Y. Maday (2000): Inf-sup conditions for the mortar spectral element method discretization of the Stokes problem. Numer. Math. **85**(2), 257–281

M. Benzi, G.H. Golub, J. Liesen (2005): Numerical solution of saddle point problems. Acta Numer. **14**, 1–37

J. Bergh and J. Löfstrom (1976): *Interpolation Spaces. An Introduction* (Springer-Verlag, Berlin)

S. Berlin, M. Wiegel, D.S. Henningson (1999): Numerical and experimental investigations of oblique boundary layer transition. J. Fluid Mech. **393**, 23–57

P.S. Bernard and J.M. Wallace (2002) *Turbulent Flow: Analysis, Measurement, and Prediction* (John Wiley & Sons, Hoboken, NJ)

C. Bernardi (1996): Indicateurs d'erreur en $h-N$ version des éléments spectraux. Math. Mod. Numer. Anal. **30**, 1–38

C. Bernardi, C. Canuto, Y. Maday (1988): Generalized inf-sup conditions for Chebyshev spectral approximation of the Stokes problem. SIAM J. Numer. Anal. **25**(6), 1237–1271

C. Bernardi, C. Canuto, Y. Maday (1991): Spectral approximations of the Stokes equations with boundary conditions on the pressure. SIAM J. Numer. Anal. **28**(2), 333–362

C. Bernardi, N. Chorfi (2006): Spectral dscretization of the vorticity, velocity, and pressure formualtion of the Stokes problem. SIAM J. Numer. Anal. **44**(2), 826–850

C. Bernardi, M. Dauge, Y. Maday (1999): *Spectral Methods for Axisymmetric Domains. Numerical algorithms and test due to Mejdi Azaiez* Series in Applied Mathematics (Gauthier-Villars, Paris)

C. Bernardi, V. Girault, Y. Maday (1992): Mixed spectral element approximation of the Navier–Stokes equations in the streamfunction and vorticity formulation. IMA J. Numer. Anal. **12**(4), 565–608

C. Bernardi, Y. Maday (1986): 'A Staggered Grid Spectral Method for the Stokes Problem'. In: *Proc. 6th Int. Symp. Finite Element Methods in Flow Problems*, ed. by M.O. Bristeau, R. Glowinski, A. Hausel, J. Periaux, pp. 33–37

C. Bernardi, Y. Maday (1988a): Spectral methods for the approximation of fourth-order problems: application to the Stokes and Navier–Stokes equations. Comput. & Structures **30**(1–2), 205–216

C. Bernardi, Y. Maday (1988b): Analysis of a staggered grid algorithm for the Stokes equation. Int. J. Num. Meth. Fluids **8**, 537–557

C. Bernardi, Y. Maday (1990): Relèvement polynomial de traces et applications. RAIRO Modél. Math. Anal. Numér. **24**, 557–611

C. Bernardi, Y. Maday (1992): *Approximations Spectrales de Problèmes aux Limites Elliptiques* (Springer-Verlag, Paris)

C. Bernardi, Y. Maday (1997): 'Spectral methods'. In: *Handbook of Numerical Analysis*, **V** Techniques of Scientific Computing, ed. by P.J. Ciarlet and J.L. Lions (North Holland, Amsterdam) pp. 209–486

C. Bernardi, Y. Maday (1999): Uniform inf-sup conditions for the spectral discretization of the Stokes problem. Math. Mod. Methods Appl. Sci. **9**, 395–414

C. Bernardi, Y. Maday, B. Métivet (1986): Une méthode directe de collocation pour le problème de Stokes. C.R. Acad. Sci. Paris **302**, serie I, 163–166

C. Bernardi, Y. Maday, B. Métivet (1987a): Spectral approximation of the periodic/nonperiodic Navier–Stokes equations. Numer. Math. **51**(6), 655–700

C. Bernardi, Y. Maday, B. Métivet (1987b): Computation of the pressure in the spectral approximation of the Stokes problem. Rech. Aerosp. (Engl. Ed.) **1987–1**, 1–21

C. Bernardi, Y. Maday, A.T. Patera (1994): 'A New Nonconforming Approach to Domain Decomposition: the Mortar Element Method'. In: *Nonlinear Partial Differential Equations and Their Applications. Collège de France Seminar*, Vol. XI, ed. by H. Brezis, J.L. Lions (Longman, Harlow) pp. 13–51

C. Bernardi, Y. Maday, F. Rapetti (2005): Basics and some applications of the mortar element method. GAMM-Mitt. **28**, 97–123

L. Berselli, T. Iliescu, W.J. Layton (2006): *The Mathematics of Large Eddy Simulation of Turbulent Flows* (Springer-Verlag, New York)

F.P. Bertolotti (1991): Linear and Nonlinear Stability of Boundary Layers with Streamwise Varying Properties' PhD thesis, Ohio State University, Columbus, OH

F.P. Bertolotti, Th. Herbert (1991): Analysis of the linear stability of compressible boundary layers using the PSE. Theor. Comput. Fluid Dyn. **3**(2), 117–124

F.P. Bertolotti, Th. Herbert, P.R. Spalart (1992): Linear and nonlinear stability of the Blasius boundary layer. J. Fluid Mech. **242**, 441–474

K. Black (2000): Spectral element approximation of convection-diffusion type problems. Appl. Numer. Math. **33**(1–4), 373–379

G.A. Blaisdell (1991): Numerical Simulation of Compressible homogeneous Turbulence. PhD thesis, Stanford University, Palo Alto, CA

G.A. Blaisdell, G.N. Coleman, N.N. Mansour (1996): Rapid distortion theory for compressible homogeneous turbulence under isotropic mean strain. Phys. Fluids A **8**(10), 2692–2705

G.A. Blaisdell, N.N. Mansour, W.C. Reynolds (1993): Compressibility effects on the growth and structure of homogeneous turbulent shear flows. J. Fluid Mech. **256**, 443–485

G.A. Blaisdell, E.T. Spryopoulos, J.H. Qin (1996): The effect of the formulation of the nonlinear terms on aliasing errors in spectral methods. Appl. Numer. Math. **21**(3), 207–219

W. Blake (1975): 'A Statistical Description of Pressure and Velocity Fields at the Trailing Edge of a Flat Strut.' David Taylor Naval Ship Research and Development Center Report 4241

J.M. Boland, R.A. Nicolaides (1983): Stability of finite elements under divergence constraints. SIAM J. Numer. Anal. **20**(4), 722–731

J.-F. Bourgat, R. Glowinski, P. Le Tallec, M. Vidrascu (1989): 'Variational Formulation and Algorithm for Trace Operator in Domain Decomposition Calculations'. In *Domain Decomposition Methods*, ed. by T.F. Chan et al. (SIAM, Philadelphia) pp. 3–16

J.P. Boyd (2001): *Chebyshev and Fourier Spectral Methods*, 2nd edition (Dover, New York)

J.P. Boyd and N. Flyer (1999): Compatibility conditions for time-dependent partial differential equations and the rate of convergence of Chebyshev and Fourier spectral methods, Comput. Meth. Appl. Mech. Engrg. **175**(3–4), 281–309

M. Braack, E. Burman (2006): Local projection stabilization for the Oseen problem and its interpretation as a variational multiscale method. SIAM J. Numer. Anal. **43**(6), 2544–2566

M. Braack, E. Burman, V. John, G. Lube (2007): Stabilized finite element methods for the generalized Oseen problem. Comput. Meth. Appl. Mech. Engrg. **196**(4–6), 853–866

M.E. Brachet (1991): Direct simulation of three-dimensional turbulence in the Taylor-Green vortex. Fluid Dyn. Res. **8**, 1–8

M.E. Brachet, D.I. Meiron, S.A. Orszag, B.G. Nickel, R.H. Morf, U. Frisch (1983): Small-scale structure of the Taylor-Green vortex. J. Fluid Mech. **130**, 411–452

J.H. Bramble, J.E. Pasciak (1997): Iterative techniques for time dependent Stokes problems. Comput. Meth. Appl. **33**(1–2), 13–30

F. Brezzi (1974): On the existence, uniqueness and approximation of saddle point problems arising from Lagrangian multipliers. RAIRO Numer. Anal. **8**, 129–151

F. Brezzi, M. Fortin (1992): *Mixed and Hybrid Finite Element Methods* (Springer-Verlag, New York)

F. Brezzi, J. Rappaz, P.A. Raviart (1980): Finite dimensional approximation of nonlinear problems, Part I: Branches of nonsingular solutions. Numer. Math. **36**(1), 1–25

F. Brezzi, J. Rappaz, P.A. Raviart (1981a): Finite dimensional approximation of nonlinear problems, Part II: Limit points. Numer. Math. **37**(1), 1–28

F. Brezzi, J. Rappaz, P.A. Raviart (1981b): Finite dimensional approximation of nonlinear problems, Part III: Simple bifurcation points. Numer. Math. **38**(1), 1–30

T.J. Bridges, P.J. Morris (1984): Differential eigenvalue problems in which the parameter appears nonlinearly. J. Comput. Phys. **55**(3), 437–460

T.J. Bridges, P.J. Morris (1987): Boundary layer stability calculations. Phys. Fluids **30**(11), pp. 3351–3358

T.J. Bridges, L.N. Vaserstein (1986): The local evaluation of the derivative of a determinant. J. Comput. Phys. **65**(1), 107–119

E.O. Brigham (1974): *The Fast Fourier Transform* (Prentice-Hall, Englewood Cliffs, NJ)

A.N. Brooks, T.J.R. Hughes (1982): Streamline upwind / Petrov-Galerkin formulations for convection dominated flows with particular emphasis on the incompressible Navier–Stokes equations. Comput. Meth. Appl. Mech. Engrg. **32**(1–3), 199–259

G.P. Brooks, J.M. Powers (2004): A Karhunenoeve least-squares technique for optimization of geometry of a blunt body in supersonic flow. J. Comput. Phys. **195**(1), 387–412

U. Brosa, S. Grossmann (2002): Hydrodynamic vector potentials. European Phys. J. B **26**, 121–132

P.N. Brown, Y. Saad (1990): Hybrid Krylov methods for nonlinear systems of equations. SIAM J. Sci. Stat. Comput. **11**(3), 450–481

J.C. Buell (1990): 'Direct simulations of Compressible Wall Bounded Turbulence'. In: *Annual Research Briefs – 1990* (Center for Turbulence Research, Stanford Univ.) pp. 347–356

E. Burman, A. Ern (2007): Continuous interior penalty hp-finite element methods for transport problems. Math. Comput. To appear. (EPFL-IACS report 02-2005.)

E. Burman, M.A. Fernández, P. Hansbo (2006): Continuous interior penalty finite element method for Oseen's equations. SIAM J. Numer. Anal. **44**(3), 1248–1274

J.C. Butcher (2003) *Numerical Methods for Ordinary Differential Equations* (John Wiley & Sons, New York)

D.S. Butler (1960): The numerical solution of hyperbolic systems of differential equations in three independent variables. Proc. R. Soc. London. Ser. A **255**, 232–252

J. Cahouet, J.-P. Chabard (1988): Some fast 3D finite element solvers for the generalized Stokes problem. Int. J. Numer. Meth. Fluids **8**(8), 869–895

A.B. Cain, J.H. Ferziger, W.C. Reynolds (1984): Discrete orthogonal function expansions for non-uniform grids using the fast Fourier transform. J. Comput. Phys. **56**(2), 272–286

G. Caloz, J. Rappaz (1997): 'Numerical Analysis for Nonlinear and Bifurcation Problems'. In: *Handbook of Numerical Analysis*, **V** Techniques of Scientific Computing, ed. by P.J. Ciarlet and J.L. Lions (North Holland, Amsterdam) pp. 487–637

C. Cambon, G.N. Coleman, N.N. Mansour (1993): Rapid distortion analysis and direct simulation of compressible homogeneous turbulence at finite Mach number. J. Fluid Mech. **257**, 641–665

C. Canuto (1988): Spectral methods and a maximum principle. Math. Comput. **51** (1988), 615–629

C. Canuto, P. Gervasio, A. Quarteroni (2007): Preconditioning G-NI spectral methods in simple- and multi-domains. Proceedings of ICOSAHOM '07

C. Canuto, M.Y. Hussaini, A. Quarteroni, T.A. Zang (1988): *Spectral Methods in Fluid Dynamics* (Springer-Verlag, New York)

C. Canuto, M.Y. Hussaini, A. Quarteroni, T.A. Zang (2006): *Spectral Methods. Fundamentals in Single Domains* (Springer-Verlag, Berlin)

C. Canuto, A. Quarteroni (1987): On the boundary treatment in spectral methods for hyperbolic systems. J. Comput. Phys. **71**(1), 100–110

C. Canuto, A. Russo, V. Van Kemenade (1998): Stabilized spectral methods for the Navier–Stokes equations: residual-free bubbles and preconditioning. Comput. Meth. Appl. Mech. Engrg. **166**(1–2), 65–83

C. Canuto, G. Sacchi-Landriani (1986): Analysis of the Kleiser-Schumann method. Numer. Math. **50**(2), 217–243

C. Canuto, V. Van Kemenade (1996): Bubble-stabilized spectral methods for the incompressible Navier–Stokes equations. Comput. Meth. Appl. Mech. Engrg. **135**(1–2), 35–61

D. Carati, G.S. Winckelmans, H.E. Jeanmart (2001): On the modelling of the subgrid-scale and filtered-scale stress tensors in large-eddy simulation. J. Fluid Mech. **441**, 119–138

M.H. Carpenter, D. Gottlieb, S. Abarbanel (1993): The stability of numerical boundary treatments for compact high-order finite-difference schemes. J. Comput. Phys. **108**(1), 272–295.

M.H. Carpenter, D. Gottlieb, S. Abarbanel (1994): Time-stable finite-difference schemes for solving hyperbolic systems: methodology and application to high-order compact schemes. J. Comput. Phys. **111**(2), 220–236

M.H. Carpenter, C. Kennedy (1994): 'Fourth-order 2N-storage Runge-Kutta Schemes'. NASA TM–109111

M.A. Casarin (1996): Schwarz Preconditioners for Spectral and Mortar Finite Element Methods with Applications to Incompressible Fluids. PhD thesis, Courant Institute of Mathematical Sciences, New York University, New York, NY

M.A. Casarin (1997): Quasi-optimal Schwarz methods for the conforming spectral element discretization. SIAM J. Numer. Anal. **34**(6), 2482–2502

M.A. Casarin (2001): Schwarz preconditioners for the spectral element discretization of the steady Stokes and Navier–Stokes equations. Numer. Math. **89**(2), 307–339

T. Cebeci, P. Bradshaw (1977): *Momentum Transport in Boundary Layers* (McGraw-Hill, New York)

T. Cebeci, K. Stewartson (1980): On stability and transition in three-dimensional flows. AIAA J. **18**(4), 398–405

S. Chandrasekhar (1961): *Hydrodynamic and Hydromagnetic Stability* (Oxford Univ. Press, London)

C.-L. Chang, M.R. Malik, G. Erlebacher. M.Y. Hussaini (1991): 'Compressible Stability of Growing Boundary Layers Using Parabolized Stability Equations'. AIAA Paper No. 91–1636

T. Chaves, E.L. Ortiz (1968): On the numerical solution of two-point boundary value problems for linear differenlial equations. Z. Angew. Math. Mech. **48**, 415–418

C.-J. Chen, S.-Y. Jaw (1998): *Fundamentals of Turbulence Modeling* (Taylor & Francis, Washington, DC)

G. Chen , M.Y.Hussaini (2006): 'Unsteady optimal boundary-control of trailing edge noise'. AIAA Paper No. 2006–2511

L. Chilton, M. Suri (2000): On the construction of stable curvilinear p version elements for mixed formulations of elasticity and Stokes flow. Numer. Math. **86**(1), 29–48

L. Chilton, M. Suri (2001): Locking-free mixed *hp* finite element methods for curvilinear domains. Comput. Meth. Appl. Mech. Engrg. **190**(26-27), 3427–3442

A.J. Chorin (1968): Numerical solution of the Navier–Stokes equations. Math. Comput. **22**, 745–762

A.J. Chorin, J.E. Marsden (1993): *A Mathematical Introduction to Fluid Mechanics*, 3rd edition (Springer-Verlag, New York)

B. Cockburn, G.E. Karniadakis, C.-W. Shu (2000): 'The Development of Discontinuous Galerkin Methods'. In: *Discontinuous Galerkin Methods*, Lecture Notes in Computational Science and Engineering **11** (Springer-Verlag, Heidelberg) pp. 3–50

G.N. Coleman, J. Kim, R.D. Moser (1995): A numerical study of turbulent supersonic isothermal-wall channel flow. J. Fluid Mech. **305**, 159–183

L. Collatz (1966): *The Numerical Treatment of Differential Equations* (Springer-Verlag, Berlin)

S.S. Collis (2002): 'The DG/VMS Method for Unified Turbulence Simulation'. AIAA Paper No. 2002–3124

J.W. Cooley, P.A.W. Lewis, P.D. Welch (1969): The fast Fourier transform and its applications. IEEE Trans. Educ. **12**, 27–34

J.W. Cooley, J.W. Tukey (1965): An algorithm for the machine calculation of complex Fourier series. Math. Comput. **19**, 297–301

P. Cornille (1982): A pseudospectral scheme for the numerical calculation of shocks. J. Comput. Phys. **47**(1), 146–159

R. Courant, K.O. Friedrichs (1976): *Supersonic Flow and Shock Waves* (Springer-Verlag, New York)

W. Couzy (1995): Spectral element discretization of the unsteady Navier–Stokes equations and its iterative solution on parallel computers. PhD thesis No. 1380, EPF Lausanne, Switzerland

W.O. Criminale, T.L. Jackson, R.D. Joslin (2003): *Theory and Computation in Hydrodynamic Stability* (Cambridge University Press, Cambridge)

J. Curry, J. Herring, J. Loncaric, S.A. Orszag (1984): Order and disorder in two- and three-dimensional Bénard convection. J. Fluid Mech. **147**, 1–38

J. le R. d'Alembert (1752): *Essai d'une nouvelle theorie de la resistance des fluides* (Opuscules Mathematiques)

G. Danabasoglu, S. Biringen and C.L. Streett (1990): 'Numerical simulation of spatially-evolving instability control in plane channel flow'. AIAA Paper No. 90–1530

O. Darrigol (2002): Between hydrodynamics and elasticity theory: the first five births of the Navier–Stokes equations. Arch. History of Exact Science **56**(2), 95–150

S.H. Davis (1976): The stability of time-periodic flows. Ann. Rev. Fluid Mech. **8**, 57–74

J.W. Deardorff (1970): A numerical study of three-dimensional turbulent channel flow at large Reynolds numbers. J. Fluid Mech. **41**, 453–480

P. Delorme (1984): Numerical simulation of homogeneous, isotropic, two-dimensional turbulence in compressible flow. Rech. Aerosp. **1984-1**, 1–13

P. Delorme (1985): Simulation numérique de turbulence homogène compressible avec ou sans cisaillement imposé. PhD thesis, University of Poitiers, Poitiers, France (available in English as Numerical Simulation of Compressible Homogeneous Turbulence, NASA Report N89–15365)

P. Demaret, M.O. Deville (1989): Chebyshev pseudo-spectral solution of the Stokes equations using finite element preconditioning. J. Comput. Phys. **83**(2), 463–484

M.O. Deville, P.F. Fischer, E.H. Mund (2002): *High-Order Methods for Incompressible Fluid Flow* (Cambridge University Press, Cambridge)

M. Deville, L. Kleiser, F. Montigny-Rannou (1984): Pressure and time treatment of Chebyshev spectral solution of a Stokes problem. Int. J. Numer. Meth. Fluids **4**, 1149–1163

W.S. Don (1994): Numerical study of pseudospectral methods in shock wave applications. J. Comput. Phys. **110**(1), 103–111

J.J. Dongarra, B. Straughan, D.W. Walker (1996): Chebyshev tau-QZ algorithm methods for calculating spectra of hydrodynamic stability problems. Appl. Numer. Math. **22**(4), 399–434

P.G. Drazin, W.M. Reid (2004): *Hydrodynamic Stability* (Cambridge Univ. Press, Cambridge)

M. Dryja, O.B. Widlund (1995): Schwarz methods of Neumann-Neumann type for three dimensional elliptic finite element problems. Comm. Pure Appl. Math. **48**, 121–155

T. Dubois, F. Jauberteau, R. Temam (1998): Incremental unknowns, multilevel methods and the numerical simulation of turbulence. Comput. Meth. Appl. Mech. Engrg. **159**(1–2), 123–189

T. Dubois, F. Jauberteau, R. Temam (1999): *Dynamic Multilevel Methods and the Numerical Simulation of Turbulence* (Cambridge Univ. Press, Cambridge)

T. Dubois, A. Maranville (1994): Existence and uniqueness results for a velocity formulation of Navier–Stokes equations in a channel. Applicable Anal. **55**, 103–138

P.W. Duck, G. Erlebacher, M.Y. Hussaini (1994): On the linear stability of compressible plane Couette flow. J. Fluid Mech. **258**, 131–165

D.W. Dunn (1953): On the Stability of the Laminar Boundary Layer in a Compressible Fluid. PhD thesis, Massachusetts Institute of Technology, Cambridge, MA

D.W. Dunn, C.C. Lin (1955): On the stability of the laminar boundary layer in a compressible fluid. J. Aero. Sci. **22**, 455–477

N.M. El-Hady, T.A. Zang, U. Piomelli (1994): Application of the dynamic subgrid-scale model to axisymmetric transitional boundary layer at high speed. Phys. Fluids A **6**(3), 1299–1309

H.C. Elman, D.J. Silvester, A.J. Wathen (2002): Performance and analysis of saddle point preconditioners for the discrete steady-state Navier–Stokes equations. Numer. Math. **90**(4), 665–688

H.W. Emmons (1949): 'The Numerical Solution of the Turbulence Problem'. In: *Proc. Symp. Applied Mathematics*, Vol. 1 (McGraw-Hill, New York) pp. 67–71

G. Erlebacher, M.Y. Hussaini (1990): Numerical experiments in supersonic boundary-layer stability. Phys. Fluids A **2**(1), 94–104

G. Erlebacher, M.Y. Hussaini (1991): Non-linear evolution of a second mode wave in supersonic boundary layers, Appl. Numer. Math. **7**(1), 73–91

G. Erlebacher, M.Y. Hussaini, H.O. Kreiss, S. Sarkar (1990): The analysis and simulation of compressible turbulence. Theor. Comput. Fluid Dyn. **2**(2), 73–75

G. Erlebacher, M.Y. Hussaini, C.G. Speziale, T.A. Zang (1992): Toward the large-eddy simulation of compressible turbulent flows. J. Fluid. Mech. **238**, 155–185

C. Eskilsson, S.J. Sherwin (2004): A triangular spectral/hp discontinuous Galerkin method for modelling 2D shallow water equations. Inter. J. Numer. Meth. Fluids **45**(6), 605–623.

L. Euler (1755): Principes generaux du mouvement des fluides. Mem. Acad. Sci. Berlin **11**, 274–315

C. Farhat (1992): 'A Saddle-point Principle Domain Decomposition Method for the Solution of Solid Mechanics Problems'. In: *Fifth International Symposium on Domain Decomposition Methods for Partial Differential Equations*, Norfolk, VA (SIAM, Philadelphia) pp. 271–292

H.F. Fasel, U. Rist, U. Konzelman (1990): Numerical investigation of the three-dimensional development in boundary-layer transition. AIAA J. **28**(1), 29–37

W.J. Feiereisen, W.J. Reynolds, J.H. Ferziger (1981): 'Numerical Simulation of Compressible, Homogeneous Turbulent Shear Flow'. Rep. TF–13 (Dept. Mechanical Engineering, Stanford Univ.)

H. Feng (2003): 'A Spectral Element Method With hp Mesh Adaptation'. PhD thesis, George Washington University, Washington, DC

L. Filippini, A. Toselli (2002): 'hp Finite Element Approximations on Non-matching Grids for the Stokes Problem'. Technical report 2002–22. ETH, Eidgenössische Technische Hochschule Zürich, Seminar für Angewandte Mathematik

T.M. Fischer (1993): A spectral Galerkin approximation of the Orr-Sommerfeld eigenvalue problem in a semi-infinite domain. Numer. Math. **66**(1), 159–179

P.F. Fischer (1997): An overlapping Schwarz method for spectral element solution of the incompressible Navier–Stokes equations. J. Comput. Phys. **133**(1), 84–101

P.F. Fischer, G.W. Kruse, F. Loth (2002): Spectral element methods for transitional flows in complex geometries. J. Sci. Comput. **17**(1), 81–98

P.F. Fischer, F. Loth, S.E. Lee, S.-W. Lee, D.S. Smith, H.S. Bassiouny (2007): Simulation of high Reynolds number vascular flows. Comput. Meth. Appl. Mech. Engrg. **196**(31-32), 3049–3060

P.F. Fischer, J.W. Lottes (2005): 'Hybrid Schwarz-multigrid Methods for the Spectral Element Method: extensions to Navier–Stokes'. In: *Domain Decomposition Methods in Science and Engineering*, Lecture Notes in Computational Science and Engineering **40** (Springer-Verlag, Berlin) pp. 35–49

P.F. Fischer, N.I. Miller, H.M. Tufo (2000): 'An Overlapping Schwarz Method for Spectral Element Simulation of Three-dimensional Incompressible Flows'. In: *Parallel Solution of Partial Differential Equations*, IMA Vol. Math. Appl. **120** (Springer-Verlag, New York) pp. 159–180

P.F. Fischer, J.S. Mullen (2001): Filter-based stabilization of spectral element methods. C.R. Acad. Sci. Sér. I – Anal. Numér. **332**, 265–270

P.F. Fischer, E.M. Rønquist (1994): Spectral element methods for large scale parallel Navier–Stokes calculations. Comput. Meth. Appl. Mech. Engrg. **116**(1–4), 69–76

J. Flores, J. Barton, T. Holst, T. Pulliam (1985): 'Comparison of the Full Potential and Euler Formulations for Computing Transonic Airfoil Flows'. In: *Proceeding of the 9th International Conference on Numerical Methods in Fluid Dynamics*. Lecture Notes in Physics **218**(5), ed. by Soubbaramayer, J. Boujot (Springer-Verlag, Heidelberg) pp. 971–974

C. Foias, O. Manley, R. Rosa, R. Temam (2001): *Navier–Stokes Equations and Turbulence* (Cambridge Univ. Press, Cambridge)

M. Fortin, R. Peyret, R. Temam (1971): Résolution numérique des équations de Navier–Stokes pour un fluide incompressible. J. Mech. **10**, 357–390

L.P. Franca, S.L. Frey (1992): Stabilized finite element methods: II. The incompressible Navier–Stokes equations. Comput. Meth. Appl. Mech. Engrg. **84**(2), 209–233

M. Frigo, S.G. Johnson (2005): The design and implementation of FFTW3. Proc. IEEE **93**(2), 216–231

W. Froude (1872): 'Experiments on the Surface-friction Experienced by a Plane Moving Through Water', 42nd Rep. Br. Assoc. Adv. Sci., 118–124

D. Funaro, D. Gottlieb (1991): Convergence results for pseudospectral approximations of hyperbolic systems by a penalty-type boundary treatment. Math. Comput. **57**, 585–596

G.P. Galdi (1994a): *An Introduction to the Mathematical Theory of the Navier–Stokes Equations.* Vol. 1. Linearized Steady Problems (Springer-Verlag, New York)

G.P. Galdi (1994b): *An Introduction to the Mathematical Theory of the Navier–Stokes Equations.* Vol. 2. Nonlinear Steady Problems (Springer-Verlag, New York)

M. Gaster (1962): A note on the relation between temporally-increasing and spatially-increasing disturbances in hydrodynamic stability. J. Fluid Mech. **14**, 222–224

T.B. Gatski, M.Y. Hussaini, J.L. Lumley (eds.) (1996): *Simulation and Modeling of Turbulent Flows* (Oxford University Press, Oxford)

C.W. Gear (1971): *Numerical Initial Value Problems in Ordinary Differential Equations* (Prentice-Hall, Englewood Cliffs, NJ)

A. Gelb, J. Tanner (2006): Robust reprojection methods for the resolution of the Gibbs phenomenon. Appl. Comput. Harmon. Anal. **20**, 3–25

M. Germano (1992): Turbulence: the filtering approach. J. Fluid Mech. **238**, 325–336

M. Germano, U. Piomelli, P. Moin, W.H. Cabot (1991): A dynamic subgrid-scale eddy viscosity model. Phys. Fluids A **3**(7), 1760–1765

P. Gervasio, F. Saleri (1998): Stabilized spectral element approximation for the Navier–Stokes equations. Numer. Meth. Partial Diff. Equat. **14**, 115–141

P. Gervasio, F. Saleri, A. Veneziani (2006): Algebraic fractional-step schemes with spectral methods for the incompressible Navier–Stokes equations. J. Comput. Phys. **214**(1), 347–365

N. Gilbert, L. Kleiser (1990): 'Near-wall Phenomena in Transition to Turbulence'. In: *Near-Wall Turbulence: 1988 Zoran Zaric Memorial Conference.* ed. by S.J. Kline, N.H. Afgan (Hemisphere, Washington) pp. 7–27

M.B. Giles (1990): Nonreflecting boundary conditions for Euler equation calculations. AIAA J. **28**(12), 2050–2058

V. Girault, P.A. Raviart (1986): *Finite Element Approximation of the Navier–Stokes Equations: Theory and Algorithms* (Springer-Verlag, Heidelberg)

J. Glimm, H.L. Li, Y. Liu, N. Zhao (2001): Conservative front tracking and level set algorithms. Proc. Nat'l Acad. Sci. **98**(25), 14198–14201

R. Glowinski (1984): *Numerical Methods for Nonlinear Variational Problems* (Springer, New York)

G.H. Golub, C.F. Van Loan (1996): *Matrix Computations* (John Hopkins Univ. Press, Baltimore)

D.L. Goodstein (1975): *States of Matter* (Prentice Hall, Englewood Cliffs)

W.J. Gordon, C.A. Hall (1973a): Construction of curvilinear coordinate systems and their applications to mesh generation. Int. J. Numer. Meth. Engrg. **7**, 461–477

W.J. Gordon, C.A. Hall (1973b): Transfinite element methods: blending-function interpolation over arbitrary curved element domains. Numer. Math. **21**(2), 109–129

H. Görtler (1957): A new series for the calculation of steady laminar boundary layer flows. J. Math. Mech. **6**(1), 1–66

D. Gottlieb, M. Gunzburger, E. Turkel (1982): On numerical boundary treatment for hyperbolic systems. SIAM J. Numer. Anal. **19**(4), 671–697

D. Gottlieb, J.S. Hesthaven (2001): Spectral methods for hyperbolic problems. J. Comput. Appl. Math. **128**(1–2), 83–131

D. Gottlieb, L. Lustman, S.A. Orszag (1981): Spectral calculations of one-dimensional inviscid compressible flow. SIAM J. Sci. Stat. Comput. **2**(3), 296–310

D. Gottlieb, L. Lustman, E. Tadmor (1987a): Stability analysis of spectral methods for hyperbolic initial-boundary value problems. SIAM J. Numer. Anal. **24**(2), 241–256

D. Gottlieb, L. Lustman, E. Tadmor (1987b): Convergence of spectral methods for hyper-bolic initial-boundary value systems. SIAM J. Numer. Anal. **24**(3), 532–537

S. Gottlieb, C.W. Shu, E. Tadmor (2001): Strong stability preserving high order time discretization methods. SIAM Review **43**, 89–112

D. Gottlieb, E. Turkel (1985): 'Topics in Spectral Methods for Time Dependent Problems'. In: *Numerical Methods in Fluid Dynamics*, ed. by F. Brezzi (Springer-Verlag, Heidelberg) pp. 115–155

I. Grattan-Guinness (1990): *Convolutions in French Mathematics: 1800–1840* (Birkhäuser, Basel)

A. Greenbaum (1997): *Iterative Methods for Solving Linear Systems*, Frontiers in Applied Mathematics **17** (SIAM, Philadelphia)

P.M. Gresho (1991): Incompressible fluid dynamics: some fundamental formulation issues. Ann. Rev. Fluid Mech. **23**, 413–453

P.M. Gresho, R.L. Sani (1998): *Incompressible Flow and the Finite Element Method: Advection-Diffusion and Isothermal Laminar Flow* (Wiley, New York)

S.E. Guarini, R.D. Moser, K. Shariff, A. Wray (2000): Direct numerical simulation of a supersonic turbulent boundary layer at Mach 2.5. J. Fluid Mech. **414**, 1–33.

J.L. Guermond, P. Minev, J. Shen (2006): An overview of projection methods for viscous incompressible flows. Comput. Meth. Applied Mech. Engrg. **195**(44-47), 6011–6045

J.L. Guermond, J. Shen (2003): A new class of truly consistent splitting schemes for incompressible flows. J. Comput. Phys. **192**(1), 262–27

B. Guo (2005): 'Recent progress in a-posteriori error analysis for the p and hp Finite Element Methods'. In *Recent Advances in Adaptive Computation* Contemp. Math. **383**, pp. 47–67

B. Guo, I. Babuška (1986): The h, p, and hp versions of the finite element method in 1 dimension. Parts I, II, III. Numer. Math. **49**(6), 577–683

B. Gustafsson, H.-O. Kreiss, A. Sundström (1972): Stability theory of difference approximations for mixed initial boundary value problems. II. Math. Comput. **26**, 649–686

B. Gustafsson, A. Sundstrom (1978): Incompletely parabolic problems in fluid dynamics. SIAM J. Appl. Math. **35**(2), 343–357

M.M. Hafez, J.C. South, E.M. Murman (1979): Artificial compressibility methods for numerical solutions of transonic full potential equation. AIAA J. **17**(8), 838–444

W.W. Hager, (2000): Runge-Kutta methods in optimal control and transformed adjoint system. Numer. Math. **87**(2), 247–282

T. Hagstrom (1999): Radiation boundary conditions for the numerical simulation of waves. Acta Numerica **8**, 47–106

E. Hairer, S.P. Norsett, G. Wanner (1993) *Solving Ordinary Differential Equations I. Nonstiff Problems* (Springer-Verlag, Heidelberg)

E. Hairer, G. Wanner (1996) *Solving Ordinary Differential Equations II. Stiff and Differential–Algebraic Problems* (Springer-Verlag, Heidelberg)

P. Hall (1983): The linear development of Gortler vortices in growing boundary layers, J. Fluid Mech. **130**, 41–58

F.R. Hama, J. Nutant (1963): 'Detailed flowfield observations in the transition process in a thick boundary layer'. In: *Proc. 1963 Heat Transfer and Fluid Mechanics Institute* (Stanford Univ. Press, Palo Alto) pp. 77–93

F. Hamba (1999): Effects of pressure fluctuations on turbulence growth in compressible homogeneous shear flow, Phys. Fluids A **11**(6), 1623–1635

A. Hanifi, P.J. Schmid, D.S. Henningson (1996): Transient growth in compressible boundary layer flow. Phys. Fluids A **8**(3), 826–837

F.H. Harlow, J.E. Welsh (1965): Numerical calculation of time-dependent viscous incompressible flow. Phys. Fluids **8**(12), 2182–2189

R.S. Heeg, B.J. Geurts (1998): Spatial instabilities of the incompressible attachment-line flow using sparse matrix Jacobi-Davidson techniques. Appl. Sci. Res. **59**(4), 315–329

Th. Herbert (1977a): 'Die Neutrale Fläche der Ebenen Poiseuille-Strömung' Habilitation, Univ. Stuttgart, Stuttgart

Th. Herbert (1977b): 'Finite Amplitude Stability of Plane Parallel Flows'. In: *Laminar-Turbulent Transition*, AGARD–CP–224

Th. Herbert (1983a): Stability of plane Poiseuille flow: theory, and experiment. Fluid Dyn. Trans. **11**, 77–126

Th. Herbert (1983b): Secondary instability of plane channel flow to subharmonic three-dimensional disturbances. Phys. Fluids **26**(4), 871–874

Th. Herbert (1988): Secondary instability of boundary layers. Ann. Rev. Fluid Mech. **20**, 487–526

Th. Herbert (1990): 'A Code for Linear Stability Analysis'. In: *Instability and Transition II*, ed. by M.Y. Hussaini, R.G. Voigt (Springer-Verlag, New York) pp. 121–144

Th. Herbert (1991): 'Boundary-Layer Transition – Analysis and Prediction Revisited'. AIAA Paper No. 91–0737

Th. Herbert (1997): Parabolized stability equations. Ann. Rev. Fluid Mech. **29**, 245–283

J.S. Hesthaven (1997): A stable penalty method for the compressible Navier–Stokes equations: II. One-dimensional domain decomposition schemes. SIAM J. Sci. Comput. **18**(3), 658–685

J.S. Hesthaven (1999): A stable penalty method for the compressible Navier–Stokes equations: III. Multidimensional domain decomposition schemes. SIAM J. Sci. Comput. **20**(1), 62–93

J.S. Hesthaven (2000): Spectral penalty methods. Appl. Numer. Math. **33**(1–4), 23–41

J.S. Hesthaven, D. Gottlieb (1996): A stable penalty method for the compressible Navier–Stokes equations: I. Open boundary conditions. SIAM J. Sci. Comput. **17**(3), 579–612

J.S. Hesthaven, S. Gottlieb, D. Gottlieb (2006); *Spectral Methods for Time-Dependent Problems* (Cambridge University Press, Cambridge)

J.C. Heywood, R. Rannacher (1986): Finite element approximation of the nonstationary Navier–Stokes problem, part II: stability of solutions and error estimates uniform in time. SIAM J. Numer. Anal. **23**(4), 750–777

J.O. Hinze (1975): *Turbulence* (McGraw-Hill, New York)

D. Hoff (2002): Dynamics of singularity surfaces for compressible, viscous flows in two space dimensions. Comm. Pure Appl. Math. **55**, 1365–1407

D. Hoff (2006): Uniqueness of weak solutions of the Navier–Stokes equations of multidimensional, compressible flow. SIAM J. Math. Anal. **37**(6), 1742–1760

T.L. Holst, W.F. Ballhaus (1979): Fast, conservative schemes for the full potential equation applied to transonic flows. AIAA J. **17**(2), 145–152

P. Houston, Ch. Schwab, E. Süli (2000): Stabilized *hp*-finite element methods for first-order hyperbolic problems. SIAM J. Numer. Anal. **37**(5), 1618–1643

P. Houston, Ch. Schwab, E. Süli (2002): Discontinuous *hp*-finite element methods for advection-diffusion-reaction problems. SIAM J. Numer. Anal. **39**(6), 2133–2163

L. Howarth (1953): *Modern Developments in Fluid Dynamics. High Speed Flow*, Vol. 1(Clarendon Press, Oxford)

F.Q. Hu (2001): A stable, perfectly matched layer for linearized Euler equations in unsplit physical variables. J. Comput. Phys. **173**(2), 455–480

F.Q. Hu, M.Y. Hussaini, P. Rasetarinera (1999): An analysis of the discontinuous Galerkin method for wave propagation problems. J. Comput. Phys. **151**(2), 921–946

N. Hu, X. Guo, I. Katz (1998): Norman bounds for eigenvalues and condition numbers in the *p* version of the finite element method. Math. Comput. **67**, 1423–1450

S. Huberson, Y. Morchoisne (1983): Large eddy simulation by spectral methods or by multilevel particle methods. AIAA Paper No. 83–1880

P. Huerre, P.A. Monkewitz (1990): Local and global instabilities in spatially developing flows, Ann. Rev. Fluid Mech. **22**, 473–537

T.J.R. Hughes, G. Scovazzi, L.P. Franca (2004): 'Multiscale and Stabilized Methods'. In: *Encyclopedia of Computational Mechanics*,
  ed. by E. Stein, R. De Borst, T.J.R. Hughes (Wiley, Chichester)

T.J.R. Hughes, L. Mazzei, K.E. Jansen (2000): Large eddy simulation
  and the variational multiscale method. Comput. Visual Sci. **3**,
  47–59

T.J.R. Hughes, L. Mazzei, A.A. Oberai, A. Wray (2001): The multiscale formulation of large eddy simulation: decay of homogeneous
  isotropic turbulence. Phys. Fluids A **13**(2), 505–512

R.R. Huilgol (1975): On the concept of the Deborah number. Trans.
  Soc. Rheology **19**(2), 297–306

M.Y. Hussaini, D.A. Kopriva, M.D. Salas, T.A. Zang (1985a): Spectral methods for the Euler equations. Part 1: Fourier methods
  and shock-capturing. AIAA J. **23**(1), 64–70

M.Y. Hussaini, D.A. Kopriva, M.D. Salas, T.A. Zang (1985b): Spectral methods for the Euler equations: Part 2. Chebyshev methods
  and shock-fitting. AIAA J. **23**(2), 234–240

G.B. Jacobs, D.A. Kopriva, F. Mashayek (2005): Validation study
  of a multidomain spectral code for simulation of turbulent flows.
  AIAA J. **43**(6), 1256–1264

A. Jameson (1978): 'Transonic Flow Calculations'. In: *Numerical
  Methods in Fluid Dynamics*, ed. by H.J. Wirz, J.J. Smolderen
  (Hemisphere) pp. 1–87

A. Jameson, H. Schmidt, E. Turkel (1981): 'Numerical Solutions of
  the Euler Equations by Finite Volume Methods Using Runge-
  Kutta Time Stepping Schemes'. AIAA Paper No. 81–1259

R.D. Joslin, C.L. Streett, C.-L. Chang (1993): Spatial direct numerical simulation of boundary-layer transition mechanisms: validation of PSE theory. Theor. Comput. Fluid Dyn. **4**(6), 271–288

Y.S. Kachanov, V.Y. Levchenko (1984): The resonant interaction of
  disturbances at laminar-turbulent transition in a boundary layer.
  J. Fluid Mech. **138**, 209–47

Y. Kaneda, T. Ishihara (2006): High-resolution direct numerical simulation of turbulence. J. Turbulence **7**(N20), 1–17

H.S. Kang, S. Chester, C. Meneveau (2003): Decaying turbulence in
  an active-grid-generated flow and comparisons with large-eddy
  simulations. J. Fluid Mech. **480**, 129–160.

G.E. Karniadakis, M. Israeli, S.A. Orszag (1991): High-order splitting
  methods for the incompressible Navier–Stokes equations. J. Comput. Phys. **97**(2), 414–443

G.E. Karniadakis, S.J. Sherwin (1995): A triangular spectral element
  method; applications to the incompressible Navier–Stokes equations. Comput. Meth. Appl. Mech. Engrg. **123**(1–4), 189–229

G.E. Karniadakis, S.J. Sherwin (1999): *Spectral/hp element methods for CFD* (Oxford University Press, New York)

G.E. Karniadakis, S.J. Sherwin (2005): *Spectral/hp Element Methods for Computational Fluid Dynamics*, 2nd edition (Oxford University Press, New York)

D.E. Keyes (1995): 'Aerodynamic Applications of Newton-Krylov-Schwarz Solvers'. In: *Proceedings of the 14th International Conference on Numerical Methods in Fluid Dynamics*, ed. by S.M. Deshpande, Lecture Notes in Physics **453** (Springer-Verlag, Berlin) pp. 1–20

M.R. Khorrami, M.R. Malik, R.L. Ash (1989): Application of spectral collocation techniques to the stability of swirling flows. J. Comput. Phys. **81**(1), 206–229

S. Kida, S.A. Orzsag (1990): Energy and spectral dynamics in forced compressible turbulence. J. Sci. Comput. **5**(2), 85–125

J. Kim, P. Moin (1985): Application of a fractional-step method to incompressible Navier–Stokes equations. J. Comput. Phys. **59**(2), 308–323

J. Kim, P. Moin, R.D. Moser (1987): Turbulent statistics in fully developed turbulent channel flow at low Reynolds number. J. Fluid Mech. **177**, 133–166.

N.P. Kirchner (2000): Computational aspects of the spectral Galerkin FEM for the Orr-Sommerfeld equation. Int. J. Numer. Meth. Fluids **32**(1), 119–137

A. Klawonn (1998a): Block-triangular preconditioners for saddle point problems with a penalty term. SIAM J. Sci. Comput. **19**(1), 172–184

A. Klawonn (1998b): An optimal preconditioner for a class of saddle point problems with a penalty term. SIAM J. Sci. Comput. **19**(2), 540–552

A. Klawonn, L.F. Pavarino (1998): Overlapping Schwarz methods for mixed linear elasticity and Stokes problems. Comput. Meth. Appl. Mech. Engrg. **165**(1–4), 233–245

A. Klawonn, L.F. Pavarino (2000): A comparison of overlapping Schwarz methods and block preconditioners for saddle point problems. Numer. Lin. Alg. Appl. **7**, 1–25

L. Kleiser (1986): private communication

L. Kleiser, U. Schumann (1980): 'Treatment of Incompressibility and Boundary Conditions in 3-D Numerical Spectral Simulations of Plane Channel Flows'. In: *Proceedings of the 3rd GAMM Conference on Numerical Methods in Fluid Mechanics*, ed. by E.H. Hirschel (Vieweg, Braunschweig) pp. 165–173

L. Kleiser, U. Schumann (1984): 'Spectral Simulation of the Laminar-Turbulent Transition Process in Plane Poiseuille Flow'. In: *Spec-

tral Mathods for Partial Differential Equations, ed. by R.G. Voigt, D. Gottlieb, M.Y. Hussaini (SIAM-CBMS, Philadelphia) pp. 141–163

W. Koch, F.P. Bertolotti, A. Stoltez, S. Hein (2000): Nonlinear equilibrium solutions in a three-dimensional boundary layer and their secondary instability. J. Fluid Mech. **406**, 131–174

D.A. Kopriva (1991): Multidomain spectral solution of the Euler gas-dynamics equations. J. Comput. Phys. **96**(2), 428–450

D.A. Kopriva (1992): Spectral solution of inviscid supersonic flows over wedges and axisymmetric cones. Comput. & Fluids **21**(2), 247–266

D.A. Kopriva (1993): Spectral solution of the viscous blunt-body problem. AIAA J. **31**(7), 1235–1242

D.A. Kopriva (1996a): A conservative staggered-grid Chebyshev multidomain method for compressible flows. II. A semi-structured method. J. Comput. Phys. **128**(2), 475–488

D.A. Kopriva (1996b): Spectral solution of the viscous blunt-body problem 2: Multidomain approximation. AIAA J. **34**(3), 560–564

D.A. Kopriva (1998): 'Euler Computations on Unstructured Quadrilateral Grids by a Staggered-Grid Chebyshev Method'. AIAA Paper No. 98–0132

D.A. Kopriva (1999): Shock-fitted multidomain solution of supersonic flows. Comput. Meth. Appl. Mech. Engrg. **175**(3–4), 383–394

D.A. Kopriva (2006): Metric identities and the discontinuous spectral element method on curvilinear meshes. J. Sci. Comput. **26**(3), 301–327

D.A. Kopriva, J.H. Kolias (1996): A conservative staggered-grid Chebyshev multidomain method for compressible flows. J. Comput. Phys. **125**(1), 244–261

D.A. Kopriva, S.L. Woodruff, M.Y. Hussaini (2002): Computation of electromagnetic scattering with a non-conforming discontinuous spectral element method. Int. J. Numer. Meth. Engrg. **53**(1), 105–122

D.A. Kopriva, T.A. Zang, M.Y. Hussaini (1991): Spectral methods for the Euler equations: the blunt body problem revisited. AIAA J. **29**(9), 1458–1462

D.A. Kopriva, T.A. Zang, M.D. Salas, M.Y. Hussaini (1984): 'Pseudospectral Solution of Two-dimensional Gas-Dynamics Problems'. In: *Proceedings of the 5th GAMM Conference on Numerical Methods in Fluid Mechanics*, ed. by M. Pandolfi, R. Piva (Vieweg, Braunschweig) pp. 185–192

V.G. Korneev, J. Flaherty, T. Oden, J. Fish (2002): Additive Schwarz algorithms for solving *hp*-version finite element systems on triangular meshes. Appl. Numer. Math. **43**(3), 399–421

V.G. Korneev, S. Jensen (1997): Preconditioning of the p-version of the finite element method. Comput. Meth. Appl. Mech. Engrg. **150**(1–4), 215–238

V.G. Korneev, S. Jensen (1999): Domain decomposition preconditioning in the hierarchical p-version of the finite element method. Appl. Numer. Math. **29**(4), 479–518

L.I.G. Kovasznay (1948): Laminar flow behind two-dimensional grid. Proc. Cambridge Philos. Soc. **44**, 58–62

A.G. Kravchenko, P. Moin (1997): On the effect of numerical errors in large eddy simulations of turbulent flows. J. Comput. Phys. **131**(2), 310–322

H.-O. Kreiss, J. Lorenz (1989): *Initial-Boundary Value Problems and the Navier–Stokes Equations* (Academic Press, San Diego)

H.-O. Kreiss, J. Lorenz, M.J. Naughton (1991): Convergence of the solutions of the compressible to the solutions of the incompressible Navier–Stokes equations. Adv. Appl. Math. **12**(2), 1876–214

H.-O. Kreiss, J. Oliger (1972): Comparison of accurate methods for the integration of hyperbolic equations. Tellus **24**, 199–215

H.-O. Kreiss, J. Oliger (1979): Stability of the Fourier method. SIAM J. Numer. Anal. **16**(3), 421–433

G. Kruse (1997): 'Parallel Nonconforming Spectral Element Solution of the Incompressible Navier–Stokes Equations in Three Dimensions'. PhD thesis, Brown University, Providence, RI

D. Kucheman (1938): Storungsbewegungen in einer Gasstromung mit Grenzschicht. Z. Angew. Math. Mech. **18**, 79–84

N.N. Kuznetsov (1976): Accuracy of some approximate methods for computing the weak solutions of a first order quasi linear equation. Ž. Vyčisl. Mat. i Mat. Fiz. **16**, 1489–1502 (In Russian)

O.A. Ladyženskaya (1969): *The Mathematical Theory of Viscous Incompressible Flow* (Gordon & Breach, New York)

P.A. Lagerstram (1989): 'Laminar Flow Theory', In: *Laminar Boundary Layers*, ed. by F. Moore (Clarendon Press, Oxford) pp. 20–285

J.D. Lambert (1991): *Numerical Methods for Ordinary Differential Systems : The Initial Value Problem* (John Wiley & Sons, New York)

L.D. Landau, E.M. Lifschitz (1987): *Fluid Mechanics*, Course of Theoretical Physics, Vol. 6 (Butterworth-Heinemann, Oxford)

E. Laurien: 'Numerische Simulation zur Aktiven Beeinflussung des Laminar-Turbulenten Übergangs in der Plattengrenzschichtslrömung'. DFVLR–FB 86–05 (1986)

P.D. Lax (1973): *Hyperbolic Systems of Conservation Laws and the Mathematical Theory of Shock Waves*, (SIAM-CBMS, Philadelphia)

L. Lees, C.C. Lin (1946): 'Investigation of the Stability of the Laminar Boundary Layer in a Compressible Fluid.' NACA TN–1115

R.B. Lehoucq, D.C. Sorensen, C. Yang (1998): *ARPACK Users' Guide: Solution of Large-Scale Eigenvalue Problems with Implicitly Restarted Arnoldi Method* (SIAM, Philadelphia)

S.K. Lele (1992): Compact finite difference schemes with spectral-like resolution. J. Comput. Phys. **103**(1), 16–42

A. Leonard (1974): 'Energy Cascade in Large-eddy Simulations of Turbulent Fluid Flows'. In: *Proceedings of the Symposium on Turbulent Diffusion in Environmental Pollution*, Advances in Geophysics **18A**, ed. by F.N. Frenkiel, R. E. Munn (Academic Press, New York) pp. 237–248

A. Leonard, A. Wray (1982): 'New Numerical Method for the Simulation of Three-Dimensional Flow in a Pipe'. In: *Proceedings of the 8th International Conference on Numerical Methods in Fluid Dynamics*, ed. by E. Krause (Springer-Verlag, Heidelberg) pp. 335–342

J. Leray (1933): Étude de diverse équations intégrales nonlinéaires et de quelques problèmes que pose l'hydrodynamique. J. Math. Pure Appl. **12**, 1–82

M. Lesieur, O. Metais (1996): New trends in large-eddy simulations of turbulence. Ann. Rev. Fluid Mech. **28**, 45–82

R.J. LeVeque (2002): *Finite Volume Methods for Hyperbolic Problems* (Cambridge University Press, New York)

F. Li, M.R. Malik (1996): On the nature of PSE approximation. Theor. Comput. Fluid Dyn. **8**(4), 253–273

F. Li, M.R. Malik (1997): Spectral analysis of parabolized stability equations. Comput. & Fluids **26**(3), 279–297

H.W. Liepmann, A. Roshko (2001): *Elements of Gas Dynamics* (Dover Publications, New York)

J. Lighthill (1989): *An Informal Introduction to Theoretical Fluid Mechanics* (Oxford University Press, Oxford)

D.K. Lilly (1992): A proposed modification of the Germano subgrid-scale closure method. Phys. Fluids A **4**(3), 633–635

C.C. Lin (1945): On the stability of two-dimensional parallel flows, Part I, II, III. Quart. Appl. Math. **3**, 117–142, 218–234, 277–301

J.-L. Lions (1969): *Quelques Methodes de Résolution des Problèmes aux Limites non Linéaires* (Dunod, Paris)

P.-L. Lions (1990): 'On the Schwarz Alternating Method III: a Variant for Non-overlapping Subdomains'. In: *Third International Symposium on Domain Decomposition Methods for Partial Differential Equations*, ed. by T.F. Chan et al. (SIAM, Philadelphia) pp. 202–231

P.-L. Lions (1996): *Mathematical Topics in Fluid Mechanics*. Vol. 1. Incompressible Models (Oxford University Press, New York)

P.-L. Lions (1998): *Mathematical Topics in Fluid Mechanics*. Vol. 2. Compressible Models (Oxford University Press, New York)

L. Lorenz (1881): Über das Leitungsvermögen der Metalle für Wärme und Electricität. Ann. Phya. Chem. **13**, 582–606

J.W. Lottes, P.F. Fischer (2005): Hybrid multigrid/Schwarz algorithms for the spectral element method. J. Sci. Comput. **24**(1), 45–78

T.S. Lund, X. Wu, K.D. Squires (1998): Generation of turbulent inflow data for spatially-developing boundary layer simulations. J. Comput. Phys. **140**(2), 233–258

A. Lundbladh, D.S. Henningson, A.V. Johansson (1992): 'An Efficient Spectral Integration Method for the Solution of the Navier–Stokes Equations'. FFA–TN 1992–28, Aeronautical Research Institute of Sweden, Bromma

A. Lundbladh, A.V. Johansson (1991): Direct simulation of turbulent spots in plane Couette flow. J. Fluid Mech. **229**, 499–516

A. Lundbladh, P.J. Schmid, S. Berlin, D.S. Henningson (1994): 'Simulations of bypass transition for spatially evolving disturbances'. In: *Application of Direct and Large-Eddy Simulation of Transition and Turbulence*, AGARD–CP–551, pp. 18.1–18.13

Y. Ma, X. Zhong (2003): Receptivity of a supersonic boundary layer over a flat plate. Part 1. Wave structures and interactions. J. Fluid Mech. **488**, 31–78

M.G. Macaraeg, C.L. Streett (1991): Linear stability of high-speed mixing layers. Appl. Numer. Math. **7**(1), 93–128

M.G. Macaraeg, C.L. Streett, M.Y. Hussaini (1988): 'A Spectral Collocation Solution to the Compressible Stability Eigenvalue Problem'. NASA TP–2858

L.M. Mack (1969): 'Boundary-layer Stability Theory'. Jet Propulsion Laboratory Doc. 900–277–REV–A (also NASA CR–131501)

L.M. Mack (1984): 'Boundary-Layer Linear Stability Theory'. In *Special Course on Stability and Transition to Turbulence*, AGARD–R-709

Y. Maday (1989): Relèvememt de traces polynomiales et interpolations hilbertiennes entre espaces de polynômes. C.R. Acad. Sci. Paris **309**, Série I, 463–468

Y. Maday, B. Métivet (1987): Chebyshev spectral approximation of Navier–Stokes equations in a two-dimensional domain. RAIRO Modél. Math. Anal. Numér. **21**, 93–123

Y. Maday, R. Muñoz (1989): 'Numerical Analysis of a Multigrid Method for Spectral Approximations'. In: *Proceedings of the 11th International Conference on Numerical Methods in Fluid Dynam-*

ics, ed. by D.L. Dwoyer, M.Y. Hussaini, R.G. Voigt (Springer-Verlag, Heidelberg) pp. 389–394

Y. Maday, A.T. Patera (1989): 'Spectral Element Methods for the Navier–Stokes Rquations'. In: *State-of-the-art Surveys in Computational Mechanics*, ed. by A.K. Noor, T.J. Oden (ASME, New York) pp. 71–143

Y. Maday, A.T. Patera, E.M. Rønquist (1987): 'A Well-posed Optimal Spectral Element Approximation for the Stokes Problem'. ICASE Report 87–48, Hampton, VA

Y. Maday, A.T. Patera, E.M. Rønquist (1990): An operator-integration-factor splitting method for time-dependent problems: Application to incompressible fluid flow. J. Sci. Comput. **5**(4), 263–292

Y. Maday, A. Quarteroni (1982): Spectral and pseudospectral approximations of Navier–Stokes equations. SIAM J. Numer. Anal. **19**(4), 761–780

Y. Maday, F. Rapetti, B. Wohlmuth (2002): 'Mortar element coupling between global scalar and local vector potentials to solve eddy current problems'. In *Numerical Mathematics and Advanced Applications*, ed. by F. Brezzi et al. (Springer-Verlag, Milano) pp. 847–865

T. Maeder, N.A. Adams, L. Kleiser (2001): Direct simulation of turbulent supersonic boundary layers by an extended temporal approach. J. Fluid Mech. **429**, 187–216

J.-F. Maitre, O. Pourquier (1996): Condition number and diagonal preconditioning: comparison of the p-version and the spectral element methods. Numer. Math. **74**(1), 69–84

A.J. Majda, A.L. Bertozzi (2002): *Vorticity and Incompressible Flow* (Cambridge University Press, New York)

A. Majda, J. McDonough, S. Osher (1978): The Fourier method for nonsmooth initial data. Math. Comput. **32**, 1041–1081

M.R. Malik (1982): 'COSAL-A Black Box Compressible Stability Analysis Code for Transition Prediction in Three-dimensional Boundary Layers'. NASA CR–165925

M.R. Malik (1989): Prediction and Control of Transition in Supersonic and Hypersonic Boundary Layers. AIAA J. **27**(11), 1487–1493

M.R. Malik (1990): Numerical methods for hypersonic boundary layer stability. J. Comput. Phys. **86**(2), 376–413

M.R. Malik (2003): Hypersonic flight transition data analysis using parabolized stability equations with chemistry effects. J. Spacecraft Rockets **40**(3), 332–344

M.R. Malik and S.A. Orszag (1987): Linear stability analysis of three-dimensional compressible boundary layers. J. Sci. Comput. **2**(1), 77–97

M.R. Malik, T.A. Zang, M.Y. Hussaini (1985): A spectral collocation method for the Navier–Stokes equations. J. Comput. Phys. **61**(1), 64–88

J. Mandel (1993): Balancing domain decomposition. Comm. Appl. Numer. Meth. **9**, 233–241

G.I. Marchuk (1975): *Methods of Numerical Mathematics* (Springer-Verlag, Heidelberg)

P.S. Marcus (1984a): Simulation of Taylor-Couette flow. Part 1. Numerical methods and comparison with experiment. J. Fluid Mech. **146**, 45–64

P.S. Marcus (1984b): Simulation of Taylor-Couette flow. Part 2. Numerical results for wavy-vortex one with one traveling wave. J. Fluid Mech. **146**, 65–113

P.S. Marcus, L.S. Tuckerman (1987a): Simulation of flow between concentric rotating spheres. Part 1. Steady states. J. Fluid Mech. **185**, 1–30

P.S. Marcus, L.S. Tuckerman (1987b): Simulation of flow between concentric rotating spheres. Part 2. Transitions. J. Fluid Mech. **185**, 31–66

A. Matsumura, T. Nishida (1979): The initial value problem for the equations of motion of compressible viscous and heat-conductive fluids. Proc. Japan Acad. Ser. A Math. Sci. **55**, 337–342

C. Mavriplis (1989): 'Nonconforming Discretization and a posteriori Error Estimations for Adaptive Spectral Element Techniques'. PhD thesis, Massachusetts Institute of Technology, Cambridge, MA

J.C. Maxwell (1872): *Theory of Heat* (D. Appleton, New York)

J.B. McLaughlin, S.A. Orszag (1982): Transition from periodic to chaotic thermal convection. J. Fluid Mech. **122**, 123–142

J.M. Melenk (2002): On condition number in hp-FEM with Gauss-Lobatto-based shape functions. J. Comput. Appl. Math. **139**, 21–48

J.M. Melenk (2003): *hp-Finite Element Methods for Singular Perturbations.* (Springer, Heidelberg)

J.M. Melenk, N.P. Kirchner, Ch. Schwab (2000): Spectral Galerkin discretization for hydrodynamic stability problems. Computing **65**, 97–118.

J.M. Melenk, B.I. Wohlmuth (2001): On residual-based a posteriori error estimation in hp-FEM. A posteriori error estimation and adaptive computational methods. Adv. Comput. Math. **15**, 311–331

A. Meseguer, L.N. Trefethen (2003): Linearized pipe flow to Reynolds number 10^7. J. Comput. Phys. **186**(1), 178–197

R.W. Metcalfe, S.A. Orszag, M.E. Brachet, S. Menon, J. Riley (1987): Secondary instability of a temporally growing mixing layer. J. Fluid Mech. **184**, 207–243

B. Métivet (1987): 'Résolution Spectrale des Équations de Navier–Stokes par une Méthodes de Sous-Domaines Courbes'. Thèse de Doctorat, Univ. Paris 6

B. Mohammadi, O. Pironneau (2001): *Applied Shape Optimization for Fluids* (Oxford University Press, New York)

A.H. Mohammadian, V. Shankar, W.F. Hall (1991): Computation of electromagnetic scattering and radiation using a time-domain finite-volume discretization procedure. Computer Phys. Comm. **68**, 175–196

P. Moin, J. Kim (1980): On the numerical solution of time-dependent viscous incompressible fluid flows involving solid boundaries. J. Comput. Phys. **35**(3), 381–392

F.K. Moore (ed.) (1989): *Laminar Boundary Layers* (Clarendon Press, Oxford)

F. Montigny-Rannou, Y. Morchoisne (1987): A spectral method with staggered grid for incompressible Navier–Stokes equations. Int. J. Numer. Meth. Fluids **7**, 175–189

Y. Morchoisne (1983): 'Résolution des Équations de Navier–Stokes par une Méthode Spectrale de Sous-Domaines'. In: *Proceedings of the 3rd International Conference on Numerical Methods in Science and Engineering* (Gamni, Paris)

G. Moretti (1987): Computation of flows with shocks. Ann. Rev. Fluid Mech. **19**, 313–337

G. Moretti (2002): Thirty-six years of shock fitting. Comput. & Fluids **31**(4–7), 719–723

G. Moretti, M.D. Salas (1969): 'The Blunt Body Problem for a Viscous Rarefied Gas Flow'. AIAA Paper No. 69–139

R.D. Moser, P. Moin (1987): The effects of curvature in wall-bounded turbulent flows. J. Fluid Mech. **175**, 479–510

R.D. Moser, P. Moin, A. Leonard (1983): A spectral numerical method for the Navier–Stokes equations with applications to Taylor-Couette flow. J. Comput. Phys. **52**(3), 524–544

R.D. Moser, M.M. Rogers, D.W. Ewing (1998): Self-similarity of time-evolving plane wakes. J. Fluid Mech. **367**, 255–289

R. Muñoz-Sola (1997): Polynomial liftings on a tetrahedron and applications to the h-p version of the finite element method in three dimensions. SIAM J. Numer. Anal. **34**(1), 282–314

J.W. Murdock (1986): 'Three-Dimensional Numerical Study of Boundary-Layer Stability'. AIAA Paper No. 86–0434

E.M. Murman, J.D. Cole (1971): Calculation of plane steady transonic flows. AIAA J. **9**(1), 114–121

C.L.M. Navier (1827): Mèmoire sur le lois du mouvement des fluides. Mèm. Acad. Roy. Sci. Paris **6**, 389–440

N. Nayar, J.M. Ortega (1993): Computation of selected eigenvalues of generalized eigenvalue problems. J Comput Phys. **108**, 8–14

A.H. Nayfeh (1980): Stability of three-dimensional boundary layers. AIAA J. **18**(4), 406–416

L.L. Ng, G. Erlebacher (1992): Secondary instabilities in compressible boundary layers. Phys. Fluids A **4**(4), 710–726

L.L. Ng, T.A. Zang (1993): Secondary instability mechanisms in compressible axisymmetric boundary layers. AIAA J. **31**(9), 1605–1610

R.A. Nicolaides (1982): Existence, uniqueness and approximation for generalized saddle point problems. SIAM J. Numer. Anal. **19**(2), 349–357

M. Nishioka, M. Asai, S. Iida (1980): 'An Experimental Investigation of the Secondary Instability in Laminar-Turbulent Transition'. In: *Laminar-Turbulent Transition*, ed. by R. Eppler, H. Fasel (Springer-Verlag, Heidelberg) pp. 37–46

J. Nordström, N. Nordin, D. Henninsgon (1999): The fringe region technique and the Fourier method used in the direct numerical simulation of spatially evolving viscous flows. SIAM J. Sci. Comput. **20**(4), 1365–1393

J.T. Oden, I. Babuška, C.E. Baumann (1998): A discontinuous *hp* finite element method for diffusion problems. J. Comput. Phys. **146**(2), 491–519

J. Oliger, A. Sundström (1978): Theoretical and practical aspects of some initial boundary value problems in fluid dynamics. SIAM J. Appl. Math. **35**(3), 419–446

S.A. Orszag (1969): Numerical methods for the simulation of turbulence. Phys. Fluids Suppl. II. **12**, 250–257

S.A. Orszag (1971a): Numerical simulation of incompressible flows within simple boundaries: I. Galerkin (spectral) representations. Stud. Appl. Math. **50**, 293–327

S.A. Orszag (1971b): Accurate solution of the Orr-Sommerfeld stability equation. J. Fluid Mech. **50**, 689–703

S.A. Orszag (1974): 'The Numerical Simulation of the Taylor-Green Vortex'. In: *Computing Methods in Applied Sciences and Engineering*, Lecture Notes in Computer Science **11**, ed. by G. Goos, J. Hartmanis (Springer-Verlag, Berlin) pp. 51–64

S.A. Orszag (1980): Spectral methods for problems in complex geometries. J. Comput. Phys. **37**(1), 70–92

S.A. Orszag, M. Israeli, M.O. Deville (1986): Boundary conditions for incompressible flows. J. Sci. Comput. **1**(1), 75–111

S.A. Orszag, L.C. Kells (1980): Transition to turbulence in plane Poiseuille flow and plane Couette flow. J. Fluid Mech. **96**, 159–205

S.A. Orszag, A.T. Patera (1980): Subcritical transition to turbulence in plane channel flows. Phys. Rev. Lett. **45**(12), 989–993

S.A. Orszag, A.T. Patera (1983): Secondary instability of wall-bounded shear flows. J. Fluid Mech. **128**, 347–385

S.A. Orszag, G.S. Patterson, Jr. (1972): Numerical simulation of three dimensional homogeneous isotropic turbulence. Phys. Rev. Lett. **28**(2), 76–79

S. Osher (1969): Systems of difference equations with general homogeneous boundary conditions. Trans. Am. Math. Soc. **137**, 177–201

S. Osher (1984): Riemann solvers, the entropy condtion, and difference approximations. SIAM J. Numer. Anal. **31**(1), 217–235

S. Osher, J.A. Sethian (1988): Fronts propagating with curvature-dependent speed: algorithms based on Hamilton-Jacobi formulations. J. Comput. Phys. **79**(1), 12–49

F. Pasquarelli, A. Quarteroni, G. Sacchi-Landriani (1987): Spectral approximations of the Stokes problem by divergence-free functions. J. Sci. Comput. **2**(3), 195–226

R. Pasquetti, L.F. Pavarino, F. Rapetti, E. Zampieri (2007a): 'Overlapping Schwarz Preconditioners for Fekete Spectral Elements'. In *Domain Decomposition Methods in Science and Engineering XVI*, Lecture Notes in Computational Science and Engineering **55**, ed. by O.B. Widlund, D.E. Keyes (Springer-Verlag, Heidelberg) pp. 715–722

R. Pasquetti, F. Rapetti, L.F. Pavarino, E. Zampieri (2007b): Overlapping Schwarz methods for Fekete and Gauss-Lobatto spectral elements. SIAM J. Sci. Comput. To appear

T. Passot, A. Pouquet (1987): Numerical simulation of compressible homogeneous flows in the turbulent regime. J. Fluid Mech. **181**, 441–466

T. Passot, A. Pouquet (1988): Hyperviscosity for compressible flows using spectral methods. J. Comput. Phys. **75**(2), 300–313

S.V. Patankar (1980): *Numerical Heat Transfer and Fluid Flow* (McGraw-Hill, New York)

A.T. Patera (1984): A spectral element method for fluid dynamics: laminar flow in a channel expansion. J. Comput. Phys. **54**(3), 468–488

G.S. Patterson, Jr., S.A. Orszag (1971): Spectral calculations of isotropic turbulence: Efficient removal of aliasing interaction. Phys. Fluids **14**(11), 2538–2541

L.F. Pavarino (1994a): Additive Schwarz methods for the *p*-version finite element method. Num. Math. **66**(4), 493–515

L.F. Pavarino (1994b): Schwarz methods with local refinement for the *p*-version finite element method. Num. Math. **69**(2), 185–211

L.F. Pavarino (1997): Neumann-Neumann algorithms for spectral elements in three dimensions. RAIRO Math. Model. Numer. Anal. **31**, 471–493

L.F. Pavarino (2007): BDDC and FETI-DP preconditioners for spectral element discretizations. Comput. Meth. Appl. Mech. Engrg. **196**(8), 1380–1388

L.F. Pavarino, T. Warburton (2000): Overlapping Schwarz methods for unstructured spectral elements. J. Comput. Phys. **160**(1), 298–317

L.F. Pavarino, O.B. Widlund (1996): A polylogarithmic bound for an iterative substructuring method for spectral elements in three dimensions. SIAM J. Numer. Anal. **33**(4), 1303–1335

L.F. Pavarino, O.B. Widlund (1999): Iterative substructuring methods for spectral element discretization of elliptic systems. II: mixed methods for linear elasticity and Stokes flow. SIAM J. Numer. Anal. **37**(2), 353–374

L.F. Pavarino, O.B. Widlund (2002): Balancing Neumann-Neumann methods for incompressible Stokes equations. Comm. Pure Appl. Math. **55**(3), 302–335

J.B. Perot (1993): An analysis of the fractional step method. J. Comput. Phys. **108**(1), 51–58

R. Peyret (2002): *Spectral Methods for Incompressible Viscous Flow* (Springer-Verlag, Heidelberg)

T.N. Phillips, G.W. Roberts (1993): The treatment of spurious pressure modes in spectral incompressible flow calculations. J. Comput. Phys. **105**(1), 150–164

U. Piomelli (2004): 'Large-eddy and Direct Simulation of Turbulent Flows'. In: *Introduction to Turbulence Modeling*, von Karman Institute for Fluid Dynamics Lecture Series 2004–06

U. Piomelli, T.A. Zang, C.G. Speziale, M. Y. Hussaini (1990): On the large eddy simulation of transitional wall-bounded flows. Phys. Fluids A **2**(2), 257–265

T.J. Poinsot, S.K. Lele (1992): Boundary conditions for direct simulations of compressible viscous flows. J. Comput. Phys. **101**(1), 104–129

S.D. Poisson (1831): Mémoire sur les équations generales de l'équilibre et du mouvement des corps solides élastiques et des fluides. J. Ecole Poly. **13**, 1–174

S.B. Pope (2000): *Turbulent Flows* (Cambridge University Press, New York)

J.O. Pralits, A. Hanifi, D.S. Henningson (2002): Adjoint-based optimization of steady suction for disturbance control in incompressible flows. J. Fluid Mech. **467**, 129–161

L. Prandtl (1904): 'Über Flüssigkeitsbewegung bei Sehr Kleiner Reibung'. In: *Proceedings of the 3rd International Mathematics Congress, Heidelberg* pp. 484–491 (available in English in NACA TM–452 (1928))

L. Prandtl (1921): Bemekungen uber die Enstehungder Turbulenz. Z. Angew. Math. Mech. **1**, 431–436

V.G. Priymak, T. Miyazaki (1998): Accurate Navier–Stokes investigation of transitional and turbulent flows in a circular pipe. J. Comput. Phys. **142**(2), 340–411

A. Prohl (1997): *Projection and Quasi-Compressibility Methods for Solving the Incompressible Navier–Stokes Equations*, Advances in Numerical Mathematics (Teubner, Stuttgart)

C.D. Pruett (1993): On the accurate prediction of the wall-normal velocity in compressible boundary-layer flow. Int. J. Numer. Meth. Fluids **16**, 133–152

C.D. Pruett (1994): 'A spectrally accurate boundary-layer code for infinite swept wings'. NASA CR–195014

C.D. Pruett (1996): A semi-implicit method for internal boundary layers in compressible flows. Comput. Meth. Appl. Mech. Engrg. **137**(3–4), pp. 379–394

C.D. Pruett, L. Ng, G. Erlebacher (1991): On the nonlinear stability of a high-speed, axisymmetric boundary layer. Phys. Fluids A **3**(12), 2910–2926

C.D. Pruett, C.L. Streett (1991): A spectral collocation method for compressible, nonsimilar boundary layers. Int. J. Num. Meth. Fluids **13**, 713–737

C.D. Pruett, T.A. Zang (1992): Direct numerical simulation of laminar breakdown in high-speed, axisymmetric boundary layers. Theor. Comput. Fluid Dyn. **3**(6), 345–367

C.D. Pruett, T.A. Zang, C.-L. Chang, M.H. Carpenter (1995): Spatial direct numerical simulation of high-speed boundary-layer flows. Part I: Algorithmic considerations and validation. Theor. Comput. Fluid Dyn. **7**(1), 49–76

J.S. Przemieniecki (1985): *Theory of Matrix Structural Analysis* (Dover, New York)

L. Quartapelle (1993): *Numerical Solution of the Incompressible Navier–Stokes Equations* (Birkhäuser, Basel)

A. Quarteroni, F. Saleri, A. Veneziani (1999): Analysis of the Yosida method for the incompressible Navier–Stokes equations. J. Math. Pures Appl. **78**(5), 473–503

A. Quarteroni, F. Saleri, A. Veneziani (2000): Factorization methods for the numerical approximation of Navier–Stokes equations. Comput. Meth. Appl. Mech. Engrg. **188**(1–3), 505–526

A. Quarteroni, A. Valli (1994): *Numerical Approximations of Partial Differential Equations* (Springer-Verlag, Heidelberg)

A. Quarteroni, A. Valli (1999): *Domain Decomposition Methods for Partial Differential Equations* (Oxford Science Publications, Oxford)

P. Rabenold (2006): 'Parallel Adaptive Mesh Refinement for the Incompressible Navier–Stokes Equations'. AMSC663/664, Univ. of Maryland, College Park, MD

W.J.M. Rankine (1864): Trans. Inst. Nay. Archit. **5**, 316–33

R. Rannacher (1992): 'On Chorin's Projection Method for the Incompressible Navier–Stokes Equations'. In: *The Navier–Stokes equations II–theory and numerical methods*, Lecture Notes in Mathematics **1530** (Springer-Verlag, Berlin) pp. 167–183

P. Rasetarinera, M.Y. Hussaini (2001): An efficient implicit discontinuous spectral Galerkin method. J. Comput. Phys. **172**(2), 718–738

P. Rasetarinera, M.Y. Hussaini, F.Q. Hu (2000): 'Some Remarks on the Accuracy of a Discontinuous Galerkin Method'. In: *Discontinuous Galerkin methods*, Lecture Notes Computational Science and Engineering **11** (Springer-Verlag, Heidelberg) pp. 407–412

L. Rayleigh (1880): 'On the Stability and Instability of Certain Fluid Motions'. In: *Scientific Papers* **1** (Cambridge University Press, Cambridge) pp. 474–487

S.C. Reddy, D.S. Henningson (1993): Energy growth in viscous channel flows. J. Fluid Mech. **252**, 209–238

S.C. Reddy, P.J. Schmid, J.S. Baggett, D.S. Henningson (1998): On stability of streamwise streaks and transition thresholds in plane channel flows. J. Fluid Mech. **365**, 269–303

D. Rempfer (2003): Low-dimensional modeling and numerical simulation of transition in simple shear flows. Ann. Rev. Fluid Mech. **35**, 229–265

L.G.M. Reyna (1982): Stability of Chebyshev Collocation. PhD thesis, Caltech, Pasadena, CA

J.J. Riley, R.W. Metcalfe (1980): 'Direct Numerical Simulation of a Perturbed, Turbulent Mixing Layer'. AIAA Paper No. 80–0274

F. Ringleb (1940): Exakte Lösungen der Differentialgleichungen einer adiabatischen Gasströmung. Z. Angew. Math. Mech. **20**, 185–198

B. Rivière, M.F. Wheeler, V. Girault (2001): A priori error estimates for finite element methods based on discontinuous approximation spaces for elliptic problems. SIAM J. Numer. Anal. **39**(3), 902–931

R.S. Rogallo (1977): 'An ILLIAC Program for the Numerical Simulation of Homogeneous, Incompressible Turbulence'. NASA TM–73203

R.S. Rogallo (1981): 'Numerical Experiments in Homogeneous Turbulence'. NASA TM–81315

M.M. Rogers, R.D. Moser (1992): The three-dimensional evolution of a plane mixing layer: the Kelvin-Helmholtz rollup. J. Fluid Mech. **243**, 183–226

E.M. Rønquist (1996): 'A Domain Decomposition Solver for the Steady Navier–Stokes Equations'. In: *Proceedings of the 3rd International Conference on Spectral and High-Order Methods*, ed. by A. Ilin and R. Scott (HJM, Houston), pp. 469–485

E.M. Rønquist (1999): 'Domain Decomposition Methods for the Steady Stokes Equations'. In: *Proceedings of the 11th International Conference on Domain Decomposition Methods*, ed. by C.-H. Lai, P.E. Bjørstad, M. Cross and O.B. Widlund (SIAM, Philadelphia), pp. 330– 340

E.M. Rønquist, A.T. Patera (1987a): A Legendre spectral element method for the Stefan problem. Int. J. Numer. Method Engrg. **24**, 2273–2299

E.M. Rønquist, A.T. Patera (1987b): Spectral element multigrid. I. Formulation and numerical results. J. Sci. Comput. **2**(4), 389–406

H. Rouse, S. Ince (1963): *History of Hydraulics* (Dover, New York)

Y. Saad (1980): Variations of Arnoldi's method for computing eigen elements of large unsymmetric matrices. Lin Algebra Appl. **34**, 269–295

Y. Saad (1996): *Iterative Methods for Sparse Linear Systems* (PWS Publishing Company, Boston)

Y. Saad, M. H. Schultz (1986): GMRES: A generalized minimal residual algorithm for solving nonsymmetric linear systems. SIAM J. Sci. Statist. Comput. **7**(3), 856–869

G. Sacchi-Landriani (1986): Convergence of the Kleiser-Schumann Method for the Navier–Stokes equations. Calcolo **23**, 383–406

G. Sacchi-Landriani, H. Vandeven (1987): Approximation polynômiale de functions à divergence nulle. C. R. Acad. Sci. Paris **304**, Serie I, 87–90

P. Sagaut (2006): *Large Eddy Simulation for Incompressible Flows. An Introduction* (Springer-Verlag, Berlin)

B. de Saint-Venant (1843): Note à joindre au mèmoire sur la dynamique des fluides. Mèm. Acad. Sci. Paris. **17**, 1240–1243

B. de Saint-Venant (1855): Mémoire sur la torsion des prismes. Mémoires de l'Academie des Sciences des Savants Etrangers. **14**, 233–560.

L. Sakell (1984): Pseudospectral solutions of one- and two-dimensional inviscid flows with shock waves. AIAA J. **22**(7), 929–934

M.D. Salas, C.R. Gumbert (1985): 'Breakdown of the Conservative Potential Equation'. AIAA Paper No. 85–0367

M.D. Salas, T.A. Zang, M.Y. Hussaini (1982): 'Shock-Fitted Euler Solutions to Shock-Vortex Interactions'. In: *Proceedings of the 8th International Conference on Numerical Methods in Fluid Dynamics*, ed. by E. Krause (Springer-Verlag, Heidelberg) pp. 461–467

F. Saleri, A. Veneziani (2005): Pressure-correction algebraic splitting methods for the incompressible Navier–Stokes equations. SIAM J. Numer. Anal. **43**(1), 174–194

N.D. Sandham, W.C. Reynolds (1991): Three-dimensional simulations of large eddies in the compressible mixing layer. J. Fluid Mech. **224**, 133–158

S. Sarkar, G. Erlebacher, M.Y. Hussaini (1991): Direct dimulation of compressible turbulence in a shear flow. Theor. Comput. Fluid Dyn. **2**(5–6), 291–305

S. Sarkar, G. Erlebacher, H.O. Kreiss, M.Y. Hussaini (1991):The analysis and modelling of dilatational terms in compressible turbulence. J. Fluid. Mech. **227**, 473–493

M. Schäfer, S. Turek (1996): 'Benchmark Computations of Laminar Flow Around a Cylinder'. Preprint SFB 359, Nummer 96-03, Universität Heidelberg

H. Schlichting (1933): Zur Entstehung der Turbulenz bei er Plattenströmung. Nachr. Ges. Wiss. Göttingen. Math. Phys. Klasse 182–208 (see also Z. Angew. Math. Mech. **13**, 171-174 (1933))

H. Schlichting, K. Gersten (1999): *Boundary-Layer Theory* (Springer-Verlag, New York)

P. Schmid (2007): Nonmodal stability theory. Ann. Rev. Fluid Mech. **39**, 139–162

P.J. Schmid and D.S. Henningson (2001): *Stability and Transition in Shear Flows* (Springer-Verlag, Heidelberg)

J. Schöberl, J.M. Melenk, C. Pechstein, S. Zaglmayr (2007): Additive Schwarz preconditioning for p-version triangular and tetrahedral finite elements. IMA J. Numer. Anal, to appear

D. Schötzau, Ch. Schwab, A. Toselli (2002): Mixed hp-DGFEM for incompressible flows. SIAM J. Numer. Anal. **40**(6), 2171–2194

D. Schötzau, Ch. Schwab, A. Toselli (2003): Stabilized hp-DGFEM for incompressible flow. Math. Models Meth. Appl. Sci. **13**, 1413–1436

G.B. Schubauer, H.K. Skramsdat (1947): Laminar boundary layer oscillations and transitions on a flat plate. J. Aero. Sci. **14**, 69–76

U. Schumann (1985): 'Algorithms for Direct Numerical Solution of Shear-Periodic Turbulence'. In: *Proceedings of the 9th Interna-*

tional Conference on Numerical Methods in Fluid Dynamics, ed. by Soubbarameyer, J. Boujot (Springer-Verlag, Heidelberg)

Ch. Schwab (1998): *p- and hp-Finite Element Methods* (Oxford Univ. Press, Oxford)

Ch. Schwab, M. Suri (1999): Mixed *hp* finite element methods for Stokes and non-Newtonian flow. Comput. Meth. Appl. Mech. Engrg. **175**(3–4), 217–241

H.A. Schwarz (1869): Über einige Abbildungsdufgaben. J. Reine Angew. Math. **70**, 105–120

A. Scotti, U. Piomelli (2001): Numerical simulation of pulsating turbulent channel flow. Phys. Fluids A **13**(5), 1367–1384

J. Serrin (1959a): 'Mathematical principles of classical fluid mechanics'. In *Handbuch der Physik, Band VIII/1*, ed. by S. Flügge, C. Truesdell (Springer-Verlag, Berlin) pp. 125–263

J. Serrin (1959b): On the uniqueness of compressible fluid motions. Arch. Rational Mech. Anal. **3**(1), 271–278

L.F. Shampine (1994): *Numerical Solution of Ordinary Differential Equations* (Chapman & Hall, New York)

S.J. Sherwin, V. Franke, J. Peiró, K. Parker (2003): One-dimensional modelling of a vascular network in space-time variables. *Mathematical modelling of the cardiovascular system*. J. Engrg. Math. **47**(3–4), 217–250.

S.J. Sherwin, J. Peiró (2003): 'Algorithms and arteries: multi-domain spectral/*hp* methods for vascular flow modelling'. In *Domain Decomposition Methods in Science and Engineering*, Natl. Auton. Univ. Mex., México, 159–170

C.-W. Shu, W.S. Don, D. Gottlieb, O. Schilling, L. Jameson (2005): Numerical convergence study of nearly-incompressible, inviscid Taylor-Green vortex flow. J. Sci. Comput. **24**(1), 1–27

C.-W. Shu, S. Osher (1988): Efficient implementation of essentially non-oscillatory shock-capturing schemes. J. Comput. Phys. **77**(2), 439–471

C.-W. Shu, S. Osher (1989): Efficient implementation of essentially non-oscillatory shock-capturing schemes. II. J. Comput. Phys. **83**(1), 32–78

C.-W. Shu, T.A. Zang, G. Erlebacher, D. Whitaker, S. Osher (1992): High-order ENO schemes applied to two- and three-dimensional compressible flow. Appl. Numer. Math. **9**(1), 45–71

V. Simoncini, M. Benzi (2004/05): Spectral properties of the Hermitian and skew-Hermitian splitting preconditioner for saddle point problems. SIAM J. Matrix Anal. Appl. **26**(2), 377–389

A. Simone, G.N. Coleman, C. Cambon (1997): The effect of compressibility on turbulent shear flow: a rapid-distortion-theory and direct-numerical-simulation study. J. Fluid Mech. **330**, 307–338

G. Sleijpen, H. van der Vorst (1996): A Jacobi-Davidson iteration method for linear eigenvalue problems. SIAM J. Matrix Anal. Appl. **17**(2), 401–425

J. Smagorinsky (1963): General circulation experiments with the primitive equations. Mon. Weather Rev. **91**, 99–164

B.F. Smith, P.E. Bjørstad, W.D. Gropp (1996): *Domain Decomposition. Parallel Multilevel Methods for Elliptic Partial Differential Equations* (Cambridge University Press, Cambridge)

A. M. O. Smith, N. Gamberoni (1956): 'Transition, Pressure Gradient and Stability Theory'. Report No. ES 26388, Douglas Aircraft Co., Inc., El Segundo, CA

F.T. Smith (1979): On the nonparallel flow stability of the Blasius boundary layer. Proc. Roy. Soc. Lond. A **366**(1724), 91–109

P.R. Spalart (1984): A spectral method for external viscous flows. Contemp. Math. **28**, 315–335

P.R. Spalart (1986a): 'Numerical Simulation of Boundary Layers, Part 1: Weak Formulation and Numerical Method'. NASA TM–88222

P.R. Spalart (1986b): Numerical study of sink-flow boundary layers. J. Fluid Mech. **172**, 307–328

P.R. Spalart (1988a): 'Direct Numerical Study of Leading Edge Contamination'. In *Fluid Dynamics of Three-Dimensional Turbulent Shear Flows and Transition*. AGARD–CP–438, pp. 5.1–5.13

P.R. Spalart (1988b): Direct numerical simulation of a turbulent boundary layer up to $R_\theta = 1400$. J. Fluid Mech. **187**, 61–98

P.R. Spalart (2000): Strategies for turbulence modelling and simulations. Int. J. Heat Fluid Flow **21**, 252–263

P.R. Spalart, R.D. Moser, M.M. Rogers (1991): Spectral methods for the Navier–Stokes equations with one infinite and two periodic directions. J. Comput. Phys. **96**(2), 297–324

P.R. Spalart, J.H. Watmuff (1993): Experimental and numerical study of a turbulent boundary layer with pressure gradients. J. Fluid Mech. **242**, 337–371

P.R. Spalart, K.-S. Yang (1987): Numerical simulation of ribbon-induced transition in Blasius flow. J. Fluid Mech. **178**, 345–365

C.G. Speziale (1991): Analytical methods for the development of Reynolds-stress closures in turbulence. Ann. Rev Fluid Mech. **23**, 107–157

C.G. Speziale, G. Erlebacher, T.A. Zang, M.Y. Hussaini (1988): The subgrid-scale modeling of compressible turbulence. Phys. Fluids **31**(4), 940–942

H.B. Squire (1933): On the stability of the three-dimensional disturbances of viscous flow between parallel walls. Proc. R. Soc. London Ser. A **142**, 621–628

D. Stanescu, D. Ait-Ali-Yahia, W.G. Habashi, M.P. Robichaud (1999): Multidomain spectral computations of sound radiation from ducted fans. AIAA J. **37**(3), 296–302.

D. Stanescu, W.G. Habashi (1998): 2N storage low dissipation and dispersion Runge-Kutta schemes for computational acoustics. J. Comput. Phys. **143**(2), 674–681

D. Stanescu, M.Y. Hussaini, F. Farassat (2003): Aircraft engine noise scattering by fuselage and wings: a computational approach. J. Sound Vibration **263**(2), 319–333

D. Stanescu, D.A. Kopriva, M.Y. Hussaini (2000): Dispersion analysis for discontinuous spectral element methods. J. Sci. Comput. **15**(2), 149–171

D. Stanescu, J. Xu, M.Y. Hussaini, F. Farassat (2002): Computation of engine noise propagation and scattering off an aircraft. Int. J. Aeroacoustics **1**(4), 403–420

D. Stefanica (2002): FETI and FETI-DP methods for spectral and mortar spectral elements: a performance comparison. J. Sci. Comput. **17**(1–4), 629–638

D. Stefanica, A. Klawonn (1999): 'The FETI Method for Mortar Finite Elements'. In: *Proceedings of the 11th International Conference on Domain Decomposition Methods* (Published on line at www.ddm.org/DD11/index.html) pp. 121–129

R. Stenberg, M. Suri (1996): Mixed *hp* finite element methods for problems in elasticity and Stokes flow. Numer. Math. **72**(3), 367–389

G.G. Stokes (1845): On the theories of the internal friction of fluids in motion. Trans. Cambridge Philos. Soc. **8**, 287–319

C.L. Streett (1983): 'A Spectral Method for the Solution of Transonic Potential Flow About an Arbitrary Two-dimensional Airfoil'. AIAA Paper No. 83–1949

C.L. Streett, M.G. Macaraeg (1989): Spectral multidomain for large-scale fluid dynamic simulations. Appl. Numer. Math. **6**(1–2), 123–139

C.L. Streett, T.A. Zang, M.Y. Hussaini (1984): 'Spectral Methods for Solution of the Boundary-Layer Equations'. AIAA Paper No. 84–0170

C.L. Streett, T.A. Zang, M.Y. Hussaini (1985): Spectral multigrid methods with applications to transonic potential flow. J. Comput. Phys. **57**(1), 43–76

J. Strikwerda (1976): Initial Boundary Value Problems for Incompletely Parabolic Systems. PhD thesis, Stanford University, Palo Alto, CA

A. Suddhoo, I.M. Hall (1985): Test cases for the plane potential flow past multi-element aerofoils. Aeronautical J. **89**, 403–414

W. Sutherland (1893): The viscosity of gases and molecular force. Philosophical Magazine **5**(36), 507–531

E. Tadmor (1989): The convergence of spectral methods for nonlinear conservation laws. SIAM J. Numer. Anal. **26**(1), 30–44

I. Tani (1977): History of boundary layer theory. Ann. Rev. Fluid Mech. **9**, 87–111

T. Tatsumi, T. Yoshimura (1990): Stability of the laminar flow in a rectangular duct. J. Fluid Mech. **212**, 437–449

G.I. Taylor (1915): Eddy motion in the atmosphere. Phil. Trans. Roy. Soc. London **215**, 1–26

G.I. Taylor, A.E. Green (1937): Mechanism of the production of small eddies from large ones. Proc Roy. Soc. A **158**, 499–521

G.I. Taylor, J.W. Maccoll (1933): The air pressure on a cone moving at high speeds. Proc. R. Soc. A **139**, 278

T.D. Taylor, R.B. Myers, J.H. Albert (1981): Pseudospectral calculations of shock waves, rarefaction waves and contact surfaces. Comput. & Fluids **9**(4), 469–473

R. Temam (1969): Sur l'approximation de la solution des équations de Navier–Stokes par la méthode des pas fractionnaires II. Arch. Rat. Mech. Anal. **33**(5), 377–385

R. Temam (1995): *Navier–Stokes Equations and Nonlinear Functional Analysis*, 2nd edition (SIAM-CBMS, Philadelphia)

R. Temam (2001): *Navier-Stokes Equations* (North-Holland, Amsterdam)

C. Temperton (1983): Self-sorting mixed-radix fast Fourier transforms. J. Comput. Phys. **52**(1), 1–23

V. Theofilis (2003): Advances in global linear instability analysis of nonparallel and three-dimensional flows, Prog. Aero. Sci. **39**, 249–315

K.W. Thompson (1987): Time dependent boundary conditions for hyperbolic systems. J. Comput. Phys. **68**(1), 1–24.

L.J.P. Timmermans, P. Minev, F.N. van de Vosse (1996): An approximate projection scheme for incompressible flow using spectral elements. Int. J. Num. Meth. Fluids **22**(7), 673–688

W. Tollmien (1929): Über die Entstehung der Turbulenz. 1. Mitt. Nachr. Gew. Wiss. Göttingen, Math. Phys. Klasse, 21–44 (available in English in NACA TM–609 (1931))

A. Toselli (2002): *hp*-finite element discontinuous Galerkin approximations for the Stokes problem. Math. Models Meth. Appl. Sci. **12**(11), 1565–1597

A. Toselli (2003): *hp*-finite element approximations on non-matching grids for partial differential equations with non-negative characteristic form. Math. Mod. Numer. Anal. **37**(1), 91–115

A. Toselli, X. Vasseur (2002): Neumann-Neumann and FETI preconditioners for *hp*-approximations on geometrically refined boundary layer meshes in two dimensions. Technical Report 2002–15, Seminar for Applied Mathematics, ETHZ

A. Toselli, X. Vasseur (2003): A numerical study on Neumann-Neumann and FETI methods for *hp* approximations on geometrically refined boundary layer meshes in two dimensions. Comput. Meth. Appl. Mech. Engrg. **192**(41–42), 4551–4579

A. Toselli, O. Widlund (2005): *Domain Decomposition Methods. Algorithms and Theory.* (Springer-Verlag, Berlin)

H. Touil, M.Y. Hussaini, M. Sussman (2007): Tracking discontinuities in hyperbolic conservation laws with spectral accuracy. J. Comput. Phys., in press

L.N. Trefethen (1983): Group velocity interpretation of the stability theory of Gustafsson, Kreiss, and Sundstrom. J. Comput. Phys. **49**(2), 199–217

L.N. Trefethen (1997): Pseudospectra of linear operators. SIAM Review **39**(3), 383–406

L.N. Trefethen (2000): *Spectral Methods in MATLAB* (SIAM, Philadelphia)

L.N. Trefethen, M. Embree (2005): *Spectra and Pseudospectra: The Behavior of Nonnormal Matrices and Operators* (Princeton University Press, Princeton)

C. Truesdell (1953): Notes on the history of the general equations of hydrodynamics. American Math. Monthly **60**(7), 445–458

J. Trujillo, G.E. Karniadakis (1999): A penalty method for the vorticity–velocity formulation. J. Comput. Phys. **149**(2), 32–58

S.V. Tsynkov (1998): Numerical solution of problems on unbounded domains, a review. Appl. Numer. Math. **27**(4), 465–532

L. Tuckerman (1989a): Divergence-free velocity fields in nonperiodic geometries. J. Comput. Phys. **80**(2), 403–441

L. Tuckerman (1989b): Transformations of matrices into banded form. J. Comput. Phys. **84**(2), 360–376

H.M. Tufo, P.F. Fischer (2001): Fast parallel direct solvers for coarse grid problems. J. Par. Dist. Comput. **61**, 151–177

A. Valli (1983): Periodic and stationary solutions for compressible Navier–Stokes equations via a stability method. Ann. Scuola Norm. Sup. Pisa Cl. Sci. (4) **10**, 607–647

H.A. van der Vorst (1992): Bi-CGSTAB: a fast and smoothly converging variant of Bi-CG for the solution of nonsymmetric linear systems. SIAM J. Sci. Statist. Comput. **13**(2), 631–644

M. Van Dyke (1969): Higher-order boundary-layer theory. Ann. Rev. Fluid Mech. **1**, 265–292

M. Van Dyke (1975): *Perturbation Methods in Fluid Mechanics*, 2nd edition (Parabolic Press, Stanford)

J.L. Van Ingen (1956): 'A Suggested Semi-Empirical Method for the Calculation of the Boundary Layer Transition Region'. Rep. VTH–74 (Depart. Aero. Engrg., University of Technology, Delft, Holland)

J.A. van Kan (1986): A second-order accurate pressure-correction scheme for viscous incompressible flow. SIAM J. Sci. Statist. Comput. **7**(3), 870–891

H. Vandeven (1991): Family of spectral filters for discontinuous problems. J. Sci. Comput. **6**(2), 159–192

H.A. van der Vorst (2003): *Iterative Krylov Methods for Large Linear Systems* (Cambridge University Press, Cambridge)

A. Veneziani (2003): Block factorized preconditioners for high order time dependent approximation of Navier–Stokes equations. Numer. Meth. Part. Diff. Eqs. **19**(4), 487–510

R. Verfürth (1996): *A Review of A Posteriori Error Estimation and Adaptive Mesh Refinement Techniques* (Wiley-Teubner, Chichester)

C.H. Von Kerczek (1982): The instability of oscillatory plane Poiseuille flow. J. Fluid Mech. **116**, 91–114

J. Von Neumann (1949): 'Recent Theories of Turbulence'. 1949 Report to ONR. In: *John von Neumann Collected Works*, Vol. 6, ed. by A.H. Taub (Macmillan, New York) pp. 437–472

T. Warburton (2000): 'Application of the Discontinuous Galerkin Method to Maxwell's Equations Using Unstructured Polymorphic hp-Finite Elements. In: *Discontinuous Galerkin methods*, Lecture Notes in Computational Science and Engineering **11** (Springer-Verlag, Heidelberg) pp. 451–458.

T. Warburton, L.F. Pavarino, J.S. Hesthaven (2000): A pseudospectral scheme for the incompressible Navier–Stokes equations using unstructured spectral elements. J. Comput. Phys. **164**(1), 1–21

R.F. Warming, R.M. Beam, B.J. Hyett (1975): Diagonalization and simultaneous symmetrization of the gas-dynamic matrices. Math. Comput. **29**, 1037–1045

W. Wasow (1948): The complex asymptotic theory of a fourth order differential equation of hydrodynamics. Ann. Math. **49**, 852–871

P. Wesseling (2001): *Principles of Computational Fluid Dynamics* (Springer-Verlag, Berlin)

T.P. Wihler, P. Frauenfelder, Ch. Schwab (2003): Exponential convergence of the hp-DGFEM for diffusion problems. Comput. Math. Appl. **46**(1), 183–205

D.C. Wilcox (1993): *Turbulence Modeling for CFD* (DCS Industries, Inc., La Canada, CA)

D. Wilhelm, L. Kleiser (2000): Stable and unstable formulations of the convection operator in spectral element simulations. Appl. Numer. Math. **33**(1–4), 275–280

J.H. Williamson (1980): Low-storage Runge-Kutta schemes. J. Comput. Phys. **35**(1), 48–56

J.H. Wilkinson (1965): *The Algebraic Eigenvalue Problem* (Clarendon Press, Oxford)

B. Wohlmuth (2001): *Discretization Techniques and Iterative Solvers Based on Domain Decomposition*, Lectures Notes in Computational Science and Engineering **17** (Springer-Verlag, Heidelberg)

P.R. Woodward, P. Colella (1984): The numerical simulation of two-dimensional fluid flow with strong shocks. J. Comput. Phys. **54**(1), 115–173

S.F. Wornom (1978): 'Critical Study of Higher Order Numerical Methods for Solving the Boundary-Layer Equations'. NASA TP–1302

T.G. Wright, L.N. Trefethen (2001): Large-scale computation of pseudospectra using ARPACK and eigs. SIAM J. Sci. Comp., **23**(2), 591–605

N.N. Yanenko (1971): *The Method of Fractional Steps for Solving Multi-dimensional Problems of Mathematical Physics in Several Variables* (English translation), ed. by M. Holt (Springer-Verlag,Heidelberg)

A. Yoshizawa (1986): Statistical theory for compressible turbulent shear flows, with the application to subgrid modeling. Phys. Fluids **29**(7), 2152–2164

T.A. Zang (1990): Spectral methods for simulations of transition and turbulence. Comput. Meth. Appl. Mech. Engrg. **80**(1–3), 209–221

T.A. Zang, R.B. Dahlburg, J.P. Dahlburg (1992): Direct and large-eddy simulations of three-dimensional compressible Navier–Stokes turbulence. Phys. Fluids A **4**(1), 127–140

T.A. Zang, M.Y. Hussaini (1981): 'Mixed Spectral/Finite Difference Approximations for Slightly Viscous Flows'. In: *Proceedings of the 7th International Conference on Numerical Methods in Fluid Dynamics*, ed. by W.C. Reynolds, R.W. MacCormack (Springer-Verlag, Heidelberg) pp. 461–466

T.A. Zang, M.Y. Hussaini (1986): On spectral multigrid methods for the time-dependent Navier–Stokes equations. Appl. Math. Comput. **19**(1–4), 359–372

T.A. Zang, M.Y. Hussaini (1987): 'Numerical Simulation of Nonlinear Interactions in Channel and Boundary-layer Transition'. In: *Nonlinear Wave Interaction, in Fluids*. AMD–87, ed. by

R.W. Miksad, T.R. Akylas, Th. Herbert, (ASME, New York) pp. 131–145

T.A. Zang, S.E. Krist (1989): Numerical experiments on stability and transition in plane channel flow. Theor. Comput. Fluid Dyn. **1**(1), 41–64

T.A. Zang, S.E. Krist, G. Erlebacher, M.Y. Hussaini (1987): 'Nonlinear Structures in the Later Stages of Transition'. AIAA Paper No. 87–1204

L. Zannetti, G. Colasurdo (1981): Unsteady compressible flow: a computational method consistent with the physical phenomena. AIAA J. **19**(7), 852–856

A. Zebib (1984): A Chebyshev method for the solution of boundary value problems. J. Comput Phys. **53**(3), 443–455

Index

Scientific Computation

springer.com

Scientific Computation